Synchrotron Radiation Research

Advances in Surface and Interface Science

Volume 1

Techniques

SYNCHROTRON RADIATION RESEARCH
Advances in Surface and Interface Science

Volume 1: Techniques
Edited by Robert Z. Bachrach

Volume 2: Issues and Technology
Edited by Robert Z. Bachrach

Synchrotron Radiation Research

Advances in Surface and
Interface Science
Volume 1

Techniques

Edited by

Robert Z. Bachrach

Applied Materials Inc.
Santa Clara, California
and Stanford University
Stanford, California

Plenum Press • New York and London

Library of Congress Cataloging-in-Publication Data

Synchrotron radiation research : advances in surface and interface
  science / edited by Robert Z. Bachrach.
        p.   cm.
    Includes bibliographical references and index.
    Contents: v. 1. Techniques -- v. 2. Issues and technology.
    ISBN 0-306-43872-0 (v. 1). -- ISBN 0-306-43873-9 (v. 2)
    1. Energy-band theory of solids. 2. Solids--Surfaces.
  3. Synchrotron radiation.   I. Bachrach, R. Z. (Robert Z.)
    [DNLM: 1. Particle Accelerators. 2. Surface Properties.
  3. Technology, Radiologic.   QC 787.S9 S992]
  QC176.8.E4S88   1992
  530.4'1--dc20
  DNLM/DLC
  for Library of Congress   92-6482
                                                                    CIP

ISBN 0-306-43872-0

©1992 Plenum Press, New York
A Division of Plenum Publishing Corporation
233 Spring Street, New York, N.Y. 10013

To Professor Frederick C. Brown whose vision of the research
opportunities and potential of synchrotron radiation in the 1960s and
development of enabling instrumentation provided impetus for some of the
significant progress described in this book

and

Dr. George E. Pake, whose appreciation of what is important in frontier
science led him to sponsor the Xerox Palo Alto Research Center's
participation in synchrotron radiation starting in 1972, sponsoring the
development at SSRL of the Grasshopper Monochromator and of
Multiundulator Beamline V

*Advances in physics follow advances in instrumentation,
once the instruments are made, many people will make the discoveries.*

— Abraham Pais,
*Inward Bound:
Of Matter and Forces in
the Physical World*

Contributors

J. W. Allen • Department of Physics, The Harrison M. Randall Laboratory of Physics, University of Michigan, Ann Arbor, Michigan 48109-1120.

Robert Z. Bachrach • Applied Materials, Inc., Santa Clara, California 95054-3299 *and* Stanford Synchrotron Radiation Laboratory, Stanford University, Stanford, California 94305. *Previous address*: Xerox Palo Alto Research Center, Palo Alto, California 94304.

A. Bianconi • Consorzio Interuniversitario di Fisica della Materia (INFM), Dipartimento di Fisica, Università "La Sapienza," 00185 Rome, Italy.

Marvin L. Cohen • Department of Physics, University of California, *and* Materials and Chemical Sciences Division, Lawrence Berkeley Laboratory, Berkeley, California 94720.

W. Eberhardt • Institut für Festkörperforschung des Forschungszentrums Jülich, D5170 Jülich, Germany.

P. Eisenberger • Princeton University, Princeton, New Jersey 08544.

Charles S. Fadley • Department of Chemistry, University of Hawaii, Honolulu, Hawaii 96822. *Present address*: Department of Physics, University of California–Davis, Davis, California 95616, *and* Materials Science Division, Lawrence Berkeley Laboratory, Berkeley, California 94720.

Anders Flodström • Department of Materials Science, Royal Institute of Technology, S-100 44 Stockholm, Sweden.

P. H. Fuoss • AT&T Bell Laboratories, Murray Hill, New Jersey 07974.

Borje Johansson • Condensed Matter Theory Group, Department of Physics, University of Uppsala, S-751 21 Uppsala, Sweden.

K. S. Liang • Exxon Corporate Research Laboratories, Annandale, New Jersey 08801.

A. Marcelli • Instituto Nazionale di Fisica Nucleare (INFN), Laboratori Nazionali di Frascati, 00044 Frascati, Italy.

Ralf Nyholm • MAX-Laboratory, University of Lund, S-221 00 Lund, Sweden.

Victor Rehn • Physics Division, Research Department, Naval Weapons Center, China Lake, California 93555. *Present address*: Office of Naval Research–Asian Office, APO AP 96337-0007.

Richard Rosenberg • Synchrotron-Radiation Center, University of Wisconsin, Stoughton, Wisconsin 53589. *Present address*: Advanced Photon Source, Argonne National Laboratory, Argonne, Illinois 60439.

J. E. Rowe • AT&T Bell Laboratories, Murray Hill, New Jersey 07974.

Foreword

In the summer of 1972, I had the privilege and responsibility of organizing a Gordon Conference on the "High-Energy Spectroscopy of Solids." The Thursday evening session focused on future directions for high-energy spectroscopy. The possibilities associated with synchrotron radiation for future research became a central issue. I was asked to choose the members of the panel and chair the session. Although all five members of the panel went on to have distinguished careers using synchrotron radiation, at the time some of them were skeptical about the future role of synchrotron radiation sources in high-energy photon spectroscopy.

The discussion became heated, and many members of the audience spoke both pro and con. One member of the panel produced a detailed argument that synchrotron radiation would never rival standard X-ray tubes. We found out that there were estimates for properties of synchrotrons that differed by orders of magnitude from those of X-ray tubes.

That much uncertainty was expressed at a meeting that took place less than twenty years ago. It is hard to believe that, even though at that time synchrotron radiation was already being used for photoemission studies of solids and surfaces and intershell excitations in solids, the potential impact and importance of this area was not fully realized even by the experts. Today synchrotron radiation is one of the primary tools for studying surfaces, and synchrotron radiation has affected many other areas of condensed-matter physics—even superconductivity. The recent observation of the superconducting energy gap by photoemission is an important breakthrough. The use of synchrotron radiation together with angular-resolved photoemission spectroscopy allows the experimental determination of surface and bulk band structures that until the 1970s were in the domain of the theorists only. These developments have come quickly and have had great impact because the new tools are significantly better than those available earlier.

Specifically for the study of surfaces, the development of techniques using synchrotron radiation came at an opportune time, when experimentalists were beginning to get reproducible results due to improved high-vacuum techniques. This was also fortunate for theorists, because their machinery had just developed to a state where both electronic structure and atomic positions could be calculated using techniques that were close to first principles. Because of this, important collaborations between experimentalists and theorists were fostered, and these led to developments that could not have been achieved by either group working alone.

In the last twenty years, the important discoveries and explanations of the properties of surfaces have convinced even the most pessimistic scientists of the power of synchrotron radiation as a tool for studying surfaces. This volume is clear evidence of the great advances in this area. Although no one knows what the next twenty years will bring, it is probably safe to predict that for at least the next decade we will rely heavily on synchrotron radiation for solving the mysteries associated with the surfaces of materials.

Marvin L. Cohen

Berkeley, California

Preface

Synchrotron Radiation Research: Advances in Surface and Interface Science presents in a focused way surface-science-related research accomplished with synchrotron radiation. Previous books on synchrotron radiation have been broader in scope or more general than this two-volume book, in that they have attempted to cover the full range of application of synchrotron radiation research. This book differs in its depth of focus, and it should be a valuable complement to other resources such as the Plenum volume edited by Winick and Doniach, *Synchrotron Radiation Research*.

I have encouraged the authors to introduce all the material they require for completeness, even though other chapters deal with some aspects in more detail. Thus individual chapters are self contained to a large degree, and this is intentional. As a result, in a book such as this which comprises collected chapters by individual authors or groups, it is not possible to avoid all overlap between chapters.

The introduction to *Synchrotron Radiation Research: Advances in Surface and Interface Science* presents an overview of the book. *Volume 1* surveys the synchrotron radiation–based techniques that have enabled a number of advances to be made in understanding issues in surface science. *Volume 2* presents individual surveys of particular areas where significant impact has occurred. *Volume 2* also describes the new undulator source and advanced monochromator technology which will be seeing wide utilization in the future. Together, these will create even further opportunities for advances in surface science.

As this book is being finished, the field of synchrotron radiation–related science is going through yet another round of facility creation on a worldwide scope. This new generation of machines is being organized to exploit periodic insertion device technology in low-emittance storage rings. The field now has its own newsletter with a worldwide distribution of around seven thousand copies. All these factors lead one to anticipate that many more exciting discoveries and advances are yet to be made.

Robert Z. Bachrach

Palo Alto, California

Contents

Chapter 3

Surface EXAFS

J. E. Rowe

PART II: Photoemission Spectroscopy

Chapter 4

Angle-Resolved Photoemission Spectroscopy

W. Eberhardt

Chapter 5

Surface Core Level Spectroscopy

Anders Flodström, Ralf Nyholm, and Börje Johansson

Chapter 6

Resonant Photoemission of Solids with Strongly Correlated Electrons

J. W. Allen

PART III: Ion Spectroscopy

Chapter 7

Photon-Stimulated Desorption

Victor Rehn and Richard A. Rosenberg

PART IV: Diffraction and Scattering

Chapter 8

Grazing-Incidence X-Ray Scattering

P. H. Fuoss, K. S. Liang, and P. Eisenberger

Chapter 9

The Study of Surface Structures by Photoelectron Diffraction and Auger Electron Diffraction

Charles S. Fadley

Introduction and Overview

Robert Z. Bachrach

1. INTRODUCTION

X-rays were discovered in 1895,[1] and in the intervening years they have been both the subject of intense research and the means of investigation in many fields. X-ray science in particular bloomed during the 1920s and 1930s with much of the scope of the field being explored with characteristic line and bremsstrahlung sources.[2] The penetrating power of hard X-rays, even with weak sources, made feasible a wide range of experiments. In contrast, science with soft X-rays, the region of the spectrum where penetrating power is weak, and which therefore requires thin samples and vacuum ambients, was much less developed. As source technology advanced, ever new paths of investigation opened. One thus sees a progression of stages in X-ray-related science, each one presaged by the advent of new source technology; e.g., static water-cooled anodes, rotating anodes, synchrotron radiation with bending magnets, pulsed plasma sources, and synchrotron radiation with insertion devices. The enormous growth in the available flux, coupled with the facility of today's vacuum technology, has changed experiments from dreams to actuality as required counting times decreased from weeks to minutes.

Surface science has been a principal focus and beneficiary of the application of synchrotron radiation coupled with other instrumentation advances. Synchrotron radiation has brought some unique advantages to the study of surfaces. The aim of this book is to assess the impact of synchrotron radiation on this particularly active area of its use. Advances in science involve extensions of technique and of knowledge, and the advances can be made both in concept and in detail. In both these ways, a number of aspects of surface physics have benefited from advances related to the development of synchrotron radiation. Several aspects of synchrotron radiation have facilitated unique experiments in

Robert Z. Bachrach • Applied Materials, Inc., Santa Clara, California 95054-3299 *and* Stanford Synchrotron Radiation Laboratory, Stanford University, Stanford, California 94305. *Previous address*: Xerox Palo Alto Research Center, Palo Alto, California 94304.

Synchrotron Radiation Research: Advances in Surface and Interface Science, Volume 1: Techniques, edited by Robert Z. Bachrach. Plenum Press, New York, 1992.

surface science for structural determinations, for microchemical analysis, and for the examination of electronic phenomena.

The application of synchrotron radiation to surface physics has developed or made possible the broad range of techniques enumerated in the table of contents. This introduction discusses some of these advances in order to prepare the reader for what follows. The book employs a three-part organization: (1) the enumeration of specific measurement techniques, (2) the elaboration of specific issues without necessarily focusing on specific techniques, and (3) the provision of an overview of source and monochromator technology. Part 3 is included because, coupled with analyzer and detector systems, source and monochromator technology is likely to impact future work. Although instrumentation is not the focus of this book, it is important to remember the enabling role that advances in instrumentation have played and will continue to play in science.

In the following section of the introduction, the generation and properties of synchrotron radiation are summarized. Section 3 discusses surface-science study techniques using synchrotron radiation, and section 4 gives examples of research advances made possible by the application of synchrotron radiation.

2. SYNCHROTRON RADIATION

Synchrotron radiation, the emission of light by accelerated relativistic electrons, is the most powerful source of soft and hard X-rays available for general research. This section presents only a descriptive overview of synchrotron radiation, its generation, and its properties because the general nature of synchrotron radiation and facilities is well presented by Winick and Doniach,[3] in other books,[2,4,5] published articles,[6] planning studies for new facilities around the world,[7] and a journal concentrating on the field.[8] George Brown discusses some general aspects of synchrotron radiation and covers insertion-device sources in detail in Chapter 8 of Volume 2 of this set.

High-energy storage rings provide high circulating currents which emit pulsed beams of light. Although synchrotron radiation was first discovered in the late 1940s,[9] access to the requisite high-energy electron machines was not readily obtained. The 1950s were therefore characterized by a few demonstration experiments, but it was not until the late 1960's that work began in earnest and exciting results led to efforts to achieve consistent machine access. Work with synchrotron machines such as those at Cornell,[10] the National Bureau of Standards,[11] Tokyo,[12] and DESY[13] was supplanted by the greater capabilities of storage rings.[14] In the United States, the Tantalus storage ring in Wisconsin was developed between 1966 and 1973,[15] and the Stanford Synchrotron Radiation Laboratory in California was created in 1973 at the Stanford Positron Electron Accelerator Ring (SPEAR). Similar efforts went forward in many other laboratories around the world.[3] At the same time that the early parasitic facilities (i.e., those shared with high-energy physics experiments) showed the value of synchrotron radiation in broad areas of research, a progression of dedicated facilities were built. Today a large number of storage-ring synchrotron radiation sources are used for scientific experimentation and technological applications. More are being built for use in a wide range of contexts. The most recent generation of storage rings for synchrotron radiation emphasizes low-emittance stored-electron beams coupled with insertion-device sources.[16] Among these new facilities is a new generation of compact storage rings,[17] which will become available in conventional laboratory or industrial settings.

For materials-related synchrotron radiation experiments, the high-energy storage ring source circulates one or more tightly packed bunches of electrons, resulting in the emission of short pulses of light, which are typically polarized. Figure 1 provides a schematic summary of the major properties of synchrotron radiation. Exhaustive treatments exist in the literature and the reader is referred to the references at the end of the chapter for detailed citations.

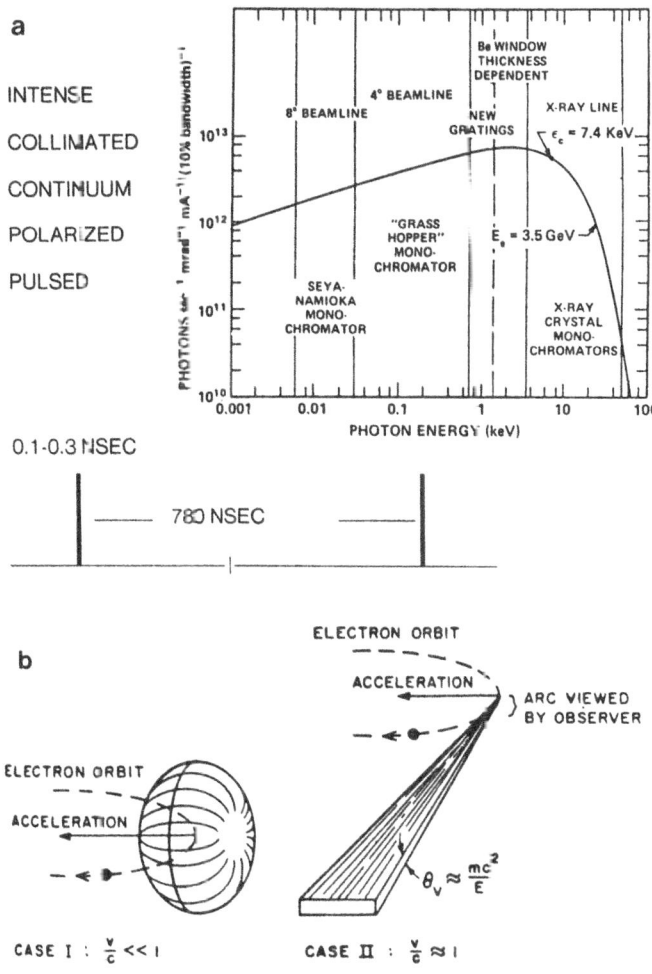

FIGURE 1. Attributes of synchrotron radiation. (a) Schematic representation of the properties of synchrotron radiation with the time structure from the SPEAR storage ring. The figure characterizes the spectrum from a bending magnet and indicates some of the monochromators used. (b) Schematic representation of the synchrotron radiation emission process. (From Ref. 3.)

Figure 1a summarizes the principal properties of synchrotron radiation, characterizes the spectral distribution from a bending magnet, and identifies some of the monochromators in use for various spectral ranges. Figure 1b depicts the emission process. The spectral and power properties of synchrotron radiation can be calculated classically, and measurements have shown that the formulation is correct to the accuracy that it can be measured. Both the spatial and temporal characteristics of synchrotron radiation lend themselves to unique applications. Some of these are described in the following chapters.

Three types of magnetic elements in the storage ring are used in the generation of synchrotron radiation: bending magnets, wigglers, and undulators. Figure 2 depicts the characteristics of the radiation from these three elements, which are situated in the magnetic lattice of the storage ring so that the radiation can be brought out into an experimental hall.[18] The elements accelerate the circulating electrons in different ways. The bending magnet effectively generates a sweeping searchlight pattern. Due to the relativistic velocity of the electrons, the radiation pulses are highly collimated with a vertical opening angle of $1/\gamma$ ($\gamma = E_0/m_e c^2$) near the synchrotron emission critical energy. Synchrotron radiation from bending-magnet sources can be used for radiometric purposes in the soft X-ray range where other types of standards are difficult to develop. The wiggler and undulator differ primarily in the magnitude of their perturbation of the circulating electron beam. Because the undulator achieves more coherence, the power P increases as the square of the number of periods N, while the divergence of the beam decreases. The new generation of insertion-device sources with very low emittances is creating opportunities to utilize their partial coherence in a number of possible experiments. The undulator is only one step removed from a laser in brightness. For example, at SPEAR, even at 20 mm from the source, the beam size is less than a centimeter.

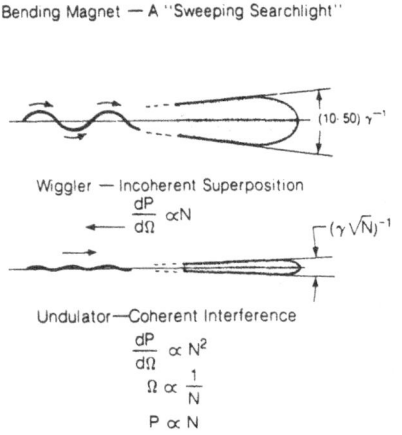

Bending Magnet — A "Sweeping Searchlight"

Wiggler — Incoherent Superposition
$$\frac{dP}{d\Omega} \propto N$$

Undulator—Coherent Interference
$$\frac{dP}{d\Omega} \propto N^2$$
$$\Omega \propto \frac{1}{N}$$
$$P \propto N$$

N = number of magnetic periods (~ 100)

$$\gamma^{-1} = \frac{m_0 c^2}{E_e} = \frac{0.511}{E_e(\text{GeV})} \text{ mrad}$$

FIGURE 2. Synchrotron radiation generation regimes from a bending magnet, a wiggler, and an undulator. (From Ref. 9.)

In a typical storage ring, the electrons can circulate with a half-life of 4–12 h. Maximum currents depend on the stored energy, but typically range from 100 to 1000 mA as the energy ranges from approximately 3 GeV to less than 1 GeV. Because the power dissipated varies as the third power of the stored energy, the total light generated in the soft X-ray range (10–3000 eV) is typically similar in the low- and high-energy machines. The primary benefit of differentiating machines into the two regimes is for optimization of the optics in the case of soft X-ray experiments.

The most recent generation of storage rings for synchrotron radiation is emphasizing low-emittance stored-electron beams coupled with insertion-device sources,[16] e.g., the Advanced Light Source being built at the Lawrence Berkeley Laboratory.[18] A forerunner of this most recent generation is Beam Line V, also known as Beam Line WUNDER, at the Stanford Synchrotron Radiation Laboratory, which uses the multiundulator insertion device shown in Fig. 3 as its source.[19] This device is currently configured with four independent permanent-magnet undulators with periods adjusted to cover the desired spectral range from 10–1000 eV. Figure 4 shows a power comparison between a bending magnet source and three of the magnetic-period elements in the multiundulator. One sees that an increase in power of at least two orders of magnitude is achieved. These

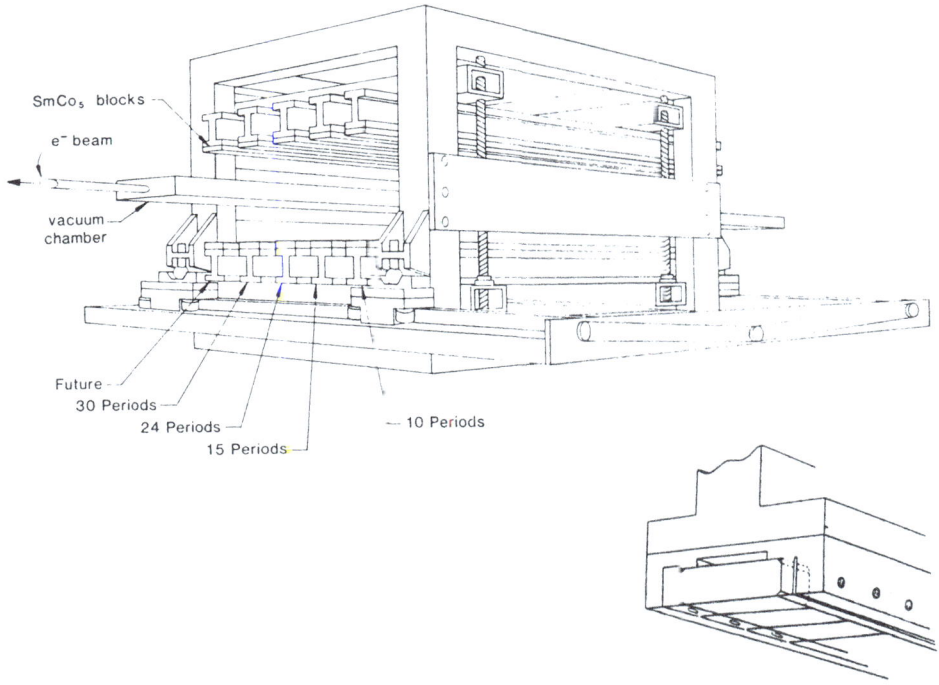

FIGURE 3. Xerox/Stanford SSRL BLV Multiundulator. The five possible insertions mounted on stainless steel I-beams are shown installed in the mover assembly surrounding the SPEAR beam pipe. The spectrum for each of the periods is changed by changing the gap between the magnets. The inset shows how the $SmCo_5$ magnet bars are mounted. (From Ref. 10.)

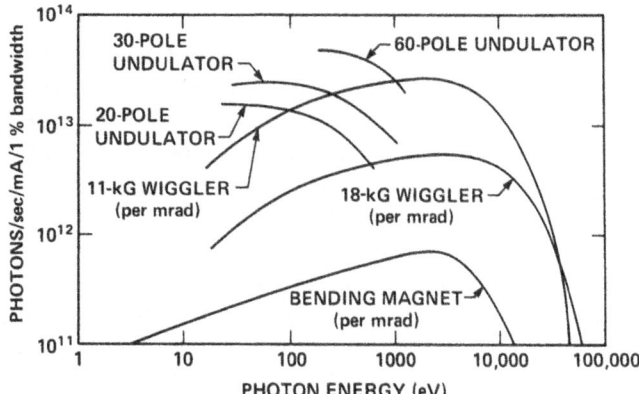

FIGURE 4. Spectral power comparison between bending magnet and undulator. The estimated first-order flux into the beamline is shown for various sources for SPEAR operating at 3 GeV under dedicated conditions. The wiggler and bending-magnet curves are a continuum. The undulator curves represent a tuning range with the low-energy cutoff determined by the minimum gap between the permanent magnets. The undulator range shifts as the square of the stored energy of the SPEAR beam.

quantitative advances in capability will create many new scientific opportunities. Although this book is principally a retrospective, it is important to keep in mind that the enormous increases in flux and brightness will open exciting new paths for the future.

Figures 5a and 5b characterize the electromagnetic spectrum and depict some of the typical application and measurement regimes.[18] Figure 5 emphasizes the breadth of possible experiments, using electromagnetic radiation as a probe, made possible by synchrotron radiation. Although the focus of investigators is different in the various energy regimes, important and exciting physics is being pursued in each of them.

Experimental access to the spectral range depicted in Fig. 5 is provided by monochromator technology. The development of sources over the last two decades has been coupled with the parallel development of monochromators, with different instruments optimized for the various broad spectral ranges. The monochromators, which disperse the radiation, are critical factors in the experiments. The monochromator is a primary transfer element in the optical system between the source and the experiment, and its design is a key element in the experimental system.

Monochromator technology is quite diverse, and within each partition of the spectral range there are many categories. One of the largest partitions is between grating and crystal monochromators. Typically gratings are most useful below 1 keV and crystals above. Surface physics work has utilized both ranges, and various reviews[20,21] provide a thorough overview of the status of this instrumentation. Chapter 9 of Volume 2 of this set, by Gwyn Williams, reviews some of the most recent advances. Examples include the Grasshopper grazing-incidence Vodar geometry-grating monochromator[22] installed at SSRL in 1974 and the Jumbo double-crystal monochromator installed in 1980.[23] These monochromators opened up the spectral range from 50 to 5000 eV for surface physics experiments.

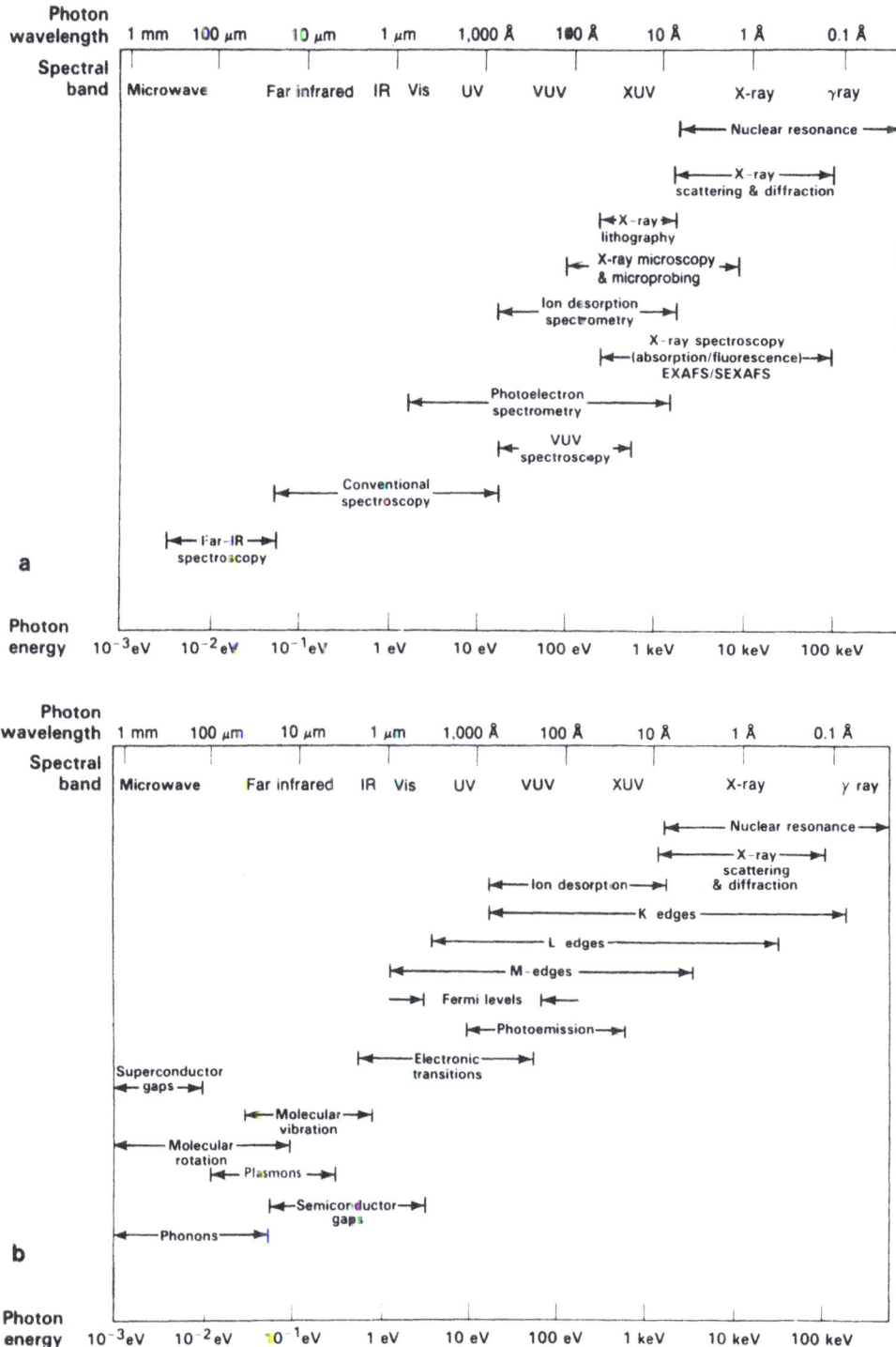

FIGURE 5. The electromagnetic spectrum accessible with synchrotron radiation. (a) Characterization techniques using synchrotron radiation. (b) Molecular, atomic, and electronic processes characteristic in various parts of the electromagnetic spectrum plotted as a function of photon energy. (From Ref. 9.)

Another example of a recent design is the Locust monochromator built for use with the multiundulator at SSRL, which became operational in 1989.[24] Because of the high power input from the multiundulator, this monochromator had to have silicon carbide optics, and the optics had to be water cooled. Figure 6a shows the primary mechanism and Fig. 6b shows the expected resolution. A complete description is given in Ref. 24.

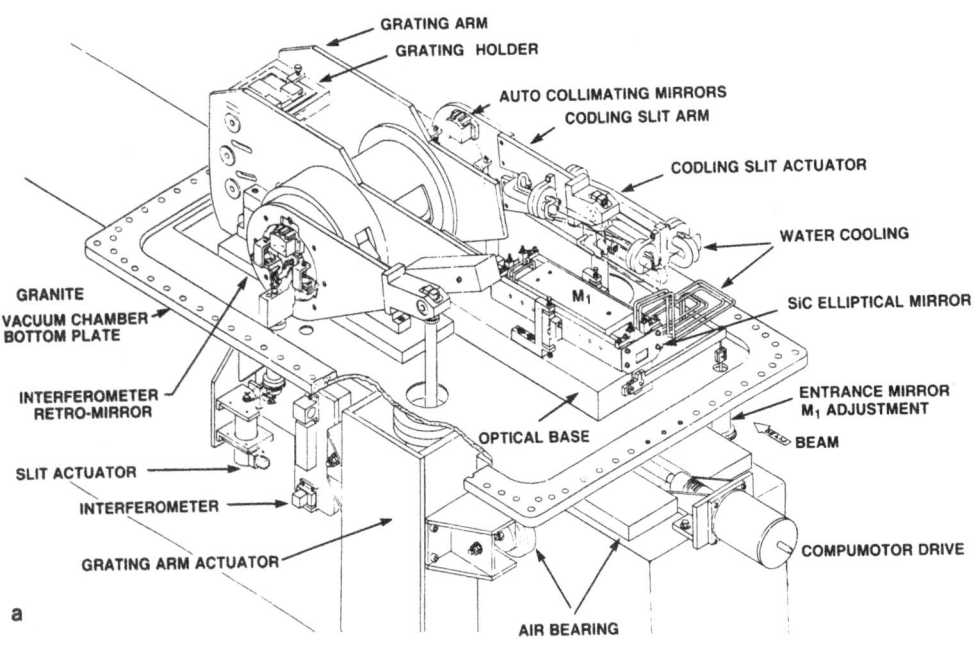

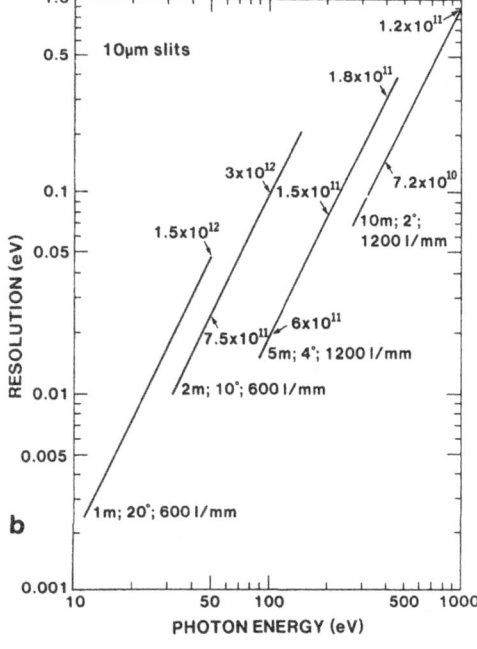

FIGURE 6. The Locust monochromator. (a) Perspective view of the main optical mechanism with the vacuum envelope removed. The monochromator was developed for multiundulator beamline V at SSRL and has water-cooled optics. All connections are introduced through the bottom plate to allow opening without disassembly of the optical mechanism. The monochromator mechanism is interferometrically positioned under computer control. (b) Estimated energy resolution *vs.* photon energy for each of the Locust gratings. The inset numbers give estimated fluxes (photons/sec) at the sample for SPEAR running 100 mA at 3 GeV (From Ref. 10.)

The Locust is a monochromator designed to provide a 1-mm beam at the sample, with high flux and high resolution over the broad energy range from 10 to 1000 eV. Using this monochromator, concurrent experiments are possible on the same sample, investigating both valence band phenomena, typically examined at low photon energies, and core-level phenomena, typically observed or examined at higher photon energies. Because of the monochromator's wide scanning range, surface EXAFS can also be performed. The optimized extended range is one of the unique aspects of the Locust monochromator. Monochromators such as the Locust at SSRL, optimized for the new sources described earlier, will operate at sufficient resolution that intrinsic linewidths will be observed.

Section 2 has provided the requisite background to synchrotron radiation generation and monochromatization and characterized the various spectral ranges. Section 3 discusses surface-science studies using synchrotron radiation in a general way and with reference to specific techniques.

3. SURFACE-SCIENCE STUDIES WITH SYNCHROTRON RADIATION

3.1. The Domain of Surface Science

Surface and interface physics[25] explores the phenomena arising on the boundaries of solid volumes in free space or at the joining of different phases. In the case of a surface, the focus is the boundary layer between a solid and free space. In the case of an interface, the focus is a phase boundary between two solid volumes. The distinction between a surface and an interface is somewhat subtle, and this definition is consistent with current usage.

The surface atomic positions for periodic condensed matter typically differ from where they would be based upon a continuation of the bulk structure of the solid. The change in surface configuration has a significant effect on the surface's electronic structure, and new surface states or interface states arise due to the truncation of a lattice at a surface or interface. Specific examples of this are the numerous spatial reconstructions or relaxations that arise on metals and on clean semiconductor surfaces. In almost all cases, as the surface or interface establishes a minimum energy configuration, surface and interface states arise in a self-consistent manner with the atomic rearrangement.

Surface studies draw together a number of threads which intertwine because of the coupled nature of the atomic and electronic structures. A major objective of surface-physics work has been to determine the atomic and electronic structure of native surfaces and then to understand the interactions with adsorbates at the various stages—physisorption, chemisorption, and chemical reaction—which lead to the development of an interface. The quantitative characterization[26] of surfaces or interfaces and their structure requires the determination of the coordinates of the near-surface atoms. In most cases, surface structures relax into the bulk structures in three to four atomic layers. A wide range of phenomena specific to surfaces occur and have been the object of detailed studies.[27–30]

The advent of functional synchrotron radiation experiments resulted in the exploration of surfaces with new techniques. These techniques allowed quantitative results to be obtained directly about both the atomic and the electronic structures of surfaces. In this section these new techniques are described and examples given of their application to surface-science studies.

A brief description of experimental equipment used to perform surface-physics experiments will help the reader understand its role in advancing the science. The experimental apparatus required to perform surface-physics experiments is typically elaborate. Figure 7 represents schematically the wide variety of probes used to study surfaces, with synchrotron radiation as one of a suite of powerful tools now available.[31] In most cases, an apparatus combines several techniques in one chamber so that a concerted analysis can be made on the same surface. The left side of Fig. 7 depicts the various sources for beams, which often pass through a monochromator to define the energy before the incident beam impinges on the sample surface. The surface is prepared in a suitable apparatus, which usually has a number of tools for characterizing the vacuum ambient, preparing an atomically clean surface and studying the surface with tools such as Auger electron spectroscopy[26] (AES) or low-energy electron diffraction[32] (LEED). The outgoing signal from the surface can consist of a variety of particles, and a broad array of detector technology is available and specialized for particular measurements. In some cases the characterization tool is also a primary experimental method. For example, LEED, in addition to displaying periodicity, can be used for quantitative structural analysis through I–V analysis of the diffracted beams.

Figure 8 depicts the various components in a state-of-the-art experimental system for a synchrotron radiation study, a system which has the objective of preparing an appropriate sample surface, delivering the required photon flux to

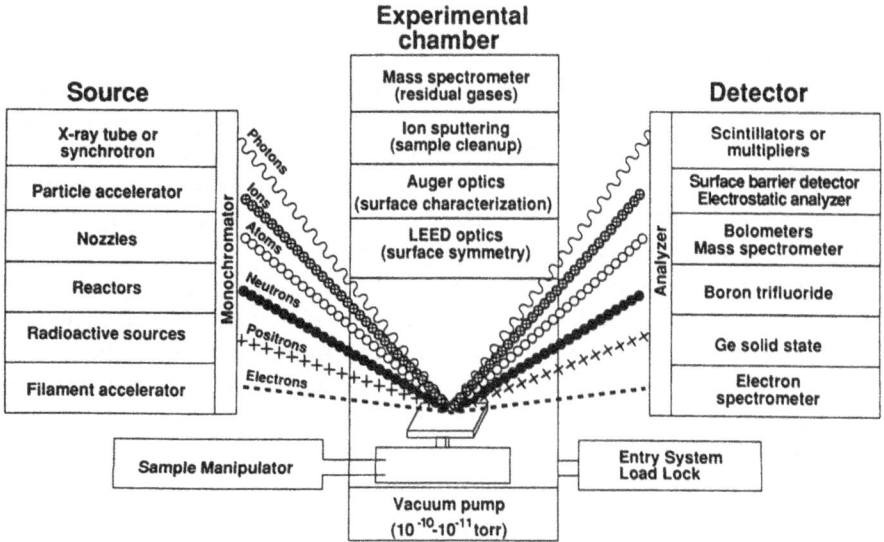

FIGURE 7. Probes used in surface-physics experiments. (Adapted from Ref. 31.)

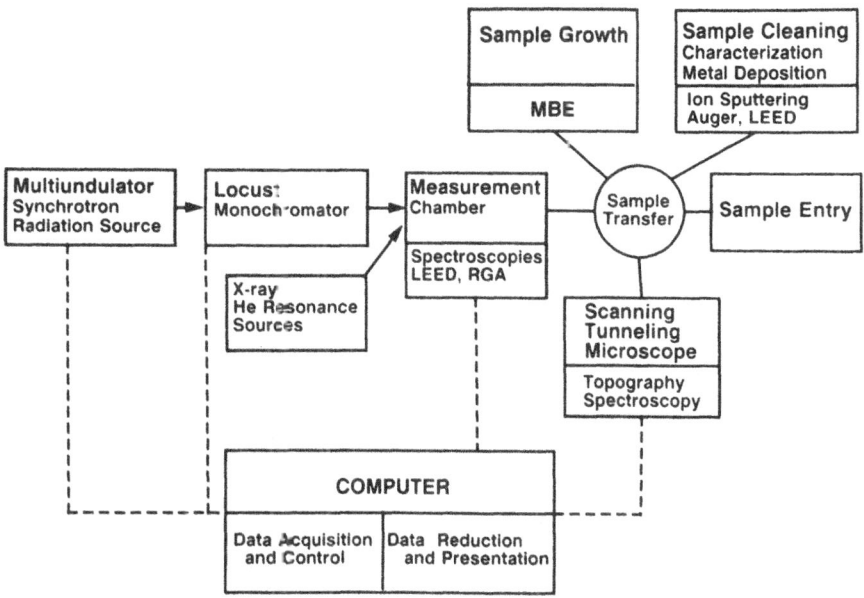

FIGURE 8. SSRL beam line V experimental system. The small-volume sample entry includes a vacuum load lock which enables samples to be brought in and out of the vacuum chamber without degrading the vacuum in the primary chambers. Although sample cleaning could be done in the growth or measurement chamber, advantages accrue from the separation. Computerization is a key component in the overall design.

the sample, and performing the analytical experiment.[33] Figure 9 shows a photograph of a modern multifunction system developed by Bringans and collaborators which embodies the components outlined in the previous figure. This particular system has a large number of *in situ* capabilities for sample preparation and for surface analysis with synchrotron radiation and conventional sources. Such ability to grow samples *in situ* is particularly useful for investigating materials where a clean surface cannot be prepared by cleavage, fracture, or other methods, e.g. the polar surfaces and interfaces of semiconductors which cannot be formed by cleavage, as discussed in Chapter 4 of Volume 2 of this set. This type of apparatus, an example of the current state-of-the-art approach to materials studies, is becoming more widely used for surface and interface studies.

3.2. Application of Synchrotron Radiation to Surface Studies

The application of synchrotron radiation to the study of surfaces has enabled a broad range of techniques to be developed. The primary surface-related measurement techniques using synchrotron radiation are angle-integrated photoemission, angle resolved photoemission, surface absorption, surface EXAFS (extended X-ray absorption fine structure), and surface diffraction measurements. These techniques, individually or in concert, reveal a variety of information about the physics and chemistry of surfaces.

There are several advantages of synchrotron radiation as compared to other radiation sources. Most of these advantages impact the various experimental

FIGURE 9. A multifunction system developed in connection with beamline V incorporates a central transfer chamber that allows samples to be inserted through a vacuum load lock (in the back and obscured) and then moved between a set of chambers. On the right is an MBE chamber in which samples can be grown and surfaces prepared. On the left is a surface-analysis chamber incorporating an angle-resolved hemispherical photoemission analyzer, a double-pass cylindrical mirror analyzer, rear-view LEED, RGA, a beam port for synchrotron radiation, an X-ray source, and a helium resonance lamp. The main vacuum pump is in the middle. A scanning tunneling microscope chamber (not shown) attaches to the valve in the middle.

modes presented in the remainder of this book. One primary advantage is the improvement in signal-to-noise ratio achieved because of the greater intensity of the synchrotron-radiation beam compared to that obtained from other radiation sources. This has facilitated working with native surfaces and low coverages of adsorbates. A second advantage due to the large photon flux is the ability to perform photoemission measurements with very high energy resolution and angular resolution. The intrinsic polarization of synchrotron radiation is often useful and has been significantly exploited in angle-resolved photoemission experiments in numerous cases.

Another significant advantage arises from the tunability of the photon energy. This has not only made possible surface absorption spectroscopies, but

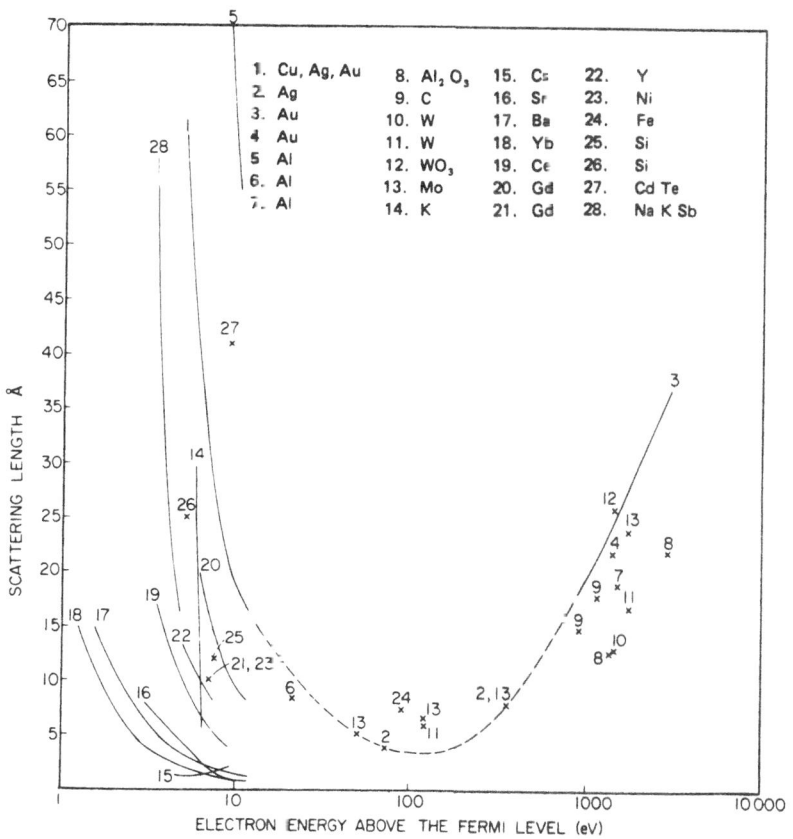

FIGURE 10. Escape depth as a function of electron final-state energy for various materials. The escape depth, in Angstroms, is shown as a function of the electron energy above the Fermi level, in electron volts. (From Ref. 34.)

has also allowed the tuning of photoemission final state energies for achieving optimal surface sensitivities by minimizing escape depths. The opportunities for escape-depth tuning were first pointed out by W. E. Spicer and I. Lindau,[34] and this effect has played a key role in many experiments. Figure 10 depicts the range of escape depths for various materials as a function of electron final-state energy. Though the measurement and interpretation of escape depths is complicated, the relevant trends are clearly seen in the figure. For final-state energies in the 100 eV range, the escape depth is usually less than 0.5 nm for a wide range of materials. An example of the use of the escape depth effect is discussed below. Tunability is also a key aspect of performing angle resolved photoemission and provides an important mode for photoelectron diffraction.

All these experimental techniques to be introduced below have in common an initial state interacting with a photon and then a detected final state containing photons, electrons, or ions. The book by Margaritondo discusses some of these techniques in further detail.[5] Most surface-related work to date has involved electron or ion final states. The major class of surface-related experiments with photon final states is grazing incidence X-ray diffraction, discussed in Chapter 8.

Very little work has been done with synchrotron-radiation-excited soft X-ray luminescence, although as discussed in Section 3.2.2, luminescence has been used to monitor the absorption process. In the following sections examples will be given of experiments with electron, photon, and atomic or ionic final states.

3.2.1. Experiments with Electron Final States

Surface Absorption. Though there are many methods of measuring photon absorption, variations of electron yield methods are the most effective in achieving surface sensitivity. Variations of electron yield techniques have been the most widely applied. Therefore surface absorption is included in the section on electron final states. Because of the wide excitation range and the consequent energy of the electron final state, surface absorption spectra divide into a number of regions. Chapters 2 and 3 expand on this subject in the context of near-edge and surface EXAFS, respectively. Chapter 4 discusses core excitons.

An absorption measurement determines the attenuation of light, which is related to the absorption coefficient and the imaginary part of the dielectric constant. These are primary parameters related to the electronic structure of matter.[35-37] The absorption coefficient $\alpha = 4\pi K/\lambda$ in cm^{-1} is related to the imaginary part of the dielectric constant by $\varepsilon_2 = 2nK = n\alpha\lambda/2\pi$. In the soft X-ray region, the real part of the dielectric constant is close to 1, so the absorption coefficient is essentially the imaginary part of the dielectric constant.

The most direct way to measure the absorption coefficient is with transmission measurements of samples of different thickness. Because α is on the order of 10^{-6} cm^{-1}, very thin samples are required. Although reflectivity measurements can be performed, this is typically difficult in the soft X-ray range because n deviates from 1 by only small values and the measurements are often dominated by scattering as discussed below under the heading "Absorption, Reflection, Scattering, and Luminescence."

Direct absorption measurements in the X-ray range do not work for surfaces, and therefore the yield techniques are used. In many cases, one is interested only in the spectral form of the absorption coefficient as opposed to its absolute value. In these cases one can use a yield technique originally described by Gudat and Kunz[38] and subsequently employed by many investigators.[39,40] This depends upon the fact that the absorption process creates an energetic core hole. The decay yield of the core hole either as photon- or electron-emission events provides a signal proportional to the absorption. The yield methods have the advantage that they can be quite surface- or interface-sensitive. The usual technique involves a yield measurement where secondary electrons, Auger electrons, or fluorescent photons are detected. A book entitled *X-ray Absorption* examines the various aspects of this technique in detail.[41]

Figure 11a, adapted from Fig. 2 in Chapter 3, depicts a typical configuration for an absorption experiment, showing the various components described in the caption. Figure 11b shows various aspects of a yield measurement and presents a schematic diagram depicting how the partial photoemission yield probes unoccupied surface states.[42,39] The top panel shows the density of states $D(E)$

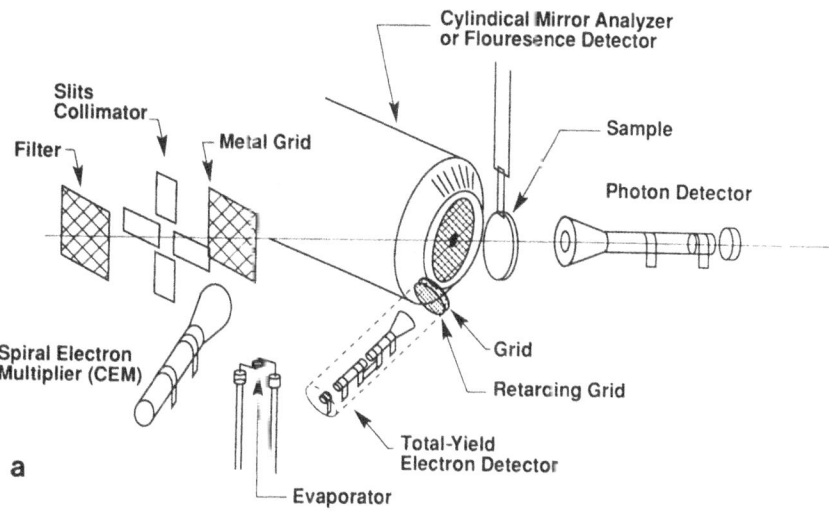

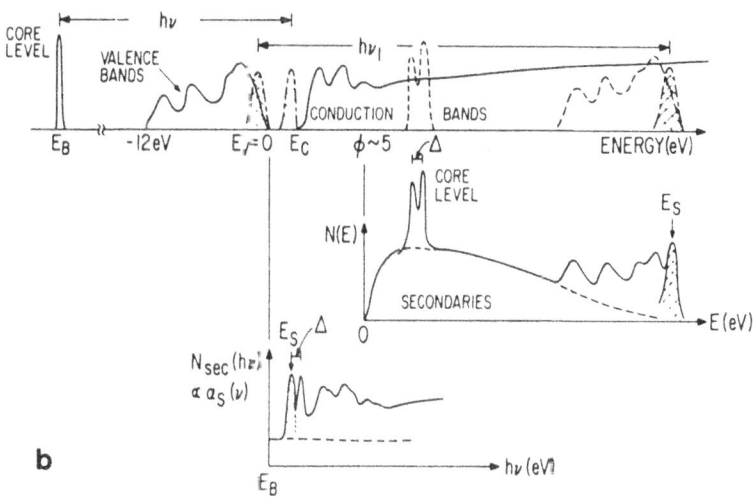

FIGURE 11. (a) Configuration of components for typical surface-absorption experiments with soft X-rays. The beam first passes through a filter and is then collimated by slits. A wire grid monitors the initial intensity either directly with absorbed current or with an electron multiplier. The grid surface can be refreshed by evaporating clean metal. The beam then impinges on the sample where, depending upon the configuration, various yield modes or transmissions are detected. Depicted are a total-yield electron detector and partial-yield detection for electrons or fluorescence. (b) How partial-yield photoemission probes unoccupied surface states. The top panel shows the density of states $D(E)$ including a core level (shown shaded) bound by energy E_B and empty surface states (shown cross-hatched) around the conduction band minimum E_C. The middle panel has the secondary-electron part of the photoemitted electron-energy distribution, which occurs at kinetic energies well below the photon energies $h\nu$ shown. The variation in the strength of this secondary emission $N_{sec}(E^*, h\nu)$ with changing $h\nu$ contains structure related to the empty surface-state density, as seen in the bottom panel. Direct core electron–emission limits the useful energy range to the work function ϕ plus the chosen constant kinetic energy E^*. (From Ref. 39.)

including a core level bound by energy E_B and empty surface states E_C (shown shaded) around the conduction band minimum. The middle panel has the secondary-electron part of the photoemitted electron-energy distribution, which occurs at kinetic energies well below the photon energy $h\nu$ shown. The variation in the strength of this secondary emission $N_{sec}(E^*, h\nu)$ with changing $h\nu$ contains structure related to the empty surface-state density, as seen in the bottom panel. Direct core-electron emission limits the useful energy range to the work function Φ plus the chosen kinetic energy E^*.

The X-ray absorption spectra have a number of spectral regions with varying degrees of structural information embedded in the signal. The structural information arises from a final-state scattering effect which modulates the transition matrix element. Some of the early surface absorption work by Bachrach and collaborators[43] exploited some of these advantages and has grown into a quite large field of endeavor which often goes under the acronym XANES or X-ray absorption near-edge structure.[44] This region of the spectra ranges from the threshold to energies typically less than 50 eV above threshold.

One clear example of surface absorption has been the extension of the EXAFS technique to surfaces, as discussed principally in Chapter 3 and in Chapter 5 of Volume 2 of this set. This has provided a tool which yields precise interatomic distances. Coupled with detailed analysis, information about bonding environments is also obtained. Other structurally sensitive techniques have also evolved. Surface EXAFS begins above the near-edge region and extends as far as possible. Surface EXAFS has proven a powerful tool for extracting bond lengths and deducing bonding sites for adsorbates.[45,46] Structural measurements with precise determination have become possible with surface EXAFS. The analysis of such data is less involved than LEED and the measurements in some specific cases have complemented the LEED structural analysis.

Figure 12 represents an example of absorption yield spectra measured in studying the oxidation of silicon,[47] which is discussed in more detail in Section 4.2. The important point to notice is that significant signatures related to the chemical structural environment are readily discerned. Figure 13 shows an application of the yield technique for determining empty surface states.[42] In this

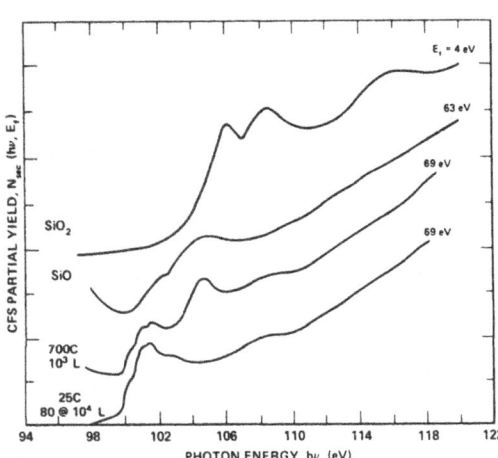

FIGURE 12. Spectral distribution for the yield of photoemitted secondary electrons measured within 3 eV at the indicated constant final-state energy E_f. Above the Si 2p absorption edge, resonances uniquely characteristic of SiO_2 (at 106, 108, and 115 eV) and SiO (at 104.5 eV) can be used as fingerprints for unknown local oxide stoichiometry (i.e., silicon monoxide containing negligible SiO_2 is present only for high temperature growth). (From Ref. 47.)

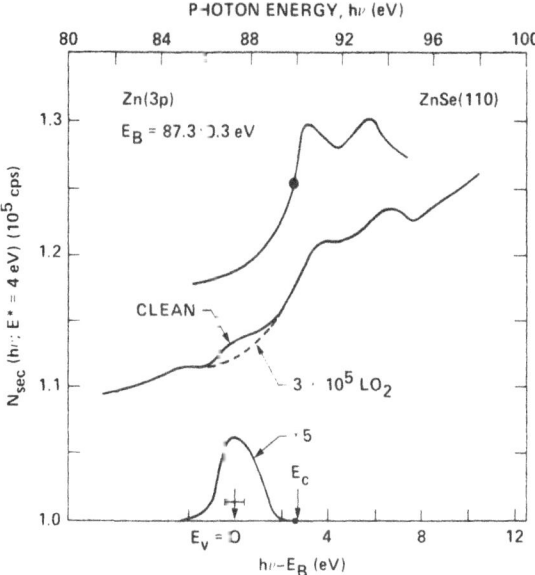

FIGURE 13. Partial yield for Zn 3p core electrons in ZnSe (110). The bulk absorption coefficient for this $M_{2,3}$ edge is shown for reference. The edge position is indicated. The empty surface-state density is shown by the difference curve between a clean and heavily oxidized ZnSe (110) surface (shown dashed where it differs from the clean spectrum). The lower scale referencing the yield to the valence band maximum E_v was set using our photoemission measurement of the Zn 3p core binding energy E_B. (From Ref. 42.)

case a differential measurement is made where the surface absorption is measured just after cleaving a surface and then after the surface is oxidized. The difference spectrum is related to surface states. Another example was the discovery of unoccupied surface resonances on single-crystal aluminum surfaces.[48] An example of these spectra for aluminum (100) and (111) surfaces is presented in Section 4.2.1.

Core and surface excitons are discussed by F. C. Brown in Chapter 2 of Volume 2. The study of surface excitons is likely to be revisited as stronger synchrotron-radiation sources become available and higher resolution monochromators eliminate some of the instrumental uncertainties arising from deconvoluting data. One novel attempt at studying surface excitons on Si used electroreflectance techniques.[49] Although the experiment yielded a null result in that no direct signal was detected for Si 2p at 100 eV while the sample was shown to be good at visible energies, a lower bound on the exciton binding energy was determined. Most theoretical analyses of the core exciton problem have found the final states to be close to effective-mass like, so there is a discrepancy between some of the experimental measurements that indicate large binding energies and the theoretical calculations.

Photoemission from Surfaces. Photoemission measures the kinetic-energy distribution of electrons emanating from a surface upon the absorption of light. Photoemission work encompasses a large and diverse body of experiments and literature, which has evolved over decades.[50] Photoemission has had a distinguished place in modern physics and is currently a widely applied tool. Photoemission work with synchrotron radiation prior to 1980 is reviewed by Lindau and Spicer.[34] Smith and Himpsel[51] have also provided a good account of work prior to 1984. The books of Ley and Cardona[52] and Feurerbacher, Fitton, and Willis[53] among others offer good descriptions of photoemission techniques and many case studies.

Theories of photoemission[54,55] have been extensively developed, but in most cases model examples invoking approximate solutions have been necessary.[56] Most of these agree that angle-integrated photoemission reveals the occupied density of states, but that cross-section, resonance effects, and other final-state effects must be taken into account. The three-step model is often used to represent the photoemission process.[57] Figure 14a schematically shows the photoemission process, which consists of the optical transition, transport to the surface, and transmission through the surface. On the left is the occupied density of states depicting the broad valence band and a narrower core level. The bands have different orbital composition, so that the photoelectron distribution represented on the right is modulated by the relative cross sections and in many circumstances by final-state effects.[58,59] The kinetic energy of the outgoing electrons is given by

$$E_{kin} = h\nu - E_b - \phi$$

where $h\nu$ is the photon energy, E_b the electron binding energy, and ϕ the work function.

Understanding the mechanisms of photoemission is the key to applying the technique to the study of materials. For example, although the three-step model

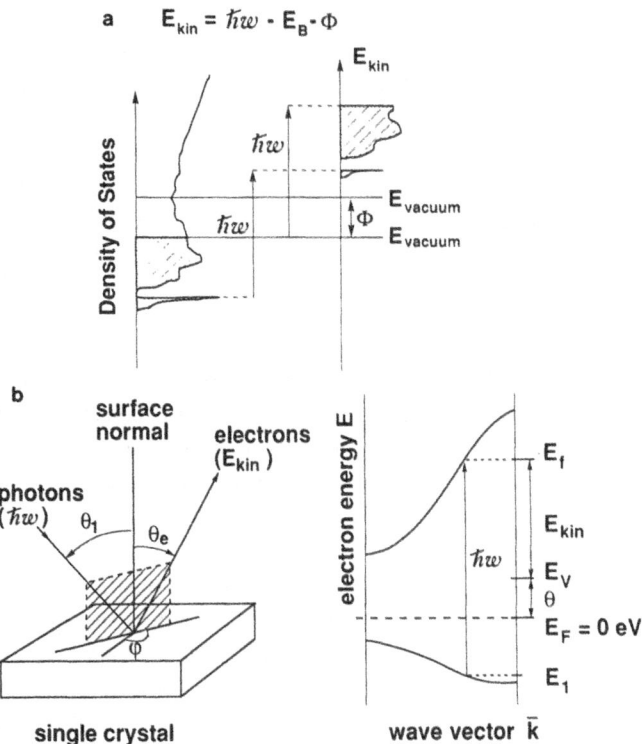

FIGURE 14. (a) Illustration of the relationship between the occupied electronic density of states (left) and the photoemitted electron kinetic-energy distribution (right) (From Ref. 53.) (b) Illustration of angle-resolved photoemission (From Ref. 141.)

is widely used for interpreting data, it has important limitations. One particularly interesting subject is that of the surface photoeffect. The polarization and variable excitation energy, have made synchrotron radiation very effective in clarifying this phenomenon.[60–62]

Within the photoemission category, the variable photon energy allows data to be collected in several modes, as described by Lapeyre,[63] which have come to be designated as the energy distribution curve (EDC), constant final state (CFS), and constant initial state (CIS). Photoemission investigations partition into valence band and core level studies. An important aspect of these works is the cross-section dependence discussed in Chapter 1 of Volume 2 by Lindau. As discussed by Rossi in Chapter 5 of Volume 2 of this set, this cross-section energy dependence allows important elemental discriminations to be made in multi-element materials.[64] In special situations, resonance effects occur which can be exploited, as discussed in Chapter 6 by Allen. The tunability of the excitation energy is a key factor in these experiments. The ability to vary the photoemission final-state energy has allowed for the adjustment of escape depths to maximize surface contrast.

A major extension of the photoemission method is measurement of the angular distribution in addition to the energy distribution.[51,65–67] Angle-resolved photoemission is discussed extensively in Chapter 4 by Eberhardt, and a particular application to semiconductors is presented in Chapter 4 of Volume 2 of this set. The possibility of using angle-resolved photoemission to gather unique information about crystalline materials and surfaces evolved slowly at first,[68–70] and then after a decade of gestation exploded in the seventies due to a number of instrumentation advances, including the utilization of synchrotron radiation.

The technique of angle-resolved photoemission depends upon the fact that the parallel component of the momentum is conserved in electron transmission through an ordered surface. The data can be analyzed directly to yield energy–momentum spectra for surface bands. Figure 14b depicts the arrangement. Even though the normal component is not strictly conserved, in many cases the work function and inner potential can be determined well enough so that quantitative use can be made, even for determining bulk bands.[71]

Several extensions of photoemission techniques have been made to obtain structural information. One such extension uses normal angle-resolved photoemission as a function of excitation energy to obtain diffraction information. Specific examples are presented in Chapter 9, by Fadley. Another method described in the chapter by Rowe obtains EXAFS information, which can be analyzed to obtain bond lengths.

A key component of photoemission measurements is the electron-energy analyzer. A vast technology now exists which optimizes for each particular combination of throughput, energy resolution, and angular resolution.[72] Additionally, instruments which provide area images have been developed, and throughput is being enhanced so that time-resolved measurements can be made.

The three principal modes of electron-energy analyzers are magnetic, electrostatic, and time of flight. Time-of-flight analysis is particularly well suited for use with pulsed synchrotron sources.[73] The primary use to date, however, has been in gas-phase spectroscopy.[74] Figure 15a shows a schematic of the simple

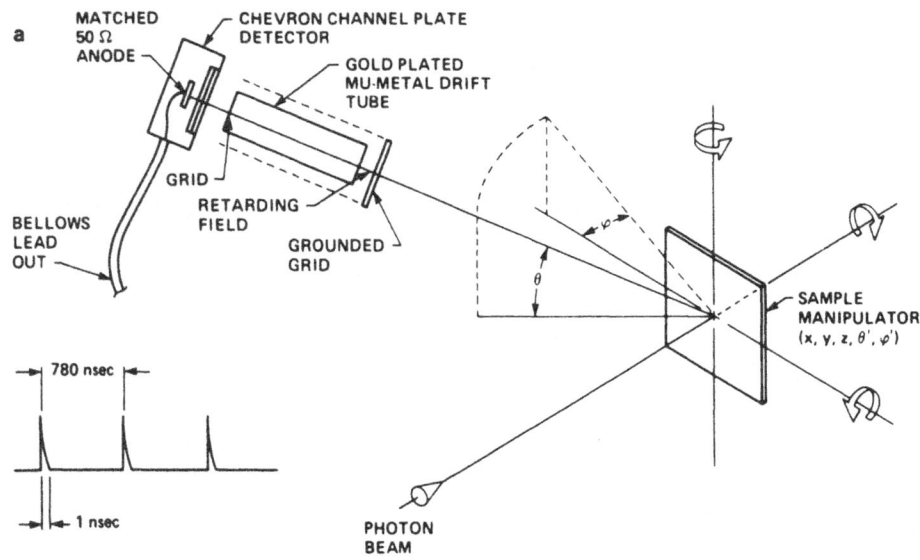

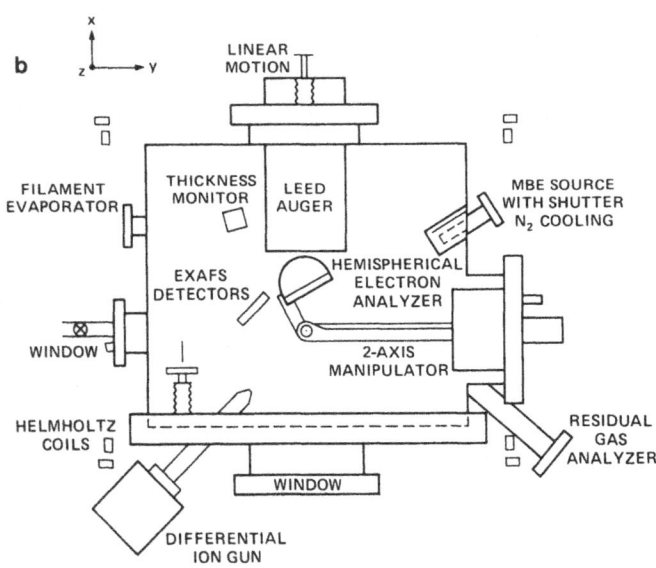

FIGURE 15. (a) Schematic of an electron time-of-flight apparatus used for measuring angle-resolved photoemission. The pulsed X-ray beam generates photoemitted electrons, which can be retarded and then traverse a field-free drift region before being detected by a fast multiplier. In the case depicted, the detector can and the bellows with the cables inside are at atmospheric pressure. The total asembly is pivoted around the sample on a two-axis manipulator. (From Ref. 73.) (b) Angle-resolved appartus using a 2-cm hemispherical electrostatic analyzer on a two-axis manipulator installed in a multifunction surface-analysis chamber. Among other functions, surfaces can be grown with molecular beam epitaxy. (From Ref. 75.)

apparatus, which is capable of very high energy and angular resolution while simultaneously detecting all energies. One would anticipate that this technique will see more use in the future at dedicated sources. Figure 15b shows an example of a typical system incorporating a small hemispherical energy analyzer for angle-resolved measurements.[75] Such measurements are discussed in detail in Chapter 5 by Eberhardt and by Bringans and Bachrach in Chapter 4 of Volume 2 of this set.

Photoelectron Diffraction. Photoelectron diffraction is discussed extensively by Fadley in Chapter 9. A major positive feature of photoelectron diffraction is the relatively simple theoretical framework which can be used to analyze the data and deduce surface structures.

In photoelectron diffraction, intensities of emitted core-level electrons are monitored as a function of either the emission direction or the photon energy. The three basic types of measurements include an azimuthal scan, a polar scan, and a scan of the photon energy in a normal or off-normal geometry. With soft X-ray excitation, scanned-angle measurements have been termed X-ray photoelectron diffraction. Scanned-energy measurements scanning the vuv-to-soft-X-ray regime have sometimes been similarly called normal or off-normal photoelectron diffraction and have sometimes been called angle-resolved photoemission fine structure, in order to emphasize their similarity to the more familiar EXAFS. The percentage modulation of intensity observed in these two classes of experiments can be as large as 50%, so the effects are easily observable.

3.2.2. Experiments with Photon Final States

Absorption, Reflection, Scattering, and Luminescence. A number of measurements make use of the forward absorption, transmission, or scattering of photons, but in the soft and hard X-ray range, these techniques are usually not very surface sensitive.[36,37]

Absorption measurements done in transmission mode, where the attenuation of the X-ray beam passing through the sample is measured, require thin samples and usually probe bulk properties. This contrasts with the yield technique, as discussed in Section 3.2.1, which can work with arbitrarily thick samples and where the surface sensitivity is determined by the escape depth of the electrons.

An exception, where X-ray absorption can probe surface properties, is the EXAFS work done on exfoliated graphite, which has an unusually large surface-to-bulk ratio. This approach has been used to look at gases physisorbed to the graphite.[76] Chapter 3, by Rowe, describes surface EXAFS in detail as well as introducing the subject of EXAFS. Other extensive reviews and books exist in the literature.[77-80]

Reflection measurements are very difficult in the X-ray range because the index of refraction is very close to 1 and signals become dominated by scattering and stray light effects. Some enhancement can be achieved by working close to the total external reflection regime.

Scattering measurements have been useful for studies related to surface roughness, but these morphological effects are outside the scope of this book.[81]

Surface-related luminescence measurements have been made principally in conjunction with a yield technique for measuring the surface absorption of an adsorbed species.[82,83] The branching ratio between photon and electron final states is small for light elements, but this is compensated for by the decreased background, so that good signal-to-noise ratio can be achieved. The luminescence yield approach has some significant advantages over electron yield in cases where measurements can be made. More details are given in the chapter by Rowe.

Grazing Incidence X-ray Diffraction. Chapter 8 by Fuoss, Liang, and Eisenberger describes in detail the scope and status of grazing-incidence diffraction and scattering techniques, which are evolving as important tools for studying surfaces. The advantage of these techniques is that in cases where the measurements can be made, the interpretation is straightforward compared with other techniques such as (LEED). The application of X-ray diffraction to surface studies requires measurements at grazing incidence where total external reflection enhances the signal. The intensity and collimation of the synchrotron radiation are then important attributes which facilitate these measurements. The use of grazing-incidence X-ray diffraction is under rapid development due in large part to the development of very high intensity X-ray beam lines. A recent review by Feidenhans'l[84] complements the chapter by Fuoss *et al.*

X-Ray Standing Wave Measurements. X-ray standing wave measurements are performed by the creation of an X-ray standing wave by Bragg scattering near the surface of the sample. This requires a highly perfect surface, but if this is possible, important crystallographic information can be achieved.[85]

The X-ray standing wave is created by interference between the incident and Bragg reflected wave, resulting in a nonpropagating wave in the direction perpendicular to the family of planes which corresponds to the Bragg reflection. The period of the standing wave is determined by the distance between the crystallographic planes. By changing either the angle of incidence or the energy of the beam, the positions of the maximum-amplitude planes of the standing wave with respect to the crystallographic planes can be varied. If the surface contains chemically unique sites or impurities, the standing wave can modulate X-ray fluorescence from these sites and the maximum in amplitudes measures the position of the unique atoms. In this way it is possible to identify positions of minority atoms with respect to the crystallographic plane.

The high collimation and energy resolution achievable with synchrotron radiation is important for facilatating standing-wave measurements. With synchrotron radiation, the standing-wave measurement is able to determine atomic positions with an accuracy on the order of 10^{-2} of the distance between crystallographic planes.

Artificially Structured Materials and Multilayer Optics. The creation of artificially structured materials has resulted from advances over the last twenty years in vacuum techniques and well-controlled deposition processes. Some examples are molecular beam epitaxy,[86] metallo-organic chemical-vapor deposition,[87] and well-controlled sputtering systems.[88] These deposition methods are able to create structures of semiconductors, metals, and insulators with interfaces often controlled to within one to two atomic layers and thicknesses on the same scale.

Artificially structured materials include metastable arrangements of material phases organized to accentuate a desired property that would better facilitate some desired application. These materials range from single-interface situations to artificially imposed periodicities ranging from atomic scale to scales on the order of optical and X-ray frequencies. Artificially structured materials exploit these imposed periodicities to create resonances that can enhance nonlinearities important to many applications. Such materials inevitably involve interfaces which might be metastable as well as the phases which are themselves metastable. The study of interface structure is important to this field.

The topic of artifically structured materials spans in two directions both because such materials are an object of study in interface science and because the structures themselves allow particular measurements to be made and optics with unique characteristics to be created.

Because the artificial periodicity can be accurately measured with an electron microscope, measurements made on these structures can be used to deduce information about the interfaces and the optical properties of the materials. A materials-study example is given below in discussing GaAs on silicon, and this is discussed in more detail by Bringans and Bachrach in Chapter 4 of Volume 2 of this set.

The multilayer optics created with artificially structured materials make use of interference.[89] These materials can also create the equivalent of artificial crystals for soft X-ray monochromators and typically consist of periods of low-Z and high-Z elements such as tungsten and carbon. The diffraction properties from such structures can be used to deduce interface properties. In this regard, they are an analogue of some of the unique characteristics of transmission gratings.

Superlattices for electronic applications are a prime example of artificially structured materials, where a vast literature now exists.[90,91] Some of these materials involve strain-layer superlattices, where suitable control of the individual layers overcomes some of the inherent problems that arise because of mismatches.

The GaAs-on-silicon system provides a representative example of an artificially structured material, although it is not used in X-ray optics. GaAs on Si is a current topic which is prototypical of the motivations for this type of effort and can provide insight into extensions into other systems, such as GaAs/ZnSe or Si/ZnSe. This extension is interesting because there is considerable interest is achieving a blue-light-emitting diode or laser and it is possible an artificially layered materials approach might provide one avenue for accomplishing this goal. GaAs on silicon is a heteroepitaxial system being actively investigated because of its technological potential for combining some of the best attributes of both materials. The 4% lattice mismatch between the two semiconductors makes formation of good-quality crystals a challenge. A principle objective in studying the GaAs-on-silicon system is combining GaAs electro-optic capability with silicon electronic capability. One hope is to be able to integrate lasers or LEDs onto a silicon chip either for interchip communication or for controlling GaAs laser arrays. Silicon substrates are also cheaper, stronger, and available in larger sizes than GaAs substrates so there are economic advantages as well.

Work from the last few years has led to a detailed understanding of Si and GaAs surfaces. Comparative studies of the low index faces of Si and GaAs show that radically different phenomena can be exhibited on different crystallographic faces. This work has then been extended to the heteroepitaxial growth of GaAs on Si and the inverse system of Si on GaAs. The GaAs-on-Si and Si-on-GaAs systems exemplify the lack of reciprocity often observed in the interface formation in strain-layered materials, depending upon the order of growth. The origin of the nonreciprocity is an important issue because it affects the ability to grow superlattices and multilayers. Progress in the GaAs-on-silicon system has been substantial, but the fundamental science is not yet well enough developed to determine whether or not the high dislocation levels are an intrinsic impediment to achieving material capable of the desired electro-optic goals. Much remains to be studied about highly strained systems.[92]

3.2.3. Experiments with Atomic or Ionic Final States

Photon-Stimulated Ion Desorption. Photon-stimulated ion desorption is discussed extensively in Chapter 7 by Rehn and Rosenberg. The process is a direct desorption phenomenon, where a single quantum photoabsorption event releases surface or adsorbed species in ionic, excited, or neutral form. PSD describes desorption induced by an electronic excitation. Thus PSD can be distinguished from excitation of molecular or crystalline vibrations, phonons, or conduction-band electrons. Among its unique characteristics, photon-stimulated ion desorption provides very good local-site selectivity and is very structurally sensitive. Provided neither the beam energy or density is too large, PSD is gentler and less disturbing than electron probes. Ions or neutral atoms desorbing from a surface at low kinetic energies are unable to displace other atoms blocking their path. Thus PSD of ions combines the unique properties of a photon beam probe with the extreme surface sensitivity of the low-energy desorbing ion.

An earlier review of photodesorption phenomena from metals and semiconductors was given by Lichtman and Shapira.[93]

4. SELECTED SURFACE-SCIENCE ADVANCES

In this section are surveyed a few selected problems that have been attacked with some of the synchrotron-radiation techniques introduced in Section 3. The selected topics relate either to developments or to particular conceptual advances. As outlined in Section 3, many of these advances were direct consequences of measurements made possible by the unique characteristics of synchrotron-radiation sources. What follows are some highlights, many of which will be described in greater detail in later chapters.

4.1. Surface Core Level Shifts

The discovery that both surface chemical shifts and native surface core level shifts could be resolved opened many opportunities for detailed surface studies. The first observations were those of Flodström, Bachrach, Bauer, and Hagstrom on oxygen-induced shifts of aluminum[94] and the clarification of native surface shifts by Eastman et al.[95] Careful high-resolution measurements of the core level binding energies complement polarization-dependent studies of the valence-band electronic structure. This type of measurement gives insight into the changes in electronic and structural properties during the growth of epitaxial layers. Chemical-shift studies of the core levels can be used to extract information on the charge transfer between the adatom and the substrate, i.e., about compound formation. The intensity of the core lines can be used to determine the stoichiometry of the formed compound and to examine the compositional gradients, the surface segregation, and the interdiffusion.[96]

Core level shifts provide an important way to monitor changes occurring during the initial stages of interface formation, whereby the surface and interface electronic structure change as the thickness of an overlayer increases from a fraction of a monolayer to several monolayers.[97] In general, the interaction between the overlayer and the substrate is very complex and depends strongly on the material system. One possibility is that a compound and/or alloy is formed between the substrate (one or both constituents in the case of a compound semiconductor) and the overlayer. In many cases a number of structure and stoichiometry-related reconstructions or surface phases occur.

Although the surface core level shift has proven to be an important signature for studying materials, in most cases it has been difficult to actually calculate the expected shifts for covalent and ionic materials,[98] and this field has required an empirical approach. One example is given below where the fluorination of silicon is discussed.

The possible examples for this type of research are now vast, so the selections presented below are meant only to be typical. Chapter 5 (by Flodström, Nyholm and Johansson) and Chapter 3 of Volume 2 of this set (by Hasse and Bradshaw) give a good projection of research problems that can be attacked by studying surface core level shifts. Spanjaard et al[99] and Egelhoff[100] have presented major reviews of the various aspects of these phenomena for both bulk and surface core level shifts in a number of contexts.

4.1.1. Al (111):O

Figure 16 shows the Al $2p$ core level shift induced by the chemisorption of oxygen on a (111) surface.[94,101] The oxidation of single-crystal aluminum surfaces is discussed in more detail in Section 4.2. The 1.4-eV-shifted core level is a signature for a distinct chemisorption stage that appears before the onset of the 2.7-eV chemical shift characteristic of the fully oxidized bulk phase. The oxidation of aluminum is discussed further in Section 4.2.1.

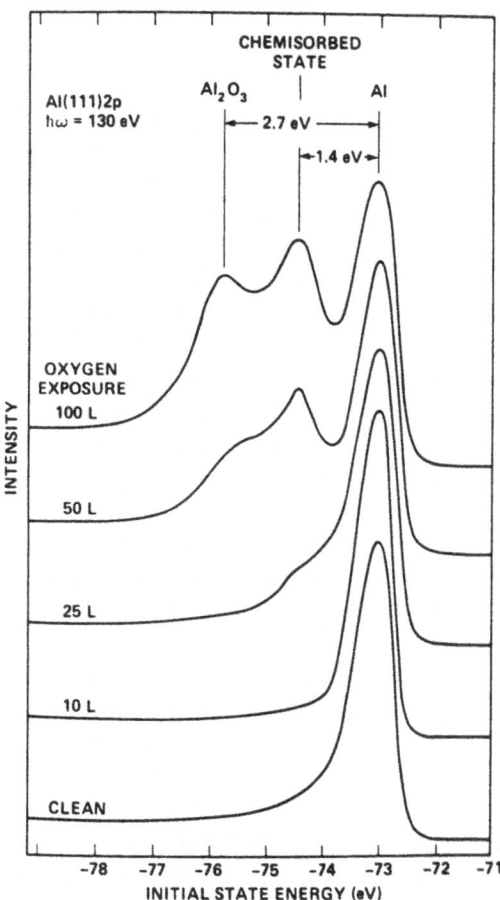

FIGURE 16. Photoemission spectra of the Al $2p$ region for an electro-polished and then sputter-annealed clean ordered Al (111) surface and surfaces exposed to different amounts of oxygen. The ability to observe the state at 1.4 eV is very sensitive to the perfection and ordering of the surface. (From Ref. 94.)

4.1.2. Si (111):As

One particularly striking example of surface core level shifts has resulted from studies of native silicon surfaces and then of the aresnic-terminated surface.[102] Figure 17 shows an example of the change in relative surface sensitivity achieved by tuning the final-state energy. Detecting a 4-eV final-state energy with 108 eV excitation energy results in a spectrum characteristic of bulk silicon. The peak shows a statistical spin–orbit splitting ratio of 2:1. Increasing the excitation energy to 130 eV so that the escape depth is near the minimum yields a radically different spectrum, as seen in Fig. 17.

Figure 17 also shows a further example. Arsenic adsorbed upon the complicated Si (111) 7 × 7 surface with the temperatures at 600 °C converts it to a near ideal 1 × 1 surface.[65,70] Analysis has shown that the As replaces the outer Si in the last double layer, resulting in a uniform site occupancy. This stable unreconstructed surface has served as a model surface for the comparison of theory and experiment. The spectrum in Fig. 17b has been deconvolved into a bulk component and the surface component. The fitting parameters obtained from this experiment have been used in analyzng the Si-on-GaAs and GaAs-on-Si experiments. The analysis has shown that the As sits in a unique site in the outer

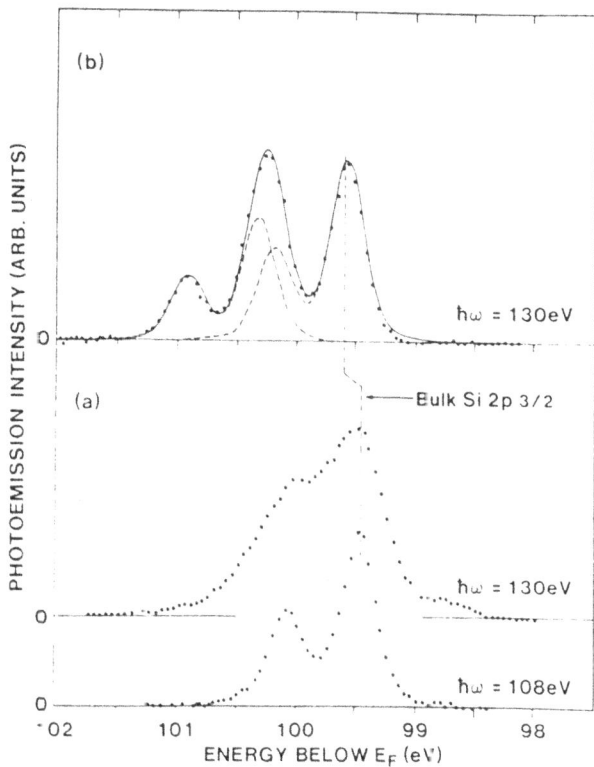

FIGURE 17. (a) Bulk-sensitive (hv = 130 eV) Si $2p$ spectra for the clean Si (111) 7 × 7 surface. (b) Surface-sensitive Si $2p$ spectrum for the Si (111):As 1 × 1 surface. The solid line is the sum of the interface and bulk components (dashed lines) obtained with fitting parameters. The spectrum intensities have been arbitrarily scaled with respect to one another after the subtraction of a quadratic background. (From Ref. 102.)

silicon double layer. The surface core level spectra are particularly compelling because the components are highly resolved.

4.1.3. GaAs (100) Surface Phases

One example of the application of surface core level phenomena to native surfaces has been the determination of the surface composition of GaAs surfaces. The possibility of variable stoichiometry results in a wide range of reconstructions on the GaAs (100) surface.[103] The large number of stable reconstructions is now known to be related to the As-to-Ga ratio in the outer atomic layers. MBE-grown GaAs[104] layers typically yields a $c(4 \times 4)$ structure when the surface is cooled to room temperature in the residual arsenic flux. Holding or annealing the surface at 300 °C will desorb any excess arsenic. Subsequent annealing of the surfaces in vacuum at temperatures ranging from 450 to 650 °C is used to prepare specific arsenic-deficient surface reconstructions observed by LEED, which follow a sequence of $c(4 \times 4)$, $c(2 \times 8)$, $c(8 \times 2)$, 1 × 6, and 4 × 6. The $c(4 \times 4)$ is stable to about 350 °C, the $c(2 \times 8)$ is obtained in the range 400–450 °C, the $c(8 \times 2)$ occurs in a very narrow range around 500 °C and is easy to miss, the 1 × 6 occurs between 500 and 550 °C, and the 4 × 6 can be obtained at temperatures above 600 °C.

Two approaches to determining the surface composition have been used. Figure 18a shows one example where the core levels were measured as a function of reconstruction. One can see both changes in the area ratio of the Ga and As 3d core levels and in the binding-energy separation. Figure 18b presents one description of the surface phase diagram as determined from core level photoemission.[103] Figure 18b plots the Ga-to-As 3d core level ratios versus As

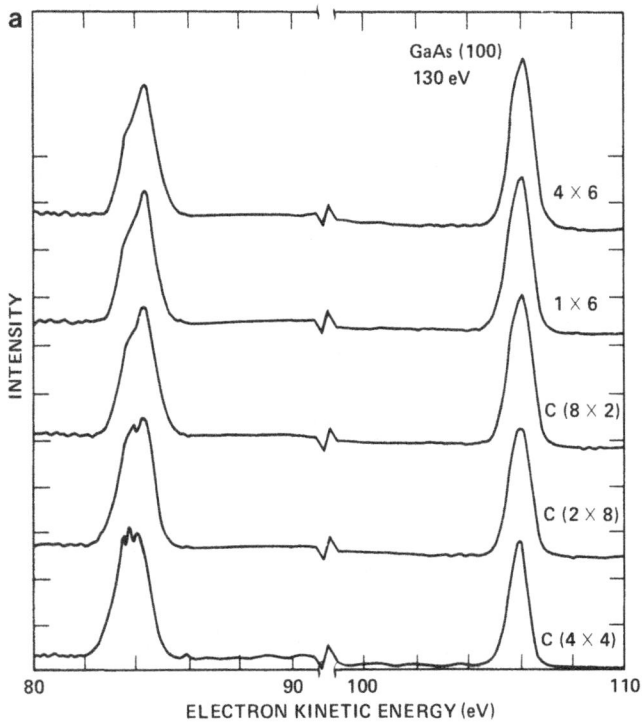

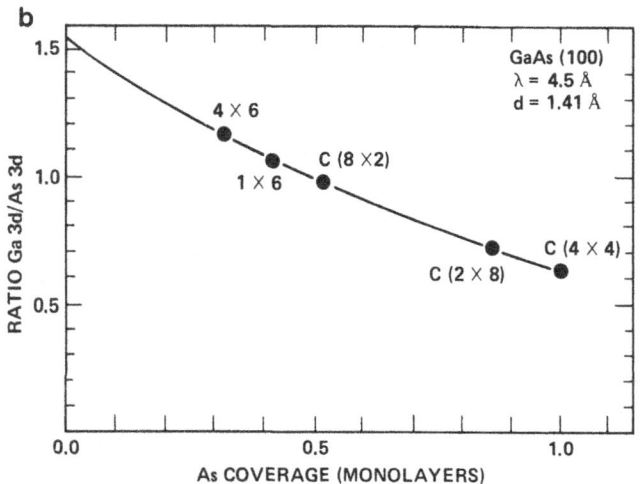

FIGURE 18. (a) GaAs (100) Ga and As 3d core level spectra measured at 130 eV as a function of reconstruction. The spectra are plotted as normalized to excitation intensity. Two effects are resolved as the reconstruction changes. One is the binding-energy difference between core lines and the other is the relative area ratio. (From Ref. 103.) (b) Surface phase diagram for GaAs (100) surface reconstructions obtained from the data in (a).

coverage. The scale was determined using an electron-yield model calibrated with respect to measurement on the stoichiometric (–10) surface. An alternative approach has resolved the surface and bulk core levels and then used a yield model to convert the ratio to a coverage scale.[105,106] Analysis of either approach requires a knowledge of the electron escape depth. Complete agreement on this phase diagram has not yet been achieved.

4.2. Insight into Oxidation and Fluorination

One of the first areas where significant application of synchrotron radiation to surface physics arose quickly and where the resolution of surface core level shifts has proved invaluable was the study of the oxidation of metals and semiconductors. The access to high-resolution core level spectroscopy with short escape depth revealed previously undetermined details of the process on a variety of surfaces. A natural first step was the investigation of adsorbate phenomena, and the methods used have evolved into an approach of broad utility. Corollary studies have examined related aspects such as fluorination. The broad range of new techniques has yielded exciting information. Chapter 3 in Volume 2 of this set, by Hasse and Bradshaw, presents in detail the adsorbate phase, and reference is made to the literature for more details on oxidation.

4.2.1. Oxidation of Aluminum Single-Crystal Surfaces

The literature on oxidation phenomena is vast,[107] but one specific example is the study of the oxidation of single-crystal aluminum[108] by the absorption and photoemission techniques introduced in the previous sections. The Al (111):O system in particular exposed a wide range of fascinating phenomena regarding the various stages of exposure to oxygen and provided a model system both for experimental exploration and for comparing theory with experiment.[109]

As discussed above, the resolution of surface core level shifts was first found in studies of the oxidation of aluminum.[94] Figure 19 shows the full Al $2p$ spectral region as a function of oxygen exposure. The inset shows the resolution of the Al $2p$ core level achieved for a clean surface with an excitation of 110 eV. The plasmon sideband is a prominent feature, and the surface plasmon attenuates even before the chemisorption state is observed on the core level. Similarly, the O $2p$ resonance in the valence region is seen at very low coverages. Detailed examination of valence-band spectra show that the oxygen is interacting with the surface even before core level shifts can be observed. As the exposure increases, a shifted core level is resolved. This is shown in more detail in Fig. 16. The 1.6-eV-shifted peak has been shown to result from an ordered chemisorption phase. The core level shift data have been extended with angle-resolved photoemission of the ordered Al (111):O chemisorbed state, and the surface energy bands have been determined.[110]

Surface absorption measurements have given details of the absorption process, an example of which is shown in Fig. 20.[111] Figure 20a compares the spectrum for clean aluminum surfaces with the spectrum for bulk aluminum, and Fig. 20b shows the effect of oxidation.

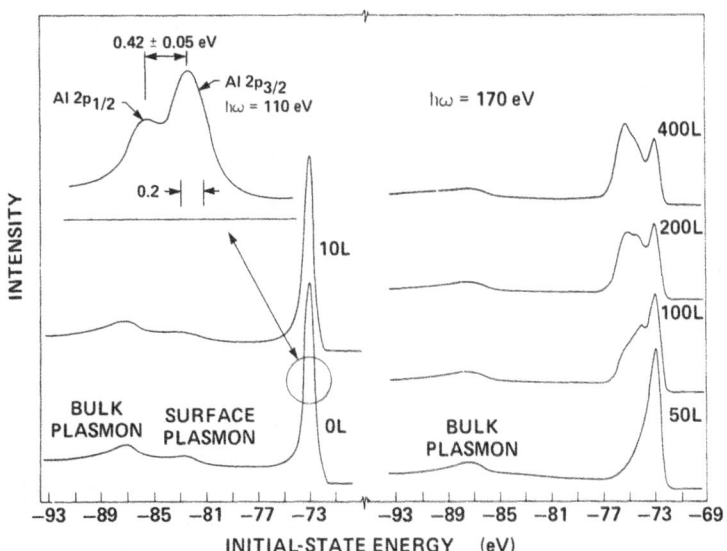

FIGURE 19. Photoelectron spectra of the Al $2p$ region for clean and oxygen-exposed aluminum films at a photon energy of 170 eV. The oxygen exposures are given in langmuirs (1 L = 10^{-6} Torr sec). The expansion shows the resolved Al $2p$ doublet obtained at a photon energy of 110 eV. (From Ref. 108.)

The investigation of this surface with absorption spectroscopies yielded additional information, and the surface was used in the development of surface EXAFS, which sought to determine the bonding height above the surface. This was done by measuring SEXAFS at the oxygen K edge.[112]

A short note by Bachrach et al[113] observing some unusual effects in the initial oxidation, as well as using surface EXAFS to measure the Al–O bond length for the chemisorption phase on the (111) surface, generated a large amount of activity on this subject. Although much critical focus was placed on what the bond length is, little attention has been paid to clarifying the oxidation stages identified. Identifying the stages better is still an area in which detailed study with probes providing higher resolution and intensity would likely yield interesting results. One aspect of this work is the identification of a surface resonance near the top of the valence band that has not yet been fully examined.[114]

4.2.2. Fluorination of Silicon

Fluorine etches silicon and is used extensively in semiconductor processing, so a fundamental understanding of the mechanisms of silicon fluorination is important. The application of synchrotron radiation techniques[115] to studying this system has created a deeper understanding of the intermediate chemical species and provided a basis for comparing theoretical modeling of the Si–F interactions.[116]

The investigation of the fluorination of silicon is another example of the application of core level surface shifts. The experiments consist of exposing Si surfaces to XeF_2 and then analyzing the surface. The spectra shown in Fig. 21

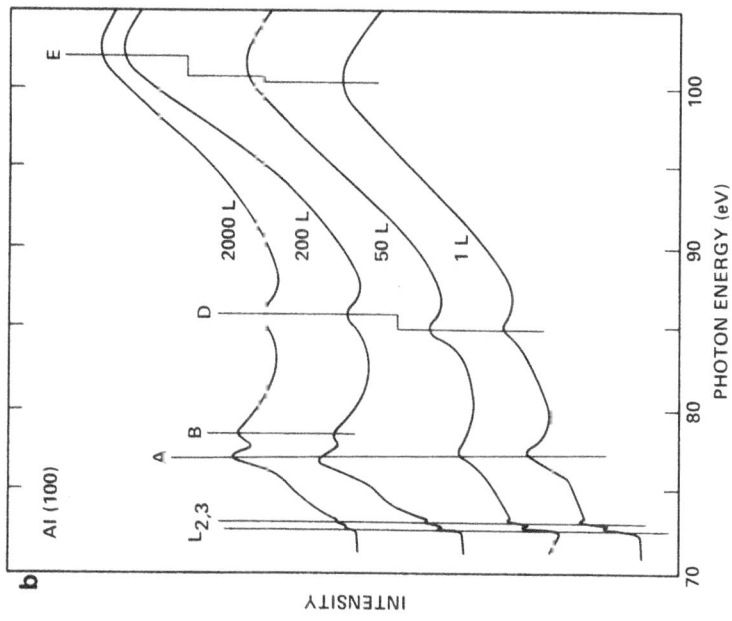

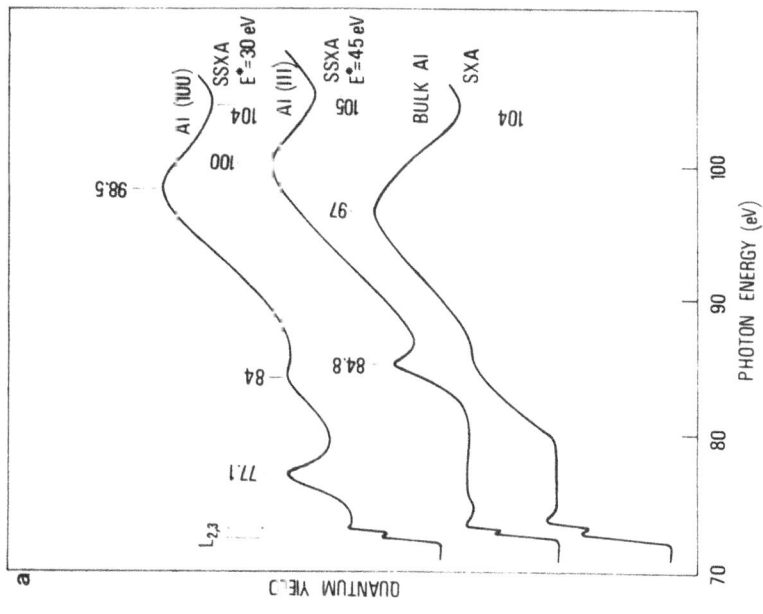

FIGURE 20. (a) Surface soft X-ray absorption spectra of Al (111) and al (100) surfaces compared in the lower part of the figure with the bulk soft X-ray absorption spectrum of a thick aluminum film for a wider spectral range. (b) Surface soft X-ray absorption spectra of Al (100) as a function of oxidation. (From Ref. 111.)

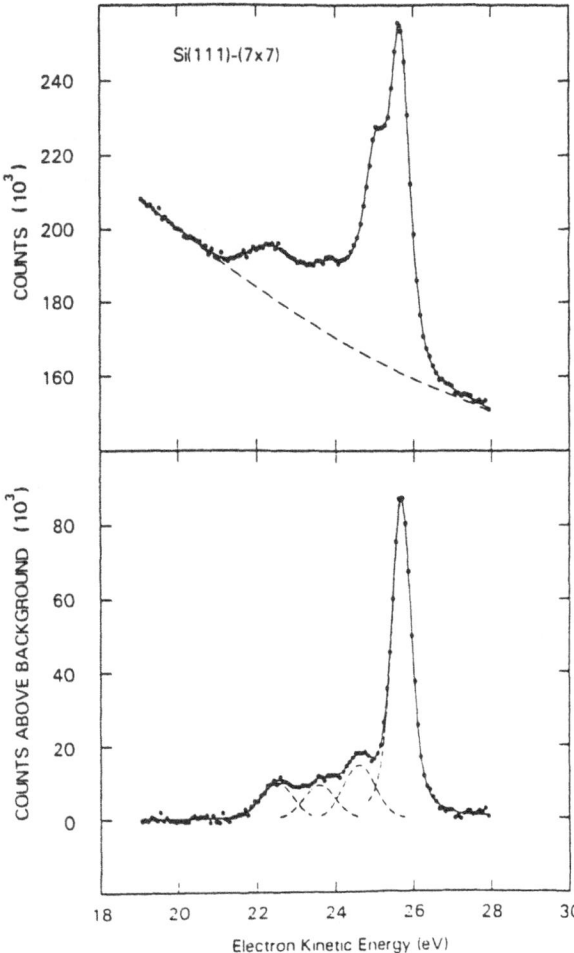

FIGURE 21. Upper panel shows a Si (111) 7×7 $2p$ photoemission spectrum after an exposure to 50 K of XeF_2. This can be compared with Fig. 17 for a clean Si (111) 7×7. The lower panel shows decomposition of the $2p^{3/2}$ portion of the spectrum into its chemical components. Note that each shifted component is associated with a Si bound to 1, 2, or 3 fluorine atoms. (From Ref. 115.)

were for particular surfaces dosed to saturation with XeF_2 and then measured in vacuum ambient. The predominant volatile species is SiF_4, but SiF and SiF_2 also evolve. The use of core level shifts has identified the presence on the surface of SiF, SiF_2, and SiF_3, as shown in Fig. 21. The magnitude of the chemical shift in this highly ionic case scales directly with the coordination of fluorine and is on the order of 1 eV per fluorine atom.

These measurements have also shown that the specific details of the attack on the surface by fluorine depend upon the structure, as well as on the electrical doping, of the surface.[117]

4.2.3. Oxidation of Silicon (111)

The oxidation of silicon has been studied extensively because of its technological importance to integrated circuits. Studies of the oxidation of silicon with synchrotron radiation have created insight into the bonding and atomic configurations at the silicon–silicon dioxide interface.[47,118,119] The specifics of oxidation depend upon the particular surface, and (100) surfaces are now used for field effect transistors while (111) surfaces are used for bipolar transistors. In each case the surface used minimizes the relevant interface states. Relating the full extent of all the work on the oxidation of silicon is beyond the scope of this section, but specific reviews exist.[120-122]

As an example, we next describe a series of experiments that led to greater understanding of the extent and composition of the transition region between bulk silicon and the silicon dioxide thermally grown *in situ* on the (111) face.[123] This work showed that immediately at the interface a suboxide exists on the (111) face. Recently work of this type has been extended to the (100) face, and similar chemistry occurs.[124,125]

In studying Si–SiO$_2$ interfaces with surface-sensitive probes, either one can begin with an oxidized silicon wafer and try to clean the surface *in situ* with chemical or plasma etching techniques, or one can prepare a clean silicon surface in the measuring system and then grow the SiO$_2$ interface of interest *in situ*. The latter approach allows, in a UHV surface-analysis system, for the oxide to be formed in a manner as similar as possible to oxidation in a furnace. With the thermally controlled *in situ* oxygen dosing of clean cleaved Si (111), samples are produced which exhibit reproducible local stoichiometry when characterized by Si (2p) CFS partial yield and photoelectron energy distribution over a wide range of conditions.

An example of the latter approach will now be discussed. The SiO$_2$ described here was formed by low-pressure, controlled O$_2$ dosing of the *in situ* cleaved Si (111) surface held at 625 °C. The transition from a Si atom in bulk silicon to one in the SiO$_2$ occurs over ~5 Å. This surface SiO$_2$ is thus representative of the thermally oxidized Si used for MOS devices since the SiO$_2$–Si transition is essentially identical to those reported for controllably etched (111) and (100) Si.

The thermally controlled *in situ* oxidation technique produces samples which exhibit uniform characteristic Si (2p) partial yield and photoelectron energy distribution over a wide range of conditions. Figure 22 shows that reproducible interfaces are obtained by cleavage in an atmospheric pressure of O$_2$, by repeated dosing with an excited oxygen plasma, and by high-temperature dosing with ground-state molecular oxygen. The evolution of Si (111) + O$_2$ into an SiO$_2$ layer is seen in both the SXPS and CFS. Comparing the CFS structure in Fig. 12 with the SiO and SiO$_2$ standards, we see that the SiO$_2$ evolves through formation of an SiO configuration. The major result to be emphasized here is that the transition from Si to SiO$_2$ is observed within one to two atomic layers, based on the surface sensitivity of our photoelectron probe.

The core photoemission of Fig. 22 shows an interface oxide characterized by a 2.5-eV Si 2p binding energy shift from the bulk Si 2p level. This same oxidation state also occurs in the spectra in Fig. 23 for an oxide formed with small repeated doses. Such oxides also exhibit a Si 2p resonance in the absorption spectrum, which is identical to that of evaporated SiO.[8] The only other explanation of this oxide state would be for its absorption fingerprint to correspond to that of Si$_2$O$_3$. For this to be consistent, the "standard" oxide formed by UHV deposition of SiO would have to form as such a Si^{+3} state. However, the slightest addition of O$_2$ to the vacuum causes SiO$_4$ quantum-well resonances characteristic of SiO$_2$ to appear

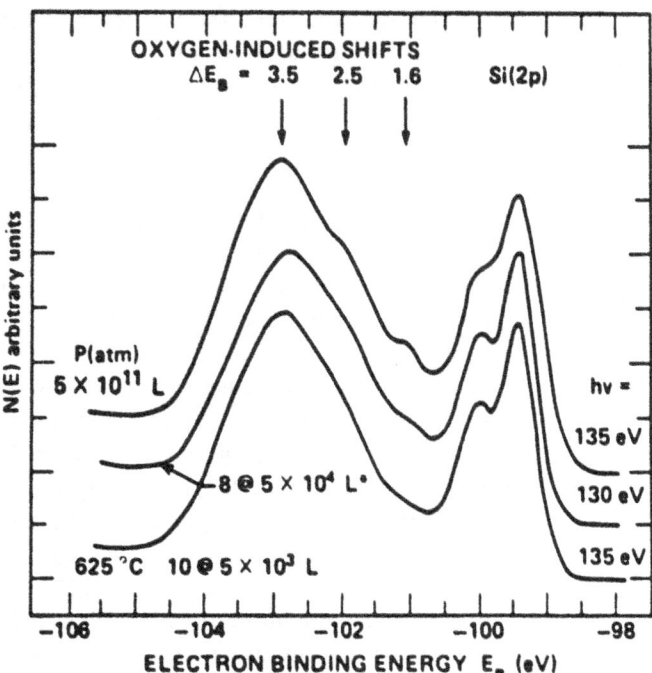

FIGURE 22. Energy-distribution curves for a cleaved Si (111) surface oxidized in three different wa
Cleavage in atmospheric pressure of O_2; (2) repeated dosing with an excited oxygen plasma; a
high-temperature dosing with ground-state molecular oxygen. Note the different oxygen-induced chemical
(From Ref. 123.)

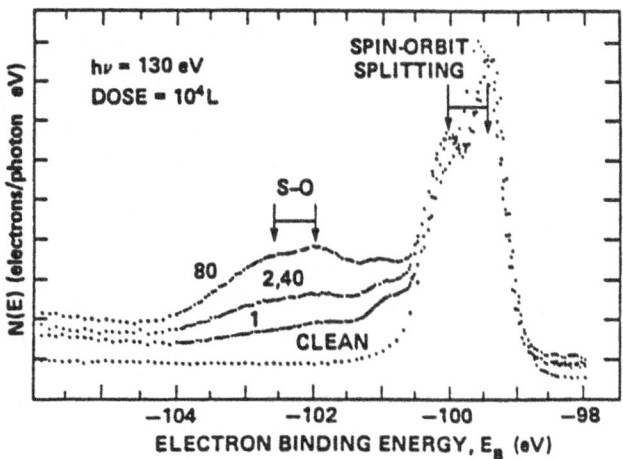

FIGURE 23. Energy-distribution curves for a cleaved Si (111) surface dosed at room temperature with un
O_2 in steps of 10^4 L. The spin–orbit splitting is indicated. (From Ref. 123.)

immediately in the CFS yield. As SiO_2 is the most thermodynamically stable oxidation state of Si, the product of the purest evaporation corresponds to the stoichiometry of the starting SiO. Thus, the absence of the SiO_4 tetrahedral signature in the partial yield of the oxide standards and of the interfacial oxide in Fig. 23 for a 700 °C oxidation allows identification of a single connective layer of unique SiO stoichiometry just beneath the SiO_2. This may be thought of as a disordered GeS-type structure. Such an SiO layer allows for the large mismatch between Si and SiO_2.

The formation of this intervening SiO layer on the underlying Si (111) is accompanied by an additional chemically shifted peak in the Si $2p$ photoemission at 1.3 ± 0.3 eV, as shown in all the SXPS data. Since the shifted binding energy is halfway between that of bulk Si and the 2.6-eV intermediate state identified as SiO above, one concludes that Si_2O is the appropriate stoichiometry describing the Si interface transition. As seen in Figs. 22 and 23, the 2.5-eV feature is always the first Si state to appear which is not 4-coordinated with Si. It therefore retains three Si nearest neighbors and has a single O bond which replaces the interface double-Si layer bond. Since this interface bonding is directed toward the (111) surface, incomplete oxidation in this first non-Si plane would lead to the P_b defect, consistent with Johnson's and Poindexter's work. From the strength of the 1.3 eV-shifted Si $2p$ emission within the very short photoelectron escape depth (7 Å for Si, ~14 Å for SiO_2), one deduces that this O is one of the atoms forming the Si^{+2} bonds of the transition layer. The conclusion is that the net distance from a Si atom in bulk Si to one in SiO_2 ~ 5.3 Å. There is no evidence in the data for the Si_2O_3 type interface bonding state.

A detailed model developed for the evolution of Si (111) + O_2 to an SiO_2 layer is summarized in Fig. 24. The Si $2p$ core electron photoemission showed that Si_2O-type bonds bridge Si (111) to SiO_2. The surface sensitivity of the constant final state (CFS) partial photoelectron yield shows that the interface is at least a couple of layers thick (>3.55 Å in extent), and contains a unique SiO layer which is characterized by negligible SiO_4 tetrahedra. Using this model, and correlating it with results from other experimental techniques such as DLTS and EPR, the possible origins of defect structure responsible for interface traps such as the P_b center can be explored.[4]

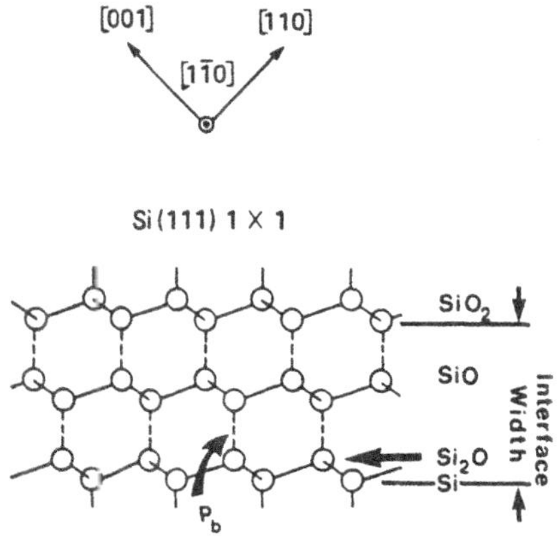

FIGURE 24. Model for the SiO_2/S (111) interface showing where Si atoms are in SiO_x environments. (From Ref. 123.)

4.3. Detection of Hydrogen with Photon-Stimulated Ion Desorption

Photon-stimulated ion desorption (PSID) is a valuable technique for investigating the oxidation of surfaces and for detecting minority species. Most important, PSID can detect hydrogen. Hydrogen plays a significant role in many materials, but it is often exceedingly difficult to detect. Adsorption of hydrogen has a strong effect upon surface reconstructions and results in clear changes in the photoemission from surfaces.[126] Photon-induced desorption detects and probes hydrogen and was first definitively demonstrated with synchrotron radiation and then turned into a powerful tool for exploring surfaces and particularly for demonstrating the presence of hydrogen. The technique is intrinsically surface-sensitive and is one of the few probes where the hydrogen bonding can be studied in detail.

Figure 25 is reproduced from Chapter 7, where a more extended discussion of PSID is presented. Acid-cleaned Si (111) was given a final rinse in D_2O before insertion into the vacuum chamber, where it was heat-cleaned to 500 K. The PSID spectra near the Si $L_{2,3}$ core threshold show the stimulated desorption of H^+, D^+, and F^+ ions. For comparison, surface absorption spectra of the underlying bulk Si and of SiO_2 are shown. The 5-eV chemical shifts from the Si core level energy indicated that all three species had issued from SiO_2-type sites.[127] The SiO_2-type site was confirmed by studying the H^+ desorption at the $O\,K$ edge and comparing the spectrum with the total photoelectron yield

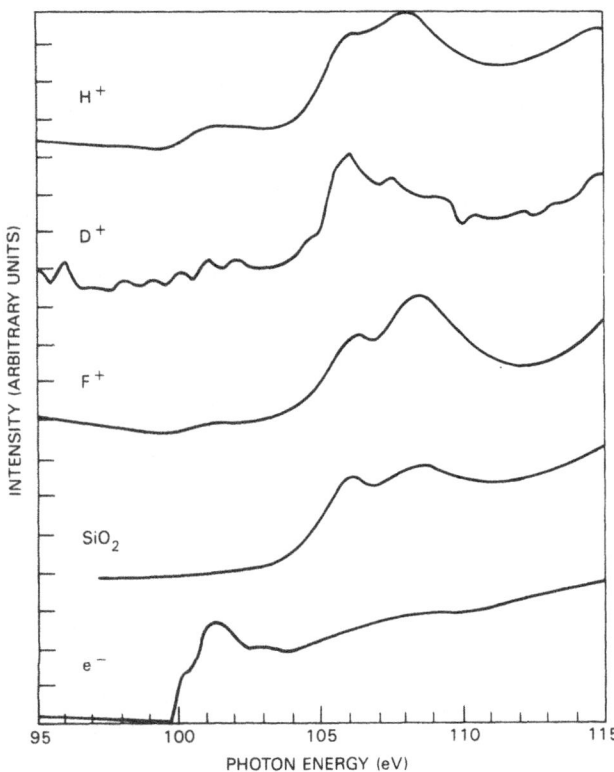

FIGURE 25. Photon-stimulated ion desorption from water-dosed Si (111). The acid-cleaned Si (111) sample was dipped in heavy water, D_2O, before introduction into the vacuum chamber. The sample was baked to 500 K, and the photon-stimulated ion desorption spectra are compared to the total photoelectron yield for Si and SiO_2. Note that the ion yield of F^+, D^+, and H^+ reflect the SiO_2 but indicate bonding in different environments. (From Chapter 7.)

spectrum from which a characteristic signature chemical shift of $+2.6\,eV$ was measured.

Another example is presented in Section 4.7.1 in the discussion of diamond surfaces.

4.4. Surface Atomic and Electronic Structure Determination

The determination of surface atomic and electronic structure requires the correlation of a wide range of physical measurements. Chapter 4 (by Eberhardt) and Chapter 4 of Volume 2 of this set by Bringans and Bachrach) present in more detail many studies which have sought to unravel aspects of surface reconstructions.[128-130]

As discussed above, the study of surface core level shifts and angle-resolved photoemission have been very effective tools for deducing structural models. Polarization-dependent, angle-resolved photoemission in particular allows the direct determination of the energy–momentum curves for surface-state bands.

The dispersion of the surface-state bands in various directions of the Brillouin zone is very sensitive to the details of the atomic arrangement. The surface energy bands can then be used as constraints in structurally sensitive model calculations and thereby lead to structural determinations. Two basic theoretical methods have been used to calculate surface bands in conjunction with the ARPES data. One is a tight-binding method developed by Chadi,[131] and the other is a first-principles pseudopotential calculation with the local density approximation as applied, for example, by Northrup[132] and others. This second method is derived from a methodology similar to that used for bulk calculations, but in this case, repetitive slabs provide the periodic boundary conditions. The tight-binding method has a certain amount of empirical input, but its predictions have been quite effective. Both approaches require testing of possible topologies to evaluate atomic structures. Both evaluate the relative total energies with respect to the atomic coordinates. For a given class of structural models, the calculation will find the best atomic coordinates, but different classes must be separately determined and evaluated.

Surface-energy band determinations with angle-resolved photoemission were first achieved with lamp sources, but synchrotron radiation added an important dimension to this work. The unique advantage of synchrotron radiation for these measurements is the polarization and ready tunability of the source, which allow the separation of surface and bulk states. Two early applications of this technique were the study of $TiSe_2$, which showed that it is a semimetal,[133] and investigation of silicon surface states, initially by Rowe et al.[134] and subsequently extended by others.[135]

4.4.1. Si (111) 2 × 1

As an example, the Si (111) 2 × 1 structure is now thought to be described by the Π-bonded chain model proposed by Pandey[136] which replaced the buckling model suggested by Haneman.[137] The first indication that the buckling model was not complete was provided by angle-resolved photoemission measurements, which found an upward-dispersing band shown in Fig. 26.[138] Tight-binding calculations attempting to fit the data to the chain model could not reproduce this form of the band with a buckling-type model, and this led Pandey[139] to propose the Π-bonded chain model, which did reproduce the observed dispersion. Subsequent extensive work of Olmstead[140] and Hansson and Uhrberg[141] among others has experimentally clarified these issues.

4.4.2. GaAs (100) c (4 × 4)

Angle resolved photoemission has been used to determine the surface-state bands on a number of GaAs surfaces.[142-144] The most detailed information for the (100) surface is available for the $c(4 \times 4)$ reconstruction, which is thought to consist of a complete arsenic layer termination. Sorting out the surface states requires great care, and a detailed analysis of the spectra is found in Chapter 4 of Volume 2 of this set. The dominant features in these spectra turn out to be direct transitions from bulk states. Analysis shows that a direct-transition photoemission

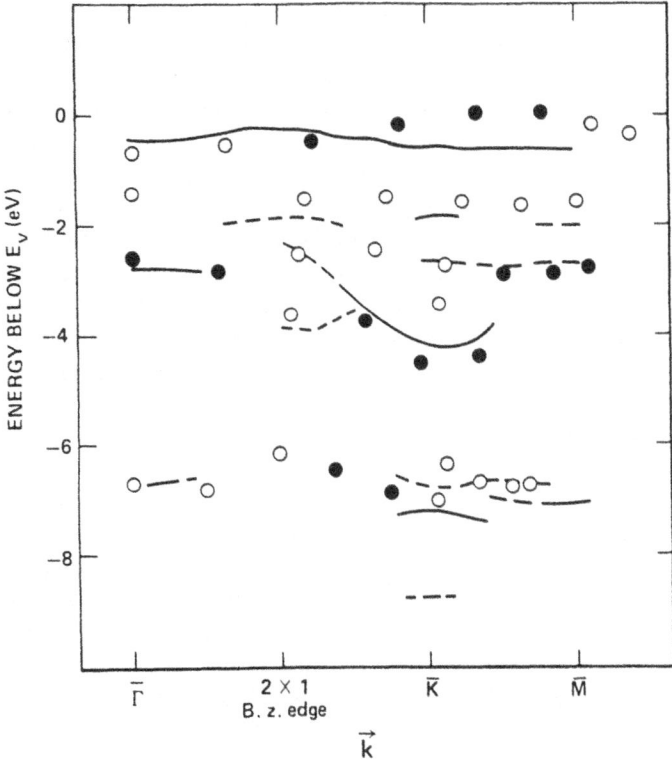

FIGURE 26. Experimental electronic-band structure along the ΓKM surface direction for Si (111) (2 × 1) compared with bands based on the Haneman chain model. Filled circles correspond to *strong peaks*, and open circles are weaker structures in the experimental spectra. The continuous curves show the energy positions of the strongest peaks in the calculated surface density of states, while the dashed curves correspond to weaker but still significant peaks. The upward dispersing band near the top of the valence band is a consequence of the Π-bonded chain structure subsequently deduced. (From Ref. 138.)

model can be used to understand which components are due to bulk transitions and thereby to sort out the surface features.

By plotting the photoemission peak positions as a function of energy position and wave vector for a particular symmetry direction, one can determine the expected points for calculated bulk transitions and thereby clearly isolate surface-related features. Figure 27 shows the derived surface-state bands for the two directions depicted in the Brillouin Zone. Interestingly, determination of the surface bands shows a 2×1 symmetry, in contrast to the 4×4 symmetry shown by LEED. Such a discrepancy between LEED and photoemission results is not uncommon. Photoemission is more sensitive to the structure within the unit cell rather than the long-range order observed by LEED. The dominant feature in the unit cell is the presence of dimers, which produce the underlying 2×1 component, as shown by the above-mentioned model developed by Chadi *et al.*[145] Surface dimerization is a unifying feature between silicon (100) and GaAs (100)

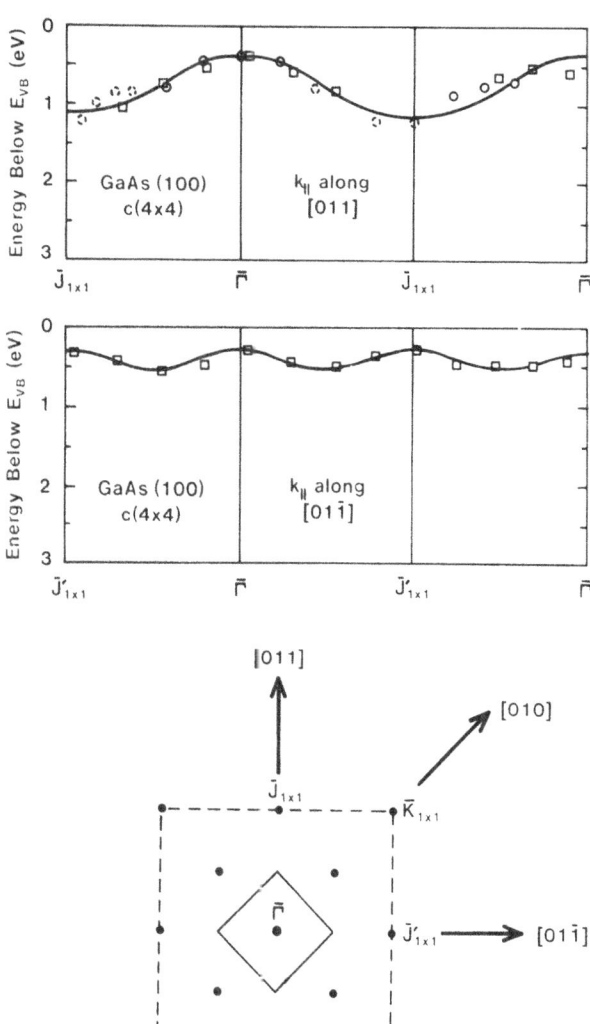

FIGURE 27. Dispersion of surface-state bands along two mirror planes of the GaAs (100) $c(4 \times 4)$ surface. The directions are shown in the representation of the Brillouin zone. The 2×1 periodicity reflects the dimerization and multidomain aspect of this surface. (From Ref. 143.)

surfaces. Secondarily, the surface origin of this peak can be seen by its sensitivity to hydrogen chemisorption.[146]

4.5. Interface Phenomena

Synchrotron radiation studies have had a major impact on the investigation of interfaces.[97,147,148] Chapter 4 of Volume 2 of this set (by Bringans and Bachrach) and Chapter 5 of that volume (by Rossi) specifically relate to these studies.

Interface phenomena form a subset of the general field of phase formation in materials systems. In some cases, particular surface phases are observed. One of the other basic aspects of interface formation is the change in the surface and interface electronic structures as the thickness of an overlayer increases from a fraction of a monolayer to several monolayers. The overlayer both introduces changes in the electronic structure of the substrate and creates new interface states. In general, the interaction between the overlayer and the substrate is very complex and depends strongly on the experimental conditions. One possibility is that a compound and/or alloy is formed between the substrate (one or both constituents in the case of a compound semiconductor). In many cases a number of structure- and stoichiometry-related reconstructions or surface phases occur.

4.5.1. Al on GaAs and on SiO_2

In studying interfaces, it is possible not only to look at static aspects, but also to measure kinetic phenomena. For example, it was found that the deposition of Al on GaAs resulted in an exchange reaction by which the aluminum replaced the surface Ga.[149-151] This reaction proceeds quite rapidly. This phenomenon, shown in Fig. 28a, has subsequently been studied extensively.

The deposition of Al on SiO_2 produces an exchange reaction which slowly reduces the SiO_2 and converts it into Al_2O_3. The kinetically constrained reaction proceeds slowly, as seen in Fig. 28b.[96] Time-resolved reaction studies will be a major area of future study.

4.5.2. Au on AlAs (100)

Spin–orbit splitting effects can be used to determine effects of metal aggregation on a surface such as the initial stages of a metal–semiconductor interface. In the case of gold deposited on Si or GaAs, there is a marked change in the spin–orbit splitting as the situation changes from low coverage, where the atoms are isolated to the actual formation of a monolayer metallic film. Figure 29 shows an example for Au on AlAs, but similar effects are seen for other systems.[152] Au on III–V semiconductors has been extensively studied because of its importance in contact metallization systems.

4.5.3. Schottky-Barrier Formation

Over the years, great effort has gone into gaining a systematic understanding of the seemingly simple metal–semiconductor system and the resulting electrical

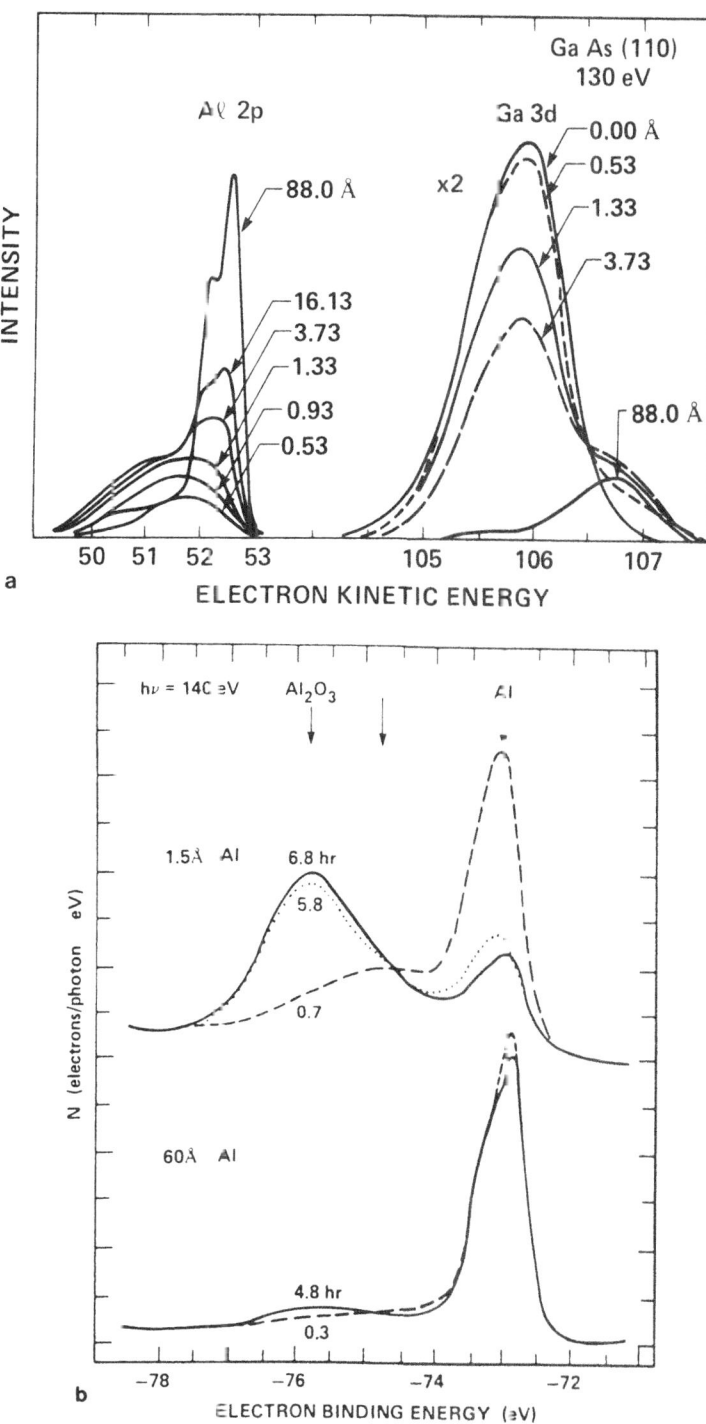

FIGURE 28. (a) Al 2p and Ga 3d core level spectra as a function of Al coverage on GaAs (110). The chemical shifts indicate that an exchange reaction takes place, creating an interfacial layer of AlAs and liberating metallic gallium. (From Ref. 149.) (b) Al 2p core level spectra as a function of time for clean Al and Al deposited on SiO₂ (From Ref. 96.)

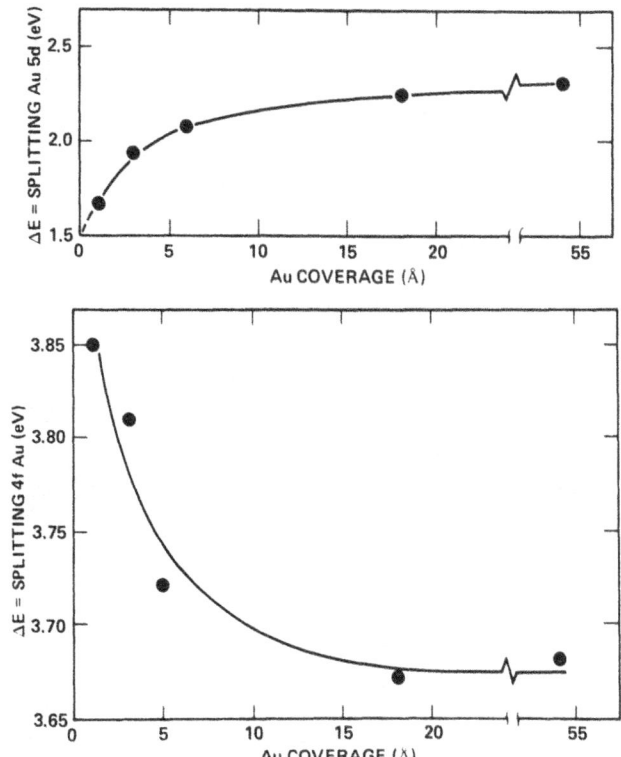

FIGURE 29. Au on AlAs (100) spin–orbit splitting of the Au 5d and Au 4f photoemission peaks as a function of Au coverage, with $hv = 130$ eV. At low coverage, the spin–orbit splitting reflects atomic gold dispersed on the surface. As the film thickens and becomes metallic, a transition in the spin–orbit splitting occurs. (From Ref. 152.)

barrier. Complete determination of the detailed mechanisms responsible for Schottky barriers has proven quite elusive, and definitive experimental determination of the mechanisms of Schottky-barrier formation is still being avidly pursued. Perhaps in fact there is not one single mechanism applicable to all the different combinations of materials encountered. There are unifying themes in the microscopic description of Schottky-barrier formation, but when the variety of metals and semiconductors are considered, there are in fact a number of mechanisms that come into play at the interface and determine the macroscopic electrical barrier.

Recent articles by Brillson[153] and by Bachrach[97] give thorough reviews of this field. The presentation here derives from those works. One particularly interesting idea resulting from the application of synchrotron radiation–excited photoemission to Schottky barriers is the unified defect model. This was proposed by Spicer[154] to explain Fermi-level pinning data obtained in studies of barrier formation on cleaved GaAs (110) surfaces. Figure 30 summarizes the data underpining the model.

Much of the impetus for renewed research work on Schottky barriers in the last ten years resulted from the advances in vacuum preparation techniques and

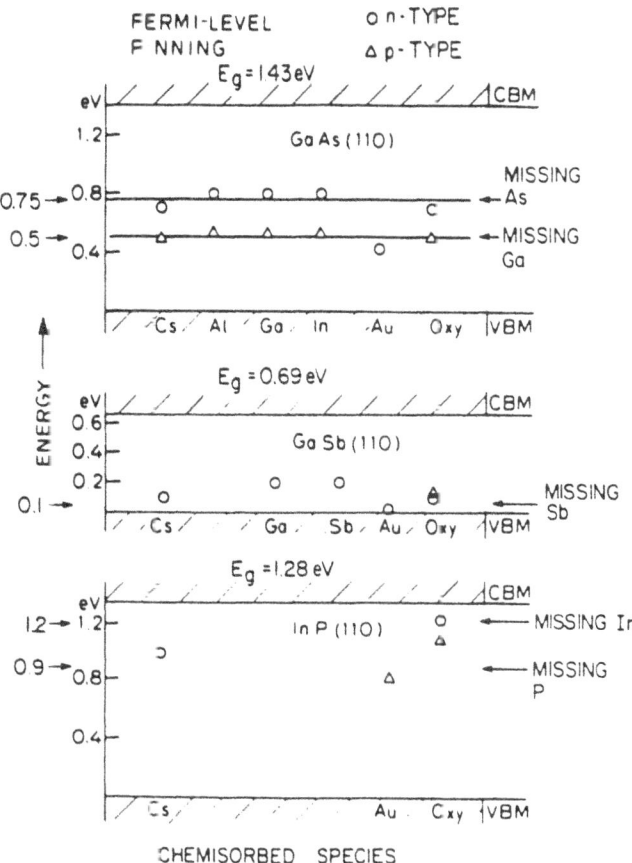

FIGURE 30. Defect-level assignments derived from Fermi-level pinning observations by Spicer *et al.* The unified-defect model is discussed in the text. (From Ref. 154.)

the development of surface-sensitive spectroscopies, both of which created new opportunities to advance the understanding of Schottky barriers. The application of these new tools over the last decade has helped to elucidate the nature and extent of the interfacial regions. Even during room-temperature formation of the metal–semiconductor interface, interdiffusion and interfacial reactions are pervasive and important. They become even more so at elevated temperatures. Therefore, specifying the extent and nature of this interfacial region is essential. An understanding of the details of the interfacial region is the major feature missing from the early work. This section will discuss some aspects of work relating to the study of surface and interface defects.

The initial condition of the clean semiconductor surface as prepared can have many different configurations, which will affect the outcome of the Schottky barrier that forms. There can be either structural or compositional disorder, reflected in structural defects such as steps, surface vacancies, or antisite arrangements. In addition, surface electronic states can be generated by the structural defects.[155,156] Many examples exist, however, in which the clean surface can be prepared in a well-ordered state, relatively free from defects, either by cleaving[157–159] or by molecular-beam epitaxial growth.[160]

In the initial metal deposition, the density of surface atoms is low enough that they do not interact. In this case, one can identify site-specific chemisorption as well as reaction or replacement.[150] As discussed in Section 4.5.1, energy released at this stage, such as the heat of condensation or reaction, may dislodge other surface atoms from their equilibrium sites. The outcome of this stage depends on the surface temperature, either through surface diffusion or through activated chemical processes.

As the surface concentration of metal atoms grows, the effect of metal–metal interactions becomes significant. Depending upon the surface mobility and the strength of the metal–metal interaction, the metal may have a tendency to form clusters. For many metals, the cluster formation or metal-film nucleation is exothermic, so this stage can also disrupt the underlying lattice.

After the formation of the initial metal overlayer, depending on the configuration of the metal, the metal film may be under significant stress (either tensile or compressional), which can be a driving force for interdiffusion and at some point for dislocation formation. Stress, dipole formation, and other phenomena drive interdiffusion, which can proceed over tens of angstroms even at room temperature. The final interface may be compositionally and spatially inhomogeneous, so that there are many types of situations that need to be evaluated in detail. These effects can be accentuated with the thermal processing that usually accompanies device processing.

Synchrotron radiation studies combined with other probes have been particularly effective in unraveling these phenomena. A wide number of techniques have evolved which have provided significant information about the microscopics of interface formation during the layer-by-layer evolution. Important for exploring interface development has been the application of core and valence-band synchrotron radiation–excited photoemission spectroscopy, along with techniques such as surface photovoltage, Kelvin probe, and Raman scattering measurements. These measurements have led to the identification of classes of localized states, which arise from the processes which disrupt the surface before or during interface formation.

For example, the various processes can possibly lead to either vacancy formation[161,162] or antisite defects[163] at the interface. If these defects have charge states associated with them, they can potentially have a strong influence on the barrier. A body of suggestive evidence for such states has accumulated from core-level photoemission measurements.[164,165]

A similar class of states arises from situations where the metal, when incorporated in the semiconductor lattice, becomes a donor or acceptor. Examples of this are Au-GaAs or Al-ZnSe. In extreme cases, if the incorporated metal atoms convert the type of the semiconductor, then a p–n junction would form and strongly influence the properties of the barrier. This doping case is analogous to the formation of a heterojunction resulting from exchange reactions and interdiffusion, such as with Al–GaAs.

The presence of surface defect states has been deduced from the observation of interdiffusion of cations and anions. Some of the first evidence for the generation of interfacial defects was provied by Bachrach and Bianconi in a study of Ga–GaAs interfaces.[166] Their demonstration with core-level photoemission of As outdiffusion was interpreted as resulting in interfacial vacancies.

Anion and cation outdiffusion has been found to be a persistent aspect of interface formation on III–Vs. Lindau and Spicer have discussed similar types of work in a review.[164] Spicer et al. have proposed that these defects are the primary mechanism for Fermi-level pinning for GaAs (110) surfaces. From these ideas, the possibility of defect levels as shown in Fig. 30 was postulated to determine barrier formation in III–V compound semiconductors.[165]

Although it is possible for vacancies to remain at the interface, the theoretical argument has been made that, in many cases, these would be associated with metal. In the case of antisite defects, the properties of isolated defects on a free surface are clear, but if the density increases, then the layer is converted into a zone of the metal. An anion antisite defect, for example a Ga on an As site, has the aspects of a metal cluster when additional monolayers develop. One can anticipate that continued studies, many of them involving synchrotron radiation, will continue to unravel and clarify these issues.

4.6. Final-State Effects in Photoemission Cross Sections

Final-state and cross-section effects are important when using photoemission spectra to deduce properties of materials. These effects are discussed extensively in Chapter 1 of Volume 2 of this set (by Lindau) and elsewhere. Chapter 5 of that volume (by Rossi) discusses some of these effects in the context of interface studies. Significant work has been done in calculating atomic subshell photoionization cross sections, and a major tabulation is presented by Yeh and Lindau[167] based upon the theoretical work of Goldberg, Fadley, and Kono.[168] Only two examples will be given in this section; Section 4.7.1 discusses carbon K-edge photoemission cross-section effects.

4.6.1. Plasmon Sideband Threshold Phenomena

The examination of final-state effects using variable energy permitted the determination of the cross-section dependence of plasmon generation. It was found that bulk plasmons show a marked threshold behavior, whereas the surface plasmon exhibited a different interaction. Figure 31 gives an example for photoemission from single-crystal aluminum.[169]

4.6.2. Cross-Section Effects in Metal Semiconductor Studies

Cross-section effects are an important component of the set of tools available to study surfaces with synchrotron radiation. For example, for transition metals, using the Cooper minimum where the cross section is low due to nodes in the wavefunction, allows discriminations to be made between the substrate and an adsorbate. Figure 32 gives an example reproduced from Chapter 5 of Volume 2 of this set.

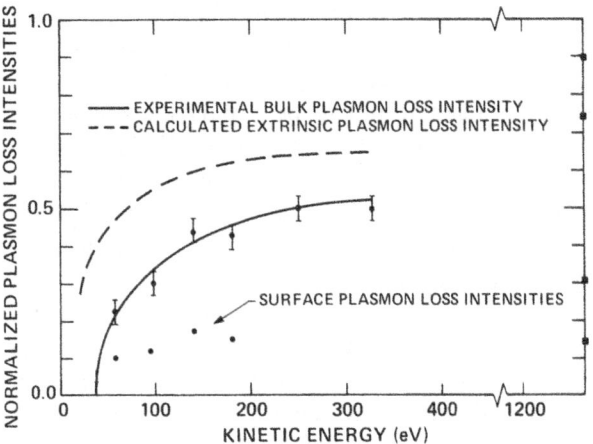

FIGURE 31. Experimental and calculated curves for the normalized bulk plasmon intensities associated with the Al 2p photoemission peak. Kinetic energies of the elastic electrons vary from about 30 to 330 eV. Also shown are experimental normalized surface-plasmon intensities. A threshold for surface plasmons was not resolved. (From Ref. 169.)

4.7. Investigations at the C, N, and O K Edges

The intensity available from the 3-GeV SPEAR storage ring at SSRL created the initial opportunity for extensive spectroscopic work studying the carbon, nitrogen, and oxygen core levels. The principal aim of the creation of the Grasshopper monochromator at SSRL was to allow experimentation in this previously inaccessible energy range. Prior to the development of the Grasshopper monochromator,[22] such studies were limited by sources and by grating and mirror technology. Conventional sources were particularly weak in the vicinity of the carbon K edge and higher energies. Significant advances have now been made

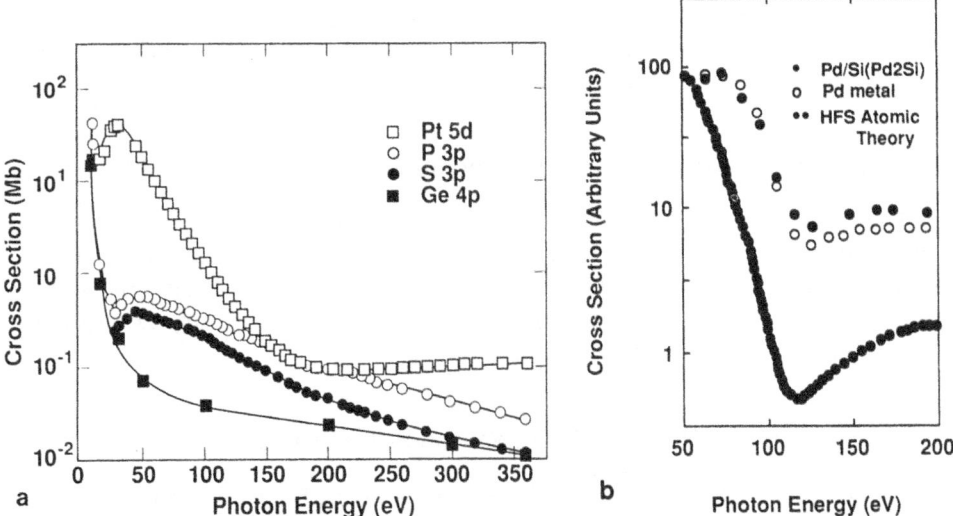

FIGURE 32. (a) Calculated atomic photoionization cross sections for Pt, a typical transition-metal 5d subshell and typical semiconductor valence-band p subshells. (b) Measured photoionization cross sections for Pd 4d states of bulk Pd, an intermixed Pd–Si (111) interface whose electronic states resemble the silicide Pd_2Si. These are compared with the calculated atomic orbital cross sections. The spectra are arbitrarily normalized at the maximum. (From Chapter 5, Vol. 2 of this set.)

in all these areas. In fact grating technology has advanced so much that many different optical mountings are successful. Some of that technology development was stimulated by the grating market created by the seventeen Grasshopper monochromators that have been built. The pioneering work done at SSRL in the period 1973–1980 developed many of the opportunities and provided the incentive for creating more advanced facilities.

Working at energies closer to threshold also has two advantages in that the absolute cross section is enhanced and final-state plasmon effects are diminished, thus simplifying spectra in some cases. Examination of the interfaces between polymers and silicon and the adsorption of small, carbon-containing molecules has also been an active area of study.

4.7.1. Carbon *K* edge

Carbon K-Edge Photoemission Cross-Section Effects. The results of a study of the graphite carbon *K* shell near threshold photoemission cross-section energy dependence of the elastic peak and plasmon sideband are an example of one system studied using the above-mentioned facilities.[170] Previously an extensive study of the valence-band region was done.[171] Experimental determination of photoionization cross sections of core levels for gases and solids in the near-threshold region is of interest because of strong final-state effects. The measurement of the energy dependence of the normalized intensity of the elastic peak gives the partial photoionization cross section between selected initial and final states and therefore makes possible the determination of final-state symmetries. In fact, in the case of graphite, it was found that the near-threshold maximum reflects the flat localized Π band in the electronic structure of the conduction band.

In addition to the elastic peak, the onset of the plasmon sideband is due to the quantum-mechanical interference term between extrinsic and intrinsic plasmon production over a kinetic-energy range of the photoelectron to $4W_p$, where W_p is the bulk plasmon energy.[172–175] Using synchrotron radiation, it is possible to tune the energy of soft X-rays in the energy range near the core-level photoionization threshold. Plasmons can be excited by the photoelectron on its way out of the metal (extrinsic process) or by the core–hole potential created in the core photoionization process (intrinsic process). In the near-edge region, the photoelectron excited on the atomic site near the surface leaves the absorbing site slowly, and both extrinsic and intrinsic processes occur on the same site. Such a negative interference effect weakening the intensity of the plasmon satellite has been measured previously in the simple isotropic metal and semiconductor crystals Al and Si,[176–178] but the observation of such an effect in the anisotropic, nearly two-dimensional electronic structure of graphite shows that this is a general phenomenon.

Analyzing photoemission data for the cross-section variation requires that one correct for several effects,[179] the principal ones being the monochromator and electron-analyzer transmission functions and the variation of probe volume due to changes in excitation and escape depth. The true spectral distribution was

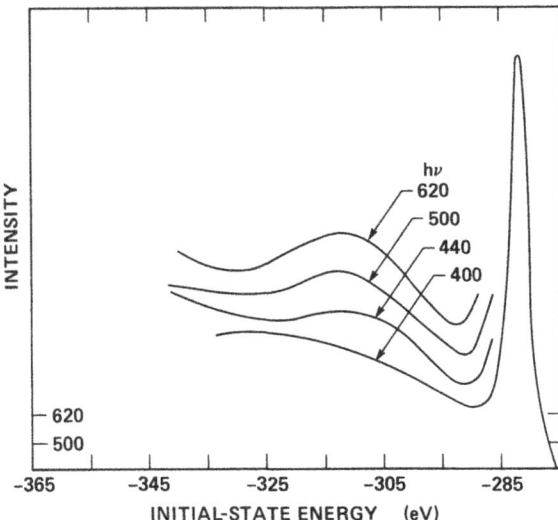

FIGURE 33. C 1*s* photoemission spectra vs. excitation energy normalized to the elastic peak, showing the variation in plasmon strength. (From Ref. 170.)

obtained both from a sodium salicylate reference and from gold photoyield, which was corrected with respect to published optical constants. The various data gave a consistent correction in the energy range used for obtaining the relative cross section.

The method used in this work for determining the partial photoemission cross sections has been to take and then analyze a series of electron energy-distribution curves as a function of the excitation energy. The spectra have been normalized to constant excitation intensity and electron-analyzer transmission. Below 400 eV, the plasmon sidebands are relatively unimportant, as seen in Fig. 33.

Figure 34 shows the relative carbon 1*s* photoemission cross section obtained by integrating the graphite data in Fig. 33. The integration used in obtaining Fig. 34 included summing over the plasmon sideband as well. At low final-state energy, the plasmon strength is small and unimportant, but it influences the result

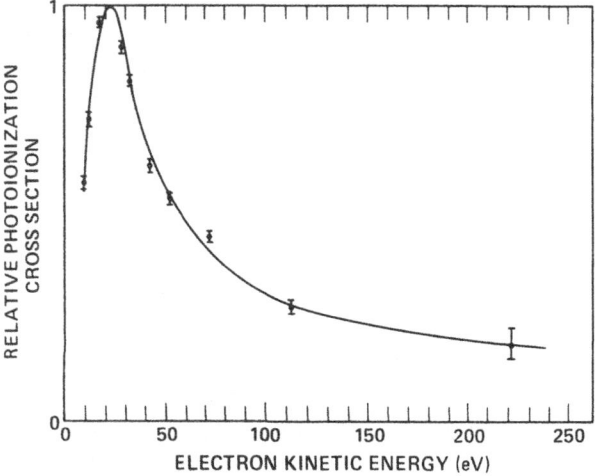

FIGURE 34. C 1*s* graphite photoemission cross section vs. electron final-state energy obtained from integrating normalized data. Note that band-structure effects change the near-threshold peak in the cross section from what would be expected for atomic carbon. (From Ref. 170.)

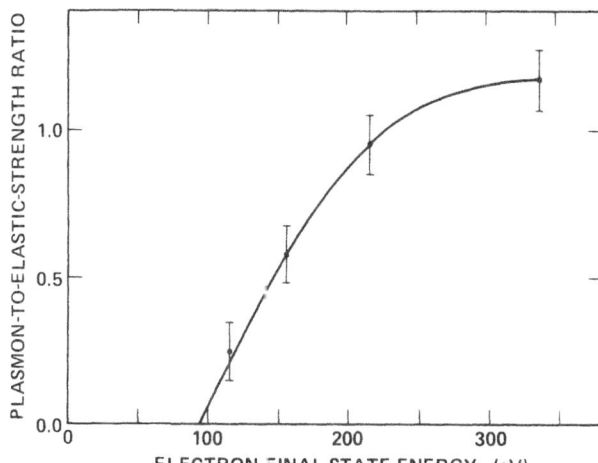

FIGURE 35. Relative plasmon strength vs. electron kinetic energy (C 1s graphite). Note the threshold behavior. (From Ref. 170.)

for the range above 100 eV. The data shown in Fig. 35 are uncorrected for the variation in escape depth, which modifies the effective sample volume. Because the light-penetration depth is large compared to the electron-escape depth, the geometrical correction is proportional to the escape depth. In a layered compound like graphite, the effects of photoemission anisotropy can be disregarded[180] because of the experimental geometry.

Whereas the plasmon sidebands are relatively unimportant below 400 eV in graphite, they become quite pronounced at higher energies, as seen in Fig. 35, where the photoemission spectra as a function of excitation energy are plotted normalized to the elastic peak. Such onset behavior was first discovered in aluminum.[176] Subsequent studies have shown these effects in other systems.[177,178] In graphite the dependence of the intensity of the bulk plasmon loss as a function of exciting photon energy has been investigated. The plasmon loss shown in Fig. 34 is dominated by the broad peak at 23 eV below the 1s direct photoemission peak. This loss is due to the bulk 2p σ-electron plasmon observed in electron-energy loss[181] and XPS[182,183] experiments. To measure the ratio between the plasmon loss and the direct 1s photoemission elastic peak, the area of the corresponding structure has been measured after subtracting the background due to secondary electrons. Accurate subtraction procedure is necessary at photon energies near the absorption threshold, where the secondary contribution is large. This procedure avoids the effect of 1s peak-width dependence on the monochromator resolution. The intensity ratio of peak areas as a function of the primary photoelectron kinetic energy is shown in Fig. 35. The photoelectron kinetic energy above the Fermi level is given by $E = h\nu - E_0$, where $E_0 = 285$ eV is the K shell photoabsorption threshold. Below $E \approx 100$ eV no resolvable plasmon loss structure appears. The detailed slope of the function of plasmon loss intensity versus photoelectron energy cannot be established. The threshold energy for plasma creation occurs at very high energy, or at about four times the bulk plasmon energy.

The photoemission 1s level cross section obtained by angular integrated photoemission primarily exhibits the shape of the atomic cross section at high energy, since line structures due to first neighbors such as EXAFS oscillations are not visible for the small number of experimental points.

The very strong peak in the $1s$ cross section near threshold can be assigned to a multiple scattering resonance. In X-ray absorption measurements, the near-edge region (often referred to as XANES[180]), extending from 8 to 70 eV above the threshold for graphite, is dominated by multiple scattering resonances localized within the first and second coordination shells.[184] This localized resonance should appear as a localized flat band in a band-structure calculation. The main contribution of the conduction band in a layer material like graphite will come from a σ band because the selection rules for polarized incident light are very strong. Under the experimental conditions, p-polarized light at grazing incidence (5°), only $\sigma \rightarrow \pi$ transitions are allowed.[171] The peak at 25 eV in Fig. 34 can then be associated with the high density of states due to the flat π_z conduction band at 27 eV above the Fermi level.[181] This experiment shows that the main peak at 25 eV is a multiple scattering resonance of π symmetry, in agreement with the band-structure calculation.[171]

The threshold for the creation of bulk $2p$ σ plasmon loss shown in Fig. 35 is higher than the expected value for a simple extrinsic plasmon creation in the bulk; thus, other effects have to be considered. A similar effect has been observed in Al and Si[176-178] using synchrotron radiation, but the effect was not so large. The extrinsic plasmon loss due to the inelastic scattering of the excited photoelectron should always be present, but the intrinsic plasmon excitation in the photoexcitation process should be taken into account. They interfere destructively[172-175,185] at low electron kinetic energy and suppress the plasmon loss sideband. Moreover, spatial effects have to be considered. The observed plasmon losses are due to hole creation near the surface. In fact the kinetic energies of the direct photoemitted electrons are in the range of minimum escape depth.

Diamond Surfaces. Diamonds have an esthetic human interest because of their beauty and are also of increasing importance in a variety of applications. Pate *et al.*[186-188] have used photoemission, photoabsorption, and photodesorption to study the surfaces of single-crystal diamond and to unravel some of the mysteries of their stability. Hydrogen plays a very important role in the stabilization of these surfaces, and it becomes bonded in the polishing process. The role of hydrogen was specifically elucidated by monitoring the H^+ PSID from excitation of the C $2s$ or C $1s$ holes, comparing it with photoemission, and correlating it with LEED. The $2 \times 2/2 \times 1$ reconstruction of the (111) surface is free of hydrogen, while the 1×1 reconstruction observed at lower temperatures and on as-polished surfaces is hydrogen terminated. The photoemission established that a surface state is formed about 2.5 eV below the valence-band maximum on the hydrogen-free $2 \times 2/2 \times 1$ surface. The surface state is removed and the H^+ yield restored with exposure to activated hydrogen. Hydrogen is seen to be strongly bonded to carbon, because heating to 600 °C has little effect on the H^+ desorption. Figure 36 compares the near edge structure for diamond with that of graphite, and Fig. 37 shows the effect of hydrogen termination on the carbon $1s$ core level.

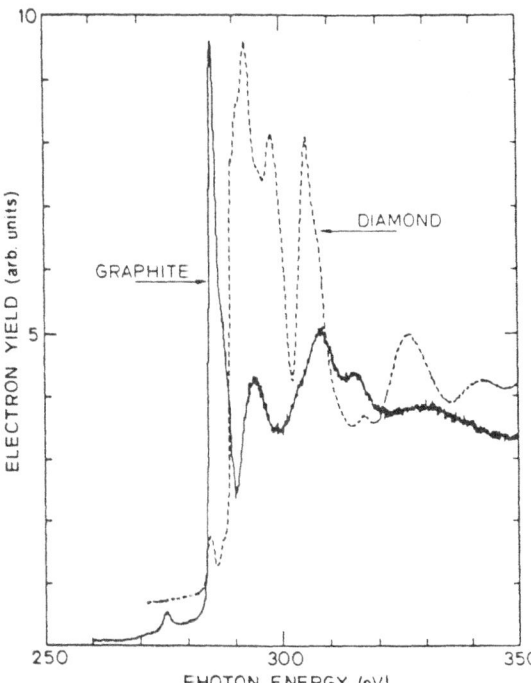

FIGURE 36. Carbon K-edge structures. A comparison of the near-edge structure of diamond, obtained by partial electron yield, with the near-edge structure of graphite, obtained by Auger electron yield. The small pre-edge feature in the graphite spectrum is due to carbon $1s$ photoelectron emission by second-order light. (From Ref. 186.)

4.7.2. Nitrogen and Oxygen K Edges

The original generation of monochromators, which tended to have 2-m dispersion lengths, did not have sufficient resolution to perform effective core-level or near-edge studies at the nitrogen and oxygen K edges. Recently completed monochromators with 10-m dispersion lengths and improved gratings and optics are changing this situation. As seen in Fig. 6b, the Locust monochro-

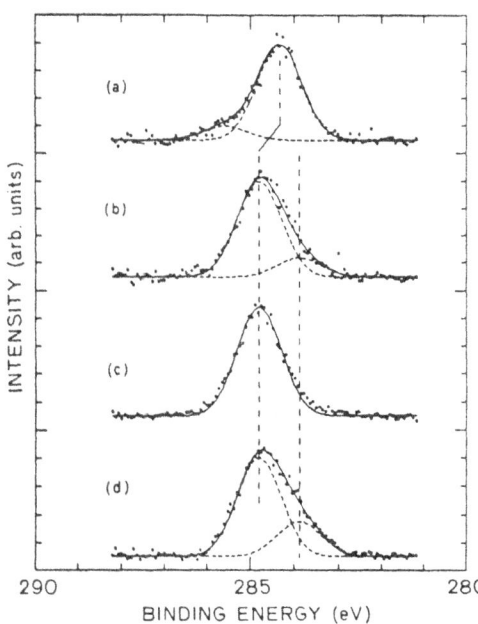

FIGURE 37. Systematics of the C $1s$ spectra in diamond as a function of hydrogen dosing and anneals ($hv = 320$ eV). The LEED remained a 1 times 1 structure after anneals, although surface states were observed with photoemission to come and go in the valence-band spectra. (a) As polished; (b) annealed (1030 °C); (c) H exposed; (d) annealed (1110 °C). (From Ref. 186.)

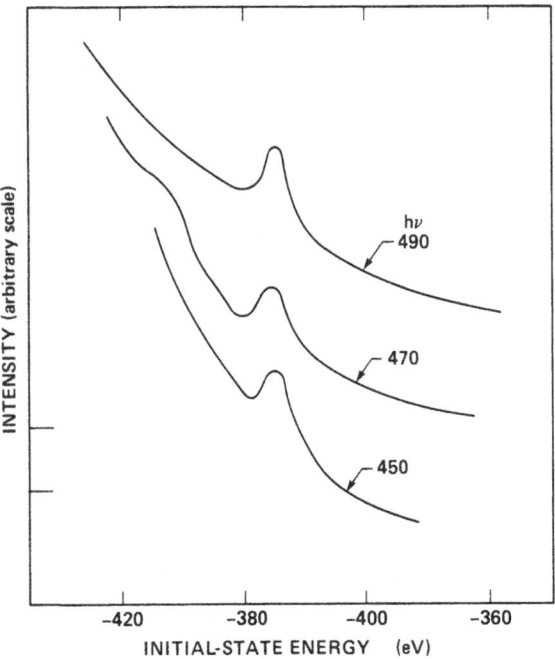

FIGURE 38. N 1s photoemission obtained from GaN. (From Ref. 170.)

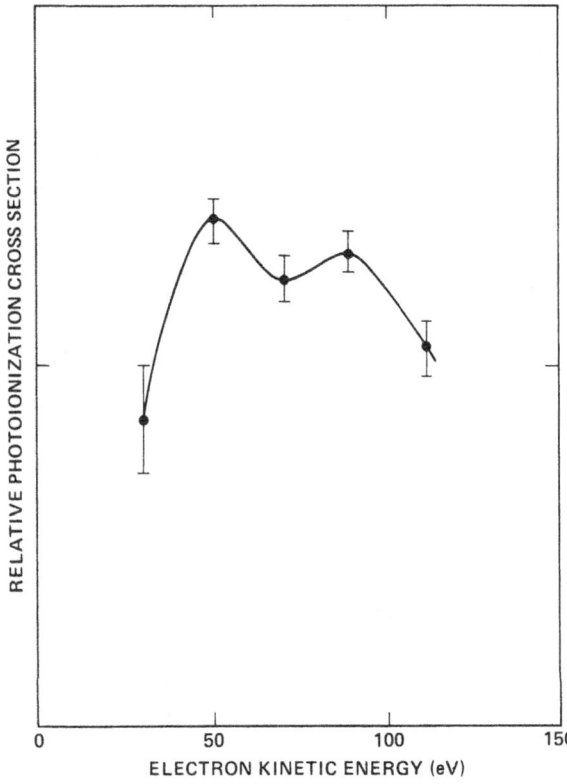

FIGURE 39. N 1s cross section derived from the data in Fig. 38. (From Ref. 170.)

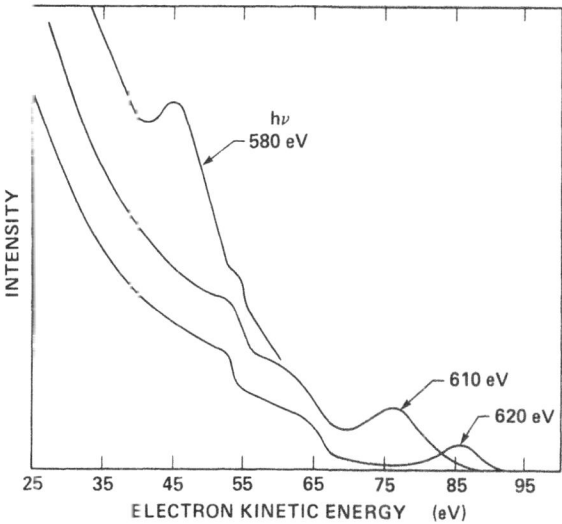

FIGURE 40. O 1s photoemission obtained from an oxidized Al (110) single-crystal surface (100 L oxygen). The resolution was not sufficient to obtain quantitative information, but was adequate for surface EXAFS, which determined the Al–O distance. (From Ref. 170.)

mator (an extension of the Grasshopper monochromator design) will make available approximately 0.1 eV resolution at the nitrogen and oxygen K edges. Experiments simultaneously exploring the chemical core level and valence aspects of condensed systems including these elements will therefore be possible. Other monochromators discussed in Chapter 9 of Volume 2 of this set present other examples.

One early investigation which did not require high resolution was the measurement of the photoemission cross section at the N and O K edge, similar to the detailed measurement of the carbon K edge presented in the previous section. Figure 38 shows the N K-edge photoemission at selected excitation energies obtained from a single crystal sample of GaN, and Fig. 39 shows the relative cross section. Figure 40 shows similar photoemission data for O on an oxidized Al (110) surface.[170] The form of the cross section for the N 1s level in the GaN presumably reflects final-state density of states effects in the solid phase compound.

5. CONCLUSIONS

In summary, many important surface and interface phenomena remain to be studied and understood, and synchrotron radiation–enabled studies will continue to open new avenues with the further enhanced capabilities that will become available. Future enhancements of beam intensity and brightness, increased energy resolution, and exploitation of polarization and time-dependent studies should provide increased opportunities for microchemical determinations, structural assignments, and exploration of kinetic phenomena in surface physics.

The chapters developed for this book provide a framework for understanding and predicting possible paths to future developments. Planning studies for new synchrotron radiation facilities are interesting guides to today's perception of future opportunities. If the past has been prologue to the present, the studies reviewed throughout this book will be prologue to the future.

REFERENCES

1. A. Pais, *Inward Bound: Of Matter and Forces in the Physical World*, Oxford, New York (1986). This book is an interesting history.
2. C. Kunz, *Synchrotron Radiation, Techniques and Applications*, Springer-Verlag, New York (1979). A short historical synopsis is provided in this book.
3. H. Winick and S. Doniach, *Synchrotron Radiation Research*, Plenum, New York (1980); see also H. Winick and A. Bienenstock, *Ann. Rev. Nucl. Part. Sci.* **28**, 33–113 (1978).
4. E. E. Koch, *Handbook of Synchrotron Radiation*, North Holland, New York (1983). See also Ref. 2.
5. G. Margaritondo, *Introduction to Synchrotron Radiation*, Oxford, New York (1988).
6. Ref. 3 (item 2); *Phys. Today*, Special Issue on Synchrotron radiation (May 1981); *Current Status of Facilities Dedicated to the Production of Synchrotron Radiation*, National Academy Press, Washington (1983). These are some representative examples.
7. "Planning Study for Advanced National Synchrotron Radiation Facilities," Department of Energy, Office of Basic Energy Sciences (March, 1984); "Synchrotron Radiation, a Perspective View for Europe," European Science Foundation, Strasbourg, France (1977). These are some representative examples.
8. *Synchrotron Radiation News*, ed. Pierre Dhez, Gordan and Breach, New York.
9. J. P. Blewett, *Nucl. Instrum. Methods* **A266**, 1 (1988).
10. D. H. Tomboulian and P. L. Hartman, *Phys. Rev.* **102**, 1423 (1956).
11. R. P. Madden and K. Codling, *J. Appl. Phys.* **36**, 380 (1965).
12. T. Sagawa, Y. Iguchi, M. Sasanuma, T. Nasu, S. Yamuguchi, S. Fujiwara, M. Nakamura, A. Ejiri, T. Masuoka, T. Sasaki, and T. Oshio, *J. Phys. Soc. Jpn.* **21**, 2587 and 2602 (1966).
13. R. Haensel and C. Kunz, *Z. Angew. Phys.* **23**, 276 (1967).
14. F. C. Brown, *Phys. Bl.* **12**, 619 (1978). A good summary history.
15. C. Gahwiller, F. C. Brown, and H. Fujita, *Rev. Sci. Instrum.* **41**, 1275 (1970); F. C. Brown, P. L. Hartman, P. Kruger, B. Lax, R. A. Smith and G. H. Vinyard, Synchrotron Radiation as a Source for the Spectroscopy of Solids, unpublished report for the Solid State Panel of the National Research Council, Washington, D.C. (March, 1966).
16. R. Tatchyn and I. Lindau, International Conference on Insertion Devices for Synchrotron Sources, Proc. of SPIE V 582 (1986).
17. G. P. Williams, *Nuclear Instrum. Methods* **A266**, 59 (1988).
18. D. Attwood, B. Hartline, and R. Johnson, The Advanced Light Source: Scientific Opportunities, Lawrence Berkeley Laboratory Publication 511 (1984).
19. R. Z. Bachrach, R. D. Bringans, B. B. Pate, and R. G. Carr, Proceeding of SPIE V 582, 251–267 (R. Tatchyn and I. Lindau, eds.) (1986).
20. J. A. R. Sampson, *Techniques of Vacuum Ultraviolet Spectroscopy*, Wiley, New York (1967).
21. R. L. Johnson, Grazing-incidence monochromators for synchrotron radiation—A review *Nucl. Instrum. Methods*, **A246**, 303–309 (1986).
22. F. C. Brown, R. Z. Bachrach, and N. Lien, *Nucl. Instrum. Methods* **152**, 73 (1978).
23. J. Cerino, J. Stohr, N. Hower, and R. Z. Bachrach, *Nucl. Instrum. Methods* **172**, 227 (1980).
24. R. Z. Bachrach, R. D. Bringans, B. B. Pate, and R. G. Carr, Proc. SPIE V 582, 251–267 (R. Tatchyn and I. Lindau, eds.) (1986); see also R. Z. Bachrach, L. E. Swartz, S. B. Hagstrom, M. H. Hecht, I. Lindau, and W. E. Spicer. *Nucl. Instrum. Methods* **208**, 105, (1983).
25. A. Zangwill, *Physics at Surfaces*, Cambridge, New York (1988).

26. P. F. Kane and G. B. Larrabee, eds, *Characterization of Solid Surfaces*, Plenum, New York (1974).

27. D. Langreth and H. Suhl, eds, *Many-Body Phenomena at Surfaces*, Academic Press, New York (1984).

28. F. Garcia-Molner and F. Flores, *Introduction to the Theory of Solid Surfaces*, Cambridge, New York (1979).

29. A. A. Madadudin, R. F. Wallis, and L. Dobrzynski, eds, *Handbook of Surfaces and Interfaces*, Vols. 1–3, Garland, New York (1978).

30. E. E. Koch, C. Kunz, and B. Sonntag, *Phys. Rep.* **29**, 153 (1977).

31. P. Eisenberger and L. C. Feldman, *Science* **214**, 300 (1981).

32. J. B. Pendry, *Low-Energy Electron Diffraction*, Academic Press, London (1974); G. A. Somorjai and M. A. van Hove, in *Structure and Bonding*, Springer-Verlag, Berlin (1979).

33. R. Z. Bachrach, R. D. Bringans, M. A. Olmstead, Proc. International School of Physics Enrico Fermi—Course CVI, "Current Trends in the Physics of Materials", Italian Physical Society, Lerici, Italy (1988).

34. I. Lindau and W. E. Spicer, Ch. 6 in Ref. 3.

35. F. Bassani and M. Altarelli, Ch. 7 in Ref. 4.

36. F. C. Brown, *Solid State Phys.* **29**, 1 (1974).

37. D. W. Lynch, Ch. 7 in Ref 2.

38. W. Gudat and C. Kunz, *Phys. Rev. Lett.* **29**, 169 (1972).

39. J. L. Freeouf and D. E. Eastman, *Crit. Rev. Solid State Sci.* **5**, 245 (1975).

40. R. Z. Bachrach, R. S. Bauer, and S. A. Flodström, Photoemission from Surfaces, 73 (R. F. Willis, B. Fitton, B. Feuerbacher, and C. Back, eds.), Ester, Noordwijk, Holland, (1976).

41. D. C. Koningsberger and R. Prins, *X-ray Absorption*, Vol. 92 in Chemical Analysis (J. D. Winefordner, ed.) Wiley, New York (1988).

42. R. S. Bauer, R. Z. Bachrach, S. A. Flodström, and J. C. McMenamin, *J. Vac. Sci. Technol.* **14**, 378 (1977).

43. R. Z. Bachrach, A. Bianconi, and S. A. Flodström, Proc. Third Int. Conference on Solid Surfaces (R. Dobrozemsky, F. Radenauer, F. P. Viehbock, A. Breth, ed.) F. Berger and Sohne, Vienna, Austria (Sept., 1977) p. 1205; A. Bianconi, R. Z. Bachrach, and S. A. Flodström, *Solid State Commun.*, **24**, 539 (1977).

44. A. Bianconi, *Appl. Surf. Sci* **6**, 392 (1981).

45. P. H. Citrin, *J. Phys. (Paris) Colloq.*, **47**, 437 (1986).

46. J. Stohr, in *Chemistry and Physics of Solid Surfaces V* (R. Vanselow and R. Howe, eds), Springer Series in Chemical Physics Vol. 35, 231, Springer, Berlin, (1984); *X-ray Absorption: Principles, Applications, Techniques of EXAFS, SEXAFS, and XANES* (D. C. Koningsberger and R. Prins, eds), Wiley, New York (1985).

47. R. S. Bauer and R. Z. Bachrach, *J. Vac. Sci. Technol.* **17**, 509 (1980).

48. R. Z. Bachrach, D. J. Chadi, and A. Bianconi, *Solid State Commun.* **28**, 93 (1978).

49. D. E. Aspnes, R. S. Bauer, R. Z. Bachrach, and J. C. McMenamin, *Phys. Rev.* **B16**, 5436 (1977); R. S. Bauer, R. Z. Bachrach, J. C. McMenamin, and D. E. Aspnes, *Nuovo Cimento* **39B**, 409 (1977).

50. T. A. Carlson, ed., *X-ray Photoelectron Spectroscopy*; *Benchmark Papers in Physical Chemistry and Chemical Physics*/2, Dowden, Hutchinson, & Ross, Stroudsburg, PA (1978). This is a collection of early papers.

51. N. V. Smith and F. J. Himpsel, in *Handbook of Synchrotron Radiation* (E. E. Koch ed.), North Holland, New York (1983).

52. L. Ley and M. Cardona, *Photoemission in Solids*, Vol. 26–27, Topics in Applied Physics, Springer Verlag, New York (1979).

53. B. Feurerbacher, B. Fitton, and R. F. Willis, *Photoemission and the Electronic Properties of Surfaces*, Wiley, New York (1978).

54. C. Caroli, B. Roulet, and D. Saint-James, "Theory of Photoemission," in *Handbook of Surfaces and Interfaces*, Vol. 1 (L. Dobrzynski, ed.) Garland, New York (1978). (Note that the chapter was submitted in 1974.)

55. P. J. Feibelman and D. E. Eastman, *Phys. Rev.* **B10**, 4932 (1974).

56. W. J. Schaich and N. W. Ashcroft, *Phys. Rev.* **B3**, 2453 (1971).

57. W. E. Spicer, *Phys. Rev.* **112**, 114 (1958).

58. C. Almbladh and L. Hedin, Ch. 8 in Ref. 4.

59. C. S. Fadley, "Basic Concepts in X-ray Photoelectron Spectroscopy," in *Electron Spectroscopy: Theory, Techniques, and Application*, Vol. 2 (C. R. Brundle and A. D. Baker, eds.), Academic Press, London (1978).

60. J. G. Endriz, *Phys. Rev.* **B7**, 3463 (1974).

61. H. Peterson, *Z. Phys.* **31**, 171 (1978).

62. H. J. Levinson, Ph.D. Thesis, U. of Pennsylvania (1980); H. J. Levinson and E. W. Plummer, *Phys. Rev.* **B24**, 628 (1981).

63. J. G. Lapeyre, A. D. Baer, J. Hermanson, J. Anderson, J. A. Knap, and p. L. Gobby, *Solid State Commun.* **15**, 1601 (1974); G. J. Lapeyre, J. Anderson, P. L. Gobby, and J. A. Knapp, *Phys. Rev. Lett.* **33**, 1290 (1977).

64. G. Rossi, *Surf. Sci. Rep.* **7** (1987).

65. E. V. Kane, *Phys. Rev. Lett.* **12**, 97 (1964).

66. J. W. Gadzuk, *Phys. Rev.* **B10**, 5030 (1974).

67. G. D. Mahan, *Phys. Rev. Lett.*, **24**, 1068 (1970).

68. G. W. Gobeli, F. G. Allen, and E. O. Kane, *Phys. Rev. Lett.* **12**, 94 (1964).

69. E. O. Kane, *Phys. Rev. Lett.* **12**, 97 (1964).

70. N. V. Smith, M. M. Traum, and F. J. DiSalvo, *Solid State Commun.* **15**, 211 (1974).

71. T. C. Chiang and D. E. Eastman, *Phys. Rev.* **B22**, 2940 (1980).

72. D. Briggs, ed., *Handbook of X-ray and Ultraviolet Photoelectron Spectroscopy*, Heyden, London (1977).

73. R. Z. Bachrach, F. C. Brown, and S. B. M. Hagstrom, *J. Vac. Sci. Technol.* **12**, 309, (1975).

74. V. Rehn, *Nucl. Instrum. Methods* **177**, 193 (1980).

75. G. V. Hansson, B. Goldberg, and R. Z. Bachrach, *Rev. Sci. Instrum.* **52**, 517 (1981).

76. S. M. Heald and E. A. Stern, *Phys. Rev.* **B17**, 4069 (1978); S. M. Heald and E. A. Stern, *J. Synth. Met.* **1**, 249 (1980).

77. E. A. Stern and S. M. Heald, Ch. 10 in Ref. 4.

78. E. A. Stern, Ch. 1 in Ref. 41.

79. T. M. Hayes and J. B. Boyce, *Solid State Phys.* **37**, 173 (1982).

80. P. A. Lee, P. H. Citrin, P. Eisenberger, and B. M. Kincaid, *Rev. Mod. Phys.* **53**, 769 (1981).

81. V. Rehn, SPIE **315**, 2 (1981); M. R. Howells, ed, *Reflecting Optics for Synchrotron Radiation*.

82. D. C. Koningsberger and R. Prins eds, *X-ray Absorption: Principles, Applications, Technique EXAFS, SEXAFS, and XANES*, Wiley, New York (1985), 450.

83. D. A. Fischer, J. B. Hastings, F. Zaera, J. Stohr, and F. Sette, *Nucl. Instrum. Methods* **A246**, 561 (1986).

84. R. Feidenhans'l, *Surf. Sci. Reps.* **10**, (1989).

85. J. A. Golovchenko, B. W. Batterman, and W. L. Brown, *Phys. Rev.* **B10**, 4239 (1974).

86. R. Z. Bachrach, "*MBE*—Molecular Beam Epitaxial Evaporative Growth" in *Crystal Growth*, 2nd edition (Brian Pamplin, ed.) Pergamon, New York 1980.

87. H. Prakash, *Prog. Cryst. Growth and Charact.* **8**, 427 (1984); see also Proc. Int. Conference on Metalorganic Vapor Phase Epitaxy, for example, *J. Cryst. Growth* **77** (1986).

88. E. Spiller and A. E. Rosenbluth, Proc. SPIE—Int. Soc. Opt. Eng. **563**, 221 (1985).

89. P. Pianetta and T. W. Barbee, Jr., *Nucl. Instrum. Methods* **A266**, 221 (1985).

90. T. W. Barbee, Jr., *Superlattices and Microstructures* **1** 311 (1985).

91. V. Narayanamurti, *Phys. Today* **37**, 24 (1984) and A. C. Gossard, *Thin Solid Films* **57**, 3 (1979).

92. J. Hwang, C. K. Shih, P. Pianetta, G. Kubiak, R. Stulen, Y. C. Pao, and J. S. Harris, *Appl. Phys. Lett.* **52**, 308–310 (1988); and J. Hwang, P. Pianetta, Y. C. Pao, C. K. Shih, Z.-X. Shen, P. A. P. Lindberg, and R. Chow, *Phys. Rev. Lett.* **6**, 1877–80 (1988).

93. D. Lichtman and Y. Shapira, "Photodesorption: A Critical Review", in *CRC Crit. Rev. Solid State Sci.*, **8**, 93 (1978).

94. S. A. Flodström, R. Z. Bachrach, R. S. Bauer, and S. B. M. Hagstrom, *Phys. Rev. Lett.* **37**, 1282 (1976).

95. D. E. Eastman, T. C. Chiang, P. Heimann, and F. J. Himpsel, *Phys. Rev. Lett.* **45**, 656 (1980).

96. R. Z. Bachrach and R. S. Bauer, *J. Vac. Sci. Technol.* **16**, 1149 (1979).

97. R. Z. Bachrach, Ch. 2 in "*Metal–Semiconductor Schottky Barrier Junctions*" (B. L. Sharma, ed.), Plenum, New York (1983).

98. F. Bechstedt, R. Enderlein, and D. Reichardt, *Physica (Utrecht)* **117B,** 825 (1983).

99. D. Spanjaard, C Guillot, M. C. Desjonquere, G. Treglia, and J. Lecante, *Surf. Sci. Rep.,* **V5,** (1985).

100. W. F. Egelhoff, Jr., *Surf. Sci. Rep.* **V6,** (1987).

101. S. A. Flodström, R. Z. Bachrach, R. S. Bauer, and S. B. M. Hagstrom, Proc. Third Int. Conference on Solid Surfaces, (R. Dobrozemsky, F. Radenauer, F. P. Viehbock, A. Breth, eds.) F. Berger and Sohne, Vienna, Austria (Sept. 1977), p. 869.

102. M. A. Olmstead, R. D. Bringans, R. I. G. Uhrberg, and R. Z. Bachrach, *Phys. Rev.* **B34,** 6401 (1986).

103. R. Z. Bachrach, R. S. Bauer P. Chiaradia, and G. V. Hansson, *J. Vac. Sci. Technol* **18,** 797 (1981).

104. R. Z. Bachrach, "MBE–Molecular Beam Epitaxial Evaporative Growth" Ch. 6 in *Crystal Growth,* 2nd edition, Brian Pamplin, ed., Pergamon, New York, (1980).

105. R. Ludeke, T. C. Chiang, and D. E. Eastman, *Physica (Utrecht)* **117B,** 819 (1983).

106. J. F. van der Veen, L. Smit, P. K. Larsen, and J. H. Neave, *Physica (Utrecht)* **117B,** 822 (1983).

107. K. Wandelt, *Surf. Sci. Rep.,* **V2,** (1982).

108. S. A. Flodström, R. Z. Bachrach, R. S. Bauer, and S. B. M. Hagstrom in *Photoemission from Surfaces,* (R. F. Willis, B. Fitton, B. Feuerbacher, and C. Backx eds.) Estec, Noordwijk, Holland (1976).

109. I. P. Batra and L. Kleinman, *J. Electron. Spectros. Rel. Phenom.* **33,** 175 (1984).

110. W. Eberhardt and F. J. Himpsel, *Phys. Rev. Lett.* **42,** 1375 (1979).

111. R. Z. Bachrach, D J. Chadi, and A. Bianconi, *Solid State Commun.* **28,** 931 (1978).

112. D. Norman, S. Brennan, R. Jaeger, and J. Stohr, *Surf. Sci.* **105,** L297 (1981).

113. R. Z. Bachrach, G V. Hansson, and R. S. Bauer, *Surf. Sci.* **109,** L560 (1981).

114. R. Z. Bachrach, S. A. Flodström, R. S. Bauer, S. B. M. Hagstrom, and D. J. Chadi, *J. Vac. Sci. Technol.* **15,** 488 (1978).

115. F. R. McFeely, J. F. Morar, N. D. Shinn, G. Landgren, and F. J. Himpsel, *Phys. Rev.* **B30,** 764, (1984).

116. C. G. van de Walle, F. R. McFeely and S. T. Pantelides, *Phys. Rev. Lett.* **61,** 1867 (1988).

117. J. A. Yarmoff and F. R. McFeely, *Phys. Rev.* **B38,** 2057 (1988).

118. R. S. Bauer, J. C. McMenamin, R. Z. Bachrach, A. Bianconi, L. Johansson, and H. Petersen in *Physics of Semiconductors* (B. L. H. Wilson, ed.) Inst. Phys Conf. Ser. **43,** 797, (1979).

119. A. Bianconi and R. S. Bauer *Surf. Sci.* **99,** 76 (1980).

120. C. R. Helms and B. E. Deal, eds, *The Physics and Chemistry of SiO₂ and the Si–SiO₂ Interface,* Plenum, New York (1988).

121. S. T. Pantelides, ed, *The Physics of SiO₂ and Its Interfaces,* Pergamon, New York (1978).

122. F. J. Grunthaner and P. J. Grunthaner, *Mat. Sci. Rep.* **1,** 65, (1986).

123. R. S. Bauer, R. Z. Bachrach, and L. J. Brillson, in *The Physics of MOS Insulators* (G. Lucovsky, S. T. Pantelides, and F. L. Galeener, eds.) Pergamon, New York, (1980), p. 221.

124. A. Ourmazd, D. W. Taylor, J A. Rentschler, and J. Bevk, *Phys. Rev. Lett.* **59,** 213 (1987).

125. P. H. Fuoss, L. J. Norton, S. Brennan, and A. Fischer-Colbrie, *Phys. Rev. Lett.* **60,** 600 (1988).

126. R. D. Bringans and R. Z. Bachrach, *Solid State Commun.* **45,** 83–86 (1983); R. Z. Bachrach and R. Bringans, *J. Phys. (Paris)* **43,** C5-145–C5-151 (1982).

127. R. A. Rosenberg, F. J. Love, Victor Rehn, I. Owen and G. Tjornton, *J. Vac. Sci. Technol.* **A4,** 1451 (1986).

128. C. B. Duke, *CRC Critical Rev. Solid State Sci.* **8,** 8 (1978).

129. A. Kahn, *Surf. Sci. Rep.* **3** (1983).

130. M. A. van Hove and S. Y. Tong, eds. *The Structure of Surfaces,* Vol. 2, Springer Series in Surface Science, Springer, Berlin (1985).

131. D. J. Chadi, *Phys. Rev.* **B29,** 785 (1984).

132. J. Northrup, *Phys. Rev. Lett.* **53,** 683 (1984); J. E. Northrup and M. L. Cohen, *Phys. Rev. Lett.* **49,** 1349 (1982).

133. R. Z. Bachrach, M. Skibowski, and F. C. Brown, *Phys. Rev. Lett.* **37,** 40 (1976).

134. J. E. Rowe, M. M. Traum, and N. V. Smith, *Phys. Rev. Lett.* **33,** 1333 (1974).

135. G. V. Hansson and R. I. G. Uhrberg, *Surf. Sci. Rep.* **9** (1988).

136. K. C. Pandey, *Phys. Rev. Lett* **47,** 1913 (1981).

137. D. Haneman, *Phys. Rev.* **121,** 1093 (1961); D. Haneman and D. L. Heron in *The Structure and Chemistry of Solid Surfaces* (G. A. Somorjai, ed.) Wiley, New York (1968).

138. G. V. Hansson, R. Z. Bachrach, R. S. Bauer, D. J. Chadi, and W. Göpel, *Surf. Sci.* **99,** 13 (1980).

139. K. C. Pandey, private communication.

140. M. A. Olmstead, *Surf. Sci. Rep.* **6,** 159 (1986).

141. G. V. Hansson and R. I. H. Uhrberg, *Surf. Sci. Rep.* **9,** 197 (1988).

142. J. F. van der Veen, P. K. Larsen, J. H. Neave, and B. A. Joyce, *Solid State Commun.* **49,** 659 (1984).

143. R. D. Bringans and R. Z. Bachrach, Proc. 17th Int. Conference on the Physics of Semiconductors (D. J. Chadi and W. A. Harrison, eds.) Springer Verlag, N.Y. (1985).

144. P. K. Larsen, J. H. Neave, and B. A. Joyce, *J. Phys. C.* **14,** 167 (1981); P. K. Larsen, J. F. van der Veen, A. Mazur, J. Pollmann, and B. H. Verbeek, *Solid State Commun.* **40,** 459 (1981).

145. D. J. Chadi, C. Tanner, and J. Ihm, *Surf. Sci.* **120,** L425 (1982); J. Ihm, D. J. Chadi, and J. D. Joannopoulos, *Phys. Rev.* **B27,** 5119 (1983).

146. R. D. Bringans and R. Z. Bachrach, *Solid State Commun.* **45,** 83 (1983).

147. R. S. Bauer, ed., *Surfaces and Interfaces: Physics and Electronics,* North Holland, Amsterdam, (1983); *Surf. Sci.* **132,** (1983).

148. R. S. Bauer, *Physics and Chemistry of Semiconductor Interfaces,* Annual Conf. Proc., *J. Vac. Sci. Technol.*

149. R. Z. Bachrach, "Metal–semiconductor surface and interface states on (110) GaAs" *J. Vac. Sci. Technol.* **15,** 1340 (1978); R. Z. Bachrach, R. S. Bauer, J. C. McMenamin, and A. Bianconi, *Physics of Semiconductors* (B. L. H. Wilson, ed.) Institute of Phys. Conf Ser **43,** 1073 (1979); L. J. Brillson, R. Z. Bachrach, R. S. Bauer, and J. C. McMenamin, *Phys. Rev. Lett.* **42,** 397 (1979).

150. D. J. Chadi and R. Z. Bachrach, Chemisorption site geometry and interface electronic structure of Ga and Al on GaAs (110), *J. Vac. Sci. Technol.* **16,** 1159 (1979).

151. R. R. Daniels, A. D. Katnani, T. Zhao, G. Margaritondo, and A. Zuriger, *Phys. Rev. Lett.* **49,** 895 (1982).

152. R. S. Bauer, R. Z. Bachrach, G. V. Hansson, and P. Chiaradia, *J. Vac. Sci. Technol.* **19,** 674, (1981).

153. L. J. Brillson, *Surf. Sci. Rep.* **2** (1982). This review contains more than 1000 references.

154. W. E. Spicer, P. W. Chye, P. R. Skeath, C. Y. Su, and i. Lindau, *J. Vac. Sci. Technol.* **16,** 1422 (1979).

155. J. van Laar and J. J. Scheer, *Surf. Sci.* **8,** 342 (1967).

156. M. Henzler, *Surf. Sci.* **36,** 109 (1973).

157. P. W. Chye, I. A. T. Sukegawa, and W. E. Spicer, *Phys. Rev. Lett.* **35,** 1602 (1975).

158. R. H. Williams, R. R. Varma, and A. McKinley, *J. Phys. C:* **10,** 4545 (1977).

159. L. J. Brillson, *Surf. Sci.* **69,** 62 (1977).

160. R. Z. Bachrach, Ch. X, in *Molecular Beam Epitaxy* (Brian Pamplin, ed.) Pergamon, New York (1980).

161. J. Ihm and J. D. Joannopoulos, *Phys. Rev.* **B26,** 4429 (1982).

162. M. S. Daw and D. L. Smith, *Phys. Rev.* **B20,** 5150 (1979); *J. Vac. Sci. Technol.* **17,** 1028 (1980).

163. R. E. Allen and J. D. Dow, *Phys. Rev.* **B25,** 1423 (1982).

164. I. Lindau and W. E. Spicer, Ch. 4 in *Electron Spectroscopy: Theory, Techniques, and Applications* Vol. 4, (C. R. Brundle and A. D. Baker, eds.) Academic Press, New York, (1981).

165. W. E. Spicer, P. W. Chye, P. R. Skeath, C. Y. Su, and I. Lindau, *J. Vac. Sci. Technol.* **16,** 1422 (1979).

166. R. Z. Bachrach and A. Bianconi, *J. Vac. Sci. Technol.* **15,** 525 (1978).

167. J. J. Yeh and I. Lindau, *At. Data Nucl. Tables* **32,** 1 (1985).

168. S. M. Goldberg, C. S. Fadley, and S. Kono, *J. Electron. Spectros. Relat. Phenom.,* **21,** 285 (1981).

169. S. A. Flodström, R. Z. Bachrach, R. S. Bauer, J. C. McMenamin, and S. B. M. Hagstorm, *J. Vac. Sci. Technol.* **14,** 303 (1977).

170. R. Z. Bachrach and A. Bianconi, Proc. Vth Int. Conf on Vacuum Ultraviolet Radiation Physics (M. C. Castex, M. Pouey, and N. Pouey, eds.) CNRS, Meudon, France (1977) VII, 213; R. Z. Bachrach and A. Bianconi, *Solid State Commun.* **V42,** 529 (1982).

171. A. Bianconi, S. B. M. Hagstrom, R. Z. Bachrach, *Phys. Rev.* **B16,** 5543 (1977).
172. J. E. Ingleslield, *Solid State Commun.* **40,** 467 (1981).
173. M. Sunjic and D. Sokevic, *Solid State Commun.* **15,** 165 (1974); **18,** 373 (1976); M. Sunjic, D. Sokevic, and A. Lucas, *J. Electron Spectrosc. Relat. Phenom.* **5,** 963 (1974).
174. D. Chastenet and P. Longe. *Phys. Rev. Lett.* **44,** 91 (1980).
175. J. W. Gadzuk in *Photoemission and the Electronic Properties of Surfaces* (B. Feuerbacher, R. Fitton, and R. F. Willis, eds.) Wiley-Interscience, New York (1978), p. 11.
176. S. A. Flodström, R. Z. Bachrach, R. S. Bauer, J. C. McMenamin, and S. B. M. Hagstrom, *J. Vac. Sci. Technol.* **14,** 303 (1977).
177. D. Norman and D. P. Woodruff, *Surf. Sci.* **79,** 76 (1979).
178. L. Johansson and I. Lindau *Solid State Commun.* **79,** 76 (1979).
179. M. H. Hecht and I. Lindau, *Nucl. Instrum. Methods* **195,** 339 (1982).
180. A. Bianconi, *Appl. Surf. Sci.* **6,** 391 (1978).
181. K. Zeppenfeld, *Z. Phys.* **211,** 391 (1978).
182. C. Webb and P. Williams, *Surf. Sci.* **53,** 110 (1975).
183. A. M. Bradshaw, S. L. Cederbaum, W. Domeke, and V. Kreupe, *J. Phys.* **C7,** 4503 (1974).
184. P. J. Durham, I. B. Pendry, and C. H. Hodges, *Solid State Commun.* **38,** 159 (1981).
185. J. J. Cheng and D. Langreth, *Phys. Rev.* **B8,** 4638 (1974).
186. B. B. Pate, P. M. Stefan, C. Binns, P. J. Jupiter, M. L. Shek, I. Lindau, and W. E. Spicer, *J. Vac. Sci. Technol.* **19,** 349 (1981).
187. B. B. Pate, M. H. Hecht, C. Binns, I. Lindau, and W. Spicer, *J. Vac. Sci. Technol.* **21,** 364 (1981).
188. B. B. Pate, Ph.D. Thesis, "The Diamond Surface: Atomic and Electronic Structure," Stanford U., January, 1984 (unpublished).

Absorption

Surface X-Ray Absorption Near-Edge Structure: XANES

A. Bianconi and A. Marcelli

1. INTRODUCTION

Surface X-ray absorption spectroscopy concerns electronic transitions from core levels of atoms at the surfaces of solids. The local character of core level excitations makes the study of atoms at the surfaces of solids attractive in investigating both the local structure and the localized electronic states. Moreover, the large energy separation between inner shells makes it possible to select a particular core level of a selected atomic species.

The knowledge of atomic arrangement of neighbor atoms around a selected atom on a surface is important, for example, in the study of chemisorption and oxidation processes. In fact, the local structure of chemisorption sites and the local structure of amorphous surface oxides are basic information not directly given by other techniques based on diffraction methods such as (LEED).[1]

Surface X-ray absorption spectroscopy was born in the middle of the 1970s when tunable X-ray synchrotron radiation emitted by intense and stable sources (electron storage rings) became available. The requirements for the X-ray source were tunability, to carry out spectroscopy, and high intensity, because of the low concentration of surface atoms.

The main experimental problem to be solved was to find experimental techniques with high surface sensitivity. Lukirskii and Brytov in 1964[2] using the continuum bremsstrahlung of a standard soft X-ray tube, and Gudat and Kunz in 1972,[3] using synchrotron radiation, demonstrated that the total electron yield (TY) of the sample surface is proportional to the bulk absorption coefficient. The sampling depth of this technique is of the order of magnitude of hundreds of angstroms, making this technique suitable for bulk investigations.

A. Bianconi • Consorzio Interuniversitario di Fisica della Materia (INFM), Dipartimento di Fisica, Università "La Sapienza," 00185 Rome, Italy. *A. Marcelli* • Istituto Nazionale di Fisica Nucleare (INFN), Laboratori Nazionali di Frascati, 00044 Frascati, Italy.

Synchrotron Radiation Research: Advances in Surface and Interface Science, Volume 1: Techniques, edited by Robert Z. Bachrach. Plenum Press, New York, 1992.

The first experiment in surface X-ray absorption with high surface sensitivity was carried out in 1977 at the Stanford synchrotron radiation facility by Bianconi, Bachrach, and Flodström[4] through detection of the Auger electron yield. The absorption spectrum of the Al surface atoms in the top monolayers of an aluminum crystal was distinguished from the aluminum bulk spectrum. (see Fig. 1). The surface contrast was increased by selecting an Auger line due to Al surface atoms interacting with chemisorbed oxygen (an interatomic aluminum–oxygen Auger transition).

The theoretical relation between Auger electron yield and surface absorption coefficient was independently predicted by Lee[5] and Landman and Adams.[6] Both the total electron yield[7] and the Auger electron yield[8] methods were used to measure the *surface extended X-ray absorption fine structure* (SEXAFS) of different atomic species chemisorbed on solids, but the absorption coefficient of the atoms on the clean surface can be detected only by selecting an Auger line with energy on the order of 50 eV, which corresponds to the minimum in the curve of the electron escape depth.[9]

SEXAFS concerns the study of the modulation of the absorption coefficient over a range of photoelectron wave vectors above about 3Å^{-1}. In this range experimental data can be analyzed in the framework of the EXAFS theory.[10] Several reviews about SEXAFS have been published over the years.[11-16]

The spectral features in the low energy range (a few tens of eV) of the X-ray absorption spectra, called *X-ray absorption near-edge structure* (XANES),[11,14] have attracted interest because they contain information about the local geometry, both bond distances and bond angles. In fact, higher-order terms of the correlation function of the atomic distribution become important in the XANES energy region, while only the pair-correlation function of the atomic distribution gives the main contribution in the EXAFS range.

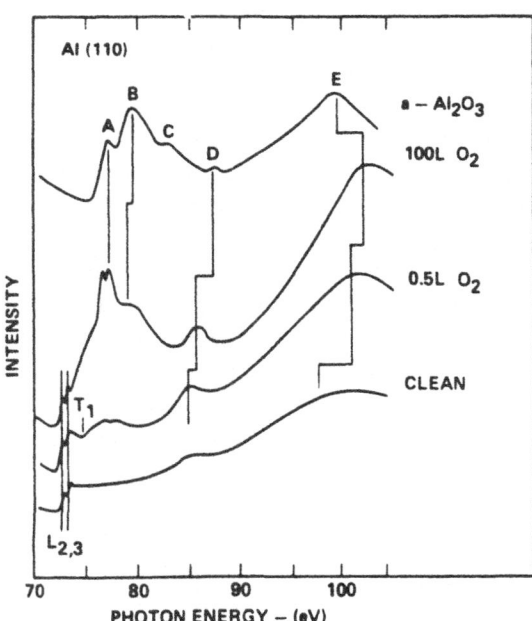

FIGURE 1. Aluminum $L_{2,3}$ surface XANES of Al ions at the Al (110) surface interacting with 100 L of chemisorbed oxygen $(1 \text{L} = 10^{-6} \text{Torr sec})$, corresponding to about one monolayer coverage, detected by Auger Al (2p) O (2s) V $(E_f = 45.5 \text{eV})$ electron yield. The lowest curve shows the minor contribution of the Al metal substrate (clean spectrum). The upper curve shows the Al L_{23} XANES of bulk amorphous Al_2O_3. The differences between the XANES curve (100 L O_2) of Al interacting with chemisorbed oxygen and that of bulk oxide (a-Al_2O_3) show the specific Al–O coordination in the chemisorbed phase. (From Ref. 4.)

In the early surface XANES experiments, the application of XANES to local geometry determination has been limited to a fingerprint approach using model compounds,[17,18] due to the lack of reliable theoretical analysis. In the last few years important advances toward quantitative theoretical analysis of the XANES data have been made.

In section 2, an overview of theoretical XANES calculations will be presented. Section 3 gives a brief panorama of the experimental methods, while section 4 reviews some applications of surface XANES.

2. CALCULATION OF XANES

X-ray absorption near-edge structure concerns the electronic transitions from atomic inner shells to unoccupied states. In classical quantum theory, the absorption cross section is given by many-body excitations of the N-electron system. Following the interaction with the photon beam of energy ω, the system is excited from the initial state i at energy E_i to a final state f at energy E_f. In the dipole approximation, the total absorption cross section is given by

$$\sigma(\omega) \sim \omega \sum_f |M_{if}|^2 \, \delta(E_i - E_f + \omega), \qquad (1)$$

where the sum is extended over all the possible final states f and M_{if} is the matrix element involving the initial Ψ_i and final Ψ_f^* many-body radial wave functions

$$M_{if} = \int \Psi_f^*(r_1, r_2, \ldots, r_n) \sum_n (\mathbf{r}_n \cdot \mathbf{e}) \Psi_i(r_1, r_2, \ldots, r_n) \, dr \qquad (2)$$

where $\mathbf{e}$ is the unitary polarization vector of the electric field and $\mathbf{r}_n$ is the vector describing the position of the nth electron.

In the one-electron approximation, X-ray absorption is described by single-particle processes. The N-electron system is separated into two parts: the single electron in the core level which is excited into an unoccupied level, and the $N - 1$ passive electrons, which do not participate directly in the electronic transition. The one-electron transition takes place in a static potential determined by a single configuration of the $N - 1$ passive electrons.

In the case of metals, the core hole in the static final-state potential is fully screened by valence electrons close to the Fermi level. Most of the XANES spectra can be interpreted within the framework of the von Barth and Grossmann final-state rule,[19,20] which states that the wave function of the excited photo-electron is determined by the final-state potential with the core hole and the relaxed $N - 1$ electrons. Therefore, in most cases, the one-electron transitions are assumed to take place in the potential of the fully relaxed configuration of the $N - 1$ passive electrons in the presence of the core hole. The fully relaxed configuration is defined as the final-state many-body configuration with the lowest energy.

Many-body effects can arise because of the presence of different many-body final-state configurations of the $N - 1$ passive electrons. Configuration interaction of many-body final states can appear because of core-hole perturbation in the final state and/or configuration interaction in the ground state as in valence fluctuating materials.[21]

2.1. The Band-Structure Approach

The absorption coefficient $\mu_c(E)$ for the transitions from the core level $c(n, l, J)$ with energy E_c and wavefunction φ_{cM}, in the one-electron approximation, can be expressed in atomic units as

$$\mu_c(E) = \frac{4\pi^2 \alpha}{\Omega/\nu} F_c(E), \qquad E > E_F, \tag{3}$$

where $\alpha^{-1} = 137.036$ is in the inverse fine-structure constant, Ω is the volume of the primitive cell, ν is the number of contributing atoms in the primitive cell, and $F_c(E)$ is the spectral distribution of oscillator strength[22]

$$F_c(E) = (\omega/3) \sum_{\mathbf{k},j} \sum_{M=-J}^{J} \mathbf{r}_{cM,\mathbf{k}j}^2 \, \delta(E - E_{\mathbf{k}j}) \tag{4}$$

$$\mathbf{r}_{cM,\mathbf{k}j} = \int \varphi_{cM}(r)\mathbf{r}\Psi_{\mathbf{k}j}(r) \, d^3r \tag{5}$$

$$= \langle \varphi_{cM} \, |\mathbf{r}| \, \Psi_{\mathbf{k}j} \rangle$$

$\Psi_{\mathbf{k}j}$ and $E_{\mathbf{k}j}$ are the wavefunction and energy of the jth conduction band at reduced vector $\mathbf{k}$, ω is the photon energy $\omega = E - E_c$, and the integration is carried out over the volume of the primitive cell. Since the dipole transitions dominate the process of photoabsorption, an electron from a core level having angular momentum l is excited into the $l \pm 1$ final states. Neglecting the spin–orbit coupling for the band states, $F_c(E)$ can be written as[23,24]

$$F_c = \frac{\omega}{3} \frac{2J + 1}{2(2l + 1)} \left[\frac{l}{2l - 1} f_{c,l-1}(E) + \frac{l + 1}{2l + 1} f_{c,l+1}(E) \right], \tag{6}$$

where the partial strength $f_{c,l}(E)$ can be factored into

$$f_{c,l}(E) = \rho_{c,l} N_l(E). \tag{7}$$

$N_l(E)$ is the angular projected density of states defined as

$$N_l(E) = 2 \sum_{\mathbf{k},j} \sum_m |\langle \mathbf{Y}_{lm} \, | \, \Psi_{\mathbf{k}j} \rangle|^2 \, \delta(E - E_{\mathbf{k}j}), \tag{8}$$

where the energy-band states are labelled by the reduced wave vector $\mathbf{k}$ and the band index j. The effective matrix element $\rho_{c,l}(E)$ is given by

$$\rho_{c,l}(E) = \frac{\langle \varphi_c \, |\mathbf{r}| \, \phi_l(E) \rangle^2}{\langle \phi_l^2(E) \rangle}, \tag{9}$$

where the wave function $\phi_l(E, r)$ is a solution of the radial Schrödinger equation inside the muffin-tin (MT) sphere of radius S $(r < S)$, and outside the MT sphere $(r \geq S)$ is given by

$$\phi_l(E, r) = [\cos \delta_l(E)]J_l[(E - V_0)^{1/2}r] - [\sin \delta_l(E)]n_l[(E - V_0)^{1/2}r], \tag{10}$$

where $\delta_l(E)$ is the l th phase shift of the muffin-tin potential, V_0 is the muffin-tin zero of the potential, and J_l and n_l are spherical Bessel functions.

The spectra of crystals are therefore interpreted in terms of the product of the partial (of selected orbital momentum $l \pm 1$) density of states and of the matrix element. Theoretical calculations of the partial and projected density of states of the crystal-band structure were reported by several authors to interpret the XANES spectra.[25-33] In Figure 2, the large differences between the K edge and the L_3 edge of the Si crystal are interpreted as arising from the difference between the $l = 0$ and the $l = 1$ partial density of unoccupied states of the conduction band.[33]

Müller et al.[23,24] have shown that the XANES spectra can be understood as the product of an atomic-like term and a solid-state term. The factorization of the partial oscillator strength into the solid-state term $\chi_l(E)$ and the atomic term $f_{c,l}^{at}(E)$ is given by

$$f_{c,l}(E) = f_{c,l}^{at}(E)\chi_l(E). \tag{11}$$

The atomic term is obtained by considering a single muffin-tin potential confined in the Wigner–Setz (WS) sphere of radius S_{WS},

$$f_{c,l}^{at}(E) = (2l + 1)N^{FE}(E)\langle \varphi_c \, |\mathbf{r}| \, \phi_l(E, r)\rangle^2, \tag{12}$$

where the integration is over the Wigner–Seitz (WS) sphere and $N^{FE}(E) = (E - V_0)^2/2\pi$ is the free electron density of states. The solid state term is

$$\chi_l(E) = \frac{N_l(E)}{N_{c,l}^{at}(E)}, \tag{13}$$

where

$$N_l^{at}(E) = \frac{1}{\Omega}(2l + 1)N^{FE}(E)\langle \phi_l^2(E) \rangle \tag{14}$$

is the projected density of states for the single-sphere problem.

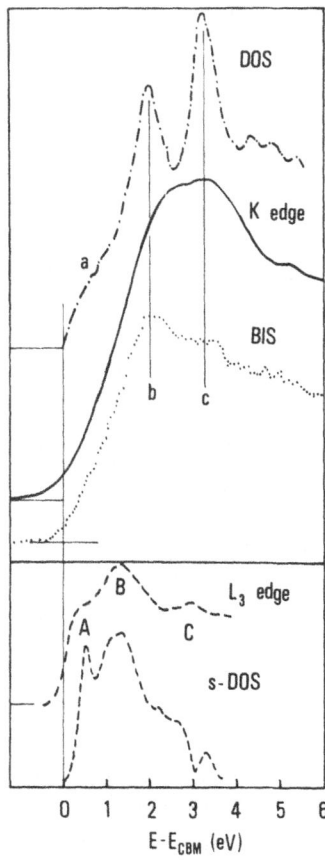

FIGURE 2. Si K-edge and Si L_3-edge absorption spectra. The Si L_3-edge spectrum is compared with the partial ($l = 0$) s density of states of the unoccupied conduction band, lower panel. The Si K-edge spectrum is compared with the total density of states (mostly p ($l = 1$)). The BIS spectrum, which is often assumed to probe the total density of states, is shown for comparison. (From Ref. 33.)

This approach leads to the following understanding of the X-ray absorption spectra: the overall magnitude and shape of a particular spectrum is determined by the corresponding atomic transition rate $f_{c,l}^{at}(E)$, and the fine structure of the spectrum is determined by the solid-state factor $\chi_l(E)$, which is proportional to the density of band states with $l \pm 1$ orbital characters.

The $l - 1$ term is often ignored because usually it exhibits a much smaller amplitude than the $l + 1$ term. However in some cases, as in the first 6 eV above threshold in the Si L_3 edge, the $l + 1$ density of states is negligible and the $l - 1$ term is the most important.

In some cases the atomic term exhibits strong resonances in a small energy range; therefore, the contribution of the solid-state factor $\chi_l(E)$ can be neglected. In these cases the spectra of solids can be interpreted in terms of atomic transitions (see for examples the $L_{2,3}$ edges of Ni and Cu in insulating systems).[34,36] In other cases the atomic portion exhibits a smooth, structureless spectrum over a large energy range, and therefore the spectral features can be assigned to the solid-state factor $\chi_l(E)$.

It is very important to include in the theory the lifetime of the photoelectron, in order to get a good agreement between calculated spectra and experiments. The lifetime of the photoelectron takes into account the inelastic scattering of the photoelectron by valence electrons, which is an essential physical aspect of the states at high energy in the conduction band, because the photoelectrons with energy E have enough energy to excite all valence electrons with binding energy smaller than E.

2.2. The Multiple Scattering Approach

Experimental evidence was found that the K-edge XANES of several systems over a range of about 50 eV depend mainly on the geometrical structure of a finite cluster of atoms around the absorbing atom.[11,37,38] An interpretation of the X-ray absorption near-edge structure was proposed in terms of multiple scattering resonances of the photoelectron in a cluster of finite size in real space. This interpretation was based on an extention to condensed systems of the shape resonances, which are localized states in the continuum. The spectra of diatomic molecules like N_2[39] show such resonances, which were interpreted according to a multiple scattering theory by Dill and Dehmer.[40]

The size of the cluster relevant for XANES was found to change in different systems ranging from a single shell to several shells around a photoabsorbing atom, depending on the electron mean free path for inelastic scattering and on the core hole lifetime in crystalline solids. In disordered systems (i.e., amorphous and liquid systems) the structural disorder reduces the number of neighbor shells to one or two.

The comparison between the calculated absorption coefficient and the experiments indicates that the theoretical spectra have to be convoluted with an intrinsic Lorentzian broadening function $\Gamma(E)$, to be added to the instrumental bandwidth, in order to obtain good agreement. The intrinsic broadening of the excited states $\Gamma(E)$ is the sum of two terms: the first is the core-hole width Γ_h and the second is the energy bandwidth $\Gamma_e(E)$ of the excited electron of energy E. These are related to τ_h, the core-hole lifetime, and to $\tau_e(E)$, the lifetime of the electron, which is a function of the mean free path of the excited electrons $\lambda(E)$. $\lambda(E)$ is determined by the inelastic scattering of the photoelectron with the passive electrons; it is energy dependent and varies for each material:

$$\Gamma_{\text{tot}}(E) = \Gamma_h + \Gamma_e(E) = \frac{h/2\pi}{\tau_h} + \frac{h/2\pi}{\tau_e(E)} = \frac{h}{2\pi}\left[\frac{1}{\tau_h} + \frac{2(2E/m)^{1/2}}{\lambda(E)}\right]. \quad (15)$$

At low kinetic energies of the electron, $\Gamma_h > \Gamma_e$ and the Γ_h term dominates, while at high energies, $\Gamma_h < \Gamma_e$, so that the $\Gamma_e(E)$ term dominates. It is possible to define an effective mean free path given by

$$\lambda_{\text{eff}}(E) = \frac{h}{2\pi}\frac{2(2E/m)^{1/2}}{\Gamma_{\text{tot}}(E)}. \quad (16)$$

In Fig. 3 the mean free path $\lambda(E)$ for a silicon crystal was obtained by photoelectron escape depth measurements. The derived $\Gamma_e(E)$ and the $\lambda_{\text{eff}}(E)$ using $\Gamma_h = 0.4$ eV for the Si K-level width are plotted. λ_{eff} is an estimate of the radius of the cluster probed by XANES. In recent years several theoretical

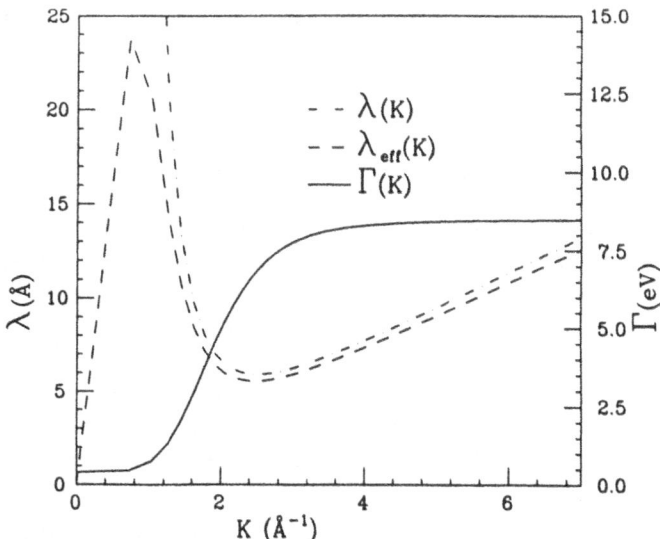

FIGURE 3. Electron mean free path $\lambda(E)$, final-state electron bandwidth $\Gamma_e(E)$, and $\lambda_{eff}(E)$ for the case of crystalline silicon.

approaches have been developed to solve the absorption cross section for core transitions in real space in the frame of multiple scattering theory.[40-53]

The generally strong scattering power of the atoms of condensed matter for low-kinetic-energy photoelectrons favors multiple scattering (MS) processes. At higher energies such that the atomic scattering power becomes small, a single scattering (SS) regime is found. In a SS regime the modulation in the absorption coefficient is substantially due to interference of the outgoing photoelectron wave from the absorbing atom with the backscattered wave from each surrounding atom, yielding extended X-ray absorption fine structure (EXAFS).[54,55] Hence EXAFS provides information about the pair-correlation function. By decreasing the photoelectron kinetic energy, a gradual transformation occurs from the EXAFS regime to the XANES regime.[56-59]

Therefore, EXAFS probes the first-order or pair-correlation function of the atomic distribution near the absorbing atom, while XANES probes the triplet and higher orders of the atomic distribution function. Interest in determination of higher-order correlation functions of local atomic distributions in complex systems has stimulated the growth of XANES.

Muliple scattering theories have been used in recent years to analyze the XANES spectra of crystals, amorphous solids, surfaces, biological molecules, liquids, catalysts and chemical compounds.[60-77] The multiple scattering method has been developed in nuclear physics to calculate nuclear scattering cross sections and in solid state physics to compute the electronic structure of solids. The extension of the bound-state molecular scattering method of Johnson and co-workers[78] to determine the one-electron wave function for continuum states was formulated first by Dill and Dehmer.[40] The continuum wave function is matched to the proper asymptotic solution of the Coulomb scattering states, and in this way the multiple scattering problem is changed from a homogeneous eigenvalue problem (bound states) to an inhomogeneous one in which the continuum wave function is determined by an asymptotic T-matrix normalization condition.[42] In this scheme the total potential is represented by a cluster of nonoverlapping spherical potentials centered on the atomic sites, and the molecule as a whole is enveloped by an "outer sphere." Three regions can be identified in this partitioning:

- atomic regions (spheres centered upon nuclei, normally called region I);
- the extramolecular region (the space beyond the outer sphere radius, region III);
- and an interstitial region of complicated geometry in which the molecular potential is approximated by a constant muffin-tin potential.

The Coulomb and exchange part of the input potential are calculated on the basis of a total charge density obtained by superimposing the atomic charge densities, calculated from the tables of Clementi and Roetti,[79] of the individual atoms constituting the cluster. For the exchange potential it is possible to use both the usual energy-independent Slater approximation[80] and the energy-dependent Hedin–Lundqvist[81] potential in order to incorporate the dynamical effect. Following this theory, the expression for the absorption coefficient for a cluster of atoms, for polarized light in the vector **e** direction, is given in the dipole approximation by

$$\mu_c(E) = N_c \sigma(E, \mathbf{e}) = n_c \sigma(k, \mathbf{e}) \tag{17}$$

$$\sigma(k, \mathbf{e}) = 4\pi\omega\alpha k \, \mathbf{Im}\left[\sum_{L,L'} \langle \varphi_c \, |\mathbf{r} \cdot \mathbf{e}| \, \phi_l Y_L \rangle \tau_{L,L'} \langle \phi_l Y_L \, |\mathbf{r} \cdot \mathbf{e}| \, \varphi_c \rangle \right], \tag{18}$$

where $\sigma(E, \mathbf{e})$ is the photoabsorbing cross section, N_c is the density of atoms, $L = (l, m)$ is the angular momentum, α is the fine-structure constant, k is the photoelectron wave number, and the spin dependence has been neglected. ϕ_l's are regular solutions of the radial Schrödinger equations in the photoabsorber muffin-tin sphere, matching the same boundary conditions as in Eq. (10).[82–84] All the structural information is contained in the quantity

$$\tau_{L,L'} = (\sin \delta_l(E) \sin \delta_{l'}(E))^{-1}[(\mathbb{T}_a^- - \mathbb{G})^{-1}]_{L,L'}, \tag{19}$$

where δ_l^0 is the lth phase shift of the absorbing atom assumed to be located at site 0, $\mathbb{G} = G_{L,L'}$ is the matrix describing the free spherical wave propagation of the photoelectron from site i and angular momentum $L = (l, m)$ to site j and angular momentum $L'(l', m')$ in the angular momentum representation, and $\mathbb{T}_a = \delta_{i,j} \delta_{L,L'}[\exp(i \delta_l^i) \sin \delta_l^i]$ is the diagonal matrix of atomic t-matrix elements describing the scattering process of the L spherical wave photoelectron by the atom located at site i with phase shift δ_l^i. Under certain conditions[56–59] it is possible to express the structural factor as an absolute convergent series and to expand the photoabsorption cross section in partial contributions as follows:

$$\sigma(E, \mathbf{e}) = \sum_{n=0} \sigma_n(E, \mathbf{e}). \tag{20}$$

The $n = 0$ term represents the smoothly varying "atomic" cross section, the $n = 1$ term is always zero, and the generic term n is the contribution to the photoabsorption cross section from processes in which the photoelectron has been scattered $n - 1$ times by the surrounding atoms before returning to the photoabsorbing site. In particular, $\sigma_2(E, \mathbf{e})$, the EXAFS term in the spherical wave representation, can be written both for the K edge and the L_1 edge as an approximation to the general polarized EXAFS formula[85]

$$\sigma_2(E, \mathbf{e}) = |M_{01}(E)|^2(-1) \sum_j \cos^2 \theta_j$$

$$\times \mathbf{Im}\left\{\exp\left[\frac{2i(\rho_j + \delta_l^0)}{\rho_j^2}\right] \sum_l (-1)^l (2l + 1) t_l^j f_i^2(\rho_j)\right\}, \tag{21}$$

where $\rho_j = kR_j$, θ_j is the angle between the polarization vector $\mathbf{e}$ and the direction R_j joining the jth atom with the absorbing one, $|M_{01}(E)|^2$ is the radial matrix element between the initial $l = 0$ state and the final dipole allowed R_1 radial wavefunction, and f_l takes into account the spherical correction to the free propagators.[84,85] The total absorption coefficient can be written as

$$\mu(E, \mathbf{e}) = \mu_0\left[1 + \sum_{n\geq 2} \chi_n(E, \mathbf{e})\right], \tag{22}$$

where $\mu_0(E)$ is the structureless absorption coefficient of a central photoabsorbing atom and $\chi_n(E)$ represents the contribution arising from all MS pathways beginning and ending at the central atom and involving $n - 1$ neighboring atoms. A pictorial view of the multiple scattering pathways contributing to the XANES is shown in Fig. 4.

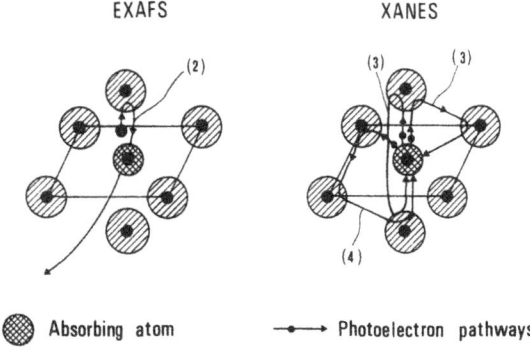

FIGURE 4. Pictorial view of the single-scattering process of the excited photoelectron giving the EXAFS signal $\chi_2(E)$ and the multiple-scattering processes giving $\chi_3(E)$ and $\chi_4(E)$ contributions.

The terms χ_n are usually plotted versus the wave vector k of the photo-electron. The EXAFS $\chi_2(k)$ term is the dominant term above wave vector values $k > 3 \sim 4 \, \text{Å}^{-1}$. The $\chi_3(k)$ term is the double scattering signal arising from all the triangular paths with one vertex on the central atom. The multiple scattering signal is defined as

$$\chi_{MS}(k) = \sum_{n \geq 2} \chi_n(k) = \frac{\mu(k) - \mu_0(k)}{\mu_0(k)} - \chi_2(k) \tag{23}$$

The general mathematical expression for $\chi_n(k)$, without taking into account the inelastic interactions of the photoelectron—i.e., its mean free path and the structural Debye–Waller factor—can be written as

$$\chi_n^l(k) = \sum_{p_n} A_n^l(k, p_n) \sin (kR_{p_n} + \phi_n^l(k, p_n) + 2\delta_l^0), \tag{24}$$

where δ_l^0 is the central atom's lth phase shift and the dependence of the amplitude and phase function A_n^l and ϕ_n^l on the particular path has been indicated symbolically by p_r. General expressions for calculating the quantities A_n^l and ϕ_n^l in terms of the atomic phase shifts and the geometry of the path p_n are provided by the MS theory. A substantial simplification of these expressions is achieved by means of a relatively simple approximation of the propagator of the photoelectron in the final state, the spherical-wave approximation (SWA),[86] which preserves the spherical-wave character of the propagation and is rather accurate even at very low wave vector values ($k \approx 1$–$2 \, \text{Å}^{-1}$).

Due to the general structure of the quantities $\chi_n^l(k)$, the amplitudes A_n^l decrease with increasing order n, so that usually $\chi_2^l(k)$ is the dominant term in the whole energy range where the series converges. Hence an analysis of the MS contribution beyond the first term is possible if the $\chi_2^l(k)$ contribution is subtracted from the experimental signal, $\chi_{MS}(k) = \chi(k) - \chi_2^l(k)$, provided good estimates for $\mu_0(k)$ and $\chi_2^l(k)$ are used.

For K edges, neglecting the angular dependence of the Hankel function in the free propagator, the usual EXAFS signal times the atomic part can be obtained for $n = 2$:

$$\mu_2 = \mu_0 \chi_2 = \mu_0 \sum_j \text{Im} \left\{ f_j(k, \pi) \exp \frac{2i \, \delta_1^0 + kr_j}{kr_j^2} \right\}, \tag{25}$$

where r_j is the distance between the central atom and the neighboring atom j and $f_j(k, \pi)$ is the usual backscattering amplitude. The first multiple scattering contribution is the μ_3 term that can be written[84]

$$\mu_3 = \mu_0 \sum_{i \neq j} \text{Im} \left\{ P_1(\cos \phi) f_i(\omega) f_j(\theta) \exp \frac{2i(\delta_1^0 + kR_{tot})}{kr_i r_{ij} r_j} \right\}. \tag{26}$$

Here r_{ij} is the distance between atoms i and j, $f_i(\omega)$ and $f_j(\theta)$ are the scattering amplitudes that now depend on the angles (by Legendre polynomials $P_1(x)$) in the triangle which joins the absorbing atom to the neighboring atoms located at sites $\mathbf{r}_i$ and $\mathbf{r}_j$, and $R_{\text{tot}} = r_i + r_{ij} + r_j$. In this expression, $\cos\phi = -\mathbf{r}_i \circ \mathbf{r}_j$, $\cos\omega = -\mathbf{r}_i \circ \mathbf{r}_{ij}$, and $\cos\theta = \mathbf{r}_j \circ \mathbf{r}_{ij}$. So the terms higher than two clearly contain information about the nth-order correlation function. To conclude, it is possible to observe that because $P_1(\cos\phi) = \cos\phi$, there is a selection rule in the pathways that contribute to the μ_3 term; in fact all the cases where $\mathbf{r}_i$ is perpendicular to $\mathbf{r}_j$ do not contribute to this term since $\cos\phi = 0$.

2.3. A Model Case: XANES of Silicon

Silicon is a particularly good material in which to test the contribution of MS in the XANES spectra for systems with noncollinear paths. The absorbing Si atom has seven shells of neighbor atoms with paths with noncollinear configurations only. In the case of collinear configurations, the high probability for forward scattering enhances MS contributions in the EXAFS,[87,88] while for noncollinear configurations the probability of large-angle scattering is very low at high kinetic energies and the EXAFS approximation has been found to work in most of these systems.

The Si K-edge XANES of crystalline silicon measured by KLL Auger yield is reported in Fig. 5, and it is compared with the spectrum of amorphous silicon.

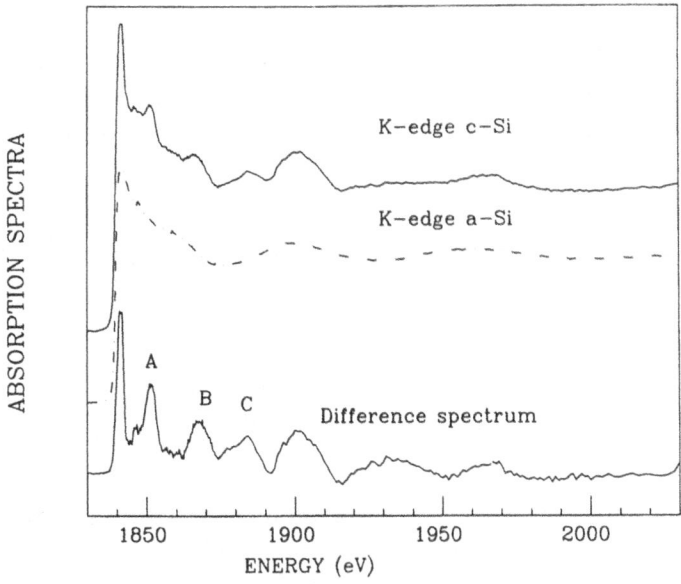

FIGURE 5. X-ray absorption spectra of crystalline (c-Si) and amorphous (a-Si) silicon determined by measuring the intensity of the KLL Auger emission as a function of the photon energy. The lower spectrum shows the difference ($\alpha_{cr} - \alpha_{am}$) multiplied by 2. The peaks A, B, and C are mainly due to the multiple-scattering contribution.

The difference spectrum shown at the bottom is determined by quenching the MS signal and EXAFS contribution of further shells. The multiple scattering effects are important in explaining the absorption spectra in the first 70 eV beyond the K threshold, producing the peaks A, B, and C in the XANES region.[59]

The multiple scattering signal was extracted from the experimental spectrum following this procedure. The modulation of the atomic absorption over the high-energy EXAFS and low-energy XANES range was determined by subtracting from the measured spectra a polynomial fitting, in order to simulate the smooth atomic absorption contribution $\alpha_0(k)$. The resulting $\chi(k)$ is shown in Fig. 6. A good fit to EXAFS oscillations in the Si K-edge spectrum from 50 to 450 eV above the absorption threshold was obtained using both the exact spherical-wave EXAFS analysis[89] and the spherical wave approximation (SWA). The $\chi_2(k)$ spherical-wave signal including the inelastic scattering and the Debye–Waller factor has been calculated in the whole energy range above threshold. A good agreement is

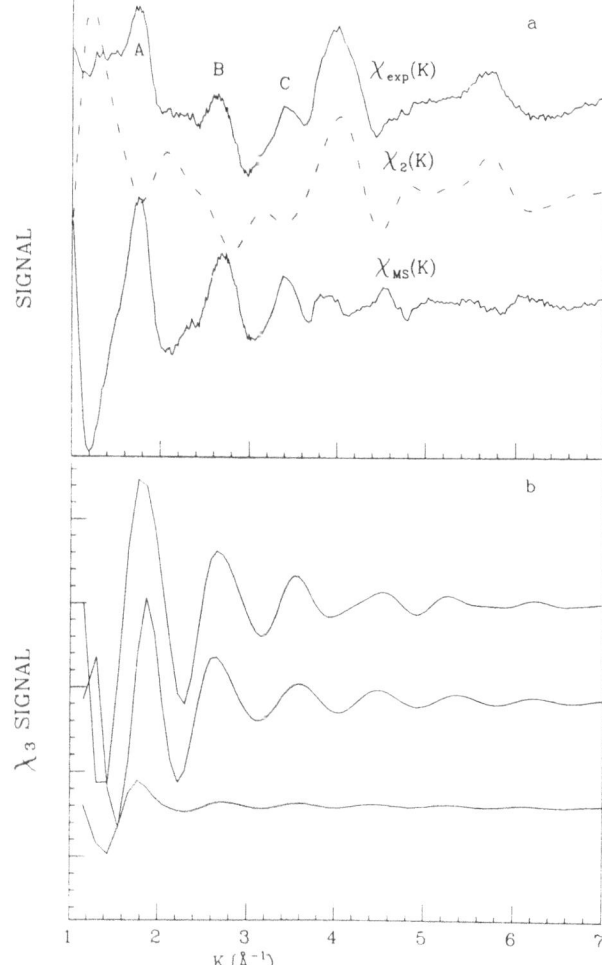

FIGURE 6. (a) Experimental $\chi(k)$ (upper curve) compared with a calculated EXAFS signal $\chi_2(k)$ using the spherical-wave formalism (dashed curve) and with the difference spectrum $\chi_{MS}(k) = \chi(k) - \chi_2(k)$ (lower curve). It is important to remark that the A, B, and C peaks are mainly due to the multiple-scattering contribution. (b) The upper curve is the total signal due to all the pathways of double scattering $\chi_3(k)$ within the first three shells, which is very close to the experimental χ_{MS} of (a) The central curve represents the contribution due to all the pathways with a length of 8.54 Å. As shown, the total χ_3 signal is dominated by the contribution of these shortest paths of double scattering. The bottom curve is the χ_3 contribution due to paths involving only the first coordination shell around the photoabsorber atom.

found between χ_2 and χ in the EXAFS range above about 100 eV from the threshold.

The experimental multiple scattering signal $\chi_{MS}(k)$ has been obtained by subtracting from the experimental oscillatory part $\chi(k)$ of the absorption spectrum the calculated EXAFS signal $\chi_2(k)$. Following this subtraction procedure over the full experimental energy range, starting from about 5 eV above threshold, a very large signal has been found in the first 70 eV above the absorption K edge, which is not possible to explain using the single scattering formalism. The result of the subtraction is plotted in Fig. 6 (lower curve), and it is compared with the $\chi(k)$ signal (upper curve). Looking at panel a of Fig. 6, it is clear that the A, B, and C peaks arise mainly from multiple scattering effects.

Since the first higher-order term in the MS series of the $\chi_3(k)$ term, a calculation of the contributions arising from the double scattering paths involving two neighbor atoms within the first three shells has been performed, introducing the mean free path damping term in the calculation of $\chi_3(k)$. The result (curve **a**) is shown in Fig. 6, panel b.

There are 756 paths of double scattering within the first three shells, and it is possible to sort them out into 41 groups according to total length and scattering angles relative to the central atom. The groups with low path degeneracy are generally negligible since they give only a weak signal. Moreover, it is possible to neglect the contribution due to the groups with a very great total length (the perimeter of the triangle), since it has been verified that the mean free path term suppresses their contribution to the spectrum in the energy range beyond 15 eV above the edge.

Panel b of Fig. 6 shows (curve **b**) the damped signal coming from 36 paths with the shortest total length ($R_{tot} = 8.54$ Å. These 36 paths can be divided into two groups, differing in the angles θ_0 between the outgoing and incoming paths directed at the photoabsorbing vertex. Figure 7 depicts the two classes of shortest double scattering paths where the difference in the angles θ_0 is emphasized.

The first of these two groups includes 12 paths contained within the first neighbor shell. This type of path, involving the photoabsorbing atom 0 and the first shell atoms 1 and 2, for example, are shown in Fig. 7 and are classified as

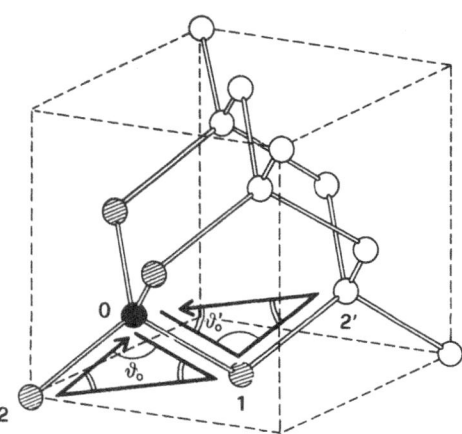

FIGURE 7. The *fcc* unit cell of the silicon structure. The photoabsorber is the black sphere, while the first nearest neighbors are shadowed. Two examples of the 36 shortest double-scattering pathways are stressed: the first, $3S1$-type, includes the atoms 0, 1, and 2, all within the first coordination shell (there are 12 paths of this type); the second, $3S2$-type, includes the atoms 0, 1 and 2' and involves an atom of the second shell (there are 24 paths of this type).

$3S1$-type paths. Their contribution is very weak as shown by curve **c** in Fig. 6. The second group includes 24 paths ($3S2$-type) involving the atoms of the second shell ($2'$ atom in Fig. 7) and gives the main contribution to the curve **b** shown in Fig. 6. Adding other damped contributions due to the other groups of pathways, the χ_3 signal does not change its shape. Therefore the most important contribution to χ_3 comes from the $3S2$-type paths with total length of 8.54 Å, as it is possible to see in Fig. 6.

The calculated χ_3 is close to the experimental oscillation χ_{MS}, as shown in Fig. 6. The differences between χ_{MS} and the theoretical χ_3 can be assigned to higher-order ($n > 3$) multiple scattering contributions.

The difference between the absorption spectra of crystalline and amorphous Si shown in Fig. 5 gives experimental evidence for multiple scattering contributions to the XANES of crystalline silicon. The EXAFS difference spectra are determined by the EXAFS contribution of further shells in the crystalline phase, but the large peaks in the difference spectrum in the low-energy XANES ranges are mostly determined by the multiple scattering signal, which is quenched in the amorphous phase.

3. EXPERIMENTAL TECHNIQUES

This section gives a brief outline of the experimental aspects of surface X-ray absorption experiments and the related detection methods.[10]

Synchrotron radiation is required for surface X-ray experiments. Indeed, the spectral brightness of this source is greater by several orders of magnitude than that of conventional X-ray tubes, providing superior signal-to-noise ratios and allowing spectral measurements in a very short time with very high resolution. In addition, ultrahigh vacuum ($\sim 10^{-10}$ Torr) is needed for experiments on well-characterized surfaces. The experimental setup for surface X-ray absorption spectroscopy requires special beam lines, dedicated sample chambers, and different detection methods, depending on several factors:

1. Energy of the X-rays involved in the experiment. It is possible to distinguish three energy ranges requiring different types of monochromators: grating monochromators for soft X-rays from 50 to 800 eV, monochromators with crystals with large spacing like beryl ($10\bar{1}0$) and InSb (111) for the range 800–3000 eV, and silicon crystal monochromators for the harder X-ray range (>3000 eV),

2. Concentration of the studied atom on the surface. The low concentration of surface atoms, about 10^{15} atoms cm^{-2} compared with the bulk concentration of 10^{19} atoms cm^{-2}, requires higher X-ray intensities to obtain a comparable signal-to-noise ratio.

3. Surface sensitivity, defined as the ratio between the signal due to the surface layer and that due to the substrate. High surface sensitivity can be achieved by limiting the penetration of the incident photon when working close to total reflection,[90-92] and by detecting particles from decay channels of the core hole with the shortest escape length.

The inner-shell photoionization process can be described by a two-step process as a first approximation. In the first step the photon excites a core hole–photoelectron pair, and in the second step the recombination process of the core hole takes place. There are many possible channels for the core-hole recombination process: the radiative type (fluorescence), producing the emission of photons, or the nonradiative type (Auger transitions), with emission of electrons or ions, which can be collected from the surface with special detectors. Several techniques can be chosen to detect surface absorption when the selected recombination channel is electronic decay.[93] A brief description follows.

The total-electron-yield technique (TY) measures the integral over the entire energy range of the electron energy distribution curves (EDC). Comparison of the absorption and electron yield has led to the conclusion that the total electron yield signal is proportional to the absorption coefficient.[3] The advantage of this method is that the counting rate is maximized, since electrons emitted over a large solid angle can be collected by providing a positive voltage on the detector. However, the surface contrast of this technique is poor because both low-energy secondary electrons and high-energy photoelectrons are collected.

Similar to TY is the partial electron yield method (PY), collecting the low-energy secondary electrons within a kinetic-energy window around the maximum of the inelastic part of the energy distribution curve (2–4 eV). Selecting only secondary electrons of low kinetic energy, the surface sensitivity is similar to that obtained by the total electron yield method.

In the energy range $h\nu < 4000\,\text{eV}$, Auger recombination has higher probability than radiative recombination, and the detection of elastically emitted Auger electrons is an efficient way to measure the surface absorption coefficient. The energy of the Auger electrons is characteristic of a particular atom; therefore, the photoabsorption cross section of a selected atomic species chemiabsorbed on a surface can be measured by monitoring the intensity of its characteristic Auger transition as a function of the photon energy. The Auger line is selected by an electron-energy analyzer operated in the constant final state (CFS) mode with an energy window of a few eV. A standard experimental set up for this type of surface X-ray absorption measurement is shown in Fig. 8.[11] The Auger electron yield (AY) technique offers the largest signal-to-background

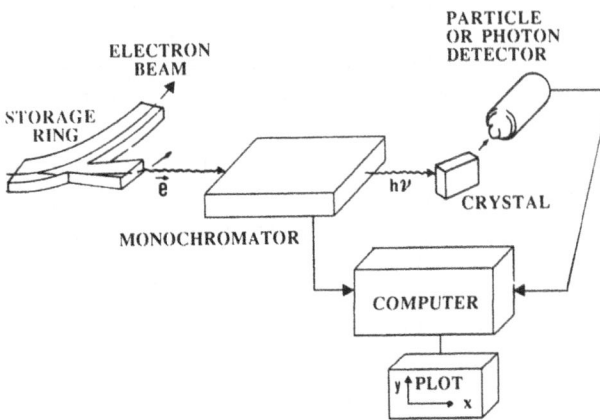

FIGURE 8. Experimental setup for surface XANES detection. In the soft X-ray region the absorption is measured by detecting the flux of the emitted electrons of energy selected by the electron-energy analyzer.

(adsorbate to substrate) ratio of all electron-yield techniques, but the smallest signal rate. This technqiue of monitoring Auger electrons at kinetic energies in the range 50–100 eV is suitable for the study of clean surfaces.[9]

The detection of the stimulated emission of ions is called photon–stimulated ion desorption (PSID).[94] The ion current due to PSID is proportional to the number of created core holes, i.e., to the photoabsorption cross section of the absorbate. It is a measure of the surface absorption with higher surface contrast than other detection methods. PSID is a surface-sensitive technique, since any ions created within the bulk are inevitably reneutralized on their way on the surface. In addition, a great advantage in principle is that surface structure of an adsorbate complex can be investigated from both the adsorbate as well as the substrate side by tuning to the appropriate absorption edge. Unfortunately, different desorption mechanisms and multielectron excitations are present in the ion yield and severely limit the applicability of PSID. The low count rates obtained with current photon flux levels provided by synchrotron radiation is also a problem. Photon-stimulated excitation, in comparison with electron excitation, offers the advantage for surface XANES that the core-hole production probability is largest at the threshold of core excitation.

In these last years, with the development of the first ultrahigh-vacuum-compatible soft X-ray detector for fluorescence photon yield (FY), detection of the XANES above the K edge of light elements has been made possible.[95] Fluorescence photon yield represents the probability of a core hole in the K or L shell being filled by a radiative process in competition with nonradiative Auger recombinations. Due to the small amount of elastically and inelastically scattered background from the sample, the FY technique gives high surface sensitivity. Using a soft X-ray proportional counter, an increase was achieved of more than a factor of 20 in surface sensitivity for chemisorbed monoloyer coverage, with signal-to-noise ratio comparable to any conventional electron-yield detector.[95,96] One of the attractive aspects of this technique is the possibility of measuring the surface X-ray absorption for in situ study of the interaction of the surface with gas with a sensitivity better than 1% of monolayer. Using window valves but with a consistent reduction of flux, pressures can range from atmospheric to 10 Torr.[97] Figure 9 shows the surface XANES of chemisorbed C_2H_2 on the surface of Cu (100) at low temperature ($T = 60$ K). The edge jump ratio J_R, which is defined as the count-rate difference above and below the edge normalized to the pre-edge background, is indicated in the figure for the different detection methods. The fluorescence photon-yield method gives the maximum jump ratio.[98,99]

A new technique for performing surface-sensitive XANES experiments is the reflection mode, or REFLEXAFS.[92] Since X-rays, on entering a substance from the air or the vacuum, are going into a medium of smaller refractive index,

$$n = 1 - \delta - i\beta \qquad (27)$$

Snell's law indicates that X-rays should be totally reflected from an optically plane surface at all glancing angles smaller than the critical angle θ_c given by $\cos \theta_c = (1 - \delta)$, so that $\theta_c \sim \sqrt{2\delta}$. Furthermore, if $\theta < \theta_c$, the incident and reflected beams interfere and give rise to standing waves above the reflecting

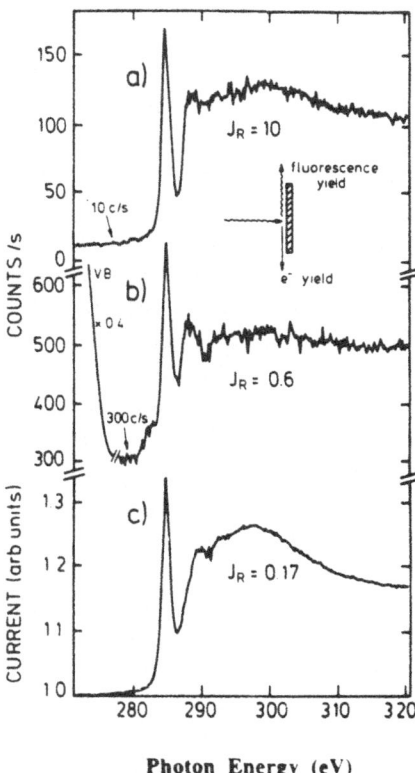

FIGURE 9. Surface X-ray absorption near-edge spectra of 10 L of C_2H_4 on Cu (100) at 60 K (about two monolayers C_2H_4) at normal X-ray incidence, recorded with different detection techniques. The detection symmetry axes are in the surface plane, along the electric vector **e**. (a) The X-ray fluorescence carbon K_α yield; (b) carbon *KVV* Auger electron yield; (c) partial electron yield (only electrons with kinetic energy larger than 220 eV were detected). J_R indicates the edge jump or signal-to-background ratio taken at 320 eV where the near-edge features fade out. (From Ref. 98.)

surface. The electrical field damps out rapidly inside the material and the penetration depth is almost unrelated to θ, provided $\theta < \theta_c$, and is on the order of 50 Å or less as a function of the material.[100] The great potentiality of this technique for surface studies can be enhanced by associating the total reflection scheme with the dispersive mode, as shown in Fig. 10. This approach opens new possibilities for *in situ* measurements of the evolution of the surface under various treatments, including nonvacuum conditions, allowing the study of real surfaces. Moreover, it can be used to investigate either top layers or epilayers.

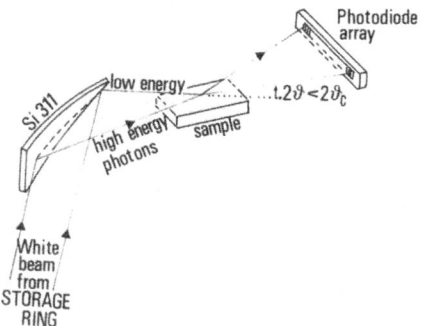

FIGURE 10. Experimental setup for surface absorption in total reflection mode combined with the dispersive X-ray optics. In the dispersive geometry the surface of the sample reflects (for grazing incident angle below the critical angle) photons, diffracted by the curved crystal, in an energy range between the low- and the high-energy limits. A position-sensitive detector is used for recording the XANES spectrum. (From Ref. 92.)

4. APPLICATIONS OF SURFACE XANES

4.1. XANES of Clean Surfaces: Al and Si

The investigation of the structure and electronic states of the surface atoms in the top monolayers of oriented crystals has attracted the interest of a large community in the last few years. The measurement of surface XANES of a single crystal requires a technique capable of distinguishing between surface and bulk atoms of the same atomic species. Surface XANES of clean surfaces have been measured by taking advantage of the short escape depth of low-energy Auger electrons with kinetic energy E_f in the range from 30 to 90 eV. By detecting the intensity of low-energy Auger electrons of kinetic energy E_f and scanning the photon energy, the surface X-ray photoabsorption spectra (with high surface contrast) for Al and Si surfaces have been measured. The upper panel of Fig. 11 shows the electron energy-distribution curve of the clean Si (111) 2×1 surface obtained by using photon excitation ($h\nu = 130$ eV) above the silicon $L_{2,3}$ absorption edge, and the inset shows the Si L_3VV Auger signal. In the lower panel the experimental escape depth λ (Å) in silicon for the corresponding kinetic energy is shown.

At the photon energy $h\nu = E_0 + E_f + \phi$, where E_0 is the absorption threshold and ϕ is the work function, the kinetic energy of direct core photoelectrons is the same as that of the selected Auger electrons E_f, and the core photoelectrons are detected by the electron analyzer, giving a spike in the Auger yield spectra. Therefore the surface XANES can be recorded in a limited

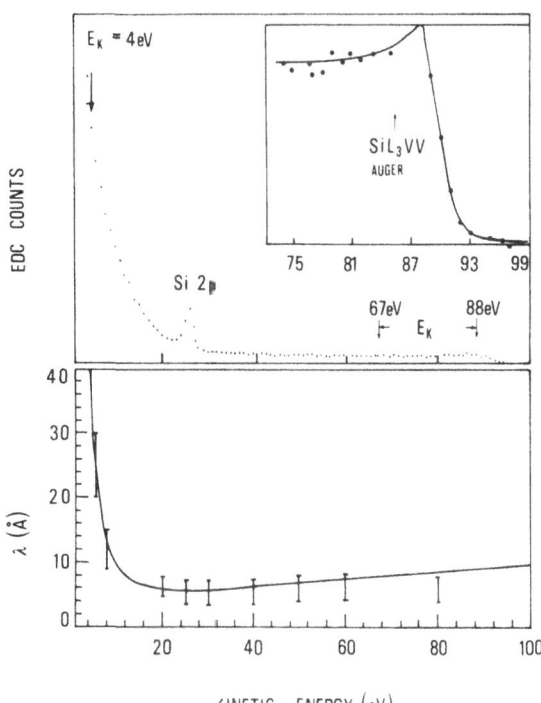

FIGURE 11. Electron energy distribution curve (EDC) of clean Si (111) 2×1 reconstructed surface obtained by using photon energy $h\nu = 130$ ev, upper panel. The inset shows the Si L_3VV Auger signal. The lower panel shows the experimental escape depth $\lambda(E)$ in silicon for emitted electrons of kinetic energy E.

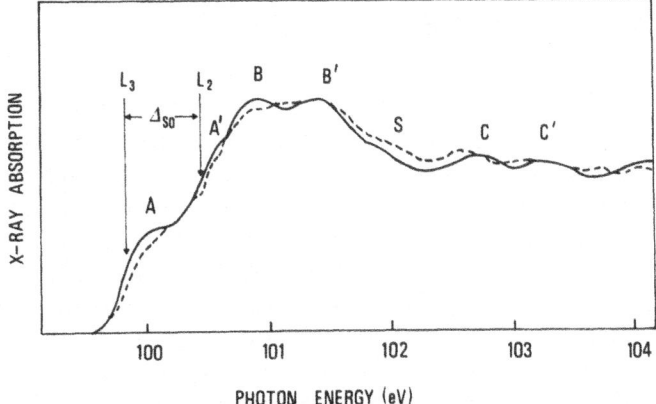

FIGURE 12. Surface (solid line) and bulk (dashed line) $L_{2,3}$ X-ray absorption spectra of a cleaved Si (111) crystal. The bulk spectrum has been obtained by selecting the final-state energy of secondary electrons at $E_f = 4$ eV. The surface X-ray absorption spectrum has been obtained by selecting the LVV Auger electrons at $E_f = 67$ eV.

photon energy range Δ above the absorption threshold $\Delta = E_f + \phi$. The bulk absorption is simply recorded by changing the final-state kinetic energy E_f from the Auger energy to the maximum of the secondary electrons at about 4 eV.

Figure 12 shows the surface X-ray absorption of the Si $L_{2,3}$ edge of a Si (111) 2 × 1 surface, recorded by selecting the LVV Auger electrons at $E_f = 67$ eV, compared with the bulk spectrum recorded at $E_f = 4$ eV. The structure in the first 10-eV energy range can be better analyzed in terms of the band-structure approach described in section 2.1. The variation between the bulk and surface spectra can be assigned to the change of the partial density of states of the conduction band induced by surface reconstruction.[33] The Si (111) 2 × 1 reconstructed surface following the Pandey model[101,102] gives a surface partial s-density reconstructed surface following the Pandey model[101,102] gives a surface partial s-density of states different from the bulk. The calculation of the partial s-density of states of the conduction band for the first few surface monolayers affected by the surface reconstruction accounts for the observed differences between the bulk and the surface absorption (see Fig. 13).

Polarized surface XANES can be performed using the linear polarization of synchrotron radiation. In Fig. 14 the surface XANES at the Si K edge obtained by

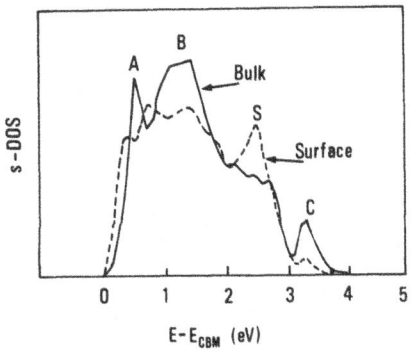

FIGURE 13. Calculated conduction-band s-density of states of bulk silicon (solid line) and of the Si (111) 2 × 1 surface (dashed line), according to Pandey's π-bonded chain model.

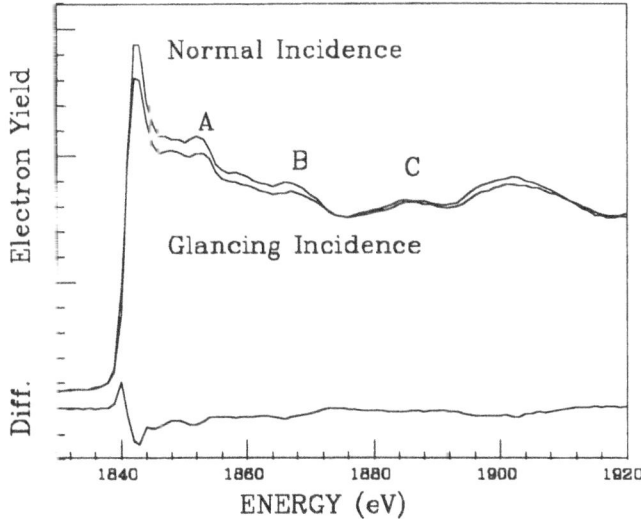

FIGURE 14. Surface silicon K-edge X-ray absorption spectra of the Si (111) 2×1 surface measured by Auger S LVV electron yield. Upper curve: polarized spectrum at normal incidence (the electric field e parallel to the surface plane); lower curve; polarized spectrum at glancing incidence (the electric field e nearly perpendicular to the surface plane). The difference curve at the bottom is the difference spectrum between the XANES of the two extreme polarizations.

recording the Si LVV Auger electrons is shown for the two extreme polarization: e ∥ n and e ⊥ n, where n is the surface normal vector and e is the electric field vector of the radiation. The difference at threshold is due to the unoccupied surface states of p-like character close to the Fermi energy, which are observed with the e ∥ n polarization at glancing incidence.[103] According to the discussion in the section 2.3, the features A, B, and C are due to higher-order multiple scattering contributions and therefore probe the atomic geometrical arrangement. The decrease of the intensity and the change of shape between normal and glancing incidence is related to the anisotropic structure of the 2×1 reconstructed surface layer.

The clean surface of Si (111) shows considerable reconstruction of the surface. On the contrary, both Al (100) and Al (111) can be considered as examples of surfaces which do not exhibit surface reconstruction. The main problem of reconstruction concerns the contraction or expansion of the spacing between the top monolayers. The first few lattice planes of ions at a metal surface can relax inward or outward because the classical Madelung forces tend to drive the first lattice plane sites inward while electronic forces usually tend to drive them outward. The spacing is calculable by direct minimization of surface energy.[104-106] Because the electronic screening is nearly perfect on the surface of simple metals like Al, Mg, and Na, the calculated face-dependent surface energies are nearly independent of small displacement of the first lattice plane.[106]

Figure 15 shows the surfaces XANES of clean aluminum surfaces (111) and (100) compared with the bulk Al XANES of the $L_{2,3}$ edge.[17,107] The aluminum $L_{2,3}$ surface XANES spectra of the Al (100) and Al (111) in the first 20 eV show differences due to the different surface partial densities of states in the conduction

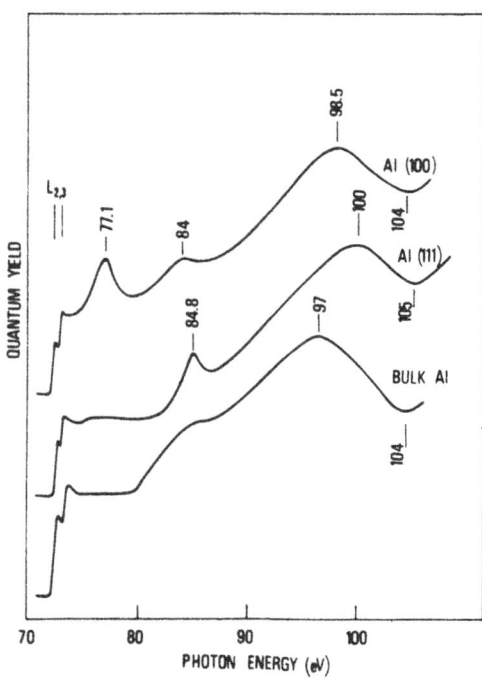

FIGURE 15. Surface soft X-ray absorption spectra of the Al (111) and Al (100) clean surfaces compared with the soft X-ray absorption spectrum of bulk aluminum. (From Ref. 107.)

band. These spectral changes are due to the presence of an anisotropic potential at the surface and to the formation of surface states in the partial gaps in the surface-projected bands. In particular, the peaks at 77.1 eV and 84 eV in the Al (100) surface have been assigned[107] to surface resonances in the partial gaps of the projected bands at 4.3 eV and 10.5 eV above the Fermi energy. The maximum at about 97 eV in the spectra is due to the delayed threshold of the $2p \rightarrow \varepsilon d$ transitions, and therefore its shift is partially due to the change of d bands at the surface. The minimum remains at the same energy ($\sim$104 eV) both for bulk aluminum and for the Al (100) surface. This minimum at 32 eV above the Fermi energy was considered to be insensitive to fine details of the electronic potential. Under this assumption and in accord with both the band structure and the multiple scattering approach discussed in sections 2.1 and 2.2, the XANES peaks follow the expansion of the volume of the crystalline cell a^3, where a is the lattice parameter. This follows the rule

$$(E - V_0)a^2 = C, \tag{28}$$

where E is the energy of the peak in the absorption spectrum, V_0 is the intersphere constant in the muffin-tin approximation, and C is a constant. As a consequence, the spacing between the top layers for the Al (100) surface is the same as in the bulk. The fact that Al (100) does not relax is confirmed by several authors using LEED methods.[108,109]

On the other hand, there is still controversy about whether the Al (111) surface contracts or expands due to the small values measured. The shift of $\sim$1 eV of the minimum in the Al (111) surface spectrum at 105 eV (see Fig. 15) is an indication of a contraction of only 1.5%, using the rule $\Delta E/(2(E - V_0)) = -\Delta R/R$, where the energy variation ΔE of the multiple scattering resonance is related to the small distance variation ΔR.[110] The various LEED data show no contraction, a slight expansion (2.2%),[111] or a slight contraction.[109]

The differing experimental results on aluminum surfaces can be affected by the different degrees of cleanliness of the surface and by the different surface sensitivities of the detection methods. When one carries out surface XANES experiments, the surface sensitivity is made very high by selecting a particular Auger energy close to the minimum of the escape depth, and the sampling depth is constant during the scan.

4.2. XANES as a Probe of Atomic Chemisorption on Crystal Surfaces

The structure of the chemisorption sites of various atomic species on the surface of oriented crystals is key information for surface science. The surface absorption spectrum of an atom chemisorbed on the surface of a crystal can be measured by Auger electron yield, selecting the particular Auger line of the chemisorbed atomic species. The XANES spectra can probe structural models of chemisorption sites via MS analysis and can provide a monitor of the structural changes at surfaces. XANES exhibits substantial advantages in comparison with other techniques because it can be measured for low adsorbate coverage ($\sim$1/100 monolayer).

Polarized SEXAFS provides direct information on bond distances between the chemisorption atom and the substrate. The joint analysis of SEXAFS and XANES for determing chemisorption site structures is very important. In fact, the bond distances obtained by SEXAFS should be used as input for XANES calculations, which gives further information on the geometrical arrangements.

The first determination of a chemisorption site by surface XANES measurements interpreted by the multiple scattering theory has been the study of oxygen chemisorption on the single-crystal Ni (100) surface.[65] This experiment shows that surface XANES is a sensitive indicator of structural changes on the surface at different oxygen coverages. With increasing oxygen coverage on the Ni (100) surface, different chemisorption geometries are produced, giving a $p(2 \times 1)$ or a $c(2 \times 1)$ LEED pattern. The XANES spectra make it possible to distinguish between different possible surface reconstructions.

The polarized oxygen K-edge XANES spectra of chemisorbed oxygen on a Ni (100) surface with increasing oxygen exposures are reported in Fig. 16 for two different angles of incidence of the photon beam. The spectra exhibit a pronounced polarization dependence for low coverage which weakens at higher O exposures. At the highest oxygen coverage, nickel oxide is formed. However the oxygen K-edge XANES spectrum of bulk NiO[74,112,113] is different from the spectrum of the saturated oxide layer formed by fewer than three layers.

The surface near-edge spectra of O/Ni(100) were calculated using a computational scheme based on a cluster method in the frame of full multiple scattering.[44] The calculation was done for bulk Ni and for O atoms in specific surface arrangements,[65] for a cluster including 30 neighboring atoms up to a distance $>5.0\,\text{Å}$ from the central O atom. Figure 17 shows the polarized

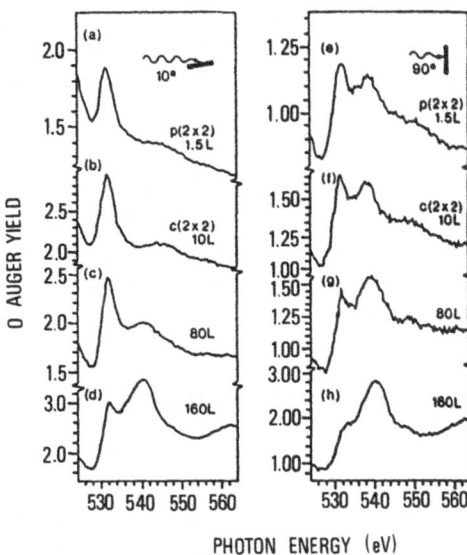

FIGURE 16. *K*-edge xanes spectra of oxygen chemisorbed on Ni (100) for increasing oxygen coverage (from 1.5 L to 160 L O_2) and two different X-ray grazing angles, $\theta = 10°$ (left) and $\theta = 90°$ (right). For $\theta = 10°$, the **e** vector makes an angle of 10° with the surface normal; it lies in the surface plane for $\theta = 90°$. (From Ref. 65.)

experimental data for a $c(2 \times 1)$ O overlayer and the calculations assuming four different chemisorption sites. Calculations have been performed for atop, bridge, and hollow sites (this last width $d_\perp = 0.9$ Å and $d_\perp = 0.2$ Å), with the bond-length constraint of 1.98 Å for the O–Ni distance as obtained by sexafs. It is possible to recognize from the figure that the calculated polarized spectra for hollow site with $d_\perp = 0.9$ Å is the one in better agreement with the experimental spectrum. This work has demonstrated not only that the xanes spectra of atomic adsorbates are dominated by nearest-neighbor atoms, but that larger clusters are necessary in the calculation.

The chemisorption of O on Cu (100) shows a 2×1 leed pattern, but different experimental measurements have yielded conflicting geometries for the

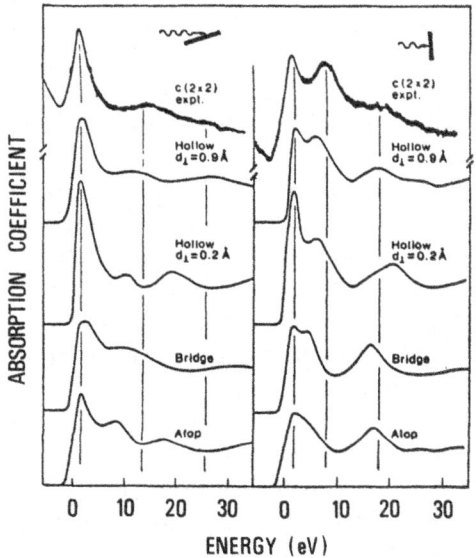

FIGURE 17. Comparison of the polarized experimental spectra (top curves in the two panels) of $c(2 \times 1)$ oxygen chemisorbed on Ni (100), and the calculated xanes for four different chemisorption sites as indicated in the figure. (From Ref. 65.)

$c(2 \times 2)$ reconstruction. This situation has been solved by quantitative XANES analysis in the framework of MS theory.[114,115] XANES spectra at the O K edge have shown that O in the $c(2 \times 2)$ overlayer occupies fourfold hollow sites with an O–Cu interlayer spacing of 0.7 ± 0.1 Å, corresponding to an O–Cu bond length of ~1.9 Å.

The spectra of the O K edge for the Cu (100) $c(2 \times 2)$ are reported in Fig. 18 for two different polar angles: $\theta = 20°$ and $\theta = 90°$, where θ is the angle between the electric field vector $\mathbf{E}$ and the surface normal. When compared with full MS calculations[115] for the fourfold hollow, bridge, and atop adsorption sites using the O–Cu bond length of 1.94 Å[116] and the truncated crystal structure for the substrate ($a = 3.615$ Å), both polarized spectra show that only a fourfold hollow site is compatible. The experimental profiles are well reproduced in both polarizations, and in the $\theta = 90°$ polarization the agreement of the peak positions is within 0.5 eV. The role of MS is shown, comparing for each adsorption site the experimental spectra with results obtained by the corresponding curved-wave single scattering calculations (dotted line in Fig. 18). The spectra for $\theta = 90°$ are insensitive to the O–Cu interlayer spacings, while the spectra for $\theta = 20°$ are sensitive to different $d_\perp$ values. In the last panel of Fig. 18 the experimental spectrum for $\theta = 20°$ (upper curve) is compared with the calculated spectra for the fourfold hollow site for different values of $d_\perp$. The best agreement between experiment and theory gives $d_\perp = 0.7 \pm 0.1$ Å.

The K-edge surface XANES of S chemisorbed on Ni (110) and on Ni (111) have been measured by Ohta et al.[117] by recording the S KLL Auger yield. A strong polarization dependence of the spectra has been found, and the chemisorption site for the $c(2 \times 2)$ sulfur on Ni (110) has been determined. The

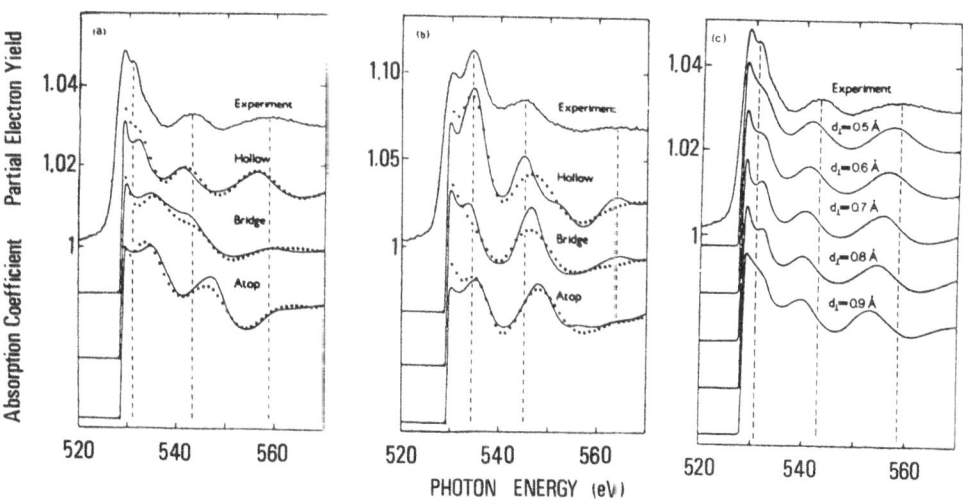

FIGURE 18. Comparison of measured (experiment) O K-edge XANES of $c(2 \times 2)$ O on Cu (100) for $\theta = 20°$ (a) and $\theta = 90°$ (b) with full multiple-scattering MS (solid lines) and single-scattering (dotted line) calculations for the fourfold hollow, bridge, and atop adsorption sites. Comparison of measured O K-edge XANES of $c(2 \times 2)$ O on Cu (100) for $\theta = 20°$ with full MS calculations for several O–Cu interlayer spacings $d_\perp$ (c). (From Ref. 116.)

analysis of the data for the reconstructed $c(2 \times 2)$ surface of S on Ni (110) shows that the sulfur atom is located on a hollow site.

4.3. XANES of Surface Oxides

The investigation of the structure of the first disordered oxide layers formed on top of surfaces exposed to oxygen gas is of interest in understanding the microscopic process of oxidation. The relevance of the structural determination of surface oxides to the technology of protective surface layers on metals and to the semiconductor–insulator or metal–insulator interfaces in electronics is well known.

The oxide layers formed on top of metals show in many cases a disordered structure, which is not possible to investigate by diffraction methods because of the lack of long-range crystalline order. However, XANES spectroscopy does not require crystalline order and it is site specific. Moreover, the disordered structure simplifies the interpretation of the experimental XANES data because it reduces the number of shells of neighbor atoms contributing to the spectrum. Therefore the XANES of amorphous surface oxide layers are determined mainly by the first shell. This last characteristic aspect of the XANES of amorphous oxides makes feasible the experimental determination of the symmetry of the coordination shell of the absorbing atom by using the empirical "fingerprint" approach. The measured spectrum of an unknown compound is compared with the spectra of model compounds. The ideal model compounds should exhibit different coordination geometries for the absorbing atom, and the number of neighbor shells contributing to their XANES spectra should be known.

The oxide formation on top of an aluminum surface following the interaction of oxygen with the metal is a classic example of oxidation processes. The study of the progression from the chemisorption of oxygen on clean Al surfaces through the oxidation phase upon oxygen exposure to an aluminum single crystal was the object of the first surface XANES experiment, as shown in Fig. 1.[17] The local structure of the first oxide layer formed on top of Al (100), Al (110), and Al (111) surfaces at oxygen exposures larger than 500 L ($1 L = 10^{-6}$ Torr sec) has been determined by XANES.[4,17,118] In Fig. 19 are shown the Al L_{23} surface X-ray absorption near-edge spectra of the Al–O complex on the Al (111) surface of the chemisorption phase at 100 L oxygen exposure (upper panel) and of saturated oxide layer at 100 L exposure at room temperature (solid line) measured by Auger quantum electron yield. By XANES it possible to follow the change of the Al local structure from the chemisorption to the oxidation phase by increasing the oxygen coverage or by heat treatment starting from the chemisorption phase. The dashed curves in Fig. 19 show the surface XANES of the oxide-like cluster formed upon heating the sample to 400 °C. By XANES it easy to see that a surface oxide-like layer with similar Al–O coordination sites are obtained by heat treatment by starting both from the chemisorption phase and from the saturated oxide phase.[118]

Information about the Al site structure in the oxide surface layer has been obtained by comparison of the Al $L_{2,3}$ XANES spectrum of the surface oxide with

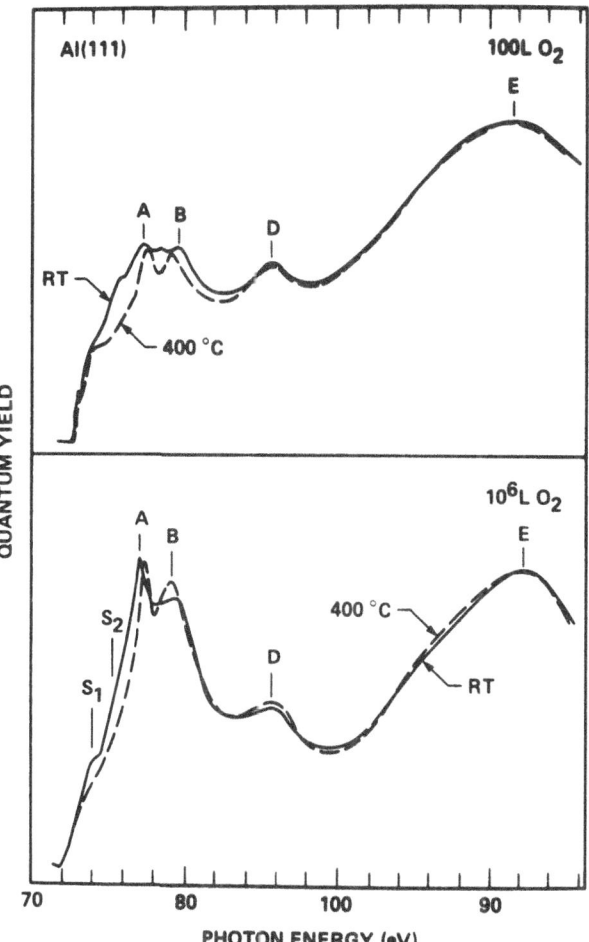

FIGURE 19. Al L_{23} surface X-ray absorption near-edge spectra measured by Auger electron yield of the Al–O complex on the Al (111) surface. Shown are the chemisorption phase at 100 L oxygen exposure (upper panel) and of the saturated oxide layer at 100 L exposure at room temperature (solid line). The dashed curves show the surface XANES of the oxide-like cluster formed upon heating the sample to 400 °C. By XANES it easy to see that a surface oxide-like layer with similar Al–O coordination sites is obtained by heat treatment by starting both from the chemisorption phase and from the saturated oxide phase. (From Ref. 118.)

the spectra of amorphous a-Al$_2$O$_3$, where the Al ion has a fourfold tetrahedral coordination, and with the spectrum of α-alumina, where the Al ion has an octahedral coordination, both shown in Fig. 20. The spectrum of the surface oxide layer does not show a similarity to the spectrum of amorphous alumina (a-Al$_2$O$_3$), thus ruling out tetrahedral coordination for the Al ion. The analogies with the XANES of octahedrally coordinated Al in α-alumina indicate an octahedral coordination for the first surface oxide layer. This result was not expected, since the Al ion has a tetrahedral coordination in the thick amorphous surface oxide layers grown by electrolysis on aluminum. These conclusions have been confirmed by Norman $et\ al.$[119] by measuring the aluminum–oxygen distance in the surface oxide layer, compared with that in a variety of aluminum oxide systems with tetrahedral and octahedral coordination, using SEXAFS at the oxygen K edge.

A similar fingerprint approach was used to identify two different chemisorption sites for H$^+$ and F$^+$ ions on α-Al$_2$O$_3$ polished parallel to the (0001) surface.[120] The XANES spectra measured by photon-stimulated desorption of H$^+$ and F$^+$ ions, compared with the bulk spectrum measured by electron yield from

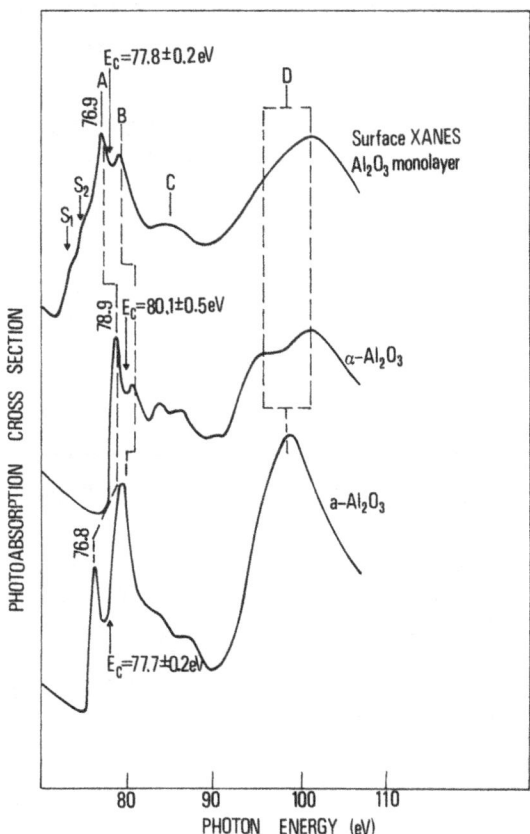

FIGURE 20. Surface XANES of the first oxide-like monolayer on Al (111), Al (100), and Al (110) surfaces (see Fig. 19) compared with the bulk X-ray absorption spectra of model compounds: amorphous a-Al$_2$O$_3$ and α-Al$_2$O$_3$ for determination of the local Al site structure. The energy E_c indicates the threshold for transition to continuum.

the α-Al$_2$O$_3$ substrate and amorphous alumina, are shown in Fig. 21. The hydrogen and fluorine bonding sites clearly differ from each other and from the octahedral bulk α-Al$_2$O$_3$ site. The comparison with the bulk spectra of model compounds shows that the hydrogen site is obviously tetrahedral, as in amorphous Al$_2$O$_3$, while the fluorine site has components similar to both tetrahedral and octahedral sites, suggesting a new site or a mixture of two.

The silicon L_{23}-edge spectrum of the first oxide on the Si (111) surface shows the formation of a SiO$_2$-like oxide (formed by SiO$_4$ units) when the surface is exposed to oxygen at room temperature and the formation of a SiO-like surface oxide on the same silicon surface when the surface temperature rises above 700 °C.[18] Figure 22 shows the L_3 spectra of surface silicon oxides. The spectrum of the oxide grown at room temperature is very similar to the bulk spectrum of silicon dioxide except for the peak named I which is due to transitions at the Si–SiO$_2$ interface. The spectrum of the surface oxide formed at high temperature shows large variation attributed to the formation of a local coordination around silicon as in SiO$_x$ ($x \sim 1$). Therefore it was possible to determine the local structures of the first amorphous silicon oxide layer grown on top of silicon crystal at room and at high temperature. Moreover, surface, XANES did prove direct experimental evidence for a specific silicon site structure in the SiO$_x$ oxide.

The local structure of the first nickel oxide layer formed on the nickel surface

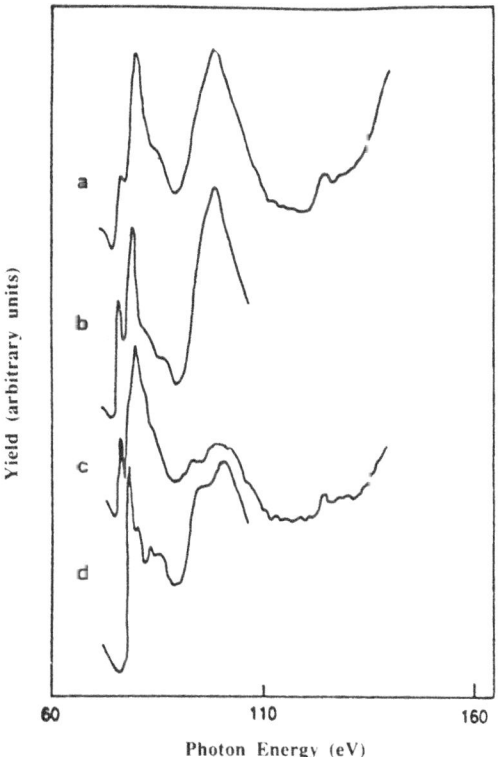

FIGURE 21. Photon-stimulated desorption XANES aluminum L_{23} spectra the desorption yield of (a) H^+ and (c) F^+ from α-Al_2O_3 and the electron yield from (b) amorphous Al_2O_3 and (d) α-Al_2O_3.

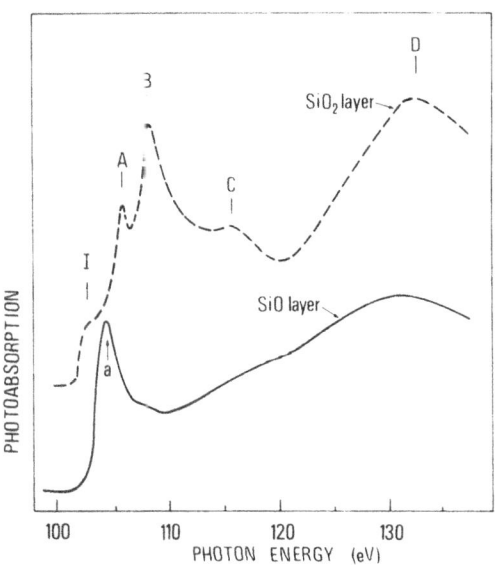

FIGURE 22. Comparison between the surface soft X-ray absorption of the saturated oxide layers formed on the Si (111) surface at room temperature (dashed line) and of the oxide layer formed at 700 °C (solid line).

was studied by O K-edge surface XANES.[113] By increasing the oxygen exposure,
the O chemisorption phase is followed by the formation of an oxide phase for
exposure larger than 150 L, as shown in Fig. 16. The intensity of the first peak A
at the oxygen absorption threshold, shown in Fig. 23, is a measure of the
covalency of the nickel–oxygen bond. In fact, the ground state of NiO should be
described by the configuration interaction between $3d^8$ and $3d^9\underline{L}$ configurations
(where $\underline{L}$ indicates a hole in the oxygen $2p$ orbital), because of the large
electronic correlation. The insulating nature of NiO is determined by the
correlation gap $3d^8 + 3d^8 \rightarrow 3d^9 + 3d^8\underline{L}$ characteristic of charge-transfer insulators. The first peak at the oxygen K threshold is due to the transition from the
ground state to the final state configuration $O(\underline{1s})Ni(3d^9)$, where the underline
indicates the core hole.[74] Therefore this peak can be described as a transition to
the upper Hubbard band, and its intensity is a measure of the probability of
finding a hole in the oxygen $2p$ orbital, i.e., of Ni–O covalency. The quenching of
this peak was found in nickel-deficient $Ni_{(1-\delta)}O$ for $\delta = 0.03$, showing that a
small amount of doping due to nickel defects is enough to suppress the
correlation gap forming new states in the gap.[113] This effect appears in the oxygen
K edge by a decrease of the intensity of the peak A and the formation of new
features at lower energy due to states induced by doping in the gap.

Using the total reflection method and dispersive X-ray optics, Dartyge et al.[92]
performed in situ time-resolved observations of surface modification of a copper
"real" metal surface under chemical or electrochemical treatments. They
followed the thermal oxidation of the copper surface by annealing a vapor-
deposited metallic film for two hours at 200 °C. The copper K edge XANES shows
without ambiguity that this annealing treatment leads to Cu_2O oxide, by
comparison with model compound Cu_2O.

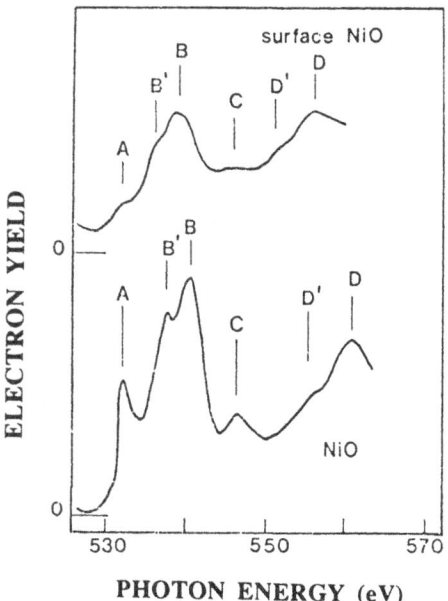

FIGURE 23. Oxygen K-edge XANES of nickel
oxide layer growth on Ni surface (top) compared
with NiO (bottom).

4. XANES Spectra of Chemisorbed Molecules

Core electron excitation spectra of gas-phase molecules have been studied for a long time by means of various experimental techniques. A complete list of available references in the field of gas-phase molecular core excitations can be found in Ref. 121.

In the molecular absorption spectra of low-Z atoms, a set of bound states is found below the ionization threshold. The most intense peaks fall within a Rydberg below the ionization threshold. These are bound states determined by transition to unfilled molecular orbitals, Rydberg states, and multielectron shake-up final states. Above the ionization threshold other, much broader resonances are observed. They were first assigned to "inner well resonances," due to an electronegative cage on the atoms surrounding the absorbing atom. Now they are better known as "shape resonances"[122,123]; in fact, they appear also at the K edge of covalent N_2 molecules. As discussed in section 2, these resonances are a particular case of multiple scattering resonances in the continuum, which are determined by the geometrical arrangement of the constituent atoms and therefore can be used to study the structure of chemisorbed molecules on surfaces.

In Fig. 24, the N K-edge spectrum of N_2 in a gas phase is shown.[39,124] Only one broad continuum shape resonance appears above the ionization threshold, and a transition to a bound π state gives the sharpest peak at the threshold.

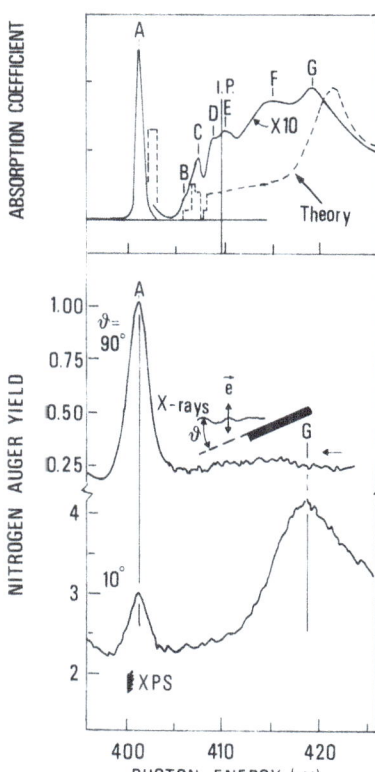

FIGURE 24. The upper panel shows the nitrogen K-edge absorption of N_2 gas (solid line) compared with Dill and Dehmer MS calculation (dashed line). (From Refs. 39, 124.) The lower panel shows the polarized N K-edge of the N_2 molecule chemisorbed on the Ni (110) surface. (From Ref. 125.)

Between these two transitions are observed a set of excitations to bound states and double electron transitions (shake-up transitions) in which, concomitant with the one-electron ionization process, another electron is promoted from an occupied valence orbital to an empty molecular orbital of the same symmetry type. We focus our discussion on the shape resonance in the continuum.

The shape resonance for nitrogen molecules was first explained by Dehmer and Dill as a relative increase, around a particular energy, in the amplitude of the final-state continuum wave function of σ symmetry near the photoionized atom in the molecule.[122,123] The final-state electron is trapped in a quasibound state decaying away with a lifetime $\tau = h\Gamma_\gamma^{-1}$. Near-edge structures arise from scattering processes within the intramolecular potential created by the atomic cores and the valence charge distribution of the molecule.

In the lower part of Fig. 24 are shown the polarized spectra of N_2 chemisorbed at 90 K on the Ni (100) surface[125] for two different angles of incidence of the synchrotron radiation with respect to the Ni surface. The $\theta = 10°$ spectrum of the chemisorbed molecule mainly shows a σ-resonance, while the $\theta = 90°$ spectrum shows only the π-resonance. The differences in the nitrogen spectra, induced by the chemisorption state, are the broadening of the σ-resonance and the disappearance of the all-Rydberg and/or multielectron excitations. In general, no new structures induced by the metal substrate are detectable.

Figure 24 shows the polarization dependence of the resonances and in particular the opposite behavior of the π and σ excitations caused by the vertical orientation of N_2 on the surface. The physical meaning of this dependence arises from the fact that the σ-shape resonance involves final states which are symmetric with respect to a reflection plane containing the molecular bond axis; on the contrary, π-resonance involves antisymmetric final states with respect to the same plane. Therefore in a photoabsorption spectrum, due to the dipole selection rules, the σ-shape resonance is maximized when the electric field e of the radiation is parallel to the bond axis, while for polarization orthogonal to the bond the spectrum is featureless. In contrast to the σ-resonances, which are present in all molecules, the π-resonances are due to transitions of $1s$ electrons into the antibonding π^* orbital. They are strongest if the e vector is parallel to the π orbital and give information on the hybridization of the bond. Accurate peak-position analysis of the resonances should provide valuable information from the two final-state resonances coupled with change in the intramolecular bond length and distortions in molecular groups containing multiple bonds. Experimental evidence for the correlation among energy, position of the σ resonances, and bond lengths in the gas phase was found in hydrocarbons.[67,126]

The absorption spectra of molecules of the C_2H_n type with $n = 2$, 4, and 6 are shown in Fig. 25. The XANES of C_2H_2 and C_2H_4 were obtained by transmission measurements as a function of the gas pressure, using synchrotron radiation.[126] The similar XANES spectra of C_2H_n have been measured by Hitchcock and Brion[127] by electron energy-loss spectroscopy.

The results of *ab initio* multiple scattering calculations for oriented N_2 and C_2H_n ($n = 2, 4, 6$) molecules are reported in Figs. 26, 27, 28, and 29. For each molecule, the z axis has been oriented along the main bond (C–C or N–N), and

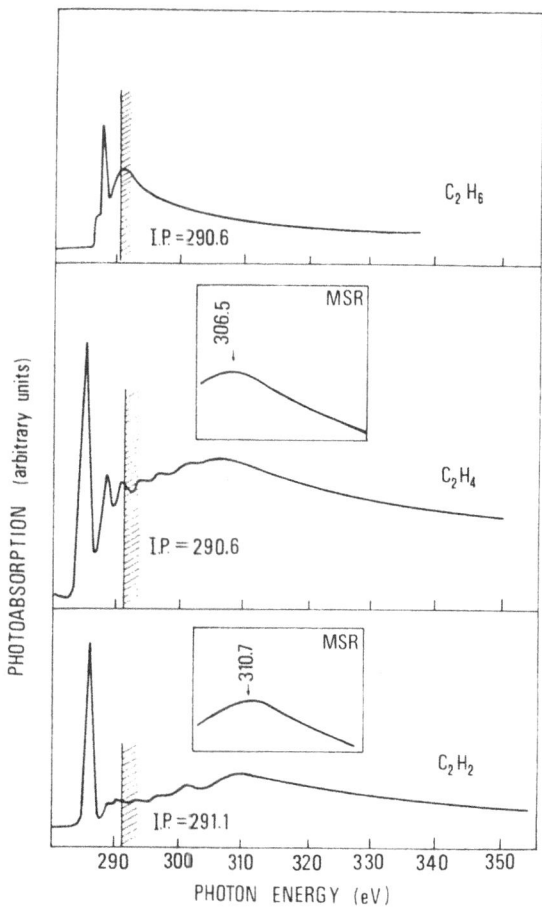

FIGURE 25. Experimental absorption spectra for C_2H_2, C_2H_4, and C_2H_6. This latter spectrum is obtained by electron energy-loss spectroscopy. (From Ref. 126.)

in the case of the planar C_2H_4 the x axis was assumed to lie in the plane of the molecule. Polarized absorption cross sections (i.e., σ_x for $\mathbf{e} \parallel \mathbf{x}$) and the total cross section (σ_{tot}) for unpolarized spectra were calculated. The absorption cross sections in the continuum above the ionization threshold are reported in Figs. 26–29.[128] The ionization threshold $E_t = (h\nu - IP) = 0$ is determined with reference to a final state potential which is not self-consistently determined. The calculated energy position of the various continuum resonances relative to the ionization threshold does not agree very well with experiments; despite this, the computed spectra clearly show the well-known $l = 3$ resonance due to the transition of a K-shell electron to a continuum state of σ-symmetry, when the polarization of the incident radiation is parallel to the axis of the molecule (z-polarization). In the calculations a resonance of $l = 2$ character due to the presence of the hydrogen atoms is present in both C_2H_4 and C_2H_6 spectra in transverse polarization (x polarization in C_2H_4 and x and y polarizations in C_2H_6). Despite their weak scattering power, the hydrogen-induced features are not entirely negligible. Recently, identification was made of the C–H resonance in K-shell excitation spectra of free ethylene and ethylene chemisorbed on various metal substrate at energies close to the ionization potential.[129]

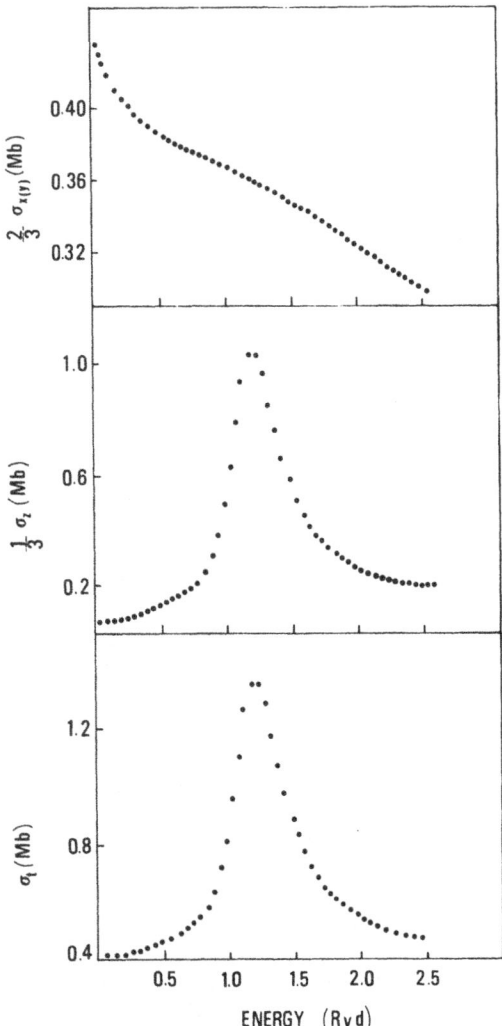

FIGURE 26. Calculated photoabsorption spectra for oriented N_2 molecule for longitudinal polarization $\sigma_z(E)$, for transverse polarization $\sigma_x(E) = \sigma_y(E)$ and for random orientation $\sigma_{tot} = (\sigma_z + \sigma_x + \sigma_y)/3$.

In the frame of multiple scattering theory, shape resonances in the cluster case are associated with singularities of the cluster K_C-matrix.[130] In the electron molecular scattering, as in the atomic case, resonances occur whenever some eigenvalue λ_m of the Hermitian K_C-matrix goes to infinity. Under the assumption that the atomic phase shifts are smooth functions of E, which is always true in the energy region where atomic resonances are located, the approximate rule can be deduced for small variation of the interatomic distances R, as given first in Ref. 67:

$$k_r R = \text{constant}, \tag{29}$$

where k_r is the resonance wave vector, R is the distance from the atomic scatterers in the cluster, and the constant is determined by the details of the atomic scattering phase shift. This rule is verified in molecules or clusters with identical angular geometrical arrangement but different bond-length scale,

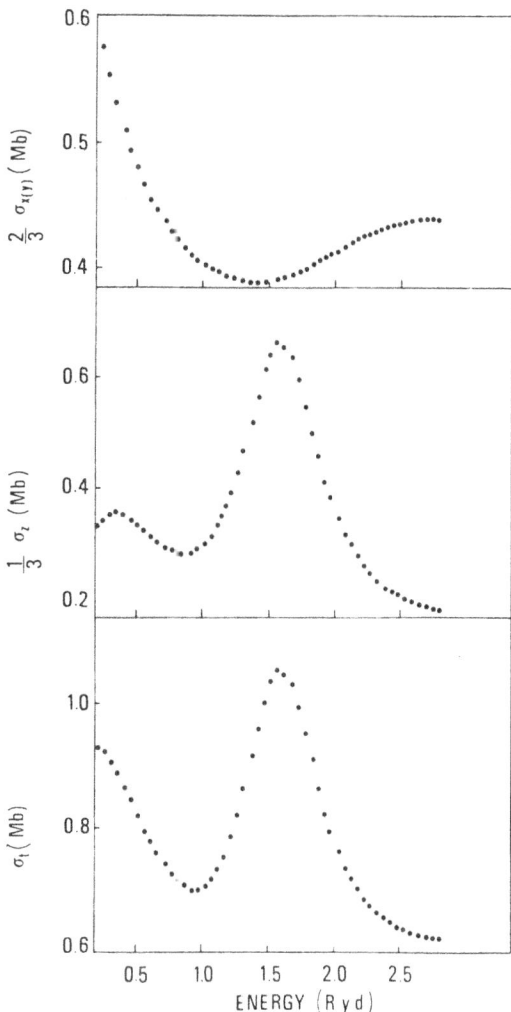

FIGURE 27. Same as Fig. 26, but for C_2H_2 molecule.

assuming that the phase shifts are "transferable" in the sense that they are functions only of the atomic species and rather insensitive to the environment. Very simple applications occur both in diatomic molecules (e.g., C_2H_n group neglecting hydrogen contributions) and in atomic clusters where the main MS resonance is due to the first coordination shell, at distance R from the photoabsorbing atom. The photoelectron wave vector k_r is referred to the average potential in the interstitial region between the ion cores V_0 (the "muffin-tin zero") so that the simple relation (29) becomes

$$\int [E - V(\mathbf{r})]^{1/2} \, d\mathbf{r} \sim [E - V_0]^{1/2} R = \text{constant}. \qquad (30)$$

The near-edge structures of the absorption spectra of molecular adsorbates are essentially dominated by intramolecular scattering, with only a small or negligible

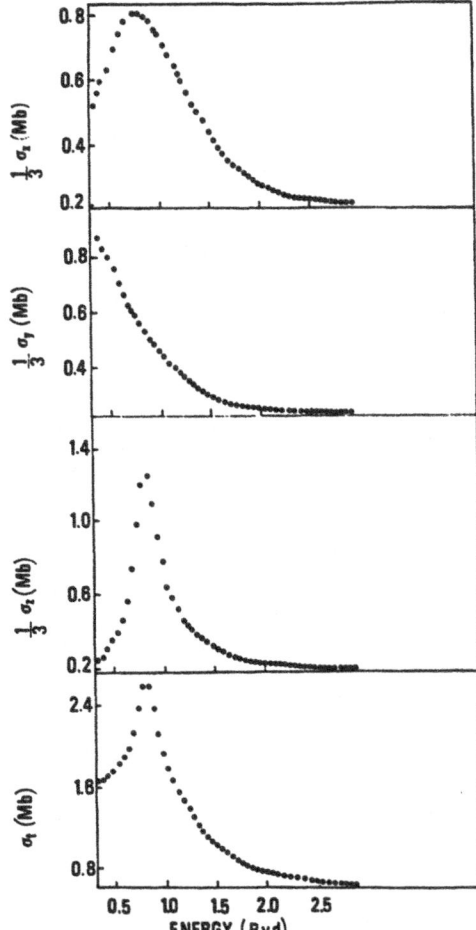

FIGURE 28. Same as Fig. 26, but for C_2H_4 Planar molecule. In this case $\sigma_x \neq \sigma_y$.

scattering contribution from the substrate atoms. A lot of chemical physics processes are based on chemisorption; for example, chemisorption on transition metals is the first step of many catalytic processes. Assuming that the interaction with the substrate is not strong enough to modify V_0, it is possible to determine the stretching of bond length in a molecule upon chemisorption. However, caution must be used in order to extract bond distances from Eq. (30), because it is strictly valid only when the molecular geometry is the same and the variation of the interatomic distances is of the order of 10% or less. Different geometries give different MS resonances at different energies, and large changes of interatomic distances induce changes in V_0 and in the phase shifts, giving a different constant in Eq. (30)

Figure 30 shows the absorption spectra at the O K edge for three molecules with carbon–oxygen bonds, chemisorbed on Cu (100).[66] The angle of the **e** vector to the surface has been chosen to maximize the intensity of the σ-shape resonance. The three molecular species are CO, with a short triple C–O bond

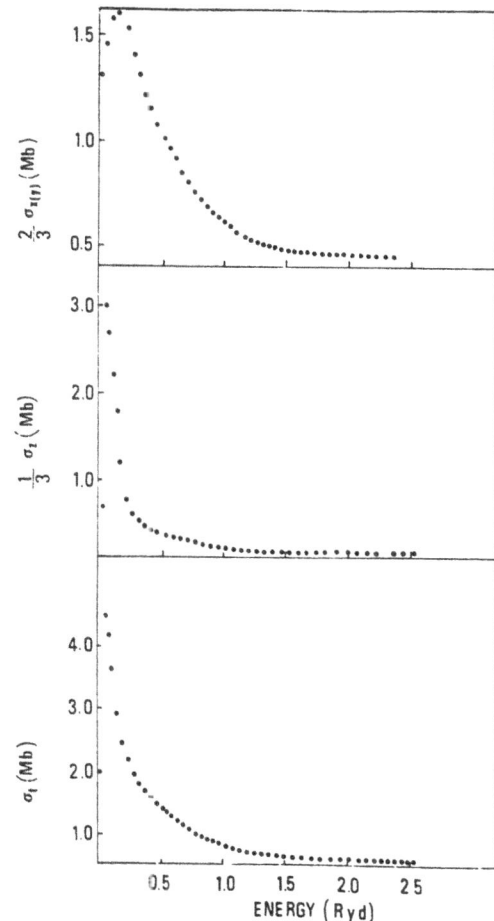

FIGURE 29. Same as Fig. 26, but for C_2H_6 molecule.

(1.13 Å), formate (HCO_2), which is a pseudodouble C–O bond; and methoxy (CH_3O), with a longer single C–O bond (1.43 Å). At first approximation, hydrogen atoms may be ignored in this work. A π resonance peak is seen both for CO and formate close to the O K-edge threshold, while no π character is detected in the single C–O bond for methoxy. The structure called X is not a weak π resonance, but rather the atomic-like absorption step at the O K edge. The energy position of the σ-shape resonance decreases with increasing C–O bond length. Using Eq. (30), Stöhr *et al.* determined the unknown bond of formate for the chemisorbed case.[66]

Because the σ resonances arise from the scattering of the photoelectron by neighboring carbons, the scattering phase shifts were supposedly identical in all three cases. Also the muffin-tin constant V_0 was assumed to be the same in the molecular and in the chemisorption phase and independent of the bond distance. Using the values of bond change from triple carbon–oxygen bond (CO/Cu (100)) to single bond (CH_3O/Cu (100)) as reference systems, the energy difference between the σ-shape resonance excitation energy E_σ and the 1s binding energy

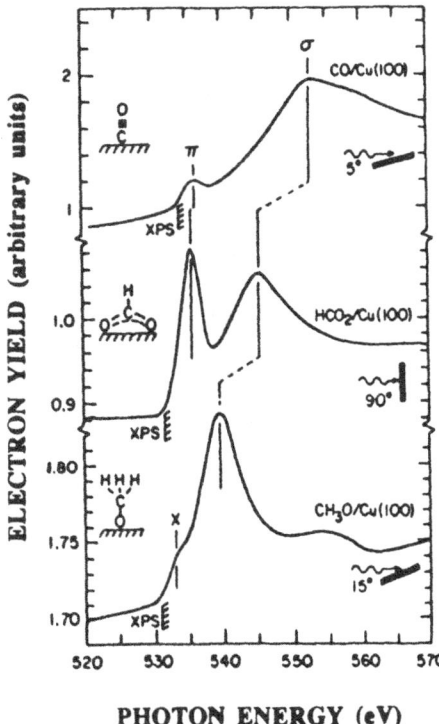

FIGURE 30. O K-edge XANES spectra of CO, HCO_2, and CH_3O on Cu (100). The angle of incidence is chosen in order to maximize the σ-shape resonance. (From Ref. 66.)

E_B relative to the Fermi level E_F was determined. The bond length extracted for the formate case was $R = 1.25 \pm 0.08$ Å, to be compared with a simple linear approximation between σ-shape resonance excitation energy and R that yields the value $R = 1.27 \pm 0.04$ Å. This value is longer than in the gas phase, but in excellent agreement with the value $R = 1.25 \pm 0.02$ Å for the C–O bond length in formate ions coordinated by a variety of metals, as a result of stronger interaction of the oxygen atoms with the substrate. This type of approach can give only a qualitative indication of the variation of bond distances. In fact the XANES spectra of formate and methoxy cannot be compared because of the large geometrical difference.

Sette *et al.*[131,132] have empirically established, for the gas phase and for chemisorbed molecules containing low-Z atoms, that a linear relation between the energy of the shape resonance and bond length holds within several classes of molecules. This relation depends on the number Z_t, defined as the sum of the atomic numbers of the absorbing and the scattering atoms (i.e., $Z_t = 12$ for C–C bonds, while $Z_t = 14$ includes molecules with N–N, B–F, or C–O bonds).

A quantitative analysis of the carbon–carbon (C–C) distance variation from the gas phase to the chemisorption phase can be carried out in the case of chemisorption of hydrocarbons on metal surfaces. The Eq. (30) has been tested for the linear molecule C_2H_2 by performing several C K-shell multiple scattering calculations (with incident light polarized along the C–C bond) with different values of the C–C distance.[67]

In Fig. 31 the experimental values of the energies of the σ resonances above the ionization potential for C_2H_2 and C_2H_4 in the gas phase and chemisorbed on

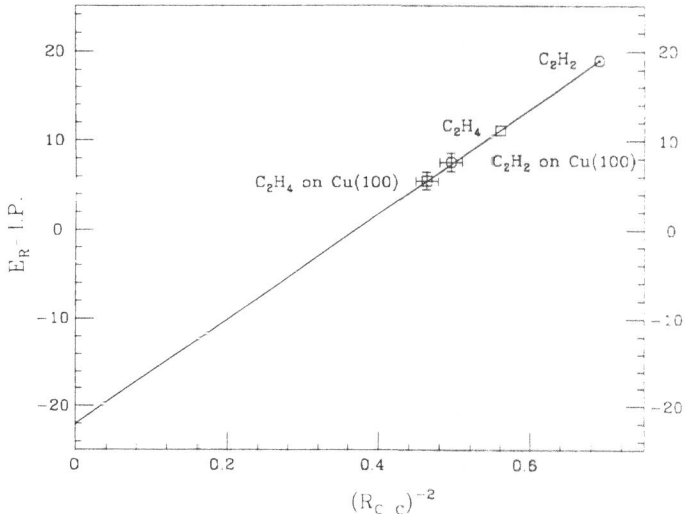

FIGURE 31. Linear relation between the σ resonance energy E_R and $(1/R)^2$, where R is the C–C distance for hydrocarbons C_2H_2 and C_2H_4 in the gas phase and in the chemisorbed phase on Cu (100).

Cu (100) at 60 K[133] are plotted versus $1/R^2$ (R is the interatomic C–C distance). R has been measured also by SEXAFS in the same chemisorption phase.[134] A linear relation is found with $V_0 \sim 22\,\text{eV}$. This value has been found by recent MS calculation for C–C distances in the range of $1.4 \pm 0.11\,\text{Å}$.

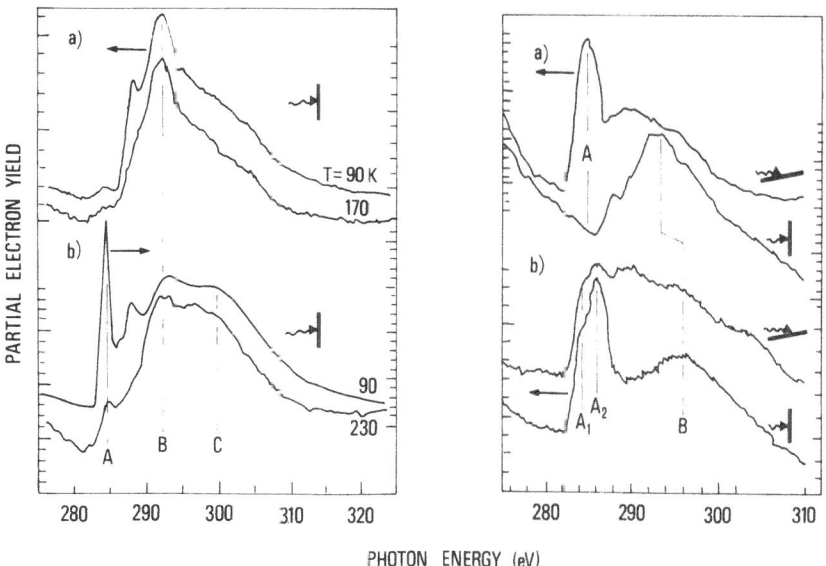

FIGURE 32. Left panel: (a) surface X-ray absorption spectra of cyclohexane (C_6D_{12}) multilayer ($T = 90$ K) and monolayer ($T = 170$ K) on Pt (111) recorded at normal X-ray incidence; (b) spectra of cycloheptatriene (C_7H_8) as a multilayer (90 K) and monolayer (230 K) on Pt (111) for normal X-ray incidence. Right panel: (a) surface X-ray absorption spectra of ethylene (C_2H_4) on Pt (111) at saturation coverage at $T = 90$ K for grazing and normal X-ray incidence; (b) spectra of deuterated acetylene (C_2D_2) on Pt (111) under same condition as (a). (From Ref. 132.)

A nice application of near-edge spectroscopy for structural investigation is the study of the reaction-intermediate states in the chemisorption of hydrocarbons on Pt (111).[132] Figure 32 shows the near-edge structure at the C K edge of condensed multilayers (90 K) of two cyclic hydrocarbons: C_6D_{12} and C_7H_8, compared with the respective chemisorbed monolayers (170 K) on Pt (111). For the cyclohexane system (C_6D_{12}) the multilayer and monolayer spectra are nearly identical, except for the Rydberg excitation at ~288 eV, which is quenched by the interaction with the Pt (111) surface in the monolayer case. In these hydrocarbons both spectra are dominated by peak B, which is clearly assigned to the σ-shape resonance produced by the single C–C bond. For weakly chemisorbed molecules, the σ resonance energy to a very good approximation remains the same as in the gas phase (≥ 1 eV) not only for simple diatomic molecules like N_2, CO, and NO, but also for the cycloheptatriene (C_7H_8) system. The multilayer spectra (90 K) is characterized by a pronounced π resonance (A), two σ resonances (B and C) and the same Rydberg-type excitation at ~288 eV that disappears in the monolayer spectra (230 K). In this last case, while both σ resonances are almost unchanged at 230 K, the peak A is strongly reduced in intensity. The presence in this ring-shaped structure of single and double C–C bonds give two σ-shape resonances, while the A peak is assigned to the transition to the corresponding π orbital. Using a linear rule between σ resonance positions and $1s$ binding energy, the estimated R values are 1.51 ± 0.03 Å for the C–C bond in C_6D_{12} and 1.37 ± 0.04 Å and 1.50 ± 0.03 Å, respectively, for the double and single C–C bond in C_7H_8. These are very close to the respective gas-phase values and in agreement with the prediction of higher stability for intramolecular bonding of ring structures.

In strong contrast with these values for ring-like structures are the R values extracted by the spectra reported in Fig. 32. The spectrum of ethylene (C_2H_4) on Pt (111) at monolayer coverage (90 K) gives in fact a C–C bond length of 1.49 ± 0.03 Å, with a stretching of 0.15 Å relative to the gas phase. Also for the acetylene (C_2D_2) the C–C distance is 1.49 ± 0.03 Å, remarkably larger (0.25 Å) relative to the gas phase. These dramatic bond stretch show the presence in these molecules of a strong interaction of the π states with the metal surface.

XANES spectroscopy is a valuable local probe to study intramolecular bonding, molecular structure, and the orientation and hybridization of chemisorbed systems; however, caution should be exercised in extending results to systems containing three or more collinear atoms such as CO_2, N_2O, NO_2, or COS, where multiple scattering effects play an important role and where the molecular orbitals are partially spread out over all three atoms, destroying the concept of resonant enhancement along a particular interatomic axis.

4.5. Molecular Orientation on Surfaces

Molecular chemisorption exhibits strong dependence of the resonance condition on the polarization of the incident radiation, as has already been shown in the case of nitrogen and chemisorbed hydrocarbons. In the frame of multiple scattering theory, the scattering matrix[130] is block diagonal, and the off-diagonal elements between basis functions belonging to different irreducible representations are zero. Each distinct polarization of incident light selects, through the Wigner–Eckart theorem, the exact diagonal subblock of the matrix involved in the expression for the cross section. Polarizations which select these particular subblocks cause a resonance structure in the spectrum, which may be completely absent for other polarizations. This has been experimentally demonstrated in the photoabsorption spectra with polarized light for oriented molecules with cylindrical symmetry. In this case for incident polarization along the bond, the absorption spectra show a strong resonance feature (the well known $l = 3$ resonance), whereas for polarization orthogonal to the bond the spectrum is featureless.[125,136–138]

The problem of molecular orientation on surfaces has been widely studied by many experimental techniques. For example, the orientation of chemisorbed carbon monoxide on the surface of metals like Ni is certainly one of the most popular topics in surface science.

Figure 33 shows the absorption spectra near the C K edge for a saturation

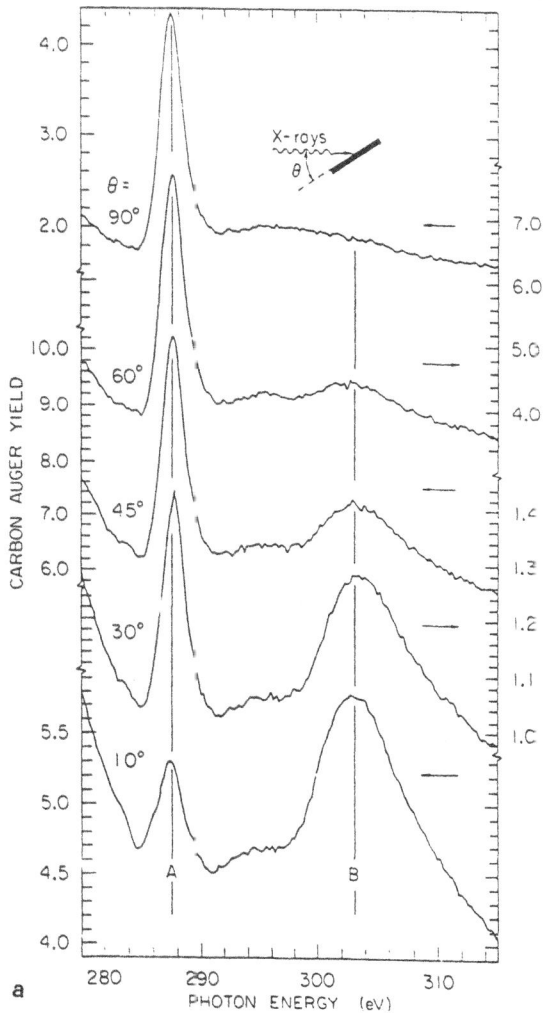

FIGURE 33. (a) Surface absorption spectra above the C K edge for CO on Ni (100) at $T = 180$ K as a function of grazing angle θ. (b) Surface absorption spectra above the O K edge for the same system as a function of incidence angle θ.

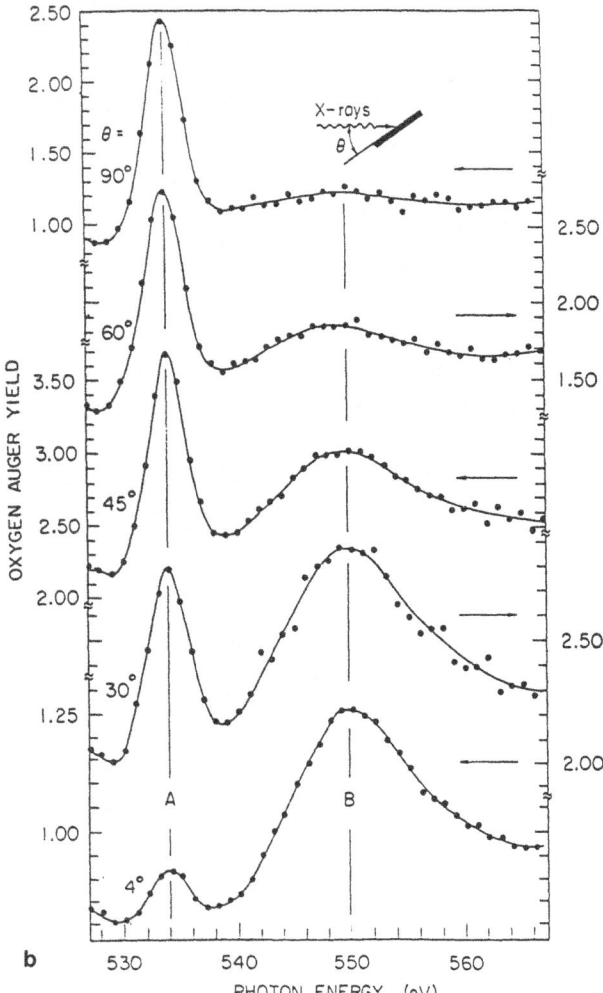

FIGURE 33. (*Continued*)

coverage of CO on Ni (100), measured at different angles of incidence.[136] At normal incidence ($\theta = 90°$), with the **e** vector parallel to the sample surface, the resonance at 287.5 eV dominates the spectrum. As θ decreases, the **e** vector rotates toward the surface normal; but while the peak A decreases, a new resonance at 303.5 eV emerges. This new resonance is maximized at grazing incidence ($\theta = 10°$) where, in contrast peak A has almost vanished. The O K-edge spectra on the same system, shown in Fig. 33, shows similar behavior, with the peak labeled A at 534 eV and the peak B at 550 eV. In both cases peaks B are readily assigned to the σ-shape resonance, because the molecular σ orbital is oriented along the molecular axis. The opposite polarization dependence identifies peak A as originating from a transition from the $1s$ state to the unfilled π^* bound molecular orbital state. As a consequence, this transition is not detectable by photoemission experiments because it lies below the continuum K-shell ionization threshold.

The intensity of these transitions can be derived from Fermi's golden rule, which links the resonance intensity I to the matrix element of the photoabsorption cross section as given in Eqs. (1) and (2). Therefore the cross section changes with the polarization angle θ as a function of the molecular orientation. Due to the dipole selections rule, and under the hypothesis of fully linearly polarized incident light, we can write for an oriented molecule with cylindrical symmetry this simple cross section:

$$\mu(\delta) = \frac{\mu_0}{4\pi}[1 + \tfrac{1}{2}\beta_m(3\cos^2\delta - 1)], \tag{31}$$

where μ_0 is the integrated photoabsorption cross section for random molecular orientation and $\beta_m(h\nu)$ is an asymmetric parameter of molecular physics. Considering that δ is the angle between the **e** vector and the intramolecular symmetry axis, due to the form of β_m the photoabsorption cross section contains no interference term between σ and π ionization amplitudes, $\beta_m = 2$ for s initial and σ final states, and $\beta_m = -1$ for s-to-π transitions giving, respectively, the simple relations $\mu \sim \cos^2\delta$ and $\mu \sim \sin^2\delta$. Thus for a molecule oriented along the surface normal, the $\sigma(\pi)$ resonance should have a maximum for $\theta = 0°(90°)$. Looking at Fig. 33, we see that the CO molecule stands upright on the Ni (100) surface. Possible deviations from the perfect angular dependence given above are due to a tilted molecular axis or to a nonlinear component of the synchrotron radiation. In fact, due to a small residual elliptical component of the polarization of the synchrotron radiation, the intensity of peak A does not vanish completely for $\theta = 0°$, but remains finite below $\theta \sim 10°$, as shown in Fig. 33.[137] A detailed analysis of the intensities of the σ and π resonances of the CO data as a function of polarization angles indicates a maximum deviation of the molecular axis from the sample normal of $10°$, which is of the same order as the vibrational amplitude.[136] Results for CO provide the basis for the determination of the orientation of NO on Ni (100). X-ray absorption measurements reported for a saturation coverage of NO give identical results compared to carbon monoxide. Only one Ni atom, the one directly bonded to the C atoms, seems to contribute to XANES spectra in molecular systems. The substrate contributions are negligible, in great contrast with atomic adsorption, as in the case of oxygen where almost 30 neighbor atoms contribute to the spectrum.[65] Analysis of the intramolecular scattering in the molecular case gives quantitative information on the molecular bond lengths and the molecular orientation, and it may be used to distinguish between molecular and atomic (dissociative) chemisorption.

The early stages of molecular dissociation are extremely important in the understanding of reaction chemistry. Oxidation has been studied on many metal surfaces, but controversial results have been obtained from various experimental techniques. Surface XANES is an ideal probe to investigate the molecular chemisorption and the detailed bonding and structure of the molecule. Stöhr *et al.*, in the case of the controversial problem of the orientation of O_2 chemisorbed on Ag (110) and Pt (111), extracted the bond length of the molecular oxygen.[138] Figure 34 illustrates the spectra of oxygen in the case of chemisorption on Ag (110) at 90 K, obtained by surface oxygen K-edge absorption spectra at various polar and azimuthal orientations of the electric field. All the spectra are

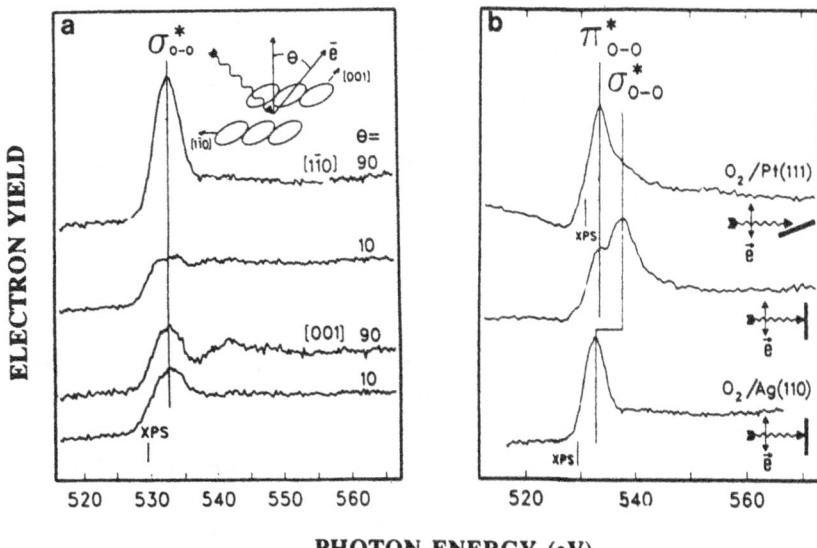

FIGURE 34. Oxygen K-edge XANES spectra for O_2 On Ag (110) at 90 K as a function of polar and azimuthal **e** orientations. The O–O σ^* peak at 532.6 eV is strongest when **E** lies along the O–O bond direction which occurs when **e** is along the $[1\bar{1}0]$ azimuth and parallel to the surface ($\theta = 90°$). The line at 529.3 eV marks the O $(1s)$ binding energy relative to the Fermi level for O_2 on Ag (110). (From Ref. 138.) (b) XANES spectra for O_2 on Pt (111) at 100 K as a function of polar **e** orientation and in comparison to O_2 on Ag (110). The two peaks in the O_2/Pt (111) spectra are assigned to the O–O π^* resonance at 533.1 eV and to the O–O σ^* resonance at 538.0 eV, and their angular dependence show that the O–O bond is parallel to the surface. Comparison with O_2/Ag (110) reveals a shift of the σ^* resonance, indicating a shorter O–O bond for O_2 on Pt (111). The line at 530 8 eV marks the O $(1s)$ binding energy of O_2 on Pt (111). (From Ref. 66.)

characterized by a peak at 532.6 eV, assigned to a transition from the O $1s$ core level to the unfilled σ^* antiboding orbital of the O–O bond. The lack of π^* resonance in the spectra indicates a complete filling of this orbital and shows the presence of a single order O–O bond. The second weak peak observed around 542 eV in the spectra with the electric field along the [001] azimuth and with $\theta = 90°$ is assigned to a scattering resonance due to the adsorbate–substrate bond. The analysis of the spectra reveals that the O_2 molecule lies approximately parallel to the surface with an uncertainty of 12° and parallel to the $[1\bar{1}0]$ azimuth. In fact, the O–O σ^* peak is maximized when the electric field is parallel to the surface ($\theta = 90°$) and to the $[1\bar{1}0]$ azimuth, giving directly the approximate orientation of the bond. Also in this case, more accurate analysis requires exact knowledge of the polarization factor of the synchrotron radiation.[125,137] Figure 34 refers to O_2 on Pt (111) at 100 K. Here the oxygen K-edge spectra exhibit two resonances: a σ^* resonance at 538.0 eV shifted by 5.4 eV if referred to the Ag (110) surface, and a new π^* resonance at 531.1 eV. The correlation established between the position of the σ^* resonance and the intramolecular bond length in free and chemisorbed molecules gives an O–O bond length of 1.32 ± 0.05 Å on Pt (111) and 1.47 ± 0.05 Å on Ag (110), with a difference of 0.15 ± 0.03 Å corresponding exactly to the 5.4 eV shift of the σ^* peak. These results show that the O–O band is a stretched bond oriented parallel to the surface; moreover, the presence of a π^* peak for O_2 on Pt (111) is an indication of a partly unfilled π^* orbital with a bond order larger than one. XANES analysis is in agreement with previous conclusions from vibrational and photoemission experiments, but in addition an accurate analysis of the bond lengths of the intensities of the σ^* resonances should give a direct measure of the degree of rehybridization in these systems.

XANES has been used to study complex molecular adsorbates. Examples are the determination of the structural transformation of the thiophene (C_4H_4S) molecule on the Pt (111) and Ni (100) surfaces as a function of temperature. XANES has also been used to study the dynamic of the hydrodesulfurization process.[96,139] Polarization-dependent spectra at the C K edge in the framework of the dipole selection rules have shown that the thiophene molecule on Pt (111) at 150 K is oriented with the ring plane tilted by about 40° from surface, while after annealing at 180 K the molecule lies down on the surface. Careful analysis of the structures and of their broadening indicate a bonding to the metal through the π^* orbitals of the ring and an interaction of the Pt surface with the σ^* orbitals near the S atom in the case of parallel bonded molecules. The thermal decomposition process of the thiophene molecule on the Pt (111) surface has been investigated in detail by comparison of the S L-XANES and the C K-XANES results as a function of increasing annealing temperatures. This is shown in Fig. 35. The S L_{23} spectrum at 180 K, corresponding to a monolayer coverage, shows two well-resolved

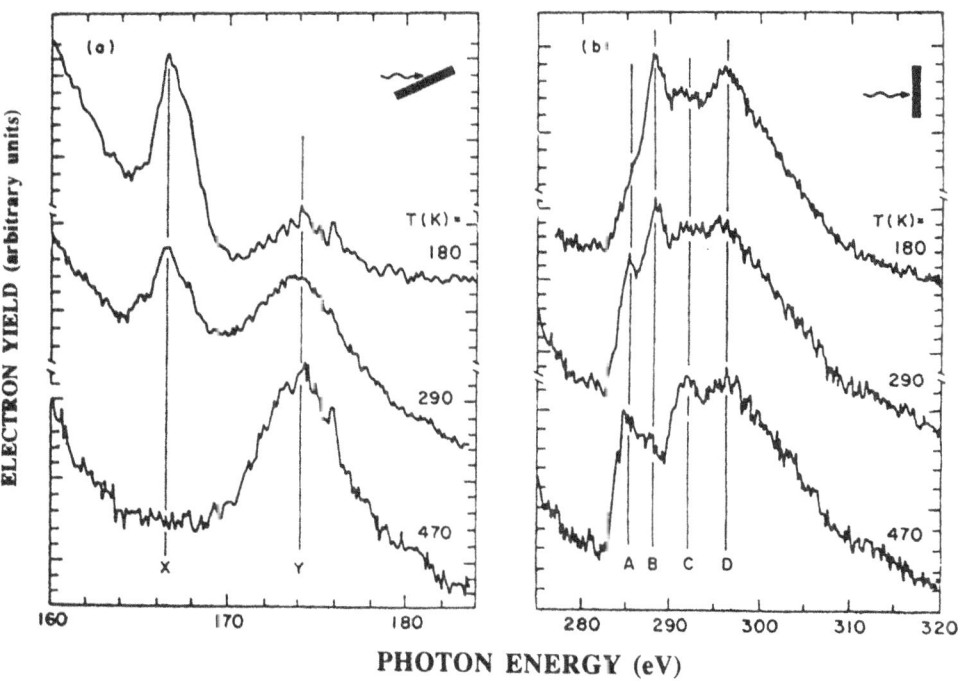

FIGURE 35. (a) XANES spectra at grazing X-ray incidence of the S $L_{2,3}$ edge for C_4H_4S on Pt (111) after annealing to different temperatures. Peak X represents a resonance characteristic of the C–S bond. (b) Spectra recorded for the C K edge under the same conditions as in (a) In this case the resonance associated with the C–S bond (peak B) disappears. (From Ref. 139.)

structures X and Y at about 166 and 174 eV. As one increases the annealing temperature, peak X decreases in intensity and is almost undetectable at ~470 K, while peak Y exhibits the opposite behavior. Peak X is a resonance corresponding to transitions of S $2p$ electrons into unfilled π^* and σ^* molecular orbitals associated with the S–C bond in thiophene. Peak Y is an atomic S resonance in the continuum assigned to excitations to d-like states.

This behavior illustrated in Fig. 35 indicates that thermal breaking of the C–S bond in the C_4H_4S molecule starts before 290 K and is completed at 470 K. The C K-edge spectrum reported in Fig. 35 strongly supports this conclusion. In addition peak B, assigned to a C–S bond, follows the same trend as peak X in the sulfur spectra. As the temperature increases, evident cleavage of the C–S bond is accompanied by a reduction of peak D associated with C–C bonds, and by the appearance of peak A. Peak A is shifted by 0.5 eV to lower energy from the π resonance of thiophene at lower temperatures. This is the result of the formation of a molecular species with a similar skeleton, but slightly inclined ($<20°$) relative to the surface, and with a new C–Pt bond, as previously observed in other similar systems.

Additional information on the C–S bond's breaking mechanism has been gained from the X-ray absorption study of thiophene adsorption on Ni (100).[96] A XANES experiment performed by fluorescence photon yield, monitoring the S K_α, made the detection of only one monolayer of thiophene, which corresponds to a sulfur sensitivity of about 0.08 monolayers. The study of dissociation of C_4H_4S on Ni (100) suggests a site-dependent C–S bond breaking on the clean Ni (100) surface, with the thiophene ring dissociated after S interaction with the Ni surface. The presence of characteristic resonances shows that S atoms are bonded in the fourfold hollow Ni sites above a temperature of about 100 K. The spectra at different coverages demonstrate that the first layer of dissociated molecules passivates the surface, and at increasing coverage the molecules remain undissociated in the upper layers. The same mechanism is activated by the oxygen atoms, so that when about half of the active fourfold hollow sites on the Ni surface are occupied (as in a $c(2 \times 2)$ O reconstructed surface), no additional thiophene dissociation occurs.

These experiments show that polarization-dependent X-ray absorption spectroscopy may probe not only the structure but also the nature of complex molecular reactions with metal surfaces such as the hydrodesulphurization, which is one important industrial process. A new possibility for investigation of chemisorption processes occurring under real catalytic conditions[96,97] is given by the fluorescence photon yield, which allows the study of samples also in nonvacuum conditions up to high pressure, and at extremely low coverage ($\ll 0.1$ monolayer). This technique has a sensitivity better than 1% of a monolayer of CO with a detector designed to be used both in ultrahigh vacuum conditions and at gas pressures up to 10 Torr. The first X-ray absorption near-edge structure measurements under atmospheric pressures were performed in the investigation of the reactivity of CO with H_2 on Ni (100) detected by C K_α fluorescence photon yield.[95,96] Figure 36 shows typical results obtained at the C K edge. Spectrum a was taken *in situ* during an exposure of the Ni 100 surface to a mixture of 10^{-6} Torr CO and 0.1 Torr H_2, while spectrum b was taken during a saturation

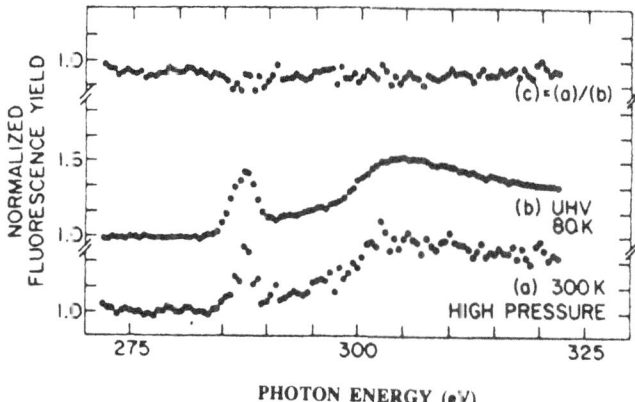

FIGURE 36. Normalized fluorescence-yield XANES taken for coadsorbed hydrogen and CO on Ni (100) at room temperature under different pressure regimes: (a) continuous exposure to 1×10^{-6} Torr CO and 0.1 Torr H_2; (b) saturation under vacuum with 10 L H_2 first, followed by 20 L CO; (c) ratio between spectra (a) and (b). (From Ref. 95.)

under vacuum conditions with 10 L of H_2 followed by 20 L of CO. No significant change in the CO chemisorption geometry upon high-pressure treatment was detected, as shown by the ratio between glancing incidence spectra under various conditions. comparison of the data reasonably excludes any formation of intermediate species in the methanation reaction, since neither new surface compounds nor changes in bond angles or bond lengths are detected by XANES in this system. A careful analysis of the data indicates that exposure of carbon monoxide to high pressures of H_2 at room temperature results in the displacement of the CO molecule from the surface, whereas when the Ni (100) is exposed to a high-pressure mixture of CO and H_2, CO molecules continue to chemisorb with their axis perpendicular to the surface, as is well known under UHV conditions.

Surface XANES studies of the chemisorption geometry of complex chemical organic conducting polymers such as poly-3-methylthiophene electrochemically deposited on Pt surfaces have been carried out.[140] The polymeric chains were found to be well ordered on the metallic surface even for thicknesses up to 50 Å. When the polymers are electrochemically doped to a conducting state, the spectra show changes as a function of thickness. These results in the field of electrochemical interfaces indicate that this technique is now ready for a large number of applications in surface science.

5. REFERENCES

1. J. P. Pendry, *Low Energy Electron Diffraction*, Academic, New York (1974).
2. A. P. Lukirskii and I. A. Brytov, Investigation of the energy structure of Be and BeO by ultra-soft X-ray spectroscopy, *Sov. Phys.-Solid State* **6**, 33–41 (1964).
3. W. Gudat and C. Kunz, Close similarity between photoelectric yield and photoabsorption spectra in the soft-X-ray range, *Phys. Rev. Lett.* **29**, 169–172 (1972).

4. A. Bianconi, R. Z. Bachrach, and S. A. Flodström, Study of the initial oxidation of single crystal aluminum by inter-atomic Auger yield spectroscopy, *Solid State Commun.* **24,** 539–542 (1977).

5. P. A. Lee, Possibility of adsorbate position determination using final-state interference effects, *Phys. Rev.* **B13,** 5261–5270 (1976).

6. U. Landman and D. L. Adams, Extended X-ray-absorption fine structure-auger process for surface structure analysis: Theoretical considerations of a proposed experiment, *Proc. Nat. Acad. Sci. USA* **73,** 2550–2553 (1976).

7. P H. Citrin, P. Eisemberger, and R. C. Hewitt, Extended X-ray absorption fine structure of surface atoms on single-crystal substrates: Iodine adsorbed on Ag (111), *Phys. Rev. Lett.* **41,** 309–312 (1978).

8. J. Stöhr, D. Denley, and P. Perfetti, Surface extended X-ray absorption fine structure in the soft-X-ray region: Study of an oxidized Al surface, *Phys. Rev.* **B18,** 4132–4135 (1978).

9. A. Bianconi and R. Z. Bachrach, Al surface relaxation using surface extended X-ray absorption fine structure, *Phys. Rev. Lett.* **42,** 104–108 (1979).

10. R. Prinz and D. Konigsberger, *X-ray Absorption: Principles and Techniques of EXAFS, SEXAFS and XANES,* Wiley, New York (1986).

11. A. Bianconi, Surface X-ray absorption spectroscopy: Surface EXAFS and surface XANES, *Appl. Surf. Sci.* **6,** 392–418 (1980).

12. J. Stöhr, EXAFS and Surface EXAFS: principles analysis and applications, in *Emission and Scattering Techniques* (P. Day, ed.), Reidel, Dordrecht (1981), pp. 213–250.

13. P. H. Citrin, An overview of SEXAFS during the past decade, in *EXAFS and Near Edge Structure IV* (P. Lagarde, D. Raoux and J. Petiau, eds.), *J. Phys. (Paris)* **47-C8,** 437–472 (1986).

14. D. Norman, X-Ray absorption spectroscopy (EXAFS and XANES) at surface, *J. Phys.* (Paris) **C19,** 3273–3311 (1986).

15. J. Stöhr, SEXAFS: everything you always wanted to know about SEXAFS but were afraid to ask, in *X-ray Absorption: Principles and Techniques of EXAFS, SEXAFS and XANES* (R. Prinz and D. Konigsberger, eds.), Wiley, New York (1986), pp. 443–571.

16. J. Rowe, "Surface EXAFS," in Chapter 3 in this volume.

17. A. Bianconi, R. Z. Bachrach, and S. A. Flodström, Oxygen chemisorption on Al: Unoccupied extrinsic surface resonances and stie-structure determination by surface soft X-ray absorption, *Phys. Rev.* **B19,** 3879–3888 (1979).

18. A. Bianconi and R. S. Bauer, Evidence of SiO at the Si-oxide interface by surface soft X-ray absorption near edge spectroscopy, *Surf. Sci.* **99,** 76–86 (1980).

19. U. von Barth and G. Grossmann, The effect of the core hole on X-ray emission spectra in simple metal, *Solid State Commun.* **32,** 645–649 (1979).

20. U. von Barth and G. Grossmann, Dynamical effects in X-ray spectra and the final-state rule, *Phys. Rev.* **B25,** 5150–5179 (1982).

21. A. Kotani, T. Jo, and J. C. Parlebas, Many-body effects in core-level spectroscopy of rare-earth compounds, *Adv. Phys.* **37,** 37–85 (1988).

22. U. Fano and J. W. Cooper, Spectral distribution of atomic oscillator strengths, *Rev. Mod. Phys.* **40,** 441–507 (1968).

23. J. E. Müller, O. Jepsen, and J. W. Wilkins, X-ray absorption spectra: K-edges of $3d$ transition metals, L-edges of $3d$ and $4d$ metals, and M-edges of palladium, *Solid State Commun.* **42,** 365–368 (1982).

24. J. E. Müller and J. W. Wilkins, Band-structure approach to the X-ray spectra of metals, *Phys. Rev.* **B29,** 4331–4348 (1984).

25. D. A. Papaconstantopoulos, Densities of states and calculated K X-ray spectra of TiFe, *Phys. Rev. Lett.* **31,** 1050–1052 (1973).

26. J. W. McCaffrey and D. A. Papaconstantopoulos, Calculated K X-ray absorption spectrum of calcium, *Solid State Commun.* **14,** 1055–1058 (1974).

27. R. P. Gupta, A. J. Freeman, and J. D. Dow, Band Theory of K-edge transitions in Li, *Phys. Lett.* **59A,** 226–228 (1976).

28. R. P. Gupta and A. J. Freeman, Role of band structure on the X-ray edge-shape in Na Metal, *Phys. Lett.* **59A,** 223–225 (1976).

29. R. P. Gupta and A. J. Freeman, Band-structure contributions to X-ray emission and absorption spectra and edges in magnesium, *Phys. Rev. Lett.* **36,** 1194–1197 (1976).

30. F. Szmulowicz and D. M. Pease, Augmented-plane-wave calculation and measurements of K and L X-ray spectra for solid Ni, *Phys. Rev.* **B17,** 3341–3355 (1978).

31. J. E. Müller, O. Jepsen, O. K. Andersen, and J. W. Wilkins, Systematic structure in the K-edge photoabsorption spectra of the $4d$ transition metals: theory, *Phys. Rev. Lett.* **40,** 720–722 (1978).

32. F. Szmulowicz and B. Segall, K X-ray absorption in aluminum, *Phys. Rev.* **B21,** 5628–5635 (1980).

33. A. Bianconi, R. Del Sole, A. Selloni, P. Chiaradia, M. Fanfoni, and I. Davoli, Partial density of unoccupied states and $L_{2,3}$-X-ray absorption spectrum of bulk silicon and of the Si (111) 2 times 1 surface, *Solid State Commun.* **64,** 1313–1316 (1987).

34. G. A. Sawatzky, Electronic structure of transition metal compounds as studied by high energy spectroscopies, in *Core Level Spectroscopy in Condensed Systems*, (J. Kanamori and A. Kotani, eds.), Springer Verlag, Berlin (1988), pp. 99–124, and reference therein.

35. J. Fink, Th. Müller-Heinzerling, B. Scheerer, W. Speier, F. U. Hillebrecht, J. C. Fuggle, J. Zaanen, and G. A. Sawatzki, $2p$ absorption spectra of the 3d elements, *Phys. Rev.* **B32,** 4899–4904 (1985).

36. J. Zaanen, G. A. Sawatzky, J. Fink, W. Speier, and J. C. Fuggle, $L_{2,3}$ absorption spectra of the lighter $3d$ transition metals. *Phys. Rev.* **B32,** 4905–4913 (1985).

37. A. Bianconi, S. Doniach, and D. Lublin, X-Ray Ca K-edge of calcium adenosine triphosphate system and of simple Ca compounds, *Chem. Phys. Lett.* **59,** 121–124 (1978).

38. M. Belli, A. Scafati, A. Bianconi, S. Mobilio, S. Palladino, A. Reale, and E. Burattini, X-Ray absorption near edge structures (XANES) in simple and complex Mn compounds, *Solid State Commun.* **35,** 355–361 (1980).

39. A. Bianconi, H. Petersen, F. C. Brown, and R. Z. Bachrach, K-shell photoabsorption spectra of N_2 and N_2O using synchrotron radiation, *Phys. Rev.* **A17,** 1907–1911 (1978).

40. D. Dill and J. L. Dehmer, Electron-molecule scattering and molecular photoionization using the multiple-scattering method, *J. Chem. Phys.* **61,** 692–699 (1974).

41. F. W. Kutzler, C. R. Natoli, D. K. Misemer, S. Doniach, and K. O. Hodgson, Use of one-electron theory for the interpretation of near edge structure in K-shell X-ray absorption spectra of transition metal complexes, *J. Chem. Phys.* **73,** 3274–3288 (1980).

42. C. R. Natoli, D. K. Misemer, S. Doniach, and F. W. Kutzler, First-principles calculation of X-ray absorption-edge structure in molecular clusters, *Phys. Rev.* **A22,** 1104–1108 (1980).

43. P. J. Durham, J. B. Pendry, and C. H. Hodges, XANES: Determination of bond angles and multi-atom correlations in order and disorder systems, *Solid State Commun.* **38,** 159–162 (1981).

44. P. J. Durham, J. B. Pendry, and C. H. Hodges, Calculation of X-ray absorption near-edge structure, XANES, *Comput. Phys. Commun.* **25,** 193–205 (1982).

45. R. V. Vedrinskii, L. A. Bugaev, I. I. Gegusin, V. L. Kraizman, A. A. Novakovich, R. E. Ruus, A. A. Maiste, and M. A. Elango, X-Ray absorption near edge structure (XANES) for KCl, *Solid State Commun.* **44,** 1401–1407 (1982).

46. T. Fujikawa, T. Matsuura, and H. Kuroda, X-Ray absorption near edge structure (XANES) studied by the short-range order multiple scattering theory, *J. Phys. Soc. Jpn.* **52,** 905–912 (1983).

47. A. Bianconi, L. Incoccia and S. Stipcich, eds. *EXAFS and Near Edge Structure*, Springer Series in Chem. Phys. Vol. 27, Springer, Berlin (1983).

48. K. O. Hodgson, B. Hedman and J. E. Penner-Hahn, eds. *EXAFS and Near Edge Structure III*, Springer Proc. Phys. Vol. 2, Springer, Berlin (1984).

49. M. Kitamura, S. Muramatsu, and C. Sugiura, Multiple-scattering approach to the X-ray absorption spectra of $3d$ transition metals, *Phys. Rev.* **B33,** 5294–5300 (1986).

50. L. T. Wille, P. J. Durham, and P. A. Sterne, X-ray absorption in ionic materials, in *EXAFS and Near Edge Structure IV* (P. Lagarde, D. Raoux, and J. Petiau, eds.), *J. Phys. (Paris)* **47-C8,** 43–47 (1986).

51. L. A. Bugaev, I. I. Gegusin, A. A. Novakovich and R. V. Vedrinskii, Crystal potential and size effects in XANES K-spectra of alkali halides, in *EXAFS and Near Edge Structure IV* (P. Lagarde, D. Raoux, and J. Petiau, eds.), *J. Phys. (Paris)* **47-C8,** 101–104 (1986).

52. R. F. Pettifer, D. L. Foulis, and C. Hermes, Multiple scattering calculations for biological catalysts, in *EXAFS and Near Edge Structure IV* (P. Lagarde, D. Raoux, and J. Petiau, eds), *J. Phys. (Paris)* **47-C8,** 545–550 (1986).

53. P. J. Durham, Theory of XANES, in *X-ray Absorption: Principles and Techniques of EXAFS, SEXAFS and XANES* (R. Prinz and D. Konigsberger, eds.), Wiley, New York (1986), pp. 53–84.

54. D. E. Sayers, E. A. Stern, and F. W. Lytle, New technique for investigating noncrystalline structures: Fourier analysis of the extended X-ray absorption fine structure, *Phys. Rev. Lett.* **27**, 1204–1207 (1971).

55. P. A. Lee and J. B. Pendry, Theory of the extended X-ray absorption fine structure. *Phys. Rev.* **B11**, 2795–2811 (1975).

56. A. Bianconi, J. Garcia, A. Marcelli, M. Benfatto, C. R. Natoli, and I. Davoli, Probing higher order correlation functions in liquids by XANES (X-Ray Absorption Near Edge Structure), *J. Phys. (Paris)* **46-C9**, 101–106 (1985).

57. J. Garcia, M. Benfatto, C. R. Natoli, A. Bianconi, I. Davoli, and A. Marcelli, Three particle correlation function of metal ions in tetrahedral coordination determined by XANES, *Solid State Commun.* **58**, 595–599 (1986).

58. M. Benfatto, C. R. Natoli, A. Bianconi, J. Garcia, A. Marcelli, M. Fanfoni, and I. Davoli, Multiple-scattering regime and higher-order correlations in X-ray-absorption spectra of liquid solutions, *Phys. Rev.* **B34**, 5774–5781 (1986).

59. A. Bianconi, A. Di Cicco, N. V. Pavel, M. Benfatto, A. Marcelli, C. R. Natoli P. Pianetta, and J. C. Woicik, Multiple-scattering effects in the *K*-edge X-ray absorption near-edge structure of crystalline and amorphous silicon, *Phys. Rev.* **B36**, 6426–6433 (1987).

60. G. N. Greaves, P. J. Durham, G. Diakun, and P. Quinn, Near-edge X-ray absorption spectra for metallic Cu and Mn, *Nature (London)* **294**, 139–142 (1981).

61. F. W. Kutzler, R. A. Scott, J. M. Berg, K. O. Hodgson, S. Doniach, S. P. Cramer, and C. H. Chang, Single-Crystal polarized X-ray absorption spectroscopy. Observation and theory for $(MoO_2S_2)^{2-}$, *J. Am. Chem. Soc.* **103**, 6083–6088 (1981).

62. A. Bianconi, M. Dell'Ariccia, P. J. Durham, and J. B. Pendry, Multiple-scattering resonances and structural effects in the X-ray-absorption near-edge spectra of Fe(II) and Fe(III) hexacyanide complexes, *Phys. Rev.* **B26**, 6502–6508 (1982).

63. J. E. Hahn, R. A. Scott, K. O. Hodgson, S. Doniach, S. R. Desjardins, and E. I. Solomon, Observation of an electric quadrupole transition in the X-ray absorption spectrum of a Cu(II) complex, *Chem. Phys. Lett.* **88**, 595–598 (1982).

64. R. A. Scott, J. E. Hahn, S. Doniach, H. C. Freeman, and K. O. Hodgson, Polarized X-ray absorption spectra of oriented plastocyanin single crystals. Investigation of methionine-copper coordination, *J. Am. Chem. Soc.* **104**, 5364–5369 (1982).

65. D. Norman, J. Stöhr, R. Jaeger, P. J. Durham, and J. B. Pendry, Determination of local atomic arrangements at surfaces from near-edge X-ray-absorption fine structure studies: O on Ni (100), *Phys. Rev. Lett.* **51**, 2052–2055 (1983).

66. J. Stöhr, J. L. Gland, W. Eberhardt, D. Outka, R. J. Madix, F. Sette, R. J. Koestner, and U. Döbler, Bonding and bond lengths of chemisorbed molecules from near-edge X-ray absorption fine-structure studies, *Phys. Rev. Lett.* **51**, 2414–2417 (1983).

67. A. Bianconi, M. Dell'Ariccia, A. Gargano, and C. R. Natoli, Bond length determination using XANES, in *EXAFS and Near Edge Structure* (A. Bianconi, L. Incoccia, and S. Stipcich, eds.), Springer Series in Chem. Phys. Vol. 27, Springer, Berlin (1983), pp. 57–61.

68. A. Bianconi, A. Congiu-Castellano, P. J. Durham, S. S. Hasnain, and S. Phillips, The CO bond angle of carboxymyoglobin determined by angular-resolved XANES spectroscopy, *Nature (London)* **318**, 685–687 (1985).

69. D. D. Vvedensky and J. B. Pendry, Comment on "Experimental study of multiple scattering in X-ray-absorption near-edge structure", *Phys. Rev. Lett.* **54**, 2725 (1985).

70. A. Bianconi, E. Fritsch, G. Calas, and J. Petiau, X-ray-absorption near-edge structure of 3*d* transition elements in tetrahedral coordination: the effect of bond-length variation, *Phys. Rev.* **B32**, 4292–4295 (1985).

71. D. Norman, K. B. Garg, and P. J. Durham, The X-Ray absorption near edge structure of transition metal oxides: A one-electron interpretation, *Solid State Commun.* **56**, 895–898 (1985).

72. H. Oizumi, J. Iizuka, H. Oyanagi, T. Fujikawa, T. Ohta, and S. Usami, *K*-edge XANES of GaP, InP and GaSb, *Jpn. J. Appl. Phys.* **24**, 1475–1478 (1985).

73. T. A. Smith, J. E. Penner-Hahn, M. A. Berding, S. Doniach, and K. O. Hodgson, Polarized

X-ray absorption edge spectroscopy of single-crystal copper(II) complexes, *J. Am. Chem. Soc.* **107,** 5945–5955 (1985).

74. I. Davoli, A. Marcelli, A. Bianconi, M. Tomellini, and M. Fanfoni, Multielectron configurations in the X-ray absorption near-edge structure of NiO at the oxygen K threshold, *Phys. Rev.* **B33,** 2979–2982 (1986).

75. M. Kitamura, C. Sugiura, and S. Muramatsu, X-ray-absorption near-edge-structure study of diamond: a multiple-scattering approach, *Solid State Commun.* **62,** 663–665 (1986).

76. S. Stizza, M. Benfatto, A. Bianconi, J. Garcia, G. Mancin., and C. R. Natoli, in *EXAFS and near Edge Structure IV* (P. Lagarde, D. Raoux, and J. Petiau, eds.), *J. Phys. (Paris)* **47-C8,** 691–696 (1986).

77. A. Bianconi and A. Congiu Castellano, eds., *Biophysics and Synchrotron Radiation,* Springer Series in Biophysics, Vol. 2, Springer, Berlin (1987).

78. K. H. Johnson, Multiple-scattering model for polyatomic molecules. *J. Chem. Phys.* **45,** 3085–3095 (1966); Scattered-wave theory of the chemical bond, *Adv. Quantum Chem.* **7,** 143–185 (1973).

79. E. Clementi and C. Roetti, Roothan–Hartree–Fock atomic wavefunctions, in *Atomic Data and Nuclear Data Tables,* Vol. 14, Academic, New York (1974), pp. 177–478.

80. K. Schwartz, Optimization on the statistical exchange parameter α for the free atoms of H through Nb, *Phys. Rev.* **B5,** 2466–2468 (1972).

81. L. Hedin and B. I. Lundqvist, Explicit local exchange-correlation potentials, *J. Phys.* **C4,** 2064–2083 (1971).

82. C. R. Natoli, M. Benfatto, and S. Doniach, Use of general potentials in multiple scattering theory, *Phys. Rev.* **A34,** 4682–4694 (1986).

83. W. L. Schaich, Derivation of single-scattering formulas for X-ray-absorption and high-energy electron-loss spectroscopies, *Phys. Rev.* **B29,** 6513–6519 (1934).

84. C. R. Natoli and M. Benfatto, A unifying scheme of interpretation of X-ray absorption spectra based on the multiple scattering theory, in *EXAFS and Near Edge Structure IV,* (P. Lagarde, D. Raoux, and J. Petiau, eds.) *J. Phys. (Paris)* **47-C8,** 11–23 (1986).

85. M. Benfatto, C. R. Natoli, C. Brouder, R. F. Pettifer, and M. F. Ruiz Lopez, Polarized curved wave EXAFS: Theory and application, *Phys. Rev.* **B39,** 1935–1939 (1989).

86. J. J. Rehr, R. C. Albers, C. R. Natoli, and E. A. Stern, New high-energy approximation for X-ray-absorption near-edge structure, *Phys. Rev.* **B34,** 4350–4353 (1986).

87. J. J. Boland, S. E. Crane, and J. D. Baldeschwieler, Theory of extended X-ray absorption fine structure: Single and multiple scattering formalisms, *J. Chem. Phys.* **77,** 142–153 (1982).

88. B. K. Teo, Bond angle determination by EXAFS: A new dimension, in *EXAFS and Near Edge Structure* (A. Bianconi, L. Incoccia, and S. Stipcich, eds.), Springer Series in Chem. Phys. Vol. 27, Springer, Berlin (1983), pp. 11–21.

89. A. Di Cicco, N. V. Pavel, and A. Bianconi, Spherical wave EXAFS analysis of the silicon K-edge X-ray absorption spectrum, *Solid State Commun.* **61,** 635–639 (1987).

90. R. Barchewitz, M. Cremonese-Visicato, and G. Onori, X-ray photoabsorption of solids by specular reflection, *J. Phys.* **C11,** 4439–4445 (1978).

91. S. M. Heald, E. Keller, and E. A. Stern, Fluorescence detection of surface EXAFS, *Phys. Lett.* **103A,** 155–158 (1984).

92. E. Dartyge, A. Fontaine, G. Tourillon, R. Cortes, and A. Jucha, X-ray absorption spectroscopy in dispersive mode and by total reflection, *Phys. Lett.* **113A,** 384–388 (1986).

93. J. Stöhr, C. Noguera, and T. Kendelewicz, Auger and photoelectron contributions to the electron-yield surface extended X-ray-absorption fine-structure signal, *Phys. Rev.* **B30,** 5571–5579 (1984).

94. V. Rehn and R. Rosenberg, Photon-Stimulated Desorption, this volume.

95. F. Zaera, D. A. Fischer, S. Shen, and J. L. Gland, Fluorescence yield near-edge X-ray absorption spectroscopy under atmospheric conditions: CO and H_2 coadsorption on Ni (100) at pressures between 10^{-9} and 0.1 Torr, *Surf. Sci.* **194,** 205–216 (1988).

96. J. Stöhr, E. B. Kollin, D. A. Fischer, J. B. Hastings, F. Zaera, and F. Sette, Surface extended X-ray absorption fine structure of low Z absorbates studied with fluorescence detection, *Phys. Rev. Lett.* **55,** 1468–1471 (1985).

97. D. A. Fischer and J. L. Gland, Soft X-ray fluorescence (UHV compatible) proportional counters

for NEXAFS and SEXAFS above the C, N, O and S K edge, in *Soft X-Ray Optics and Technology, SPIE 733*, (1986), pp. 504–507.

98. D. A. Fischer, U. Döbler, D. Arvanitis, L. Wenzel, K. Baberschke, and J. Stöhr, Carbon K-edge structure of chemisorbed molecules by means of fluorescence detection, *Surf. Sci.* **177**, 114–120 (1986).

99. D. Arvanitis, U. Döbler, L. Wenzel, K. Baberschke, and J. Stöhr, A new technique for submonolayer NEXAFS: Fluorescence yield at the carbon K edge, in *EXAFS and Near Edge Structure IV* (P. Lagarde, D. Raoux, and J. Petiau, eds.), *J. Phys. (Paris)* **47-C8**, 173–178 (1986).

100. R. P. Phizackerley, Z. U. Rek, G. B. Stephenson, S. D. Conradson, K. O. Hodgson, T. Matsushita, and H. Oyanagi, An energy-dispersive spectrometer for the rapid measurement of X-ray absorption spectra using synchrotron radiation, *J. Appl. Crystallog.* **16**, 220–232 (1983).

101. K. C. Pandey, New π-bonded chain model for Si (111) (2 times 1) surface, *Phys. Rev. Lett.* **47**, 1913–1917 (1981).

102. K. C. Pandey, Reconstruction of semiconductor surfaces: buckling, ionicity, and π-bonded chains, *Phys. Rev. Lett.* **49**, 223–226 (1982).

103. J. C. Woicik, B. B. Pate, and P. Pianetta, Silicon (111) 2×1 Surface States: K-edge transitions and surface selective $L_{2,3}VV$ Auger lineshape, *Physica* **B158**, 576–577 (1989).

104. M. Hietschold, G. Paash, and I. Bartos, Adiabatic variational calculation of the lattice relaxation at metal surfaces, *Phys. Status Solidi* **B101**, 239–252 (1980).

105. U. Landman, R. N. Hill and M. Mostoller, Lattice relaxation at metal surfaces: An electrostatic model, *Phys. Rev.* **B21**, 448–457 (1980).

106. J. P. Perdew, Physics of lattice relaxation at surfaces of simple metals, *Phys. Rev.* **B25**, 6291–6299 (1982).

107. R. Z. Bachrach, D. J. Chadi, and A. Bianconi, Unoccupied surface resonances on aluminum single crystal surfaces, *Solid State Commun.* **28**, 931–934 (1978).

108. D. W. Jepsen, P. M. Marcus, and F. Jona, Accurate calculation of the low-energy electron-diffraction spectra of Al by the Layer–Korringa–Kohn–Rostoker method, *Phys. Rev. Lett.* **26**, 1365–368 (1971).

109. P. E. Vilijoen, B. J. Wessels, G. L. P. Berning, and J. P. Roux, Temperature dependent low energy electron diffraction from aluminum, *J. Vac. Sci. Technol.* **20**, 204–212 (1982), and references therein.

110. A. Bianconi, XANES spectroscopy for local structures in complex systems, in *EXAFS and Near Edge Structure* (A. Bianconi, L. Incoccia and S. Stipcich, eds.), Springer Series in Chem. Phys. vol. 27, Springer, Berlin (1983), pp. 118–129.

111. F. Jona, D. Sondericker, and P. M. Marcus, Al (111) revisited, *J. Phys.* **C13**, L155–L158 (1980).

112. L. A. Grunes, R. D. Leapman, C. N. Wilker, R. Hoffman, and A. B. Kunz, Oxygen K near-edge fine structure: an electron-energy-loss investigation with comparison to new theory for selected $3d$ transition-metal oxides, *Phys. Rev.* **B25**, 7157–7173 (1982).

113. I. Davoli, M. Tomellini, and M. Fanfoni, Surface Ni oxide studied by oxygen K-XANES, in *EXAFS and Near Edge Structure IV* (P. Lagarde, D. Raoux, and J. Petiau, eds.), *J. Phys. (Paris)* **47-C8**, 517–520 (1986).

114. D. D. Vvedensky, D. K. Saldin, and J. B. Pendry, An update of DLXANES, the calculation of X-ray absorption near edge structure, *Comput. Phys. Commun.* **40**, 421–424 (1986).

115. D. D. Vvedensky, J. B. Pendry, U. Döbler, and K. Baberschke, Quantitative multiple scattering analysis of near-edge X-ray-absorption fine structure: $c(2 \times 2)O$ on Cu (100), *Phys. Rev.* **B35**, 7756–7759 (1987).

116. U. Döbler, K. Baberschke, J. Stöhr, and D. A. Outka, Structure of $c(2 \times 2)$ oxygen on Cu (100): A surface extended X-ray absorption fine-structure study, *Phys. Rev.* **B31**, 2532–2534 (1985).

117. T. Ohta, Y. Kitajima, P. M. Stefan, M. L. Shek Stefan, N. Kosugi, and H. Kuroda, Surface EXAFS and XANES studies of S/Ni (110) and S/Ni (111), in *EXAFS and Near Edge Structure IV* (P. Lagarde, D. Raoux, and J. Petiau, eds.), *J. Phys. (Paris)* **47-C8**, 503–508 (1986).

118. A. Bianconi, R. Z. Bachrach, S. B. M. Hagstrom, and S. A. Flodstrom, Al–Al_2O_3 interface study using surface X-ray absorption and photoemission spectroscopy, *Phys. Rev.* **B19**, 2837–2843 (1979).

119. D. Norman, S. Brennan, R. Jaeger and J. Stöhr, Structure models for the interaction of oxygen with Al (111) and Al implied by photoemission and surface EXAFS, *Surf. Sci.* **105**, L297–L306 (1981).

120. M. L. Knotek, R. H. Stulen, G. M. Loubriel, V. Rehn, R. A. Rosemberg, and C. C. Parks, Photon stimulated desorption of H^+ and F^+ from BeO, Al_2O_3 and SiO_2: comparison of near edge structure to photoelectron yield, *Surf. Sci.* **133**, 291–304 (1983).

121. A. P. Hitchcock, An update of bibliography of atomic and molecular inner-shell excitation studies, *J. Electron. Spectrosc. & Relat. Phenom.* **25**, 245–275 (1982).

122. J. L. Dehmer and D. Dill, Shape resonances in K-Shell photoionization of diatomic molecules, *Phys. Rev. Lett.* **35**, 213–215 (1975).

123. J. L. Dehmer and D. Dill, Molecular effects on inner-shell photoabsorption. K-shell spectrum of N_2, *J. Chem. Phys.* **65**, 5327–5334 (1976).

124. H. Petersen, A. Bianconi, F. C. Brown and R. Z. Bachrach, The absolute N_2 K-photoabsorption cross section up to $h\mu = 450\,eV$, *Chem. Phys. Lett.* **58**, 263–266 (1978).

125. J. Stöhr and R. Jaeger, Absorption-edge resonances, core-hole screening, and orientation of chemisorbed molecules: CO, NO, and N_2 on Ni (100), *Phys. Rev.* **B26**, 4111–4131 (1982).

126. A. Bianconi, F. C. Brown, and R. Z. Bachrach (unpublished); A. Bianconi, X-Ray absorption near edge structure (XANES) and their applications to local structure determination, in *EXAFS for Inorganic Systems*, (C. D. Garner and S. S. Hasnain, eds.), Daresbury Report DL/SCI/R17, 13–22 (1981).

127. A. P. Hitchcock and C. E. Brion, Carbon K-shell excitation of C_2H_2, C_2H_4, C_2H_6 and C_6H_6 by 2.5 keV, *J. Electron. Spectrosc. Relat. Phenom.* **10**, 317–330 (1977).

128. M. Dell'Ariccia, A. Gargano, C. R. Natoli, and A. Bianconi, A calculation of C K-shell X-ray absorption spectra of C_2H_n ($n = 2, 4, 6$) oriented molecules: correlation between position of the multiple scattering resonance in the continuum and the C–C bond length, LNF Report 84/51 (P).

129. S. Della Longa and A. Marcelli (unpublished).

130. C. R. Natoli, Inner shell X-ray photoabsorption as a structural and electronic probe of matter, Lectures at NATO Advanced Study Institute Vimeiro, Portugal (1987); LNF Report 87/83 (PT) and references therein.

131. F. Sette, J. Stöhr, and A. P. Hitchcock, Determination of intramolecular bond lengths in gas phase molecules from K shell shape resonances, *J. Chem. Phys.* **81**, 4906–4914 (1984).

132. J. Stöhr, F. Sette, and A. L. Johnson, Near-edge X-ray-absorption fine-structures studies of chemisorbed hydrocarbons: Bond lengths with a ruler, *Phys. Rev. Lett.* **53**, 1684–1687 (1984).

133. D. Arvanitis, K. Baberschke, L. Wenzel, and U. Döbbler, Experimental study of the chemisorbed state of C_2H_2, C_2H_4 and C_2H_6 on noble-metal surfaces, *Phys. Rev. Lett.* **57**, 3175–3178 (1986).

134. D. Arvanitis, L. Wenzel, and K. Baberschke, Direct evidence of a stretched C–C distance for C_2H_2 and C_2H_2 on Cu (100) at 60 K, *Phys. Rev. Lett.* **59**, 2435–2438 (1987).

135. J. Stöhr, D. A. Outka, K. Baberschke, D. Arvanitis, and J. A. Horsley, Identification of C–H resonances in the K-shell excitation spectra of gas-phase, chemisorbed, and polymeric hydrocarbons, *Phys. Rev.* **B36**, 2976–2979 (1987).

136. J. Stöhr, K. Baberschke, R. Jaeger, R. Treichler, and S. Brennan, Orientation of chemisorbed molecules from surface-absorption fine-structure measurements: CO and NO on Ni (100), *Phys. Rev. Lett.* **47**, 381–384 (1981).

137. J. Stöhr and D. A. Outka, Determination of molecular orientations on surfaces from the angular dependence of near-edge X-ray absorption fine-structure spectra, *Phys. Rev.* **B36**, 7891–7905 (1987).

138. D. A. Outka, J. Stöhr, W. Jark, P. Stevens, J. Solomon, and R. J. Madix, Orientation and bond length of molecular oxygen on Ag (110) and Pt (111): A near-edge X-ray-absorption fine-structure study, *Phys. Rev.* **B35**, 4119–4122 (1987).

139. J. Stöhr, J. L. Gland, E. B. Kollin, R. J. Koestner, A. L. Johnson, E. L. Muetterties and F. Sette, Desulfurization and structural transformation of thiophene on the Pt (111) surface, *Phys. Rev. Lett.* **53**, 2161–2164 (1984).

140. G. Tourillon, E. Dartyge, A. Fontaine, R. Garrett, M. Sagurton, P. Xu, and G. P. Williams, Chemisorption geometry of poly-3-methylthiophene electrochemically deposited on Pt as observed by NEXAFS, *Europhys. Lett.* **4**, 1391–1396 (1987).

3

Surface EXAFS

J. E. Rowe

1. INTRODUCTION

In the last decade there has been an increased wealth of information on surface properties of solids, both for atomically pure (or clean) surfaces of metals and semiconductors and for surfaces with adsorbed atoms and molecules. The latter systems have many fundamental properties, such as 2-dimensional phase transitions and electronic instabilities. In order to describe these dynamical properties of surfaces, it is necessary to know accurately the atomic structural coordinates of the first one or two layers at the surface. Since 1977–78 studies[1,2] of surfaces using synchrotron radiation have provided new structural information with a precision of ± 0.03 Å, which is better than the accuracy achieved using most other surface-structure probes. In this chapter, we shall describe the essential physics of two experimental methods which are explicitly designed to determine surface structures by exploiting the properties of elemental specificity and sensitivity to local structure. The two methods—surface extended X-ray absorption fine structure (SEXAFS) and angle-resolved photoemission extended fine structure (ARPEFS) are closely related, as shown below. In a heuristic sense, it is useful to think of SEXAFS and ARPEFS, respectively, as the integral and differential manifestations of the same phenomenon, by analogy with the broader relationship between absorption and angle-resolved photoemission. This phenomenon is the spectral extended fine structure due to the elastic scattering of photoelectrons originating from core level photoexcitation of surface atoms.

In both SEXAFS and ARPEFS the sample is irradiated with monochromatized synchrotron radiation of energy $h\nu$. To achieve elemental specificity, $h\nu$ is varied across the edge energy E_0 of a characteristic cross core level of the element of interest. The signal from this element is usually identified through an enhanced electron flux from the sample above the edge energy, $h\nu \approx E_0$. Surface sensitivity is attained because the short (≈ 10 Å) mean free path of electrons in solids with energies of 10–1000 eV assures a strong signal from electrons that

---

J. E. Rowe • AT&T Bell Laboratories, Murray Hill, New Jersey 07974.

Synchrotron Radiation Research: Advances in Surface and Interface Science, Volume 1: Techniques, edited by Robert Z. Bachrach. Plenum Press, New York, 1992.

originated near the surface. Electron-energy analysis can be used to increase both the elemental and surface sensitivities of the electron signal, as described below.

Several books[3-7] and review articles on EXAFS[8,9] in general and surface EXAFS[10-12] in particular are available, and these works constitute an extensive bibliography.

2. THEORY OF SEXAFS AND ARPEFS

Extended fine structure appears as an oscillatory modulation of the electron signal $I(h\nu)$, extending for hundreds of electron volts above the edge energy. This structure arises through interference between the primary photoelectron wave $\Psi_0(k)$ and a particular subset $\Psi_s^1(k)$ of the waves created by elastic scattering of $\Psi_0(k)$ from neighboring atoms. The particular subset $\Psi_s^1(k)$ relevant to a given experiment is determined by the method of detection. In SEXAFS, $\Psi_s^1(k)$ consists of waves scattered back to the source atom: the SEXAFS signal arises from modulations due to constructive or destructive interference between $\Psi_0(k)$ and $\Psi1_s(k)$ at the source. In an oversimplified model calculation, we could imagine elastic backscattering of a spherical photoelectron wave

$$\Psi_0(k) = \frac{e^{ikr}}{r} \tag{1}$$

from a spherical shell of radius R with scattering amplitude $f(k)/R$, yielding

$$\Psi_s^1 = \frac{f_\pi(k)}{R} e^{ik(r+2R)}. \tag{2}$$

The normalized SEXAFS signal $\chi_E(k)$ is given by

$$\chi_E(k) = |\Psi_0(k) + \Psi_s^1(k)|^2 \approx \frac{f_\pi(k)}{kR^2} \sin 2kR, \tag{3}$$

exclusive of atomic phase shifts. Clearly $k\chi_E(k)$ is directly related to the extra distance $2R$ traveled by the scattered wave, and R can be readily obtained from the measured quantity $\chi_E(k)$. In real surface–adsorbate systems, R is, of course, replaced by the distances to the nearest, next-nearest, etc., neighbor atoms.

The ARPEFS method differs in that a direction is inherent in the detection of the angle-resolved photoelectron signal. The primary wave has a given amplitude $\Psi_0(k, d)$ in the direction of the detector d. Interfering with this wave is the set of waves $\Psi_s(k, d)$ scattered off neighboring atoms j, through scattering angles α_j, and propagating in direction d toward the detector. The ARPEFS signal $\Psi_A(k)$ is made up of terms of the form

$$\frac{f(\alpha_j, k)}{kr_j} \cos kr_j(1 - \cos \alpha_j). \tag{4}$$

The high angular sensitivity that this form implies is both an advantage and a burden.

To understand the similarities between SEXAFS and ARPEFS, let us explore in more detail how they can be derived from a common theory due to Lee.[13] Consider a simple two-atom configuration with the atom absorbing the X-ray at the origin and the nearest neighbor (NN) atom at a distance r_1. The intensity of photoelectrons in direction k is

$$I(k) = \text{const} \left| \hat{\varepsilon} \cdot \hat{k} + \frac{f(\alpha_1, k)}{r_1} e^{ikr_1(1-\cos\alpha_1)} \hat{\varepsilon} \cdot \hat{r}_1 \right|^2, \qquad (5)$$

where $\hat{\varepsilon}$ is the polarization direction. The two terms in Eq. (5) describe the interference of the direct electron wave function with the wave function that propagates a distance r_1 to the nearest neighbor and is then scattered through the angle α_1 to reach the detector. Figure 1 shows this process schematically. The difference in the path length traveled is $r_1(1 - \cos\alpha_1)$, and this form of Eq. (5) is used directly in analyzing ARPEFS data. If one integrates Eq. (5) over all directions of photoelectron direction $\hat{k}$, then the total absorption probability is obtained: i.e., the EXAFS or SEXAFS function given in Eq. (3). A detailed discussion of this integration procedure is not appropriate here. The extension of Eq. (5) to more realistic surface configurations involves a summation over atomic neighbors at distances $\hat{r}_j$ and implies an independent (or single-scattering) contribution to the total intensity of each atom. As discussed by Lee and coworkers, this single-scattering picture is justified in EXAFS by some cancellation effects in the angular integration of Eq. (5) and by the fact that the most important multiple-scattering terms have the effect of replacing the path-length phase $2kR$ by $2kR + 2\delta_1(k) + \Phi(k)$. The additional terms $\delta_1(k)$ and $\Phi(k)$ are the central absorbing atom phase shift and backscattering phase shift, respectively. These terms have an approximately linear dependence on the wave vector k and can be obtained from theoretical calculations or from known model compound systems and by fitting the experimental data.

Figure 2 shows the outline for the iterative procedure used in analyzing EXAFS data. In most cases to date the preferred method is to compare experimental data

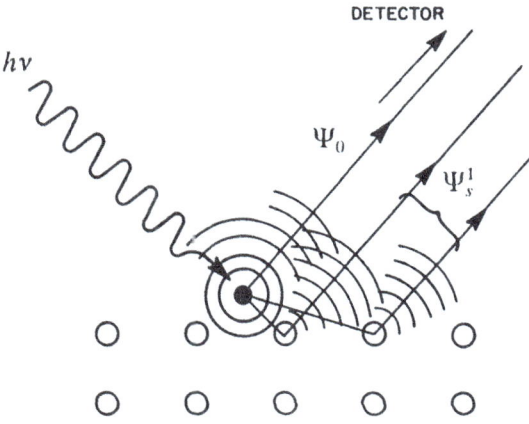

FIGURE 1. Schematic diagram showing the wave-function interference common to both EXAFS and photoemission or ARPEFS.

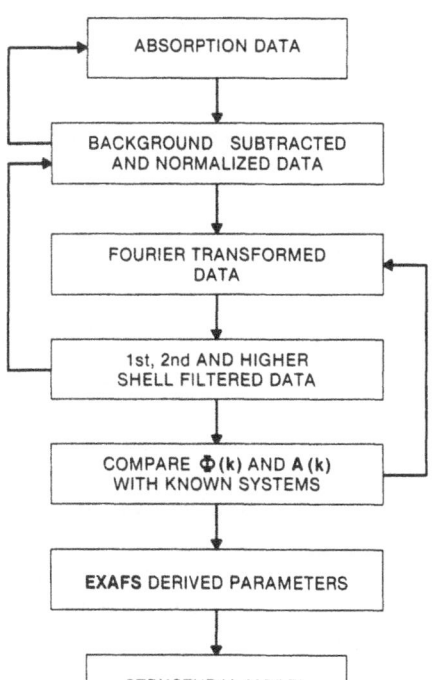

FIGURE 2. Block diagram showing the analysis procedure most commonly used for determining structural parameters from EXAFS. Empirical phase and
amplitude parameters from known systems are used
to increase the precision of the derived EXAFS
parameters.

for model compound systems with known geometry and data for an unknown
surface system. Some representative examples of such comparisons are given
below. An extensive discussion of EXAFS data analysis is given by Sayers and
Bunker.[14]

3. EXPERIMENTAL TECHNIQUES

Surface-structure determinations using the SEXAFS and ARPEFS methods
require synchrotron radiation with photon beam line optics including monochromators and detectors including those for incident intensity and I_0, and monitors.
The beam line optics typically consists of grazing incidence mirrors, adjustable
slits, and an energy-defining element such as a grating or a crystal. The coating
material of the reflecting mirrors and grating and the monochromator crystals
need to be chosen carefully in order to minimize the introduction of structures
into the transmitted intensity. All absorption edges and EXAFS modulations of the
coating and crystal materials will be directly imposed onto the energy dependence
of the reflected intensity because the absorption coefficient and reflectivity are
linked through Fresnel's equations.

3.1. Beam-Line Considerations

The monochromator is the heart of the beam line and is mainly responsible
for the important quantities of photon flux, spectral resolution, spectral purity,
and spatial beam stability. High photon flux levels are needed for SEXAFS and

ARPEFS: thus crystals should be chosen with high efficiencies or Bragg reflectivities. For SEXAFS and ARPEFS only moderate spectral resolution (<5 eV) is necessary. The details of the near-edge absorption fine structure contain complementary information, and one may also wish to separate chemically shifted spectral components; thus resolution on the order of 1 eV is desirable.

Both gratings and crystals reflect higher orders of the first-order or principal energy. These can amount to several percent of the principal intensity. The harmonics can be suppressed by filters with an absorption edge above the first and below the appropriate higher-harmonic energy or more efficiently by using a mirror reflection with a suitable cutoff energy determined by the incidence angle and coating material. For a double-crystal monochromator the rocking curve (i.e., resolution) for a higher-order reflection is narrower and usually not centered on the first-order reflection. Thus, detuning of the two crystals will preferentially suppress the higher-order intensity. The problem of higher-order contamination is most difficult to deal with in the soft X-ray region (250–1500 eV) since the separation of harmonics is comparable to the length of a typical SEXAFS scan. For example, in order to carry out SEXAFS or ARPEFS measurements above the C (285 eV), N (400 eV) and O (530 eV) K edges, a spectrally pure energy from 250 to 850 eV is needed. If we choose a filter or mirror with a 850-eV cutoff energy, the energy range 250–425 eV will still suffer from higher-order contamination.

Another source of nonmonochromatic spectral contamination is scattered light. For grating monochromators, the grating itself is usually the main source of scattered light. For crystal monochromators we have to remember that radiation up to about 3 keV is reflected by any smooth surface. For example, in order to avoid having the specular-reflected and Bragg-reflected radiation superimposed, one crystal in a double crystal monochromator needs to be cut slightly asymmetrically (Θ_A < 1 percent). The specular beam is then reflected differently by Θ_A and can be eliminated by a collimator. For crystal monochromators, sudden intensity changes can occur at certain energies due to the fact that several Bragg reflections are satisfied simultaneously. If these Bragg "glitches" are too strong they cannot be accurately normalized and the resulting data's analysis is severely limited.

3.2. Surface EXAFS Yield Measurements

A postmonochromator experimental arrangement to monitor the incident intensity, I_0, is essential for SEXAFS and ARPEFS as shown in Fig. 3. The monochromatic beam is first trimmed to an appropriate size by an arrangement of collimators. The collimator and a beam-intensity monitor are usually in a small chamber in front of the actual sample and detector chamber. The beam-intensity monitor is used to normalize the signal from the sample to any fluctuations, modulations, or structures of the incident X-ray beam intensity. Thus this reference monitor should provide an output signal which is proportional to the number of incident photons and which is not signal-to-noise limited, and it should have a constant or at least smoothly varying quantum efficiency over the investigated energy range.

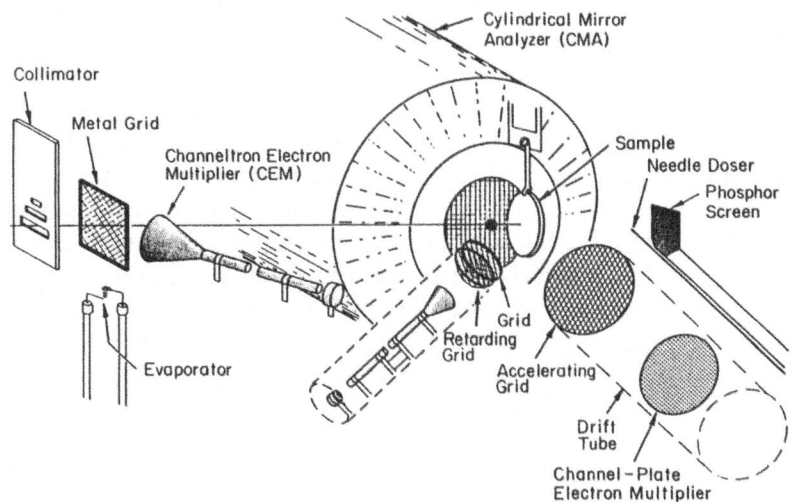

FIGURE 3. Experimental detector arrangement for surface EXAFS or angle-resolved photoemission with synchrotron radiation. The metal grid and CEM monitor the relative incident-light intensity. The C detector is used for Auger yield or photoemission (angle-resolved with the addition of an aperture).

A typical reference monitor scheme consists of a high transmission 80–90% metal grid which can be coated *in situ*. The total current signal from this grid amplified by a high current channeltron electron multiplier serves as a dynamic intensity monitor. The coating material of the grid applied by *in situ* evaporation is chosen so as not to exhibit any absorption structures in the photon energy range of interest.[10]

The SEXAFS chamber is equipped with the usual surface physics preparation and characterization tools such as a sputter gun, sample cleaver, residual gas analyzer, LEED optics, gas-dosing system, and evaporators. A phosphor screen which can be positioned in place of the sample allows one to properly focus and align the X-ray beam with respect to the focal spot of the electron energy analyzer, usually a cylindrical mirror analyzer (CMA).

In many experimental arrangements, a CMA is used for Auger yield measurements, while total-yield and partial-yield measurements are carried out with a simple retarding grid detector consisting of two hemispherical grids and a high-gain electron multiplier. The output signal of the electron multipliers is high enough to employ current-measurement techniques using a floating high-voltage battery box and a current amplifier. Previous work[10] has shown that the total yield and partial yield of secondary photoelectrons versus photon energy gives the X-ray absorption spectrum. Since the Auger yield is proportional to the total number density of core holes excited in the sample, Auger yield is also proportional to the X-ray absorption coefficient. Thus for SEXAFS the quantity typically measured is the Auger yield, total electron or partial electron yield versus photon energy. It is also possible to measure SEXAFS by monitoring the desorbing ion signal, as discussed in Chapter 9.

Another very successful form of yield measurement detects X-ray fluorescence rather than electrons. Although the efficiency is sometimes low, the signal-

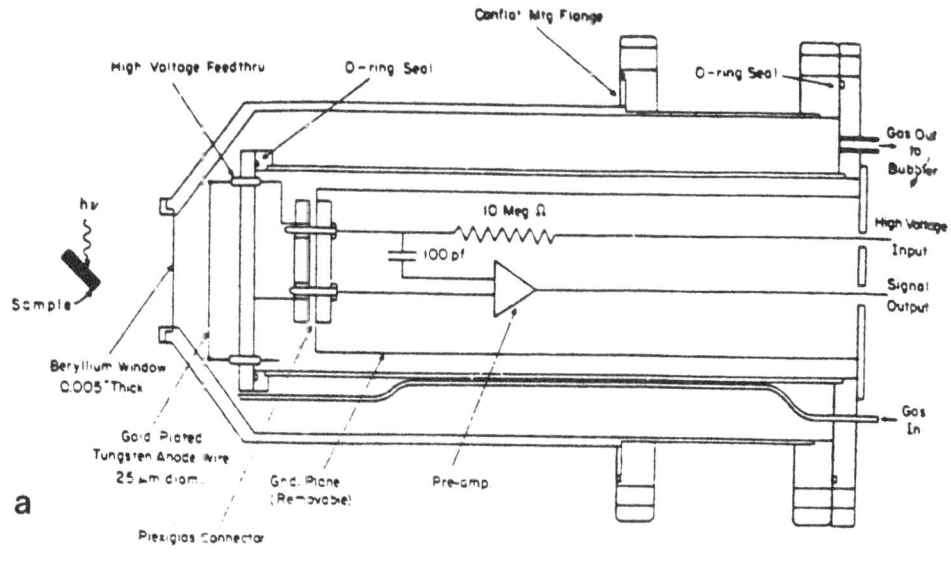

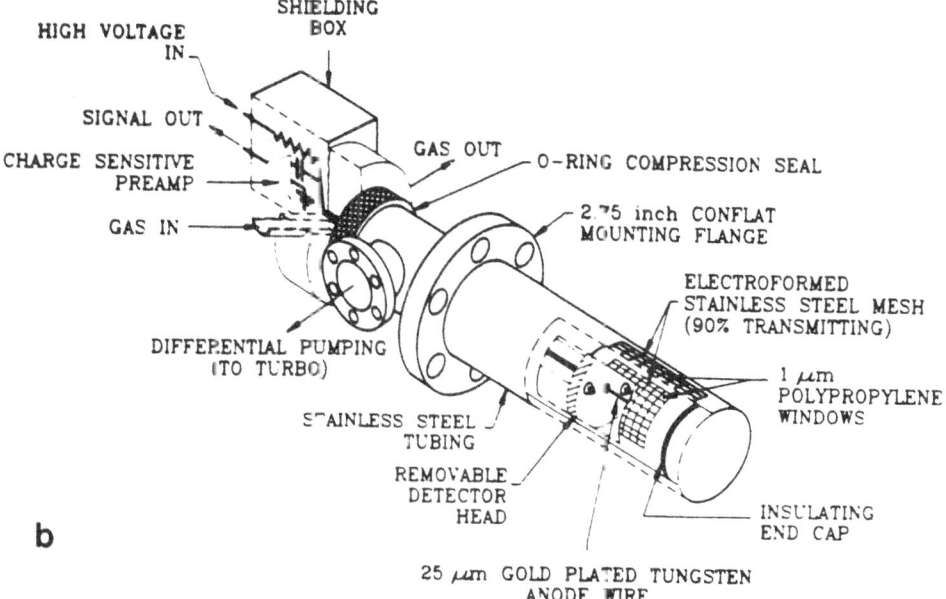

FIGURE 4. Fluorescence yield detector schematic.

to-noise ratio is good because of the absence of the secondary electron background. Figure 4 shows examples of fluorescence yield detectors.

3.3. Angle-Resolved Photoemission Fine Structure

For ARPEFS studies an angle-resolved electron detector is required. It must be capable of fairly high resolution energy analysis because only the elastic photoelectron peak carries the signal. A multichannel energy analyzer is desirable to permit simultaneous recording of peak (signal) and baseline (background) data, thereby facilitating efficient data reduction.

To establish the essential criteria for angular variability in an ARPEFS detector, we can replace the full sample–detector system by a single primitive scattering event. Three vectors are needed to characterize the experimental geometry of this event: the momentum vectors describing photoelectron propagation toward the detector and toward the scattering atom, respectively, and the polarization vector of the exciting radiation, $\mathbf{E}$. Three angles are therefore important in ARPEFS:

$$\alpha = \angle(\mathbf{p}, \mathbf{p}'), \qquad \beta = \angle(\boldsymbol{\varepsilon}, \mathbf{p}'), \quad \text{and} \quad \nu = \angle(\boldsymbol{\varepsilon}, \mathbf{p}).$$

In SEXAFS, by contrast, only one angle is variable; it is the analogue of β.

Present-day synchrotron radiation sources provide linearly polarized radiation with $\mathbf{E}$ fixed in the horizontal plane. For ARPEFS studies, the detector should be rotatable through $\pi/2$ around the beam direction to cover the range $0 < \nu < \pi/2$. A versatile sample manipulator will then permit variation of the other angles through $0 < \alpha < \pi$, $0 < \beta < \pi/2$.

Angular resolution in ARPEFS is another matter. J. J. Barton has shown that, while an accurate knowledge of the angles—especially α—is important, a relatively large detector acceptance angle is permissible. In fact, large acceptance angles may be advantageous, as they would emphasize near neighbors in the $X(k)$ curves by averaging out distant-atom contributions.

4. ATOMIC ADSORBATES ON METALS

4.1. Iodine on Ag and Cu

The choice of iodine as an adsorbate for the initial SEXAFS experiments[1] was governed in part by its suitability for study with the available synchrotron X-ray monochromator and focusing mirror and in part by the fact that the $(\sqrt{3} \times \sqrt{3})$ $R30°$ I:Ag(111) structure had been previously characterized with LEED by Forstmann et al., thereby providing a basis for comparison with the new technique. A simple system was sought so that the phenomenology of the SEXAFS technique could be assessed. The SEXAFS data on the $(\sqrt{3} \times \sqrt{3})$ 30° structures of atomic I adsorbed on Ag(111) and Cu(111) surfaces are shown in Figs. 5 and 6. Bond lengths were directly directly determined from this data. Both of the adsorbate–metal bond lengths at 110 K are larger than those in AgI and CuI at 300 K by ±0.07 Å and that the iodine atom occupies the threefold coordination site on these surfaces. The 1/3 monolayer system has a M–I–M (M = metal) bond angle of ~60° with coordination number of 3, while the bulk system has a M–I–M bond angle of ~109° and coordination number of 4; thus it should not be surprising that the I–M bond lengths are also different. The earlier LEED study for the I–Ag system maintained that the I–Ag(111) (300 K) distance was the same as that in AgI (300 K) to within about ±0.14 Å, while several LEED studies of other

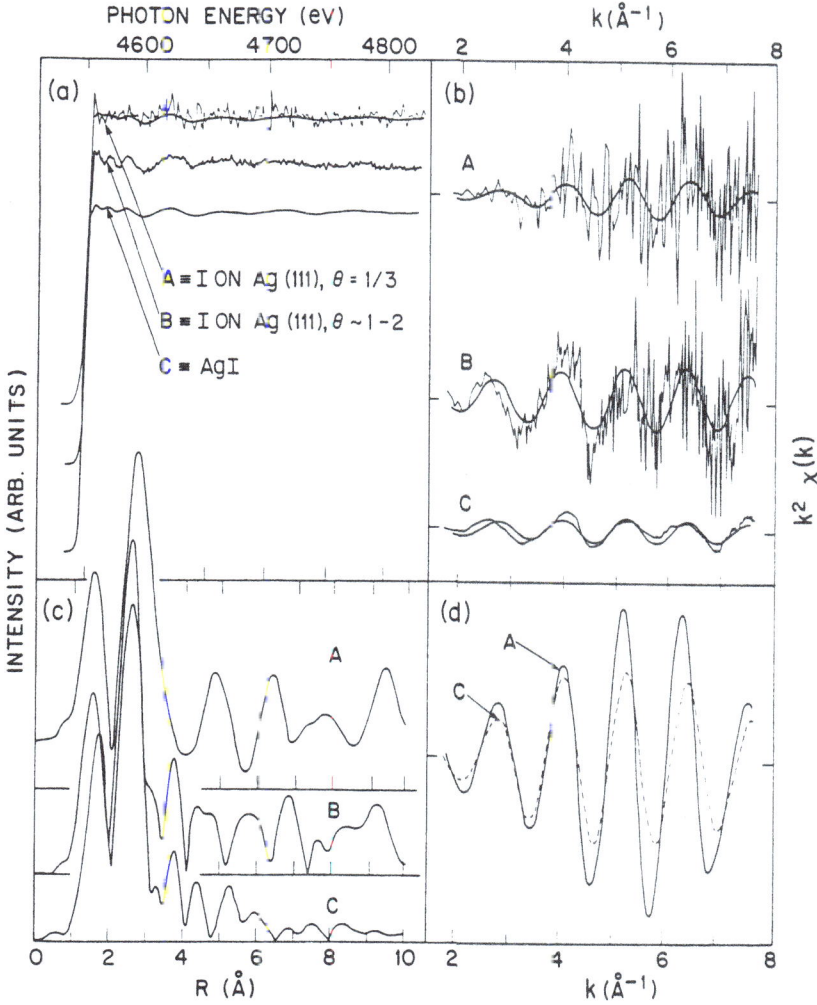

FIGURE 5. (a) Early X-ray absorption data for three different I/Ag systems taken under nondedicated running conditions. Curves in panels (b), (c), and (d) correspond to the total EXAFS function, to the Fourier-transformed data, and to the first-neighbor filtered data, respectively.

systems involving more electronegative adsorbates on transition-metal surfaces found the adsorbate–substrate bond lengths were shorter than those in analogous stable bulk compounds. The SEXAFS bond lengths reported here are therefore not at all obvious to interpret unambiguously.

4.2. Chlorine on Cu (100)

The geometry and electronic structure of $c(2 \times 2)$ Cl on Cu (100) was investigated in 1982 by Citrin *et al.* in a combined study utilizing SEXAFS, angle-resolved photoemission spectroscopy (ARPES) and a self-consistent electronic-structure calculation. This combined scheme allowed the structural parameters derived from the SEXAFS data to be taken as input parameters for a

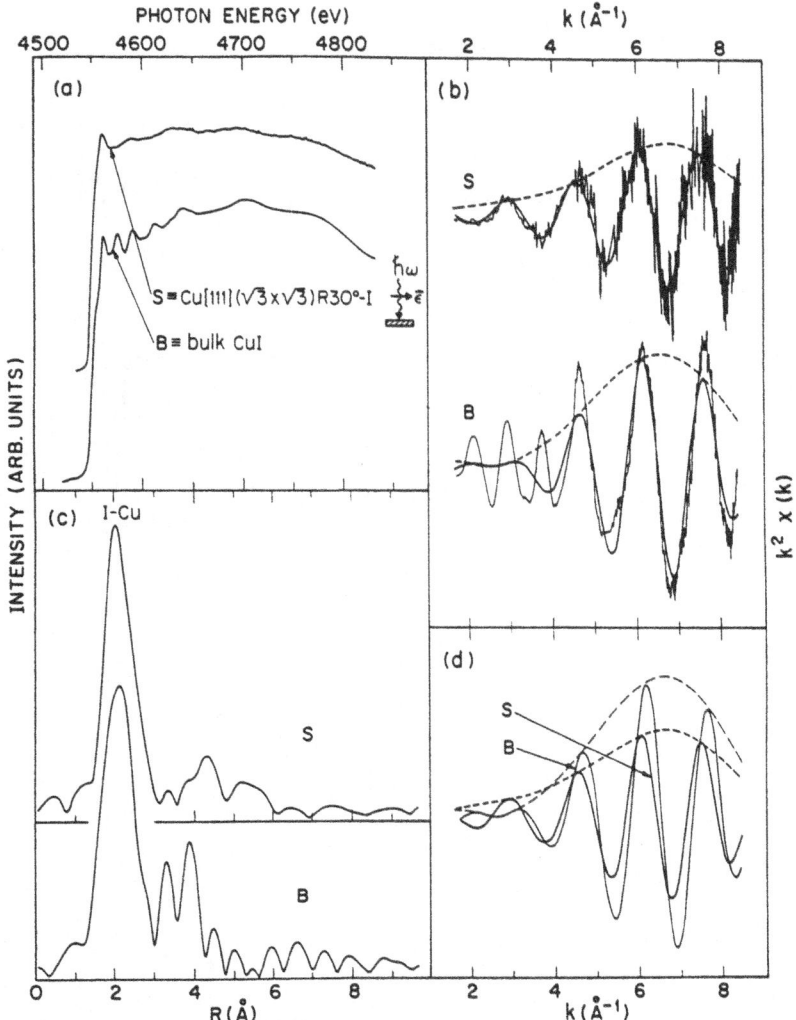

FIGURE 6. (a) X-ray absorption data for bulk CuI and for Cu (111)-I with the incident angle of the X-ray beam at 90°. Curves (b), (c), and (d) correspond to the total EXAFS function, the Fourier transform function, and the first-neighbor data. The first-neighbor data is superimposed on the total EXAFS function in panel (b) for comparison.

surface linear augmented plane wave calculation of a layered slab. The calculated dispersion of Cl-induced surface states and resonances was then compared to those from previously published and newly measured ARPES spectra.

Figure 7 shows the fluorescence yield spectra for Cu (100) $c(2 \times 2)$–Cl for two different polarizations.[15] Figure 8 shows the radial function for the data described in Fig. 7. The chemisorption geometry for $c(2 \times 2)$ Cl on Cu (100) as derived by SEXAFS consists of a simple Cl overlayer in fourfold hollow sites with a bond length of 2.37 ± 0.02 Å corresponding to $Z = 0.53 \pm 0.03$ Å. The SEXAFS measurements above the Cl K edge (2820 eV) were carried out by elastic Auger yield detection using the Cl KLL Auger line and CuCl with $N = 4$ and

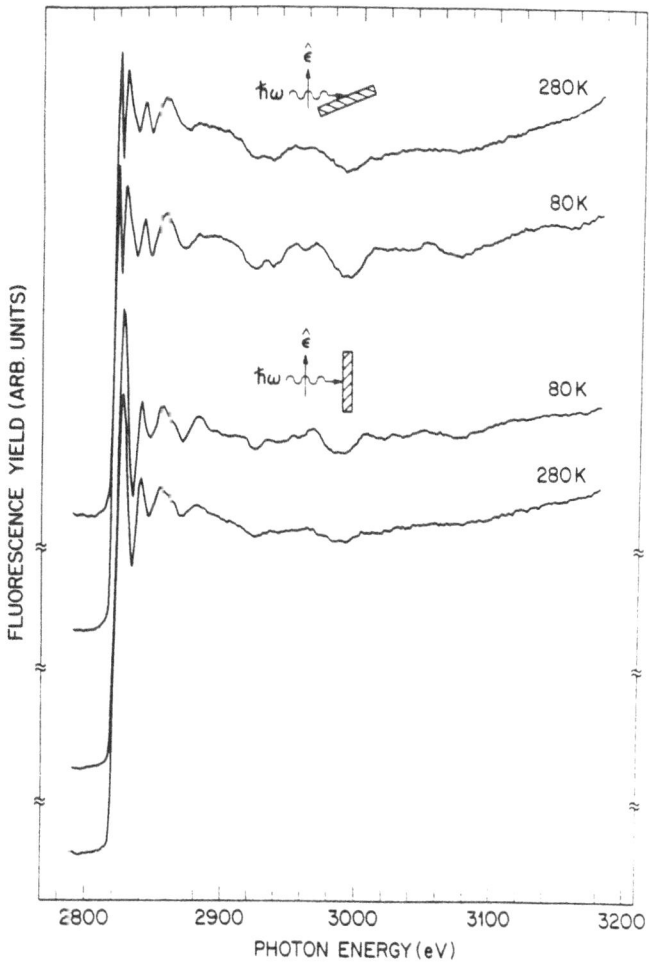

FIGURE 7. Fluorescence yield spectra for Cu (100) $c(2 \times 2)$–Cl for two different polarizations.

$R = 2.341$ Å served as a model compound. The Fourier transform revealed the first NN Cl–Cu distance (2.37 Å) and a weaker polarization-dependent distance corresponding to a combination of four third NN Cu atoms in the second layer and to eight fourth NN Cu atoms in the surface plane. This second observed distance falls at 4.31 ± 0.04 Å for $\Theta = 90°$ and 4.26 ± 0.05 Å for $\Theta = 0.05°$ and confirms the chemisorption site determined from the SEXAFS amplitude. With the structural parameters determined by SEXAFS, good agreement was obtained between the measured and calculated dispersion of Cl-induced valence band features. This study is a good example of how the structural parameters determined by SEXAFS serve as input for first-principles theoretical calculations of surface electronic structure and other theoretical studies such as adsorption energies, interaction potential curves, and surface vibrational modes.

4.3. Sulfur on Ni (100)

The initial chemisorption of S on Ni (100) is characterized by $p(2 \times 2)$ (0.25 ML) and $c(2 \times 2)$ (0.5 ML) overlayer structures which have been well

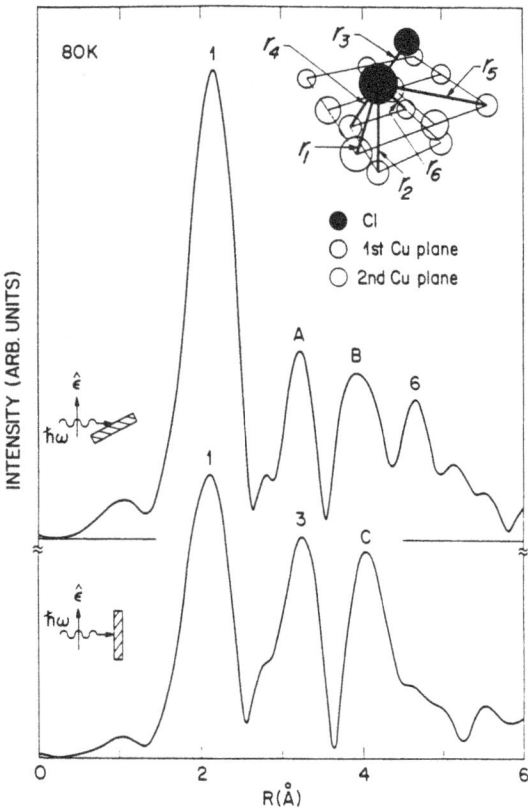

FIGURE 8. Radial function for data described in Fig. 7.

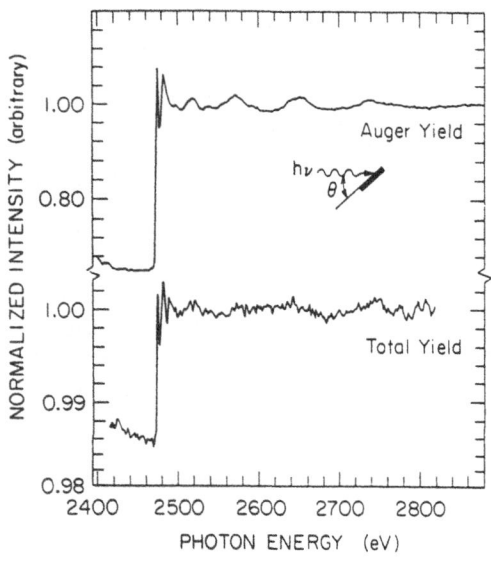

FIGURE 9. Auger electron yield and total electron yield EXAFS K-edge spectra for $c(2 \times 2)$ S on Ni(100), corresponding to half a S monolayer. Both spectra were recorded at 45° X-ray incidence. Note that the edge jump is only 1–5% for the total yield signal; this leads to an incomplete removal of monochromator noise.

characterized over the last 10–15 years by ion-neutralization spectroscopy photoemission, EELS, photoelectron diffraction, low-energy ion scattering, and theoretical electronic, and crystallographic structure calculations. In 1981, Brennan, Stöhr, and Jaeger investigated the $c(2 \times 2)$ S on Ni (100) system by SEXAFS. The system was chosen for the very reason that it appeared to be one of the best understood ones in surface science, and thus allowed one to test the SEXAFS technique for K edges and assess its reliability and accuracy as compared to other surface structural techniques applied previously.

Figure 9 shows the Auger electron yield and total electron yield EXAFS K-edge spectra for $c(2 \times 2)$ S on Ni (100) corresponding to half a S monolayer. Both

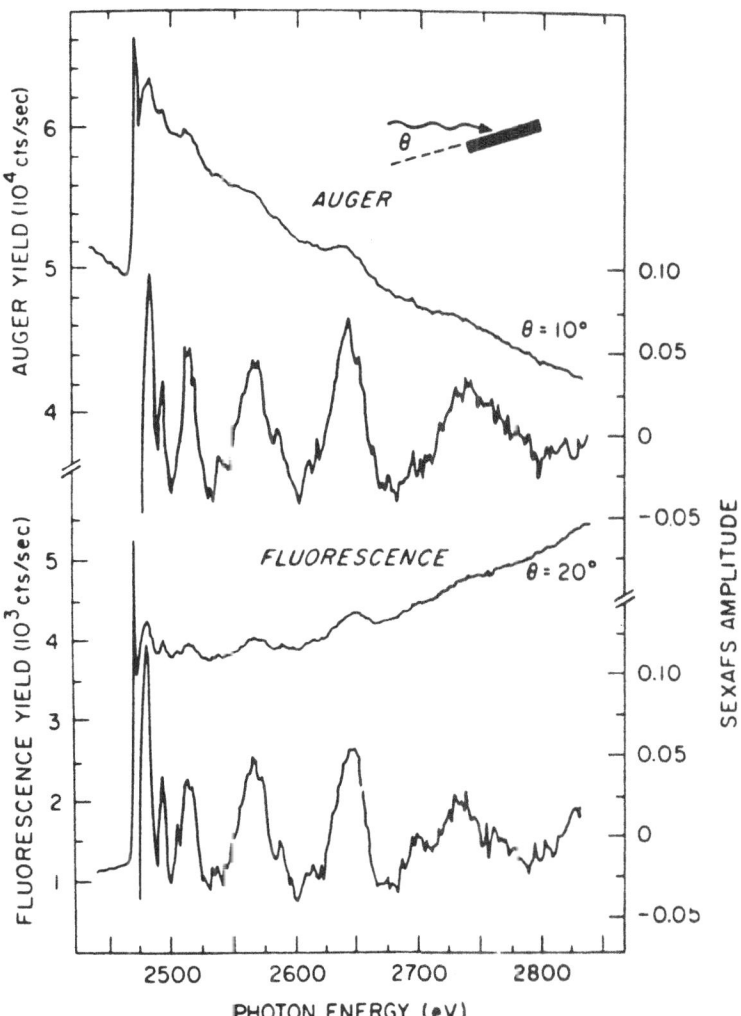

FIGURE 10. Auger electron yield and X-ray fluorescence yield SEXAFS spectra above the S K edge for $c(2 \times 2)$ S on Ni (100), corresponding to half a S monolayer. Both spectra were recorded at grazing X-ray incidence. Underneath each spectrum the SEXAFS oscillations after background subtraction are shown enlarged.

spectra were recorded at 45° X-ray incidence. Note that the edge jump is only 1–5% for the total yield signal; this leads to an incomplete removal of monochromator noise. Figure 10 shows the Auger electron yield and X-ray fluorescence yield SEXAFS spectra above the S K edge for $c(2 \times 2)$ S on Ni (100) corresponding to half a S monolayer. Both spectra were recorded at grazing X-ray incidence. Underneath each spectrum the SEXAFS oscillations after background subtraction are shown enlarged.

The SEXAFS spectra recorded for three polarization directions were analyzed by comparison to the total yield EXAFS spectrum of bulk NiS which served as a model compound. The Fourier transforms shown in Fig. 11 are dominated by the nearest neighbor (Ni) S–Ni peak. The $\Theta = 90°$ spectrum clearly reveals a second weaker peak around ~3.8 Å. This peak is also observed for the two other polarization directions, although it is close to the noise level for $\Theta = 10°$.

By detailed analysis of the phase and amplitude of the first NN peak one can accurately determine the local structure of $c(2 \times 2)$ S on Ni. In agreement with the structure determination by LEED, sulfur was found to chemisorb in the fourfold hollow site with a S–Ni bond length of $R = 2.23 \pm 0.02$ Å corresponding to $Z = 1.37 \pm 0.03$ Å. In addition to the analysis of the first NN SEXAFS amplitude, the chemisorption site was independently derived by the second-neighbor distance for the $\Theta = 90°$ spectrum. This peak corresponds

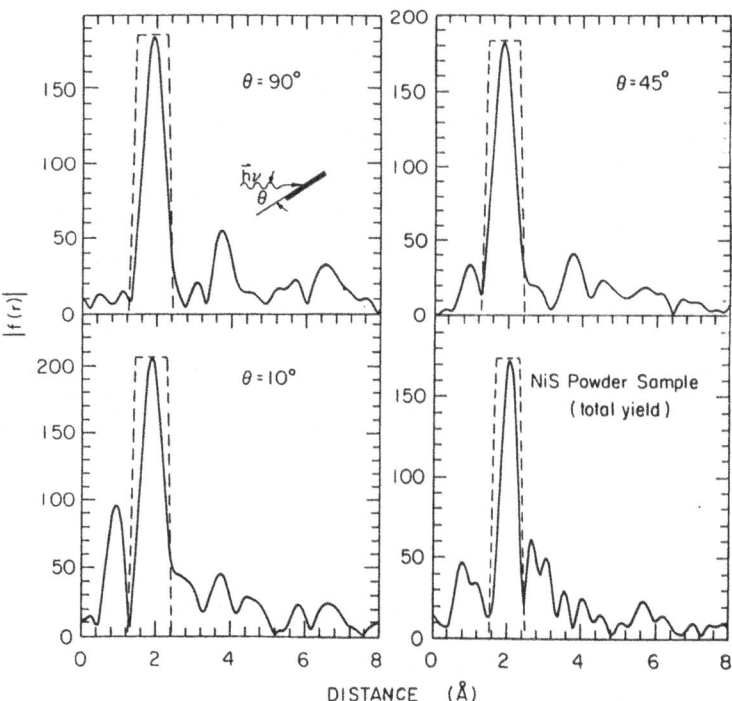

FIGURE 11. Fourier transforms of the EXAFS signals for $c(2 \times 2)$ S on Ni (100) recorded for different angles of incidence, showing the first-neighbor peak at ~2 Å and the second-neighbor peak at ~3.8 Å. The Fourier transform of the bulk model compound NiS is also shown for comparison.

preferentially to the fourth-nearest-neighbor Ni shell consisting of the eight second-nearest-neighbors in the surface plane.

Turning to ARPEFS, the most thoroughly studied prototype system is $c(2 \times 2)$ S–Ni (100), in which sulfur is known to occupy a fourfold hollow site. This system was in fact used by Liebsch in his early papers predicting the photoelectron diffraction phenomenon. Photoelectron diffraction has been observed with electrons excited from the both S $1s$ and the S $2p$ shells. The fourfold hollow site geometry was independently confirmed by comparison of the observed S $2p$ excitation curve with theory. The derived perpendicular distance of sulfur atoms above the surface Ni layer was $d_\perp = 1.30 \pm 0.04$ Å, somewhat more accurate than typical LEED limits of ± 0.1 Å. The ARPEFS data obtained for S $1s$ electron excitation along the [011] direction showed structure in $\chi(k)$ similar to SEXAFS on the same system, but with very large oscillations (up to 50%), more frequencies, and slower attenuation with k. All these differences are predictable from Eqs. (3–5). Fourier transformation of $\Psi(k)$ yielded peaks at several path lengths although the extent to which these peaks can be identified with specific scattering centers is not yet settled.

Analysis of the main Fourier transform peak in the ARPEFS $\chi(k)$ data for $c(2 \times 2)$ S/Ni (100)[011] yields a S–Ni bond distance of 2.24 Å in good agreement with the SEXAFS value [10] of 2.23 Å, and consistent with $d_\perp = 1.35$ Å. The 3% discrepancy between this result and the earlier $d_\perp$ value may arise in part from multiple scattering. It may also reflect surface-layer expansion of the underlying nickel by a few percent. ARPEFS studies have confirmed lattice expansions of several percent in two other cases.

In summary, the $c(2 \times 2)$ S/Ni (100) results show that ARPEFS $\chi(k)$ data25460 for appropriate directions can yield distances to an accuracy of ± 0.03 Å or better. In addition, the capability of ARPEFS to measure substrate lattice expansion appears to be established.

5. SEMICONDUCTOR SURFACES AND INTERFACES

Semiconductor surfaces differ in many ways from metal surfaces since they are not close-packed structures but consist of tetrahedrally bonded bulk lattices. This results in a surface with broken bonds, usually referred to as dangling bonds. In contrast to metals, the arrangement of atoms near a semiconductor surface is modified considerably, with a long-range superlattice such as the 2×1 and 7×7 structures formed for Si (111) surfaces. From a chemisorption point of view, it may be expected that the directional bonding in semiconductors may lead to simple overlayer geometries which result from saturation of the dangling surface bonds. For ordered overlayers, techniques which probe long-range order may be overwhelmed by the structural complexities introduced by the underlying substrate. SEXAFS and ARPEFS offer the advantage that they are local probes, and as such sample the structure within about 5 Å of an adsorbed atom.

In the following we shall discuss SEXAFS studies of chemisorption phenomena on Si (111) and Ge (111). Some of these systems are also discussed in Chapter 16 of Volume 2 of this set by Woicik and Pianetta.

5.1. Chlorine and Iodine on Si (111) and Ge (111)

Figure 12 shows the local structure of the bulk Si (111) plane as well as several high-symmetry adsorbate bonding configurations. Although the clean surfaces have more complex geometries than are shown in Fig. 12, one expects the chemisorbed overlayer to saturate the surface broken bonds and in some favorable cases to remove the reconstruction. A particularly simple overlayer system is Cl on Si (111) and Ge (111) surfaces. Early polarization-dependent valence-band studies on cleaved Si and Ge suggested that the two systems have very different chemisorption geometries. Chlorine was proposed to chemisorb in the atop ionic site on Si (111) and in the threefold covalent site on Ge (111). Both surfaces have previously been investigated by Citrin, Eisenberger, and Rowe by means of SEXAFS. For both Si (111) 7×7 and Ge (111) 2×8 starting surfaces, the Cl adsorption results in SEXAFS oscillations, which are only pronounced at grazing X-ray incidence ($\Theta = 10°$) and essentially nonexistent at normal incidence. This clearly indicates that for both Si (111) and Ge (111), Cl chemisorbs in the onefold atop site, a finding confirmed by the strong polarization dependence of the near-edge fine structure. These results are a beautiful example of the strong polarization dependence of the K-edge EXAFS and fine structure for anisotropic chemisorption geometries.

The SEXAFS data and Fourier transforms for two cases are shown in Fig. 13, as well as the case of Te, discussed in the next section. Note the differences in first NN amplitudes and phases as a function of polarization for the three cases. For I/Si (111) a 7×7 LEED pattern corresponding to ~ 1 ML was observed. Using a vapor SiI(CH$_3$)$_3$ model compound with an I–Si distance of 2.46 ± 0.02 Å yielded the Iodine–Si first NN distance on the surface of 2.44 ± 0.03 Å. The data

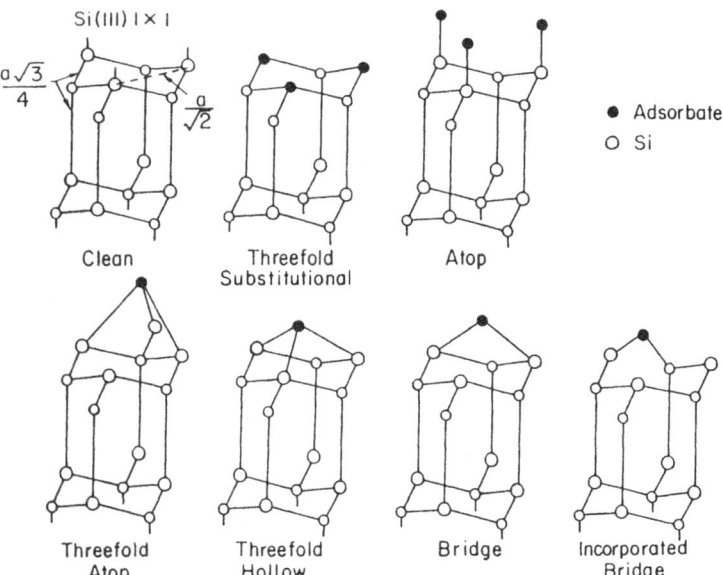

FIGURE 12. Models of adsorption sites for Si (111) surfaces showing the double layer staclang of the (111) planes of the diamond structure.

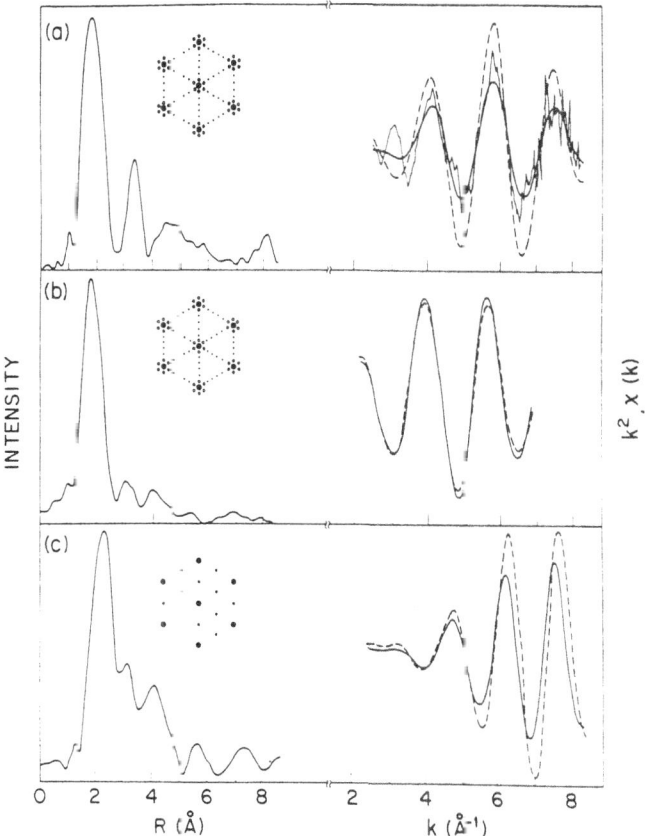

FIGURE 13. Fourier transforms of EXAFS data taken at 90° incidence, LEED patterns observed and first-neighbor EXAFS functions for several semiconductor–adsorbate systems (a) Si (111) (7 × 7)–I; (b) Si (111) (7 × 7)–Te; (c) Ge (111) (2 × 2)–Te. Note the strong polarization dependence of the EXS functions where the dashed lines correspond to grazing incidence ~ 20°, compared to normal 90° incidence.

is shown in Fig. 14. The authors found that iodine chemisorbs in the atop site as expected for a monovalent atom. The results for I/Ge (111) at 1 ML coverage and a 1 × 1 LEED pattern is similar to Si (111). Here GeI_4 vapor served as a model compound with an I–Ge bond length of 2.50 ± 0.03 Å. I is also found to chemisorb on top of a surface Ge atom with a bond length of 2.50 ± 0.04 Å as in the vapor model.

5.2. Tellurium on Si (111) and Ge (111)

The short-range structure of Te and I on Si (111) 7 × 7 and Ge (111) 2 × 8 surfaces was investigated by Citrin, Eisenberger, and Rowe using the L_3 X-ray edge. The goal of this study was to test the ideas of bonding trends expected from simple chemical-bond saturation arguments for monovalent iodine and divalent Te. Three of the studied cases were found to be consistent with such arguments, while Te on Ge (111) occupying a threefold site was not. The SEXAFS study made use of polarization-dependent first and second nearest-neighbor

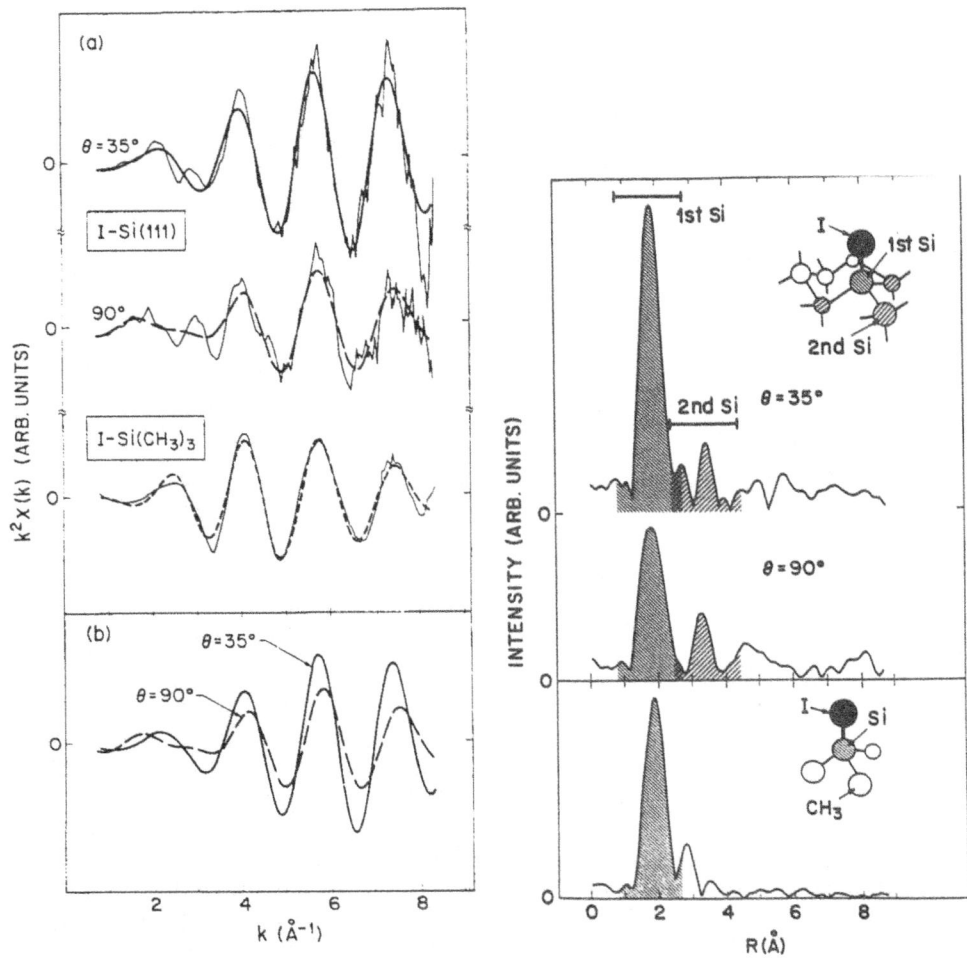

FIGURE 14. SiI(CH₃)₃ reference data used in analysis.

distances and relative and absolute coordination numbers to establish the local chemisorption geometry. The data are, in general, consistent with a locally unreconstructed substrate in the presence of the adsorbates. For I/Si (111) the second NN distance to the Si atoms in the second layer is found to be greater by $0.1\,\text{Å}$ than for an unrelaxed Si (111) 1×1 substrate, suggesting outward relaxation of the surface Si atoms.

Figure 13 indicates a different chemisorption geometry for Te/Si (111) characterized by $\sim 1\,\text{ML}$ coverage and a 7×7 LEED pattern. Analysis using SiI(CH₃)₃ as a model, which gives small bond length ($\Delta R \sim 0.01\,\text{Å}$) and amplitude ($<10\%$) errors, results in an experimental Te–Si surface bond length of $2.47 \pm 0.03\,\text{Å}$. Amplitude analysis excludes all but the bridge site, although the threefold atop site can only be excluded by comparison of the measured ($N^* = 2.9 \pm 0.04$) and calculated ($N^* = 3.8$) absolute N^* values for $\Theta = 10°$. The distinction between these sites is supported by bond-length considerations.

For both the bridge and threefold atop sites, the experimentally determined distance is a superposition of first and second NN distances. The measured distance of 2.47 ± 0.03 Å would imply that for the threefold atop site the Te atom is unphysically close $(R \sim 1.8$ Å$)$ to the Si atom directly underneath. Therefore Te on Si (111) around monolayer coverage occupies the twofold bridge site with a first NN distance of 2.44 ± 0.03 Å. Te on Si (111) has also been studied by Comin, Citrin, and Rowe in 1983 at lower coverage. At 0.5 ML coverage a 3×1 and 2×2 LEED pattern is observed which changes to a $(\sqrt{3} \times \sqrt{3}) R30°$ pattern at 0.25 ML coverage. For these two cases a larger Te–Si bond length of 2.51 ± 0.04 Å has been determined. The identification of the bridge site is the first of its kind and can be explained by the divalent nature of Te.

Te on Ge (111) forms a 2×2 LEED pattern at about 0.5 ML coverage. The SEXAFS signal (Fig. 13) of the first NN shell exhibits a larger amplitude at grazing than normal X-ray incidence and a polarization-dependent NN distance with a 0.08 Å larger distance at normal incidence $(R = 2.75 \pm 0.04$ Å$)$. Using GeI$_4$ as a model, amplitude analysis establishes the threefold atop chemisorption

TABLE 1. Summary of Surface EXAFS Results for Si (111) and Ge (111) Surfaces

System	Bondlength (Å)	Site	Coverage	Comments
Si (111) (7 × 7)–I	2.44 ± 0.23	Atop	Sat.	I–Si (2nd nn) distance = 4.01 ± 0.04 Å, different from calculated 3.91 ± 0.03 assuming ideal Si (111) (1 × 1)
Si (111) (7 × 7)–Te	2.44 ± 0.04	Bridge	Sat.	Distance of 2.47 ± 0.3 Å reflects average of two 1st and one 2nd nn atom distances.
Si (111) ($\sqrt{3} \times \sqrt{3}$) (R30°)–Te	2.52 ± 0.05	Bridge	1/3	Measured distance of 2.59 ± 0.3 Å averages 1st and 2nd nn atoms as above
Ge (111) (1 × 1)–I	2.50 ± 0.04	Atop	Sat	I–Ge (2nd nn) distance = 4.06 ± 0.04 Å assuming ideal Ge (111) (1 × 1).
Ge (111) (2 × 2)–Te	2.45 ± 0.05	*hcp* hollow	1/2	Measured distances are polarization-dependent.
Si (111) (7 × 7)–Cl	2.03 ± 0.03	Atop	Sat.	Cl–Si (2nd nn) distance = 3.51 ± 0.06 Å, slightly smaller than calculated 3.59 Å, assuming Si (111) (1 × 1).
Si (111) ($\sqrt{19} \times \sqrt{19}$)–Cl	1.98 ± 0.04	Atop	Sat.	Shorter bond lengths than Si (11) (7 × 7)–Cl
Ge (111) (1 × 1)–Cl	2.07 ± 0.03	Atop	Sat.	
Si (111) (7 × 7) + Ni	2.37 ± 0.03	Imbedded hollow	1/2	

site. The Te is found to be above the Ge atom in the second plane at a distance of 2.45 ± 0.04 Å, with three Ge atoms in the surface plane as second *NN* at 2.78 Å. This explains the large measured average distance and the polarization dependence. This is a rather unusual chemisorption site with complex bonding, which had never been observed.

Table 1 summarizes the various cases discussed in this section.

REFERENCES

1. P. M. Citrin, P. Eisenberger, and R. C. Hewitt, *Phys. Rev. Lett.* **41,** 309 (1978).
2. P. M. Citrin, P. Eisenberger, and R. C. Hewitt, *Phys. Rev. Lett.* **45,** 1948 (1980).
3. H. Winick and S. Doniach, *Synchrotron Radiation Research,* Plenum, New York (1980); H. Winick and A. Bienenstock, *Ann. Rev. Nucl. Part. Sci.,* **28,** 33–113 (1978).
4. E. E. Koch, *Handbook of Synchrotron Radiation,* North Holland, New York (1983).
5. G. Margaritondo, *Introduction to Synchrotron Radiation,* Oxford, New York (1988).
6. D. C. Koningsberger and R. Prins, eds. *X-ray Absorption, Principles, Applications, Techniques of EXAFS, SEXAFS, and XANES,* Wiley, New York (1988).
7. B. K. Teo and D. C. Joy, eds. *EXAFS Spectroscopy, Techniques and Applications,* Plenum, New York (1981).
8. E. A. Stern in Ref. 6.
9. T. M. Hayes and J. B. Joyce, in *Solid State Physics,* Vol. 37 (H. Ehrenreich, F. Seitz, and D. Turnbull, eds.), Academic Press, New York, (1982) p. 173.
10. J. Stöhr in Chemistry and Physics of Solid Surfaces V (R. Vanselow and R. Howe, eds.), Springer Series in Chemical Physics Vol. 35, Springer, Berlin, (1984), p. 231; *X-ray Absorption: Principles, Applications, Techniques of EXAFS, SEXAFS, and XANES,* (R. Prins and D. C. Koningsberger, eds.), Wiley, New York, (1985).
11. P. H. Citrin, *Journal de Physique (Paris) Colloq* **47,** 437, (1986).
12. See Chapter 6 of Volume 2 of this set, by Woicik and Pianetta.
13. P. A. Lee, P. H. Citrin, P. Eisenberger, and B. M. Kincaid, *Rev. Mod. Phys.* **53,** 769 (1981).
14. D. E. Sayers and B. A. Bunker in Ref. 6.
15. P. H. Citrin, D. R. Harnann, L. F. Mattheiss, and J. E. Rowe, *Phys. Rev. Lett.* **49,** 1712 (1982); F. Sette, C. T. Chen, J. E. Rowe, and P. H. Citrin, *Phys. Rev. Lett.* **59,** 311 (1987).

Photoemission Spectroscopy

Angle-Resolved Photoelectron Spectroscopy

W. Eberhardt

1. INTRODUCTION

Over the past decade, angle-resolved photoemission spectroscopy has matured from an exotic technique into a very powerful tool to determine the electronic structure of solids, surfaces, and interfaces.[1,2] Since not only the energy but also the direction of the photoexcited electrons and thus the components of the electron momentum vector are determined by the experiment, all relevant quantum numbers of the electronic states of a solid or a surface can be determined. Angle-resolved photoemission spectroscopy is indeed the only existing technique that allows a direct and unambiguous determination of a two- or three-dimensional band structure $E(k)$ of a solid or a surface. These band structures have been calculated for many decades and the concepts can be found in any elementary solid-state physics textbook. It is therefore very fascinating to have an experimental technique at hand to verify these calculations. Even more important than this connection with theory is the potential technological impact generated by these measurements. The electron interactions among the individual atoms forming a solid or a surface, as shown by the electronic structure, determine all the properties of the material, as for example its mechanical stability, corrosion resistance, chemical reactivity, electrical and thermal conductivity, or magnetism.

Two-dimensional systems like the interior and exterior surfaces of a solid have their own electronic structures, which are different from the bulk electronic structures because of the discontinuity introduced by the surface and the different coordination of the surface atoms. Corrosion or embrittlement as well as the catalytic activity of a material are related to the electronic structure of internal and external surfaces. In semiconductor technology, metal–semiconductor or semiconductor–semiconductor interfaces largely determine the electrical charac-

W. Eberhardt • Institut für Festkörperforschung des Forschungszentrums Jülich, D5170 Jülich, Germany.

Synchrotron Radiation Research: Advances in Surface and Interface Science, Volume 1: Techniques, edited by Robert Z. Bachrach. Plenum Press, New York, 1992.

teristics of a device. Even though these systems are very complex and often do not have the two- or three-dimensional periodicity required for a strict application of angle-resolved photoemission, very valuable information can be obtained by studying ideal model systems. For example, the involvement of individual surface states in the Schottky-barrier formation at metal–semiconductor interfaces is relevant to electronic devices. Epitaxial metal overlayers on various substrates give insight into the magnetism of thin films, and the chemisorption bond of molecules on single-crystal transition-metal surfaces dan serve as a model system for complex catalytic reactions used by industry.

The full power of angle-resolved photoemission as a technique unfolds only in connection with a tunable excitation source like synchrotron radiation. Only by continuously varying the energy of the photons used to excite the photoelectrons can we separate the two-dimensional electronic structure of a surface or interface from the underlying electronic structure of the bulk. In addition, we can also make use of the polarization of the synchrotron radiation source and the selection rules inherent in this mode of excitation to identify the symmetries of the system. For example, using polarization selection rules we can determine the orientation of a molecular axis with respect to the surface in a chemisorbed molecule.

In the following sections the basic theoretical concepts used in the application of angle-resolved photoemission are outlined, followed by discussion of some examples where this technique has been used successfully to study interesting problems in materials science. Because of the success of the technique and the large number of publications resulting from it, this article is not intended to be a review of all this work. Accordingly, the selection of examples presented here reflects the possibilities in materials science–related studies opened up by the technique, rather than the extent of work actually performed. Many people and research groups have done excellent work developing the technique and finding interesting technological applications for it.

Note Added in Proof. This article was written in 1986 and early 1987 and accordingly reflects the state of the field and the views of the author at that particular time. Since then the field has developed most rapidly in the area of angle- and spin-resolved photoemission related to thin-film magnetism, but also with respect to high-resolution studies and the involvement of surface states in surface reconstructions. Since the major emphasis of this article is to give a textbook-like introduction to the technique of angle-resolved photoemission and its application to the study of two-dimensional systems, I trust that there is still some value in this work even though by now some of the examples used are somewhat dated.

2. THEORETICAL AND EXPERIMENTAL CONCEPTS

2.1. Energy Conservation

The concept inherent to all photoemission studies is that the photon and its energy has to be absorbed as a whole in the process, as orginally formulated by

Einstein[3] in his theory of the photoelectric effect. Therefore, if we measure the kinetic energy E_{kin} of a photoejected electron and subtract from this energy the energy of the photon $h\nu$ used to excite the electron, we can determine the binding energy, E_B inside the solid or atom prior to its excitation according to the equation

$$E_B = E_{kin} - h\nu. \tag{1}$$

This is illustrated for CO in Fig. 1, where the kinetic energy distribution of the photoelectrons is shown.[4] The peaks in this spectrum reflect the shell structure of the electrons. The peaks at $E_B = 543\,eV$ and at $E_B = 295\,eV$ correspond to the ionization of an O or C $1s$ electron, respectively. The other peaks at lower binding energies are due to the emission from the valence electrons as marked. The additional weak structures in the range from $E_B = 20\,eV$ to $40\,eV$ will be discussed below.

When two or more atoms are brought together to form a molecule or a solid, the electrons interact and the binding energy of the orbitals will change, as indicated schematically in Fig. 1. The valence electrons will form new, combined molecular orbitals and the core electrons will also change in energy because of charge transfer and a difference in screening of the nucleus in the new environment. This simple concept has led to the development of ESCA[4] (*electron spectroscopy for chemical analysis*), in which binding energy shifts are correlated with changes in the chemical environment of the individual atoms. Even though this concept is widely used because of its intriguing simplicity, care has to be taken in the interpretation of the binding-energy shifts. Secondary effects, which have nothing to do with charge transfer, will affect the apparent measured

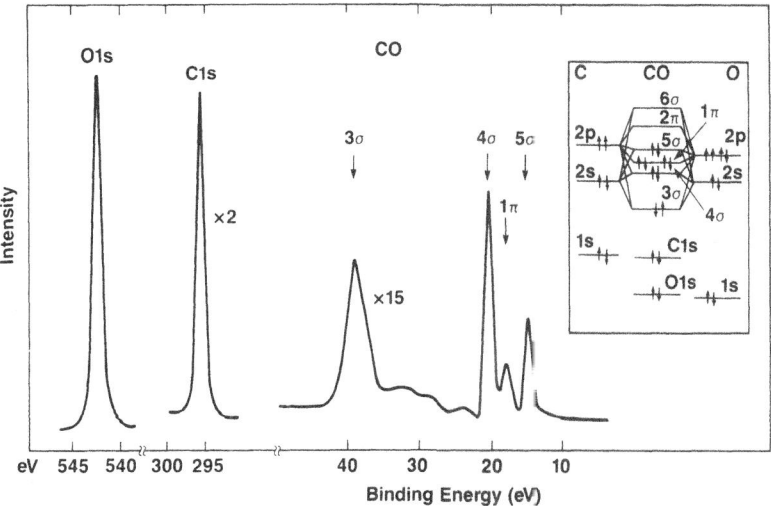

FIGURE 1. Energy Distribution Curve (EDC) of photoelectrons ejected from isolated CO molecules in gas phase. (From Ref 4.) The electrons were excited with Mg K_α X-rays. The insert shows schematically how the electronic states of the individual atoms are combined to form the electronic states of the molecule.

binding energies (see section 4). This is probably the reason why in the literature this type of photoelectron spectroscopy is more often referred to as XPS (*X*-ray *p*hotoelectron *s*pectroscopy), especially in studies dealing with the mechanism itself.

Not all peaks or structures in a photoelectron spectrum are caused by these direct photoemission processes. Secondary processes, like the Auger decay of an ionic configuration created in the initial photoemission event, also cause sharp lines to appear in the spectra. However, these Auger lines appear at a fixed kinetic energy, independent of the energy of the incident photons, and therefore can be readily identified by recording photoelectron spectra at different excitation energies.

The EDCs (*energy distribution curves*) of electrons emitted from solids rather than isolated atoms or molecules exhibit a large fraction of scattered electrons. This results in a smooth background of scattered electrons in the EDCs which rises toward low energies. Since, with a few exceptions, the scattering processes do not introduce specific energy losses to the electrons, the scattered electrons cannot be identified anymore with respect to their initial excitation process. This makes photoemission a surface-sensitive probe, because the mean free path for an electron inside a solid is on the order of 10 to 30 Å, dependent on its kinetic energy. The electrons that we can clearly identify originate within this near-surface region of the solid. Discrete energy-loss mechanisms, associated with, for example, the excitation of a plasmon or phonon, do also exist inside a solid, and these scattering events are responsible for the appearance of satellite lines on the low-kinetic-energy side of photoemission peaks. These features will be discussed under the heading "Complications and Special Effects" later in this chapter.

2.2. Photoemission Intensity and Matrix Elements

The transition probability per unit time between two eigenfunctions $|i\rangle$ and $\langle f|$ of the same Hamiltonian H_0 under a small perturbation H' is given by Fermi's golden rule as

$$\frac{d\omega}{dt} = \left(\frac{2\pi}{\hbar}\right)^* |\langle i| H' |f\rangle|^{2*} \delta(E_f - E_i - h\nu). \tag{2}$$

The perturbation H' to the system in a photon field is found by replacing the momentum operator $\mathbf{P}$ by $\mathbf{P} + (e/c)\mathbf{A}$ in the original Hamiltonian H_0. Thus H' has the form

$$H' = (e/2mc)(\mathbf{A} \cdot \mathbf{P} + \mathbf{P} \cdot \mathbf{A}) - e\Phi + (e^2/2mc^2) |\mathbf{A}|^2, \tag{3}$$

where $\mathbf{A}$ and Φ are the vector and scalar potentials of the incident light field. It is always possible to choose a gauge where $\Phi = 0$. Neglecting the diamagnetic term $|\mathbf{A}|^2$ because it is small, and using the commutator $[\mathbf{P}, \mathbf{A}] = -i\hbar\nabla \cdot \mathbf{A}$, the differential photoionization cross section can be written as

$$d\sigma/d\omega \sim |\langle f| 2\mathbf{A} \cdot \mathbf{P} - i\hbar\nabla \cdot \mathbf{A} |i\rangle|^{2*} \delta(E_f - E_i - h\nu). \tag{4}$$

The $-i\hbar\nabla \cdot \mathbf{A}$ term is zero for a transverse wave. Near the surface of a crystal, however, longitudinal field components are induced, depending on the frequency of the incident light and its interaction with the electronic charge at the surface.[5,6] These induced fields are especially strong when the frequency of the incident light corresponds to the plasma frequency of the solid, and they cause large enhancements in the photoemission cross section at these energies. These effects have been experimentally verified by studying the angle resolved photoemission of Al.[7] Well above the plasma energy these effects play a minor role so that they are in general neglected and the cross section is written in the form

$$d\sigma/d\omega \sim |\langle f| \mathbf{P} |i\rangle \cdot \mathbf{A}|^{2*} \delta(E_f - E_i - h\nu). \tag{5a}$$

This electric dipole matrix element can also be written in two alternate representations by using the commutation relation of H_0 with $\mathbf{P}$ and $\mathbf{r}$, yielding

$$d\sigma/d\omega \sim |\langle f| \mathbf{r} |i\rangle \cdot \mathbf{A}|^{2*} \delta(E_f - E_i - h\nu) \tag{5b}$$

$$d\sigma/d\omega \sim |\langle f| \nabla V |i\rangle \cdot \mathbf{A}|^{2*} \delta(E_f - E_i - h\nu). \tag{5c}$$

These equations are the ones most commonly used for the description of the photoemission cross section. The δ-function assures the conservation of energy between the initial state of the system and its final state, as discussed in section 2.1, and all symmetry-related aspects are derived from the dipole matrix element $\langle f| \mathbf{A} \cdot \mathbf{P} |i\rangle$. So far nothing has entered about the exact nature of the wave functions $\langle f|$ and $|i\rangle$. These wave functions should be considered the total wave functions of the system, which has the following direct implications.

First, the golden rule as applied above is not strictly valid, because $|i\rangle$ and $\langle f|$ are not eigenfunctions of the same Hamiltonian. $|i\rangle$ is the wave function of the neutral system, whereas $\langle f|$ is the wave function of the singly ionized system. *Photoemission is a spectroscopy of ionized states and does not probe the ground state of the system.* Since the ejection of an electron causes all other electronic orbitals to react to the increase in electrostatic potential, different multiparticle configurations, with different energies, can result from the ejection of the same electron from the neutral system. This causes the appearance of additional lines, so-called shake-up satellites, in the EDCs of the ejected electrons. Thus any binding energy extracted from a photoemission experiment according to Eq. (1) is in the true sense the energy of a specific hole excited state configuration of the system. These effects sometimes play a role, not only for isolated atoms or molecules but even in the case of photoemission out of a delocalized valence band of a solid, where intuition leads one to believe that one hole spread over all atoms of the solid does not make a difference. This will be discussed in more detail in section 4.

Second, in a solid the number of existing electronic states is restricted in energy and momentum space, as given by the band structure. Thus we can excite an electron only if the photon energy matches the energy difference between the initial and final states. This has led to the "direct transition" model in solid-state photoemission since at photon energies less than 1 keV the momentum of the photon can usually be neglected. This model implies that the electron momentum does not change, modulo a vector of the reciprocal lattice, when the electron is excited. In the standard band structure schema, which is plotted in the reduced Brillouin zone, the excitation corresponds to a vertical transition between occupied and unoccupied bands at the same reduced **k**. This process is generally referred to as a direct transition and is illustrated by the vertical arrows in Fig. 2.

As will be discussed in the next section, in angle-resolved photoemission we can position our detector such that we observe only transitions along selected lines in **k**-space. If we measure, for example, the normal emission from the (100) surface of Cu at a photon energy of 26 eV, we will observe a spectrum with two peaks, as indicated in Fig. 2. Due to emission into exponentially decaying evanescent final states in the near-surface region, we will also see some emission from other states, for example at the Fermi level E_F. These transitions, however, are in general weaker than the direct interband transitions. Increasing the photon energy will cause, corresponding to the dispersion of the final-state band, the direct transitions to occur at a **k**-point closer to the center of the Brillouin zone. Thus, as the total kinetic energy of the photoemission peaks will increase, the

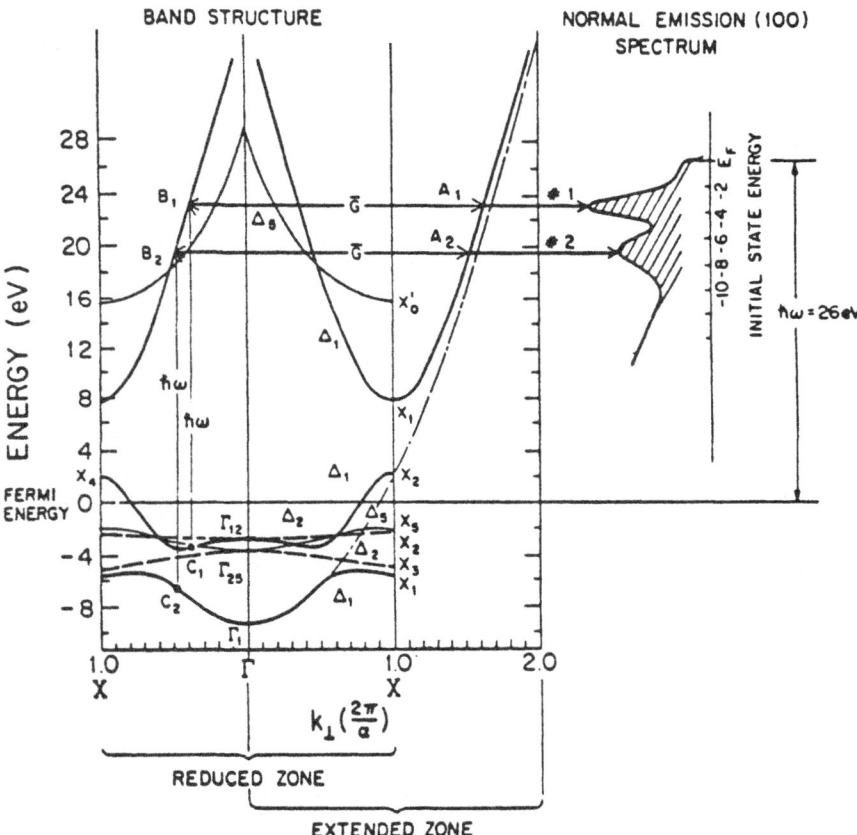

FIGURE 2. Band structure of Cu along the (100) direction plotted in the reduced and extended zone scheme. The direct interband transitions occurring at a photon energy of $h\nu = 26$ eV in this crystallographic direction are indicated. This results in an EDC with two major peaks, as shown schematically on the right. The free-electron-like band is shown by the dash-dotted curve.

binding energy of the strong peaks in the EDCs, measured relative to the Fermi level E_F, will increase also. This is the basis for a band-structure determination using angle-resolved photoemission with synchrotron radiation.

2.3. Momentum Conservation

Unlike the orbitals of isolated atoms or molecules, the valence electrons of a solid form bands, in which the electrons are described by two quantum numbers, the energy E and the wavevector $\mathbf{k}$. Whereas the bandwith of the valence bands is on the order of 10 eV, core levels in a solid are essentially atomic-like because of the negligible overlap between core level wavefunctions in neighboring atoms. Consequently, just as in atoms or molecules, a simple energy analysis of the photoelectrons is sufficient to characterize the core electrons in a solid. In order to determine the valence electronic structure of a solid, however, we need to measure both the energy and the momentum components of the electrons with respect to the crystallographic directions. This is achieved by measuring not only the energy of the emitted photoelectrons, but also their direction with respect to the crystallographic axis. Thus we determine the components of the electron momentum after it has left the crystal.

In order to determine the momentum of the electron initial state inside the crystal, we have to make use of some concepts developed during the last decade to guide in the interpretation of angle-resolved photoemission spectra in the determination of bulk band structures. In the direct-transition model, the momentum components of the excited electron inside the crystal are identical, modulo a reciprocal lattice vector, to the momentum of the initial state. Therefore we have to establish only the relationship between the measured momentum components in the vacuum and the momentum components of the electron after the excitation, but inside the crystal.

As the electron leaves the crystal it has to overcome the work-function barrier at the surface. This results in a renormalization of the electron energy and reduces the momentum component normal to the surface of the crystal. The momentum components parallel to the surface, on the other hand, are conserved for a system that is periodical in two dimensions. Thus we can measure the electronic structure of a surface or any two-dimensional periodic system exactly, without any approximations or interpolations. The magnitude of the momentum component parallel to the surface $\mathbf{k}_{\parallel}$ is readily calculated according to

$$\hbar k_{\parallel} = \sqrt{(2mE_{\text{KIN}})} \cdot \sin \beta \tag{6}$$

where E_{KIN} is the kinetic energy of the electron in vacuum and β is the angle of the electron detector with respect to the normal of the crystal. This relationship makes it obvious why the normal emission geometry is so convenient for the experimental determination of bulk band structures. Since the parallel momentum component is conserved, we know that an electron leaving the crystal in the direction of the surface normal, propagated, independent of its kinetic energy, normal to the surface inside the crystal. In addition, because of the direct interband transition, the initial state is to be found along the line in **k**-space normal to the surface also. Thus by keeping $k_\parallel$ fixed and scanning the excitation energy, we can probe the electronic structure of solids along "rods" in **k**-space oriented normal to the crystal surface. Umklapp processes, where an additional reciprocal lattice vector is added when the electron is excited or when it leaves the solid, add a slight complication to this picture, but in general these processes are much weaker and do not dominate the spectra.

2.4. Experimental Methods for Determining Bulk Band Structures

One possible procedure to measure the bulk electronic structure offers itself immediately. We align our detector along the surface normal and take a sequence of spectra at a series of photon energies as shown in Fig. 3 for Ni (100).[8] The left panel shows the actual EDCs, whereas the right panel illustrates the band structure. Taking the *sp*-band of Ni as an example, which is the feature indicated by tic marks in Fig. 3, we observe that with increasing photon energy we approach

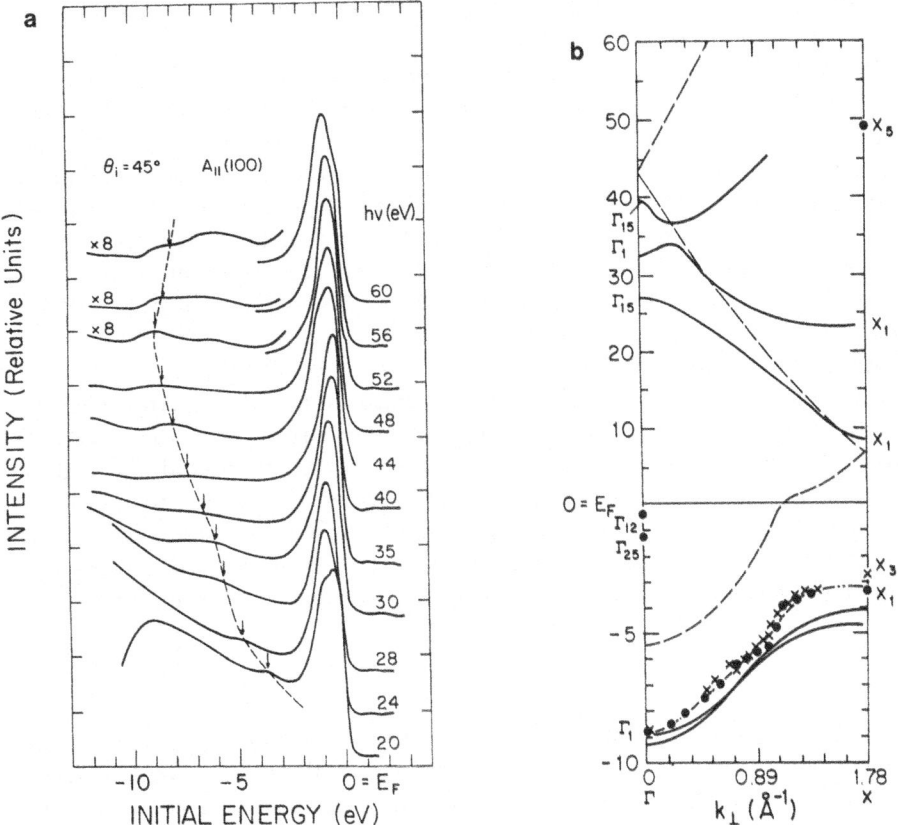

FIGURE 3. (a) Normal emission EDCs of Ni (100) as a function of excitation energy and (b) bandstructure of Ni along the (100) direction. (From Ref. 8.) The emission from the Ni *sp* band is indicated by the ticmarks and the dashed line in (a). The measured initial- and final-state bands are compared in (b) to the calculated band structure.

the maximum of the binding energy at about 52 eV. At higher photon energies the *sp*-band feature in the EDCs starts dispersing towards E_F. This reversal of the initial-state dispersion indicates that at a photon energy of 52 eV the transitions occur at the Γ-point of the bulk band structure. Because of the direct transition, the final-state band is located at Γ too. This method can be applied at other critical points also. Usually a symmetry point has at least one band that exhibits an extremal behavior as the transitions approach this point. Thus, determining where the extrema occur in the dispersion of the initial state bands, we are able to measure essentially all symmetry points of a bulk band structure without any assumption or approximation for the final states of the transition.

Obviously the method just described works for off-normal photoemission too, if we scan the photon energy and take spectra at fixed $k_\parallel$, as is achieved by a simultaneous change of excitation energy and detection angle according to Eq. (6). After determining the high symmetry points of a band structure, we obtain the dispersion of the bands throughout the Brillouin zone by interpolating the final states between these points. Usually a free electron–like final state band fitted to both accurately determined extremal points gives adequate results.

2.5. Identification of Surface States

Even though nature provides us with some inherently two-dimensional systems, like graphite or layered compounds, the most interesting and common two-dimensional systems are surfaces and interfaces, which by their nature are in contact with a three-dimensional substrate. In order to study these systems, we have to separate the emission from intrinsic or extrinsic surface states and the photoemission coming from the underlying bulk. In the early days this was done by putting some adsorbate onto the surface and monitoring changes in the photoemission. This procedure was nicknamed the "crud test" for surface states. Whatever changed or disappeared in the EDCs had to have been correlated with the emission from the clean surface. Obviously this is not the most accurate way to measure these states. Fortunately, by now we do not have to rely on the "crud test" anymore and have developed some more exact criteria.

First, any surface state or resonance is a truly two-dimensional state, even if it extends a few lattice planes deep into the solid. This means that these states do not exhibit any dispersion with a change in normal momentum of the photoelectron. Contrary to the behavior of bulk-band photoemission, discussed above, the energy position of these surface states measured relative to the Fermi level E_F is not going to change if we change the photon energy and keep $k_\parallel$ fixed. The intensity of these states will change with excitation energy, which will be discussed below. Obviously, a continuously tunable excitation esource, like synchrotron radiation, is one of the key ingredients for this test.

Second, a true surface state is located in a gap or at least a symmetry gap of the bulk band structure projected onto the surface. In other words, a surface state exists in a region of the energy and momentum space of the solid where there are no bulk states of the same $k_\parallel$ and symmetry in existence. Bulk states themselves may have a slightly modified wavefunction near the surface. This modification of the wave function in the near-surface region will make these states sensitive to the

surface conditions. These modified states are commonly called surface resonances. Surface resonances are difficult to identify experimentally, because
even the true bulk photoemission signal changes when the surface conditions
change.

The third and last criterion is the old "crud test," with some refinements.
Surface states and resonances are indeed sensitive to the potential peculiarities at
the surface. Changes in the surface potential or surroundings will, in general,
have an effect on the electronic states of the surface. The changes introduced at
the surface by the addition of atoms or molecules could mean merely a change in
the electrostatic potential at the location of the surface atoms. Another
alternative could be a change in the symmetry or reconstruction of the surface
caused by the presence of an adsorbate. Last but not least, the surface states can
also be directly involved in the formation of the chemisorption bond. However it
is important to note that not all changes in the surface conditions will have an
effect on all surface states. Moreover, surface states do not really "disappear" if
we put an adsorbate onto the surface. They become interface states, and their
charge is redistributed in the energy and momentum space near the surface. An
example is shown below in which a surface state "happily" exists, apparently
unchanged, with a saturated layer of adsorbate on the surface.

The applications of these criteria can be illustrated using the *sp*-like surface
state on Cu (111) as an example. The low-index Cu surfaces have been studied
quite extensively in angle-resolved photoemission, and the current status is
summarized in a recent publication by S. D. Kevan and coworkers.[9] The *sp*

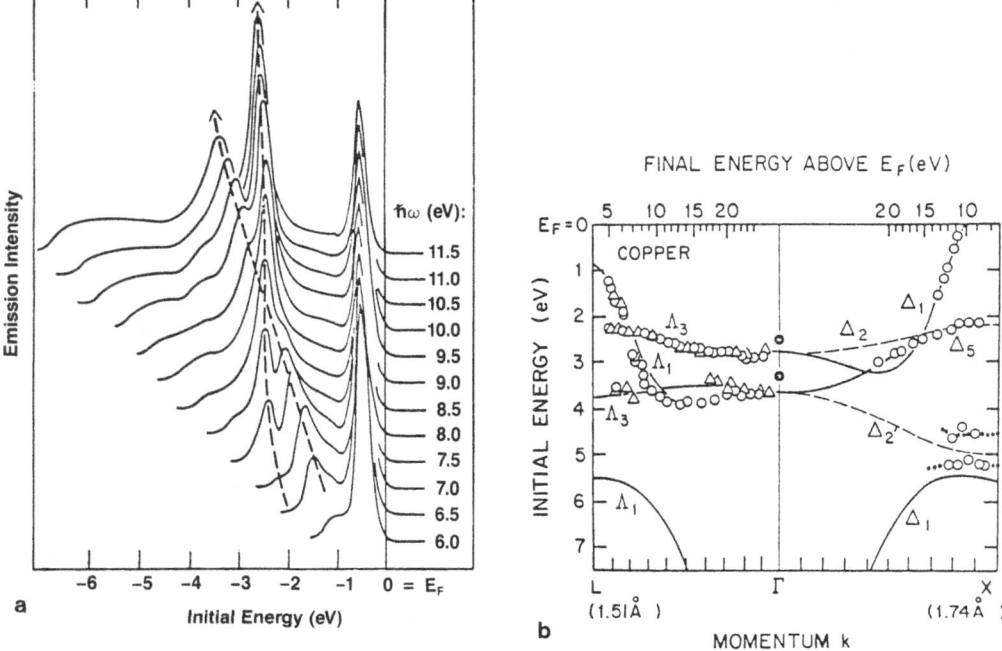

FIGURE 4. (a) Normal emission *p*-polarization EDCs of Cu (111) and (b) comparison between
experimental and theoretical band structure. (From Ref. 11.) The peak in the photoemission signal
near E_F clearly comes from a region of energy and momentum space where there is a gap in the
band structure. Therefore this state is a true surface state.

surface state on Cu (111) was first discovered by P. O. Gartland and B. J. Slagsvold.[10] The data in Fig. 4 is taken from the publication of J. Knapp and coworkers.[11] In Fig. 4a the normal-emission photoemission EDCs are shown for a variety of photon energies between 6 eV and 11.5 eV. We clearly recognize three different features in these EDCs. The two peaks showing dispersion with a change in photon energy, which corresponds to a change in the normal momentum component of the direct transition, are due to the emission from the Λ_1 and Λ_3 initial-state bulk bands. The sharp peak at a binding energy of 0.4 eV is caused by the emission from a surface state. We can see from the comparison of the measured bands[11] with the calculated band structure[12] in Fig. 4b that there exists a gap in the bulk band structure along the Λ-axis near E_F. All states with $\mathbf{k}_\parallel = 0$ on the (111) surface have a normal momentum component located somewhere along the Λ-axis. Since the state at -0.4 eV is located in a gap of the bulk band structure of the solid, it has to be a surface state. So far these data show that this state fulfills the first two criteria described above.

How does this state react to the presence of an adsorbate on the surface? In Fig. 5, angle-resolved EDCs from the clean Cu (111) crystal are shown in comparison with the same spectra taken after the adsorption of CO and N_2.[13] After CO exposure the sp-surface state has clearly disappeared, but after N_2 adsorption it is attenuated in intensity but remains visible in the photoemission. The conclusion of this observation is that the sp surface state takes part in the weak chemisorption bond of CO on the surface,[13,14] whereas physisorbed N_2 only attenuates its photoemission signal, due to an increase in surface scattering.

We also notice from the spectra shown in Figs. 4 and 5 that this surface state exhibits a dramatic decrease in cross section at higher photon energies. In the

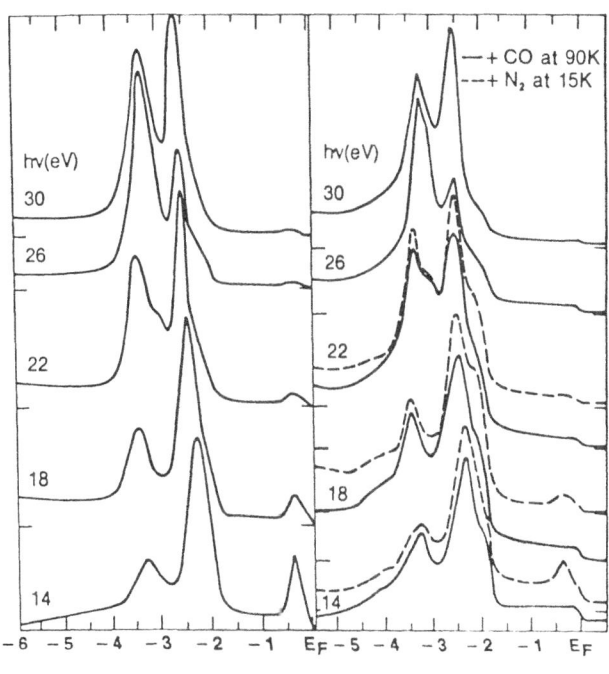

FIGURE 5. Normal emission EDC of the clean and adsorbate-covered Cu (111) surface showing how the sp surface state reacts to the presence of CO or N_2 molecules on the surface. (From Ref. 13.)

Inital-State Energy (eV)

EDCs taken at 30 eV and higher it has essentially vanished. At higher photon energies around 70 eV, however, the same surface state becomes clearly visible again. This oscillatory behavior of the surface state cross section was first reported and explained by S. G. Louie et al.[15] Any surface state that extends several layers into the bulk, like the sp surface state on Cu (111), will couple preferentially to final states, which have a wavelength about equal to the lattice spacing normal to the surface. Under these conditions the contributions from the various crystal layers add up constructively to the cross section. D-like surface states, which in general are more localized in the outermost lattice plane, consequently do not exhibit these cross section oscillations.

2.6. Symmetry and Polarization Selection Rules

Angle-resolved photoemission, coupled with a polarized light source like synchrotron radiation, not only enables us to determine the energy levels of an atom or a molecule as well as the bandstructure $E(\mathbf{k})$ of a solid or a surface, but also serves as a tool to identify directly the symmetry of the initial-state electronic wave functions.

Starting with the emission of an isolated atom and linearly polarized light, the angular-dependent cross section is given by[16]

$$\frac{d\sigma}{d\Omega} = I_0 \left[1 + \frac{\beta}{2} (3 \cos^2 \delta - 1) \right], \tag{7}$$

where I_0 is the total intensity, δ is the angle between the polarization vector and the direction of the emitted electron, and β is the angular asymmetry parameter, which depends on the initial- and final-state wavefunctions. β can range from -1 to $+2$ and $\beta = 2$ is valid for a plane-wave final state. This results in the classical picture of a $\cos^2 \delta$ distribution, with no emission perpendicular to the direction of polarization. Unfortunately, simple intuitive rules, e.g., $\beta = 2$ for an s state, are almost not applicable. Final-state contributions and peculiarities seem to destroy the simple picture. For a state-of-the-art experiment, a rather sophisticated calculational scheme is necessary to explain the changes of the β parameters with photon energy for the individual orbitals of a molecule. In several cases, however, these studies yield the additional benefit to resolve ambiguities in the orbital assignment.

The next step is to consider the emission from molecules oriented by the presence of a surface. Molecules chemisorbed on a surface are, with the exception of vibrational motion, essentially frozen in their orientation with respect to the surface. This orientation might change upon a change in substrate temperature or coverage, but within each of these states the molecules are more or less aligned with respect to the substrate. This general orientation largely enhances the angular variations of the emission pattern and may be used to determine the adsorption geometry.

The geometric structure of the adsorbate is certainly one important input parameter in understanding the chemisorption mechanism. The second important piece of information is the identification of the molecular orbitals and their relative shifts upon the formation of the chemisorption bond with the surface. Being able to answer both of these questions was one of the highlights of the early days of angle-resolved photoemission.

The classical results for CO on Ni (100) demonstrate the capabilities of angle-resolved photoemission in this area and the application of the technique. In general, symmetry selection rules can be derived from the symmetry properties of the dipole matrix element $\langle f| \mathbf{A} \cdot \mathbf{P} |i \rangle$. As introduced above, $\langle f|$ and $|i \rangle$ are the final and initial state, respectively, and $\mathbf{A} \cdot \mathbf{P}$ is the dipole operator. All these functions, including the matrix element itself, are continuous in space. In the experiment, we place the detector into any particular mirror plane of the system. This includes the direction along the molecular axis of CO or one of the mirror planes normal to the surface of the Ni substrate. This test verifies whether CO is bound in such a way that the mirror symmetry of the substrate is maintained for the combined system. In order for us to detect any signal in this geometry, both the final state $\langle f|$ and the total dipole matrix element have to be even with respect to this mirror plane, since any odd continuous function is identical to zero in the mirror plane itself.

In performing this test, we have to distinguish between two cases, which can be selected by choosing the appropriate polarization direction of the light. If the polarization vector is located in the same mirror plane as the detector, then the dipole operator $\mathbf{A} \cdot \mathbf{P}$ is even with respect to this mirror plane. Since the final state $\langle f|$ and the whole matrix element have to be even, we will get an emission only if the initial state $|i \rangle$ is also even. Conversely, if the polarization vector is oriented perpendicular to the mirror plane picked out by the detector, the dipole operator is odd, and for the whole matrix element to be even, only odd initial states will contribute.

In 1976, R. J. Smith et al.[17] were the first ones to explore these symmetry selection rules to determine the orientation of CO absorbed in a saturated monolayer on Ni (100). These results are shown in Fig. 6. Spectrum a is taken in

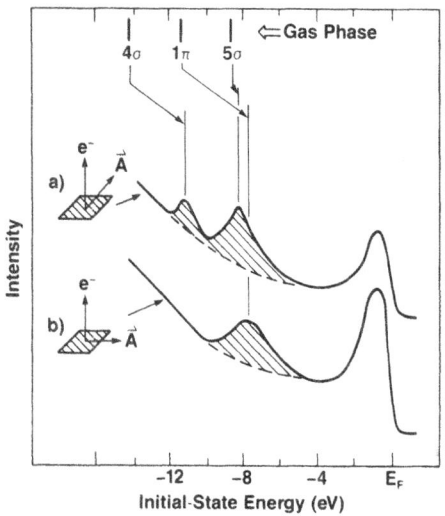

FIGURE 6. Photoemission EDCs from CO adsorbed onto Ni (100), illustrating the use of polarization selection rules to identify the symmetry of adsorbate induced states. (From Ref. 17.) In geometry (a), π and σ-type orbitals contribute to the photoemission, whereas in geometry (b) only the π symmetry states show up. The corresponding gas-phase ionization energies of the three CO outer valence orbitals, corrected for the work function of the system, are shown as bars on the top of the figure.

a geometry where all states will show up in the detector, whereas in spectrum *b* only odd initial states are excited into the direction of the detector. The shaded peaks are caused by the adsorbate. Compared to the three-peak structure observed for isolated CO, as indicated by the bar diagram at the top of Fig. 6, we find only two distinct features in the top spectrum and one in the bottom curve. The peak at the highest binding energy, at 11.2 eV below E_F, is due to emission from the 4σ orbital of CO. Since this peak disappears in curve *b*, we conclude that the mirror symmetry of the substrate is not broken by the adsorbate. This is consistent with CO adsorbing with its molecular axis normal to the surface. Because of this symmetry, the peak at -7.9 eV in curve *b* is exclusively due to the 1π emission of the chemisorbed CO. In curve *a* both the 1π and 5σ emission contribute to the adsorbate peak at about -8 eV. However, this peak is observed at a larger binding energy than the 1π peak in curve *b*. Thus Smith *et al.*[17] concluded that the level ordering between the 5σ and 1π orbitals is inverted in the chemisorbed phase. The large relative shift of the 5σ level is a consequence of the 5σ interaction with the substrate, which is described as 5σ donation and constitutes part of the chemisorption bond of CO.

This symmetry test with polarized light is a very powerful tool in the identification of adsorption configurations or the orbital assignments in the chemisorbed phase. The test can even be applied repeatedly, first to a deeper-lying orbital like the 3σ or 4σ in CO to determine the orientation of the molecular axis and the total symmetry of the system. Afterwards, once the total symmetry is established, the same selection rules can be applied to identify the orbitals that interact with the substrate.

Obviously, the polarization selection rules govern the emission not only from atomic or molecular orbitals but from band-like states as well. This greatly facilitates any band structure determination. For the same reasons as were just discussed for molecules, only the totally symmetric final states and their extensions into the vacuum will have nonvanishing amplitude at the detector. This was first pointed out by J. Hermanson for the case of the normal emission geometry on the low index planes of *fcc* and *bcc* surfaces.[18] The extension to the mirror plane emission is straightforward. Additionally selection rules for the optical excitation, which can be derived using group theory, come into play. This limits the number of possible transitions between the occupied initial-state bands and the final-state bands leading to the detector. Today, tables can be found in the literature containing these selection rules for optical transitions at all symmetry points and lines of the *fcc*, *bcc*,[19] and *hcp*[20] lattices. Exploitation of these selection rules, using a polarized excitation source like synchrotron radiation, supplies the knowledgeable researcher with an extremely powerful tool to probe the band structures of solids and surfaces.

2.7. Final-State Effects

So far the discussion has concentrated mainly on initial-state properties and how they manifest themselves in the angle-resolved photoemission spectra. Often the final states are just approximated by plane waves, and cross sections are calculated in this approximation. Any deviation from a purely plane-wave final state is observed as an additional structure in the photoemission signal. Next will be discussed two classes of final-state effects. The first group are localized resonances, so called shape resonances that show up several eV above the ionization threshold in molecules. The second class of effects deals with photoelectron diffraction. Both of these phenomena are observed as intensity modulations in the photoemission signal either under variation of the excitation energy or by varying the detector angle. These effects are also of interest from a materials-science point of view because they reveal insight into adsorption geometries and bond lengths of atoms or molecules on surfaces.

Shape resonances were first theoretically described by J. Dehmer and D. Dill[21] and by J. Davenport.[22] These are short-lived, virtually bound states well above the ionization continuum, trapped by a large angular-momentum barrier. The presence of these states in the continuum causes an enhancement of the transition rate which can be easily observed in the partial absorption cross section of the valence orbitals that couple to these resonances. Figure 7a shows the partial ionization cross section of the 4σ orbital of CO.[23] In CO the shape

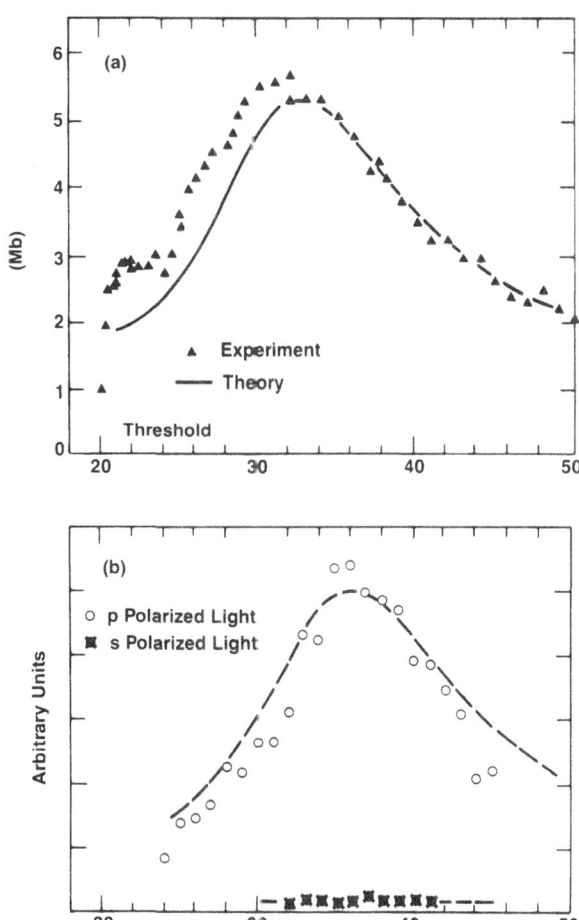

FIGURE 7. Partial photoionization cross section of the 4σ orbital as a function of photon energy for (a) gas phase CO. (From Ref. 23.) (b) The same for CO adsorbed onto Ni (100). (From Ref. 24.) The shape resonance is maintained in the chemisorbed state and can be excited only by a polarization component normal to the surface.

resonance has σ-symmetry; consequently, only the 4σ and 5σ orbitals couple to it, but only with the polarization component along the molecular axis. In isolated gas-phase CO the 4σ resonance transition occurs at a photon energy of 32 eV, or about 12 eV above the ionization threshold. When CO is chemisorbed onto a surface, the resonance shifts slightly in energy, as seen in Fig. 7b, but persists essentially as strongly as in the isolated molecule.[24]

In angle-resolved photoemission the shape resonance in these linear molecules leads to a strong directional emission along the molecular axis.[25] The half-width of this emission is about 50 degrees FWHM. This can be used to identify the direction of the molecular axis when the molecule is chemisorbed.

The shape resonances are also observed for transitions involving core levels as initial states and are strong enough to show up as features in the total absorption cross section, again several eV above the ionization threshold.[26,27] Recently Natoli[28] suggested that the shape resonance position is sensitive to the internuclear bond distance in these molecules. In a simple scattering model he found the wavelength of the final-state wave function at the shape resonance energy to match the internuclear spacing. This relationship between the bond length and shape resonance position was verified empirically by Hitchcock and coworkers[29] for a quite extensive series of small, isolated molecules in gas phase. Even for chemisorbed molecules the same relationship seemed to hold, upon proper treatment of the reference level problem.[28,30] This opened up the exciting possibility of studying changes in bond lengths of chemisorbed molecules in previously unmatched detail. Recently, however, a study of the 4σ valence shape resonance position in CO on a K-promoted Ru (001) surface[31] revealed no difference in the resonance position compared to the CO adsorption on the unpromoted Ru (001) surface, whereas by different methods a severe CO bond weakening was established for CO on the K-promoted surface. This questions the unambiguity of the shape-resonance position as an indicator of the bond length in chemisorbed molecules. Obviously we need to develop a better theoretical understanding of the effects on the shape resonance caused by the chemisorption bond and the presence of the surface atoms.

The second class of final-state effects to be mentioned here are best generalized by the term photoelectron diffraction. Again, the true final state of the photoelectrons is not a plane-wave final state, but for this purpose is best described by a plane wave with scattering contributions from the neighboring atoms. Depending on the final-state wavelength these backscattered portions add in or out of phase and thus cause a characteristic oscillation of the photoemission intensity. These studies have been mostly performed on core levels. Since core level photoemission is discussed in another chapter of this book,[32] this section will be rather short. The effects are especially strong in the normal emission geometry because of the high symmetry and the resulting degeneracy of the scattering contributions.[33] Intensity modulations up to 40% of the signal have been observed in the normal emission photoelectron diffraction studies of chalcogens adsorbed onto transition-metal surfaces. A theoretical analysis of these spectra yields the bonding site and bond length of the adsorbate atoms.

The same diffraction effects are observed under an arbitrary polar angle at a fixed excitation energy by changing the azimuth of the electron detector.[34] These "flower patterns" contain the same information as the normal photoelectron diffraction data; however, the magnitude of the total modulations is much smaller because of the lack of degeneracy of the scattering contributions.

For grazing exit angles and X-ray photon energies, C. S. Fadley[35] has shown that the theoretical analysis becomes simpler, because only single scattering events have to be included in the calculations to give satisfactory results for adsorbate geometries. In all other geometries and conditions, a full multiple scattering LEED-type calculation is necessary to obtain any structural information.

3. SPECIFIC CASE STUDIES AND EXAMPLES

3.1. Layered Compounds

The layered compounds such as graphite or transition-metal dichalcogenites can be viewed as intrinsically two-dimensional systems, because the layers are held together by a van der Waals type bond and accordingly the interaction between the layers or sandwiches is quite small. For two-dimensional periodic systems there is no ambiguity about the electron **k** vector, as explained above. Consequently, layered compounds were some of the first materials studied by angle-resolved photoemission.[36,37] The work of N. V. Smith, M. M. Traum, and F. J. DiSalvo[36] on $1T$-TaS_2 was the first demonstration that the band structure of a solid could be mapped using angle-resolved photoemission. In addition to mapping the band structure, N. V. Smith et al.[37] found rather pronounced intensity variations in the emission of the Ta $5d$-derived band upon changing the azimuth of the detector at a constant polar angle. The measured "flower pattern" did exhibit a very pronounced threefold symmetry, which was explained originally as due to shadowing of the Ta emission by the S atoms in the top layer.

More important, this study clearly demonstrated that the simple plane-wave final-state approximation is insufficient to describe the azimuthal angular dependence in general. In the plane-wave final-state approximation there would be no photoemission in a direction as long as its electron momentum is perpendicular to the polarization of the light ($\mathbf{A} \cdot \mathbf{P} = 0$), whereas the experiment found appreciable intensity even in these directions. A. Liebsch later was able to theoretically reproduce the angular pattern of the Ta d-bands rather well using a time-reversed LEED wave function for the final state.[38]

3.1.1. Charge-Density Waves in Layered Compounds

The current interest in these systems[39–42] comes from the occurence of phase transitions involving charge-density waves (CDWs) in these layered compounds. Often the CDW is incommensurate with the lattice; however, in $1T$-$TiSe_2$ the phase transition occurs around 200 K directly between a normal phase and a commensurate CDW phase. The CDW has a periodicity of $2a_0 \times 2a_0 \times 2c_0$, which means that in the reciprocal lattice the Γ, A, L, and M points all become equivalent. The relationship of the Brillouin zone in the normal phase to the Brillouin zone of the CDW phase is shown in Fig. 8. Because of the

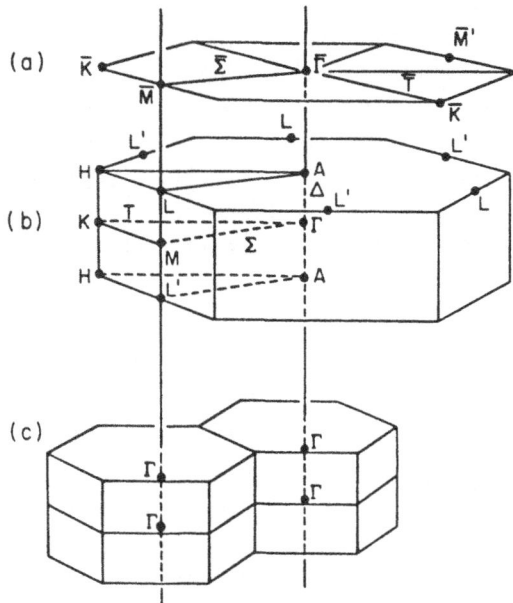

FIGURE 8. Surface (a) and bulk (b) Brillouin zone of transition metal dichalcogenites in the 1T phase. In the presence of a commensurate $2a_0$ times $2a_0$ times $2c_0$ CDW phase, the Γ, A, L, and M points of the original Brillouin zone become degenerate, and the zone is reduced as shown in (c).

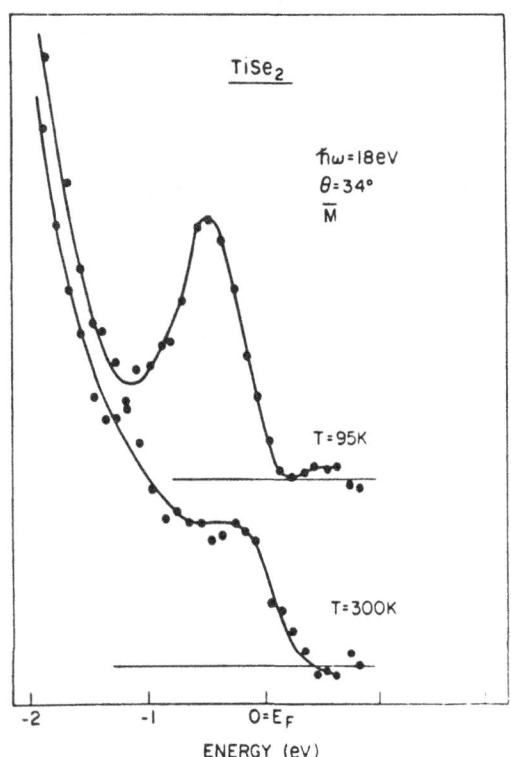

FIGURE 9. EDCs of 1T TiSe$_2$ taken at $T = 300$ K (bottom) in the normal phase and (top) at $T = 95$ K in the CDW phase (From Ref. 39.) In the CDW phase an additional peak is observed in the EDCs near E_F in the curves taken close to the SBZ boundary at $\bar{M}$.

additional vector introduced by the CDW, the Brillouin zone of the CDW state is smaller and we expect to see the Γ-point "backfolded" into the M-point of the normal phase. This is strictly true in the theoretical picture. In practice, these might be only small effects in the measured angle-resolved photoemission spectra, since the perturbation of the crystal potential by the CDW is relatively small also.

In Fig. 9 two EDCs are shown corresponding to the emission from states near the $\bar{M}$ point of the SBZ (surface *Brillouin* zone). In the CDW phase a new peak is observed near E_F. This peak corresponds to the emission from the folded Se 4p bands, which have a band maximum near Γ. N. Stoffel et al.[41] verified this interpretation by measuring the dispersion of this peak and comparing it to the dispersion of the equivalent peak observed near normal emission.

In EDCs corresponding to the emission from states near the L point (not shown here), a triangular peak is observed at $\bar{E}_F$, which clearly does not correspond to a Fermi function. This triangular shape is caused by a small electron pocket of Ti 3d derived bands, which dip below E_F just near the zone boundary. The interaction of these Ti 3d states with the Se 4p states near Γ is supposed to drive the transition into the CDW phase.[43,44] From this theoretical concept the following two questions arise: first, whether the Ti 3d and Se 4p bands overlap in energy in the normal phase, and second, whether a gap opens up in the CDW phase.

Before trying to answer these questions, we have to briefly discuss what one would expect to observe in the photoemission of the Se 4p bands. Se has a sufficiently large Z such that the spin–orbit splitting in the 4p levels has to be taken into account. As conveniently illustrated by a tight-binding approach, in a two-dimensional layer the atomic-like p states form three bands, which are commonly labeled p_x, p_y, and p_z according to their atomic origin. At the center of the zone, at Γ, p_x and p_y are degenerate, while the p_z-type band is split off by the crystal-field splitting. When relativistic effects are introduced, the p levels are each split into a $p_{3/2}$ and a $p_{1/2}$ component, just as in the atom. The crystal field will split the $p_{3/2}$ level into two components according to $m_j = 3/2$ and $m_j = 1/2$. The $p_{3/2}$ $m_j \pm 3/2$ band is of "pure" p_{xy} symmetry, whereas the $p_{3/2}$ $m_j \pm 1/2$ band and the $p_{1/2}$ $m_j \pm 1/2$ both are of mixed p_{xyz} symmetry. Consequently, even at the center of the Brillouin zone we expect to see three p-symmetry bands rather than two as in the nonrelativistic case.

In order to determine the band positions at Γ as accurately as possible, one has to take into account even small effects of the dispersion along the axis normal to the surface. This is done by recording spectra in normal emission geometry with different excitation energies, as shown in Fig. 10. Contrary to our expectations, we observe a strongly dispersing feature in these spectra. The energy of this band varies between 0.35 eV and 2.15 eV and it is assigned to the Se 4p p_z band.[42] Thus even in layered compounds there are bands showing a strong dispersion with momentum normal to the planes. This demonstrates that the interplanar bonding is not purely of the van der Waals type but rather mediated by the chalcogen p-orbitals normal to the planes. Additionally we see a peak at about 0.2 eV below E_F and in the spectra taken near the A point we see a sharp spike at E_F. O. Anderson et al.[42] assign these features to the spin–orbit split uppermost p band. Assuming a constant spin–orbit splitting, these authors

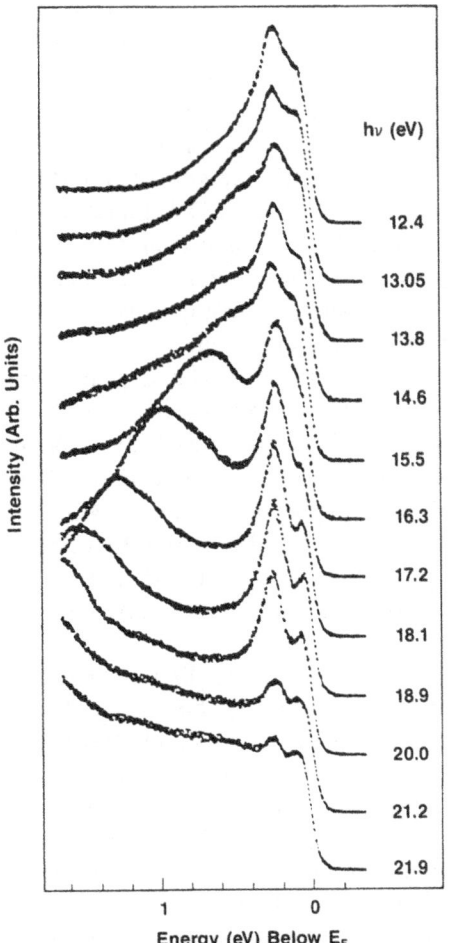

hν (eV)

12.4

13.05

13.8

14.6

15.5

16.3

17.2

18.1

18.9

20.0

21.2

21.9

Intensity (Arb. Units)

1   0

Energy (eV) Below E_F

FIGURE 10. Normal emission EDCs of 1T TiSe$_2$ taken as a function of photon energy as indicated next to each curve. (From Ref. 42.)

come to the conclusion that the lower-binding-energy component of this spin orbit "doublet" crosses the Fermi level between A and Γ. This implies that there is a small pocket of holes right at Γ. In a similar manner the bottom of the Ti $3d$ bands is found near the L point and located at 0.07 eV below E_F. Thus, according to this study, there is an overlap between the Se $4p$ bands and the Ti $3d$ bands in energy.

In contrast to these results, N. Stoffel et al.[41] previously came to the conclusion that the Ti $3d$ bands are located at 0.23 eV below E_F and that the maximum of the Se $4p$-bands is located at 0.29 eV. Thus these researchers concluded that the material is a semiconductor with a rather small band gap. Both studies[41,42] agree that there are no large gaps opening up in the CDW phase.

It is important to note here that two groups of experimentalists have come up with quite different conclusions about the same material based on spectra that to an unbiased observer look almost alike. This also touches on the question of how useful angle-resolved photoemission is as a technique for answering such important questions in materials science. Therefore, some comments should be added at this point.

The first unresolved question is whether both groups really studied the same material. Depending on growth conditions, samples off in stoichiometry are easily produced. Any change in the amount of Ti in the samples causes a large shift in the occupancy of the Ti $3d$ states near the L point, because the electron pocket around this point is very small.

Second, the exact position of the Ti $3d$ L point depends on how accurately the line-shape function of the photoelectron analyzer is known. The emission of the Ti $3d$ states is never clearly split off from the Fermi level, and the analyzer and the Fermi function have to be folded to determine the exact location of these states.

Third, the assumption that the spin–orbit splitting is constant along the Δ-axis is certainly not justified. If the spin–orbit splitting is not constant across the zone, then it is not totally obvious whether the upper Se p band really crosses the Fermi level. Again this would result in different conclusions about the nature of the material.

Even though these discrepancies are present in the literature, it is nevertheless very exciting to have a technique at hand which is powerful enough to address these questions experimentally at all. Only recent developments of instruments with about 1° angular resolution and an energy resolution far better than 100 meV make it possible to even attempt to study these questions. The issue of TiSe$_2$ is especially of interest because theory cannot predict with any certainty the existence of a band gap. TiS$_2$ undoubtedly is a semiconductor, whereas TiTe$_2$ is a metal.[45] The CDW phase has been observed only for TiSe$_2$ but not for TiS$_2$,[43] and even the exact mechanism leading to this instability is not completely understood. It is an open question whether the CDW phase is caused mainly by an electronic interaction between the Ti d bands and the Se p bands[46] or by the softening of a phonon mode leading to the observed lattice distortion accompanying the CDW.[47]

3.1.2. Charge Transfer in Intercalated Graphite

Graphite was one of the first compounds studied in the early days of angle-resolved photoemission,[48] just as were the transition-metal dichalcogenites. Graphite is a semimetal with rather poor conductivity. However, intercalating it with alkali metals or other dopants results in a conductivity similar to that of Cu.[49] Upon intercalation, the graphite host lattice expands slightly along the c-axis to incorporate ordered layers of, for example, alkali-metal atoms. The periodicity of the superlattice structure is given by the stage of the intercalation. Stage 1 means that graphite and alkali-metal layers alternate, whereas for a stage n intercalation compound n layers of graphite are separated by one layer of alkali-metal atoms. The changes in the electronic properties are caused by charge transfer from the alkali-metal atoms into the graphite bands. Since graphite is a semimetal with a vanishing density of states at the Fermi level, even a relatively small charge transfer can thus cause major changes in the electronic properties.

As a model system for these intercalation compounds, we studied the stage 1 Li intercalated LiC$_6$.[50] The Li atoms in this compound are in registry with the C lattice and form a $\sqrt{3} \times \sqrt{3}$ superstructure. Accordingly, the Brillouin zone of

LiC_6 is smaller than that of graphite, and the bands are folded back into the first zone. Again the question arises as to how strong this superstructure effect is in practice. This is answered by the spectra shown in Fig. 11, which contains a series of EDCs of LiC_6, taken at a photon energy of 40 eV, for various polar angles of emission. In normal emission the strongest peak is due to the emission from the π bands at 9.3 eV below E_F. The bottom of the σ band shows up at 22.5 eV below E_F. The weak structure at -5.5 eV is the top of the second σ-band, whereas the other structures at -0.5 eV, -13 eV, and -15.2 eV are due to backfolded bands. A comparison of this result with the band structure of graphite shows that the band shifts are not rigid, as expected from a simple charge-transfer model. For example, the top of the second σ band shifts only by 1 eV, whereas the other bands shift by about 2 eV. This means that there is an appreciable interaction between the Li atoms and even the sp_2 hybridized in-plane graphite bands, rather than just a charge transfer.

Upon variation of the polar angle of the electron detector, we observe the expected dispersion of all bands. The most interesting observation is that at a polar angle of 30° the previously weak feature near E_F becomes rather strong. Now we are no longer detecting not a backfolded hand, but the original unfolded band. Thus we can directly see in these curves that the band folding due to the superstructure shows up as a second-order effect in the photoemission spectra. The perturbation of the total potential is noticeable, but not strong enough to make this emission comparable in intensity to the primary emission.

3.2. Adsorbed Atoms and Molecules

Adsorbed atoms and molecules have been widely studied with a whole variety of techniques, angle-resolved photoemission being a very important one. These adsorbate systems can be viewed as model systems for the understanding of

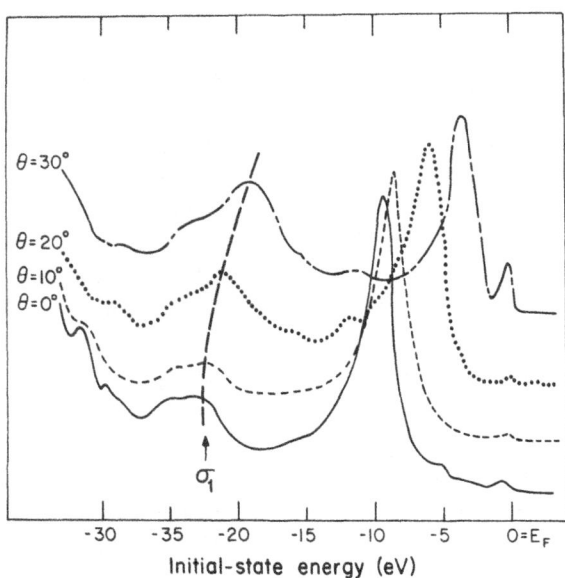

FIGURE 11. Angle-resolved photo-emission of LiC_6 stage I intercalated graphite as a function of polar angle taken at a photon energy of $h\nu = 40$ eV. (From Ref. 50.)

catalysis, corrosion, or embrittlement, even if the technique employed requires that the studies are undertaken on single-crystal surfaces. The individual molecular orbitals in the adsorbed species are identified using symmetry-selection rules, as explained above. Relative energy shifts reveals which orbitals are involved in the chemisorption bond. In addition, the angular emission characteristics reveal the orientation of the molecular symmetry axis. However, we can do even more; we can also determine the electronic structure of the adsorbate-covered surface and compare it to the electronic structure of the clean surface. Thus we are able to deduce which substrate states are active in the chemisorption bond and why the bonding within the top metal layers is possibly disturbed when an adsorbate is present, explaining embrittlement cf the original solid material. Following are a few examples illustrating these various aspects.

3.2.1. Adsorption Geometries and Chemical Transformations

Originally[17,24] the orientation of CO on Ni (100) was determined successfully by angle-resolved photoemission to be upright, parallel to the surface normal, with the carbon end towards the surface. The same identical adsorption geometry was confirmed for CO on almost any other transition-metal surface, and the same seemed to be true for other diatomic molecules like N_2 or NO. Afterwards the search was on to find a CO molecule that was tilted with respect to the surface normal,[51] and nowadays it seems to be fashionable to find CO molecules that are bonded parallel to the surface.[52] The special interest in these latter systems originates from catalysis research. The first step in catalytic reactions involving the CO molecule has to be the breaking or weakening of the C–O bond. This only occurs in the more exotic adsorption geometries, and thus these species are viewed as precursors or intermediates for these reactions. Upright, terminally bound CO was found to be almost as strongly bound as the isolated gas-phase CO molecule.

At this point these results will not be discussed in detail, but it should be pointed out, that an accurate ($\pm 5°$) bond-angle determination by angle-resolved photoemission for a diatomic molecule that is not upright on the surface is rather difficult. In order to calculate the emission intensity of a given molecular orbital as a function of the detector angle, one has to take into account, nonlocal field effects, which can significantly alter the relative polarization components of the electric field at the surface compared to a macroscopic estimate.[5-7] Thus this calculation becomes a major effort, and one might consider combining these measurements with different techniques such as NEXAFS, vibrational spectroscopy, or ESDIAD in order to obtain the best possible bond-angle information.

Rather than dwelling on the studies of diatomic adsorbates, here are some examples of slightly larger molecules whose molecular orientation was determined or confirmed by angle-resolved photoemission. The first one is ammonia (NH_3) on Ni (111). Ammonia is bound to the surface through the nitrogen lone-pair orbital with the H atoms pointing away from the surface.[53] W. M. Kang et al.[54] measured and calculated the anisotropy of the photoemission from the degenerate 1e orbitals. These orbitals are the N–H bonding orbitals and have a threefold symmetry. In this experiment, the polar angle of the detector and the angle of incidence and the direction of the incoming light were held fixed. The intensity of the emission from the 1e orbitals was measured as a function of the rotation angle of the crystal as shown in Fig. 12b. The measured intensity of the 1e orbital emission, as shown in Fig. 12c, clearly displays a threefold symmetry.

The comparison with calculations,[54] Fig. 12d, leads to the conclusion that the experimentally observed emission is dominated by the molecular features. Scattering within the ordered adsorbate layer produces quite strong modulations, whereas the substrate has a rather small effect on the emission pattern. Actually the experimental curves seem to agree best with the calculations for the oriented molecule alone. The observation of a threefold symmetry of the emission pattern also implies that the molecules are azimuthally oriented with respect to the substrate and do not rotate freely. At a temperature of 80 K, this is quite surprising. According to Netzer and Madey,[55] the freezing of the rotation is caused by trace amounts (less than 0.05 monolayer) of impurities, presumably O, on the surface.

Intuitively, one is inclined to attribute the maxima in the emission intensity

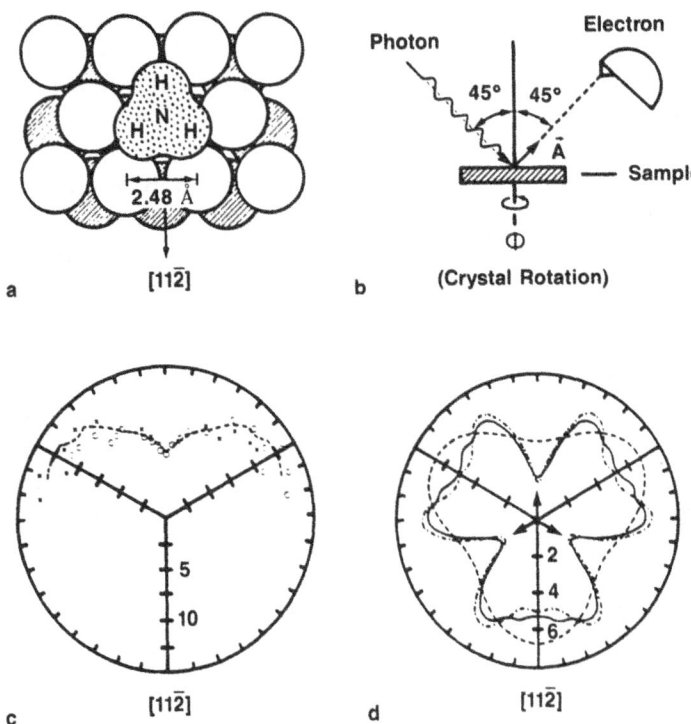

FIGURE 12. NH_3 adsorption on Ni (111) (a) adsorption geometry. The adsorption site is chosen arbitrarily to be on top of a substrate atom. (b) Schematic of the geometry of the experiment. (c) *Measured intensity from the 1e orbital of* NH_3 taken at a photon energy of $h\nu = 42$ eV, as a function of the rotation angle of the crystal as shown in (b); (d) Calculated emission pattern from an oriented NH_3 molecule (dashed curve), a (1 × 1) NH_3 layer (solid line) and a (1 × 1) NH_3 layer in contact with the Ni (111) substrate (dash-dotted line). The direction of the N–H bonds is indicated by the arrows. (From Ref. 54.)

from the $1e$ orbitals with the direction of the N–H bond since the $1e$ orbitals are attributed to this bond. The calculations[54] however show that the maxima of the emission occur in the direction just in the center, between the N–H bond directions, which are indicated by the arrows in Fig. 12d. Thus NH_3 on Ni (111) is an interesting example, demonstrating that intuition and angle-resolved photo-emission are not always compatible.

The second example deals with the adsorption of C_2H_2 and C_2H_4 on Pt (111). These molecules undergo chemical transformations on this surface depending on the surface temperature. LEED, thermal desorption, photoemission and LEELS (*low-energy electron-loss spectroscopy*) studies have led to suggestions of several different configurations for these adsorbates in the two temperature regimes, at 150 K and after heating the surface to about 400 K (see sources listed in Ref. 56). The most probable configurations are shown at the top of Fig. 13. In the low-temperature state it is generally accepted that the molecules are adsorbed in

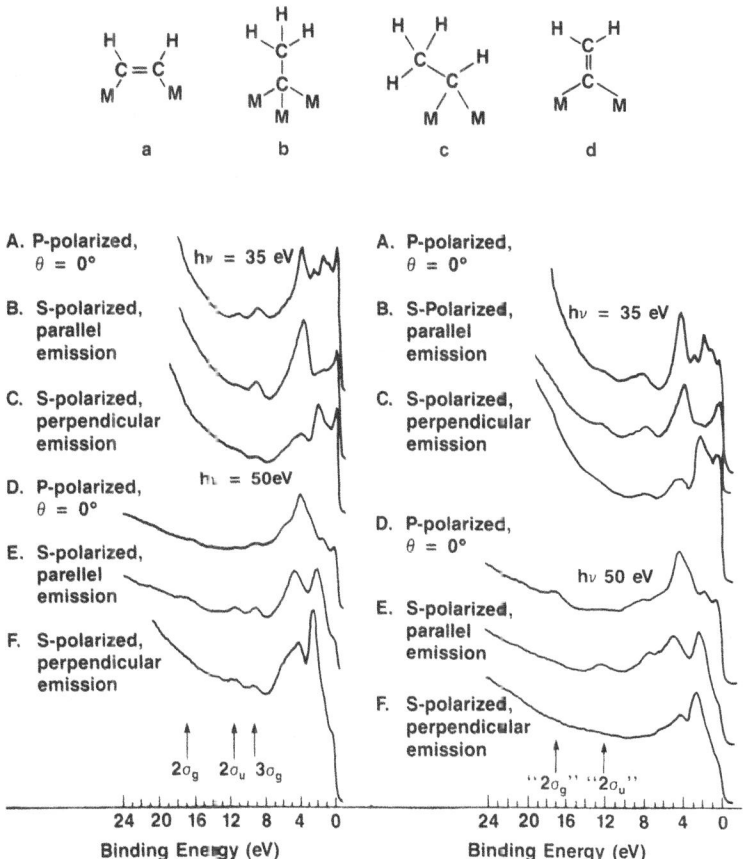

FIGURE 13. Angle-resolved photoemission of C_2H_2 adsorbed onto Pt (111). At the top the various proposed adsorbate configurations are shown. All the spectra were accumulated with the sample at a temperature of $T = 150$ K. The curves in the right panel were taken directly after adsorption of C_2H_2 at $T = 150$ K, whereas for the curves in the left panel the adsorbate was annealed at a temperature of about $T = 400$ K.

a configuration where the C–C bond axis is parallel to the surface (configuration *a*), whereas for the high-temperature state several models are proposed in which the C–C bond axis is either tilted or normal to the surface (configurations *b* through *d*). Another open question about the high-temperature phase is whether the C–C bond of this surface species is a single or a double bond.

In the bottom panels of Fig. 13, the angle-resolved photoemission of C_2H_2 adsorbed onto Pt (111) at 150 K and after annealing the surface at 400 K is shown under various light and detector geometries. The adsorbate induced features are relatively weak compared to the strong emission of the Pt substrate. Nevertheless, at least three different adsorbate "bands," indicated by the arrows, can be found in the spectra between 7 and 20 eV binding energy. In the low-temperature phase the bands are assigned as follows: The band at 17 eV originates from the $2\sigma_g$ states, the band at 11.5 eV from the $2\sigma_u$ states, and the band at 9.3 eV from the $3\sigma_g$ states. The $1\pi_u$-derived features fall within the range of the strong Pt *d*-band emission and are not discernible. Both the $2\sigma_u$ and the $3\sigma_g$ bands are observed upon excitation with *s*-polarized light in both detector geometries, parallel and also perpendicular to the polarization vector of the light. Therefore these orbitals must be oriented in such a way that they are not symmetric with respect to the mirror plane *the detector is located in*. A π-bonded species with the C–C bond axis parallel to the surface, as shown in configuration *a*, can be expected to be present on the surface in three different domains. In this case, 2/3 of the molecules present on the surface fulfill the symmetry condition shown by the angle-resolved photoemission, even if the remaining 1/3 is adsorbed symmetric to the mirror plane of the detector. Therefore the picture obtained by angle-resolved photoemission is certainly consistent with the models proposed by the studies using other surface-science techniques.

Whereas everybody seems to be in agreement about the nature of the low-temperature adsorbed species, the models for the high-temperature phase are more controversial. The first observation is that the photoemission signal is clearly different from that of the low-temperature phase. Furthermore, we realize that the exact nature of the adsorbate states will be different for the various proposed species, which are shown as Fig. 13b through d for C_2H_2. However, for the more tightly bound orbitals we are discussing here, one can safely assumed that at least their gross symmetry remains unchanged. The molecular structure of the surface species will thus be characterized by the orientation of the C–C bond axis and the hybridization of the C–C bond. The polarization test in this case unambiguously shows that the C–C bond axis is oriented normal or close to normal to the surface. This clearly rules out the existence of an ethylidene species c. The orientation of the C–C bond axis is derived from the disappearance of the $2\sigma_u$ derived orbital at 12.3 eV binding energy in a detector geometry normal to the polarization vector of the light. Independent of the exact nature of the surface species, the respective orbital at this binding energy and in this order with respect to the other orbitals will always be a σ-type orbital, which is symmetric around the molecular axis. The same is true for the orbital at a binding energy of 17 eV. However, this orbital does not show up in either geometry with *s*-polarized light, and therefore cannot be used to identify the orientation.

After the orientation of the high-temperature C_2H_2 derived species has been identified, we still have to deal with the bond order or hybridization. This is done by studying the emission band at a binding energy of 8 eV. Whereas a triple-bonded species can be positively ruled out, no absolute distinction can be made between the single- or double-bonded species. Following the detailed discussion by M. Albert et al.[56] the single bound ethylidyne species b is by far the most likely candidate.

3.2.2. Lateral Interactions and Band Formation

So far the photoemission from chemisorbed molecules has been discussed from the viewpoint of an isolated single molecule, which is oriented and modified by the presence of the surface. However, the adsorbed molecules in general are not isolated from each other but form an overlayer, which often is ordered and in registry with the substrate. Consequently, the adsorbate states are two-dimensional band-like states, just as the states of the underlying substrate. The first band structure and dispersion measurements were reported for atomic adsorbates such as Xe on Pd (100),[57] Cl on Si (111),[58] or O on Al (111).[59] Later, these studies were followed by detailed studies of chalcogen adsorption on a whole variety of transition-metal surfaces (see summary in Ref. 2). Physisorbed Xe exhibits hardly any dispersion at all,[57] even though it is adsorbed in a close-packed structure. O on the other hand, shows a dispersion of a couple of eV in the observed $p_{x,y}$ bands,[59] much more than the calculations[60] predict for an isolated monolayer. In this case the interaction between the adsorbate atoms is mediated through the substrate, and this enlarges the bandwidth appreciably compared to the direct overlap interactions between neighboring O atoms.

Naturally the adsorbate bands have also been studied for CO on various substrates. Most of these studies have been summarized in Refs. 2 and 61. As originally discovered by Tracy and Palmberg,[62] CO undergoes a number of structural phase transitions involving a compression of the CO layer with increasing coverage, when the substrate is cooled with LN_2. Thus, the nearest-neighbor distance of the adsorbed molecules can be almost continuously varied between the room-temperature adsorption phases and the low-temperature phase, which is a close-packed hexagonal phase. On Co (0001) both of these phases are hexagonal, and the nearest-neighbor distance between these two phases varies from 4.35 Å in the low-coverage commensurate phase at room temperature to 3.29 Å in the high-coverage, incommensurate, close-packed hexagonal, low-temperature phase.[61] Thus this adsorbate system is again a model system for the study of electronic interactions between neighboring chemisorbed molecules.

Adsorbate bandwidth measurements for various substrates are summarized in Fig. 14. The experimentally obtained CO 4σ bandwidth is plotted in Fig. 14 for several CO/substrate combinations as a function of the nearest-neighbor distance. A simple tight binding calculation predicts this bandwidth to scale exponentially with the distance between the molecules. This behavior is verified by the results shown in Fig. 14. In the semilog plot all the adsorbate substrate combinations fall resonably well onto a straight line. The results for the nonhexagonal phases were scaled in Fig. 14, as indicated by the dashed lines, to correct for the reduced

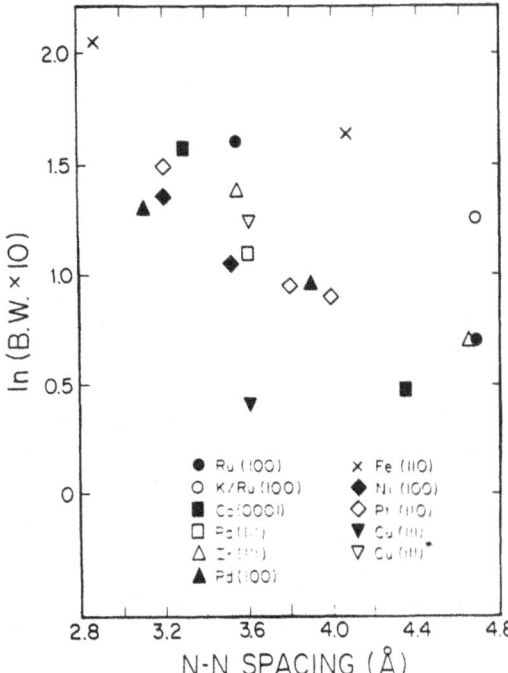

FIGURE 14. Plot of the CO 4σ bandwidth [In (bandwidth times 10)] vs. nearest-neighbor distances for various CO adsorption systems. (From Ref. 63.) The data points are scaled to CO overlayers having hexagonal structure with six nearest neighbors. Corrections have been made for overlayers with a different structure. The CO/Cu (111) value has been adjusted with respect to the position and intensity distribution between shakeup satellite and 4σ "main line" as discussed in Ref. 142.

number of nearest neighbors in these systems. As far as we know, the 5σ bandwidth is about twice as large as the 4σ bandwidth in these systems, and it also scales in a similar way with the nearest-neighbor distance. This indicates that the dispersion is a result of the direct overlap between the adsorbate molecular wavefunctions, and it is not grossly modified by the substrate, contrary to the atomic adsorbates mentioned above.

Whereas the σ-bands are hardly influenced by the substrate, the 1π derived bands are clearly dependent on the coordination of the overlayer with respect to the substrate.[61,63,64] This is illustrated by the studies of the adsorption of CO on Fe (110) by Jensen and Rhodin.[64] Needless to add here, the symmetry classification of 5σ or 1π is relevant only at the center of the surface Brillouin zone $\bar{\Gamma}$. Away from $\bar{\Gamma}$ the even 1π component mixes with the σ-bands, mostly the 5σ. The 4σ can usually be regarded as more pure, because of the larger energy separation from the 1π band.

For most CO adsorbate systems the 5σ and 1π band at $\bar{\Gamma}$ are so close in energy that they are difficult to separate. In order to separate the two components by making use of the polarization selection rules, the measurement has to be taken in the second Brillouin zone. For CO on Fe (110), the separation between the 5σ and 1π bands is unusually large and can be measured directly.[64] The large energy difference between the 4σ and 1π bands can also be taken as an indication of a stretched CO bond in a precursor state to dissociation. Dissociation occurs spontaneously on this surface near room temperature. Rosen et al.[65] have calculated the energy of the CO-derived orbitals and found that the $4\sigma–1\pi$ energy spacing is quite sensitive to the CO bond length. This will be discussed more in the section about alkali-promoted surfaces (3.3.3)

Another interesting result of the study of CO on Fe (110) is that the bands are much closer to isotropic in periodicity than expected from the simplest explanation of the $p(1 \times 2)$ LEED structure. The observed isotropy in the band dispersions has led Jensen and Rhodin[64] to propose a new adsorption geometry for the CO molecules, which is considerably more complicated than the simple $p(1 \times 2)$ structure and would be better described by a (2×2) structure. In order to reconcile the missing spots in the LEED pattern and the new structure, domains are invoked which destroy the long-range order such that it becomes again a (1×2) system.[64] This also implies that the coherence length in LEED and photoemission has to be quite different.

The last example of molecular adsorbates discussed in this section is the observation of the effects of azimuthal orientational ordering on the bandstructure in physisorbed layers of N_2 on graphite and CO on Ag (111) by Schmeisser et $al.$[66] From LEED studies,[67] which were later refined by X-ray scattering,[68] it is known that at low temperatures (20 K), N_2 adsorbs on graphite with its bond axis parallel to the surface. In addition to that, alternating rows of adsorbed molecules form a $2(\sqrt{3} \times 3)$ herringbone structure below a temperature of 27 K. Upon warming above the transition temperature of 27 K, the orientational ordering is lost and the molecules become free rotators centered at the same adsorption sites.[67]

In Fig. 15, the normal emission from N_2 on graphite below and above the phase transition is shown for different photon energies. The emission from the $3\sigma_g$ orbital in N_2 exhibits a splitting, which disappears when the molecules are not rotationally fixed anymore in the herringbone structure. This splitting is due to the formation of a bonding–antibonding orbital pair, when there are two molecules in the unit cell. From a simple tight binding calculation[66] it can be seen that this splitting is most prominent at the center of the Brillouin zone and disappears at the zone boundaries. The features at the Fermi level and at a fixed kinetic energy of 7 eV and 9.5 eV are due to umklapp scattering of graphite

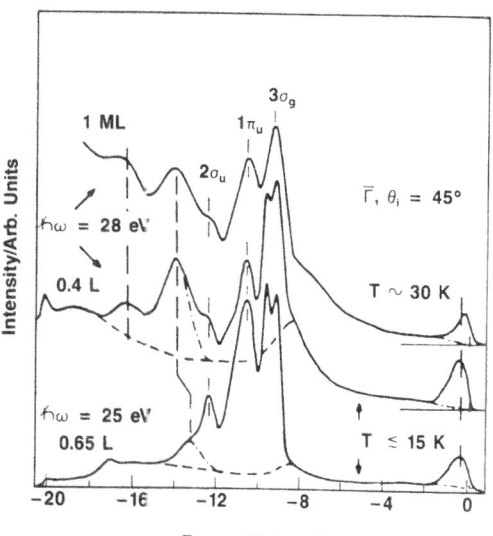

FIGURE 15. Normal emission photoemission from physisorbed N_2 on graphite. The $3\sigma_g$ emission peak exhibits a double structure at Γ in the low-temperature phase due to the lateral interactions in the herringbone configuration of the adsorbate layer. Upon warming, this doublet structure disappears when the overlayer enters the disordered phase. (From Ref. 66.).

substrate emission by the ordered overlayer. Since the molecular centers remain unchanged by the phase transition, these features are present in all spectra.

The center of gravity of the three N_2 derived bands, determined for each one individually by integration over all $\mathbf{k}_\parallel$, shows approximately the same splitting as in the isolated gas-phase N_2 molecule. This is characteristic for a physisorbed molecule and is in contrast to the chemisorbed state of N_2 on Ni (110).[69] In the chemisorbed state the N–N bond axis is perpendicular to the surface, not parallel as in the physisorbed state, and the center of gravity of the $3\sigma_g$ derived band is shifted relative to the other N_2 derived bands, indicating the interaction of the $3\sigma_g$ states with the substrate.

The same behavior is observed for CO on Ag (111).[66] Again in the physisorbed state the relative spacing of the CO derived orbitals is the same as for the gas-phase molecule, and the bond axis is parallel to the surface. On the other hand, in the chemisorbed state, observed on almost all other metal surfaces, the molecular axis is oriented perpendicular to the surface, and the 5σ orbital interacts strongly with the substrate.

3.3. Surface Electronic Structure of Transition Metals

So far we have dealt with adsorbate systems from the point of view of the adsorbed atoms or molecules. Obviously this presents only half the information. The other side of the problem is contained in the electronic structure of the surface of the substrate. In order to really understand chemisorption or catalysis, one has to pay special attention to the surface states of the substrate and their interaction with the adsorbed atoms or molecules. In the early days of photoemission the structures that disappeared out of the photoemission signal of the clean surface upon exposure to an adsorbate were interpreted as surface states. Obviously this is only a very crude test and not entirely true, as demonstrated above. The surface states of the clean surface become interface states when the surface is covered with an adsorbate, and as a consequence they exist, in general, in different pockets of energy and momentum space. Comparing the surface electronic structure of the chemisorption system with the electronic structure of the clean surface prior to exposure to the adsorbate, one may deduce which states of the surface are involved in forming the chemisorption bond and why certain surfaces are more reactive than others. This is a formidable task, but also one of the great achievements of angle-resolved photoemission, and some of these results are highlighted in the following. Starting with a discussion of some interesting results for clean surface, I follow up with some detailed studies of adsorption systems.

3.3.1. Surface States: Reconstructions and Stepped Surfaces

Surface states are a direct consequence of the disruption of the perfect symmetry of the infinite-three dimensional solid. At the surface, the electrons can exist in regions of energy and momentum space forbidden to bulk electrons. These surface states are two-dimensional, they are localized with respect to the extent of their wave functions into the bulk of the solid or into the vacuum, but within the surface region they are periodic and in general delocalized states. Apart from these true surface states, which are not degenerate to any electronic states of the bulk having the same quantum numbers, there are surface resonances observed which best can be described as bulk states having a clearly enhanced wave function amplitude in the near-surface region. These surface resonances are not readily identified experimentally in an unambigous way. The only indication of the existance of a surface resonance comes from the sensitivity of the photoemission signal, which is assigned to a bulk interband transition, to surface conditions such as the presence of an adsorbate.

One inherent question one might ask about surface states is how far they extend into the bulk. As a rule of thumb, there are two classes of surface states, which have different depths and can be characterized by the atomic character of the wave functions. In general, *sp*-like surface states penetrate several layers into the bulk solid. The less the *sp*-like surface state is split off in energy from the bulk bands, the farther it penetrates into the bulk of the crystal. On the other hand, *d*-like surface states are usually concentrated within the first lattice plane.

The different penetration results in a characteristically different behavior of the cross section of the surface states with changes in photon energy. The cross section of the *d*-like surface states commonly is rather strong at low photon energies and drops fairly rapidly and continuosly with increasing photon energy. The *sp*-like surface states, on the contrary, have an oscillating cross section. These oscillations in the cross section of an *sp*-like surface state were first observed by S. Louie et al.[15] These authors found that the surface state on the (111) surface of Cu has a rather strong cross section at about 70 eV photon energy, as shown in Fig. 16. This state also emits rather strongly at low photon energies, but becomes almost invisible in EDCs taken at around 30 eV photon energy (see also Fig. 4). The maxima in the cross section occur when the wavelength of the final state wave function corresponds to the spacing of the layers normal to the surface. Under these conditions all contributions of the surface state that are spread out over several lattice planes add in phase and thus cause a strong signal. The minima are caused, obviously, when these contributions cancel each other at the intermediate electron wavelengths or excitation energies.

Surface states are also intimately related to the phenomenon of surface reconstructions. The best example is the 2×1 reconstruction of the Si (111)

FIGURE 16. Measured intensity of the *sp*-like surface state on Cu (111), showing a resonance at $h\nu = 70$ eV. (From Ref 15.) The calculated cross section is included as a solid curve.

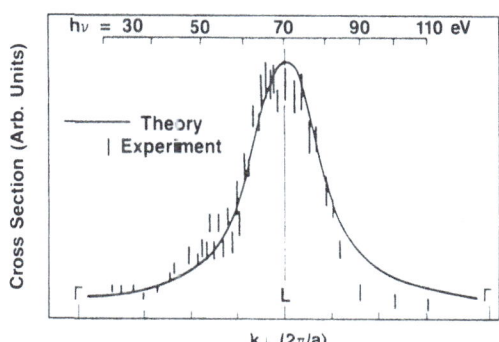

surface. The unreconstructed Si (111) surface should have a half-filled dangling-bond surface-state band, which would make the surface metallic. The (2 × 1) reconstruction actually lowers the total energy of the surface by making it semiconducting. Together with other surface techniques like ion scattering, angle-resolved photoemission studies of the Si (111) (2 × 1) reconstructed surface[70] actually helped to promote the π-bonded chain model[71] over various "buckling" models. Prior to these studies, for almost two decades the Si (111) (2 × 1) surface had been described by the "buckling" models in various stages of refinement. However, the measured electronic structure of this surface could not be reconciled with these models, and this has led to the development of the π-bonded chain model, which is now generally accepted.

With respect to the involvement of surface states in reconstructions on metal surfaces, the situation is not as clear. One example of a surface state supposed to be the driving force for a reconstruction is the W (100) surface. This surface undergoes a completely reversible phase transition from a 1 × 1 surface to a 2 × 2 reconstructed surface on cooling below 370 K.[72,73] Posternak et al.[74] calculated the electronic structure of this surface and proposed a mechanism which involved a surface state as the driving force for the reconstruction. This surface state was calculated to cross the Fermi level at the midpoint between $\bar{\Gamma}$ and $\bar{M}$. On the reconstructed surface, this point would be located at the Brillouin-zero boundary, where a gap would open up in the surface-state band and thus lower the total energy of the surface. In the subsequent experimental studies[75,76] the existence of this surface state was verified approximately as calculated. However upon a closer look it was ruled out as the driving force for the reconstruction, because the Fermi-level crossing occurs at an appreciable distance from the midpoint between $\bar{\Gamma}$ and $\bar{M}$ (see footnote in Ref. 76).

From a materials-science point of view, surfaces with steps or reconstructed surfaces often reveal very interesting chemisorption properties.[77-81] The structure sensitivity of surface reactions is a widely discussed issue in catalysis research. So far, most of the surface-science studies on model systems of the effect of steps on adsorbate bonding and dissociation have been carried out on Ni and Pt surfaces. In general, in these studies the adsorbed species were identified by low-energy electron-loss spectroscopy. Several molecules dissociate in the presence of steps, whereas they adsorb intact on the perfectly cut surface. Ethylene, for example, remains molecularly adsorbed on Ni (111) at 200 K, but dissociates at 150 K on a 5 (111) × (110) Ni surface.[77] Acetylene dehydrogenates completely on this stepped Ni surface at 150 K,[78] whereas it remains intact up to 400 K on Ni (111). Even CO dissociates on a stepped Ni (111) surface,[79] and once the step sites are blocked, only molecular adsorption is observed. Similar effects are observed on a stepped Pt (111) surface.[80] On the Pt (100) surface it is not so much the presence of steps, but rather the reconstruction of the surface itself that is held responsible for special chemisorption properties.[81,82]

Even though the effects of steps or reconstructions on the surface chemistry are known, relatively few studies have been published in which an attempt was made to correlate the electronic structure of a surface with its special chemical behavior. In part, this might be related to the fact that the two communities, catalysis science and surface physics, do not have a well-established communication link. Some of the few existing examples of angle-resolved photoemission experiments in which a connection between the electronic structure of the surface and a special catalytic behavior has been made, deal with composite surfaces such as alkali-modified surfaces and carbon overlayers on Ni, both of which systems are discussed below. Nothing equivalent has been done for pure metallic surfaces. One example of where an angle-resolved photoemission study is needed to shed light on the structure sensitivity of the chemisorption behavior is the Pt (100) surface.[81,82] There even exist angle-integrated photoemission studies which clearly show a strong peak close to E_F on the (1 × 1) surface, which is absent on the reconstructed surface.[83,84]

In angle-resolved photoemission, however, only the equivalent reconstruction on the Au (100) surface has been studied.[85] The (5 × 20) reconstruction of this surface has been described as the formation of an *hcp* overlayer, in which six rows of surface atoms extend over 5 rows of substrate atoms. Thus the topmost layer actually resembles the top layer of the (111) surface. Both of these surfaces exhibit a surface state, marked a and c, respectively, in Fig. 17, which is sensitive to the surface structure. These states are located close to the top of the *d* bands near the boundary of the respective SBZ. Note that the SBZ, as shown in Fig. 17, is different for the two surfaces because of the reconstruction. The surface state on the (1 × 1) surface is analogous to the ones observed on the Ni (100)[86] and Cu (100)[87] surfaces, both of which do not reconstruct. The surface-state band on the (5 × 20) reconstructed surface, peak c, resembles a surface state observed on Au (111),[88,89] which is explained by the presence of the

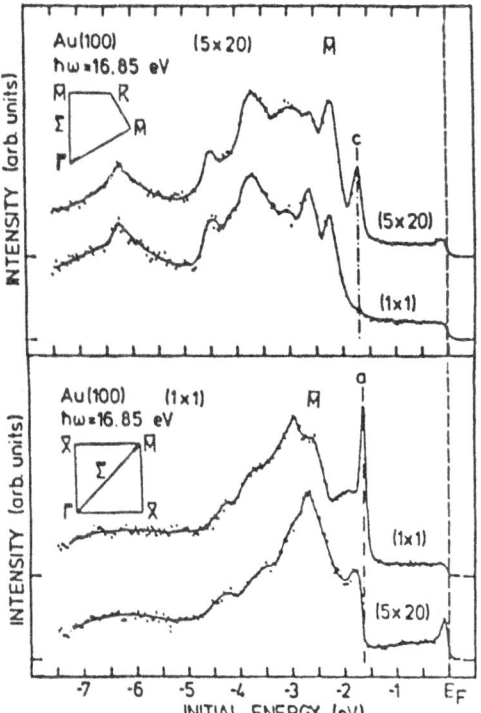

FIGURE 17. Angle-resolved photoemission EDCs from normal (1 × 1) and reconstructed (5 × 20) Au (100). The curves in the top [bottom] panel are taken near the $\bar{M}$ point of the (5 × 20) [(1 × 1)] SBZ. The two different SBZs are shown schematically. Two surface states, labelled a and c, are found in the spectra, which are sensitive to the reconstruction of the surface. (From Ref. 85.).

hcp overlayer on the (5 × 20) reconstructed surface. Whether or not these results can be transferred to the Pt surface is an open question. Even if the Pt (100) surface shows the analogous surface states, then their role in the catalysis needs to be explored in a different set of experiments. Again, as in the case of the W surface, it is also questionable whether the change in these surface states is the driving force for the reconstruction.

Regarding experiments on stepped surfaces, the only results presented here are the two studies on Au (211)[90] and Cu (211).[91] These surfaces actually represent three rows of atoms with an arrangement comparable to the (111) surface followed by a monoatomic step, where the atoms are arranged as on the (100) surface. Therefore these surfaces are also described by the notation [3(111) × (100)]. The general observation made in these two studies is that the bulk photoemission can be explained by the bulk band structure projected onto the direction of the macroscopic surface direction. Essentially, the peaks observed correspond to **k**-conserving transitions along the macroscopic directions in **k**-space. Even the polarization selection rules apply according to the macroscopic picture. Even the surface states found on the Au (211) surface[90] are located in the gaps of the bulk band structure projected onto the macroscopic surface direction. No surface states corresponding to the microscopic atomic arrangement, either of the (111) terraces or of the (100) steps, were found, in contrast to the results for the reconstructed Au surfaces discussed in the previous paragraph. An explanation of this difference might lie in the different extinction lengths of the various surface states.

As a general summary of this section about surface states on clean metal surfaces, it can be said that the surface states of the low-index surfaces are quite well understood. They are located in gaps in the projected band structure, and the excitation cross section can be understood in terms of the extension of these states into the bulk of the material. The situation for reconstructed or stepped surfaces is not as clear. The role of surface states as a driving force for the surface reconstructions and the special chemisorption properties of reconstructed or stepped surfaces constitute a wide open field for future studies.

3.3.2. A Detailed Picture of the Chemisorption Bond

In order to understand the formation of the chemisorption bond, one has to study not only the electronic states of the adsorbate, but also the surface counterpart. Surface states are intimately involved in the formation of the chemical bond with adsorbed molecules. In the early days the disappearance of a surface state in the presence of an adsorbate was taken as a proof for the existence of surface states. A more accurate analysis reveals that the surface states do not disappear, but rather get relocated to a different region in energy–momentum space. This is caused either by a pure electrostatic interaction or by the direct involvement of the surface state in forming the bond with the adsorbate. Even if no chemical bond is formed with the surface, the surface states nevertheless might shift in energy, because of the addition of another layer of atoms to the crystal. This is what I refer to as a "pure electrostatic" interaction.

It would be intriguing to measure the strength of the adsorbate bond or the heat of adsorption by measuring the shifts in all the electronic states involved. Even though these shifts are clearly related to the bond strength, the value of the bond strength cannot be obtained this way. The formulas show that the total energy of the surface differs from the sum of the single-particle energies because of the double counting of the electron–electron interaction terms. Thus, even if the surface does not reconstruct (i.e., the ion–ion interaction does not change), the heat of adsorption cannot be extracted from these measurements. A second reason this concept does not work lies in the fact that photoemission is not a ground-state spectroscopy, even though the results are generally interpreted in this framework. The shortcomings resulting from this simplification will be discussed in more detail at the end of this review.

It is a formidable effort to map the electronic structure for a clean and adsorbate-covered surface, even if this is restricted to the high symmetry lines and points of the SBZ. The first example of this actually being done is the hydrogen-covered Pd (111) surface.[91,92] The interaction of hydrogen with Pd is technologically very important, and hydrogen is also the simplest possible adsorption system from a theoretical point of view. The H_2 molecule dissociates upon contact with the surface, and hydrogen is present in atomic form on the surface.

The normal emission EDCs of the H/Pd (111) surface are shown in Fig. 18.

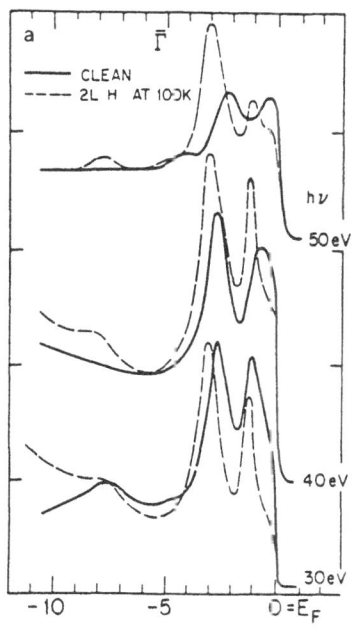

FIGURE 18. (a) Normal emission EDCs from a clean and a hydrogen-covered Pd (111) surface taken at different photon energies as indicated. (From Ref. 92.) (b) Difference curves between the spectra of the adsorbate-covered surface and the clean surface. The surface states of the clean (adsorbate covered) surface show up as negative (positive) peaks in these difference curves, as indicated.

There is a lot of structure in the first 4 eV below E_F, and this structure changes with H adsorption. Additionally we see a hydrogen-induced feature at -7.9 eV, well below the Pd d bands. The interpretation of this latter peak is quite straightforward. It is the H $1s$-derived "band", and it exhibits a 1×1 periodicity with respect to the SBZ, as shown in Fig. 19. The d-band emission and the other surface states can be derived only by taking a whole series of EDCs over a large interval of excitation energies. This results in the picture indicated by the difference curves in Fig. 18. The negative features in the difference curves are surface states of the clean surface, whereas the positive features are the surface states of the hydrogen-covered surface. Performing this type of analysis for different $\mathbf{k}_{\parallel}$ points along the symmetry lines of the SBZ, we come up with the results shown in Fig. 19. The measured surface states[93] are shown as points, and the calculated[94] surface states are shown as lines. Except for the states near the $\bar{M}$ point of the SBZ, the agreement between experiment and theory is quite good. This seeming disagreement between experiment and theory is partially due to the

(a)

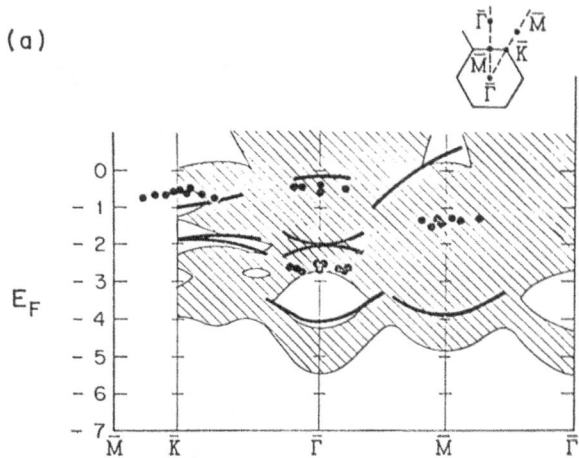

(b)

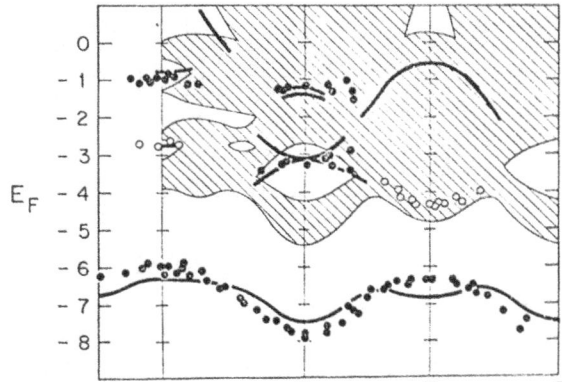

FIGURE 19. Measured surface states, indicated by solid and open circles (weak features), of the clean (a) and hydrogen-covered (b) Pd (111) surface. (From Ref. 93.) The calculated surface states are given by the solid lines. (From Ref. 94.) The projected bulk band structure is shown as the shaded area.

acceptance criteria for a surface state in the slab calculation. The $\mathbf{k}_\parallel$-resolved calculated density of states at the $\bar{M}$ point clearly shows a peak at -4 eV for the H-covered surface comparable to the observation of the experiment (for more details see Ref. 93). Taking these "experimental" errors into account, this study actually demonstrates that a state-of-the-art calculation and experiment produce acceptable results even for a composite surface. Moreover, comparing the experimental results with calculations for various high-symmetry adsorption sites, we can safely rule out an on-top adsorption geometry. Only a threefold hollow position of the adsorbed H-atoms is consistent with these results.

The experiments just discussed were done for saturation coverage, which corresponds to one monolayer. Therefore, it is not surprising that the H-induced split-off band exhibits a (1×1) dispersion. For a coverage less than one monolayer, down to about 0.5 on Ni (111) or 0.3 on Pd (111), the total bandwidth of the split-off state is reduced, but it still exhibits a (1×1) dispersion.[95] At the same time we observe a continuous shift of the intrinsic surface states back toward the position of the clean surface. Both the reduced width of the split-off state and the continuous shift of the intrinsic surface states cause us to rule out the formation of adsorbate islands as the explanation for the (1×1) dispersion of the split-off band. The (1×1) periodicity can be explained only by the large admixture of substrate d-character as suggested by the theoretical analysis of the orbital content in this state.[93,94] Both at the zone center and at the boundary, this state is largely characterized by substrate d states "pulled down" by the presence of the H atoms on the surface. This also explains the large cross section of this state compared to the emission of the substrate sp band, which is barely visible in the spectra.

Another puzzle is the difference in the bandwidth of the H $1s$ band between the Pd (111) and Ni (111) surfaces. The bandwidth of the H-induced band on Ni is about twice as large (4.2 eV) as on Pd (2 eV), a finding which cannot be explained by the 10% decrease in lattice constant between Pd and Ni. Moreover, an isolated monolayer of H atoms, arranged in the correct geometry to match the Pd (111) surface, has a calculated bandwidth of 4 eV.[96] We also note that, in the experiments, the measured H-induced split-off band on Pd deviates strongly from the normal parabolic bandshape and exhibits an almost flat dispersion from the halfway point to the zone boundary. The explanation for both the unusual dispersion and the small bandwidth of H on Pd again lies in the interaction with the Pd d electrons.

The exact nature of the H adsorbate layer for submonolayer coverage is still a mystery. We have reconciled the observation of a (1×1) periodicity in the H-induced band by the strong admixture of substrate d states and ruled out island formation because of the continuous, almost linear shift of the intrinsic surface states with coverage. Whether the H atoms actually form a uniform-density lattice gas in which the average distance varies with concentration, or whether the atoms are adsorbed in fixed threefold hollow positions with a random distribution of vacancies, is as yet unresolved. Also, the H atoms repel each other rather strongly and may move into a subsurface position,[97,98] which is calculated to have a higher binding energy.[99] On the other hand, the subsurface hydrogen is supposed to change the symmetry of the surface state near E_F at the center of the

SBZ.[100] We have experimentally checked the symmetry of this state and did not observe a change in symmetry at low hydrogen coverages.[95]

3.3.3. Composite Surfaces: Modification by Alkali Promotion

Alkali atoms are used in catalysis to promote bond breaking, for example of CO in the Fischer–Tropsch reaction or N_2 in ammonia synthesis. As a model system, the coadsorption of CO and alkali atoms has been studied on single-crystal transition-metal surfaces.[101] Potassium is the most commonly used alkali metal, and the changes in CO chemisorption are quite common for a variety of transition-metal surfaces and almost independent of the substrate. The effects of the alkali promotion can be generalized as:

1. a weakening of the CO bond, as evidenced by a large decrease in the CO stretch frequency combined with an increase in the overtone anharmonicities;
2. an increase in the metal–CO bond strength; and
3. an increase in the dissociation rate.

Generally these effects are explained on the basis of the Anderson model for alkali adsorption and the Blyholder model for CO chemisorption. In this framework the alkali atoms donate their charge to the surface. An increased charge density at the surface leads to a larger backdonation into the 2π orbital of CO, which is antibonding with respect to the CO bond.

This model may be qualitatively correct and appropriate to explain adsorption site–induced changes in the CO chemisorption on unpromoted surfaces. However, it is certainly inadequate to explain the full magnitude of the changes observed in alkali-promoted systems, because it would involve an unrealistically high charge transfer into the 2π orbital. This statement is easily substantiated by experimental data for the CO^- negative ion, which has one extra electron in the 2π orbital. The stretch frequency of CO^-, which exits as an isolated molecule, is lowered by $300 \, cm^{-1}$ compared to the neutral CO molecule. Extrapolating this result linearly, we would need a charge of about two electrons in the 2π orbital in order to explain the $600 \, cm^{-1}$ shift found for CO on K-promoted Ru (0001).[102] It is unrealistic to assume that CO is present on the alkali-promoted surface in the equivalent form of CO^{2-}. Therefore the question is: what is the mechanism that causes this drastic bond weakening without the creation of this unlikely, highly charged species, which does not even exist in isolated form?

CO has been found to exhibit a similar change in stretch frequency under other special conditions, such as in cluster compounds,[103] in deep hollow sites on Fe (111),[104] and on Cr (110).[52] In general the bonding under these circumstances is described as π-bonding in contrast to the 5σ-donation, 2π-backdonation model of Blyholder. π bonding means a more active involvement of the CO π orbitals, 1π and 2π, in the formation of the chemisorption bond. In extreme cases, as in the carbonyl clusters, this is also reflected by the chemisorption geometry, where CO is bonded side-on to the metal. A similar configuration with the molecular axis parallel to the surface has also been proposed for CO on Cr (110)[52] in order to explain the low stretch frequency and the absence of an ESDIAD (electron-stimulated *d*esorption *i*on *a*ngular *d*istribution) signal.

The adsorption geometry can be easily checked by angle-resolved photo-emission, as described above, making use of the selection rules under excitation with polarized light. The results are shown in Fig. 20 for a saturated coverage of CO on Ru (0001), and CO on Ru (0001) covered with a monolayer of K atoms.[105] The results for the unpromoted surface basically show a vanishing 4σ orbital, when the polarization vector is oriented perpendicular to the plane described by the detector and the surface normal ("forbidden geometry"). The small remnant of the 4σ signal can be attributed to the incomplete polarization of the light. This

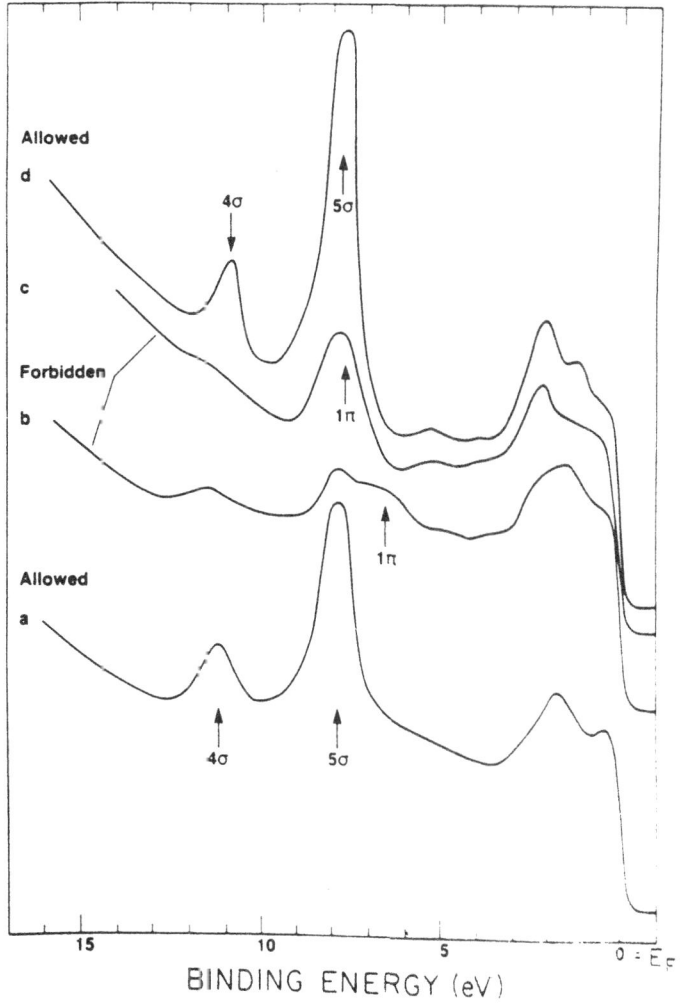

FIGURE 20. Angle-resolved photoemission EDCs of CO chemisorbed on the clean (curves c and d) and alkali-promoted ($\theta_K = 0.33$) Ru (0001) surface (curves a and b). (From Ref. 105.) The light is incident normal to the surface (*s* polarized) and the detector is placed at a polar angle of 20° from the normal into a mirror plane of the crystal. For the curves marked allowed (forbidden) the polarization vector is located in the same mirror plane (perpendicular to the mirror plane) the detector is located in. $h\nu = 31$ eV.

behavior is characteristic of CO adsorbed onto almost any transition-metal surface and is consistent with CO adsorbed with its axis normal to the surface.

The results for the K-promoted surface are distinctly different. Here an appreciable photoemission signal of the 4σ orbital is observed under all polarization conditions. This could be caused by CO not being oriented perpendicular to the surface. However, the more general and correct interpretation is that the symmetry of the adsorbed molecule is broken such that there exists no mirror plane symmetry. Measurements of the angular emission profile of the 4σ shape resonance[31] actually demonstrate that the molecule is still oriented with its axis perpendicular to the surface.

Apart from the break of symmetry, the spectra in Fig. 20 also show that there is a difference in the energy position of the 1π-derived orbital between CO on the bare Ru and on the promoted surface. For the CO/K/Ru system, the 1π orbital is shifted by more than 1 eV closer to E_F. This shift is consistent with the weakening and lengthening of the C–O molecular bond. In this emission geometry this is not a pure 1π state, but rather band formation and mixing are occurring between even π and σ components. Therefore, the 1π classification is not absolutely correct. Examining the cross section of this peak with variation in photon energy, we independently verified that the peak labelled 1π has largely π character.[105]

The important remaining issue with respect to the chemical behavior of this surface is to establish the difference between the electronic structure of the clean and of the alkali-promoted surface. Figure 21 shows a comparison of the normal emission features for clean Ru (0001) and K-promoted Ru (0001). Apart from a shift in the existing surface resonance, a new peak is observed at about 7 eV below E_F. This state has almost no dispersion and exists throughout the SBZ.

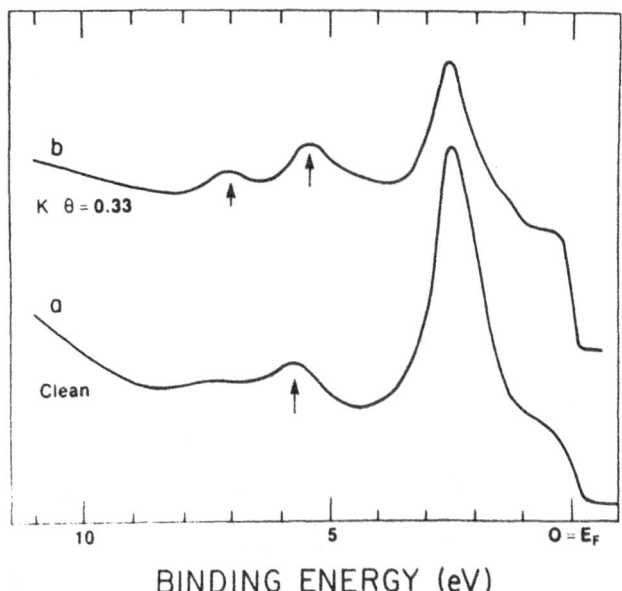

BINDING ENERGY (eV)

FIGURE 21. Normal emission EDCs of clean Ru (0001) and of the alkali-promoted surface. (From Ref. 105.) The surface states and resonances are indicated by arrows. $h\nu = 30$ eV.

Similar alkali metal–induced states below the substrate d bands have also been found for K/Cu (100),[105] K/Fe (110),[106] K/Ni (100),[107] and Na/Ag (110).[108] These features went largely unnoticed in these studies probably because they did not correspond to the generally expected Anderson–Newns picture for alkali adsorption. They are definitely not due to the alkali s electron, but must contain very large admixtures of substrate states. The s electron should contribute to the states located near E_F, but even on Cu (100)[105] no significant change in the intrinsically weak emission near E_F is found prior to the adsorption of CO.

Having reported all these experimental observations, we are still left with the open question of the real cause for the bond weakening of CO on the promoted surface. Obviously the change in the 1π orbital is related to it. The question whether the 1π orbital changes because the bond is changed or the bond changes because of a special interaction involving the 1π orbital is almost academic, but cannot be resolved by these data. What we learn from these spectra, however, is that there is more going on than just an increased 2π backdonation. Altogether, the old picture of an increased 2π backdonation is not wrong, but it just does not describe the situation completely. This is obvious from the change in symmetry of the total system. It is possible that the K-derived surface features are directly involved in an interaction with the 1π orbital especially since these states are close to each other in energy. The possibility of an interaction of this nature has to be investigated by theory.

Consequently, the description of CO as a π-bonded system should be understood in a way that recognizes that both sets of π electrons are involved. This does not necessarily reflect the specific adsorption geometry, but rather the mechanism involved in the adsorbate substrate interaction. The proposed adsorption structure, according to which CO is adsorbed upright but deep inside the threefold hollow sites of the K adlayer, can intuitively be described as side-on bonding with all its implications. Thus, all the experimental observations would be reconciled.

The studies of multicomponent coadsorption systems by angle-resolved photoemission is presently at the forefront of the field. The motivation for these studies is that an improvement in the understanding of these systems gets us a step closer to the real world of catalysis. The study of model systems is a worthwhile undertaking, dictated mostly by the limitation in being able to apply angle-resolved photoemission only to well-ordered and periodic systems. Nevertheless, the ability to make use of the special selection rules in these systems is what makes angle-resolved photoemission still a powerful tool.

Recently a few more two-component systems have been studied by angle-resolved photoemission. These are mentioned here in a general way without going into too much detail. One is the system of Cu on Ru (0001),[109] and the other one is $c(2 \times 2)$ C on Ni (100).[110] The Cu/Ru system exhibits a true interface state that exists at the interface between the Ru substrate and the first Cu adlayer, which is stressed by a 5% lattice mismatch. The most interesting observation is that the interface state is not changed by the addition of more Cu layers. The only effect observed is an attenuation in the photoemission signal, which can be correlated with the increase in escape depth for the electrons originating at the interface.

The last system to be mentioned here is the $c(2 \times 2)$ C overlayer on Ni

(100). The carbidic phase of C on Ni (100) is the steady-state active phase of a Ni (100) single-crystal catalyst surface during a methanation reaction. McConville *et al.*[110] measured the electronic structure of this system and found reasonable agreement with their own calculated results. However, despite the microscopic knowledge of the electronic structure, the question of why this phase is active in catalysis unfortunately remains unresolved.

3.4. Semiconductor Surfaces and Interfaces

Semiconductor surfaces and interfaces are of great technological interest. Surface states are directly involved in the Schottky barrier formation at the metal–semiconductor interface. With respect to the application of angle-resolved photoemission to the measurement of the band structure of these systems, there is no principal difference between semiconductor and metal surfaces. The difference, at least in the past, lay in the difficulty in preparing and characterizing these surfaces reproducibly. Most of the intrinsic semiconductor surfaces are reconstructed in a variety of ways depending on the surface preparation. These surface reconstructions involve displacements of the atoms from their ideal lattice positions up to four layers deep. In contrast, reconstructions on metal surfaces mostly involve the top layer of atoms only. In addition, the reconstructions on the semiconductor surfaces and interfaces often are lower in symmetry than the bulk, and thus lead to the formation of domains. The angle-resolved photoemission from a multidomain surface may give ambiguous results.

The publication of a number of contradictory experimental results in the past (see Refs. 2 and 111) are attributed largely to these difficulties in sample preparation and characterization. Fortunately, the situation has considerably improved since these past reviews.[2,111] Today, there is general agreement about the intrinsic surfaces of Ge and Si to the extent that even detailed models of the surface geometry are confirmed or ruled out by comparing the calculated and measured surface states. Also, angle-resolved photoemission studies of samples grown by MBE and characterized *in situ* have yielded nice results on composite semiconductor surfaces and metal–semiconductor interfaces. Following is a brief summary of the present understanding of the intrinsic semiconductor surfaces. Special attention is given to cases where angle-resolved photoemission has made important contributions to finding the correct structural model for the surface.

3.4.1. The Intrinsic Semiconductor Surfaces

Most of the studies on clean Si surfaces deal with the Si (100) 2 × 1 surface and the Si (111) surface in the 2 × 1 and 7 × 7 reconstructions. For the Si (100) 2 × 1 reconstructed surface, several models were proposed. These include a chain-type reconstruction[112] as well as vacancy and pairing models.[113] However, these calculations predicted the surface to be metallic, whereas the subsequent photoemission studies[114] showed that the surface was semiconducting. Following these experimental studies, Chadi[115] presented an asymmetric dimer model for the Si (100) 2 × 1 surface based on calculations that also minimized the total surface energy. The calculated electronic structure at least was qualitatively in agreement with the photoemission results, even though the surface-state bandwidth was about twice as large as measured. In the following years, the experimental results were expanded to different symmetry directions of the SBZ[116] and the theory was refined.[117] Now there is, in general, good agreement between experiment and theory for the asymmetric dimer model of the surface reconstruction.

The Si (111) 2×1 surface is the natural structure obtained after cleavage of the sample. Again, this is a multidomain surface, and that has led to some misunderstanding and confusion in the interpretation of the early angle-resolved photoemission results. The currently accepted model for the reconstruction of the surface is the π-bonded chain model.[71,118] The earlier "buckling" models were found to be absolutely inconsistent with the measured surface-state dispersion.[70,119,120] In the course of the recent experiments, it was also established that the photoemission structure, which was for a long time interpreted as a "back bond" surface state, is actually due to a bulk direct interband transition.[121] The π-bonded chain models do not incorporate any "back bonds," and this discrepancy had to be resolved in order to have consistency between theory and experiment.

The Si (111) 7×7 reconstruction is considerably more complex than the 2×1 reconstruction. Moreover, the surface has such a large unit cell that it is practically impossible to calculate the electronic structure for the proposed structural models. Therefore, a detailed comparison between photoemission experiments and electronic-structure calculations cannot be made as yet. Other surface probes, such as LEED, face similar problems with calculations. The best structural information on this reconstructed surface comes from such experimental techniques as ion scattering or scanning tunneling microscopy, which do not require complicated calculations. As far as the electronic structure is concerned, there is fairly good agreement about the existence of three surface-state bands. The highest surface state exhibits a Fermi-like edge, such that the surface itself is metallic.[122]

Reports of the existence of a metallic surface on semiconductors seem to come up every so often in the literature. Sometimes the metallic behavior is due to impurities on the surface. On the Ge (100) surface, however, S. D. Kevan[123] found a metal-to-insulator transition to occur with change in temperature. At room temperature there exists a surface state, which exhibits a Fermi edge–like cutoff. If the crystal is cooled to 77 K, the surface state gradually shifts away from E_F, such that at 77 K the surface is semiconducting. At the same time a sharp $c (4 \times 2)$ LEED pattern forms. No discrete phase transition is observed, but rather the changes occur in a continuous fashion. This has led the authors to propose a model for this surface in which the driving force for the reconstruction is the interaction between dangling bonds located on neighboring dimers. The elementary excitation in the disorder transition then corresponds to the flipping of a single surface dimer, and the room temperature surface reconstructs as a "disordered" 4×2 surface rather than 2×1.

3.4.2. Metal Overlayers on Semiconductors

Metal–semiconductor interfaces are of special technological interest. Contact and Schottky-barrier formation determines the electronic properties of semiconductor devices. Many of these interfaces are very complex because of defect formation and interdiffusion. Using MBE techniques, however, various abrupt and well-ordered overlayer systems have also been prepared and characterized. For example, Al, In, and Ga all form stable overlayers on Si (111) at 1/3 monolayer coverage. Monolayers of As form ordered overlayers on Si (111) 1×1, Ge (111) 1×1, and Si (100) 2×1, when the clean surfaces are exposed to As at elevated temperatures, and GaAs (110) and Sb form another stable surface configuration. All these results are reviewed in a recent paper by G. V. Hansson.[124] Here only the study of the 1×1 phase of As on Ge (111)[125] is presented as an example of how the development of *in situ* MBE growth-preparation methods for these ordered overlayers has opened up the field for the application of angle-resolved photoemission.

Ge is tetrahedrally coordinated in the bulk, and the clean Ge (111) surface has dangling bonds and is unstable with respect to reconstruction into a 1×2 or 2×8 geometry. By replacing the outer layer of a Ge crystal with a layer of As atoms, one obtains a system in which all the atoms are in their optimum bonding configuration. All the Ge atoms find themselves in a fourfold coordination, and all As atoms are threefold coordinated, as they are found in bulk Ge or As. Thus the composite surface forms an "ideal" (111) surface. R. D. Bringans *et al.*[125] prepared this surface by evaporating As onto a Ge (111) 2×8 reconstructed surface at a substrate temperature of 400 °C. A sharp 1×1 LEED pattern was observed for the composite surface.

A strong new surface state is observed for the composite surface, as the photoemission spectra in Fig. 22 (top) show. At the bottom of Fig. 22 the measured surface-state dispersion is compared to a calculation.[125] The position of the As atoms was determined by energy minimization in a pseudopotential total-energy calculation.[126] The resulting Ge–As bond length is 2.52 Å, which is fairly close to the normal Ge–Ge (As–As) bond length of 2.45 Å (2.51 Å). These atomic positions were used to calculate the electronic structure in a local density approximation. The calculated band disperses just like a dangling bond band on an "ideal" Ge (111) surface, but is shifted in energy by the substitution of As for Ge. The bandwidth and binding energy for the surface state are underestimated by the calculation, but similar discrepancies between theory and experiment have been observed for the Ge (111) 2×1 surface.[126] Recently the band-structure calculations for As on Ge (111) have been refined by including self-energy corrections.[127] In the new theory the surface-state band is shifted to higher binding energies and the calculated bandwidth has increased, resulting in much better agreement with experiment. In summary, the comparison between experiment and theory confirms the picture of As on Ge (111) as an almost ideal (111) semiconductor surface with respect to its topology and dangling bond surface state.

3.5. Spin Resolved Studies and Magnetic Systems

Detecting not only the energy and momentum but also the spin of the emitted photoelectrons opens up a new dimension. Experimentally this adds the

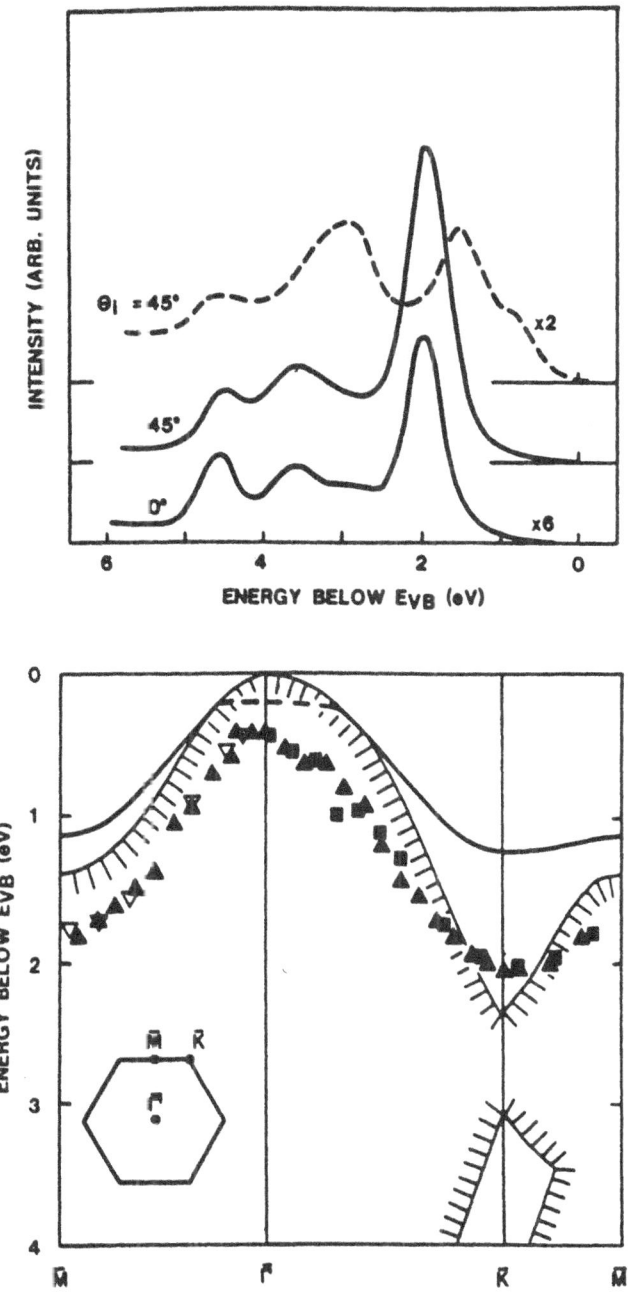

FIGURE 22. Angle-resolved photoemission spectra (top, solid curves) for As on Ge (111) 1 × 1 taken near the $\bar{K}$ point of the SBZ (From Ref. 125.) The As dangling-bond surface state is located at a binding energy of 2 eV below the valence-band maximum. The emission of the clean Ge (111) 2 × 8 surface in the same detector geometry is shown as dashed curve. The bottom panel of the figure shows a comparison between the experimental surface-state dispersion for As on Ge (111) (triangles and squares) and the calculated dispersion (solid line). The shaded regions show the edge of the projected bulk band structure.

complication that spin-resolving detectors have very low efficiency. Conventionally, the asymmetry in the Mott scattering is used to detect the spin alignment of
an electron. Such a detector has an efficiency of only about 10^{-4} to 10^{-5}. Over the
past years, new spin-resolving detectors have been designed and built that have
an efficiency about one to two orders of magnitude higher than the Mott detector
and which are so compact that they can be included in a movable angle-resolving
electron spectrometer.[128] Together with the new undulator sources, this will make
it a lot easier to make such measurements in the future.

One inherent question to be asked about any new experimental technique is,
what type of information can we get from it which is not obtainable otherwise? At
first glance, the answer to this question for spin-resolved photoemission is that
one can study magnetism and the spin–orbit and magnetic exchange interaction.
In general, the splittings in the electronic structure resulting from these
interactions can also be resolved by a conventional detector. However, the
additional information is about the alignment of the spin relative to an externally
applied field or a macroscopic magnetization of the sample. This new information
nevertheless does not come easy. The price one currently has to pay for it is much
lower counting statistics, which for all practical purposes results in a deteriorated
electron energy and angular resolution compared to state-of-the-art conventional
angle-resolved photoelectron spectroscopy.

Historically, the first measurements of the magnetic exchange splitting were
reported by E. Dietz et al.[129] for Ni. These results were confirmed shortly
thereafter by Eastman et al.[130] Different values for the exchange splitting
measured at various points of the Brillouin zone were added later by other
groups.[8,131] completing the picture. This confirmed the theoretical predictions of
A. Liebsch,[132] who had calculated that the exchange splitting is different for
different symmetry (e_g vs. t_{2g}) bands. Another important result of these studies
was that the magnetic exchange splitting was reduced, but did persist above the
Curie temperature. Later these measurements were repeated with simultaneous
spin detection.[133]

The observation of a persistent exchange splitting does not contradict the fact
that the macroscopic magnetization of the sample has vanished. As pointed out
by M. B. Stearns,[134] the spin polarization is measured in the laboratory frame and
obviously has to go to zero above the Curie temperature just as the magnetic
moment does. The measurement of the exchange splitting in photoemission,
however, is a microscopic probe of the magnetic interactions. This splitting is
observed in these measurements as long as the time for the photoemission process
(10^{-15} sec) is shorter than the spin–flip time (typically 10^{-13} sec). This just means
that, on the timescale of photoemission, the system exhibits local magnetic
interactions which fluctuate on a much longer timescale to average out any
resulting magnetic moment.

Using the circularly polarized light emitted above and below the plane of the
synchrotron radiation, spin–polarized photoemission can be applied to nonmagnetic materials. The spin polarization of the photoelectrons is induced by the
excitation using circularly polarized light coupling with specific selection rules to
the various components of the band structure. Thus the relativistic band structure

of Pt has been measured by Eyers *et al.*[135] These bulk measurements are not discussed here in any more detail, since the emphasis of this review is on two- and low-dimensional systems.

Surface magnetism and the question of the existence of magnetically dead layers on surfaces have been discussed for a long time in the literature. The first magnetic surface states were also observed on Ni surfaces.[86,136] As for the bulk magnetic exchange splitting, the polarization of these states was not actually observed, but deduced from a comparison with the calculated bands. How this is done can be illustrated using the Ni (100) surface as example. The (100) surface has a square SBZ with two mirror planes along the [10] and [11] directions. Figure 23 shows the photoemission signal taken at different $k_{\parallel}$ values along these lines for the clean surface and after contamination. A strong emission feature, observed very close to E_F in both crystallographic directions, is sensitive to contamination. The feature observed along the [11] direction exhibits even character under the usual polarization test, whereas the feature observed in the [10] direction has odd symmetry with respect to the mirror plane it is located in. Changing the photon energy and keeping $k_{\parallel}$ fixed by a simultaneous change of the detector angle, we find that both of these states do not show any dispersion with change in momentum normal to the surface, as expected for surface states.

All these observations indicate that these features are surface states. The equivalent surface states, at least for the $\bar{M}$ point ([11] direction), have also been

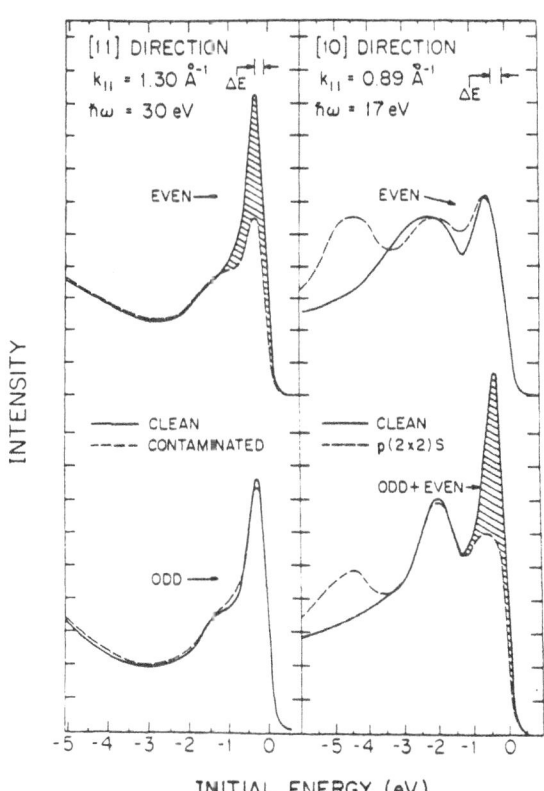

FIGURE 23. Angle-resolved photo-emission of Ni (100). The curves on the left top (bottom) panel are taken at a polar angle of 30° in the [11] azimuth and with s-polarized light within (normal to) the mirror plane the detector is placed in. The curves at the right are taken under the equivalent conditions in the [10] direction of the SBZ. (From Ref. 86.)

observed for the Cu (100)[87] and Au (100)[85] surfaces, as has already been discussed above. The next and final step in the analysis is the comparison with the bulk band structure projected onto the two symmetry direction of the SBZ. These projections are shown in Fig. 24, and the regions of $k_\parallel$-space where the surface-state emission is observed are also indicated. The projected bands are plotted not only according to symmetry, but also according to the spin of the bands. The projected band structure was derived from calculated bands and corrected slightly using the experimental data points available at the time. There is a gap in the projected band structure along these symmetry directions, but only in one spin orientation. Consequently, if these states are true surface states and not surface resonances, then they are spin polarized; i.e., they contain only electrons of majority (minority) spin in the state observed along the [10] ([11]) direction, as shown in Fig. 24. The alternate explanation is that these states are surface resonances, but then the question remains why these states are so narrow. At the time these results initially were published there was no other support available for this observation. Later, however, a calculation by Zhu et al.[137] confirmed the odd majority spin state around the $\bar{M}$ point of the SBZ.

Recently, thin films of Fe and Co grown epitaxially on GaAs were studied with spin-resolved photoemission.[138,139] The most interesting result of these studies was that very thin Fe films showed an anomalous decrease in magnetization. At the same time the easy axis for magnetization also changed. The spin polarization of the scattered secondary electrons was taken as a measure for the magnetization of the sample. The anomalous behavior of the thinnest films has been attributed to compressive strain in the first layer due to the large lattice mismatch.

At the time this review is being written, a substantial fraction of the studies of spin-resolved electronic structure were actually carried out without explicitly detecting the spin of the particle. In the future more of the truly spin-resolved

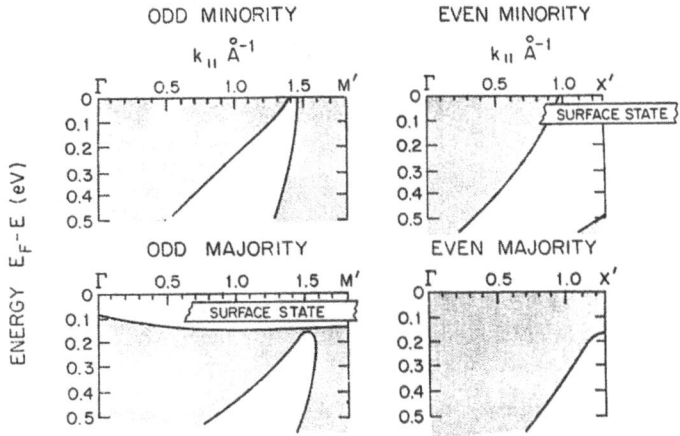

FIGURE 24. Projection of the Ni bulk bands according to spin and symmetry onto the (left) [10] and (right) [11] symmetry axis of the Ni (100) SBZ. (From Ref. 86.) The surface states shown in Fig. 23 are located, as indicated, in the gaps of the projected band structure.

studies will be performed, and surely these studies will lead to the discovery of new and unexpected phenomena which will justify the large amount of effort going into these experiments.

4. COMPLICATIONS AND SPECIAL EFFECTS

4.1. Multielectron Effects

Photoelectron spectroscopy is an excited-state spectroscopy, even though the results are commonly interpreted to reflect a picture of the ground state. The difference can be most easily seen by considering the photoemission from an isolated atom or molecule. Upon removal of one electron from the system, all other electrons relax and become more tightly bound to the nucleus. This relaxation actually makes it easier to remove the photoejected electron from the system. However, strictly speaking, the measured binding energy is not the energy of the photoelectron before it got excited and ejected, but rather the energy it takes to remove this electron. The difference between the two energy values is accounted for by the change in energy of all the other remaining electrons. In the case of electron spectroscopy from solids, this might sound like more semantics in view of the large number of electrons in a solid. It certainly does not make a noticeable difference to the electronic structure of a solid to remove one of about 10^{23} electrons. However, quantum mechanics tells us that the ground state of a system with 10^{23} electrons and the ground state of the same system with $(10^{23} - 1)$ electrons are orthogonal.[40] As a consequence of this orthogonality, the shape of core electron absorption edges in metals is modified due to secondary excitations of electrons near E_F. These many-body effects also influence the binding energy and the line shape of the core electron photoemission in solids. With respect to the valence-band photoemission in solids, these effects are not so obvious and not so widely discussed in the literature. Therefore, in the next section these effects are illustrated in more detail, using as an example the photoemission of CO in the gas phase and adsorbed in a monolayer on a surface.

4.1.1. Binding Energies in the Presence of Shakeup

Several peaks are observed in the photoemission spectrum of CO in Fig. 1 in the binding energy range between 21 eV and 50 eV. According to the single-particle picture, only one level of the CO molecule, the 3σ level, falls within that range, at a binding energy of approximately 35 eV. Theory tells us that not even this peak is due to a single-hole final-state configuration. All structures in the photoemission spectrum in this range correspond to multiple excited states of the CO ion. As the photoexcitation occurs, all the wave functions have to adjust to the change in the electronic interaction. The wave functions of the ion are not orthogonal to the wave functions of the ground states. Accordingly, there is a small but finite probability that, for example, a 1π electron will find itself in the 2π orbital of the ion rather than the 1π as expected. The additional energy remaining in the ion is taken from the photoelectron so that satellite lines appear in the energy-distribution curves corresponding to these discrete excitations in the ions. The energy of the lines observed in the photoemission spectrum thus is given by $E_{kin} = h\nu - E_{ion}$, where $h\nu$ is the photon energy and E_{ion} is the energy of the remaining ionic configuration. This ionic configuration can correspond to a single-hole state (main line) or a two-hole, one-electron configuration (shakeup satellite), where in addition to the ionized electron another electron has been promoted into an excited state.

Theoretically, several sum rules have been derived for these processes.[141] The first is that the total intensity of the excitation has to be shared between all the ionic configurations; i.e., the satellite lines "borrow" intensity from the main line. The second is that the first moment in energy taken over the main line and all shakeup configurations corresponds to the single-particle binding energy (Koopman's theorem). Thus the main line will be shifted to a lower binding energy if there is any shakeup satellite associated with it. This shift increases with the intensity of the satellite line and its separation in energy.

4.1.2. Band-Structure Measurements and Shakeup

Most of these multielectron effects have been studied using core level photoemission from isolated molecules because the shakeup intensity is rather small and core electrons have sharp single-hole-state lines. One classic example where these many-body effects distort the photoemission from a solid is the valence-band photoemission of Ni. Whereas the $3d$ band mapping by angle-resolved photoemission for Cu produces results that are in agreement with the calculated band structure, for Ni the measured $3d$ bands are about 30% narrower than calculated[8] and the observed exchange splitting is about 50% smaller. The difference between the photoemission from Cu and Ni is that there exists a broad shakeup satellite about 6 eV below the Fermi level in the EDCs of Ni. Since the center of gravity for the peaks in the spectral function is maintained, the appearance of this satellite causes an upward shift of all the d-band peaks closer to E_F. This shakeup satellite also affects the measurement of the exchange splitting, because the hole states near E_F, and therefore the hole–hole interaction, is different for different spin states. A. Liebsch[132] has actually calculated the spectral function for the majority and minority electronic states in Ni. Using only one adjustable parameter, the hole–hole interaction U, he is able to reconcile the differences between the experimental and theoretical band structure and exchange-splitting data for Ni.

The question of whether these effects alter or maintain the **k**-resolution of angle-resolved photoemission cannot be answered by the examination of the Ni spectra. It is impossible to determine the intensity of the shakeup satellite accurately enough to extract a single-particle band structure measurement from the spectra. This can be done, however, for the band structure of CO adsorbed on Cu,[142] where a strong shakeup satellite is observed correlated with the 4σ emission. As discussed above, for an adsorbate overlayer we do expect to see dispersion in the adsorbate states with change of parallel momentum. The experimental curves[142] for a saturated layer of CO on Cu (100) are shown in Fig.

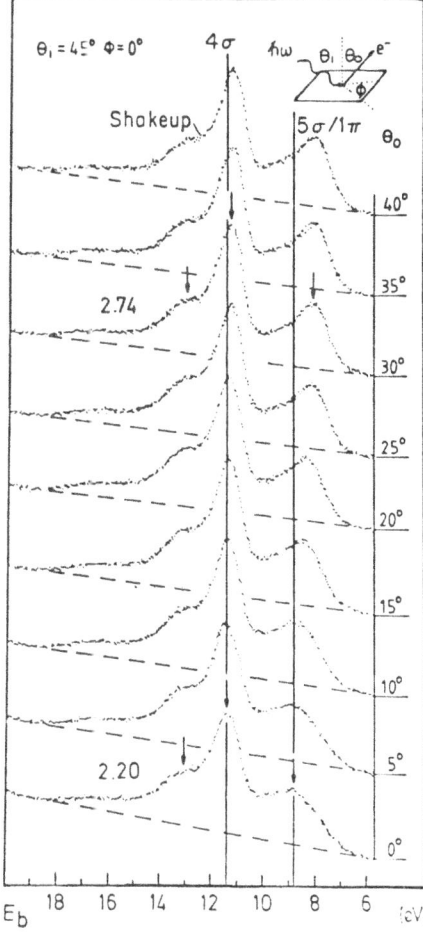

FIGURE 25. Adsorbate-derived peaks in the photo-emission of CO adsorbed at 80 K onto Cu (111). The polar angle the spectra are taken at is marked next to each curve. The curve taken at a polar angle of 30° corresponds to emission from states near the boundary of the SBZ. The ratio of the intensity of the 4σ main line relative to the shakeup satellite is indicated for the center and the boundary of the SBZ. $\hbar\omega = 40$ eV. (From Ref. 142.).

25, and Fig. 26 shows the observed experimental band structure extracted from the data shown in Fig. 25. The 5σ band exhibits a dispersion and bandwidth as expected for the saturated overlayer. The width of the 4σ band is about half the width of the 5σ band in all systems measured so far[63] where no shakeup is present. For CO on Cu (100), the position of both the 4σ main line and of the satellite associated with it does not change with momentum parallel to the surface. However, the relative intensities of the main line and the satellite change in such a way that the center of gravity of both peaks exhibits the expected dispersion as shown in Fig. 26. This means that these many-body effects do not destroy the momentum resolution of the angle-resolved photoemission technique.

The shakeup excitation picture and the band structure picture can be unified by using a **k**-dependent spectral function to describe the excited state spectrum.[143] The only remaining difficulty is to carefully measure the intensities of all states in the spectrum. This gets very complicated if more than one primary excited state is present in the spectrum, as is generally the case in angle-resolved photoemission band structure studies of solids. As discussed for the Ni case, it may be impossible to resolve how much of the satellite intensity has to be attributed to each of the

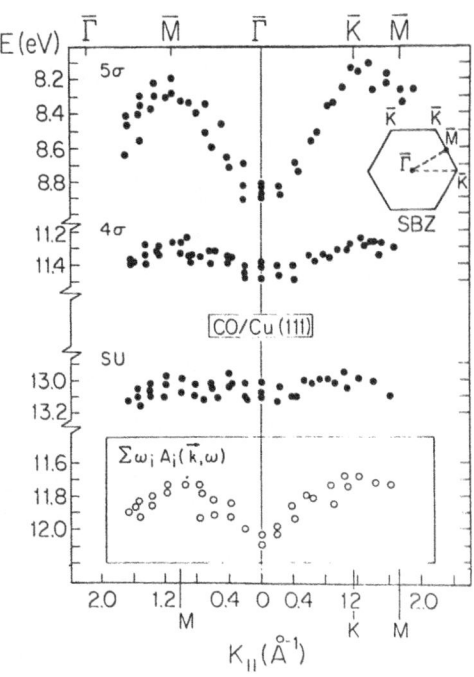

FIGURE 26. Peak position as a function of parallel momentum of the adsorbate-induced features of CO on Cu (111) in two perpendicular directions of the SBZ (solid circles). The open circles show the center of gravity between the 4σ main line and its satellite as a function of parallel momentum. (From Ref. 142.)

main lines present in the spectrum in order to get the correct single-particle values for each of the main lines.

4.2. Photoemission Linewidth and Broadening Mechanisms

The energy and momentum resolution achievable in angle-resolved photoemission is ultimately not determined by the electron spectrometer, but rather directly correlated with the intrinsic width of the observed structures. With today's electron spectrometers, an energy resolution of 50 meV at an angular acceptance of ±1 degree is readily obtainable. It is important to understand the various mechanisms leading to the intrinsic width of the observed structures, because once we understand the width, and especially the lineshape, we are able to extract the peak position more accurately. The broadening effects can be generalized into two categories. One broadening mechanism is associated with the lifetimes of the various particles generated in the photoemission process. The other is caused by the nuclear motion and phonon scattering processes in general. Both of these will be discussed here.

4.2.1. Lifetime Broadening

The energy of a quantum-mechanical state is accurately defined only if the state has infinite lifetime. Otherwise, the energy and the lifetime of the state are correlated via the uncertainty principle, which relates the energetic width Γ (full width half maximum) of a state to its lifetime τ: $\Gamma = \hbar/\tau = 6.6 \times 10^{-16}$ eV sec$/\tau$. In general both the electron and the hole generated in the photoemission process contribute to the total lifetime of the state. The electron is subject to scattering processes, and the hole gets filled via a radiationless Auger decay. Consequently the peak width Γ is given by the Lorentzian convolution of the inverse hole and electron lifetimes (Γ_h, Γ_e):

$$\Gamma = \frac{\Gamma_h + [v_h/v_e]\Gamma_e}{1 - v_h/v_e},$$

where v_h and v_e are the group velocities of the hole and the electron normal to the surface. In off-normal-emission geometries, obviously, the other velocity components also enter into the lifetime.

Two extreme conditions are possible which allow the experimentalist to separately determine electron or hole lifetimes. If $v_h \ll v_e$, i.e., in the case of flat initial (hole) state bands and high-energy (steep) final-state bands as in the d-band emission from transition or noble metals, then the peak width essentially reflects the initial state or hole lifetime. For transitions from the Fermi level, on the other hand, the hole lifetime is infinite, and the width of the transition directly reflects the lifetime broadening in the final state.[144,145]

Any hole state separated in energy from E_F in a metal can decay via an Auger process creating two holes at E_F and an excited electron. Thus the hole lifetime decreases approximately quadratically with separation from the Fermi level as the phase space available for the Auger decay increases. The lifetime of the excited electron is largely determined by electron–electron scattering and also increases with separation from E_F. The experimentally determined[146] electron and hole lifetimes for Al and Zn are shown in Fig. 27.

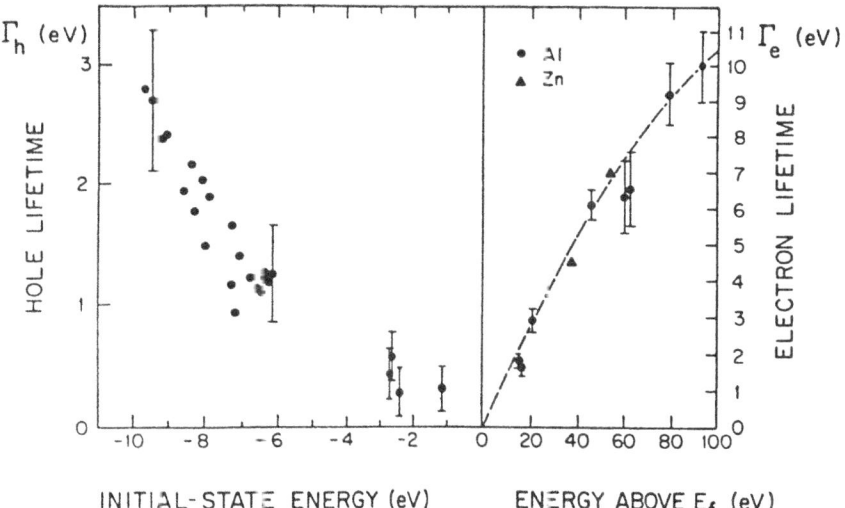

FIGURE 27. Lifetime broadening. Measured electron and hole lifetimes as a function of energy relative to E_F for Al and Zn. (From Ref. 146.)

4.2.2. Electron–Phonon Coupling

At any finite temperature the atoms in a solid move about their equilibrium positions. This results in a broadening of the electronic states from the ideal zero-temperature limit. At elevated temperatures, additionally, the lattice expands, which causes the initial- and final-state wavefunctions and the matrix elements between them to change with temperature. Moreover, phonon-assisted nondirect transitions contribute to the observed EDC. This apparently weakens the momentum selectivity of the angle-resolved photoemission technique. Any measured EDC always contains a mostly featureless background emission which cannot be attributed to direct interband transitions. One example of this is the photoemission from simple metals where some intensity is always observed at the Fermi level. Thus, because of momentum conservation rules, these electrons have to originate either from nondirect phonon-assisted transitions or, alternately, from transitions into evanescent final states.

N. J. Shevchik[147] originally came up with a model in which the photoemission signal was approximated by a direct transition component and a nondirect density-of-states component, both of which were independent of temperature. The temperature dependence was introduced by modulating the relative intensity of these components with a Debye–Waller type exponential function of temperature. In this description the direct transitions become attenuated as the temperature increases and the spectrum approaches more and more a density-of-states distribution.

Several experimental studies were carried out along these lines, and they more or less seemed to match these expectations.[147,148] Recently this model was refined by R. C. White *et al.*,[149] showing that the phonon scattering preferentially involves processes with very small momentum change ΔQ. Thus transitions close to the direct transitions will preferentially contribute to the spectrum, resulting in a more gradual transition to the density-of-states picture. This coincides with an apparent smearing and broadening of the structures observed in the spectra.

Phonon scattering also results in energy loss of the photoelectron. Whereas the phonon spectrum of a solid is almost continuous, adsorbates are known to exhibit discrete vibrational modes in the energy range between 50 meV and 250 meV. These modes are very convenient for identifying specific adsorbate types and sites. Conventionally this is done by high-resolution electron energy-loss spectroscopy or by infrared absorption spectroscopy. These modes should also show up as discrete energy-loss sidebands in the photoemission spectra of adsorbate layers. However, usually the adsorbate peaks are broadened so much by the lifetime broadening due to interatomic Auger decay involving substrate electrons that it is impossible to resolve these sidebands. So far, vibrational substructure in the photoemission from adsorbates has been observed only for O_2 and other small molecules adsorbed onto graphite.[150] The interaction with graphite is very weak, and therefore the lifetime of the adsorbate hole states is longer than on a metal surface. From these data the change in internuclear separation upon the removal of an electron can be determined, as well as the dissocation energy of the molecule on the surface, both of which are different than in the gas phase.

5. SUMMARY AND OUTLOOK

Angle-resolved photoemission is a very powerful technique to completely determine the electronic structure of a two- or three-dimensional system. Synchrotron radiation is essential to make use of the full power of this technique, because the photon energy has to be tuned to observe direct transitions at selected points in momentum space. The polarization of the radiation can be used to apply simple selection rules, which give insight into the symmetry of the system. The outstanding feature of angle-resolved photoemission is that it is the only available experimental technique which yields a direct determination of the band structure of a solid or a surface. The band structure concept of solids had been established for a long time, but prior to the application of angle-resolved photoemission, only indirect proof of the existence of the band structure could be obtained experimentally, where transitions between states were observed. With angle-resolved photoemission it has been possible for the first time to measure independently the initial electronic states.

To a limited extent the results obtained with this technique also give insight into the geometric structure of the system studied. This information includes the total symmetry of the system. The most specific information about the geometry can be obtained from a comparison with calculations, if different geometries result in a measurable change in the electronic structure.

The full power of angle-resolved photoemission as a technique is restricted to periodic systems. This restriction has different consequences for the areas of applied science and technology, which may benefit from these studies. Single crystalline materials are being used in semiconductor devices; therefore, the study of single-crystal semiconductor surfaces and interfaces couples directly into these applications. With respect to materials science, the areas of thin-film magnetism and magnetic materials show great promise for the application of angle-resolved photoemission coupled with simultaneous spin detection. In surface-science and catalysis research, another major area of application for this technique, in general polycrystalline materials are being used technologically. Here the understanding of the relevant catalysis and corrosion processes has to be derived from studies of a carefully selected series of model systems.

In my opinion, at the time this article is being written, with respect to all these applied areas we have just scratched the surface. Angle-resolved photoemission as a technique has not yet been applied broadly to these areas, because in the past the major interest was in the development of the technique itself. Now angle-resolved photoemission has matured as a technique, and because of its unique powers and capabilities some very exciting research lies ahead of us.

ACKNOWLEDGMENT

I want to thank Dr. J. Davenport for his valuable comments and a careful reading of this manuscript.

REFERENCES

1. F. J. Himpsel, *Advances in Physics* **32,** 1–51 (1983).
2. E. W. Plummer and W. Eberhardt, in *Advances in Chemical Physics,* Vol 49, (I. Prigogine and S. A. Rice, eds.) Wiley, New York (1982), pp. 533–656.
3. A. Einstein, *Ann. Phys. (Leipzig)* **17,** 132 (1905).
4. K. Siegbahn, C. Nordling, G. Johansson, J. Hedman, P. F. Hadin, K. Hamrin, U. Gelius, T. Bergmark, L. O. Werme, R. Manne, and Y. Baer, *ESCA Applied to Free Molecules,* North Holland, Amsterdam (1969).
5. P. J. Feibelman, *Phys. Rev. Lett.* **34,** 1092 (1975); *Phys. Rev.* **B12,** 1319 (1975).
6. K. L. Kliewer, *Phys. Rev.* **B12,** 1319 (1975); **B14,** 1412 (1976).
7. H. J. Levinson, E. W. Plummer, and P. J. Feibelman, *Phys. Rev. Lett.* **43,** 952 (1979).
8. W. Eberhardt and E. W. Plummer, *Phys. Rev.* **b21,** 3245 (1980).
9. S. D. Kevan, N. G. Stoffel, and N. V. Smith, *Phys. Rev.* **B31,** 3348 (1985).
10. P. O. Gartland and B. J. Slagsvold, *Phys. Rev.* **B12,** 4047 (1975).
11. J. A. Knapp, F. J. Himpsel, and D. E. Eastman, *Phys. Rev.* **B19,** 4952 (1979).
12. J. F. Janak, A. R. Williams, and V. L. Moruzzi, *Phys. Rev.* **B11,** 1522 (1975).
13. W. Eberhardt and E. W. Plummer, *Phys. Rev.* **B28,** 3605 (1983).
14. J. Paul, S. A. Lindgren, and L. Walden, *Solid State Commun.* **40,** 395 (1981).
15. S. G. Louie, P. Thiry, R. Pinchaux, Y. Petroff, D. Chandesris, and J. Lecante, *Phys. Rev. Lett.* **44,** 549 (1980).
16. J. Cooper and R. N. Zare, *J. Chem. Phys.* **48,** 942 (1968).
17. R. J. Smith, J. Anderson, and G. J. Lapeyre, *Phys. Rev. Lett.* **37,** 1081 (1976).
18. J. Hermanson, *Solid State Commun.* **22,** 9 (1977).
19. W. Eberhardt and F. J. Himpsel, *Phys. Rev.* **B21,** 5572 (1980).
20. R. Benbow, *Phys. Rev.* **B22,** 3775 (1980).
21. J. Dehmer and D. Dill, *Phys. Rev. Lett.* **35,** 213 (1975).
22. J. Davenport, *Phys. Rev. Lett.* **36,** 945 (1976).
23. E. W. Plummer, T. Gustafsson, W. Gudat, and D. E. Eastman, *Phys. Rev.* **A15,** 2339 (1977).
24. C. L. Allyn, T. Gustafsson, E. W. Plummer, *Chem. Phys. Lett.* **47,** 127 (1977).
25. F. Greuter, D. Heskett, E. W. Plummer, and H. J. Freund, *Phys. Rev.* **B27,** 7117 (1983).
26. A. P. Hitchcock and C. E. Brion *J. Electron. Spectros. Relat. Phenom.* **10,** 3 (1983).
27. J. Dehmer, D. Dill, and A. C. Parr in *Photophysics and Photochemistry in the Vacuum Ultraviolet,* (S. P. McGlynn, G. Findley, and R. Haebner., eds.) D. Reidel Dordrecht (1985), pp. 341–408.
28. C. R. Natoli in *EXAFS and Near Edge Structure,* Springer Series in Chemical Physics, Vol. 27, A. Bianconi, L. Incoccia, and S. Stipcich, eds., Springer, Berlin (1983), p. 43.
29. A. P. Hitchcock, S. Beaulieu, T. Seel, J. Stöhr, and F. Sette, *J. Chem. Phys.* **80,** 3927 (1984).
30. J. Stöhr, J. L. Gland, W. Eberhardt, D. Outka, R. J. Madix, F. Sette, R. J. Koestner and U. Döbler, *Phys. Rev. Lett.* **51,** 2414 (1983).
31. W. Eberhardt, F. M. Hoffmann, R. DePaola, D. Heskett, E. W. Plummer, and H. R. Moser, *Chem. Phys. Lett.* **124,** 237 (1986).
32. A. Flodström, R. Nyholm, and B. Johansson, Chapter 5 of this volume.
33. S. D. Kevan, D. H. Rosenblatt, D. Denley, B. C. Lu, and D. A. Shirley, *Phys. Rev. Lett.* **41,** 1565 (1978).
34. D. P. Woodruff, D. Norman, B. W. Holland, N. V. Smith, H. H. Farrell, and M. M. Traum, *Phys. Rev. Lett.* **41,** 1130 (1978).
35. C. S. Fadley, *Prog. Surf. Sci.* **16,** 275 (1984) and Chapter 9 in this volume.
36. N. V. Smith, M. M. Traum, and F. J. DiSalvo, *Solid State Commun.* **15,** 211 (1974); N. V. Smith, M. M. Traum, J. A. Knapp, J. Anderson, and G. J. Lapeyre, *Phys. Rev.* **B13,** 4462 (1976).
37. R. Z. Bachrach, N. Skibowski, and F. C. Brown, *Phys. Rev. Lett.* **37,** 40 (1976).
38. A. Liebsch, *Solid State Commun.* **19,** 1193 (1976).
39. N. G. Stoffel, F. Levy, C. M. Bertoni, and G. Margaritondo, *Solid State Commun.* **41,** 53 (1982).

40. O. Anderson, G. Karschnick, R. Manzke, and M. Skibowski, *Solid State Commun.* **53**, 339 (1985).
41. N. G. Stoffel, S. Kevan, and N. V. Smith, *Phys. Rev.* **B31**, 8049 (1985).
42. O. Anderson, R. Manzke, and M. Skibowski, *Phys. Rev. Lett.* **55**, 2188 (1985).
43. A. Zunger and A. J. Freeman, *Phys. Rev.* **B17**, 1839 (1978).
44. N. Suzuki, A. Yamamoto, and K. Motizuki, *Solid State Commun.* **49**, 1039 (1984).
45. D. K. G. deBoer, C. F. van Bruggen, G. W. Bus, R. Coehorn, C. Haas, G. Sawatzky, H. W. Myron, D. Norman, and H. Padmore, *Phys. Rev.* **B29**, 6797 (1984).
46. J. A. Wilson, *Solid State Commun* **22**, 551 (1976).
47. K. Motizuki, N. Suzuki, and Y. Takoaka, *Solid State Commun.* **46**, 995 (1981).
48. P. M. Williams, *Nuovo Cimento* **38B**, 216 (1977).
49. J. E. Fischer and T. E. Thompson, *Phys. Today* **31**, 36 (1978).
50. W. Eberhardt, I. T. McGovern, E. W. Plummer, and J. E. Fischer, *Phys. Rev. Lett.* **44**, 200 (1980).
51. P. Hofmann, S. R. Bare, N. V. Richardson, and D. A. King, *Solid State Commun.* **42**, 645 (1982).
52. N. D. Shinn, and T. E. Madey, *J. Vac. Sci. Technol.* **A3**, 1673 (1985).
53. C. W. Seabury, T. N. Rhodin, R. J. Purtell, and R. P. Merrill. *Surf. Sci.* **93**, 117 (1980).
54. W. M. Kang, C. H. Li, S. Y. Tong, C. W. Seabury, K. Jacobi, T. N. Rhodin, R. J. Purtell, and R. P. Merrill, *Phys. Rev. Lett.* **47**, 931 (1981).
55. F. P. Netzer, and T. E. Madey, *Phys. Rev. Lett.* **47**, 928 (1981).
56. M. R. Albert, L. G. Sneddon, W. Eberhardt, F. Greuter, T. Gustafsson, and E. W. Plummer, *Surf. Sci.* **120**, 19 (1982).
57. M. Scheffler, K. Horn, A. M. Bradshaw, and K. Kambe, *Surf. Sci.* **80**, 69 (1978).
58. P. K. Larsen, N. V. Smith, M. Schlüter, H. Farrel, K. Ho, and M. Cohen, *Phys. Rev.* **B17**, 2612 (1978).
59. W. Eberhardt, and F. J. Himpsel, *Phys. Rev. Lett.* **42**, 1375 (1979).
60. I. P. Batra, and S. Ciraci, *Phys. Rev. Lett.* **39**, 774 (1977).
61. F. Greuter, D. Heskett, E. W. Plummer, and H. J. Freund, *Phys. Rev.* **B27**, 7117 (1983).
62. J. C. Tracy, and P. W. Palmberg, *J. Chem. Phys.* **51**, 4852 (1969).
63. D. Heskett, E. W. Plummer, R. DePaola, W. Eberhardt, F. M. Hoffmann, and H. R. Moser, *Surf. Sci.* **164**, 490 (1985).
64. E. S. Jensen, and T. N. Rhodin, *Phys. Rev.* **B27**, 3338 (1983).
65. A. Rosen, P. Grundevik, and T. Morovic, *Surf. Sci.* **95**, 477 (1980).
66. D. Schmeisser, F. Greuter, E. W. Plummer, and H. J. Freund, *Phys. Rev. Lett.* **54**, 2095 (1985).
67. R. D. Diehl, M. F. Toney, and S. C. Fain, *Phys. Rev. Lett.* **48**, 177 (1982); R. D. Diehl and S. C. Fain, *Surf. Sci.* **125**, 116 (1983).
68. P. A. Heiney, P. W. Stephens, S. G. J. Mochrie, J. Akimitsu, and R. J. Birgeneau, *Surf. Sci.* **125**, 539 (1983).
69. K. Horn, J. DiNardo, W. Eberhardt, H. J. Freund, and E. W. Plummer, *Surf. Sci.* **118**, 465 (1982).
70. R. I. G. Uhrberg, G. V. Hansson, J. M. Nicholls, and S. A. Flodström, *Phys. Rev. Lett.* **48**, 1032 (1982) and references therein.
71. K. C. Pandey, *Phys. Rev. Lett.* **47**, 1913 (1981).
72. T. E. Felter, R. A. Barker, and P. J. Estrup, *Phys. Rev. Lett.* **38**, 1138 (1977).
73. M. K. Debe and D. A. King, *J. Phys.* (Paris) **C10**, L303 (1977).
74. M. Posternak, H. Krakauer. A. J. Freeman, and D. D. Koelling, *Phys. Rev.* **B21**, 5601 (1980).
75. J. C. Campuzano, D. A. King, C. Somerton, and J. E. Inglesfield, *Phys. Rev. Lett.* **45**, 1649 (1980).
76. M. I. Holmes, and T. Gustafsson, *Phys. Rev. Lett.* **47**, 443 (1981).
77. S. Lehwald, and H. Ibach, *Surf. Sci.* **89**, 425 (1979).
78. S. Lehwald, W. Erley, H. Ibach, and H. Wagner, *Chem. Phys. Lett.* **42**, 360 (1979).
79. W. Erley, H. Ibach, S. Lehwald, and H. Wagner, *Surf. Sci.* **83**, 585 (1979).
80. D. W. Blakeley, and G. A. Somorjai, *J. Catal.* **42**, 181 (1979).
81. H. P. Bonzel, G. Broden, and G. Pirug, *J. Catal.* **53**, 96 (1978).
82. C. R. Helms, H. P. Bonzel, and S. Kelemen, *J. Chem. Phys.* **65**, 1773 (1976).

83. H. P. Bonzel, C. R. Helms, and S. Kelemen, *Phys. Rev. Lett.* **35,** 1237 (1975).

84. G. Thornton, R. F. Davis, K. A. Mills, and D. A. Shirley, *Solid State Commun.* **34,** 87 (1980).

85. P. Heimann, J. Hermanson, H. Miosga, and H. Neddermeyer, *Phys. Rev. Lett.* **43,** 1757 (1979).

86. E. W. Plummer and W. Eberhardt, *Phys. Rev.* **B20,** 1444 (1979).

87. P. Heimann, J. Hermanson, H. Miosga, and H. Neddermeyer, *Phys. Rev. Lett.* **42,** 1782 (1979).

88. P. Heimann and H. Neddermeyer, *J. Phys.* (Paris) **F7,** L37 (1977).

89. P. Heimann, H. Neddermeyer, and H. F. Roloff, *J. Phys.* (Paris) **C10,** L17 (1977).

90. P. Heimann, H. Miosga, and H. Neddermeyer, *Phys. Rev. Lett.* **42,** 801 (1979).

91. R. F. Davis, R. S. Williams, S. D. Kevan, P. S. Wehner, and D. A. Shirley, *Phys. Rev.* **B31,** 1997 (1985).

92. Eberhardt, F. Greuter, and E. W. Plummer, *Phys. Rev. Lett.* **46,** 1085 (1981).

93. W. Eberhardt, S. G. Louie, and E. W. Plummer, *Phys. Rev.* **B28,** 465 (1983).

94. S. G. Louie, *Phys. Rev. Lett.* **42,** 476 (1979).

95. F. Greuter, I. Strathy, E. W. Plummer, and W. Eberhardt, *Phys. Rev.* **B33,** 736 (1986).

96. S. Chubb and J. Davenport, *Phys. Rev.* **B31,** 3278 (1985).

97. R. J. Behm, V. Penka, M. G. Cattania. K. Christmann, and G. Ertl, *J. Chem. Phys.* **78,** 7486 (1983).

98. T. E. Felter and R. H. Stulen, *J. Vac. Sci. Technol.* **A3,** 1566 (1985).

99. M. S. Daw and M. I. Baskes, *Phys. Rev.* **B29,** 6443 (1984).

100. C. T. Chan and S. G. Louie, *Solid State Commun.* **48,** 417 (1983).

101. H. P. Bonzel, *J. Vac. Sci. Technol.* **A2,** 866 (1984).

102. F. M. Hoffmann and R. A. dePaola, *Phys. Rev. Lett.* **52,** 1697 (1984).

103. W. A. Herrmann, H. Biersack, M. L. Ziegler, K. Weidenhammer, R. Siegel, and D. Rehder, *J. Am. Chem. Soc.* **103,** 1692 (1981).

104. U. Seip, M. C. Tsi, K. Christmann, J. Küppers, and G. Ertl, *Surf. Sci.* **139,** 29 (1984).

105. W. Eberhardt, F. M. Hoffmann, R. A. dePaola, D. Heskett, I. Strathy, E. W. Plummer, and H. R. Moser, *Phys. Rev. Lett.* **54,** 1856 (1985).

106. M. P. Kiskinova, G. Pirug, and H. P. Bonzel, *Surf. Sci.* **133,** 321 (1983).

107. Y. M. Sun, H. S. Luftman, and J. M. White, *Surf. Sci.* **139,** 379 (1984); *Surf. Sci.* **141,** 82 (1984).

108. D. Briggs, R. A. Marbrow, and R. M. Lambert, *Surf. Sci.* **65,** 314 (1977).

109. J. E. Houston, C. H. F. Peden, P. J. Feibelman, and D. R. Hamann, *Phys. Rev. Lett.* **56,** 375 (1986).

110. C. F. McConville, D. P. Woodruff, S. D. Kevan, M. Weinert, and J. W. Davenport, *Phys. Rev.* **B34,** 2199 (1986).

111. D. E. Eastman, *J. Vac. Sci. Technol.* **17,** 492 (1980).

112. G. P. Kerker, S. G. Louie, and M. L. Cohen, *Phys. Rev.* **B17,** 7066 (1978).

113. J. A. Appelbaum, G. A. Baraff, and D. R. Hamann, *Phys. Rev.* **B14,** 588 (1976); **B15,** 2408 (1977).

114. F. J. Himpsel and D. E. Eastman, *J. Vac. Sci. Technol.* **16,** 1297 (1979).

115. D. J. Chadi, *Phys. Rev. Lett.* **43,** 43 (1979).

116. R. I. G. Uhrberg, G. V. Hansson, J. M. Nicholls, and S. A. Flodström, *Phys. Rev.* **B24,** 4684 (1981).

117. A. Mazur and J. Pollmann, *Phys. Rev.* **B26,** 7086 (1982).

118. J. E. Northrup and M. L. Cohen, *Phys. Rev. Lett.* **49,** 1349 (1982).

119. G. V. Hansson, R. Z. Bachrach, R. S. Bauer, D. J. Chadi, and W. Göpel, *Surf. Sci.* **99,** 13 (1980).

120. F. J. Himpsel, P. Heimann, and D. E. Eastman, *Phys. Rev.* **B24,** 2003 (1981).

121. R. I. G. Uhrberg, G. V. Hansson, U. O. Karlsson, J. M. Micholls, P. E. S. Persson, S. A. Flodström, R. Engelhardt, and E. E. Koch, *Phys. Rev. Lett.* **52,** 2265 (1984).

122. D. E. Eastman, F. J. Himpsel, J. A. Knapp, and K. C. Pandey in *Physics of Semiconductors* (B. L. H. Wilson, ed.) Inst. of Physics, Bristol and London (1978), p. 1059.

123. S. D. Kevan, *Phys. Rev.* **B32,** 2344 (1985).

124. G. V. Hansson, *Phys. Scr.* **t17,** 70 (1987) and references therein.

125. R. D. Bringans, R. I. G. Uhrberg, R. Z. Bachrach, and J. E. Northrup, *Phys. Rev. Lett.* **55,** 533 (1985).

126. J. E. Northrup and M. L. Cohen, *Phys. Rev.* **B27,** 6553 (1983).

127. M. S. Hybertsen and S. G. Louie, *Phys. Rev. Lett.* **55**, 1418 (1985).
128. J. Unguris, D. T. Pierce, and R. J. Celotta, *Rev. Sci. Instrum.* **57**, 1314 (1986).
129. E. Dietz, U. Gerhardt, and C. J. Maetz, *Phys. Rev. Lett.* **40**, 892 (1978).
130. D. E. Eastman, F. J. Himpsel, and J. A. Knapp, *Phys. Rev. Lett.* **40**, 1514 (1978).
131. P. Heimann, F. J. Himpsel, and D. E. Eastman, *Solid State Commun.* **39**, 219 (1981).
132. A. Liebsch, *Phys. Rev. Lett.* **43**, 1431 (1979).
133. H. Hopster, R. Raue, G. Güntherodt, E. Kisker, R. Clauberg, and M. Campagna, *Phys. Rev. Lett.* **51**, 829 (1983).
134. M. B. Stearns, *J. Appl. Phys.* **57**, 3030 (1985).
135. A. Eyers, F. Schäfers, G. Schönhense, U. Heinzmann, H. P. Oepen, K. Hünlich, J. Kirschner, and G. Borstel, *Phys. Rev. Lett.* **52**, 1559 (1984).
136. W. Eberhardt, E. W. Plummer, K. Horn, and J. Erskine, *Phys. Rev. Lett.* **45**, 273 (1980).
137. X. Zhu, J. Hermanson, F. J. Arlinghaus, J. G. Gay, R. Richter, and J. R. Smith, *Phys. Rev.* **B29**, 4426 (1984).
138. K. Schröder, G. A. Prinz, K. H. Walker, and E. Kisker, *J. Appl. Phys.* **57**, 3669 (1985).
139. G. A. Prinz, E. Kisker, K. B. Hathaway, K. Schröder, and K. H. Walker, *J. Appl. Phys.* **57**, 3024 (1985).
140. P. W. Anderson, *Phys. Rev. Lett.* **18**, 1049 (1967).
141. R. E. Watson and M. L. Perlman in *Structure and Bonding* Vol. 24, (J. P. Dunitz, P. Hemmerich, R. H. Holm, J. A. Ibers, C. K. Jørgensen, J. B. Neilands, D. Reinen, and R. J. P. Williams eds.) Springer Berlin (1975), pp. 83–132 and references therein.
142. H. J. Freund, W. Eberhardt, D. Heskett, and E. W. Plummer, *Phys. Rev. Lett.* **50**, 768 (1983).
143. B. L. Lundquist, *Phys. Kondens. Mater.* **6**, 193 (1967); **7**, 117 (1968); **9**, 2236 (1969).
144. D. E. Eastman, J. A. Knapp, and F. J. Himpsel, *Phys. Rev. Lett.* **41**, 825 (1978).
145. F. J. Himpsel and W. Eberhardt, *Solid State Commun.* **31**, 747 (1979).
146. H. J. Levinson, F. Greuter, and E. W. Plummer, *Phys. Rev.* **B27**, 727 (1983).
147. N. J. Shevchik, *J. Phys. (Paris)* **C10**, L555 (1977) and *Phys. Rev.* **B16**, 3428 (1977) and *Phys. Rev.* **B20**, 3020 (1979).
148. Z. Hussain, C. S. Fadley, S. Kono, and L. F. Wagner, *Phys. Rev.* **B22**, 3750 (1980).
149. R. C. White, C. S. Fadley, M. Sagurton, P. Roubin, D. Chandesris, J. Lecante, C. Guillot, and Z. Hussain, *Phys. Rev.* **B35**, 1147 (1987).
150. W. Eberhardt and E. W. Plummer, *Phys. Rev. Lett.* **47**, 1476 (1981).

5

Surface Core Level Spectroscopy

Anders Flodström, Ralf Nyholm, and Börje Johansson

1. INTRODUCTION

After the discovery that core level binding energies are sensitive to the charge distribution of the valence electrons, core level photoelectron spectroscopy has become one of the main analytical techniques in physics and chemistry. For example, an increase in oxidation number is accompanied by stepwise changes in the core level binding energy. The core level chemical shifts observed for atoms in different charge states usually range from 0.5 to 5 eV. Under the acronym ESCA (*electron spectroscopy for chemical analysis*), Siegbahn and co-workers developed the technique with extensive studies of chemical shifts in solid compounds,[1] and later in molecules.[2] This work was performed with fixed photon energy sources, MgKα and AlKα primarily. The main objective driving the field was to develop excitation sources and kinetic-energy electron analyzers with improved resolution to detect smaller chemical shifts with increased precision.

The concept of surface-sensitive core level spectroscopy for solids was still dormant. Early theoretical predictions of the inelastic scattering lengths of the photoexcited electrons in solids[3] were first appreciated in ultraviolet valence-band photoemission studies.[4-6] The calculated scattering lengths were found to be in qualitative agreement with experimental determinations. It became, for example, clear that photoemission yielding electrons with kinetic energies in the range of 10 eV to 1000 eV is a most surface-sensitive technique with probe depths of a few atomic layers. The implementation of synchrotron radiation as a new excitation source opened up the possibility of investigating some shallow core levels of important metals and semiconductors with extreme surface sensitivity (see Fig. 1). It was still argued, however, that the initial (chemisorption) oxidation states of the surface-layer atoms could perhaps not be followed separately. The reason for this skepticism was the idea that all the atoms of the solid served as a charge

Anders Flodström • Department of Materials Science, Royal Institute of Technology, S-100 44 Stockholm, Sweden *Ralf Nyholm* • MAX-Laboratory, University of Lund, S-221 00 Lund, Sweden *Börje Johansson* • Condensed Matter Theory Group, Department of Physics, University of Uppsala, S-751 21 Uppsala, Sweden.

Synchrotron Radiation Research: Advances in Surface and Interface Science, Volume 1: Techniques, edited by Robert Z. Bachrach. Plenum Press, New York, 1992.

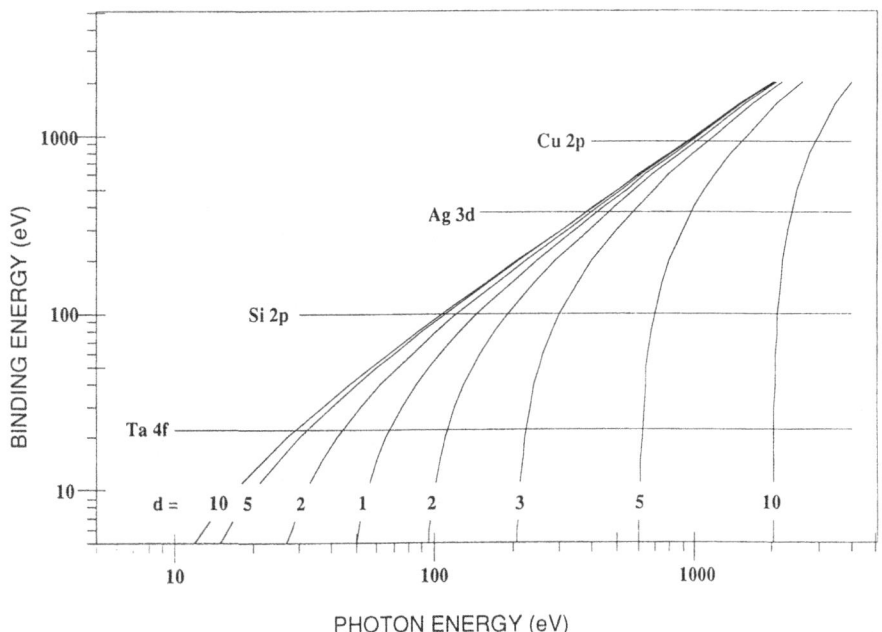

FIGURE 1. Escape depth of the photoelectrons given in number of atomic layers (d) as function of excitation energy and core level binding energy. A few typical core levels are displayed explicitly.

reservoir and that the resulting chemical shift of the surface-layer atoms should be minute. This view is, of course, appealing for metals, but should not be valid, e.g., for covalent semiconductor surfaces. In reality it turned out that the idea of a surface with its own characteristic electronic structure is applicable to both metals and semiconductors. But to discover the surface core level shifts, both high-energy resolution and extreme surface sensistivity are simultaneously needed, since the surface core level shifts are only of the order of a few tenths of an electron volt. The first surface core level shift was reported by Flodström et al.[7] for the Al (111) 1 × 1 O chemisorption system, where a surface core level shift of 1.4 eV toward higher binding energy in the Al $2p$ core level was detected (see Fig. 2). The more fundamental surface atom binding energy shift for a clean surface, due to the reduced coordination number of the surface atoms as compared to the bulk atoms, was first observed in an ordinary XPS (X-ray photoelectron spectroscopy) experiment. By detecting emitted photoelectrons at glancing angles, Citrin et al.[8] observed the surface core level shift of the Au surface atoms. Shortly afterward, Tran Minh Duc et al.[9] beautifully demonstrated the concept of a clearly separated surface core level peak in a synchrotron radiation–excited photoemission experiment on the clean W (100) surface (see Fig. 3). These experiments started a new field of surface physics and chemistry. During the last years many studies of surface core level shifts for clean metals and semiconductors, of chemisorption systems, of alloys, and of metal-on-semiconductor systems have appeared.[10–14] The field has matured, and surface core level spectroscopy is nowadays a widely used tool, almost as common as ordinary XPS.

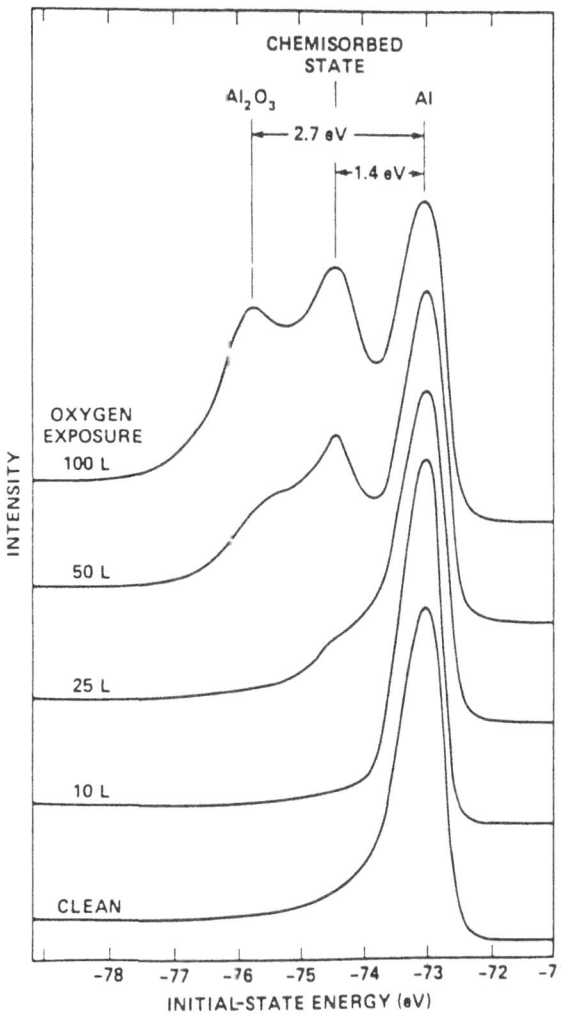

FIGURE 2. Al 2p core evel spectra from an oxygen-exposed Al surface recorded at 130 eV photon energy. Observe the well-resolved +1.4 eV surface peak of the chemisorbed oxygen state. (From Ref. 7.)

As in ordinary XPS experiments, the observed chemical shift is a combination of initial- and final-state effects, i.e., the measured core level binding energy is the difference in total energy of the system with and without a core hole. Thus we cannot argue that a measured core level binding energy directly compares with a ground state calculation for the initial state.[15–17] In a series of papers Johansson, Mårtensson, and Rosengren[10,13,16–19] have shown how this ambiguity in comparing theoretical and measured core level binding energies can be avoided. The first treatment of a surface core level shift in terms of a total energy difference was made by Johansson,[18] and later Johansson, Mårtensson, and Rosengren[17,19] extended this treatment by using a thermodynamic Born–Haber cycle. In this way both initial and final-state effects are included, and the measured surface core level shifts are related to the differences in surface energies of the element investigated and the next element in the periodic table. Similar approaches have been developed for chemisorption systems[11] and alloys.[20–24] The

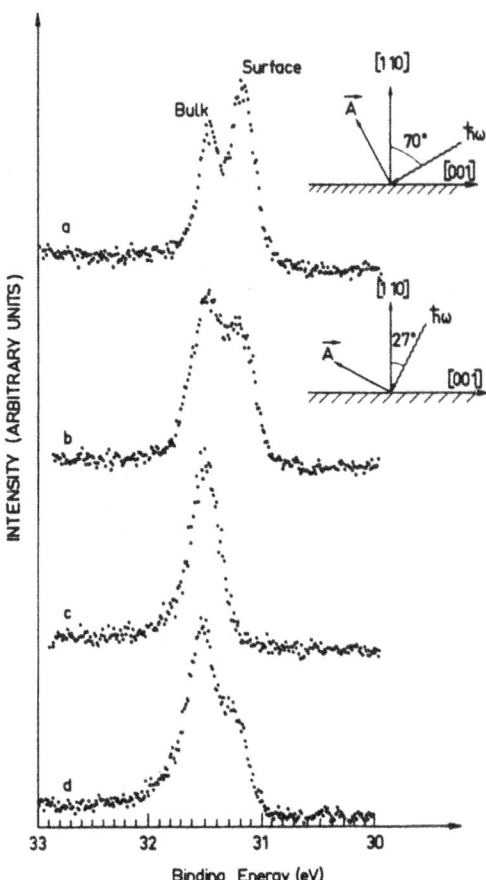

FIGURE 3. W 4f core level spectra from clean and gas-exposed W surfaces recorded at 70 eV photon energy. Clearly shown is the separate −0.28 eV surface peak of the clean metal surface. (From Ref. 9, reproduced with permission.)

thermodynamical model and its extensions have often been applied in the development of surface core level spectroscopy.

Because of the very high experimental resolution in surface core level spectroscopy experiments and the need to separate bulk and surface core-level peaks that are close together in binding energy, new demands have been put on curve fitting of the core level line shapes. For the first time in core level spectroscopy the full machinery of lifetime broadening, core level peak asymmetry, phonon-induced broadening, and accurate control of the experimental resolution has to be involved. Nowadays one routinely uses computer codes developed for this purpose, and relatively little attention is paid to the physics of the various broadening mechanisms.

The number of usable core levels for surface core level spectroscopy is limited by their intrinsic widths and by the experimental access to photons with highly accurate excitation energies to yield photoelectrons having kinetic energies of 25–100 eV. Interesting cases such as the C, N, and O 1s core levels in important chemisorption systems, and the 2p and 3d core levels of the 3d and 4d transition elements and their associated alloys have binding energies that have not yet become readily accessible for surface core level spectroscopy due to limitations of the experimental techniques. Still, surface core level spectroscopy is a mature surface-analysis tool with many applications in surface physics and chemistry.

2. THEORETICAL MODEL FOR THE SURFACE CORE LEVEL SHIFTS

2.1. Metallic Systems

In the following theoretical treatment of the binding energy of a core electron in a metallic system, we follow closely the account given by Johansson and Mårtensson.[10] The most essential part of the theory is the assumption of a completely screened final state.[15–18,25–28] This means that in the final state the conduction electrons are assumed to have fully relaxed in the presence of a core hole on the core-ionized atom. The core-electron binding energy is then the difference in total energy between the final state and the initial, unperturbed state. Thus we define the core level binding energy for an elemental metal as the total energy difference

$$E_c^M(X) = E_{N-1}^M(X) - E_N^M. \tag{2.1}$$

M is here used to denote that the system is a metal, and X is the core level from which an electron has been removed and brought to infinity. N is the total number of electrons in the metal. For the corresponding free atom, which will be denoted by A, the atomic core level binding energy is written in an analogous way:

$$E_c^A(X) = E_{n-1}^A(X) - E_n^A, \tag{2.2}$$

where n is the number of electrons in the atom. The binding-energy shift of the core level between the free atom and the metal is of particular interest. The shift, δE_c, can be expressed as

$$\delta E_c = E_c^A(X) - E_c^M(X) = E_{n-1}^A(X) - E_{N-1}^M(X) - (E_n^A - E_N^M). \tag{2.3}$$

The total number of atoms in the metal is N/n. Therefore we can write the total energy of the initial metal state as

$$E_N^M = \frac{N}{n}(E_n^A - E_{\text{coh}}), \tag{2.4}$$

where the positive quantity E_{coh} is the cohesive energy of the metal. The shift δE_c can now be expressed as

$$\begin{aligned} \delta E_c &= E_{n-1}^A(X) - E_{N-1}^M(X) + ((N/n) - 1)(E_n^A - E_{\text{coh}}) - E_{\text{coh}} \\ &= E_{n-1}^A(X) - E_{N-1}^M(X) + E_{N-n}^M - E_{\text{coh}}. \end{aligned} \tag{2.5}$$

Here E_{N-n}^M is the total energy of a metal with $(N/n) - 1$ atoms, i.e., one fewer than for E_N^M.

In Eq. (2.5) we have isolated one contribution to the core level binding energy shift, namely E_{coh}. This term describes exactly the change of the initial-state energy between the free atom and the metal.

It is now useful to decompose the term $E_{N-1}^M(X)$ into two parts: first the term $E_N^M(X)$, where a core electron corresponding to X has been brought to the Fermi level, and second the term ϕ, where an electron from the Fermi level has been brought to the vacuum level. This last term is the work function. Also, the atomic term $E_{n-1}^A(X)$ can be decomposed into two terms: $E_n^A(X)$, where the core electron has been brought from the level X into the lowest unoccupied valence orbital of the atom (in the presence of the core hole), and $I^*(X)$, where this extra valence electron has been ionized away. Equation (2.5) can thereby be written as

$$\delta E_c = E_n^A(X) + I^*(X) - E_N^M(X) - \phi + E_{N-n}^M - E_{\text{coh}}. \qquad (2.6)$$

The total energy $E_N^M(X)$ corresponds to a metallic system with $(N/n) - 1$ atoms without a core hole and one atom where an electron has been excited from the core level X to the Fermi level. Thus $E_N^M(X)$ refers to an electrically neutral system.

The total energy $E_N^M(X)$ can be conveniently decomposed into three terms; one describing the total energy of a system with $(N/n) - 1$ (nonexcited) metal atoms, E_{N-n}^M; one describing the total energy for a condensed metal atom with a core hole but with an extra valence electron, $E_n^M(X)$; and finally a term describing the interaction between this core hole metal atom and the surrounding metal, $E_{Z^*}^{\text{imp}}(Z)$, where we use a notation which will be convenient later. Thus the term $E_N^M(X)$ can be written as

$$E_N^M(X) = E_{N-n}^M + E_n^M(X) + E_{Z^*}^{\text{imp}}(Z). \qquad (2.7)$$

Equations (2.6) and (2.7) give

$$\delta E_c = I^*(X) - \phi + E_n^A(X) - E_n^M(X) - E_{Z^*}^{\text{imp}}(Z) - E_{\text{coh}}. \qquad (2.8)$$

However, the difference $E_n^A(X) - E_n^M(X)$ defines a generalized cohesive energy, $E_{\text{coh}}(X)$, since it is the energy difference between a core-hole atom and a core-hole metal. Thus Eq. (2.8) takes the form

$$\delta E_c = I^*(X) - \phi + E_{\text{coh}}(X) - E_{\text{coh}} - E_{Z^*}^{\text{imp}}(Z). \qquad (2.9)$$

Defining a new core level binding-energy shift, ΔE_c, as

$$\Delta E_c = \delta E_c + \phi = E_c^A(X) - (E_c^M(X) - \phi), \qquad (2.10)$$

we obtain a more practical expression for the shift, since experimentally the metallic core level binding energies are normally determined relative to the Fermi energy. This is exactly the meaning of $(E_c^M(X) - \phi)$.

The energy-shift expression in Eq. (2.10) can be written as

$$\Delta E_c = I^*(X) + E_{coh}(X) - E_{coh} - E_{Z^*}^{imp}(Z), \qquad (2.11)$$

which is the form derived by Johansson and Mårtensson.[17] The charge neutralization (screening) of the final state is to a large extent accounted for by the $I^*(X)$ term. This expresses, however, only an atomic screening, while the contribution to the shift $(E_{coh}(X) - E_{Z^*}^{imp}(Z))$ describes how the energy of this screening charge is modified within the metallic host.

From the given analysis, the metallic core level binding energy relative to the Fermi energy, $E_{c,F}^M$, can be expressed as

$$E_{c,F}^M = E_c^A + E_{coh}^Z - E_{coh}^{Z^*} - I_1^{Z^*} - E_{Z^*}^{imp}(Z). \qquad (2.12)$$

Here the notation has been slightly changed. Z denotes the atomic number of the metal in consideration, and Z^* stands for the core-ionized Z-atom. $I_1^{Z^*}$ is the first ionization potential for a neutralized Z atom with a core hole. As before, $E_{Z^*}^{imp}(Z)$ is the heat of solution for a substitutional metallic Z^*-atom impurity in the Z metal, with no relaxation of the atomic positions of the host-metal atoms. The reason for this is the vertical nature of the photoionization process.

The obtained decomposition of the core level binding energy, expressed by Eq. (2.12), refers to an atom in the bulk. The same analysis can be repeated for a surface atom, and the core level binding energy for a surface atom takes the form

$$E_{c,F}^{surf} = E_c^A + E_{coh,surf}^E - E_{coh,surf}^{Z^*} - I_1^{Z^*} + E_{Z^*}^{imp,surf}(Z). \qquad (2.13)$$

Here the surface cohesive energy, $E_{coh,surf}$, enters both for the original Z metal and for the hypothetical Z^* core-hole metal. Similarly, there is also a surface impurity solution energy, $E_{Z^*}^{imp,surf}(Z)$.

It is now of considerable interest to consider the difference Δ_c in core level binding energy between a surface and a bulk atom, what is called the surface core level shift:

$$\Delta_c = E_{c,F}^{surf} - E_{c,F}^M = (E_{coh}^{Z^*} - E_{coh,surf}^{Z^*}) - (E_{coh}^Z - E_{coh,surf}^Z)$$
$$- (E_{Z^*}^{imp}(Z) - E_{Z^*}^{imp,surf}(Z)). \qquad (2.14)$$

Since the difference $E_{coh} - E_{coh,surf}$ is the surface energy E_S, the surface core level shift may be expressed as

$$\Delta_c = E_S^{Z^*} - E_S^Z - (E_{Z^*}^{imp}(Z) - E^{imp,surf}Z^*(Z)). \qquad (2.15)$$

This expression for Δ_c shows that there is a direct connection between the surface core level shift and the surface energies, but with a modification from the difference in heat of solution for a bulk and a surface substitutional impurity.

In Fig. 4 we illustrate a most direct way to obtain the surface core level shift, based on the definition of the core level binding energy as an energy difference between the final and initial states. Thus $E_{c,F}^M$ is the energy difference between a Z^* impurity atom in the bulk of the Z metal host and the original nonexcited Z

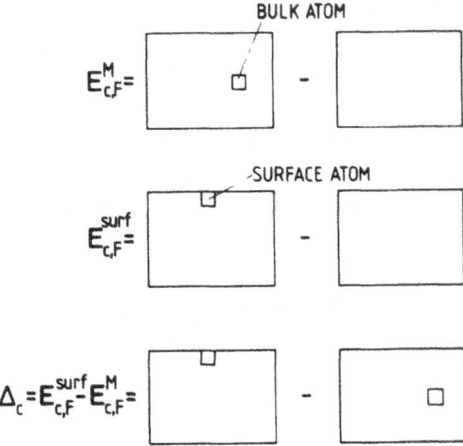

FIGURE 4. Schematic picture of the final and initial state at the surface and in the bulk of the core level photoemission process. Observe that the surface shift Δ_c directly yields the surface segregation energy of a $Z + 1$ atom in a Z metal host.

metal. $E_{c,F}^{surf}$ is the corresponding difference for the case of a core-hole atom at the surface. From Fig. 4 we notice immediately that the surface shift is the energy difference between a Z^*-atom impurity in a substitutional surface position and in a substitutional bulk position. This means that the surface core level shift Δ_c is nothing but the surface segregation energy for a Z^* impurity as shown by Johansson and Mårtensson[17] and further developed by Rosengren and Johansson.[29] With the $(Z + 1)$ equivalent core approximation, which will be introduced below, the surface core level shift gives important information about the segregation potential for a $(Z + 1)$ substitutional impurity in a Z metal host.

In the actual calculation of the core level binding-energy shift, it is most useful to use the equivalent core approximation, namely the replacement of the core-ionized atom Z^* by a $(Z + 1)$ atom. This is most accurate since the screening charge involves only the valence electrons, and they are very similar for the Z^* and $(Z + 1)$ atoms. By means of this approximation, the core level binding-energy shift between the free atom and the condensed phase has been calculated for all the elemental metals,[17] and a very good agreement with experiment was obtained.

Introducing the equivalent core approximation, the surface core level shift takes the simple form

$$\Delta_c = E_S^{Z+1} - E_S^Z - (E_{Z+1}^{imp}(Z) - E_{Z+1}^{imp,surf}(Z)). \tag{2.16}$$

In the $Z + 1$ approximation, the surface core level shift can be interpreted as the heat of surface segregation of a $(Z + 1)$ substitutional impurity in the Z metal. Since many technologically important alloy systems are of the type $Z_x(Z + 1)_{1-x}$, the surface core level shift provides useful data.

From empirical studies it has been found that the surface energy can be approximately related to the cohesive energy by [30]

$$E_S \cong 0.2 E_{coh}. \tag{2.17}$$

A similar relation should also be approximately valid for the impurity term[17]

$$E_{Z+1}^{\text{imp,surf}}(Z) \cong 0.8 E_{Z+1}^{\text{imp}}(Z). \tag{2.18}$$

With these approximations the surface core level shift can be written

$$\Delta_c \cong 0.2[E_{\text{coh}}^{Z+1} - E_{\text{coh}}^{Z} - E_{Z+1}^{\text{imp}}(Z)]. \tag{2.19}$$

To the first order it seems possible to neglect the impurity term. Therefore the surface shift can be approximated by the simple expression

$$\Delta_c \cong 0.2[E_{\text{coh}}^{Z+1} - E_{\text{coh}}^{Z}]. \tag{2.20}$$

This shows clearly that the surface core level shift depends on both initial- and final-state effects, since both indices, Z and $(Z + 1)$, enter into the expression for the shift.

Since the surface energy is different for different crystallographic surfaces, the expression in Eq. (2.16) shows that the surface core level shift will depend on the specific crystal character of the surface. Theoretical calculations[19] give in general a rather good agreement with experiments on single crystals and demonstrate that surface core level shifts can be used as an experimental tool for structural surface studies.

2.2. The 4f Levels in Rare-Earth Systems

With the assumption that the $4f$ electrons behave essentially as core electrons for the lanthanide metals,[5,16] Eq. (2.12) can be used for the $4f$ binding energy relative to the Fermi level:

$$E_{4f,F}^{M} = E_{4f}^{A} + E_{\text{coh}}^{Z} - E_{\text{coh}}^{Z^*} - I_1^{Z^*} + E_{Z^*}^{\text{imp}}(Z). \tag{2.21}$$

If the system is a divalent metal such as Eu ($4f^7$) it is most interesting to notice that Z^*, which then corresponds to $4f^6$, actually is nothing but trivalent Eu. Thus the quantity $E_{\text{coh}}^{Z^*}$ refers to the cohesive energy of a hypothetical trivalent Eu metal. We introduce the quantity $\Delta E_{\text{II,III}}$ which is the energy needed to create a trivalent Eu metal from the divalent metallic ground state. Then it is obvious that the first four terms in Eq. (2.21) constitute just this energy. Therefore the $4f$ binding energy can be expressed as[28]

$$E_{4f,F}^{M} = \Delta E_{\text{II,III}} + E_{\text{III}}^{\text{imp}}(\text{II}), \tag{2.22}$$

where $E_{\text{III}}^{\text{imp}}(\text{II})$ is the heat of solution of a trivalent impurity in the divalent host metal. If the impurity term could be neglected, the $4f$ binding energy would therefore be a direct measure of the energy difference between the trivalent and divalent metallic states.

In exactly the same way, the $4f$ binding energy for a trivalent rare-earth metal can be expressed as

$$E_{4f,F}^{M} = \Delta E_{\text{III,IV}} + E_{\text{IV}}^{\text{imp}}(\text{III}), \tag{2.23}$$

where $\Delta E_{\text{III,IV}}$ Is the energy difference between the tetravalent and trivalent metallic states and $E_{\text{IV}}^{\text{imp}}(\text{III})$ is the heat of solution of a tetravalent impurity in the trivalent host metal.

It should be remarked that with this formulation the treatment does not actually rely on the $(Z + 1)$ approximation for the screening state, but refers to the next-higher valence state of the element under investigation. This valence state might actually be realized in certain alloys and compounds or at very high pressures. The energy differences $\Delta E_{\text{II,III}}$ and $\Delta E_{\text{III,IV}}$ can be very accurately derived in a totally independent way by means of a thermodynamical treatment.[15,16,31] The remaining contribution in Eqs. (2.22) and (2.23), the impurity terms $E_{\text{III}}^{\text{imp}}(\text{II})$ and $E_{\text{IV}}^{\text{imp}}(\text{III})$, can be obtained from the Miedema scheme.[32] A comparison between the theoretical and the experimental binding energies for the lanthanide metals shows a very good agreement.[16] Thus it can be claimed that the $4f$ binding energies, as determined by photoelectron spectroscopy, are now well understood for the lanthanides.

In the same manner as above, the surface shift of the $4f$ binding energy, $\Delta_{c,4f}^{\text{II}}$ for a divalent metal can be written as[18]

$$\Delta_{c,4f}^{\text{II}} = E_{S}^{\text{III}} - E_{S}^{\text{II}} - [E_{\text{III}}^{\text{imp}}(\text{II}) - E_{\text{III}}^{\text{imp,surf}}(\text{II})]. \tag{2.24}$$

E_{S}^{III} (E_{S}^{II}) is the surface energy for the trivalent (divalent) metallic state of the lanthanide element. The impurity terms have a corresponding meaning to those in Eq. (2.16).

With the same type of approximations as before, we can write the following simple expression for the $4f$ surface shift for a divalent lanthanide metal:

$$\Delta_{c,4f}^{\text{II}} \cong 0.2(E_{\text{coh}}^{\text{III}} - E_{\text{coh}}^{\text{II}}), \tag{2.25}$$

and similarily the surface shift for a trivalent lanthanide metal becomes,

$$\Delta_{c,4f}^{\text{III}} \cong 0.2(E_{\text{coh}}^{\text{IV}} - E_{\text{coh}}^{\text{III}}). \tag{2.26}$$

Since in general $E_{\text{coh}}^{\text{II}} \cong 40\,\text{kcal/mol}$, $E_{\text{coh}}^{\text{III}} \cong 100\,\text{kcal/mol}$, and $E_{\text{coh}}^{\text{IV}} \cong 145\,\text{kcal/mol}$ (1 eV/atom = 23 kcal/mol), we obtain, to the first order of approximation, the result that a divalent rare-earth metal should have a surface $4f$ binding energy shift of about 0.5 eV (toward higher binding energies), and a trivalent metal a shift of about 0.4 eV. This then means that the $4f$ electrons for the lanthanides always will be more bound at the surface than in the bulk (provided the surface and bulk atoms have the same valence). A divalent $4f^{n+1}$ atom is therefore more stable relative to a trivalent $4f^{n}$ configuration at a surface position than for a bulk position.[18]

3. CURVE FITTING

The purpose of performing a curve fitting to the surface-sensitive experimental core level spectra is to deduce accurate numbers for the surface core level shifts. This information can then be used to improve the understanding of important physical and chemical properties of surfaces. The curve fitting is far from trivial and requires an understanding of the various contributions to the core level peak line shape arising from many fundamental physical processes. A complete description of the core level photoemission process is presently not available. Therefore we have to rely on some *ad hoc* assumptions in fitting the experimental spectra. As an example, we show in Fig. 5 a survey spectrum for Mg recorded at 130 eV photon energy.[33] We recognize some typical features in the photoelectron spectrum: the sharp and strong Mg $2p$ core level peak, with its associated surface and bulk plasmon losses, the *LVV* Auger structure; and the Mg $2s$ core level peak. This last feature has a low intensity compared to the Mg $2p$ peak because of its low cross section for photon energies relatively close to the photoionization threshold. All these features rest on a structureless secondary-electron background. The objective is to curve-fit the energy region near the core level peak of interest. It is therefore not necessary to characterize the plasmon losses in terms of intrinsic and extrinsic contributions and their varying intensities as a function of photon energy. Furthermore, if we limit our choice of exciting photon energies, we can hopefully avoid interference with the Auger structures, since they are constant kinetic-energy features. But, even so, a peak fitting of the Mg $2p$ core level peak in a narrow, say 5 eV, energy region centered at the Mg $2p$ peak poses problems.

The increase in background intensity toward lower kinetic energies in Fig. 5 is due to primary photoelectrons suffering energy losses from inelastic scattering events during transport through the solid. It is still an unsolved problem how to correctly distinguish the spectral contribution from these inelastically scattered

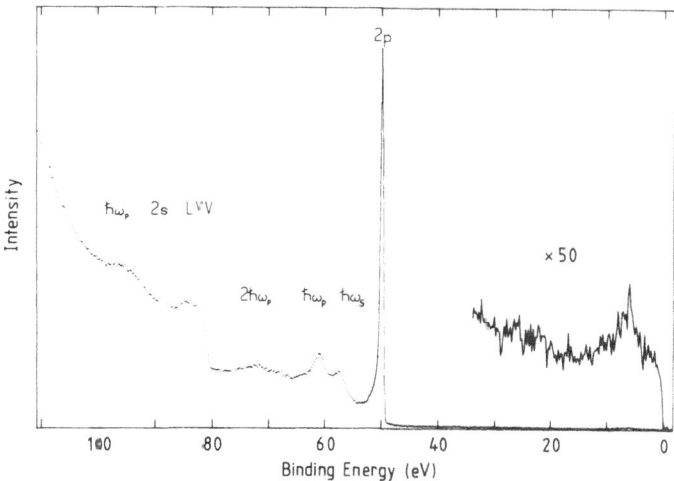

FIGURE 5. Survey photoelectron spectrum of Mg recorded at 130 eV photon energy. A typical energy range to be peak fitted is about 5 eV centered around the Mg $2p$ peak. (From Ref. 33.)

electrons. Several models have been employed in curve-fitting procedures. Perhaps the most frequently used background model is the so called Shirley background,[34] where the inelastic intensity at a certain point is considered to be proportional to the integrated intensity at higher kinetic energies. Also, calculated background distributions based on measured electron energy-loss curves have been proposed.[35] Theoretical models applicable in the XPS region (high kinetic energies) have been given (see, e.g., Refs. 36–39), but their use requires knowledge of the cross section for inelastic electron scattering.

It should be noted that the inelastic intensity not only emanates from the actual core level studied, but the more shallow electronic levels also contribute to this intensity. Toward low kinetic energies (<10 eV), inelastic electrons give rise to a steeply increasing background intensity (see, e.g., the Mg $2s$ region in Fig. 5). It is difficult to perform accurate curve fits to a core level peak superimposed on such an inelastic background. Experimentally, one could get an estimate of the intensity distribution of this inelastic tail by recording constant initial-state (CIS) spectra over the kinetic-energy range of interest and by chosing the initial state at a slightly lower binding (higher kinetic) energy than that of the core level investigated.

At higher binding energies but close to the core level peak, another contribution to the background is due to the following, intrinsic physical process. The sudden change in the core potential caused by the core electron emission will shake up electron–hole pair excitations in the primary excitation process. These electron–hole pairs are the low-energy equivalents of the intrinsic shakeup plasmons in the core-excitation process. The electron–hole pairs take their energy from the primary photoelectron and give rise to a tailing off of the core level peak toward lower kinetic energies. It has been shown theoretically that the cross section for this intrinsic low-energy loss process has a $E^{\alpha-1}$ energy dependence, where E is the energy distance to lower kinetic energies from the main peak (zero loss) and α ($0 < \alpha < 1$) is the so-called asymmetry index.[40–44] Calculations show that the loss tail extends from the main peak to an energy comparable to the valence-band width, i.e., 5–10 eV. It should be stressed that the calculations of this intrinsic loss process and its energy dependence are based on free-electron-gas models of metals. Several analytical expressions can be implemented in cure-fitting procedures in order to describe an asymmetric core level line shape. The most common yields the so called Doniach–Šunjić line shape,[40] but the models of Mahan[41] and Wertheim and Walker[42] are also sometimes used. In the latter model, the valence-band density of states is taken directly into account.

Since the extrinsic background already has a nonvanishing intensity in close vicinity to the main peak, it will cause an apparent distortion of the line shape. In practice it is difficult to distinguish between the asymmetry caused by electron–hole pair excitations and the effect of extrinsic losses. Thus the derived asymmetry index α will often depend on the assumed functional form and intensity of the extrinsic background. This is especially troublesome in spectra recorded at low kinetic electron energies where the core level peak is superimposed on the steeply rising tail of inelastically scattered electrons. In the ordinary, bulk-sensitive high-resolution XPS spectra this is not a severe problem, since for high kinetic energy of the primary electrons the cross section for scattering against the valence electrons is small. The asymmetry due to intrinsic electron–hole pair production can then be obtained much more reliably. In fitting surface-sensitive high-resolution core level spectra, one normally uses the asymmetry index x obtained from high-resolution XPS spectra. One also uses the lifetime widths obtained from the XPS spectra for the bulk core level peaks. Here one can obtain independent information by recording the synchrotron radiation excited core level spectra close to the photoionization threshold, where the bulk contribution dominates (see Fig. 1).

Two separate broadening mechanisms for the width of the core level peak have to be considered. The sudden change in the core potential in the photoionization process also shakes up phonons due to the different bonding to the surroundings for the atom in the ground state and the core-ionized atom. In the limit of many phonons the resulting broadening is Gaussian.[43,45] This process has been shown to be the dominating broadening mechanism in core level photoemission from maximum valence compounds such as MgO, Al_2O_3, and LiF. Here the screening of a created core hole is very inefficient because of a lack of conduction electrons with low excitation energies. On the other hand, for most metals the phonon broadening is small and often difficult to separate from the Gaussian instrumental broadening. However, we note two exceptions, Li and Be, where the phonon broadening is important in fitting the experimental spectra.

The other broadening mechanism we have to consider is due to the limited lifetime of the core hole, before it is filled by an Auger event. The limited lifetime-induced line shape is Lorentzian; the width can vary from a few meV up to several eV. Often the lifetime width determines if a given core level can be used for surface core level spectroscopy. For example, the shallow $4p$ core levels of the fifth-period elements are readily accessible to monochromatized synchrotron radiation with high energy resolution, but their lifetime widths are, except for the very first elements in the series, too broad to be of use in the present context.[46] The large lifetime widths are due to the very efficient super Coster–Kronig $N_{23}N_{45}N_{45}$ decay channel.[46,47] For the elements following Pd a breakdown of the one-electron picture occurs, and the spectral strength of the $4p$ ionization becomes smeared out over a large energy range.[46,48] Also, in many other cases Coster–Kronig and super-Coster–Kronig transitions make the lifetime for a core level short and the energy precision low.

A broadening mechanism of a different nature is observed for the surface core level peaks. Much surface core level spectroscopy has been performed on polycrystalline films, where one expects the surface to be made up of low-index single-crystal grains. These crystallites will exhibit different surface core level shifts, the differences being perhaps of the order of 0.2–0.3 eV.[19] The distribution of low-index surfaces between different grains is not known, and when fitting the experimental data one simply chooses to broaden the surface core level peak with an extra Gaussian contribution. This is illustrated in Fig. 6 where we show the Au $4f_{7/2}$ peak fitted with a bulk and an extra broadened surface core level. We also show a possible resolution of the surface core level peak into contributions from the three low-index single-crystal surfaces with their different surface core level shifts obtained from independent measurements on single crystals.[49]

The curve fitting of high-resolution surface-sensitive core level spectra can be

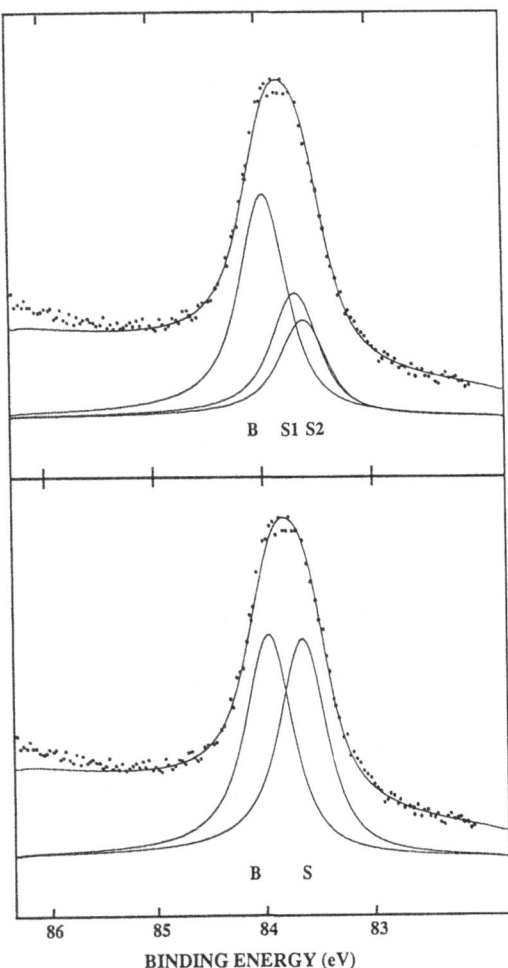

B   S1 S2

B   S

86   85   84   83

BINDING ENERGY (eV)

FIGURE 6. Au $4f_{7/2}$ core level spectra fitted with (top) bulk and separate surface peaks $S1$ [$\Delta_c = -0.28$ eV (100) surface contribution] and $S2$ [$\Delta_c = -0.35$ eV (110) and (111) surface contributions], (bottom) bulk and a single broadened surface peak. $h\nu = 120$ eV. (Surface shifts from Ref. 49.)

summarized as follows: For the bulk and surface core level peaks, a Doniach–Šunjić model line shape with an asymmetry and a lifetime width given by XPS data is chosen. A Gaussian broadening due to the phonon and instrumental contributions is added. If necessary, an additional broadening of the surface core level peak is included in order to account for the polycrystalline character of the sample. These model line shapes for the bulk and surface contributions are then added together with the anticipated intensity ratio and energy shift. A background intensity is added with a given shape, often the Shirley background[34] or simply a polynomial or exponential background. The best agreement between this model spectrum and the measured spectrum is then sought, primarily by varying the bulk-to-surface-intensity ratio, the surface shifts, and the background intensity. If necessary, adjustments of the line shape parameters are also made. It should be emphasized that, although the actual background model used has an influence on the derived core level line shape parameters, it usually has a negligible influence on the derived peak positions. Thus reliable surface core level shifts can be determined even with rather crude background models.

The experimental resolution should be minimized in order to make the broadening contributions from the physical processes as easily observable as possible. At present, for photon energies below 200 eV, one routinely obtains 0.1-to-0.3 eV photon energy resolution at photoemission beamlines of almost all synchrotron radiation facilities. At many facilities, this energy resolution is now also becoming available for photon energies up to 1000 eV, which will make a number of new, interesting systems available for high-resolution surface-sensitive core level spectroscopy. In Fig. 7, we summarize the best candidates for surface core level spectroscopy, having an experimental resolution of 0.2 eV, available up to 200 eV photon energy or up to 1000 eV photon energy (italicized). Here we assume that the intrinsic lifetime width of a core level should be below 0.5 to 0.6 eV for the core level to be of use in a high-resolution surface core level spectroscopy experiment. Perhaps the main advantage of a 0.1-to-0.3-eV resolution at higher photon energies is that one can study adsorbate systems for

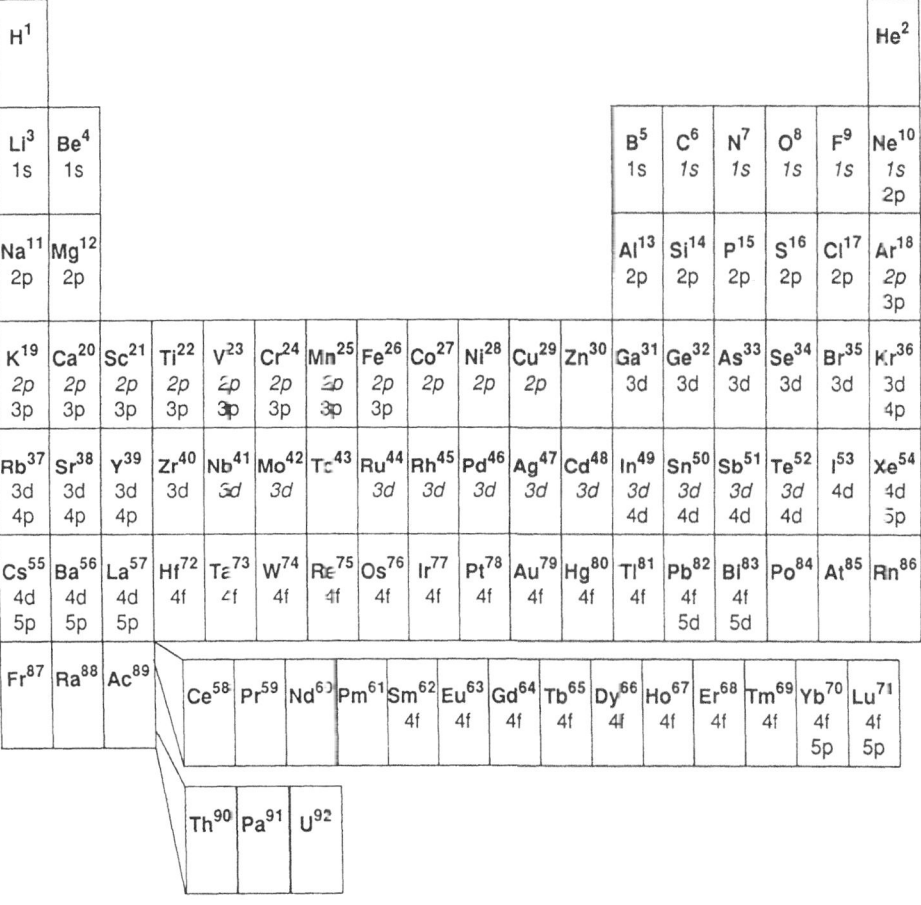

FIGURE 7. The periodic table, displaying the possible core levels available for surface core level spectroscopy with 0.2 eV energy resolution up to 200 eV photon energy or (*italic letters*) 0.2 eV energy resolution up to 1000 eV photon energy. Only core levels with a lifetime width less than 0.5–0.6 eV have been included.

molecules containing C, N, and O. But the also interesting $Z/Z + 1$ alloy systems of $3d$ and $4d$ transition metals will thereby become available for high-resolution surface core level spectroscopy, e.g., CuNi and AgPd for surface-segregation studies.

4. THE LANTHANIDES AND THE 5d TRANSITION METALS

4.1. Surface Core Level Shifts

Surface core level shifts have been extensively studied for the lanthanides[50] and the $5d$ transition metals.[10] Unfortunately, experimental single-crystal surface core level shifts exist only for some of the $5d$ transition metal surfaces, and experimental data from ordered Sm and Yb films have only recently become available. These metals have narrow $4f$ levels at binding energies below 100 eV that are readily accessible for high-resolution synchrotron radiation excitation. The $4f$ lifetime widths vary with $5d$ band occupancy for these elements but are below 400 meV.[51] The phonon broadening is small because of the relatively small change in atomic size between a Z and $Z + 1$ metal atom.[52] Thus in the most favorable cases we will find separate surface core level peaks and be able to resolve surface core level shifts below 0.1 eV. In other cases the surface core level peaks will appear only as pronounced shoulders in the $4f$ photoelectron spectra, and curve fitting is needed in order to separate the bulk and surface contributions.

Experimental $4f$ photoemission spectra from the lanthanides and the $5d$ transition metals appear very different. The $4f$ core level binding energies of the lanthanides are very shallow and energetically located in the valence-band region; even so, the $4f$ electrons may to a very good approximation be looked on as core electrons. Except for some Ce systems, the $4f$ electrons show no sign of participation in the bonding.[31] The photoionization of a lanthanide $4f$ electron in a metallic system may therefore be treated in the same way as an ordinary core–electron ionization. The complete screening picture for the lanthanide $4f$ ionization was suggested early,[25] and a thermodynamical approach using this principle was also introduced to account for the $4f$ binding energies in the pure elements.[15,16] See also section 2.2.

The final state for the $4f^n \rightarrow 4f^{n-1}$ excitation consists of a distribution of multiplet levels of the $4f^{n-1}$ configuration, and their intensities are well described by fractional parentage coefficients.[53-55] In the present context we will refer only to the position of the lowest-lying multiplet level of the final state, i.e., the level closest to the Fermi energy, when we discuss the $4f$ binding energy. A great number of photoelectron spectroscopy studies have been performed for the lanthanide systems. One reason for this interest has undoubtedly been the occurrence of the so-called mixed-valence phenomenon for quite a number of lanthanide compounds, where actually the $4f^{n+1} \rightarrow 4f^n$ process is directly involved in the formation of the ground state.

The complexity of the final-state multiplet structure varies with the $4f$ occupation number. Of the pure elements, only Ce ($4f^1$), Yb ($4f^{14}$), and Lu ($4f^{14}$) represent simple cases. All the other $4f$ elements exhibit complex final-state

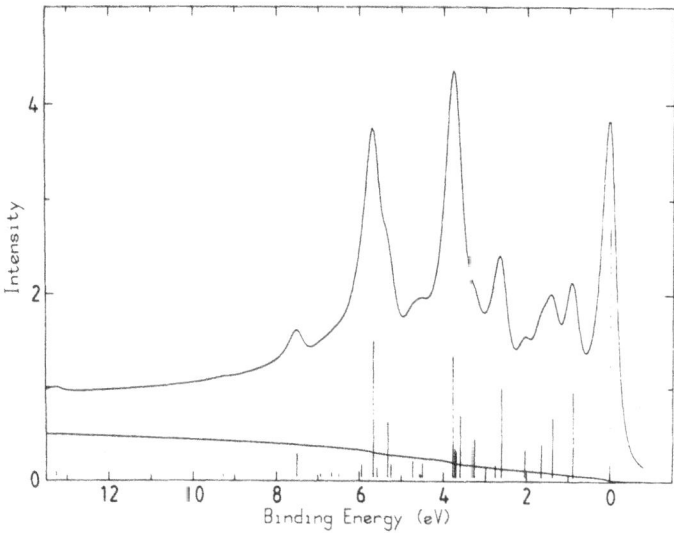

FIGURE 8. Bar diagram of the $4f^{12}$ to $4f^{11}$ multiplet structure (fractional parentage coefficients). Also shown is the Tm theoretical $4f$ photoelectron spectrum and the background intensity utilized. (From Ref. 55, reproduced with permission.)

multiplet structures when the $4f$ level is excited. These structures can be used as fingerprints of the initial $4f^n$ configuration. In Fig. 8 we exemplify this by displaying the theoretical $4f^{12} \rightarrow 4f^{11}$ multiplet structure of Tm.[55] Also shown is the theoretical $4f$ photoelectron spectrum of Tm obtained by broadening the calculated multiplet lines with a Doniach–Šunjić line shape[40] ($2\gamma = 0.24$ eV and $\alpha = 0.21$). A background and an experimental Gaussian broadening of 0.2 eV have also been added. Lang, Baer, and Cox,[56] using ordinary XPS, experimentally verified these final-state multiplet structures for all the stable lanthanide metals. To fit an experimental $4f$ surface-sensitive photoemission spectrum we need to add the bulk final-state multiplet structure and the surface-shifted final-state multiplet structure to mimic the experimental spectrum and from such an analysis obtain the experimental surface core level shift. Figure 9 exemplifies the curve fitting for Ho $4f$ spectra recorded at various photon energies.[50] The only change in the curve fitting between the spectra recorded at the different photon energies is the surface-to-bulk emission intensity ratio, which reflects the change in the inelastic electron-scattering length with electron kinetic energy. The experimental resolution also varies with photon energy, which has been incorporated in the curve fitting. The surface core level shifts obtained for the $4f$ levels of the lanthanide metals are displayed in Table 1.

The $5d$ transition metals all have filled $4f$ levels and exhibit the normal $4f_{7/2,5/2}$ doublet with the statistical intensity ratio of 4/3. The surface-sensitive $4f$ spectra of the $5d$ transition metals are simply fitted by adding a bulk and a surface $4f_{7/2,5/2}$ doublet with an appropriate intensity ratio. The surface core level shifts obtained for the $4f$ core levels of the close-packed surfaces of the $5d$ transition metals are also listed in Table 1.

Applying Eq. (2.20) to the $5d$ transition metals, we can rather easily derive

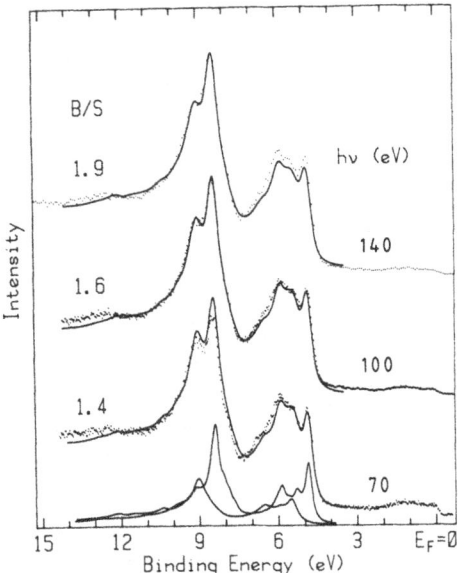

FIGURE 9. Ho 4*f* photoelectron spectra recorded at different photon energies (dotted curves). Shown are also the calculated Ho 4*f* spectra based on the fractional parentage coefficients for the f^{10} to f^9 multiplet (solid curves). At the bottom, the bulk and surface multiplets for the 70-eV photon energy spectrum are displayed. (From Ref. 55, reproduced with permission.)

TABLE 1. Experimental and Calculated Surface Core Level Shifts (in eV) of the 5*d* Transition Metals and the 4*f* Lanthanide Metals

	Experimental	Calculated
La	+0.48 5p^a	
Ce	+0.3,^{b} <+0.4^c	
Pr	+0.4^c	
Nd	+0.5^c	
Sm	+0.46−+1.23^{c,d}	
Eu	+0.63,^{c} +0.63,^{e} +0.60^f	
Gd	+0.48,^{e} +0.50^c	
Tb	+0.55^c	
Dy	+0.55^c	
Ho	+0.63^c	
Er	+0.65^c	
Tm	+0.70^c	
Yb	+0.60,^{g} +0.60,^{h} +0.60,^{c}, +0.62,^{f}	
	+0.56,^{a} +0.56 5p^a	
Yb (111)	+0.53^i	
Lu	+0.77,^{c} +0.83,^{f} +0.74 5$p,^f$ +0.70^j	+0.39^k
Hf	+0.42^l	+0.48^k
Ta (110)	+0.28,^{m}	+0.31,^{k} +0.4^n
W (110)	−0.30,^{o} −0.30^m	−0.18,^{k} −0.28^p
Re (0001)	0.0^q	−0.27,^{k} −0.18^r
Ir (111)	−0.50^s	−0.50,^{k} −0.43^p
Pt (111)	−0.39,^{t} −0.33^u	−0.40,^{k} −0.22^v
Au (111)	−0.35^w	

^{a} From Ref. 57. b From Ref. 58. ^{c} From Ref. 50. ^{d} From Ref. 59. ^{e} From Ref. 60. ^{f} From Ref. 61. ^{g} From Ref. 62. ^{h} From Ref. 63. ^{i} From Ref. 64. ^{j} From Ref. 65. ^{k} From Ref. 19. ^{l} From Ref. 71. ^{m} From Ref. 70. ^{n} From Ref. 72. ^{o} From Ref. 9. ^{p} From Ref. 73. ^{q} From Ref. 69. ^{r} From Ref. 11. ^{s} From Ref. 66. ^{t} From Ref. 67. ^{u} From Ref. 68. ^{v} From Ref. 74. ^{w} From Ref. 49.

several interesting features. For the $5d$ transition metals, the $5d$ band is successively filled through the series, first the bonding part of the $5d$ band and then the antibonding part. This d-band filling reflects itself in the parabolic variation of the cohesive energy as a function of atomic number. From this variation in the cohesive energy, one can from Eq. (2.20) conclude that the surface shift will vary essentially linearly through the series. Furthermore, there will be a change of sign of the surface shift in the middle of the series.[17,19] This prediction of a change of sign of the surface shift has been verified experimentally, as can be seen in Fig. 10. The physical origin of the change of sign can be understood as follows. We notice that the final-state valence charge distribution is essentially that of the $(Z + 1)$ element. For the $5d$ elements in the beginning of the series, this means that the bonding due to the conduction electrons is stronger in the final state than in the initial state. The reason is that, for an early $5d$ element, the screening takes place by electrons filling the bonding part of the d

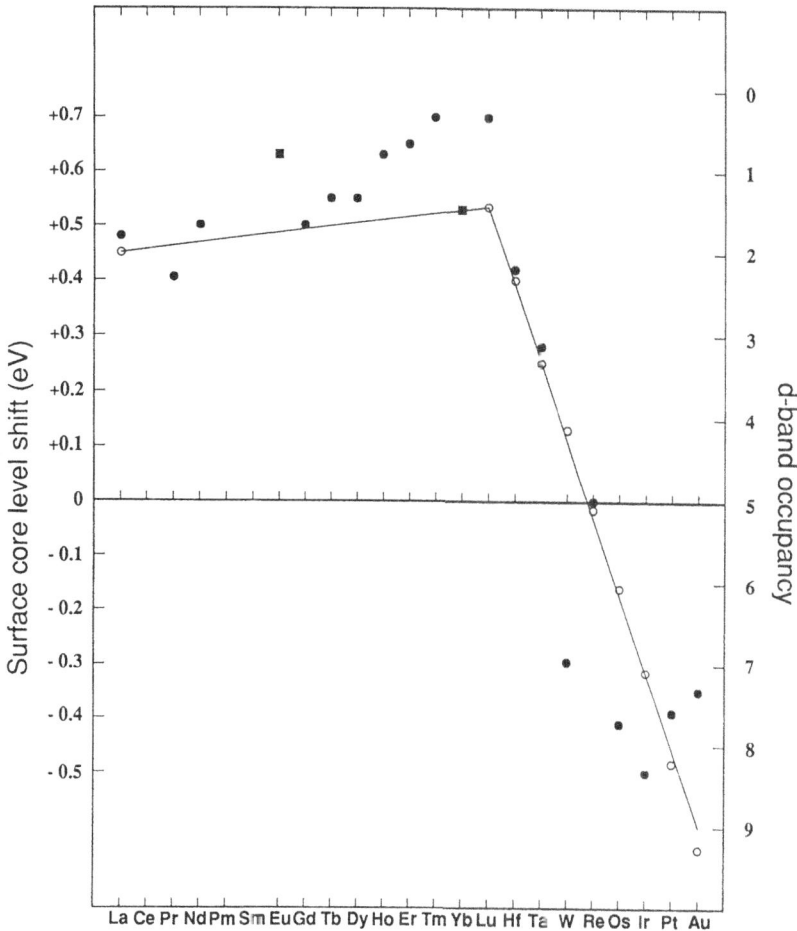

FIGURE 10. Correlation between surface core level shifts (filled symbols) and $5d$ band occupancy (open circles) for the lanthanides and the $5d$ transition metals. When available, experimental shifts for the most-close-packed surfaces have been used.

band. The same also holds for a surface atom. Due to the reduced coordination number for a surface atom compared to a bulk atom, the gain of bonding in the final state relative to the bonding in the initial state is less for a surface atom than for a bulk atom. This immediately explains the higher core level binding energy for a surface atom in the earlier transition metals. For a late $5d$ transition metal with a half-filled or more than half-filled d band, the situation is the opposite, since the $(Z + 1)$ screening now takes place within the antibonding part of the d band.

All the surface core level shifts for the lanthanide $4f$ levels are positive, which is in agreement with a less than half-filled d band (see Fig. 10). The observed increase in the surface core level shifts along the lanthanide series is in accordance with the simultaneous decrease in the number of $5d$ conduction electrons. There is a similar increase in the surface core level shift when we proceed from Hf to Lu, whereby the number of $5d$ conduction electrons is decreased by one. This comparison between the lanthanides and the early $5d$ transition metals (Lu, Hf) is based upon the assumption that the d electrons are the main source of bonding (cohesive energy) for both the lanthanides and the $5d$ transition metals. It is satisfying to notice that the surface core level shift reported for La of 0.48 eV[57] falls in between the surface core level shifts reported for Lu and Hf, 0.70 eV[65] and 0.42 eV,[71] respectively, which is in accordance with calculated d-band occupancies.[75]

In Fig. 10 we display the experimental surface core level shifts of the lanthanides and the $5d$ transition metals together with their calculated d-band occupation number.[75–77] In the case of the $5d$ transition metals, we have used the experimental surface core level shifts from the most closely packed surfaces. We notice that the surface $4f$ binding-energy shifts fit nicely into the general behavior of the d-band occupancy.

Among the lanthanides, only Eu and Yb are divalent, and they each have essentially an $(sd)^2$ valence configuration with less than one d electron. This means that the d bonding is not as dominant for Eu and Yb as it is for the trivalent lanthanides and the $5d$ transition metals. Therefore the surface energies for Eu and Yb are not strongly correlated with the d-occupation number. This is also observed in the surface core level shifts shown in Fig. 10, where Eu and Yb fall outside the trend seen for the trivalent lanthanides.

Theoretical estimates of the surface core level shifts for the lanthanides and the $5d$ transition metals can be obtained from Eq. (2.16). For this, we need the surface energies of the lanthanides and the $5d$ transition metals. For the $5d$ transition metals, the surface energies have been calculated for the various single-crystal surfaces using a tight-binding approach,[19] and the resulting surface core level shifts are included in Table 1.

For the lanthanides, an uncritical use of Eq. (2.16) poses problems. First, when we excite a $4f$ electron the screening electron will be in the $5d6s$ conduction band at the Fermi level, not an added $4f$ electron as would follow from a direct application of the $Z + 1$ approximation. Depending on the initial valence state (trivalent or divalent), this means a tetravalent metallic $(5d6s)^4$ or a trivalent metallic $(5d6s)^3$ valence configuration of the photoionized atom. Thus the $Z + 1$ screening charge is not comparable to the next element in the periodic table.

Second, there are no calculations of the surface energies of the lanthanides as a function of d-band occupancy similar to the ones reported for the $5d$ transition metals. If we use Eqs. (2.25) and (2.26), we need the cohesive energies for the lanthanides and for the corresponding final state. Such an analysis was presented in section 2.2, where we made a first-order estimate of the surface $4f$ binding energy shifts. We obtained surface $4f$ shifts of 0.50 eV and 0.40 eV for the divalent and the trivalent lanthanides, respectively.

A problem in the analysis above is the lack of correlation between the derived metallic bulk cohesive energies and the calculated d-band occupancy. In the derivation of the metallic bulk cohesive energies for the lanthanides made in section 2.2, we found the metallic cohesive energies for the divalent, trivalent, and tetravalent lanthanides to be 40, 100, and 145 kcal/mol, respectively, if the change in valency between the free atom and the metal was taken into account. We cannot, however, correct for the exact distribution between the s and d states in the valence band for the metal. In order to estimate the surface energies for the lanthanides a more microscopic approach is needed, in which the change of d-band occupation through the series should play a central role. As we noticed, the variation of the surface $4f$ binding-energy shift can be directly correlated with the calculated change of the d-occupation number from La to Lu, which shows a monotonic decrease from 2.0 to 1.5.[75,77] This variation originates from the lanthanide contraction, which forces the $5d$-band to rise in energy compared to the $6s$ band. It has been shown that the trend in the $5d$-band occupancy nicely explains the sequence of different crystal structures observed along the lanthanide series.[75,78] The calculated change of d-band occupancy is supported by the experimentally observed lifetime widths for the $4f$ core levels, both in the lanthanides and in the $5d$ transition metals. This is shown in Fig. 11 where we plot the experimentally determined $4f$ lifetime widths.[50,51,56] We observe that an increased d-band occupancy enhances the Auger process of filling the $4f$ core level, and hence the lifetime width. Thus there are strong theoretical and experimental evidences for a decreasing d-band occupation along the lanthanide series, and it is clear that in order to obtain a good quantitative agreement between the calculated and experimental surface core level shifts, calculated surface energies based on the d-band occupation are required.

The experimental surface core level shifts for the lanthanides discussed above were all measured on polycrystalline films. For all elements, a significant broadening of the surface peaks as compared to the bulk peaks was observed. When we use these experimental surface core level shifts, we rely on the assumption that the polycrystalline films are made up of low-index crystal grains (close-packed surfaces) and accept the uncertainty given by the broad surface core level peak.

The influence of surface structure on the surface core level shift was demonstrated by Schneider et al.[79] for polycrystalline Yb films grown on substrates at various temperatures. They were able to show that the broad surface peak is composed of several components, each associated with a given coordination number for the surface atoms. The measured surface core level shifts ranged from 0.51 eV to 0.86 eV.

Recently, Mårtensson et al.[64,80] have succeeded in growing ordered films of

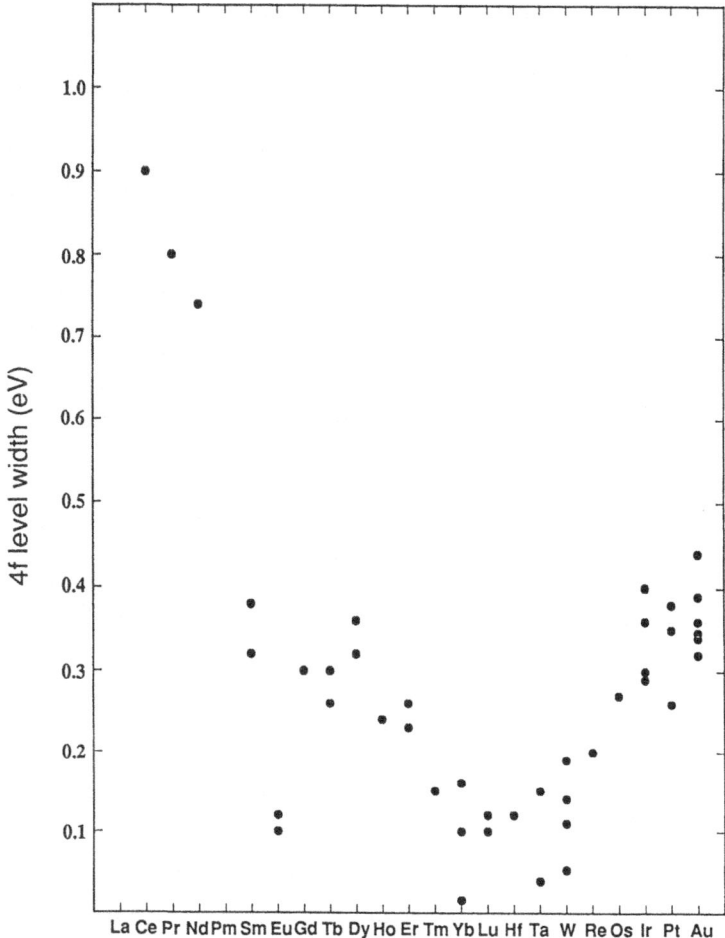

FIGURE 11. 4*f* core level lifetime widths for the lanthanides and the 5*d* transition metals. (Data compiled from Refs. 50, 51, and 56.)

Sm and Yb. The 4*f* spectrum from an ordered Sm film is shown in Fig. 12 together with the spectrum from a disordered polycrystalline film. For both Sm and Yb a significant decrease in the width of the surface core levels is observed (for Sm the outermost multiplets are surface-related; see section 4.2). For Yb, a surface core level shift of 0.53 eV was measured, which can be compared to the 0.6 eV commonly found for polycrystalline surfaces. For the other lanthanides no experiments of this kind have been performed.

4.2. Surface Valence Changes

For some of the trivalent lanthanides there is a rather small energy difference between the ground state and the hypothetical divalent metal state. Since the divalent state is more favored at a surface than in the bulk, there might be lanthanide metals for which the bulk material is trivalent but the surface is divalent. This is the case with Sm, where photoelectron spectroscopy has

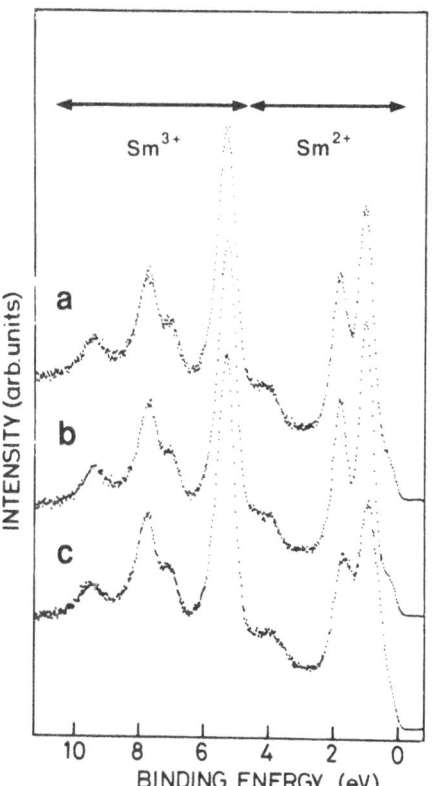

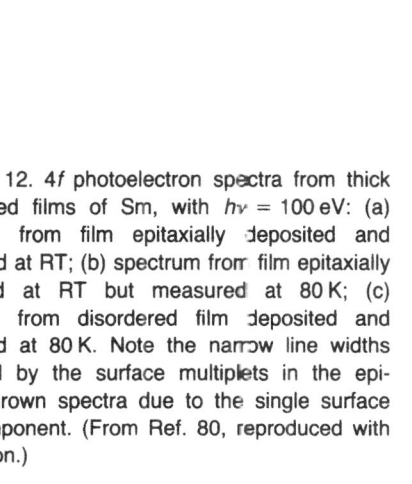

FIGURE 12. $4f$ photoelectron spectra from thick evaporated films of Sm, with $h\nu = 100\,eV$: (a) spectrum from film epitaxially deposited and measured at RT; (b) spectrum from film epitaxially deposited at RT but measured at 80 K; (c) spectrum from disordered film deposited and measured at 80 K. Note the narrow line widths displayed by the surface multiplets in the epitaxially grown spectra due to the single surface shift component. (From Ref. 80, reproduced with permission.)

demonstrated the presence of a divalent surface.[81] The energetics for this situation has been investigated theoretically in some detail[18,82] and give strong support to the experimental evidence of a divalent surface.

The $4f^6$ level for the divalent surface layer of Sm was measured by Lang and Baer[59] to be 0.77 eV below the Fermi level. Allen *et al.*[83] find an experimental $4f^6$ level binding energy of 0.65 eV. The binding energy of a $4f$ level is taken as the energy position of the lowest-lying multiplet level. The rather high $4f^6$ level binding energy suggests that the divalent state at the surface is very stable. Rosengren and Johansson[82] showed that the main cause for this nominally high stability is that the whole Sm surface layer has transformed to the divalent state. The following considerations by Johansson and Mårtensson[13] give support to this explanation. We first assume that the whole Sm metal, the surface included, is divalent. This is thus a system which must be very similar to its divalent neighbor, Eu. For Eu, the measured bulk $4f$ binding energy is 1.5 eV.[56] Since $\Delta E_{II,III}$ for Eu is 0.9 eV, the impurity term $E_{III}^{imp}(II)$ is 0.6 eV. For Sm, the corresponding energy for the impurity term should be very similar to that of Eu. The energy difference $\Delta E_{II,III}$ is -0.27 eV for Sm (the negative sign expresses the fact that Sm in the bulk is a trivalent metal). Therefore it follows that for hypothetical divalent Sm the bulk $4f^6$ binding energy is $(-0.27 + 0.6) = 0.33$ eV. This is quite remarkable and means that the divalent Sm metal is stable locally. Therefore the valence change in bulk Sm is of a collective type, since the single sites in divalent Sm are *locally stable* against a valence change. At the surface of the hypothetical divalent

Sm metal there should be a surface shift similar to that for Eu (0.6 eV). Thus the divalent $4f^6$ Sm should be expected to have a binding energy of $(0.33 + 0.6) = 0.93$ eV. This is a high binding energy, as is in agreement with experimental observations. The actually observed binding energy is somewhat less than 0.9 eV, reflecting the fact that for the real Sm system there is a trivalent layer beneath the topmost divalent surface layer.

Tm has also a tendency to attain a divalent configuration in certain compounds. To discuss the possibility of divalent Tm, it is instructive to compare it with its neighbor Yb, a divalent metal.[13] The $4f^{14}$ level in Yb has a bulk binding energy of 1.17 eV, and the surface shift for an evaporated film is 0.6 eV. Since the valence energy difference, $\Delta E_{\text{II,III}}$, is 0.5 eV for Yb and -0.78 eV for Tm, for a hypothetical divalent Tm metal the binding energy of the $4f^{13}$ level at the surface should be $1.17 + 0.6 - (0.5 + 0.78) \cong 0.5$ eV, i.e., stable, in the bulk its position will be at -0.1 eV, which means that Tm is not only globally unstable toward the trivalent state but also locally unstable. Schneider et al.[79] performed experiments on Yb surfaces evaporated at low temperatures. This experimental condition leads to an increased surface roughness, and the $4f$ surface signals become dominated by contributions from surface atoms with relatively low coordination numbers. The corresponding surface shift was measured to be 0.86 eV, i.e., considerably larger than for a film evaporated at room temperature. A corresponding geometrical configuration for hypothetical divalent Tm metal would give a surface binding energy of $1.17 + 0.86 - (0.5 + 0.78) = 0.75$ eV, which is quite high. Therefore a Tm surface with a low coordination number could be stable in a divalent configuration. A Tm surface prepared in the same way as the Yb surface was found to have a divalent surface with an f^{13} level binding energy of 0.5 eV.[84] The reduction from the value 0.75 eV, derived above, can be explained as due to the trivalent Tm atoms in the layers beneath the surface atoms. From pair-bonding calculations,[82] it can be shown that when a divalent surface on top of a trivalent bulk is just stable, then the surface $4f$ binding energy is 0.4 eV. Therefore the measured value of 0.5 eV shows that the divalent state is stable by only 0.1 eV for the Tm surface prepared at low temperatures.

5. THE FREE-ELECTRON METALS

5.1. Theoretical Considerations

Free-electron-like or simple metals are considered to have physical properties which can be closely approximated by those calculated for a free-electron gas with the appropriate electron density. They have long attracted interest, since model calculations of their physical properties should predict experimental results rather accurately. One should, however, remember that the free-electron gas at the high electron densities, corresponding to the electron densities of such important simple metals as Mg and Al, is nonbonding, and other contributions to the cohesive energy make the bonding possible. Thus, directly estimating such properties as work functions and bulk and surface cohesive energies from free-electron-gas calculations is not possible.[85] However, by including the electron–ion interaction through realistic pseudopotentials collective properties such as work functions and bulk and surface cohesive energies can also be accurately calculated for all simple metals.[85,86]

We compare the experimental surface core level shifts for the free-electron metals with shifts calculated from models based on the free-electron-gas approximation of solids. The electron density profiles at the surfaces of simple metals have been calculated using the electron-density functional formalism.[85] In Fig. 13 the calculated electron-density profiles are shown together with the corresponding electrostatic potential variations at the surfaces of Na, Mg, and Al.

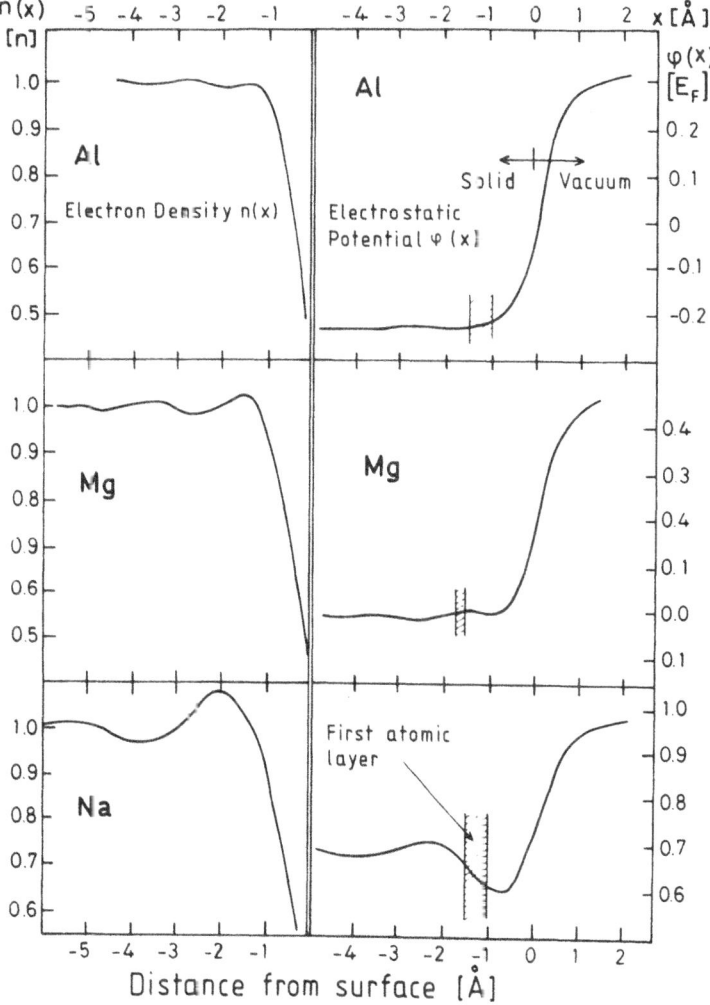

FIGURE 13. Friedel oscillations of the electronic charge density at free-electron metal surfaces and the associated variation in the electrostatic potential. Electronic charge density $n(x)$ is given in units of the bulk electronic charge density. Electrostatic potential $\varphi(\chi)$, is given in units of the Fermi energy. (From tabulated data given in Ref. 85.)

We observed that the lower the bulk electron density, the more pronounced are the oscillations in the electron density at the surface and, accordingly, in the electrostatic potential at the surface. This is due to the smaller range in reciprocal space for the Fourier expansion of the decreasing charge density at the metal surface for metals with low electron density, i.e., small Fermi wave vector, which results in larger spatial variations. The electrostatic potential variations are of the order 100–150 meV and should easily be observed as core level binding-energy shifts for the surface atoms.

For the Al (100), surface the initial state surface core level shift of the $2p$ level has been calculated self-consistently[87] to be 120 meV to lower binding energy. In addition, a surface-induced crystal field splitting of 38 meV was calculated for the $2p_{3/2}$ level. Similar initial-state self-consistent surface-electronic-structure calculations have been performed for single-crystal surfaces of a number of metals, and their accuracy is well established. We conclude that sizeable surface core level shifts are predicted for the simple metals, and that the initial-state contribution to this shift can be accurately calculated. Recently Feibelman[88] calculated self-consistently the Si atom point-defect energies at the surface and in the bulk for an Al (100) single crystal. Using the thermodynamic model he could then for the first time in an *ab initio* manner calculate a surface core level shift including both initial and final-state contributions. The calculation predicts a surface shift of -0.097 eV. The calculations should be extended to Mg and Na single-crystal surfaces where experimental data indicate larger surface shifts and a comparison with experiment would be more feasible.

We interpret the experimental surface core level shifts for the free-electron metals Li, Be, Na, Mg, and Al using the thermodynamical model reviewed in section 2.1. This model includes, as stated earlier, the electronic differences between the surface and the bulk atoms, both in the initial and final state. Using Eq. (2.16) and neglecting the difference in surface and bulk impurity solution energies, the surface core level shift is given as

$$\Delta_c = E_S^{Z+1} - E_S^Z, \tag{5.1}$$

where E_S^{Z+1} and E_S^Z are the surface energies for two consecutive elements in the periodic table. To use Eq. (5.1), the surface energies for the free-electron metals both in their initial and screened final states are needed. In Table 2 these are summarized for the free-electron metals of interest. The theoretical surface energies shown in Table 2 should be lowered about 10% to allow for the inward relaxation of the surface atoms taking place in reality. However, for estimating the surface shifts only differences between surface energies are involved and relative errors are of little importance.

5.2. The Surface Core Level Shifts of Na, Mg, and Al

The $2p$ core level binding energies of Na, Mg, and Al are 30.6 eV, 49.6 eV, and 72.7 eV, respectively. Therefore they are readily accessible for high-resolution synchrotron radiation excitation. The lifetime widths of the $2p$ levels

TABLE 2. Experimental and Calculated Surface Shifts and Calculated Surface Energies at Different Temperatures of the Free-Electron-like Metals

	Experimental surface shifts (eV)	Calculated surface shifts (eV)	Surface energies (meV/atom)
Li	$+0.56^a$	$+0.17^b$	$251,^e$ 277 100 K,^{d} 224 300 K^d
Be (0001)	-0.50^e	$-0.02/-0.15^b$	$530,^f$ 450 100 K,^{d} 406 300 K^d
B			$380,^f$ 387 100 K,^{d} 382 300 K^d
Na	$+0.22^g$	$+0.23^b$	$205,^c$ 205 100 K,^{d} 194 300 K^d
Mg	$+0.14^g$	$+0.07^b$	$404,^c$ 433 100 K,^{d} 425 300 K^d
Al	-0.12^g		
Al (100)	-0.12^g	$-0.12,^h$ $-0.097,^i$ -0.06^b	$505,^c$ 579 100 K,^{d} 571 300 K^d
Al (111)	0.0^g	$+0.02^b$	$372,^c$ 502 100 K,^{d} 495 300 K^d
Si			574 100 K,^{d} 512 300 K,^{d}

^{a} From Refs. 33, 90. ^{b} Calculated from the surface energies. ^{c} From Refs. 85, 92. ^{d} From Ref. 93. ^{e} From Ref. 91. ^{f} From Refs. 94, 95. ^{g} From Ref. 89. ^{h} From Ref. 87. ^{i} From Ref. 88.

are less than 100 meV, and the phonon-induced broadenings at RT are also small. Surface shifts below a tenth of an eV should be observable. Surface core level spectroscopy data from Na, Mg, and Al polycrystalline films and Mg (0001), Al (100), and Al (111) single crystals have been reported.[33,89,96,97] In Fig. 14, the Na $2p$ core level photoelectron spectra recorded for a polycrystalline film at four different photon energies are shown. The corresponding excited electron kinetic energies range from 5 eV to 70 eV, yielding spectra from bulk sensitive to surface sensitive (see Fig. 1). The $2p$ core level spectra change their appearance completely in the surface-sensitive part of the photon-energy range. For example, at 100 eV photon energy, the Na $2p_{3/2,1/2}$ doublet is no longer resolved. Accordingly, we use curve fitting to separate the $2p$ peak into a bulk and a surface $2p_{3/2,1/2}$ doublet. The solid curves in Fig. 14 show the results from the curve fitting, and for the 100 eV spectrum the bulk and the surface $2p$ doublets are displayed separately. It should be noted that the surface peak is considerably broader than the bulk peak, 90 meV and 20 meV, respectively. For a polycrystalline film the surface peak is a mixture of surface peaks from the surfaces of different single-crystal grains. The asymmetry and the width of the surface peak should consequently be considered as fit parameters, their physical significance possibly being related to the surface morphology of the polycrystalline film. The surface core level shift obtained for Na is +0.22 eV.[33] A similar analysis of experimental spectra for a polycrystalline Mg film yields a surface core level shift of +0.14 eV.[33,89] For a polycrystalline Al film no surface core level shift could unambiguously be detected. The surface-sensitive Al $2p$ core level spectra could be fitted equally well with a surface-induced symmetrical broadening of 50 meV of the Al $2p$ peak or with a bulk $2p$ and a surface $2p$ peak separated by 0.12 eV, the surface peak located at lower binding energy.[33]

A resolution of this ambiguity for Al should be reached by examining spectra recorded for single-crystal Al surfaces. The Al (100) single-crystal surface shows

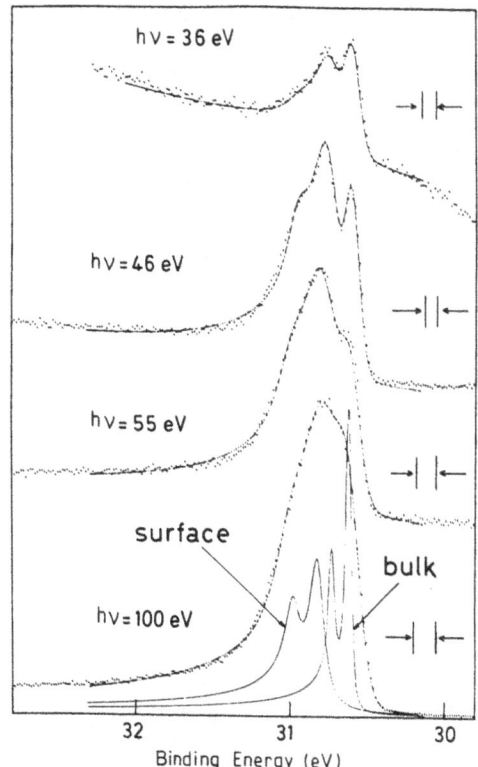

FIGURE 14. Na 2p core level spectra recorded at 100 K for various photon energies (dotted curves). Calculated fits are shown (solid curves). At the bottom, the bulk and surface components for the 100-eV photon energy spectrum are displayed. (From Ref. 33, reproduced with permission.)

the same behavior as the polycrystalline Al film. Here too, the spectra could either be fitted with a 50-meV symmetrically broadened single Al $2p$ peak or a bulk and a surface $2p$ peak separated by 0.12 eV. Possibly it can be stated that the surface-shift fit is marginally better than the surface-broadening fit. For the Al (111) single-crystal surface, no surface core level shift was observed for the $2p$ peak. Also, there was no observation of a surface-induced broadening of the $2p$ peak. It is in fact possible to fit all the Al $2p$ spectra independent of the exciting photon energy using a single peak with exactly the same line-shape parameters. The surface-crystal field splitting should be of the same magnitude for the Al (111) as for the Al (100) surface-layer atoms. We thus rule out the surface-broadening explanation for the Al (100) surface as well. We conclude that the Al (100) and Al (111) single-crystal surfaces exhibit surface shifts of −0.12 eV and 0.0 eV, respectively. We also state that the polycrystalline Al surfaces studied in these experiments exposed not only (111) but also (100) single-crystal grains.

In Table 2 we summarize the surface core level shifts of various Na, Mg, and Al surfaces. Using Eq. (5.1) and the surface energies given in Table 2, we can compare experimental and calculated surface core level shifts. The agreement is satisfying, considering the uncertainties in the surface energies used. Since the experimental surface core level shifts are mostly recorded for polycrystalline films, it is of limited value to compare them with initial-state binding-energy shifts calculated for single-crystal surfaces. We note that the initial-state surface core level shift calculated by Wimmer et al.[87] for the Al (100) surface agrees well with the corresponding experimental surface core level shift, both being 0.12 eV to lower binding energy. More important, however, are the recent *ab initio* calculations of the surface shift of Al (100) made by Feibelman.[88] These calculations, which include both initial- and final-state contributions, yield a value of -0.097 eV for the surface shift, in good agreement with experiments. This is a most promising result, and further calculations for other free-electron-like metals would be highly interesting.

5.3. The Surface Core Level Shifts of Li and Be

Li and Be display, based on the experience obtained from the Na, Mg, and Al core level spectra, very different surface-sensitive spectra. For Li and Be, the 1s core levels are excited and the screening orbitals have $2s2p$ character.

The Li 1s ($E_B = 50.0$ eV) core level spectrum recorded at 100 eV photon energy is shown in Fig. 15.[33] It displays a single peak with a shoulder to higher binding energy. It is known[43] from XPS that the Li 1s bulk peak is severely phonon-broadened due to the inefficient electronic screening of the 1s core hole. Attempts have been made to measure the corresponding bulk peak using synchrotron radiation photon energies near the 1s core level photoionization threshold, where the excited electron inelastic scattering length is even larger than in the XPS spectra. This failed, because of the very small photoexcitation cross section for the Li 1s core level at photon energies below 100 eV. A slight oxygen contamination of the Li film raised the cross section drastically, making it possible to record core level spectra near threshold. This may be due to the introduction of empty O $2p$ orbitals into the conduction band, whose matrix elements for excitation from the Li 1s core level are large. However, a clean Li 1s bulk spectrum could not be recorded, and one has to rely on the XPS spectra to obtain the Li 1s bulk peak line shape. Using the XPS bulk peak parameters, the surface-sensitive Li 1s core level spectrum was fitted and a surface core level shift of +0.7 eV was deduced. It is not very satisfactory to fit the Li 1s spectrum with two broad Gaussians and not to be able to observe the relative change in intensity between the bulk and surface emission by varying the exciting photon energy. Any possible contamination origin of the shoulder structure was checked by

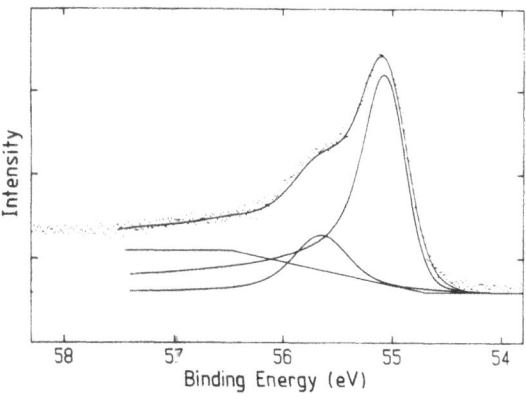

FIGURE 15. Li 1s photoelectron spectrum (dotted curve). Calculated fit composed of bulk peak, surface peak, and background intensity also displayed (solid curves). $h\nu = 100$ eV. (From Ref. 33, reproduced with permission.)

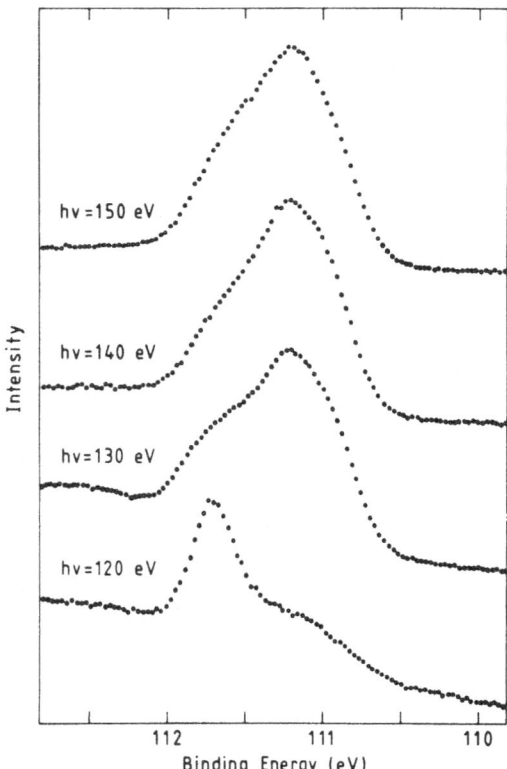

FIGURE 16. Be 1s core level spectra recorded for various photon energies. Note the change in surface sensitivity upon increasing the photon energy from the photoemission threshold. (From Ref. 91.)

exposing the Li film to oxygen. It was observed that the first oxygen-induced structure in the Li 1s core level spectrum had a chemical shift of +1.3 eV to higher binding energy, which is substantially larger than the +0.7 eV surface shift deduced from the shoulder in the clean Li 1s spectrum.

The Be (0001) 1s (E_B = 111.7 eV) core level spectra recorded at four different photon energies are shown in Fig. 16.[91] Here the cross section for the 1s core level is relatively high even near the photoionization threshold, due to the more 2p-like character of the Be conduction band. It is easy to resolve the Be 1s peak into a bulk and a surface peak using the spectra shown in Fig. 16. A surface core level shift of −0.50 eV was deduced.

As for Li, the Be 1s bulk and surface peaks are severely phonon broadened. In Table 3 we give the phonon broadenings for the Li 1s and the Be 1s surface and bulk peaks. Almbladh and Morales[99] calculated the phonon-induced width for the Li 1s core level and found it to be in good agreement with the experimental value of Citrin et al.[43,98] (see Table 3). The phonon broadening observed in Fig. 15 for the Li 1s core level is substantially larger. For the Be 1s core level no calculated phonon broadening has been reported. A significant feature of the phonon broadening of the Be 1s core level is the difference between the bulk and the surface peaks, the surface peak being significantly broader. This indicates a less effective electronic screening of the Be 1s core hole at the surface as compared to the bulk. The covalent character of the Be metal bonding can make the surface and bulk electronic structure very different and

TABLE 3. Phonon Broadenings (in eV) for the Free-Electron-like Metals at Different Temperatures

	Measured	Calculated
Na 2p	0.17 300 K,[a] 0.025 100 K[b]	0.11 300 K,[d] 0.10 80 K,[e] 0.18 300 K[e]
Mg 2p	0.14 90 K,[a] 0.16 300 K,[a] 0.02 90 K,[b] 0.075 300 K[b]	0.17 300 K[c]
Al 2p	0.05 300 K,[a] 0.02 100 K,[b] 0.05 300 K[b]	0.09 300 K[c]
Li 1s	0.23 90 K,[a] 0.32 300 K,[a] 0.37 100 K[b]	0.21 300 K,[c] 0.23 80 K,[d] 0.33 300 K,[d] 0.39 440 K[d]
Be 1s	0.24 300 K,[e] 0.60 (surface peak) 300 K[e]	0.49 300 K[c]

[a] From Refs. 43, 98. [b] From Ref. 33. [c] From Ref. 52. [d] From Ref. 99. [e] From Ref. 91.

cause large differences in the core hole screening. Furthermore, it is interesting to note that in the Be 1s surface peak one can observe additional structure. Be has a high Debye temperature and consequently high-energy phonon modes. The structure observed could be due to discrete phonon shakeup losses. We believe that further experimental and theoretical work is needed before the phonon-induced broadenings of the Li and the Be 1s core levels are fully understood. Single-crystal surface core level spectra recorded at low temperatures would especially be of great value.

To interpret the Li and Be surface core level shifts using Eq. (5.1), the surface energies of metallic Li, Be, and B are needed. For Be and B we use, due to lack of reliable experimental data, a scheme introduced by Miedema[94,95] to calculate the metallic surface energies for these elements. As an alternative, one could use the bulk cohesive energies of Be and B to estimate their surface energies; compare Eq. (2.17). For Be, a change in the valence configuration from the free atom s^2 to an sp valence state in the metal takes place.[100] The energy of the s^2-to-sp transition for the free atom can be obtained from optical data.[101] For the Be screened final state, the $(sp)^3$ of B can be used. The experimental and calculated surface energies of Li, Be, and B are summarized in Table 2. The calculated surface core level shifts for Li and Be are also shown in Table 2; they yield the correct sign but are far off numerically compared to the experimental ones. Using bulk cohesive energies and the scheme proposed above, a better agreement can be reached.[91]

6. ALLOYS AND INTERMETALLIC COMPOUNDS

6.1. Z/Z + 1 Alloys

Alloys represent particularly attractive systems to study using surface core level spectroscopy. The surface and bulk composition of alloys often differ drastically. To understand these segregation phenomenona, it is necessary to investigate the physical mechanisms causing the composition difference at the surface relative to the bulk. The surface segregation is often of utmost importance in technological applications of surface physics and chemistry, since the surface properties can be drastically different for a segregated surface as compared to a surface maintaining the bulk composition. This is often the case for surfaces used in heterogeneous catalysis.[102] It is true that ordinary XPS is already a surface-sensitive technique with a probe depth of only a few atomic layers, the degree of surface sensitivity depending on the exciting photon energy and on the binding energy of the core level in question (see Fig. 1). XPS was therefore utilized long before the discovery of surface core level shifts to investigate the surface properties of alloys. The introduction of synchrotron radiation excited photoelectron spectroscopy permitted a development toward more surface-sensitive experiments in which only the topmost layers of the surface was probed, based on the possibility of varying the photon energy and thus the excited electron kinetic energy to obtain the shortest possible electron inelastic scattering length.[103] This development also preceded the concept of surface core level shifts. An example of the surface sensitivity of synchrotron-excited photoemission is given in Fig. 17, where a spectrum from a Li film with a bulk impurity content of less than 200 ppm Na is displayed.[33] In spite of the very low abundance of Na in the bulk, we easily observe the Na $2p$ peak in the 100-eV photon energy spectrum. Note that the loss structure associated with the Na $2p$ peak is the Li surface plasmon loss at 4.2 eV energy, proving the location in the surface layer of the detected Na. Clearly in the LiNa alloy system there is a strong enrichment of Na at the surface.

Surface core-level spectroscopy is unique in the sense that the surface peaks represent emissions only from the surface-layer atoms. If the cross sections of the various core levels involved are properly taken into account, the intensity ratios obtained between the surface peaks of the two components in a binary alloy yield the surface-layer composition. The ratio of the bulk peaks does not necessarily

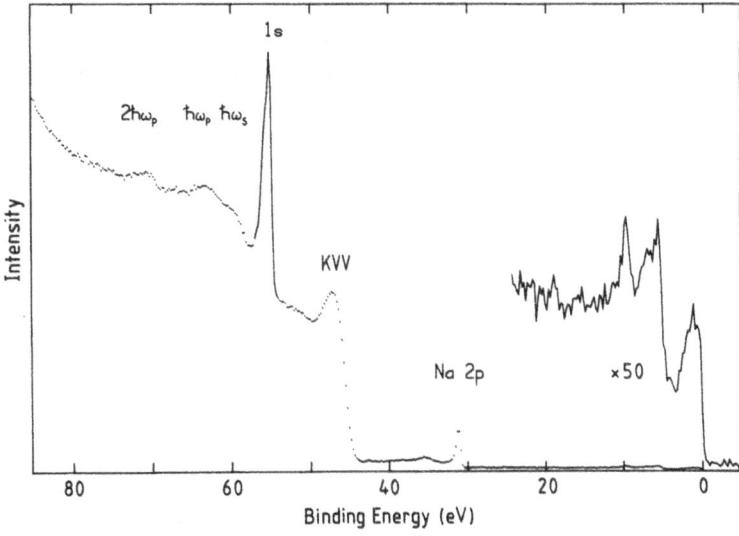

FIGURE 17. Survey photoelectron spectrum from an evaporated Li film. Na is present as an impurity with less than 0.02% bulk abundance. The relatively strong Na $2p$ signal is due to the strong surface segregation of Na in Li. $h\nu$ = 100 eV. (From Ref. 33, reproduced with permission.)

represent the bulk composition of the alloy, since the spatial extension of the surface segregation into the alloy may extend several atomic layers below the surface. Therefore the observed bulk core level peaks, often corresponding to probe depths of only 5–10 Å, are also surface probes and influenced by the surface segregation. The chemical shifts of the surface peaks of the alloy are also important to deduce and to understand. If we assume no surface segregation in an alloy, the chemical shifts of the surface peaks should be, for a close-packed surface, approximately 70–80% of the chemical shifts for the alloy bulk peaks.[61,104,105] This is due to the lower atomic coordination number at the surface, which means less influence from a change in the chemical bonding. Any deviation from this expected behavior is a sign of surface segregation due to a different bonding situation at the segregated surface as compared to the bulk.

It was shown in section 2 that the heat of segregation $Q_{segr}(Z + 1)$ in a $Z/Z + 1$ alloy system dilute in the $Z + 1$ component is equal to the surface core level shift of the Z component:

$$\Delta_c(Z) = E_S((Z + 1); Z_x Z + 1_{1-x}) - E_S(Z; Z_x Z + 1_{1-x}) = Q_{segr}(Z + 1).$$

$$(6.1)$$

This yields the following expression for the surface layer composition X_s for the $Z + 1$ component in an alloy with a bulk composition X_B

$$X_S(1 - X_S)^{-1} = X_B(1 - X_B)^{-1} \exp \frac{-\Delta_c}{kT}.$$

$$(6.2)$$

A number of technically important alloy systems are of the $Z/Z + 1$ type, e.g., the Mg/Al, Ni/Cu, Pd/Ag, Ir/Pt, and Pt/Au alloys. The surface core level shifts give information about Q_{segr}, a quantity which is difficult to measure thermochemically.

In Fig. 18 we show the Au 4f core level spectra from elemental Au and 2% Au diluted in Pt.[106] It is obvious that there is a strong surface segregation of Au in the dilute alloy system. The surface peak completely dominates the Au 4f spectrum of the $Pt_{0.98}Au_{0.02}$ alloy as compared to elemental Au, where the bulk and surface peaks are of comparable intensity. Upon alloying both the bulk and the surface Au 4f peaks exhibit chemical shifts, which are summarized in Fig. 19. We observe a bulk peak chemical shift of -0.39 eV, which is in good agreement with the chemical shift for dilute Au in Pt observed in XPS.[107] The chemical shift of the surface peak is -0.24 eV, which is significantly less than the bulk peak chemical shift even if the different coordination numbers in the bulk and at the surface are considered. This is consistent with the strong segregation indicated by the relative intensities between the surface and bulk peaks observed for the Au 4f core level spectrum in the $Pt_{0.98}Au_{0.02}$ alloy. The Au atoms at the surface of the alloy are in a more Au-like surrounding than is an Au atom in the bulk of the alloy. The surface Au atoms thus exhibit a smaller alloy chemical shift than the bulk atoms. We also observe a surface core level shift of -0.30 eV for the Pt 4f core level that can be interpreted as meaning that an Au impurity atom in the

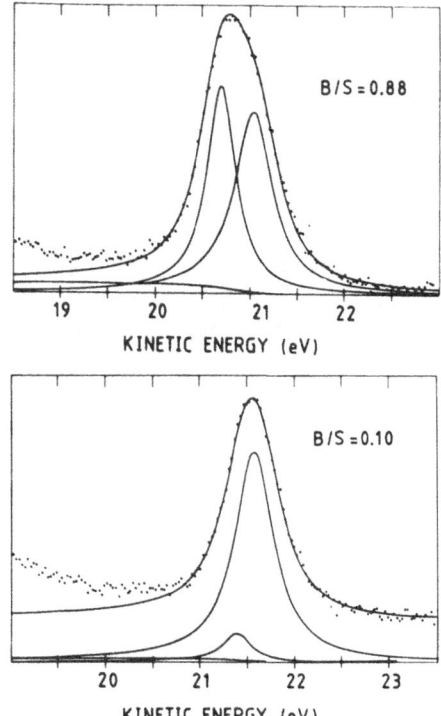

FIGURE 18. Au 4f core level spectra from (top) elemental Au and (bottom) Pt–2% Au alloy. Surface peak chemically shifted to higher kinetic energies in the alloy. $h\nu = 110$ eV. (Data from Ref. 106.)

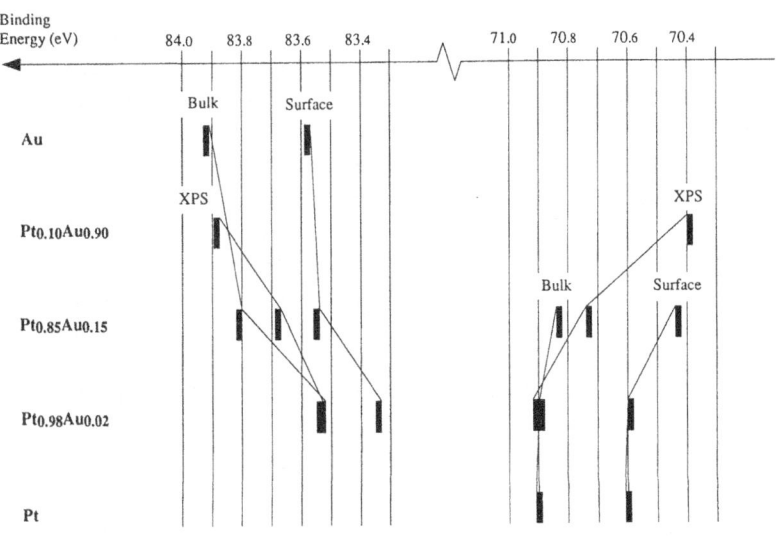

FIGURE 19. Summary of the Pt and Au 4f bulk and surface core level chemical shifts for the PtAu alloy system. (Data from Ref. 106.)

bulk will gain 0.30 eV in energy by segregating to the surface. For an alloy, the surface-layer composition can be obtained in two independent ways, directly from the surface peak intensities or indirectly from the surface core level shift by using Eq. (6.2) and the known bulk composition.

A number of alloys have been studied using surface core level spectroscopy. However, surface core level spectroscopy has so far not proved to be as important a tool in alloy-surface studies as one might have expected. Surface core level spectroscopy yields the surface layer composition but apart from that gives the same information on the surface segregation profile as any fixed photon energy or narrow-range photon energy photoemission experiment. In most synchrotron radiation photoemission studies, the rather narrow photon-energy range available does not allow a large enough variation of the electron inelastic scattering length to give reliable information on the spatial extent and shape of the surface segregation profile. The obvious suggestion would be to use photon energies close to the core level photoionization threshold, where the photoemission probe depth varies from bulk sensitive to surface sensitive over a narrow photon energy range. Unfortunately, this is often difficult because of the low photoexcitation cross section near threshold for the core level of interest, e.g., the delayed onset of the $4f$ core level photo excitation in the $5d$ transition metals. The high and steeply raising background intensity from the scattered secondary electrons also makes the signal-to-background intensity ratio small and rapidly changing. Another possibility would be to use photon energies in the X-ray region, i.e., a few keV, where the excited electron scattering length becomes large again. At present no synchrotron radiation facility is equipped with a photoemission beamline with adequate resolution for such studies. Considering the experimental problems mentioned and the practical difficulties of working at a synchrotron radiation facility, most applied electron spectroscopy studies on alloy surface segregation are still performed using Auger electron spectroscopy. Many elements have three or four useful Auger transitions spanning kinetic energies from 50 eV to 2000 eV, a probe depth change of a factor of five.[108] The easily obtained lateral resolution in electron-excited Auger electron spectroscopy is important in investigating often inhomogeneous alloys, where perhaps one has to select a special single-crystal grain or a grain boundary to obtain the correct information. In the future, however, there will be many more single-crystal surface experiments on alloys using surface core level spectroscopy. Even more important, the technique will be used to study reactions on alloy surfaces. The chemical shift of the surface peak will give information, which presently cannot be obtained in any other way, on, e.g., catalytic reactions on transition-metal alloys.

6.2. Lanthanide Compounds and Alloys

A large number of investigations have been directed toward lanthanide alloys and compounds using surface core level spectroscopy (see, e.g., Ref. 13). We will here discuss only metallic systems, which for the lanthanides often are found as intermetallic compounds with well-defined stoichiometries. One most interesting property of lanthanide compounds is the possibility of a valence change as compared to the pure metal. For instance, Eu and Yb, which are divalent as pure metals, are often observed to change to a mixed-valence or trivalent state upon alloying.

The valency is determined by the balance between the energy required to promote a $4f$ electron to the valence band and the resulting gain in metallic binding energy. Such energy considerations clearly explain why Eu and Yb retain their divalent state when going from the atomic to the metallic state, while many of the other lanthanides change into a trivalent state.[109,110] In an alloy or compound the situation can be altered by the formation of chemical bonds with the other constituent element(s).[111-113] Furthermore, the conditions at the surface may be quite different from those in the bulk. It is not unusual to find a different valence for the bulk and the surface atoms in a lanthanide compound, essentially the same phenomena as observed for Sm metal.

To illustrate this behavior we show in Fig. 20 the Eu $4f$ spectra from different $EuPd_x$ compounds.[104] In EuPd we recognize a $4f$ spectrum characteristic of divalent Eu and closely resembling the $4f$ spectrum from the pure metal. A bulk and a surface peak can be distinguished having a surface core level shift of $+0.8\,eV$. Both peaks consist of a number of closely spaced (and unresolved) multiplets from the f^6 final-state configuration. Similarly, the $EuPd_2$ system is divalent both in the bulk and at the surface. In $EuPd_2Si_2$, however, new $4f$ emission features are found between $6\,eV$ and $10\,eV$ binding energy. These are characteristic of the f^5 final-state configuration in trivalent Eu. Still, both a divalent surface peak and a divalent bulk peak, shifted close to the Fermi level, are observed. These findings lead to the conclusion that Eu is in a mixed-valence state in the bulk, whereas the surface layer remains in a pure divalent state.[13] Considerations of the energy stability of a divalent surface layer on top of a mixed-valence bulk point to the general conclusion that all mixed-valence lanthanide compounds have a stable divalent surface layer. Experimentally, this has been confirmed for several Sm, Tm, Eu, and Yb compounds. The next two compounds in Fig. 20, $EuPd_3$ and $EuPd_5$, show a purely trivalent bulk while the surface layer is still divalent.

From photoemission spectra of the $4f$ levels, as in Fig. 20, important information is obtained concerning the stability of the various valence states both at the surface and in the bulk. For instance, the photoionization of the $4f$ level for a divalent lanthanide system (ground state $f^{n+1}[ds]^2$) leads to a final state $f^n[ds]^3$ corresponding to the trivalent state of the same element. Thus the binding energy of divalent $4f$ peaks can be used to determine the energy required to transform the divalent lanthanide into a trivalent state. In a mixed-valence lanthanide system, the binding energy of the divalent peaks can also help to tell us whether the system is in a heterogeneous or homogeneous mixed-valence state.

The general treatment of bulk and surface core level shifts for lanthanide metals in section 2.2 can be extended to lanthanide compounds. The problem concerning valence stability can also be treated within the framework of this model.[13] Predictions on the stability of surface and bulk valencies for a particular type of compound (LnX, Ln = lanthanide element), can be made for the whole lanthanide series on the basis of measurements on only a few compounds.[16] Stability diagrams have been determined for compounds such as LnPd, $LnPd_3$, $LnAl_2$, and LnS.[114,115]

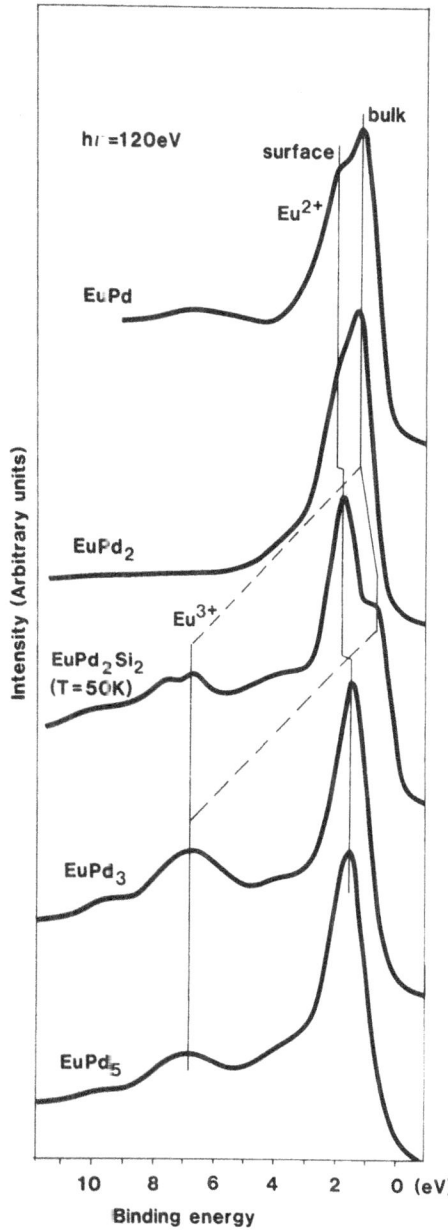

FIGURE 20. Eu $4f$ photoelectron spectra from EuPd$_x$ compounds. Note the energy positions of the $4f^6$ to $4f^5$ (divalent) and the $4f^5$ to $4f^4$ (trivalent) final-state multiplets. (From Ref. 104, reproduced with permission.)

In addition to the experiments on bulk alloys (or compounds), investigations of thin overlayers of lanthanides on metals and semiconductors have gained increasing interest. These experiments contribute to the understanding of interface properties, compound formation, surface segregation and Schottky-barrier formation. Experiments made on thin lanthanide overlayers on well-defined single-crystal substrates provide valuable information on the surface properties of lanthanide compounds. Important thermodynamic data can be obtained from these experiments using the theoretical model reviewed in section

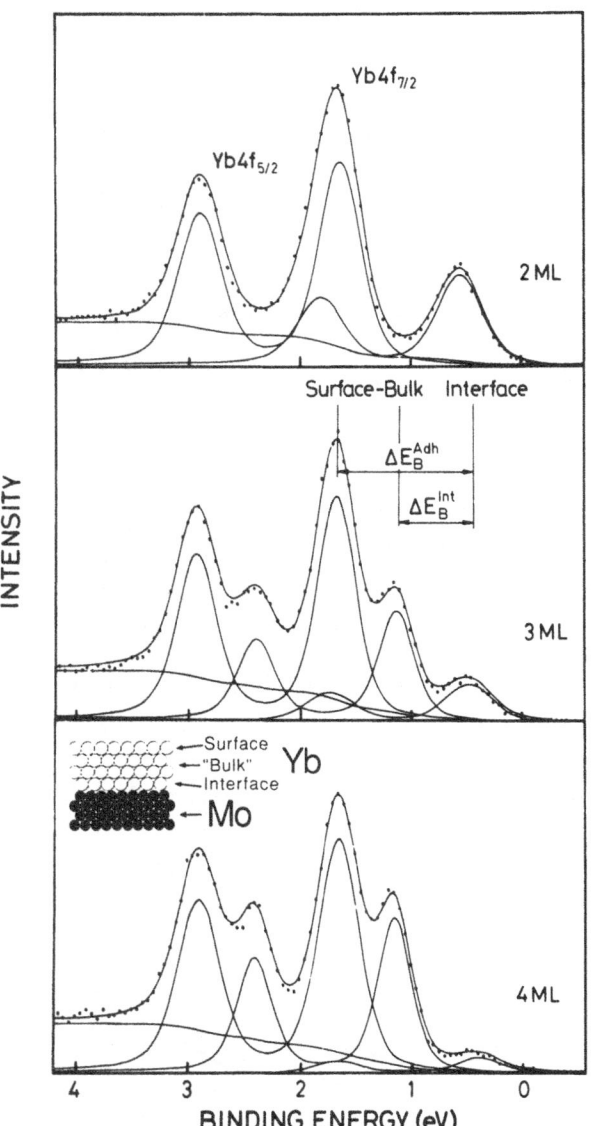

FIGURE 21. Yb 4f photoelctron spectra from an ordered Yb film epitaxially grown on a Mo (110) surface. Shown are surface, interface, and bulk Yb peaks. $hv = 100$ eV. (From Ref. 64, reproduced with permission.)

2. To illustrate this we show, in Fig. 21, photoemission spectra from the Yb/Mo (110) system.[64] In this special case Yb does not form any compound with the Mo substrate. Instead an epitaxial overlayer is formed. For thin overlayers, three doublets from the Yb 4f level are seen (see Fig. 21). These correspond to the emission from the interface, the intermediate [bulk layer(s)], and the surface layer, respectively. From the chemical shift between the surface and interface-related peaks the adhesion energy of Yb to Mo is obtained. Similarly, the chemical shift between the bulk and interface-related peaks gives the interface segregation energy. It should be noted that this kind of experiment can be applied to any combination of elements having measurable surface core level shifts.

A most interesting extension of these kind of measurements is to study compound formation in thin overlayers on single-crystal substrates. Recently Andersen *et al.*[116] have beautifully demonstrated the possibilities of characterizing such systems using surface core level spectroscopy. They studied the formation of two-dimensional (2D) Yb–Ni compounds by co-evaporation of Yb and Ni onto an Mo (110) single-crystal substrate. Using high-resolution surface core level spectroscopy in combination with LEED measurements, they were able to determine the phase diagram for the 2D Yb–Ni system. The phase diagram is determined by the heats of formation of the various 2D compounds *and* the surface energy of the substrate. For instance, it was observed that after the initial formation of YbNi$_2$ an increase in Yb coverage leads to an additional pure Yb phase, prior to the formation of Yb$_2$Ni. The reason for this behavior is the lowering of the surface energy of Mo (110) by adsorbed Yb. The detection of this pure Yb phase can be accomplished only by photoelectron spectroscopy.

7. ADSORBATE-INDUCED SURFACE CORE LEVEL SHIFTS

7.1. Chemisorption

The first unambiguous observation of an adsorbate-induced substrate surface core level shift was one reported for the $2p$ core level peak of the Al (111) surface layer atoms, which is shifted 1.4 eV toward higher binding energy upon chemisorption of a monolayer of O atoms.[7] As shown in Fig. 2, one clearly observes both the oxygen-induced surface core level shifted Al $2p$ peak and the Al $2p$ bulk peak. From a simple extension of the treatment in section 2.1, the 1.4-eV shift can, to a first order of approximation, be interpreted as the difference in chemisorption energies for O atoms on Al (111) and on metallic Si (111).[117] This is in qualitative agreement with bond strengths reported for the Al–O and the Si–O bonds. These bond strengths are estimated from the dissociation energies of Al$_2$O$_3$ and SiO$_2$.

Due to the large bonding energies gained upon oxygen interaction with most metals and semiconductors, the adsorption often causes severe substrate atom rearrangements and direct oxide formation. The complex changes in the substrate core level spectra have been investigated in detail for Ta and W.[11,73,118–120]

From a large number of measurements on adsorbate-induced substrate core level shifts, we will here give the explicit example of CO adsorbed on Pt. CO chemisorbs undissociated on a number of transition metal surfaces and, for many metals, chemisorption energies are available from temperature-stimulated desorption experiments.[121] Surface core level shifts have been reported for CO chemisorbed on Pt (110) and Pt (111) single-crystal surfaces.[68,122–125] The clean Pt (111) surface exhibits a surface shift of −0.4 eV. Upon CO chemisorption, the surface peak is shifted 1.2 eV to a higher binding energy, yielding a +0.8-eV surface shift. Thus the position of the surface peak with respect to the bulk peak changes side upon chemisorption of CO. Analogous to the case of O chemisorption on the Al (111) surface, this change indicates that the chemisorption energy for CO on the Pt (111) surface is larger than for CO on Au (111) surface. Taking the difference between the experimental chemisorption energies for CO on Au and CO on Pt,[121] one obtains 1.2 eV, in agreement with the change in the Pt

(111) surface core level shift upon CO chemisorption. The change in sign of the surface shift upon CO chemisorption on both Pt (111) and Pt (110) shows that for a PtAu alloy the heat of segregation for Au changes sign and that after CO chemisorption Pt segregates to the surface and forms the Pt–CO chemisorption bonds.

7.2. Oxidation of Lu and Hf

As mentioned, the interaction energies of oxygen with most metals are of the same order as the metal-cohesive energies. This implies that major surface-atom rearrangements are possible upon oxygen adsorption, and often the oxidation process proceeds directly from a clean surface to an bulk oxide-covered surface without passing through any intermediate chemisorption phases. This is the case for almost all lanthanide metals and some $5d$ transition metals, as for example Hf.

In Fig. 22 we observe in the clean Hf $4f$ spectrum a surface peak shifted +0.42 eV.[71] Upon small oxygen exposures, the surface peak broadens and disappears. For higher oxygen exposures the growth of the maximum-valence HfO_2 $4f$ peak shifted +3.8 eV is observed. Apart from remaining ambiguities, as a proper explanation of the broadening of the surface peak upon the initial

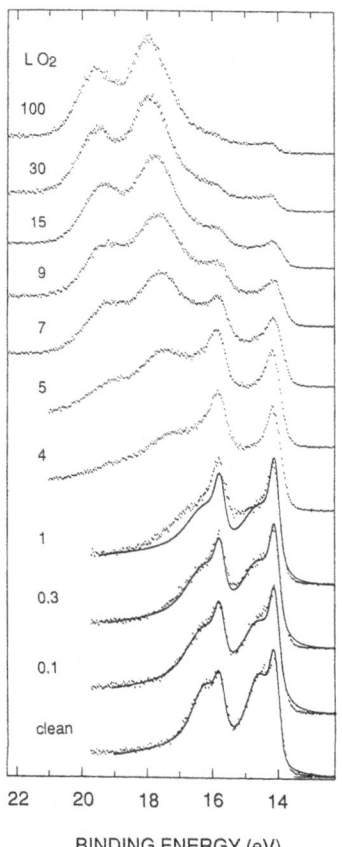

FIGURE 22. Hf $4f$ photoelectron spectra from oxygen-exposed evapoarated Hf films. Oxidation proceeds directly into the maximum valence HfO_2 oxide signified by the +3.8 eV core level chemical shift. $h\nu = 100$ eV. (From Ref. 65.)

oxygen interaction and the HfO_2 $4f$ peak not immediately reaching its final binding energy upon oxidation, the Hf metal oxidation process as observed in Fig. 22 is typical of many metals.

In Fig. 23 we observe in the clean Lu $4f$ spectrum, a surface peak shifted +0.73 eV.[65] Upon small oxygen exposures, the surface peak moves to higher binding energies but retains its intensity and line shape. This seems to contradict the local oxygen–metal bond picture since seemingly all Lu surface atoms are collectively influenced upon oxygen adsorption. Similar behavior has been observed for Ta and W single-crystal surfaces.[118] The explanation given was that the first stage of oxygen interaction with these single-crystal surfaces was a quenching of surface states in the valence band. Since the surface states are delocalized over rather large surface areas, a collective behavior in the change of the potential for the surface core levels is expected. However, the Lu surface is polycrystalline and has no preferentially oriented single-crystal grains, as proved by the large width of the surface peak. Thus a unique chemisorption phase quenching a surface state specific to a single-crystal surface seems rather unlikely.

The total change in the surface core level shift upon oxygen exposure is only 0.2 eV, implying a small charge transfer typical more of the chemical shifts observed upon alloying metals and not of the strong charge transfer expected in oxidation. There is a possibility that the surface Lu atoms form a stable metallic monoxide, LuO. The continuous binding-energy shift of the Lu surface peak upon oxygen exposure is interpreted as being due to the formation of LuO_x at the surface, yielding the final 0.2 eV chemical shift when LuO has been formed. Since the system is metallic, the oxygen interaction in this case would be of extended character, and it would be independent of the polycrystalline nature of the surface. After LuO has been formed, further oxygen exposure transforms the surface into the maximum-valence oxide Lu_2O_3, which is signified by the observed oxide chemical shift of +2.4 eV.

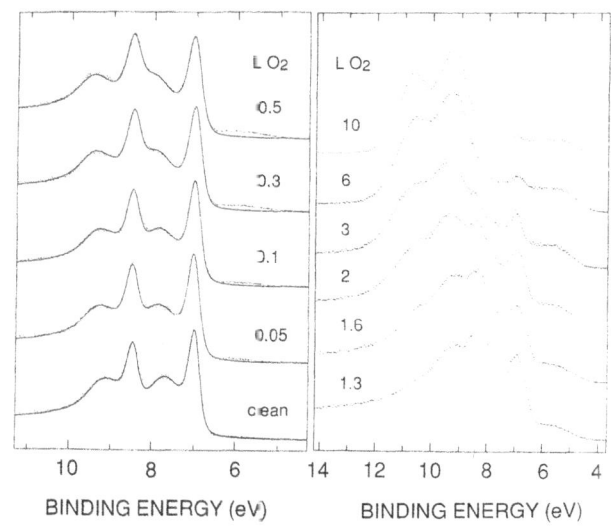

FIGURE 23. Lu $4f$ photoelectron spectra recorded from oxygen-exposed evaporated Lu films. Oxidation proceeds through two phases. The first is signified by a small, +0.2-eV chemical shift of the surface atoms. The second is the formation of the LuO_2 maximum-valence oxide signified by the +2.3-eV core level chemical shift. $hv = 100$ eV. (From Ref. 65.)

8. SURFACE CORE LEVEL SHIFTS AT SEMICONDUCTOR SURFACES

8.1. III–V and II–VI Compound Semiconductors

The screening of a core hole is less efficient in semiconductors and insulators than it is for metallic systems, the Fermi screening length in metals being less than or comparable to interatomic distances because of the high density of conduction electrons that can move freely and completely screen the positive core hole. This yields a well-defined final state of the photoemission process. As we have stressed, this fully screened core hole atom is an essential assumption for using the thermodynamic model in calculating the surface core level shifts of metal surfaces. In semiconductors and insulators, the final-state screening is incomplete, leaving a partially charged core-hole atom in the final state of the photoemission process.[126]

The dielectric screening of a core hole in a semiconductor or an insulator is reduced at the surface compared to the bulk, which should lead to a higher appearent surface core level binding energy. This is different compared to a metal where the final-state screening charge can be located either in a bonding or in an antibonding part of the conduction band and the resulting contribution to the surface core level shift could be either positive or negative. The final-state relaxation energies for semiconductors and insulators are smaller than for metals. Charge neutrality in a core excitation in semiconductors and insulators can be obtained only with the core hole and the excited electron forming a bound state, i.e., a bulk or a surface excition. Surface excitons should have lower excitation energies due to the reduced screening at the surface. It is possible that the thermodynamic model could be applied to the binding-energy difference between a surface and a bulk exciton.[126] However, for such common semiconductors as Si, Ge, and GaAs, the dielectric screening is very effective and the excitonic binding energies are on the order of tens of meV (30 meV for Si)[127] and thus considerably smaller than the core level lifetime widths. This makes it difficult to resolve any photoabsorption features that are due to either bulk or surface excitons at the core level photoionization thresholds. The existence of unoccupied surface states in the bandgap close to the conduction-band minimum further complicates the photoabsorption spectra. To obtain easily resolved excitonic structures, one has to resort to wide-bandgap semiconductors or insulators.[128]

It has been shown that the charge in surface states at semiconductor surfaces is screened over a few atomic layers.[129] The nonpolar surfaces of the III–V and II–VI compound semiconductors, i.e., the (110) surface of the zincblende structure and the ($11\bar{2}0$) and ($10\bar{1}0$) surfaces of the wurtzite structure exhibit a surface-symmetry-conserving reconstruction giving rise to anion- and cation-derived surface states. The generally accepted picture of the reconstructions of these surfaces involves a surface bond rotation, where the nonmetallic anions move outward and accumulate negative charge, the metallic cations move inward and become positively charged, as compared to bulk anions and cations, respectively.[129,130] In a simple picture one therefore expects the anion surface core level shift to be negative and the cation surface core level shift to be positive.[131] In Fig. 24 we display the Cd $4d$ core level of the $(11\bar{2}0)$ surface of wurtzite CdS.[132] A surface shift of $+0.40\,\text{eV}$ is clearly seen. In the fitting procedure Voigt line shapes are assumed. The shoulder on the high-kinetic-energy side of the $4d$ doublet has been interpreted as emission from the S $3s$ level.

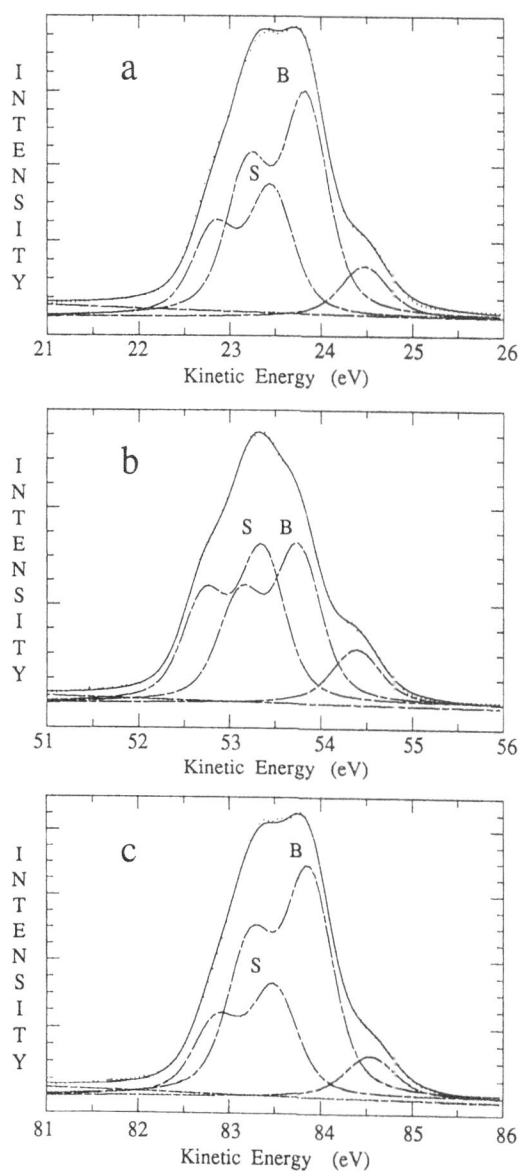

FIGURE 24. Surface-sensitive core level spectrum of the Cd $4d$ doublet in CdS $(10\bar{1}0)$ recorded at (a) 40 eV, (b) 70 eV, and (c) 100 eV photon energy. Experimental spectrum fitted with (B) bulk Cd $4d$ doublet, (S) surface Cd $4d$ doublet displaced 0.40 eV to lower kinetic energy (higher binding energy) and a S $3s$ peak at high kinetic energy. (From Ref. 132.)

TABLE 4. Surface Core Level Shifts of the Non Polar Surfaces of III–V and II–VI Compound Semiconductors

	Experimental		Calculated[a]	
	Cation shift (eV)	Anion shift (eV)	Cation shift (eV)	Anion shift (eV)
AlP			0.30	−0.69
AlAs			0.33	−0.40
AlSb	0.38[b]	−0.38[b]		
GaP	0.28,[c] 0.31[d]	−0.41[d]	0.41	−0.43
GaAs	0.28[c]	−0.37[c]	0.38	−0.40
GaSb	0.30[c]	−0.36[c]	0.35	−0.41
InP	0.30,[e] 0.35,[f] 0.28,[g] 0.33[h]	−0.24,[g] −0.31[h]	0.34	0.00
InAs	0.26,[e] 0.27[j]	−0.30[j]	0.31	−0.28
InSb	0.24,[g] 0.22[j]	−0.29[b]	0.28	−0.22
CdS	0.39[k]	−0.43[k]		
CdTe	0.24[l]	−0.26[l]		

[a] From Ref. 129. [b] From Ref. 141. [c] From Ref. 131. [d] From Ref. 137. [e] From Ref. 134. [f] From Ref. 136. [g] From Ref. 138. [h] From Ref. 140. [i] From Ref. 139. [j] From Ref. 133. [k] From Ref. 132. [l] From Ref. 135.

In Table 4 we summarize the experimental surface core level shifts reported for the III–V and II–VI semiconductors. We observe that the notion about symmetric surface shifts with respect to the bulk core levels of the surface anions and cations of the III–V and II–VI semiconductors seems to hold. The agreement with the simple surface-bond charge-transfer model is probably fortuitous. There are many contributions to the surface core level binding energies for the surface anions and cations besides the surface-bond charge transfer.[142] The Madelung potential is different for the surface atoms because of the reduced coordination at the surface. The charge transfer between the anions and the cations at the surface also changes the surface Madelung potential and creates a surface dipole layer. The reduced dielectric screening at the surface changes the final-state relaxation energy and contributes to the surface shift.

    The possible contributions to the surface shifts at the (110) III–V and II–VI semiconductor surfaces have been discussed in the literature, and different ways to interpret the observed surface core level shifts of the anions and the cations have been suggested.[129,131,135,138,143,144] A common view has been that for the III–V semiconductors the reduced-surface Madelung potential due to the lower coordination number at the surface explains the observed surface core level shifts without invoking extra surface-bond charge transfer.[138,143,144] Priester et al.[129] performed a full initial-state calculation of the surface core level shifts for the (110) surfaces of the III–V semiconductors, neglecting only possible differences in the final-state relaxation energy between the surface and the bulk core levels. Self-consistency in the calculations was assured by using a local charge neutrality condition which somewhat overestimates the calculated surface shifts. Their predictions fit the available experimental data quite well, but more interestingly they predict discrepancies to the rule of symmetric but opposite binding-energy shifts of the surface anion and cation. For example, the anion $2p$ core level of P in InP should exhibit no surface shift and the corresponding $2p$ core level in AlP a shift of -0.69 eV, while the cation surface shifts should remain close to their usual value, approximately $+0.3$ eV. Extensive experimental data on the surface core level shifts for the (110) surface of the III–V and II–VI semiconductors are being collected and compared to this full initial-state calculation (see Table 4). This will possibly yield information on the mechanisms behind the surface shifts and on the role of the core-hole final-state relaxation in semiconductors.

In the spirit of Yin and Tosatti,[126] Prince et al.[135] applied the thermodynamic model to the surface core level shifts of semiconductors. Strictly speaking, if we naively use the simple model depicted in Fig. 4, the surface segregation energy for a substitutional $Z + 1$ impurity in a semiconductor is given by the energy difference between the Z-element absorption edges for the surface and bulk core excitons. However, the dielectric screening in semiconductors is strong, as can be noted by the small binding energies with respect to the conduction-band minimum observed for core excitons (tens of meV).[127] Thus, using the surface shifts observed in photoemission to obtain the surface segregation energies should not introduce major errors. In Table 4 we observe that all experimental anion surface shifts reported for the III–V semiconductors are negative, implying that n-doping impurities should have a tendency to segregate to the surface. Contrariwise, experimental surface-segregation energies for the substitutional $Z + 1$ impurities in a semiconductor should give a first-order estimate of the corresponding surface shifts. This approach is most easily applied to semiconductors, where the surface reconstructions do not introduce different types of surface sites.

8.2. Si and Ge Reconstructed Surfaces

Semiconductor surface-reconstruction models can be tested indirectly by comparing calculated and measured surface electronic structures. One can use the calculated energy positions and dispersions of surface states and resonances for a semiconductor surface and compare them with the corresponding features as measured by means of angle-resolved photoemission or angle-resolved inverse photoemission. From scanning tunneling microscopy, surface X-ray diffraction, and low-energy electron diffraction experiments, one usually obtains fair knowledge of the surface geometry more directly before performing the electronic structure studies.

The main use of the surface core level shifts for studying semiconductor surfaces is their chemical signature. Different surface reconstructions of clean elemental semiconductor surfaces and their different adsorbate-induced surface reconstructions have their own specific set of surface core level shifts, the signature being the number of surface shifts, their binding energies, and their intensities. The number of surface peaks for a reconstructed surface is determined by the number of different surface sites. The possibility of differentiating among all surface sites is limited only by the experimental resolution and the intrinsic core level widths. The magnitude and direction of the surface shifts give a qualitative picture of the character of the surface bonds, and the intensities should yield the number of surface atoms to be associated with a specific type of site. The experimental information about these quantities provides an important test of the suggested surface reconstruction.

We exemplify surface core level spectroscopy applied to clean elemental semiconductor surfaces by spectra recorded for the Ge (111) $c(2 \times 8)$ and Ge (100) 2×1 reconstructed surfaces.[145] The 55-eV photon energy spectra shown in Fig. 25 represents superpositions of about equal amounts of surface and bulk emission. We observe that the Ge (100) 2×1 $3d$ core level spectrum can be fitted with only one surface core level peak, having a surface shift of -0.43 eV, while the Ge (111) $c(2 \times 8)$ $3d$ core level spectrum requires two surface peaks having -0.22-eV and -0.75-eV surface shifts respectively. This suggests one type of surface atoms for the Ge (100) 2×1 surface and two types of surface atoms for the Ge (111) $c(2 \times 8)$ surface. The large -0.75-eV surface shift for some surface atoms at the Ge (111) $c(2 \times 8)$ surface could be interpreted as Ge atoms located at special surface sites. The surface shift would then in a simple picture be due to the adatoms yielding the Ge (111) $c(2 \times 8)$ surface reconstruction.[146] A similar large surface shift of -0.76 eV is observed for the adatoms of the Si (111) 7×7 surface.[147] Recently this view has been challenged.[148,149] Using arguments based on charge transfer calculations,[150] the large surface shifts to lower binding energies, -0.75 eV and -0.76 eV, were attributed to the restatoms for the Ge (111) $c(2 \times 8)$ and Si (111) 7×7 surfaces, respectively.[148,149]

To compare with probable reconstruction models, one uses the number of surface atoms in the different surface sites. The observed surface peak intensities I_S can be related to a certain fraction of monolayer coverage Θ using the relation

$$\frac{I_S}{I_T} = R = \theta(1 - e^{-d_\perp/\lambda}), \qquad (8.1)$$

where I_T is the total intensity, i.e., the intensity in the surface peak under consideration plus the bulk peak intensity, $d_\perp$ is the atomic layer distance

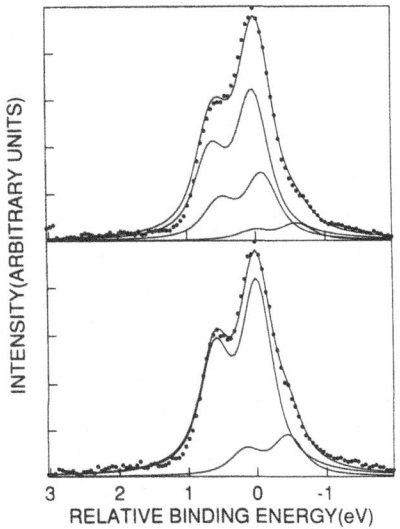

FIGURE 25. Surface-sensitive spectra of the Ge $3d$ doublet for (top) Ge (111) $c(2 \times 8)$ and (bottom) Ge (100) (2×1) surfaces. Ge (111) $c(2 \times 8)$ spectrum fitted with two surface peaks at -0.76 eV and -0.25 eV, respectively. Ge (100) (2×1) peak fitted with one surface peak at -0.43 eV. (From Ref. 145.)

perpendicular to the surface, and λ is the excited electron inelastic scattering length. We need to know λ in order to calculate Θ from the experimentally measured intensity ratio R. A simple way is to use the universal curve for excited electron inelastic scattering lengths as a function of the electron kinetic energy. This curve is based on a calculated shape and fitted to experimental results obtained in different ways. This is a dubious procedure, where scientific intuition about assumed reasonable reconstruction models often influences the interpretation of the experimental data.

More accurate ways involve adsorbing exactly a monolayer of gas atoms onto the reconstructed semiconductor surface, the monolayer criterion being the observation of a new 1×1 reconstruction pattern in LEED, the use of RHEED (reflection high-energy electron diffraction) spot intensity variations to obtain calibrated coverages during molecular-beam epitaxial growth of the surface to be investigated or the use of a "known" surface reconstruction to calibrate the surface peak intensities.[151-153] In Table 5 we give the surface core level shifts and the corresponding surface coverages for a number of important surface reconstructions of Si and Ge surfaces. The surface shifts and the corresponding coverages can be related to and interpreted in terms of various reconstruction models for these surfaces. It should be remembered that the observed surface shifts and their intensities have been taken as evidence for other reconstruction models before the ones now believed to be correct. Surface core level spectroscopy is certainly not the best tool for *a priori* surface-structure determination. Surface core level spectroscopy data reported from the cleaved Si (111) 2×1 reconstructed surface show two surface-shifted peaks, each with an estimated intensity corresponding to 0.5 ML coverage, which is consistent both with the earlier suggested buckling model and the nowadays established π-bonded chain model of this surface. Another ambiguity concerns the Si (100) 2×1 and Ge (100) 2×1 surfaces, where all surface core level spectroscopy data agree on one single-surface peak shifted about 0.5 eV to lower binding energies (see Fig. 25). The derived coverages vary between 0.5 ML and 1.0 ML. The building blocks for these surfaces are dimers, but it is not clear if the dimer atoms are equivalent (symmetric dimer) or inequivalent (asymmetric dimer). If the dimer is asymmetric, we expect a charge transfer between the dimer atoms, i.e., different surface shifts for the two dimer atoms. Recent scanning tunneling microscopy data seems to establish for the Ge (100) 2×1 surface the dominance of asymmetric dimers for well-prepared surfaces.[164] Either the charge transfer between the dimer atoms is minute and the dimer atoms are equivalent within the combined intrinsic and experimental energy resolution, so that then the correct coverage is 1.0 ML, or we are missing a surface peak and the correct coverage is 0.5 ML.

Perhaps the most promising case to which to apply the thermodynamic model are the surface shifts observed for the rest atoms of the Si (111) 7×7 and Ge (111) $c(2 \times 8)$ reconstructions, i.e., the surface atoms making up an almost perfect bulk termination of the Si and Ge crystals in the (111) direction. They interestingly show shifts in opposite directions with respect to the bulk peak (see Table 5), an experimental observation that possibly can be explained from thermodynamic data as impurity segregation-energy differences in Si and Ge.

TABLE 5. Experimental Surface Core Level Shifts and Corresponding Surface Coverages for Various Si and Ge Surfaces

	7×7 (eV/ML)	2×1 (eV/ML)	1×1 (eV/ML)	$c(2 \times 8)$ (ev/ML)
Si (111)	$-0.70/0.13^a$	$-0.37/0.5^a$	$-0.80/0.23^b$	
	$-0.7/0.16^b$	-0.59^c	$+0.75\text{-As}^d$	
	$-0.77/0.24^e$	$+0.30^c$	$+0.26\text{-H}^a$	
	$+0.36/1.2^e$	-0.41^f	$+0.16\text{-H}^a$	
	-0.76^g	$+0.24^f$		
	$+0.36^g$			
Si (100)		$-0.52/0.5^a$		
		-0.53^d		
		$-0.52/0.92^h$		
		-0.52^i		
		$+0.39^i$		
Ge (111)		-0.65^f	$-0.60/0.37^b$	$-0.77/0.33^h$
		-0.26^f	$+0.38\text{-As}^d$	$-0.27/1.9^h$
			$+0.47/1\text{-Br}^d$	$-0.70/0.32^k$
			$+0.57/1\text{-Cl}^j$	$-0.20/1.33^k$
				$-0.75/0.28^b$
				$-0.35/0.25^b$
				-0.72^d
				-0.23^d
				$-0.77/0.37^g$
				$-0.26/2.1^g$
				$-0.76/0.25^j$
				$-0.26/0.93^j$
Ge (100)		$-0.43/0.87^h$	$+0.67/1\text{-S}^l$	
		-0.42^m		
		-0.41^n		
		$-0.43/0.62^j$		
		-0.44^l		
		$-0.43/0.78^o$		

^{a} From Ref. 154. ^{b} From Ref. 155. ^{c} From Ref. 157. ^{d} From Ref. 156. ^{e} From Ref. 153.
^{f} From Ref. 158. ^{g} From Ref. 147. ^{h} From Ref. 156. ^{i} From Ref. 163. ^{j} From Ref. 151.
^{k} From Ref. 156. ^{l} From Ref. 162. ^{m} From Ref. 163. ^{n} From Ref. 151. ^{o} From Ref. 145.

The most common use of semiconductor surface core level shifts has been in surface-reaction studies. Two of the most frequently studied systems are metal-overlayer-induced reconstructions of semiconductor surfaces and oxidation of semicondcutor surfaces. In oxidation studies the possibility of observing small substrate surface core level shifts has made it possible to closely follow the oxidation of a semiconductor surface step by step and not merely observe the maximum valence bulk like oxide after an oxide layer has been formed.[165] This has led to a deeper understanding of the oxidation process of semiconductor surfaces.

O, O_2, and H_2O have been used as the oxidizing agents. The Si (100) 2×1 surface displays a very special behavior in comparison with the other low-index Si and Ge surfaces that have been investigated. When exposing the Si (100) 2×1 surface to H_2O at RT, a unity sticking coefficient is observed and

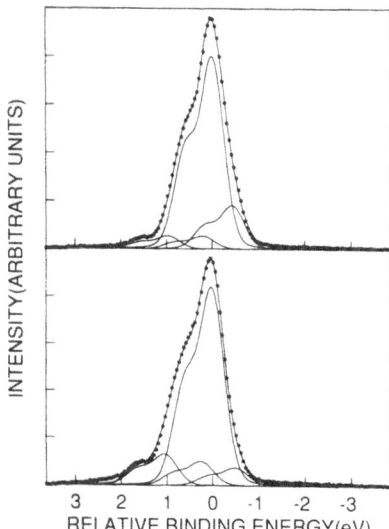

FIGURE 26. Surface-sensitive spectrum of the Si $2p$ doublet for a (top) clean and (bottom) H_2O-dosed Si (100) (2 × 1) surface. Surface peaks induced by chemisorbed OH and H at 1.0 eV and 0.25 eV, respectively, are used to fit the experimental spectrum. (From Ref. 166.)

immediate dissociation into OH and H takes place.[166] After monolayer formation, the surface becomes inert to further oxidation. In Fig. 26 we show the Si $2p$ core level spectra after H_2O exposure. We observe two surface peaks of equal intensity.[166] The larger surface shift is characteristic of an ionic Si–O or Si–OH bond and the smaller surface shift of a Si–H bond. The surface core level shifts of Si–O and Si–H have been determined in independent experiments and agree well with those observed for the dissociated H_2O on Si (100) 2 × 1. Useful information from the surface shifts is that their equal intensities yield equal coverage of OH and H, i.e., it is very probable that OH and H occupy one dimer atom each. In the case of compound semiconductors, surface core level spectroscopy is even more powerful in following the initial oxidation because one can observe the roles both of the anion and the cation atoms at the surface in the initial oxidation process.

ACKNOWLEDGMENT

Technical assistance was given by Mrs. Birgitta Norman-Ebendal. The authors would like to express their sincere gratitude to The Swedish Natural Science Research Council for financial support.

REFERENCES

1. K. Siegbahn, C. Nordling, A. Fahlman, R. Nordberg, K. Hamrin, J. Hedman, G. Johansson, T. Bergmark, S. E. Karlsson, J. Lindgren, and B. Lindberg, *Nova Acta Regiae Soc. Sci. Ups.*, **20**, (1967).
2. K. Siegbahn, C. Nordling, G. Johansson, J. Hedman, P. F. Hedén, K. Hamrin, U. Gelius, T. Bergmark, L. O. Werme, R. Manne, and Y. Baer, *ESCA Applied to Free Molecules*, North-Holland, Amsterdam (1969).
3. R. H. Rithie, F. W. Garber, M. Y. Nakai, and R. D. Birkhoff, *Adv. Radiat. Biol.* **3**, 1 (1969).

4. W. F. Krolikowski and W. E. Spicer, *Phys. Rev.* **185**, 882 (1969).
5. W. F. Krolikowski and W. E. Spicer, *Phys. Rev.* **B1**, 478 (1970).
6. D. E. Eastman, *Solid State Commun.* **8**, 41 (1970).
7. S. A. Flodström, C. W. B. Martinsson, R. Z. Bachrach, S. B. M. Hagström, and R. S. Bauer, *Phys. Rev. Lett.* **40**, 907 (1978).
8. P. H. Citrin, G. K. Wertheim, and Y. Baer, *Phys. Rev. Lett.* **41**, 1425 (1978).
9. T. M. Duc, C. Guillot, Y. Lassailly, J. Lecante, Y. Jugnet, and J. C. Vedrine, *Phys. Rev. Lett.* **43**, 789 (1979).
10. B. Johansson and N. Mårtensson, *Helv. Phys. Acta* **56**, 405 (1983).
11. D. Spanjaard, C. Guillot, M. C. Desjonquères, G. Tréglia, and J. Lecante, *Surf. Sci. Rep.*, **5** (1985).
12. W. F. Egelhoff, Jr., *Surf. Sci. Rep.*, **6** (1986).
13. B. Johansson and N. Mårtensson in *Handbook on the Physics and Chemistry of Rare Earths* (K.A. Gschneidener, L. Eyring, and S. Hüfner, eds.) North-Holland, Amsterdam (1988), p. 361.
14. Y. Jugnet, G. Grenet, and T. M. Duc in *Handbook on Synchrotron Radiation*, Vol. 2, (G. V. Marr, ed.), North-Holland, Amsterdam (1987), p. 663.
15. B. Johansson, *J. Phys.* **F4**, L169 (1974).
16. B. Johansson, *Phys. Rev.* **B20**, 1315 (1979).
17. B. Johansson and N. Mårtensson, *Phys. Rev.* **B21**, 4427 (1980).
18. B. Johansson, *Phys. Rev.* **B19**, 6615 (1979).
19. A. Rosengren and B. Johansson, *Phys. Rev.* **B22**, 3706 (1980).
20. P. Steiner, S. Hüfner, N. Mårtensson, and B. Johansson, *Solid State Commun.* **37**, 73 (1981).
21. P. Steiner and S. Hüfner, *Solid State Commun.* **37**, 79 (1981).
22. P. Steiner and S. Hüfner, *Solid State Commun.* **37**, 279 (1981).
23. N. Mårtensson, R. Nyholm, H. Calén, J. Hedman, and B. Johansson, *Phys. Rev.* **B24**, 1725 (1981).
24. P. F. de Châtel and F. R. de Boer, in *Valence Fluctuations in Solids* (L. M. Falicov, W. Hanke, and M. B. Maple, eds.) North-Holland, Amsterdam (1981), p. 377.
25. J. F. Herbst, D. N. Lowy, and R. E. Watson, *Phys. Rev.* **B6**, 1913 (1972).
26. J. F. Herbst, R. E. Watson, and J. W. Wilkins, *Phys. Rev.* **B13**, 1439 (1976).
27. J. F. Herbst, R. E. Watson, and J. W. Wilkins, *Phys. Rev.* **B17**, 3089 (1978).
28. B. Johansson and A. Rosengren, *Phys. Rev.* **B14**, 361 (1976).
29. A. Rosengren and B. Johansson, *Phys. Rev.* **B23**, 3852 (1981).
30. B. C. Allen, in *Liquid Metals, Chemistry and Physics* (S. Z. Beer, ed.), Dekker, New York (1972), p. 161.
31. B. Johansson and P. Munck, *J. Less-Common Met.* **100**, 49 (1984).
32. A. R. Miedema, *J. Less-common Met.* **46**, 67 (1976).
33. R. Kammerer, HASYLAB Internal Report 82-05, (1982).
34. D. A. Shirley, *Phys. Rev.* **B5**, 4709 (1972).
35. H. H. Madden and J. E. Houston, *J. Appl. Phys.* **47**, 3071 (1976).
36. S. Tougaard and P. Sigmund, *Phys. Rev.* **B25**, 4452 (1982).
37. S. Tougaard, *Surf. Sci.* **139**, 208 (1984).
38. S. Tougaard and B. Jörgensen, *Surf. Sci.* **143**, 482 (1984).
39. A. L. Tofterup, *Surf. Sci.* **167**, 70 (1986).
40. S. Doniach and M. Šunjić, *J. Phys.* **C3**, 285 (1970).
41. G. D. Mahan, *Phys. Rev.* **B11**, 4814 (1975).
42. G. K. Wertheim and L. R. Walker, *J. Phys.* **F6**, 2297 (1976).
43. P. H. Citrin, G. K. Wertheim, and Y. Baer, *Phys. Rev.* **B16**, 4256 (1977).
44. G. K. Wertheim and P. H. Citrin, in *Photoemission in Solids I,* Vol. 26 of Topics in Applied Physics (M. Cardona and L. Ley, eds.) Springer, Berlin (1978), p. 197.
45. L. Hedin and A. Rosengren, *J. Phys.* **F7**, 1339 (1977).
46. N. Mårtensson and R. Nyholm, *Phys. Rev.* **B24**, 7121 (1981).
47. R. Nyholm and N. Mårtensson, *Chem. Phys. Lett.* **45**, 754 (1980).
48. M. Ohno and G. Wendin, *Solid State Commun.* **39**, 875 (1981).
49. P. Heimann, J. F. van der Veen, and D. E. Eastman, *Solid State Commun.* **38**, 595 (1981).

50. F. Gerken, A. S. Flodström. J. Barth, L. I. Johansson, anc C. Kunz, *Phys. Scr.* **32,** 43 (1985).
51. R. Nyholm and N. Mårtensscn, *Phys. Rev.* **B36,** 20 (1987).
52. C. P. Flynn, *Phys. Rev. Lett.* **37,** 1445 (1976).
53. P. A. Cox, *Struct. Bonding (Berlin),* **24,** 59 (1975).
54. P. A. Cox, J. K. Lang, and Y. Baer, *J. Phys.* **F11,** 113 (1981).
55. F. Gerken, *J. Phys.* **F13,** 703 (1983); Ph.D. Thesis, University of Hamburg (1983).
56. J. K. Lang, Y. Baer, and P. A. Cox, *J. Phys.* **F11,** 121 (1981).
57. A. Nilsson, N. Mårtensson, J. Hedman, B. Eriksson, R. Bergman, and U. Gelius, *Surf. Sci.* **162,** 51 (1985).
58. R. D. Parks, N. Mårtensson. and B. Reihl, in *Valence Instabilities* (P. Wachter and H. Boppart, eds.) North Holland, Amsterdam (1982), p. 239.
59. J. K. Lang and Y. Baer, *Solid State Commun.* **31,** 945 (1979).
60. R. Kammerer, J. Barth, F. Gerken, C. Kunz, S. A. Flodström, and L. I. Johansson, *Solid State Commun.* **41,** 435 (1982).
61. G. Kaindl, W. D. Schneider, C. Laubschat, B. Reihl, and N. Mårtensson, *Surf. Sci.* **126,** 105 (1983).
62. S. F. Alvarado, M. Campagna, and W. Gudat, *J. Electron. Spectros. Relat. Phenom* **18,** 43 (1980).
63. M. Hecht, L. I. Johansson. J. W. Allen, S. J. Oh, anc I. Lindau, Proc. VI VUV-Conf., Charlottesville (1980) pp. 1–64.
64. N. Mårtensson, A. Stenborg. O. Björneholm, A. Nilsson, and J. N. Andersen, *Phys. Rev. Lett.* **60,** 1731 (1988).
65. J. Schmidt-May and R. Nyhelm, in *VUV-Radiation Physics-* (A. Weinreb and A. Ron, eds.), *An. Is. Phys. Soc.* **6,** 348 (1983); J. Schmidt-May, Ph.D. Thesis, University of Hamburg (1985).
66. J. F. van der Veen, F. J. Himpsel, and D. E. Eastman, *Phys. Rev. Lett.* **44,** 189 (1980).
67. G. Apai, R. C. Baetzold, E. Shustorovitch, and R. Jaeger, *Surf. Sci.* **116,** L191 (1982).
68. G. Apai, R. C. Baetzold, P. J. Jupiter, A. J. Viescas, and I Lindau, *Surf. Sci.* **134,** 122 (1983).
69. N. Mårtensson, H. B. Saalfeld, H. Kuhlenbeck, and M. Neumann, *Phys. Rev.* **B39,** 8181 (1989).
70. C. Guillot, D. Chauveau, P Robin, J. Lecante, M. C. Desjonqueres, G. Treglia, and D. Spanjaard, *Surf. Sci.* **162,** 46 (1985).
71. R. Nyholm and J. Schmidt-May, *J. Phys.* **C17,** L113 (1984).
72. G. Treglia, M. C. Desjonqueres, D. Spanjaard, Y. Lasailly C. Guillot, Y. Jugnet, T. M. Duc, and J. Lecante, *J. Phys.* **C14,** 3463 (1981).
73. M. C. Desjonqueres, D. Spanjaard, Y. Lasailly, and C. Guillot, *Solid State Commun.* **34,** 807 (1980).
74. M. C. Desjonqueres and F. Cyrot-Lackmann, *Solid State Commun.* **18,** 1127 (1976).
75. H. L. Skriver, in *Systematics and the Properties of the Lanthanides* (S. P. Sinha, ed.) Reidel, Dordrecht (1983), p. 239.
76. O. K. Andersen, O. Jepsen, and D. Glötzel, in *Highlighs in Condensed Matter Theory* (F. Bassani, F. Fumi, and M. P. Tosi, eds.), North-Holland, New York (1985).
77. H. L. Skriver, *Phys. Rev.* **B31,** 1909 (1985).
78. J. C. Duthie and D. G. Pettifor, *Phys. Rev. Lett.* **38,** 564 (1977).
79. W.-D. Schneider, C. Laubschat, and B. Reihl, *Phys. Rev.* **B27,** 6538 (1983).
80. A. Stenborg, J. N. Andersen O. Björneholm, A. Nilsson, and N. Mårtensson, *Phys. Rev. Lett.* **63,** 187 (1989).
81. G. K. Wertheim and G. Crecelius, *Phys. Rev. Lett.* **40,** 813 (1978).
82. A. Rosengren and B. Johansson, *Phys. Rev.* **B26,** 3068 (1982).
83. J. W. Allen, L. I. Johansson, I. Lindau, and S. B. M. Hagström, *Phys. Rev.* **B21,** 1335 (1980).
84. M. Domcke, C. Laubschat, M. Prietsch, T. Mandel, G. Kaindl, and W.-D. Schneider, *Phys. Rev. Lett.* **56,** 1287 (1986).
85. N. D. Lang and W. Kohn, *Phys. Rev.* **B1,** 4555 (1970).
86. K. H. Lau and W. Kohn, *J. Phys. Chem. Solids* **37,** 99 (1976).
87. E. Wimmer, M. Weinert, A. J. Freeman and H. Krakauer, *Phys. Rev.* **B24,** 2292 (1981).
88. P. J. Feibelman, *Phys. Rev.* **B39,** 4866 (1989).
89. R. Kammerer, J. Barth, F. Gerken, C. Kunz, S. A. Flodström, and L. I. Johansson, *Phys. Rev.* **B26,** 3491 (1982).

90. R. Kammerer, A. S. Flodström, and C. Kunz, unpublished data.

91. R. Nyholm, A. S. Flodström, L. I. Johansson, S.-E. Hörnström, and J. Schmidt-May, *Surf. Sci.* **149**, 449 (1985).

92. J. H. Rose, H. B. Shore, D. Geldart, and M. Rasolt, *Solid State Commun.* **19**, 619 (1976).

93. W. Tyson and N. Miller, *Surf. Sci.* **62**, 267 (1977).

94. A. R. Miedema, *Z. Metallkd.* **69**, 287 (1978).

95. A. R. Miedema, *Physica* **100B**, 1 (1980).

96. W. Eberhardt, G. Kalkoffen, and C. Kunz, *Solid State Commun.* **32**, 901 (1979).

97. T.-C. Chiang and D. E. Eastman, *Phys. Rev.* **B23**, 6836 (1981).

98. P. H. Citrin, G. K. Wertheim, and M. Schluter, *Phys. Rev.* **B20**, 3067 (1979).

99. C.-O. Almbladh and A. L. Morales, *J. Phys.* **F15**, 991 (1985).

100. G. Augonoa, J. Koutecky, and C. Pisani, *Surf. Sci.* **121**, 355 (1982).

101. C. E. Moore, Atomic Energy Levels, US Natl. Bur. Std. Circ. No. 467, Vol. I, US Natl. Bur. Std., Washington, D.C. (1949).

102. V. Ponec, *Catal. Rev.* **11**, 41 (1975).

103. I. Lindau and W. E. Spicer, *J. Electron Spectros. Relat. Phenom.* **3**, 409 (1974).

104. V. Murgai, L. C. Gupta, R. D. Parks, N. Mårtensson, and B. Reihl, in *Valence Instabilities* (P. Wachter and P. Boppart, eds.) North Holland, Amsterdam (1982), p. 299.

105. L. I. Johansson, A. Flodström, S.-E. Hörnström, B. Johansson, J. Barth, and F. Gerken, *Solid State Commun.* **41**, 427 (1982).

106. S.-E. Hörnström, L. Johansson, A. Flodström, R. Nyholm, and J. Schmidt-May, *Surf. Sci.*, **160**, 561 (1985).

107. T. S. Chou, M. L. Perlman, and R. E. Watson, *Phys. Rev.* **B14**, 3248 (1976).

108. S.-E. Hörnström, L. I. Johansson, and A. Flodström, *Appl. Surf. Sci.* **26**, 27 (1986).

109. B. Johansson and A. Rosengren, *Phys. Rev.* **B11**, 2836 (1975).

110. B. Johansson, *J. Phys.* **F5**, 1241 (1975).

111. K. A. Gschneidner, *J. Less-Common Met.* **17**, 13 (1969).

112. A. R. Miedema, *J. Less-Common Met.* **46**, 167 (1976).

113. F. R. de Boer, W. H. Dijkman, W. C. M. Mattens, and A. R. Miedema, *J. Less-Common Met.* **64**, 241 (1979).

114. C. Laubschat, G. Kaindl, W.-D. Schneider, B. Reihl, and N. Mårtensson, *Phys. Rev.* **B33**, 6675 (1986).

115. N. Mårtensson, B. Reihl, R. A. Pollak, F. Holtzberg, and G. Kaindl, *Phys. Rev.* **B25**, 6522 (1982).

116. J. N. Andersen, O. Björneholm, M. Christiansen, A. Nilsson, C. Wigren, J. Onsgaard, A. Stenborg, and N. Mårtensson, *Surf. Sci.* **232**, 63 (1990).

117. D. Tomanek, V. Kumar, S. Holloway, and K. H. Bennemann, *Solid State Commun.* **41**, 273 (1982).

118. J. F. van der Veen, F. J. Himpsel, and D. E. Eastman, *Phys. Rev.* **B25**, 7388 (1982).

119. F. J. Himpsel, J. F. Morar, F. R. McFeely, R. A. Pollak, and G. Hollinger, *Phys. Rev.* **B30**, 7236 (1984).

120. C. Guillot, P. Roubin, J. Lecante, M. C. Desjonqueres, G. Treglia, D. Spanjaard, and Y. Jugnet, *Phys. Rev.* **B30**, 5487 (1984).

121. J. T. Yates, T. E. Madey and J. C. Campuzano, The adsorption of carbon monoxide by the transition metals, in *The Chemical Physics of Solid Surfaces and Heterogeneous Catalysis*, Vol 3b, Chemisorption Systems, (D. A. King and D. P. Woodruff eds.), Elsevier, Amsterdam (1987).

122. M. L. Shek, P. M. Stefan, C. Binns, I. Lindau, and W. E. Spicer, *Surf. Sci.* **115**, L81 (1982).

123. K. Duckers, K. C. Prince, H. P. Bonzel, V. Chab, and K. Horn, *Phys. Rev.* **B36**, 6292 (1987).

124. R. C. Baetzold, G. Apai, E. Shustrorovich, and R. Jaeger, *Phys. Rev.* **B26**, 4022 (1982).

125. G. Apai, R. C. Baetzold, E. Shustrorovich, and R. Jaeger, *Surf. Sci.* **116**, L191 (1982).

126. S. Yin and E. Tosatti, *J. Phys. Soc. Jpn* **49**, Suppl. A. 105 (1980).

127. F. Bechstedt, *Phys. Status Solidi* **112**, 9 (1982).

128. J. F. Morar, F. J. Himpsel, G. Hollinger, G. Hughes, and J. L. Jordan, *Phys. Rev. Lett.* **54**, 1960 (1985).

129. C. Priester, G. Allan, and M. Lannoo, *Phys. Rev. Lett.* **58**, 1989 (1987).

130. A. Kahn, *Surf. Sci. Rep.* **3**, 193 (1983).
131. D. E. Eastman, T.-C. Chiang, P. Heimann, and F. J. Himpsel, *Phys. Rev. Lett.* **45**, 656 (1980).
132. S. Wiklund, K. O. Magnusson, and A. S. Flodström, *Surf. Sci.* **238**, 187 (1990).
133. M. Taniguchi, S. Suga, M. Seki, B. Stin, K. L. T. Kobayashi, and M. Kanzaki, *J. Phys.* **C16**, L45 (1983).
134. H.-U. Baier, L. Koenders, and W. Monch, *Surf. Sci.* **184**, 345 (1987).
135. K. C. Prince, G. Paolucci, V. Chab, M. Surman, and A. M. Bradshaw, *Surf. Sci.* **206**, L871 (1988).
136. R. Cao, K. Miyano, T. Kendelewicz, I. Lindau, and W. E. Spicer, *Phys. Rev.* **B39**, 11146 (1989).
137. A. B. Lean and R. Ludeke, *Phys. Rev.* **B39**, 6223 (1989).
138. V. Hinkel, L. Scrba, and K. Horn, *Surf. Sci.* **194**, 597 (1988).
139. J. N. Andersen and U. O. Karlsson, *Phys. Rev.* **B41**, 3844 (1990).
140. W. G. Wilke, V. Hinkel, W. Theis, and K. Horn, *Phys. Rev.* **B40**, 9824 (1989).
141. D. H. Ehlers, F. U. Hillebrecht, C. T. Lin, E. Schönerr, and L. Ley, *Phys. Rev.* **B40**, 3812 (1989).
142. R. E. Watson, J. W. Davenport, M. L. Perlman, and T. K. Sham, *Phys. Rev.* **B24**, 1791 (1981).
143. J. W. Davenport, R. E. Watson, M. L. Perlman, and T. K. Sham, *Solid State Commun.* **40**, 999 (1981).
144. W. Mönch, *Solid State Commun.* **58**, 215 (1986).
145. C. U. S. Larsson, A. Flodström, U. O. Karlsson, and Y. Yang, *J. Vac. Sci. Technol.* **A7**, 2044 (1989).
146. R. S. Becker, J. A. Golovchenko, and B. S. Swartzenruber, *Phys. Rev. Lett.* **54**, 2678 (1985).
147. A. L. Wachs, T. Miller, A. P. Shapiro, and T.-C. Chiang *Phys. Rev.* **B35**, 5514 (1987).
148. J. Aarts, A.-J. Hoeven, and P. K. Larsen, *Phys. Rev.* **B38**, 3925 (1988).
149. C. J. Karlsson, E. Landemark, L. S. O. Johansson, U. O. Karlsson, and R. I. G. Uhrberg, *Phys. Rev.* **B41**, 1521 (1990).
150. J. E. Northrup, *Phys. Rev. Lett.* **57**, 154 (1986).
151. R. D. Schnell, F. J. Himpsel, A. Bogen, D. Rieger, and W. Steinmann, *Phys. Rev.* **B32**, 8052 (1985).
152. D. T. Rich, T. Miller, and T.-C. Chiang, *Phys. Rev. Lett.* **60**, 357 (1988).
153. T. Miller, T. C. Hsieh, and T.-C. Chiang, *Phys. Rev.* **B33**, 6983 (1986).
154. F. J. Himpsel, P. Heimann, T.-C. Chiang, and D. E. Eastman, *Phys. Rev. Lett.* **45**, 1112 (1980).
155. F. J. Himpsel, D. E. Eastman, P. Heimann, B. Reihl, C. W. White, and D. M. Zehner, *Phys. Rev.* **B24**, 1120 (1981).
156. R. D. Bringans, M. A. Olmstead, R. I. G. Uhrberg, and R. Z. Bachrach, *Phys. Rev.* **B36**, 9569 (1987).
157. S. Brennan, J. Stohr, R. Jaeger, and J. E. Rowe, *Phys. Rev. Lett.* **45**, 1414 (1980).
158. J. C Woicik, Ph.D. Dissertation, Stanford University (1989).
159. S. B. DiCenzo, P. A. Bennett, D. Tribula, P. Thiry, G. K. Wertheim, and J. E. Rowe, *Phys. Rev.* **B31**, 2330 (1985).
160. T. Miller, A. P. Shapiro, and T.-C. Chiang, *Phys. Rev.* **B31**, 7915 (1985).
161. T. Miller, E. Rosenwinkel, and T.-C. Chiang, *Solid State Commun.* **47**, 935 (1983).
162. T. Weser, A. Bogen, B. Konrad, R. D. Schnell, C. A. Schug, and W. Steinmann, *Phys. Rev.* **B35**, 8184 (1987).
163. R. McGrath, R. Cimino, W. W. Braun, G. Thornton, and I. T. McGovern, *Vacuum* **38**, 251 (1988).
164. J. A. Kubby, J. E. Griffith, R. S. Becker, and J. S. Vickers, *Phys. Rev.* **B36**, 6079 (1987).
165. G. Hollinger and F. J. Himpsel, *J. Vac. Sci. Technol.* **A1**, 640 (1983).
166. C. U. S. Larsson, A. Flodström, R. Nyhom, L. Incoccia, and F. Senf, *J. Vac. Sci. Technol.* **A5**, 3321 (1987).

6

Resonant Photoemission of Solids with Strongly Correlated Electrons

J. W. Allen

1. INTRODUCTION

This chapter describes a set of synchrotron studies in which there occurs the happy combination of a powerful spectroscopic technique and a problem of fundamental and enduring interest. The technique is known as resonant photoemission (RESPES) and the problem is that of strongly interacting electrons in solids. That the technique and the problem go together so well is not entirely a coincidence, as will be seen below.

Resonant photoemission is an aspect of a more general spectroscopic phenomenon in atoms and solids known as the giant resonance. The proceedings[1] of a recent NATO Advanced Study Institute provide a detailed description of such resonances. Here it suffices to say that a giant resonance is said to occur if essentially all the oscillator strength of some X-ray absorption edge appears in a series of relatively narrow lines. Typically this occurs for transitions in which the angular-momentum selection rule $l \rightarrow (l + 1)$ permits the transition to terminate in a final state associated with a partially filled inner atomic shell. Thus giant resonances are found, e.g., for the $3p \rightarrow 3d$, $4d \rightarrow 4f$, and $5d \rightarrow 5f$ absorption edges of transition metals, rare earths, and actinides, respectively. The final states of these resonances decay by various channels, and a full study of the resonance phenomenon involves delineating all of the decay channels.

Resonant photoemission takes advantage of the fact that an important decay channel is one which proceeds by an Auger-like matrix element in which the core hole is refilled by an electron from the final-state shell and another electron is ejected into the continuum. The ejected electron is detected as a photoelectron, and the entire process then appears as a contribution to the photon energy ($h\nu$) dependence of the photoemission cross section for electrons of the type detected.

J. W. Allen • Department of Physics, The Harrison M. Randall Laboratory of Physics, University of Michigan, Ann Arbor, Michigan 48109-1120

Synchrotron Radiation Research: Advances in Surface and Interface Science, Volume 1: Techniques, edited by Robert Z. Bachrach. Plenum Press, New York, 1992.

For the case of emission of a $3d$ electron from a transition-metal atom, this contribution can be written

$$3p^6 3d^n \xrightarrow[h\nu]{} 3p^5 3d^{n+1} \xrightarrow[e^2/r]{} 3p^6 3d^{n-1} \varepsilon_k.$$

Photoemission also occurs by the direct process

$$3p^6 3d^n \xrightarrow[h\nu]{} 3p^6 3d^{n-1} \varepsilon_k,$$

and the interference between the two processes implies that a resonance occurs in the $3d$ photoemission cross section. The simplest useful description of this effect is provided by the Fano theory of photon absorption, summarized below in section 2. The earliest work focusing on the RESPES aspect of giant resonances in solids was that done in 1977 for Ni metal by Guillot et al.[2] and in 1978 for certain rare-earth materials by Lenth et al.,[3] Allen et al.,[4] Johansson et al.,[5] and Gudat et al.,[6]

If a series of photoemission spectra are taken at various photon energies, it is found that as $h\nu$ passes through the cross section resonance, the emission from the resonating state is substantially modulated. Two other modes of measurement have proved especially useful for exploiting the resonance. The first is the constant final state (CFS) spectrum in which $h\nu$ is varied while the detected electron kinetic energy E_{KE} is fixed at a small value in the inelastic tail of the E_{KE} distribution. This spectrum is widely taken to give essentially the photon-absorption spectrum and is thus the spectrum of the giant resonance. The second is the constant initial state (CIS) spectrum, in which the photon energy and E_{KE} are simultaneously varied so as to maintain the ionization energy $E_I = h\nu - E_{KE}$ as a constant. This spectrum measures the cross section RESPES lineshape.

The electronic structure of the $3d$, $4f$, and $5f$ electrons of transition-metal, rare-earth, and actinide materials, to which RESPES is naturally applicable, has for at least 50 years provided a challenge on the way to the goal of achieving a unified theory of the electronic structure of solids. Physically this is because the electrons of partially filled inner shells are considerably, but not completely, shielded from the solid-state environment and therefore retain many atomic-like properties while at the same time participating in solid-state bonding. The resulting situation is often not well described either by a traditional band theory or by a model of weakly interacting atoms. In the band model, the states of the partially filled shell must lie at the Fermi energy of the solid and hence influence the solid's low-energy properties in an important way. The added atomic ingredient is that of strong intrasite Coulomb interactions among the electrons. These interactions tend to suppress the site-charge fluctuations inherent in the band model and thus to alter profoundly the ground-state properties from those expected in the band model. An extreme example is the formation of Mott–Hubbard insulators, insulating materials that have an odd number of electrons per unit cell and hence violate the Wilson rules. Other examples include mixed valence, the Kondo effect, heavy-fermion behavior, the interplay between magnetism and superconductivity, and, one may speculate, the occurrence of high-temperature superconductivity.

Typically, these phenomena are described by model many-body Hamiltonians, which are summarized in section 2. Ideally a unified interpretation of experimental spectroscopic and ground-state properties can be achieved for some single choice of parameters in the model Hamiltonian. An elegant example with considerable success is that of cerium materials, described in section 4. There is also recent success in calculating these parameters from first principles. A guide to the frontiers of research on this program is provided by the articles in the proceedings of a recent NATO Advanced Research Workshop.[7]

2. THEORY OVERVIEW

2.1. Introduction

The generic process involved in resonant photoemission can be thought of as having two steps. In the first step the photon field excites an intermediate state having a core hole. In the second step the core hole decays via the Coulomb interaction in an Auger-like process in which the core hole is filled by one electron and a second electron is ejected from the solid to produce a state identical to that obtained by a direct photoemission process for the ejected electron. Typically, the intermediate state is modeled as discrete, while the final state has a continuum character associated with the range of kinetic energies of the ejected electron. An explicit example[8,9] is resonant photoemission of $4f$ electrons in triply ionized Yb for photon energies near the Yb $4d \rightarrow 4f$ absorption edge. In this case the initial state is $4d^{10}4f^{13}$, the intermediate state is $4d^9 4f^{14}$, and the final continuum state is $4d^{10}4f^{12}\varepsilon_k$, where ε_k is the outgoing photoelectron state. The photon excitation of the intermediate state occurs by the $4d \rightarrow 4f$ electric dipole transition, and for the second step the Auger matrix element is $\langle 4f, 4f| e^2/r |4d, \varepsilon_k \rangle$. The final continuum can also be reached from the initial state by the direct photoemission process $4d^{10}4f^{13} \rightarrow 4d^{10}4f^{12}\varepsilon_k$. In this example the intermediate state $4d^9 4f^{14}$ is split due to the $4d$ core hole spin–orbit interaction, but electric dipole absorption is allowed from the $4d^{10}4f^{13}(^2F_{7/2})$ initial state to only one of these states, the $4d^9 4f^{14}(^2D_{5/2})$. There are also several continua $4d^{10}4f^{12}\varepsilon_k$ due to multiplet splitting of the $4f^{12}$ state.

2.2. Fano Line Shape Theory

The theory of the Fano line shape[10] is the paradigm for resonant photoemission. The simplest version of the theory diagonalizes the Hamiltonian in a subspace consisting of one sharp state ϕ with energy E_ϕ and one continuum ψ_E with states of energy E. The only off-diagonal matrix elements are V_E between ϕ and the continuum. For resonant photoemission, ϕ is the intermediate state with a core hole, excited in the first step of the process, and V_E is the Auger matrix element for the second step of the process. The exact eigenfunctions Ψ_E have the form

$$\Psi = a_E \phi + \int dE' b_{EE'} \psi_{E'} \tag{2.2.1}$$

and Fano obtained explicit expressions for a_E and $b_{EE'}$. He then calculated the matrix element of a transition operator T for exciting the state Ψ_E from the ground state Φ_g:

$$\langle \Psi_E | T | \Phi_g \rangle = a_E^* \langle \phi | T | \Phi_g \rangle + \int dE' b_{EE'}^* \langle \psi_{E'} | T | \Phi_g \rangle. \tag{2.2.2}$$

The ratio f of the transition probability $|\langle \Psi_E | T | \Phi_g \rangle|^2$ to the probability $|\langle \psi_E | T | \Phi_g \rangle|^2$ of the transition to the unperturbed continuum can be represented by a family of curves

$$f(\varepsilon) = \frac{(q + \varepsilon)^2}{1 + \varepsilon^2} = 1 + \frac{q^2 - 1 + 2q\varepsilon}{1 + \varepsilon^2}, \tag{2.2.3}$$

which gives the famous Fano resonance lineshape. In Eq. (2.2.3) the reduced energy variable ε is

$$\varepsilon = \frac{E - E_\phi - F(E)}{\pi |V_E|^2}, \tag{2.2.4}$$

where

$$F(E) = P \int dE' \frac{|V_{E'}|^2}{E - E'} \tag{2.2.5}$$

and P denotes the principal value. The definition of ε accounts for the fact that the resonance associated with the state ϕ is shifted from E_ϕ by a self-energy $F(E)$ arising from the interaction with the continuum, which also causes the state ϕ to be diffused through the continuum with a Lorentzian (virtual bound state) line shape given by the square of the coefficient a_E as

$$|a_E|^2 = \frac{|V_E|^2}{[E - E_\phi - F(E)]^2 + \pi^2 |V_E|^4}, \tag{2.2.6}$$

for which the half-width is clearly $\pi |V_E|^2$. Thus if the system is prepared in the state ϕ it will "autoionize" with a mean lifetime $\hbar / |V_E|^2$

In Eq. (2.2.3), the lineshape parameter q is defined by

$$q = \frac{\langle \Phi | T | \Phi_g \rangle}{\pi V_E^* \langle \psi_E | T | \Phi_g \rangle}, \tag{2.2.7}$$

where

$$\Phi = \phi + P \int dE' \frac{V_{E'} \psi_{E'}}{E - E'} \tag{2.2.8}$$

is the sharp state modified by the admixture of the continuum states. The line shape of Eq. (2.2.3) is plotted for several values of q in Fig. 1. The asymmetric shape reflects the phases of the wave function coefficients a_E and $b_{EE'}$ in Eq. (2.2.2), which are such that the contributions $\langle \phi | T | \Phi_g \rangle$ and $\langle \psi_E | T | \Phi_g \rangle$ to $\langle \Psi_E | T | \Phi_g \rangle$ interfere with opposite phases for energies below and above the resonance. In this case of one sharp state interacting with one continuum, there is perfect cancellation on one side of the resonance, so that f is zero for a particular value of ε. This perfect cancellation does not occur if there is more than one continuum, a case discussed further below.

2.3. Theoretical Treatments of Resonant Photoemission

The standard Fano results summarized above apply to a process such as photoabsorption rather than to photoemission. As discussed by Davis and Feldcamp,[11] by considering the limit in which the photoelectron spatial coordinate becomes infinite, the states Ψ_E can be written in a form that is essentially the product of the photoelectron wave function with a many-electron wave function of the ionized system, so that the integrated photoelectron current at infinity can be calculated. As might be expected, the result is that the photocurrent equals the

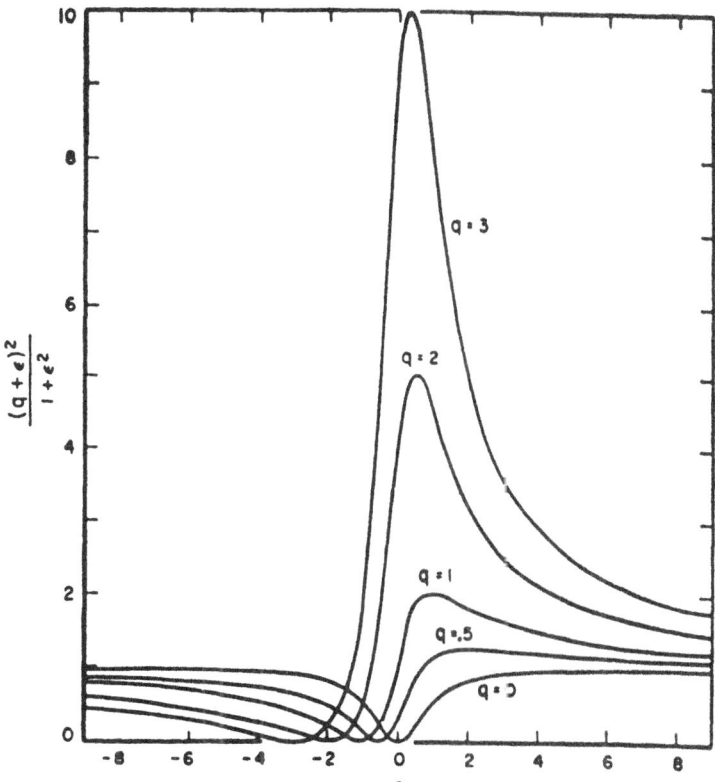

FIGURE 1. Fano lineshapes for various values of the parameter q. Reverse the scale of the abscissas for negative q (From Ref. 10.)

photoabsorption; i.e., for each photon absorbed a photoelectron is emitted when there is no other decay mechanism for the intermediate state.

A more general Fano situation is that of several continua,[11] corresponding to several different photoionized states of the system, and also several intermediate states, as occurs when Coulomb interactions between a core hole and an unfilled valence shell produce multiplet splittings of the state ϕ. If no other decay mechanisms of the intermediate states are considered, it remains true that the total photocurrent summed over all photoionization channels is equal to the photoabsorption and has the same photon-energy dependence. However, the photon-energy dependence of the photocurrent of any particular photoionization channel is in general different from that of another channel and from that of the photoabsorption. In addition, with several discrete states, coupled to one another through the continua, the resonance line shape associated with each state is not in general the simple Fano one. If there are other decay channels of the intermediate states, such as real Auger emission or radiative decay, the total photoabsorption will exceed the total photoemission current. Yafet[12,13] has shown that for the case of including an Auger channel, with only one discrete state and only one photoemission (continuum) channel, the photon-energy dependences of the photoabsorption and the photoemission both have the Fano form, but have different values of the parameter q.

A number of other theoretical treatments of resonant photoemission have been given. Some of these are directed at general aspects of the resonance[14–20] and others are specialized to particular physical systems, including semiconductors,[21] metallic transition-metal systems,[22–29] insulating transition-metal or rare-earth compounds,[21,30–34] metallic cerium or other rare-earth materials,[17,18,35,36] and insulating or metallic uranium materials.[37] The various theories can also be distinguished according to whether the physical system is described by calculated potentials and wavefunctions or by a model Hamiltonian and according to the degree of sophistication in treating the dynamics of the various parts of the total process. There is interaction among all these differences. It is often difficult to carry through a fully general formulation for a specific system, and a careful treatment of one aspect often complicates the treatment of another part of the problem. One tradeoff that often occurs is that theories with first-principles potentials and calculated wave functions usually treat many-body dynamics in some average way, while detailed treatments of many-body dynamics generally proceed from model Hamiltonians in which the underlying states, energy levels, and interactions are postulated and parametrized, rather than being calculated *a priori*. Wendin[20] has given a good overview of the first approach and discusses a number of important aspects of giant dipole resonances, including their relation to one-electron shape resonances. Gunnarsson and Schönhammer[17] and Gunnarsson and Li[18] have employed the second approach to give a very general treatment of dynamical effects in both the resonance and normal photoemission channels and have then applied their results to the important problem of resonant photoemission in cerium systems, discussed in section 4, where dynamical effects involving the Kondo problem can be observed and described by the impurity Anderson Hamiltonian. Other work[22–29] using the second approach has focused on the Hubbard Hamiltonian.

2.4. Model Many-Body Hamiltonians

The Hubbard and the Anderson Hamiltonians just mentioned constitute the two models most commonly in use to describe strong Coulomb correlation effects. Since these models are encountered frequently in discussions of materials where resonant photoemission is observed, it is useful to give here a brief description of their basic features. The Hubbard Hamiltonian[38] introduces an s band, characterized by one-electron matrix elements t_{ij} for electrons to hop from a Wannier orbital on site i to a Wannier orbital on site j, and an on-site Coulomb repulsion U. U gives the increase in energy required to put two electrons on a site. When $U \gg t_{ij}$, the s band is split by U, giving an insulator for half-filling, i.e., for one electron per site. In this situation, the ground state has one electron in the Wannier orbital on each site. The addition of another electron produces a doubly occupied site and hence costs energy U, giving rise to the splitting of the s band. The resulting insulator is called a Mott–Hubbard insulator, and systems close to this condition are said to have "narrow" bands. In the impurity Anderson Hamiltonian,[39] a local s orbital, again with on-site Coulomb repulsion U, is hybridized with one-electron matrix element V to a continuum band characterized by width W or Fermi energy E_F. The local orbital binding energy relative to E_F is generally denoted ε_d or ε_f. For $U \gg \varepsilon_f > \rho V^2$, where ρ is the continuum density of states, there is nearly integer occupation of the local orbital, and a local magnetic moment is stable. The periodic version, with such an s orbital at every site of a lattice, is sometimes regarded as an extended Hubbard model having two bands, one with $t_{ij} = 0$ and nonzero U and the other with nonzero t_{ij} and $U = 0$. The two bands are then hybridized to each other. Both the Anderson and Hubbard models can be augmented to include orbital degeneracy in the narrow bands and off-diagonal site-Coulomb interactions among the narrow-band electrons, leading to atomic multiplet formation. In addition the Anderson model can include a Coulomb repulsion between the narrow-band and the continuum electrons, or even a nonzero Coulomb repulsion for the continuum electrons. Many aspects of these models are described and discussed in the poceedings of a recent NATO workshop.[7]

3. TRANSITION-METAL COMPOUNDS

3.1. Introduction

In 1977 Guillot *et al.*[2] reported that for photon energies near the $3p$ edge of nickel metal, there was a strong resonance enhancement of a weak valence-band structure about 6 eV below E_F, and there has been much subsequent related work.[29,40–52] This structure came to be called the 6-eV satellite and has been interpreted[2,22,23,25,28,53–59] as an electron-removal process resulting in two d holes, i.e., a $3d^8$ configuration, on a Ni site. Theoretical treatments employ the Hubbard model, sometimes including[60,61] multiplet splittings of the local d^8 configuration. The Coulomb interaction U is comparable to the one-electron bandwidth B, and the satellite energy is determined by both $2B$ and U. The emission nearer E_F, called the main band, is in reasonable agreement with band theory, except that B is reduced[23] from the band-theory value. This band narrowing is a Coulomb correlation effect, and it is the major consequence of transferring some of the d-spectral weight into the satellite. In these descriptions, Ni is not a strongly correlated metal because the renormalization of the bands[62] is not very large. The major problem has been a difficulty in fitting both the position and the weight of the satellite simultaneously. There have been conflicting reports as to the occurrence of valence-band satellites for the other $3d$ transition metals, and the current consensus is that they are not observed.[63–69] So it seems that the correlation effects occurring in the atomic states[70–75] of these elements are lost in the solid. There are also studies of correlation effects in other elemental metals and atoms.[29,46,56,68,76]

The work on resonantly enhanced valence-band satellites in transition metals stimulated studies of transition-metal compounds, and by now a large number have been measured.[77–101] Intrinsically interesting in these materials is their wide variety of electrical and magnetic properties, with related questions involving the $3d$ states, such as their bandwidth, their location relative to the Fermi level, and whether to model the effects of Coulomb interactions from an itinerant or local starting point. The electronic structure of these materials is also important in relation to studies of compound formation at the interfaces between metals and semiconductors and in connection with metallic glass formation and other metallurgical issues. Most recently the discovery of high T_c superconductivity in some Cu oxides has provided even greater impetus for studies of transition-metal compounds. There is at least a superficial similarity between the valence-band spectra of the pure transition metals and of their compounds. Typically there is a "main" d band and a satellite to higher binding energies. Both structures resonate for photon energies around the $3p$ edge, the satellite with a resonance shape and the main band with an antiresonance shape. For metallic compounds, the data have been discussed qualitatively with the same model as described above for nickel metal, while for some insulating oxides, especially NiO, a strongly correlated description based on the impurity Anderson Hamiltonian, described in section 2, has been semiquantitatively successful. The discussion below focuses mostly on the latter work and its possible application to the superconducting copper oxides.

3.2. Nickel Oxide

NiO is an antiferromagnetic insulator with the rocksalt crystal structure. It has for 50 years[102] served as a focus for arguments as to the origin of its insulating gap. Two very different views are that the gap can be obtained from spin-polarized band theory[103] and that the gap directly reflects a splitting of the Ni $3d$ states due to Coulomb interactions,[38,104–106] as in the Hubbard model described in section 2. Recent Bremsstrahlung isochromat spectroscopy (BIS) and resonant photoemission work show[88] that the former view is not correct, but that the latter view requires modification to an Anderson model, as is now described.

A band calculation performed for paramagnetic NiO yields a metal, as might be expected since the Ni $3d$ states are partially occupied. Nonetheless NiO is a good insulator above its Neel temperature, $T_N = 520$ K. It is found that spin-polarized density functional calculations,[103] which account for the magnetic ordering, are able to predict the antiferromagnetism of NiO and also its insulating ground state, with a band gap of 0.3 eV. The band-theoretic view would argue that above T_N the local moments continue to exist and to polarize the band structure to preserve the gap. Ultraviolet BIS studies of oxidized Ni surfaces[107,108] observed a small feature just above E_F, which was assigned as the $3d$ states of NiO and taken to support the small gap obtained in band theory. Figure 2 shows combined resonant photoemission and X-ray BIS spectra obtained from a cleaved single crystal of NiO.[88] The band gap is directly obtained from these spectra to be 4.3 eV, an order of magnitude larger than the band theory prediction. Except for tailing into the band gap due to the experimental resolution, there is no detectable structure in the gap, showing that the structure found near to E_F for oxidized Ni is not intrinsic to NiO. Such structure appears only after intense argon-ion bombardment and is accompanied by the appearance of extra O $1s$ structure and a change in the Ni $2p$ line shape, suggesting defect formation.

Thus the measured intrinsic band gap of NiO is an order of magnitude larger than predicted by the local density functional calculation. The main reason is that the effective one-electron potential of the density functional theory utilizes the charge distribution of the ground state in determining the energies of all excited states. This procedure then assumes at the outset that there is no change in the d–d Coulomb repulsion energy for the charge distributions of the unbound hole–electron pair excitations required for d-charge transport, whereas in the Hubbard or Anderson model Hamiltonians an extra d electron at one site and a missing d electron at another site give an increase in the energy equal to U, the Coulomb repulsion between pairs of d electrons on a site.

The bar diagram at the top of Fig. 2 is a qualitative picture of the electronic

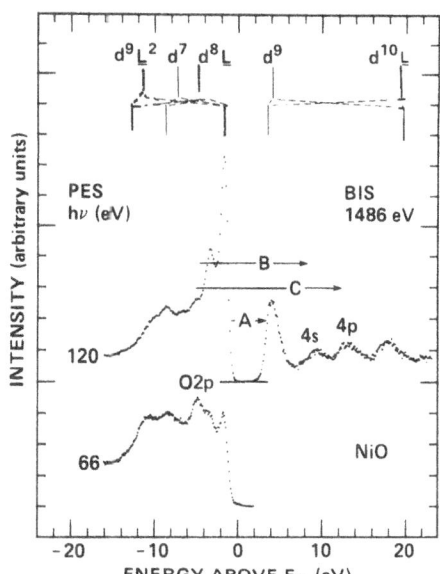

FIGURE 2. Complete valence- and conduction-band spectrum of NiO (100). The bar diagram shows the assignment of the peaks according to the local-cluster calculation. Also indicated are the positions of the $4s$ and $4p$ bands. The arrows indicate the expected energies of the nonexcitonic optical transitions. (From Ref. 88.)

structure expected from a local cluster model for NiO, as proposed by Fujimori *et al.*[32,33] In this approach an $(NiO_6)^{10-}$ cluster is regarded as a separable unit and its electronic structure is described by configuration interaction. In essence it is an impurity Anderson Hamiltonian description of the Ni $3d$ electrons hybridized to the oxygen $2p$ states, but neglecting the $2p$ bandwidth. Figure 3 shows an energy-level diagram of the configurations included, as well as the effect of hybridization. The ground state is a linear combination of properly symmetrized d^8 and d^9L configurations, where L denotes a hole relative to filled ligand O $2p$ states. In order to hybridize with the $3d$ state, L must be a linear combination of orbitals on the six oxygen sites with d symmetry relative to the central Ni site. Similarly the final states reached in photoemission are described as linear combinations of symmetrized d^7, d^8L, and d^9L^2 configurations, and the final states reached in BIS as linear combinations of d^9 and $d^{10}L$ configurations. Not shown, but included in the calculation of Ref. 33, are multiplet spreadings of the configurations, involving crystal field and term splittings of the d manifold. The arrows in Fig. 3 show the PES and BIS transitions.

To analyze the cluster model energies, it is convenient to define the following:

$$E_L = E(d^nL) - E(d^n), \qquad (3.2.1)$$

$$E_d = E(d^9) - E(d^8) \qquad (3.2.2)$$

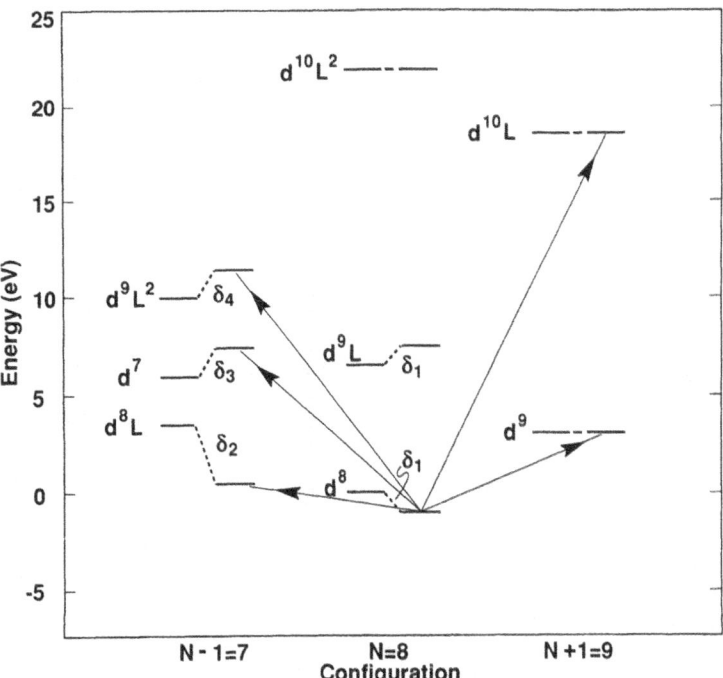

FIGURE 3. Cluster-model energy-level diagram for NiO.

and to take $E(d^8) = 0$. The site energies $E(d^n)$ are given by the relation[38]

$$E(d^n) = n\varepsilon_d + (\tfrac{1}{2})n(n-1)U. \tag{3.2.3}$$

Here ε_d is the one-electron orbital energy, and the second term counts the pairs of electrons to obtain the total Coulomb repulsion energy. Using Eq. (3.2.3), the energies of the various configurations are found to be, for
$N = 9$ electrons:

$$E(d^9) = E_d = -\varepsilon_d + 8U \tag{3.2.4}$$

$$E(d^{10}L) = E_L + 2E_d + U \tag{3.2.5}$$

$N = 8$ electrons:

$$E(d^8) = 0 \tag{3.2.6}$$

$$E(d^9L) = \Delta = E_d + E_L \tag{3.2.7}$$

$N = 7$ electrons:

$$E(d^8L) = E_L \tag{3.2.8}$$

$$E(d^7) = -E_d + U. \tag{3.2.9}$$

Figure 3 has been drawn for $E_L = 3.5\,\text{eV}$, $E_d = 3\,\text{eV}$, and $U = 9\,\text{eV}$. The quantity Δ in Eq. (3.2.7) is the energy required to move an electron from an oxygen atom to a nickel atom and is called[88] the charge-transfer energy. Nickel–oxygen hybridization matrix elements exist between d^8L and d^7, between d^8L and d^9L^2, between d^9L and d^8, between d^9L and $d^{10}L^2$, and between d^9 and $d^{10}L$. These mixings induce shifts of the levels, which are shown by the dashed lines and are labeled by δ_i. In drawing the figure, the effects of $d^{10}L^2$ and $d^{10}L$ have been neglected because their energies are very high, which also implies no shift in the d^9 level. The other shifts have been treated very crudely as follows. It is easily shown from Eqs. (3.2.4–3.2.9) that d^9L lies above d^8 by the same energy as d^9L^2 lies above d^8L. The separation of d^7 and d^9L^2 is neglected. It is also observed that because d^7 and d^8 have 3 and 2 holes, respectively, there are 3 ways for the mixing of d^8L with d^7 and with d^9L^2, and 2 ways for the mixing of d^8 with d^9L. Using second-order perturbation theory, it is then expected that $2\delta_2 = 6\delta_1$, $2\delta_3 = 3\delta_1$, and $2\delta_4 = 3\delta_1$, i.e., the shifts are weighted relatively by the number of channels repelling a level. With the choices above for E_L, E_d, and U, $\delta_1 = 1\,\text{eV}$ leads to reasonable agreement with experiment, as shown in Fig. 2. Parameters in the range between these values and $E_L = 4.5\,\text{eV}$, $E_d = 4.75\,\text{eV}$, $U = 7.25\,\text{eV}$, $\delta_1 = 1.5\,\text{eV}$ provide a similar quality of overall agreement. In Fig. 2, the separation between the bare (i.e., without hybridization) d^7 and d^9 configurations is $U + 2\delta_1$ because the zero of energy is the experimental one, which includes the ground-state hybridization shift.

In this model for the electronic structure of NiO, the large band gap does not directly reflect the even larger Coulomb interaction U, but is caused by U in combination with the one-electron energy Δ. Thus NiO is not a Mott–Hubbard insulator in the simplest sense. In the more traditional Mott–Hubbard picture, the gap is a direct measure of U and the ground state is essentially d^8. Then the strongest of the low-binding-energy PES structures are ascribed to d^7 final states, and various shakeup mechanisms are proposed for the higher-binding-energy peaks near 10 eV. This Mott–Hubbard picture would obtain in the present Anderson impurity model only for $U \ll \Delta$. Gap formation has been explored over the entire range $U \gg \Delta$ to $U \ll \Delta$ for the impurity Anderson model, with the ligand states treated more realistically as a continuum, by Zaanen et al.[109] The results yield a kind of phase diagram which can be used for classifying various types of insulating and metallic transition-metal compounds. Zaanen has also modeled the PES/BIS spectrum and the Ni $2p$ core-level spectrum for NiO using the Anderson impurity model, finding parameter values not greatly different from those of the simple cluster model.

A challenging and fundamental problem for spectroscopy and for the theory is to relate spectroscopically determined parameters to near–ground state quantities, such as the superexchange interaction J for NiO. A simple attempt at this has been given, as follows. Anderson[105] has derived an expression for J in terms of the σ-bonding covalent contribution to the cubic crystal field splitting of the d orbitals. This quantity can be recognized as $\delta_1/2$, remembering that the ground state hybridization shift includes two d holes, and neglecting the π-bonding contribution, which is expected to be much smaller. Then Eq. (33) of Ref. 105 can be evaluated as

$$J = 2(\delta_1/6)U_{dd}^{-1}. \tag{3.2.10}$$

This relation is derived from a Hubbard model based on d Wannier functions obtained from a band structure. Thus the underlying one-electron functions are hybridized nickel $3d$–oxygen $2p$ orbitals, so the Coulomb energy U_{dd} is not the U of the cluster model. It is plausible[110,111] to identify as U_{dd} the smallest ionization energy–electron affinity energy difference of the d electrons, i.e., the value given by the 4.3 eV gap. For the range of values $1.5\,\text{eV} > \delta_1 > 1.0\,\text{eV}$ given above, this yields[111] $160\,\text{K} < J < 345\,\text{K}$, in good agreement with the value of 260 K inferred from the experimental Neel temperature 520 K by using molecular-field theory, but smaller than the value $J = 440\,\text{K}$ obtained by neutron scattering from magnons. The agreement is probably fortuitously good, but the exercise serves to make the points[110] that the relation of the present approach to that employed in the Anderson superexchange theory requires interpretation of U_{dd} and that hybridization probably acts to reduce, i.e., screen, the cluster-model Coulomb interaction U to a lower value of roughly the gap by the relaxation of d-spectral weight from the d^7 to the d^8L configurations. Zaanen and Sawatzky[112] have discussed these issues for superexchange in more detail. As discussed further in the next paragraph, for the eigenstate associated with the d^8L configuration, the local orbital occupation is much closer to its ground-state value than for the d^7

configuration, and thereby this relaxed peak has some conceptual commonality with the density functional eigenvalues.

The validity of this new picture for NiO depends greatly on the validity of the interpretation of its valence-band PES spectrum, especially the assignment of the $3d$ peaks. The most direct evidence is provided by the analysis of the valence-band RESPES behavior. Figure 4 shows this behavior[79,83] for photon energies in the vicinity of the Ni $3p$ edge. Peak A and its shoulder and peak C resonate and therefore are assigned as $3d$. Peak B, which does not resonate, is assigned to the nonbonding oxygen $2p$ states, i.e., linear combinations of oxygen states in the cluster model, which do not have d symmetry and hence do not mix with the d functions of the central Ni site. The assignment of peak A and its shoulder to d^8L final states and of peak C to d^7 final states permits a theoretical interpretation of their differing photon energy dependences, shown in Fig. 5, in which the former has a resonance dip while the latter has a resonance peak. Fujimori and Minami[33] have given a detailed theoretical treatment of the cluster model, including crystal field and multiplet splittings, and find that the structure of the XPS spectrum is accounted for at least as well as in the more traditional Mott–Hubbard assignments, as shown in Fig. 6, and that the differing RESPES behavior of the peaks is also reproduced, at least qualitatively, as shown in Fig. 7.

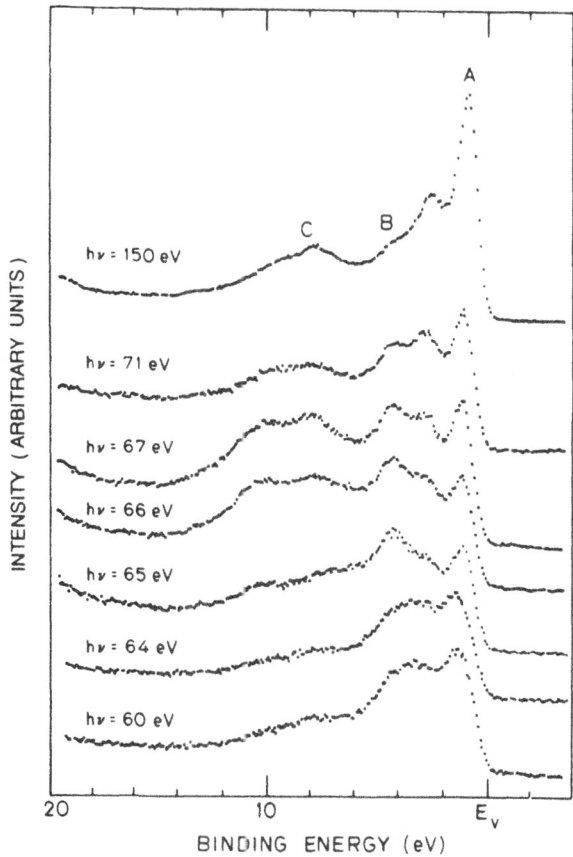

FIGURE 4. Electron energy distribution curves for the valence band of NiO (100) at various incident photon energies around the Ni $3p$ edge, showing RESPES behavior. The binding energy is measured from the top of the valence band. (From Ref. 79.)

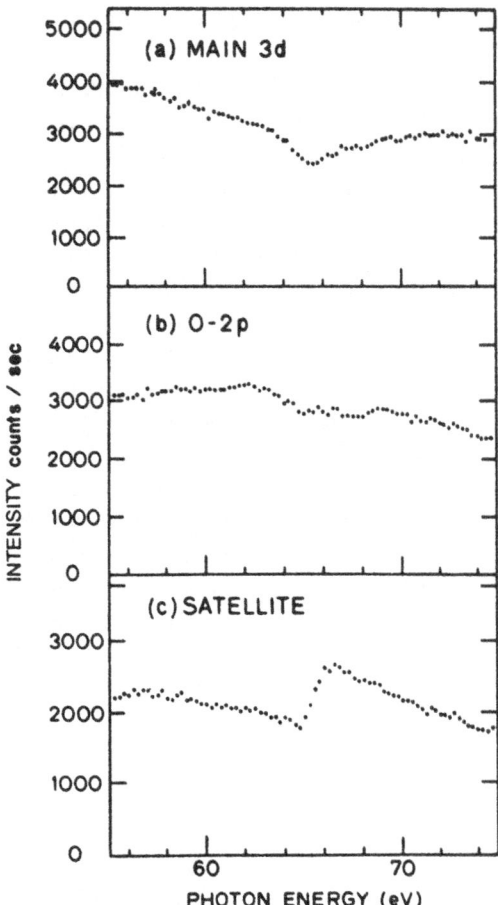

FIGURE 5. Constant initial-state spectra of various peaks in the NiO valence-band photoemission of Fig. 4. (a) d^8L emission, peak A; (b) Nonbonding O $2p$ emission, peak B; (c) d^7 final state, peak C. (From Ref. 79.)

An important point illustrated in the decomposition of the theoretical spectrum in Fig. 8 is that much more than 50% of the total $3d$ weight occurs in peak A even though the initial state is 70% d^8 and the d^7 component of the final state of peak A is only 30%. This counterintuitive result can be traced[113] to the fact that the phasings in the wave functions for the ground states of the 8- and 7-electron systems are the same, which leads to constructive interference for the transition of peak A and destructive interference for that of peak C. Because of this quantum interference effect, peak A can be described quite correctly as a fermion excitation having most of the $3d$ weight, but having largely ligand hole character in the wave function.

3.3. Other Compounds

The cluster model described above for NiO was originally developed and applied in detail to the Cu dihalides[114] by Sawatzky and his collaborators. The study concluded that Cu $2p$ core level spectra and LVV Auger spectra could be interpreted only with the same valence-band assignments, but analogous for Cu, namely that d^9L final states have a smaller binding energy than the d^8 final states,

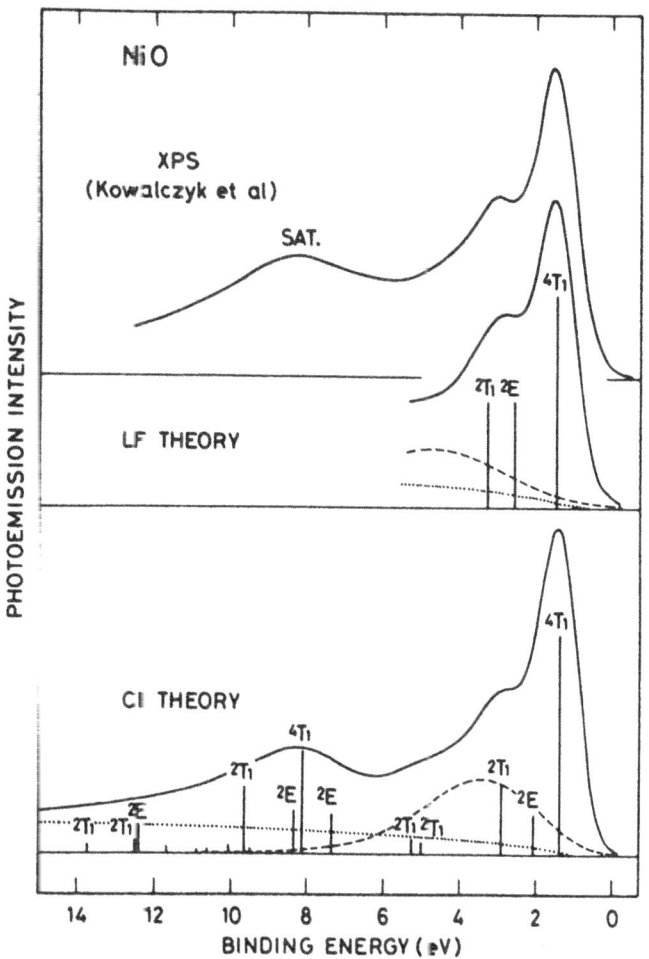

FIGURE 6. Comparison of the experimental XPS valence-band spectrum of NiO (top), the conventional theoretical spectrum calculated as d^7 final states (middle), and the theoretical spectrum calculated with the cluster model (bottom). The dashed line shows emission from the nonbonding oxygen $2p$ states, and the dotted line is a background. (From Ref. 33.)

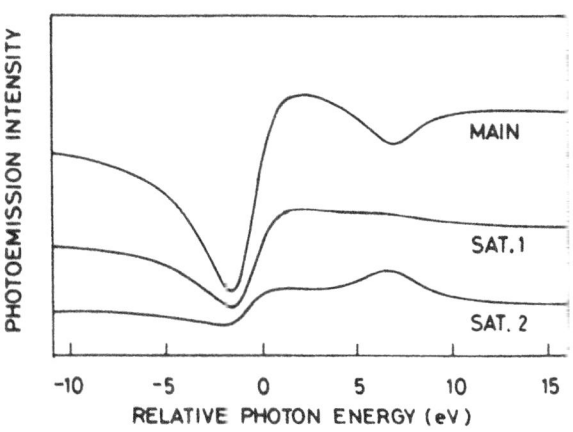

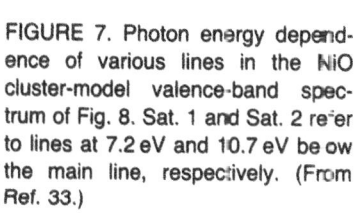

FIGURE 7. Photon energy dependence of various lines in the NiO cluster-model valence-band spectrum of Fig. 8. Sat. 1 and Sat. 2 refer to lines at 7.2 eV and 10.7 eV below the main line, respectively. (From Ref. 33.)

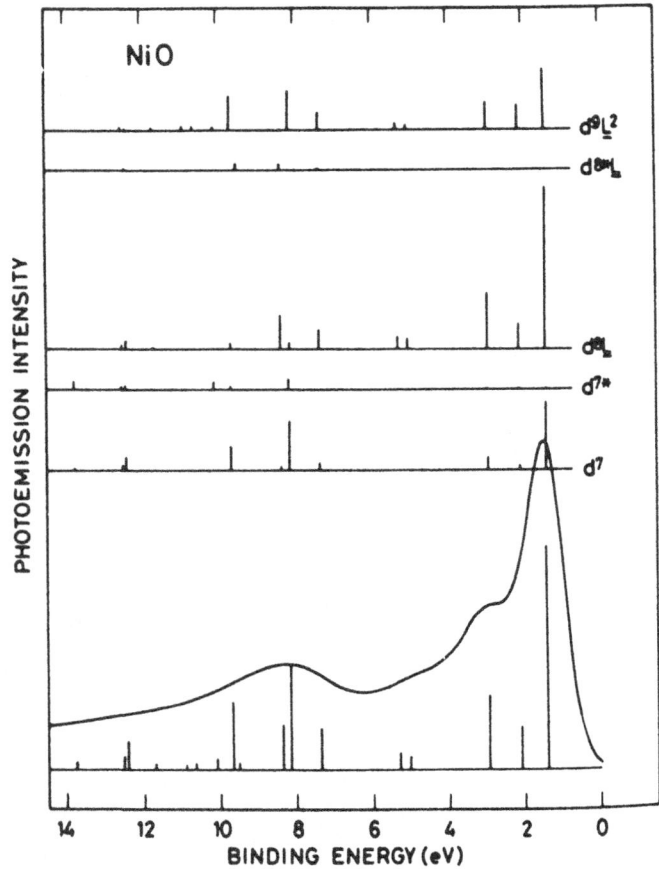

FIGURE 8. Final state components of the valence band spectrum of NiO in the cluster calculation. d^{7*} denotes $t_{2g}^4 e_g^3$ and $t_{2g}^3 e_g^4$ configurations, and d^{8*} denotes $t_{2g}^4 e_g^4$ configurations, which result from configurational interactions with the d^7 and d^8 manifolds, respectively. Note the predominant $d^8 L$ character in the main lines (0–4 eV). (From Ref. 33.)

which lie about 10 eV further from E_F. Theoretical predictions of the RESPES behavior of Cu dihalides have been made,[31,34] but there are no published data as yet. The cluster model and Anderson Hamiltonian calculations have also been applied to core level spectra of Ni dihalides.[115,116] For these materials, and other transition-metal dihalides,[83,84,93] there exist valence-band RESPES data for photon energies near the transition-metal $3p$ edge, as illustrated in Figs. 9 and 10 for NiCl$_2$ and CoCl$_2$, respectively. Presumably these data are susceptible to an interpretation like that for NiO, but a detailed analysis has not been made. In these spectra it is again true that the small binding-energy feature has a resonance dip, while the large binding-energy feature has a resonance peak.

Valence-band RESPES for photon energies around the transition-metal $3p$ edge have been made for a number of metallic transition metal compounds.[77–79,81,87,89,91,94,96,100] For these materials there is no detailed interpretation of the data, reflecting the generally difficult and still unsolved problem of obtaining a unified picture of the rich variety of spectroscopic and ground-state

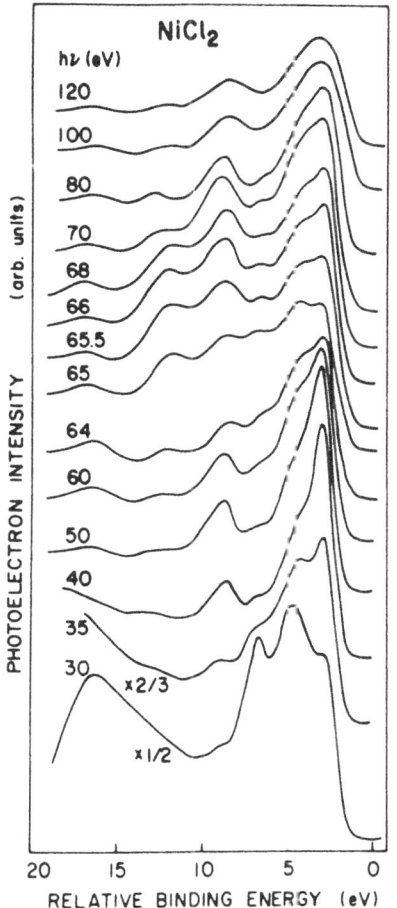

FIGURE 9. Resonant photoelectron energy-distribution curves for the valence band of $NiCl_2$ for various photon energies around the Ni $3p$ edge. The original curves for 30- and 35-eV photon energies are scaled by the factors shown. (From Ref. 84.)

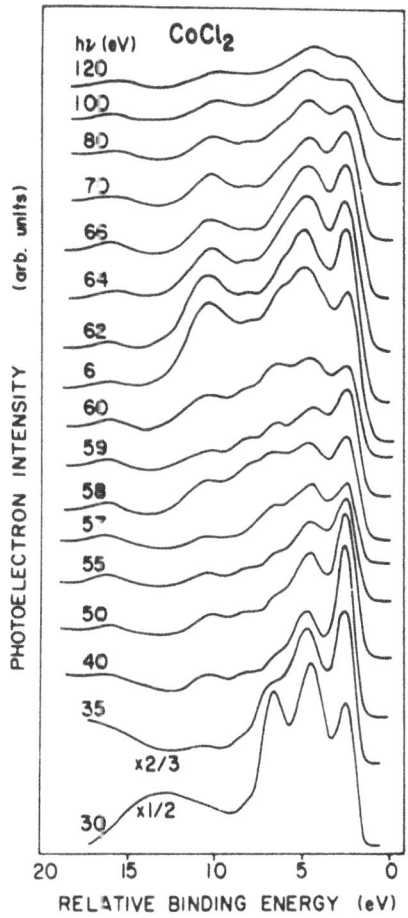

FIGURE 10. Resonant photoelectron energy-distribution curves for the valence band of $CoCl_2$ for various photon energies around the Co $3p$ edge. The original curves for 30- and 35-eV photon energies are scaled by the factors shown. (From Ref. 84.)

behavior in transition-metal compounds. Figure 11 shows, for example, valence-band RESPES data for NiAs,[96] a nonmagnetic metal. There is a weak feature at 8 eV with a resonance peak, and the main band shows a small resonance dip. The 8-eV feature is not predicted by a band calculation, implying the existence of $d-d$ Coulomb interactions, so band structure effects have been invoked to give a qualitative explanation for the lack of magnetism. Another example[81] is provided by the spectra of MnSi and CoSi, shown in Figs. 12 and 13, respectively. For both materials, the main band shows weak resonance effects and LVV Auger emission of the type observed in transition-metal spectra. For MnSi, which is ferromagnetic below 29 K, there is a weak feature at 7 eV, but it has been argued that such a feature is found in band calculations, arising from the bonding state of Si $3p$ and Mn $3d$ orbitals, and that it does not appear to resonate, so that it should not be taken as an indication of correlation effects. A similar band-structure feature

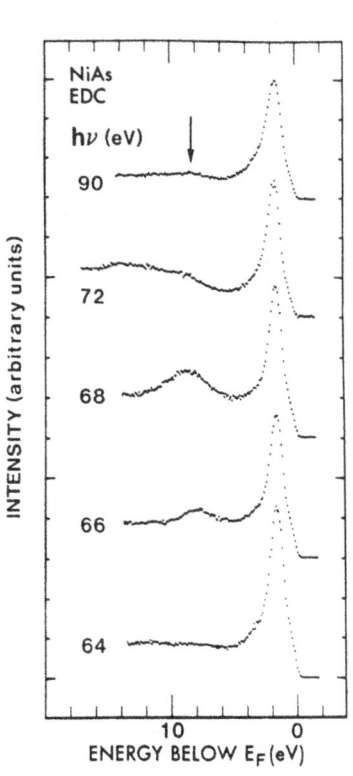

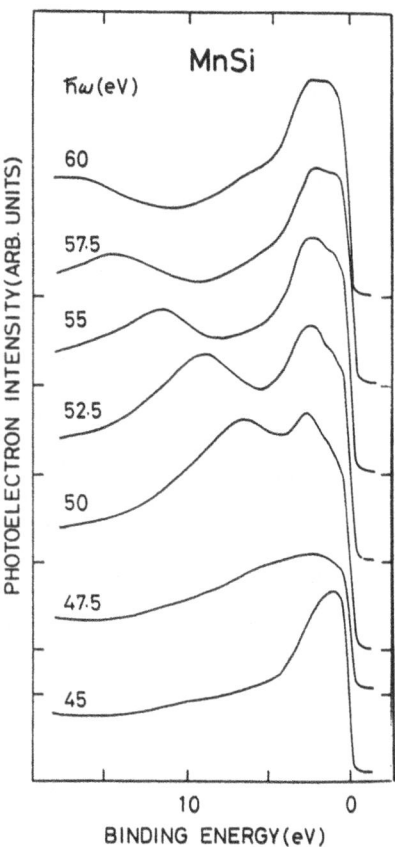

FIGURE 11. Resonant photoelectron energy-distribution curves for the valence band of NiAs for various photon energies around the Ni $3p$ edge. The arrow marks a weak satellite at 6.7 eV below the main band. $M_{23}VV$ Auger emission can also be observed. (From Ref. 96.)

FIGURE 12. Resonant photoelectron energy-distribution curves for the valence band of MnSi for various photon energies around the Mn $3p$ edge. (From Ref. 81.)

might then be expected for nonmagnetic CoSi, but is not observed. For NiSi,[89] differing opinions exist as to the presence of correlation effects in the spectra. In spite of these uncertainties, it does appear that the effect of bonding with the silicon $3p$ orbitals strongly suppresses the correlation effects normally found for transition-metal materials.

3.4. High-Temperature Superconductors

The discovery of high-temperature superconductivity in certain copper oxide compounds has raised a new set of basic conceptual issues. These revolve around the fact that the normal metallic state of the superconductors is achieved by doping "parent" insulating antiferromagnetic materials. Part of the research effort to date has been devoted to establishing that the parent materials belong to the same general class of insulators as described above for NiO and the copper dihalides. PES and RESPES spectra for superconducting $(La_{1-x}Sr_x)_2CuO_4$ and

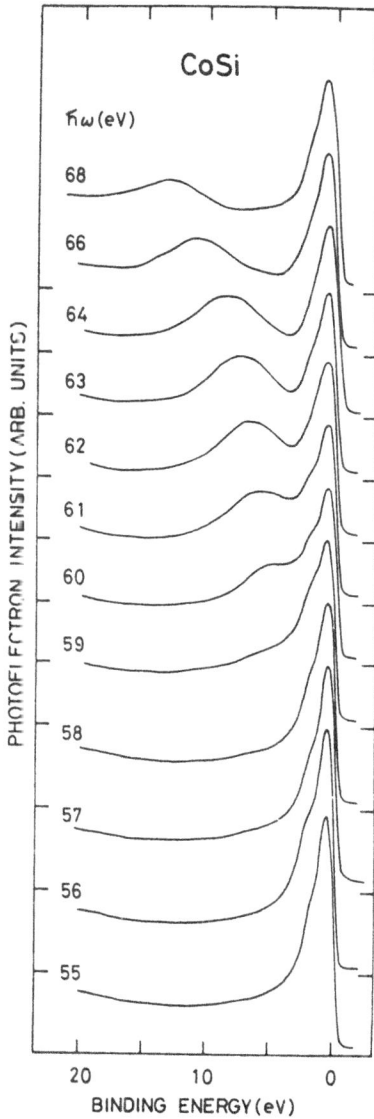

FIGURE 13. Resonant photoelectron energy-distribution curves for the valence band of CoSi for various photon energies around the Co $3p$ edge (From Ref. 81.)

$YBa_2Cu_3O_{7-\delta}$ have been obtained by several groups. Common to all the spectra are a main band and satellite features associated with the Cu $3d$ states, which resonate for photon energies around the Cu $3p$ edge in the usual way. Figure 14 shows such RESPES spectra[117] for $La_{1.8}Sr_{0.2}CuO_4$.

Feature C is almost certainly of extrinsic origin, feature D is the $3d$ satellite assigned as d^8 in the cluster picture, and peaks A and B are a mix of Cu $3d$ and O $2p$ states. The resonance dip of the main-band peaks cannot be seen because the spectra are normalized in the A–B region. The dip for peak B is greater than for peak A. These spectra and those for $YBa_2Cu_3O_{7-\delta}$ have been analyzed using the cluster model employed for the Cu dihalides, and it appears that $U > \Delta$, so that these materials are in the charge-transfer regime of the model.

What sort of electronic structure is produced by adding holes or electrons to

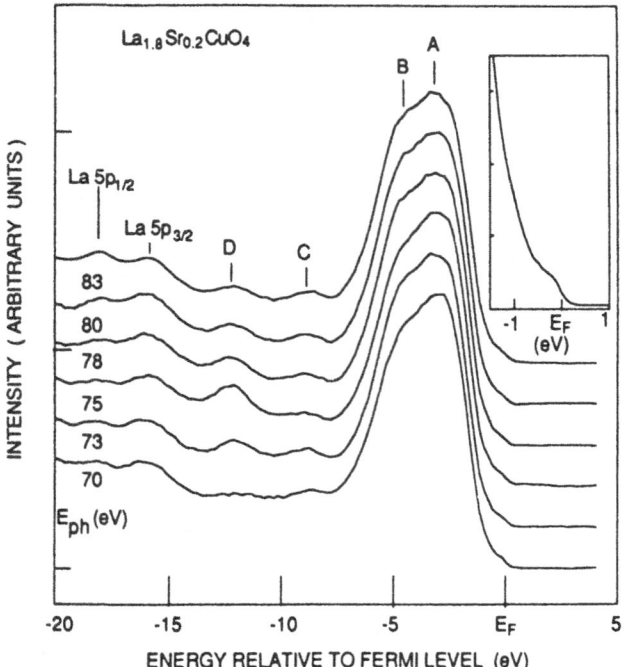

FIGURE 14. Resonant photoelectron energy-distribution curves for the valence band of $La_{1.8}Sr_{0.2}CuO_4$ for various photon energies around the Cu $3p$ edge. Inset shows the detail of the Fermi-level region at a photon energy of 70 eV. (From Ref. 117.)

such an insulator? How are the states within 0.5 eV of E_F in the spectra of Fig. 14 to be described? Does the BCS theory apply to the superconducting state? Readers interested in these fascinating questions should consult the recent literature of this fast-moving field. Articles and papers of particular interest at the time of the writing of this chapter are listed in Ref. 118.

4. RARE EARTHS

4.1. Introduction

In 1978, four papers[3-6] were published reporting the photon energy dependence of photoemission cross sections in rare-earth solids for the photon energy in the vicinity of the rare-earth $4d$ absorption edge. It was found that for these energies there occurs a resonance that is clearly associated with the $4d \rightarrow 4f$ absorption. These papers raised many of the questions that have occupied the field ever since: the determination of the $4f$ emission in Ce materials,[5] the study of surface valence changes,[4] and the behavior of the resonance itself.[3,6] It has evolved that resonant photoemission is useful as a spectroscopic tool for identifying particular components of a PES spectrum because the phenomenon is atom-specific and sometimes orbital- and valence-specific.

The RESPES work built upon XPS studies,[119] which had established a picture for the $4f$ states of rare-earth metals and compounds as follows. For materials with stable $4f$ valence $4f^n$, i.e., nearly integer occupation n of the $4f$ shell, the photoemission spectrum shows a set of peaks corresponding to the various final-state multiplets of $4f^{n-1}$. Similarly, the BIS shows a set of peaks corresponding to the various final-state multiplets of $4f^{n+1}$. The intensities of the peaks can be calculated from overlap factors obtained using coefficients of fractional parentage.[120] The energies of the lowest-lying $4f^{n-1}$ and $4f^{n+1}$ peaks relative to the Fermi energy E_F are large compared to the widths of the peaks, showing the energy barriers against fluctuations $4f^n \rightarrow 4f^{n-1}$ and $4f^n \rightarrow 4f^{n+1}$, which stabilizes the $4f^n$ ground state.

Using the site energy expression of Eq. (3.2.3) with ε_f replacing ε_d, these two excitation energies are

$$E(f^{n+1}) - E(f^n) = \varepsilon_f + nU = -E_f + U \qquad (4.1.1)$$

and

$$E(f^{n-1}) - E(f^n) = -\varepsilon_f - (n-1)U = E_f. \qquad (4.1.2)$$

The sum of the two excitation energies is U. Neglecting the ground-state hybridization shift discussed in connection with Fig. 2, U can then be determined experimentally from combined PES and BIS spectra, as done first by Lang and Baer[119] in their classic work on the rare-earth metals. They found generally good agreement between the experimental values of XPS/BIS excitation energies and values from two quite different theories.[121] Figure 15 shows their combined XPS/BIS spectra for Tm, Er, and Gd. The vertical lines below the spectra show the predicted multiplet structure of the f^{n-1} and f^{n+1} final states, and the solid line through the spectra is the sum of broadened peaks centered on the various multiplet lines. This procedure produces a good fit to the spectra. U is generally taken as the separation of the $4f$ BIS and PES lines nearest to E_F. In addition to the $4f$ features, the PES and BIS spectra also show, near E_F, band states which arise from rare-earth $6s$ and $5d$ orbitals, as well as the relevant valence electrons of other elements in the case of rare-earth compounds.

Research thinking around 1978 was dominated by the ideas of mixed valence,[122] the name given to the situation found in some materials in which the excitation energy of Eq. (4.1.2) is zero relative to E_F. The PES spectra of such materials, described further in subsection 4.4, display two sets of spectral features, $4f^n \rightarrow 4f^{n-1}$ and $4f^{n-1} \rightarrow 4f^{n-2}$, signaling a noninteger occupation of the $4f$ shell and the presence of two $4f$ valence states, $4f^n$ and $4f^{n-1}$. From Eqs. (4.1.1) and (4.1.2) it follows that the two structures are separated by U, with the $f^n \rightarrow f^{n-1}$ weight lying at E_F. It would be expected that in this situation the BIS spectrum would show $f^{n-1} \rightarrow f^n$ weight at E_F with $f^n \rightarrow f^{n+1}$ weight at U above E_F, and this has been confirmed in more recent BIS studies described in subsection 4.4. One question concerned the possibility of an altered- or mixed-valence state on the surface of Sm metal. This possibility was suggested by core level spectra but had not been verified for the $4f$ states themselves. It is discussed further in subsection 4.4. The most outstanding problem was to

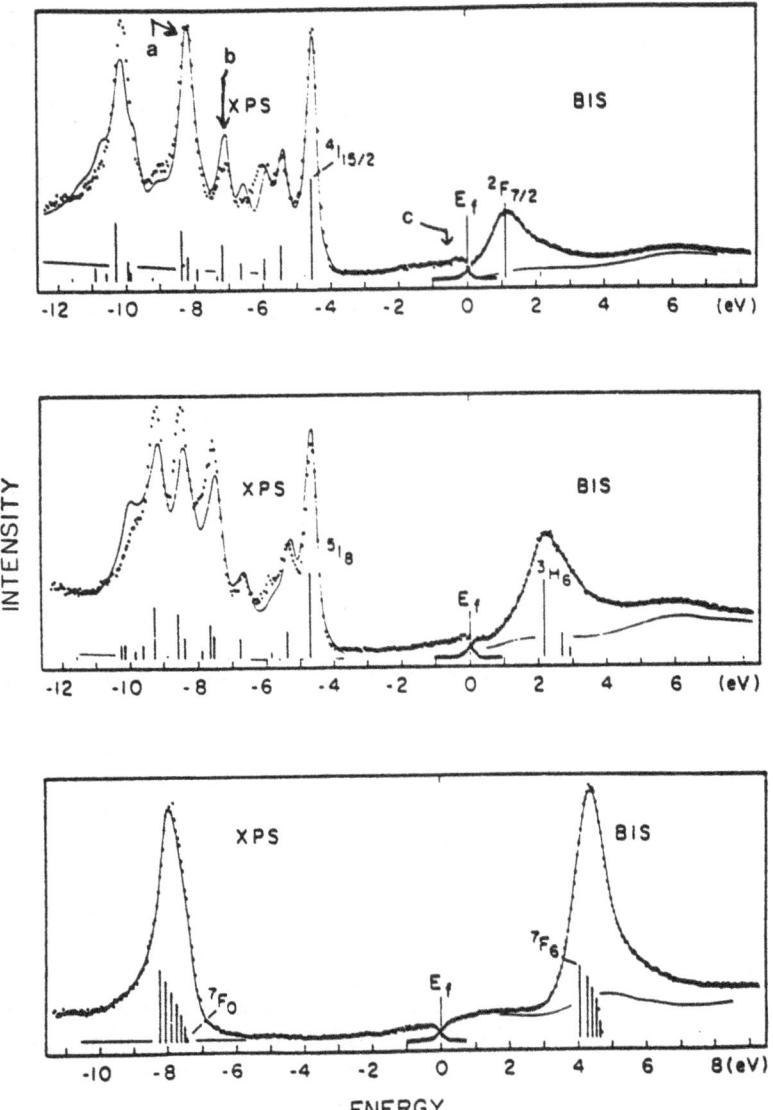

FIGURE 15. Combined XPS and BIS spectra (points) for three rare-earth metals: (top) Tm, (middle) Er, (bottom) Gd. The continuous lines are fitted curves based on the final-state line spectra, shown as vertical bars. The labels *a, b, c* in the Tm spectrum refer to positions of CIS curves in Fig. 18. (From Ref. 19.)

determine the $4f^1 \rightarrow 4f^0$ transition in γ-Ce, of great interest in connection with the unusual properties of Ce and Ce compounds, and the Ce α-γ phase transition. This is experimentally difficult for XPS because the oscillator strength for the $4f^1 \rightarrow 4f^0$ transition is about the same in XPS as that for the $5d$ emission. This topic is discussed in subsection 4.5.

4.2. Single-Valent Materials

The largest amount of work on rare-earth materials of all kinds has been done on the resonance enhancements that occur at the $4d$ edges.[3–6,8,9,123–154] In these studies it has been very useful to refer to experimental[155–160] and theoretical[161–165] results for 4d edge absorption spectra. The resonance behavior itself is most easily studied in materials where the electronic structure is simplest, the single-valent materials. Figure 16 shows as an example the energy distribution curves (EDCs) for the 4f emission of Er for photon energies around its 4d edge.[142] The resonance involves processes like the following:

$$4d^{10}5p^{6}4f^{n}5d^{m} \xrightarrow{hv} 4d^{9}5p^{6}4f^{n+1}5d^{m} \xrightarrow{e^{2}/r} 4d^{9}5p^{6}4f^{n}5d^{m}\varepsilon_{k}$$

$$\longrightarrow 4d^{10}5p^{5}4f^{n}5d^{m}\varepsilon_{k} \qquad (4.2.1)$$

$$\longrightarrow 4d^{10}5p^{6}4f^{n-1}5d^{m}\varepsilon_{k}$$

$$\longrightarrow 4d^{10}5p^{6}4f^{n}5d^{m-1}\varepsilon_{k}.$$

The four final states are, respectively, 4d, 5p, 4f, and 5d photoemission. For the last three of these, there are also the additional direct photoemission processes,

$$4d^{10}5p^{6}4f^{n}5d^{m} \xrightarrow{hv} 4d^{10}5p^{5}4f^{n}5d^{m}\varepsilon_{k}$$

$$\longrightarrow 4d^{10}5p^{6}4f^{n-1}5d^{m}\varepsilon_{k} \qquad (4.2.2)$$

$$\longrightarrow 4d^{10}5p^{6}4f^{n}5d^{m-1}\varepsilon_{k},$$

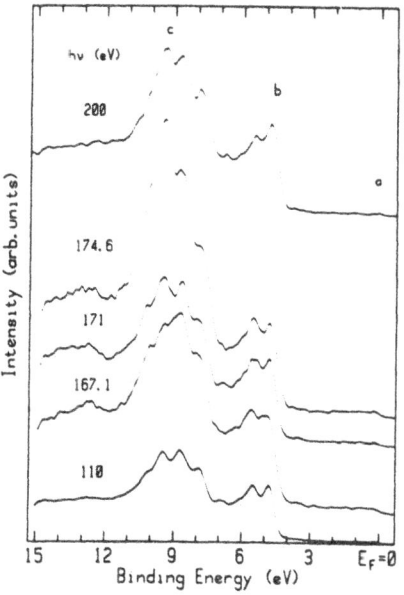

FIGURE 16. Sequence of EDCs of Er metal for photon energies in the region of the 4d edge. A large change in the relative intensities of the different $4f^{10}$ multiplets occurs for various photon energies of the resonance. The labels a, b, c refer to positions of CIS curves of Fig. 17. (From Ref. 142.)

so there is the general character of the situation leading to a Fano lineshape, described in section 2, that of a sharp state degenerate with, and coupled to, a continuum, and photon absorption to both the sharp state and the continuum. The actual situation is quite complex because there are several sharp states due to the multiplets of $4d^9 5f^{n+1}$ and several continua due to the different angular-momentum channels listed above and to multiplet and spin–orbit splittings of the $5p^5$ and $4f^{n-1}$ states. For the $4d$ photoemission process, it is less clear what description to give, and one can find shape-resonance and other treatments in the literature.[154,166]

Figures 17 and 18 show the CFS spectrum and CIS spectra for the $4f$ and valence-band ($5d/6s, p$) emission of Er and Tm, respectively.[142] The positions of the initial states of the CIS spectra are labeled in the valence-band spectra of Figs. 16 and 15, respectively. The spectra illustrate a common finding, which is that the magnitude of the enhancement of any particular $4f^{n-1}$ final state is different for each particular $4d^9 4f^{n+1}$ intermediate state. Presumably, this implies that the Auger matrix element coupling the states depends on the details of the multiplet state. The same result has been reported for the various final states of $5p^5 4f^n$. Figure 19 shows the CFS spectrum and CIS spectra for the $5p$, $4f$, and valence band of Gd.[142] The Gd CIS spectra have been fit with Fano profiles, characterized by the Fano q values given in the figure. There is considerable difference among the three lineshapes, as shown by the different q values. These q values are typical of those for other rare earths, negative for the $5p$, positive for

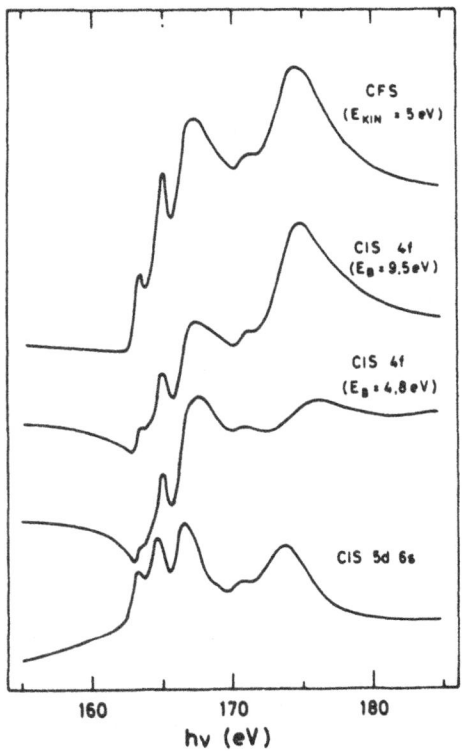

FIGURE 17. The CIS spectra of two $4f$ multiplet lines (b and c) and of the $5d/6s$ valence band (a) of the Er spectra of Fig. 16 and a CFS spectrum. (From Ref. 142.)

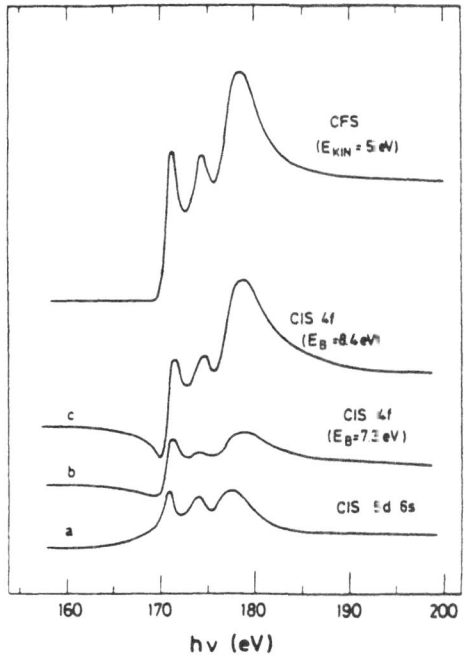

FIGURE 18. The CIS spectra of two 4*f* multiplet lines (*b* and *c*) and of the 5*d*/6*s*-valence band (a) of the Tm spectrum of Fig. 15 and a CFS spectrum. (From Ref. 142.)

FIGURE 19. Dots show the C=S and partial photoemission cross sections extracted from 4*f*, 5*p*, and valence-band CIS spectra for Gd in the region of the 4*d* edge. The solid lines show fitted Fano profiles with the given *q* values. The given E_0 value is the same for all curves. The relatively poor statistics in the 5*p* cross section above 150 eV are due to the very low count rate of the 5*p* and preceding background CIS spectra subtracted to give the cross section (From Ref. 142.)

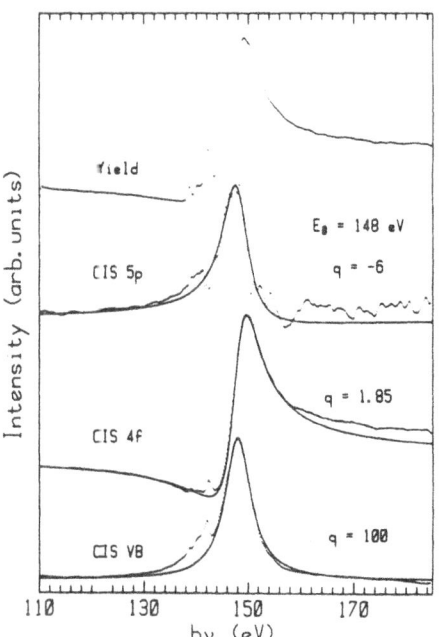

4*f*, and tending toward a Lorentzian shape for the valence band. Attention should be drawn to the fact that the resonance of the valence bands can definitely be observed in these systems where the 4*f* binding energy is large so that the valence-band emission is unobscured for several eV. For Gd, it was reported[132] that the multiplicative enhancement of the valence band is as large as that of the 4*f*s, although for a heavy rare earth the absolute magnitude of the former is much larger than for the latter because there are many 4*f* electrons. As will be discussed further in section 4.5, it appears that the 5*d* resonance becomes weaker as one moves to the left across the rare earth series, perhaps because the 5*d* orbitals are less tightly bound so that the Auger matrix element is weaker.

Resonant photoemission also has been observed for photon energies near rare-earth 3*d* and 5*p* absorption edges. The former involves $3d \rightarrow 4f$ absorption[1,167,168] and results in resonant enhancement[154,169,170] of 4*d*, 5*p* and 4*f*, and 5*d* emission. A difference[154,166] between the 3*d* and 4*d* resonance is that the 3*d* absorption edge typically lies at lower energy than the 3*d* photoemission threshold, while the 4*d* edge typically lies at higher energy than the 4*d* photoemission threshold. Thus resonant decay of the $3d^9 4f^{n+1}$ state into the 3*d* photoemission continuum is not possible. The 5*p* resonance involves $5p \rightarrow 5d$ absorption and results primarily in resonant enhancement of 5*d* conduction-band states, with little effect on the 4*f* states.[171–176]

4.3. Surface Shifts of the 4f Energies

Surface shifts are not intrinsically related to resonant photoemission, but they must be dealt with to fully understand the 4*f* binding energies. Surface shifts are important because the elastic escape depth tends to be small in the energy range 50–150 eV,[177] so that 4*f* photoemission using photon energies near the 4*d* edges is very surface sensitive. Yb metal is the simplest case.[178–183] Figure 20

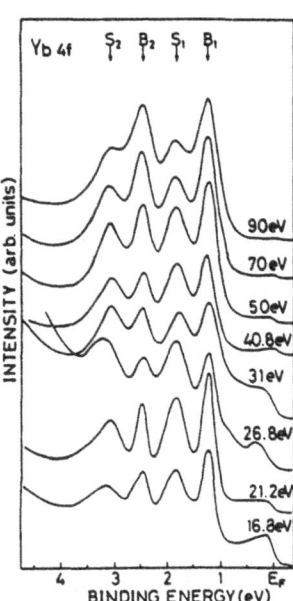

FIGURE 20. Photoelectron spectra of the 4*f* states of Yb metal with photon energies from 16.8 eV to 90 eV. The peaks labeled *S* and *B* are surface and bulk, respectively. (From Ref. 181.)

shows the $4f$ emission of Yb metal at several photon energies.[181] Since Yb metal is single-valent ($4f^{14}$) its $4f$ spectrum is expected to show two peaks corresponding to the $j = 5/2$, $7/2$ states of $4f$.[13] Instead, one observes two sets of peaks, one of which decreases in intensity relative to the other as the photon energy increases. These are the surface $4f$ peaks, shifted to higher binding energy and labeled S_1 and S_2 in Fig. 20. That the shift is to higher binding energy is theoretically predicted.[184–185] Surface shifts to higher binding energy have now been identified for nearly all the rare-earth metals.[186–188] For many of them, including Yb metal, the shift does not result in a valence difference between the surface and the bulk, but for cases where a $4f$ excitation energy, Eqs. (4.1.1) and (4.1.2), is just above or at E_F, the situation is different, as discussed in the next section.

4.4. Mixed Valence—Charge Fluctuations

When the $4f$ weight to remove (or add) an f electron has substantial weight at E_F, the f shell has noninteger occupancy and the material is said to be mixed-valent.[189,190] Traditional views[122,190] of mixed-valent materials attributed their various interesting low-energy properties to charge fluctuations of the type $4f^n \rightarrow 4f^{n-1}5d$, since the presence of $4f^n \rightarrow 4f^{n-1}$ weight at E_F implies a low energy for such fluctuations. As will be discussed further in the next section, Ce materials and possibly certain Yb materials are now known to be dominated by spin fluctuations instead, with charge fluctuations being largely virtual. In this section, we present results typical of those systems, which, for the moment at least, are still considered to be traditional mixed-valent or charge-fluctuation systems. The example to be used is TmSe.

The ground state of TmSe is a mix of $4f^{12}$ and $4f^{13}$, and it has been studied by many workers using many techniques.[190] In photoemission, its mixed valence is shown by the presence of both $4f^{13} \rightarrow 4f^{12}$ and $4f^{12} \rightarrow 4f^{11}$ peaks, first observed in XPS studies.[191] Figure 21 shows the complete $4f$ electronic structure in a combined XPS/BIS spectrum.[192] The weight for transitions back and forth between $4f^{13}$ and $4f^{12}$ crosses E_F. The PES $4f^{12}$ final-state multiplets are labeled $2+$, and the BIS $4f^{13}$ peak is labeled A. Further below and further above E_F,

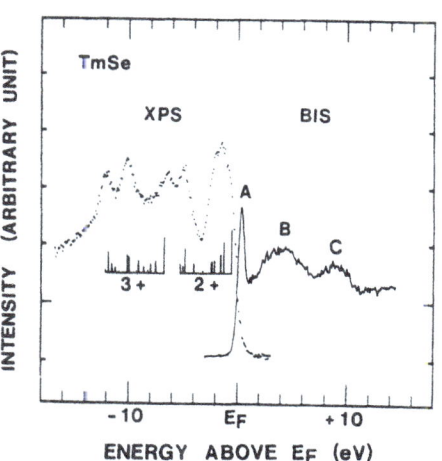

FIGURE 21. XPS and BIS spectra of the (100) surface of TmSe. The assignments of various features of the mixed-valence spectrum are given in the text. The ratio of XPS to BIS weight was chosen to approximate the ratio of electrons to holes in the $4f$ shell. (From Ref. 192.)

respectively, are the $4f^{12} \rightarrow 4f^{11}$ and $4f^{13} \rightarrow 4f^{14}$ peaks. By comparison to the BIS spectrum of YS, which shows peak B but not peaks A or C, peak B of the BIS spectrum is known to be the $s-d$ conduction band. The PES multiplet structure is not well resolved in this XPS spectrum, but has been seen in higher-resolution XPS spectra[191] and in the synchrotron-excited spectra[193] presented below. In the BIS spectrum the $4f^{14}$ state is a singlet, and a calculation of coefficients of fractional parentage has shown that only one line, the $^2F_{7/2}$, of the spin–orbit doublet of the $4f^{13}$ state has significant intensity. Figure 21 can be contrasted with Fig. 15 for single-valent Tm metal, and one can visualize that the $4f^{12} \rightarrow 4f^{13}$ BIS peak in the latter has moved to E_F in TmSe to produce the mixed-valent situation. The valences determined in these XPS and BIS spectra are 2.6 and 2.55, respectively.

In the soft X-ray photon energy range[150,193] there occurs a mix of effects due to resonant enhancement of the Tm $4f$ emission and due to a surface shift of the $4f$ binding energies. The surface shift is discussed first, and is shown in Fig. 22 by spectra[193] taken at 45, 70, and 100 eV. The $4f^{13} \rightarrow 4f^{12}$ emission is seen to be composed of two sets of $4f^{12}$ multiplets, and as with Yb, the higher-binding-energy set decreases in relative intensity with increasing photon energy. The top panel of the figure shows that oxidizing the surface leaves only the bulk peak. The surface-shifted peak lies off the Fermi level, indicating that the surface of the

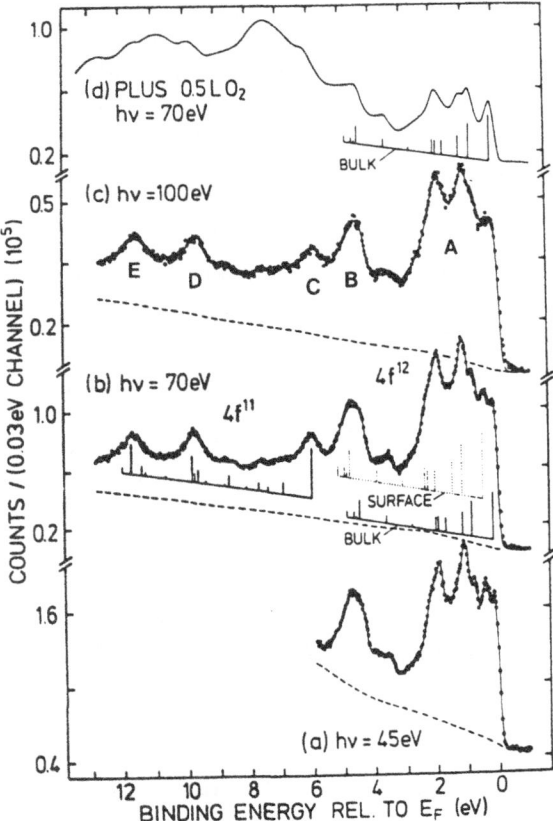

FIGURE 22. Photoemission spectra of the (100) surface of TmSe excited with the given photon energies: (a)–(c) freshly cleaved, and (d) after oxygen exposure. The bar diagrams represent the positions and relative intensities of the final-state multiplet components originating from the bulk (solid bars) and from a surface layer (dotted bars). The solid lines in (a)–(c) represent the results of fits based on the bar diagrams, with the resulting integral background plotted separately as a dashed line. The labels A–E refer to the energies of the CIS spectra of Fig. 23. (From Ref. 193.)

material is divalent, $4f^{13}$. Thus in this spectrum as compared to the XPS spectrum, the total $4f^{13} \rightarrow 4f^{12}$ intensity, relative to the trivalent part, is increased considerably. By fitting the various components of the $4f^{13} \rightarrow 4f^{12}$ emission, Kaindl *et al.*[193] have extracted the bulk portion and found a valence in good agreement with the XPS value.

The resonance behavior[150] of the $4f$ emission appears to be a linear superposition of that for each valence state. Figure 23 shows the CIS curves for various peaks in the valence band of TmSe, labeled *A* through *E* in Fig. 22, and Fig. 24 shows the CFS spectrum. The CIS curves of the divalent and trivalent parts are quite different and correspond to photon absorption to the states of $4d^9 4f^{13}$ and $4d^9 4f^{14}$. For the former, the three multiplets $^3H_6$, $^3H_5$, and $^3G_5$ occur, as seen also in the spectra of Fig. 18 for trivalent Tm metal, and for the latter, absorption only to the line $^2D_{5/2}$ is allowed. Thus one sees that the CIS spectrum can serve as a fingerprint for a certain $4f^n$ valence state. The CFS spectrum of TmSe shows a mix of the two sets of absorption spectra. Since the CFS spectrum, as opposed to the CIS spectra, arises from electrons that have suffered many collisions, it is expected to be bulk sensitive, permitting the bulk valence to be determined from this spectrum also. The valence obtained is 2.62 ± 0.15, in good agreement with other spectroscopic values.

Another system which displays effects of this type is Sm metal.[4,124,194–196]

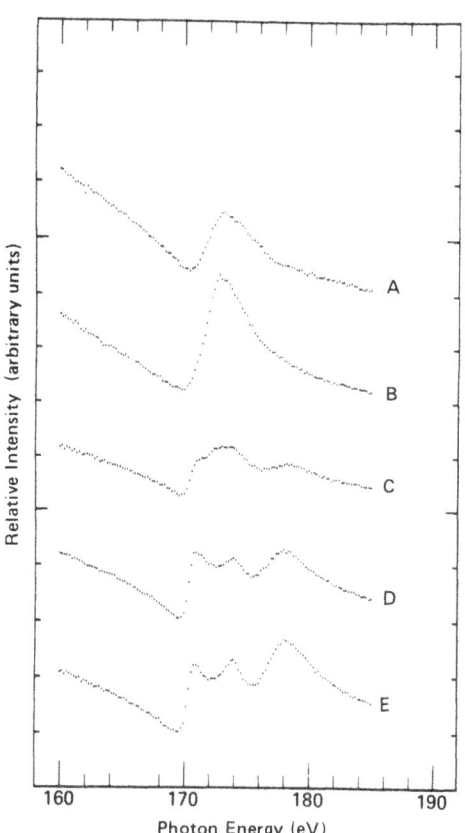

FIGURE 23. The CIS spectra of the $4f$ peaks of TmSe labeled in Fig. 22 for photon energies in the region of the Tm $4d$ edge. Peaks *A* and *B* belong to the divalent structures, while peaks *C–E* belong to the trivalent structures. (From Ref. 150.)

Photon Energy (eV)

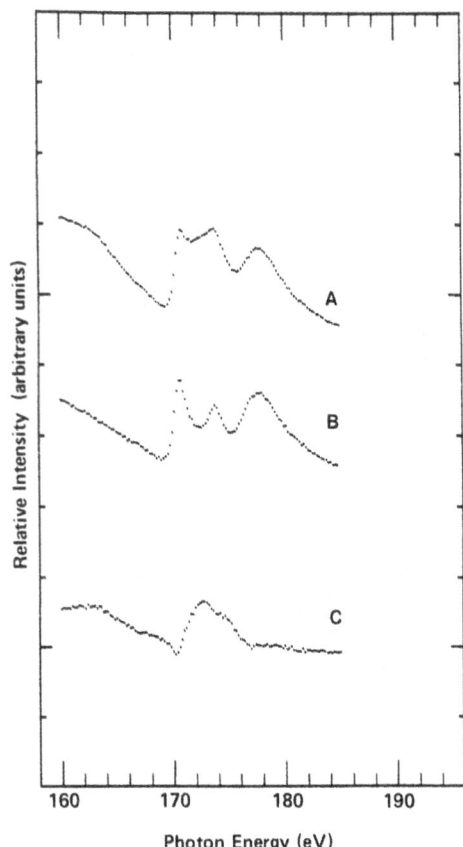

FIGURE 24. The CFS spectra of (a) TmSe, (b) Tm metal, and (c) their difference for photon energies in the region of the Tm 4d edge. (From Ref. 150.)

Studies of Sm were actually the first[4] in which the fingerprint aspects of resonant photoemission were coupled with the surface sensitivity of valence-band photoemission in this photon energy range to provide evidence of the occurrence of divalent $4f^6$ states for the top layer of evaporated films of Sm, which is trivalent in the bulk. It was initially throught[124] that the surface layer is inhomogeneously mixed-valent, but subsequent experiments gave evidence that it is purely divalent.[196] Studies of surface-induced valence change effects have been undertaken for Sm[197–204] and other[205–213] rare-earth atoms in a variety of environments, and theoretical treatments[214–217] directed at this phenomenon have been given.

4.5. Mixed Valence—Spin Fluctuations: Cerium

By far, the largest amount of resonant photoemission work in the rare earths has been done for cerium.[3,5,126–130,135–141,144–149,151,153] Several reasons combine to account for this. Cerium metal and its compounds have diverse and fascinating low-temperature properties traditionally associated with mixed valence.[190] Therefore, there was much interest in determining the binding energy of the single $4f$ electron. However, the XPS cross section for one $4f$ electron is nominally the same as that of the 3 $5d/6s$ valence electrons, so that it is not obvious in an XPS spectrum which peak is due to $4f$. This can be seen in Fig. 25, which shows both

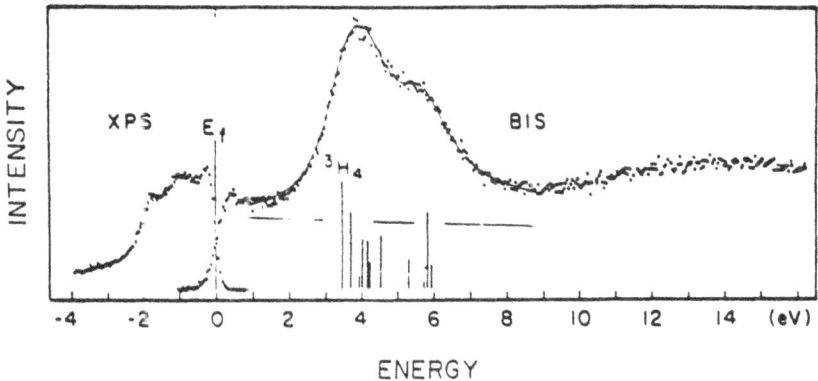

FIGURE 25. Combined XPS and BIS spectra (points) for γ-Ce. The continuous line in the BIS spectrum is a fitted curve based on the final-state $4f^2$ line spectrum, shown as vertical bars. (From Ref. 119.)

XPS and BIS.[119] While the $f^1 \rightarrow f^2$ BIS peaks are easily identified, it is not possible to determine which XPS peak is the $4f^1 \rightarrow 4f^0$ transition. Ce metal[5] and CePO$_3$[3] were among the first rare earths to be studied by resonant photoemission, and it was quickly realized that resonant photoemission might provide the spectroscopic tool to extract the $4f$ part of the spectrum.

Between 1981 and 1983, synchrotron-excited PES and BIS experimental work was combined with new theoretical results to produce a new paradigm for the electronic structure of cerium materials.[129,130,139,141,146,218–222] The theoretical framework is provided by the impurity Anderson Hamiltonian, and it has been shown that all known metallic cerium materials lie in the Kondo regime of the Hamiltonian, so that the low-energy properties are controlled by spin fluctuations rather than charge fluctuations. The distinction between these is elucidated further below. A more detailed discussion of the material of this subsection can be found in a recent review.[223]

Metallic cerium materials can be classified as being γ-like or α-like, depending on whether their properties mimic those of the γ or the α phase of cerium metal.[190] The cerium phase diagram[224] is shown in Fig. 26. The α and γ phases have the same crystal structure and are separated by a first-order phase boundary which terminates in a critical point. The γ phase has a much increased atomic volume, 15% relative to the α phase, and the magnetic properties of the two phases suggest that in the γ phase, the Ce atoms have local magnetic moments that are lost in the α phase. Compounds like CeAl$_3$, whose atomic volume and magnetic properties are γ-like at all temperatures, show one more characteristic property, a very large low-temperature T-linear specific heat coefficient, which places these materials in the group which has recently acquired the name "heavy-fermion."[189,225] The α-like materials have specific heat coefficients that are much smaller than for γ-like materials but are still somewhat larger than is typical for other solids. Neither α- nor γ-Ce is superconducting (although the α' phase is), but some α-like cerium materials such as CeCo$_2$ and CeRu$_2$ are,[140] and γ-like CeCu$_2$Si$_2$ is the original heavy-fermion superconductor.[226]

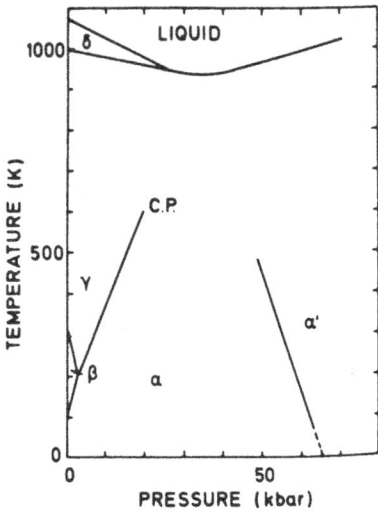

FIGURE 26. Phase diagram of Ce. (From Ref. 224.)

Making a model for the cerium α-γ phase transition is a 40-year-old problem.[122] It has generally been agreed that in the magnetic γ phase there is one electron in the $4f$ shell, providing a local Ce magnetic moment, and that the α phase involves some sort of change in the $4f$ electron. In the period 1949–1950, Zachariasen[227] and Pauling[228] independently proposed what is known as the promotional model. In this picture, the $4f$ electron leaves the $4f$ shell to become a conduction electron in the α phase, so that the local moment is lost and the increase in conduction electrons leads to a decrease (collapse) in the volume through an increased cohesive energy. Theoretical elucidation of this[229] and a related[230] model showed that the temperature and pressures involved in the transition required the $4f$ electron binding energy ε_f relative to E_F to be small (0.1 eV), and consequently that the hybridization width of this $4f$ level be even smaller (0.02 eV). In 1974, Johansson[231] employed thermodynamic arguments to predict that ε_f is about 2 eV. He argued that this precludes the promotion model and that the volume decrease in the α phase then implies that the $4f$ electrons contribute to the cohesive energy in this phase. He proposed that the mechanism by which this is accomplished is a Mott transition, i.e., that the balance between Coulomb energy U and $4f$ bandwidth is sufficiently close to permit the $4f$ electrons to change from localized to itinerant in the phase transition, thereby quenching the magnetic moments and increasing the cohesive energy to collapse the volume.

The $4f$ spectrum obtained from combined synchrotron-excited photoemission and BIS measurements shows that neither the promotional model nor the Mott transition model are correct.[146] Figure 27 shows the $4f$ spectrum for α and γ Ce[232,233] and Fig. 28 shows the $4f$ spectrum for γ-like CeAl and α-like CeNi$_2$.[146,223] Obtaining the PES part of these $4f$ spectra is discussed further below. The PES spectra of both α and γ materials show a peak 2 eV or 3 eV below E_F, confirming Johansson's prediction[231] that the binding energy of the $4f$ electron is too large for the promotional model to work. However, the $4f^1 \to 4f^2$ peak occurring in each of the BIS spectra shows that the Coulomb interaction U has essentially the same value for γ and α materials and is always much larger

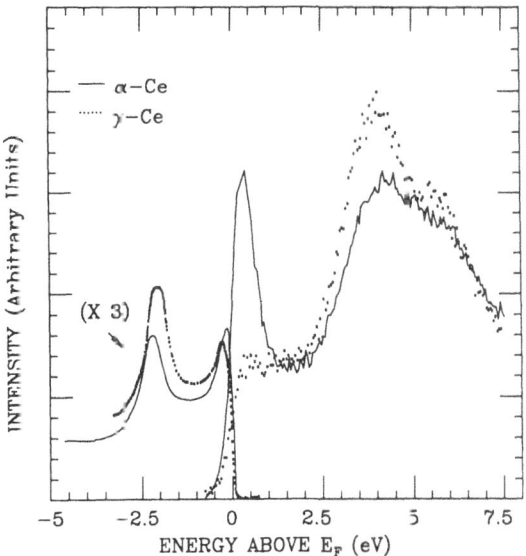

FIGURE 27. Combined PES and BIS 4f spectra of α-Ce (solid) and γ-Ce (dotted). PES spectra are for $h\nu = 60$ eV. (From Ref. 233; BIS spectra from Ref. 232.) The relative intensities of the BIS and PES portions are roughly for thirteen and one 4f electron, respectively.

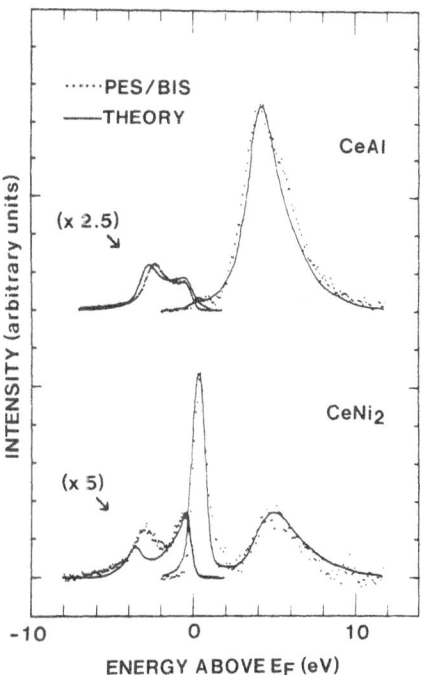

FIGURE 28. Combined theoretical and experimental 4f PES and BIS spectra for CeAl and CeNi$_2$, obtained as described in the text. The PES portion is increased by a scale factor for clarity. The Kondo temperature for CeAl is about 1–10 K, and for the experimental resolution employed, the spin–orbit sideband portion of the Kondo resonance, combined with the "finite-U" effects, gives a broad nonzero weight at E_F. The Kondo temperature for CeNi$_2$ is about 900 K and the large peak appearing in both BIS and PES, centered slightly above E_F, is the Kondo resonance in this material. (From Ref 223.)

than the widths of any of the peaks. Thus, the criterion for the Mott transition model is not fulfilled either. In fact, something intermediate between the two pictures takes place, and in changing from γ- to α-like materials, spectral weight is transferred from the peaks above and below E_F to a peak which is slightly above E_F in the BIS spectrum and has a tail extending across E_F into the PES spectrum.

A theoretical description which includes the $4f$ spectra and many of the low-energy properties can be obtained using the degenerate impurity Anderson Hamiltonian,[39] which models a local orbital of combined spin and orbital degeneracy N_f, having a binding energy ε_f, Coulomb interaction U, and hybridization $V(\varepsilon)$ to a conduction electron density of states $\rho(\varepsilon)$. The past nine years have seen enormous progress with the theory of this Hamiltonian using the techniques of the renormalization group, the Bethe ansatz, and the inverse degeneracy $(1/N_f)$ expansion.[234] Essentially exact results have been achieved for the case appropriate to Ce where the local orbital valence states are f^0, f^1, or f^2. The $(1/N_f)$ expansion has been especially important because BIS/PES spectra can be calculated, in addition to low-energy properties.[221,222,235,236] The important property of the Hamiltonian is that the ground state is a singlet and there is a low energy scale, the Kondo temperature T_K, which characterizes spin (i.e., degeneracy) fluctuations out of the singlet ground state. The nature of the relation of the high-energy parameters of the Hamiltonian to the low-energy properties is illustrated by simple relations[222,237] valid to lowest order in $(1/N_f)$, for infinite U, constant $\rho(\varepsilon)$, and $n_f < 1$. The Kondo temperature T_K is given by

$$T_K = D \exp \frac{-1}{J} \qquad (4.5.1)$$

where J, the Kondo coupling constant, is

$$J = \frac{N_f \rho V^2}{\pi \varepsilon_f} \qquad (4.5.2)$$

and D is the width of the occupied part of the conduction band. The binding energy of the singlet ground state is

$$E_K = -k_B T_K, \qquad (4.5.3)$$

and the T-linear specific heat coefficient γ and $T = 0$ static magnetic susceptibility $\chi(0)$ are given by

$$\gamma = \frac{\pi^2 k_B^2 n_f}{3 k_B T_K} \qquad (4.5.4)$$

$$\chi(0) = \frac{(N_f^2 - 1)(g\mu_B)^2 n_f}{12 k_B T_K}. \qquad (4.5.5)$$

The expression for γ reflects the fact that the spin entropy evolves over the temperature range T_K. If T_K is small, both $\chi(0)$ and γ are large. In this framework, the heavy-fermion materials can be characterized as having very small values of T_K. From Eqs. (4.5.1) and (4.5.2), it is apparent that these small values of T_K can occur even if the underlying Hamiltonian parameters, such as ε_f and V, are large. The realization[129,130,139,141,218,222,238–240] that the Kondo properties of the Anderson Hamiltonian provide the basic way to reconcile the large energy scales observed in PES/BIS spectroscopy with the small energy scales observed in transport properties was an important step forward in the cerium problem. It is further apparent that very large changes in T_K can occur for small changes in J, because of the exponential relation between them.

The direct manifestation of the low-energy scale in the PES/BIS spectrum is the occurence of a sharp peak, the Kondo resonance, near E_F, in addition to the $4f^1 \rightarrow 4f^0$ and $4f^1 \rightarrow 4f^2$ PES/BIS peaks. The resonance is centered at T_K, above E_F, with a width in BIS of T_K/N_f and a tail in PES extending T_K below E_F. The fraction of weight transferred to the resonance in the PES spectrum is proportional to $k_B T_K/N_f \rho V^2$, i.e., the ratio of the spin and charge fluctuation energy scales. Hybridization enters the PES spectrum as $N_f \rho V^2$, so the degeneracy enhances its effect. The existence of the Kondo resonance had been inferred many years ago[241,242] from approximate theories of the spin $\frac{1}{2}$ ($N_f = 2$) Hamiltonian, but the first rigorous theory was provided by Gunnarsson and Schönhammer,[221,222] using a $(1/N_f)$ expansion, and the first direct experimental identification was by Allen et al.[146] The $1/N_f$ theory also provides a physical picture of the resonance as a kind of dynamical relaxation process in which hybridization allows the local orbital to exchange electrons and holes with the conduction band and thereby return to its ground-state occupation after an electron is removed or added. To order $(1/N_f)^0$, it can be shown that the local orbital occupancy of the states reached in the resonance is exactly the same, n_f, as in the ground state, and that an electron has actually been removed from or added to the conduction band in the range of energy T_K near E_F. The relaxational picture provides a link to the results of density functional calculations,[243] as discussed further in section 5.4, on heavy–fermion actinide materials.

Figure 28 also shows theoretical fits[223] to the PES/BIS spectrum of CeAl and CeNi$_2$ using the $1/N_f$ theory, and it is apparent that the agreement is very good. For γ-like CeAl, the value of T_K is 1 to 10 K and the weight near E_F is much larger than would be expected from the Kondo resonance as described in the preceding paragraph. This increased weight has two origins. For finite U the ground state has an f^2 component which provides an $f^2 \rightarrow f^1$ PES channel that enhances the weight in the general vicinity of E_F. In the case of spin–orbit splitting (or also crystal-field splitting) of the N_f-fold degeneracy, extra weight occurs as a sideband structure[222,244,245] on the resonance in both PES and BIS at the energy of the spin–orbit splitting. The resolution of the PES spectra of Fig. 27 is just adequate to show that for γ-Ce the $4f$ spectrum peaks slightly below E_F in the spin–orbit sideband at 0.25 eV, while for α-Ce the peak is at E_F because its T_K is larger than for γ-Ce. The spin–orbit sideband and the analogous crystal-field sideband, and the temperature dependence of the resonance, have been fully and beautifully observed in recent high-resolution laboratory PES experiments using a helium lamp.[246–248] These spin–orbit effects also occur for α-like CeNi$_2$ but are numerically less important because T_K is much larger, 1000 K. The difference between large and small T_K is due largely to a difference in hybridization V. For the CeNi$_2$ spectrum, the effect of large V can be seen not only in the formation of the large Kondo resonance crossing E_F, but also in the very pronounced dip in both the experimental and theoretical PES spectra. In the calculations, this dip is easily traced to the fact that the bare f level is located in a large peak in the Ni $3d$ conduction-band density of states at this energy. The large hybridization yields bonding and antibonding states and thereby greatly contributes to splitting the f spectrum into two peaks. One has added confidence in the description because equally good fits (not shown) of $3d$ XPS[249] and X-ray absorption (XAS)[250] spectra can be achieved by the theory with essentially the same values of the Hamiltonian parameters, if the Coulomb interaction U_{fd} between the $3d$ core hole and the $4f$ electron is included in the Hamiltonian. Because of U_{fd} and the hybridization, the XAS and XPS spectra display structures corresponding to f^0, f^1, and f^2 final states, the splittings and intensities of which are determined by the Hamiltonian parameters. In this connection, it is important to point out that concurrent studies with core hole[249–251] and other spectroscopies[252,253] were also very important in leading to the new Ce picture presented here. Finally, but of equal importance for the total picture, if the spectroscopic parameters are used to calculate $\chi(0)$ or T_K, values in good agreement with experiment are obtained. The values of n_f, the f-orbital occupation, are always greater than 0.75 for metallic cerium materials, even the most α-like ones.

The Kondo volume-collapse model[220,240,254] of the cerium α–γ transition draws on the picture that emerges from electron spectroscopy studies that γ- and α-Ce differ primarily in having small and large values of T_K, respectively, and that this large variation in T_K occurs because of a small variation in the coupling constant J, primarily by a change in hybridization V. When T_K is large, as in α-Ce and α-like materials, the binding energy of Eq. (4.5.3) constitutes an important contribution to the cohesive energy. It is large enough to account for the reduced lattice constants, owing to the volume dependence of J, which for Ce is due largely to that of ρV^2. The volume dependence of ρV^2 arises in part from that of V, as follows. For materials where Ce atoms are at sites with inversion symmetry, as is often the case, $4f$–$5d$ hybridization on a site is forbidden, so that V is due to hybridization with orbitals on neighboring sites, either Ce $5d$ in Ce metal or ligand orbitals in Ce compounds. For hybridization between orbitals of angular momentum 1 and 1' on different sites, V depends on the site separation R as[255]

$$V \sim \left(\frac{1}{R}\right)^{l+l'+1}. \tag{4.5.6}$$

Thus for f–d hybridization, this contribution to ρV^2 is expected to vary inversely with the fourth power of the volume. This dependence will be reduced somewhat due to the volume dependence of ρ, but it is reasonable to expect that for small changes of volume the dependence of ρV^2 is at least linear with a negative slope.

The volume then collapses as much as is permitted by the quadratic volume dependence of the normal bulk-modulus contributions to the free energy, in order to increase ρV^2 and hence J, and thus E_K. The importance of the $4f$ electrons for the cohesion was emphasized by Johansson[231] although the specific mechanism is altogether different than in his proposal of a Mott transition. Because of the exponential form of Eq. (4.5.1), modest changes in ρV^2 produce large changes in T_K, leading to the possibility of a first-order phase transition in Ce. As shown in detail elsewhere,[220,240] in the α to γ phase transition in Ce, the value of T_K changes abruptly from being much larger to being much smaller than the temperature, and the material gives up the binding energy E_K in order to gain the spin entropy which exists for $T > T_K$. This Kondo-volume collapse model of the transition is supported semiquantitatively by the parameter values found in the electron spectroscopy studies described in this chapter. Figure 29 compares the experimental α–γ phase boundary with that calculated from a simple version of the Kondo volume-collapse model.[240] The lower critical point is a prediction of the theory, subsequently observed in alloys,[256] where it occurs at positive pressure.

Obtaining experimentally the $4f$ spectrum of cerium materials has played a large role in developing the new picture for cerium. As described in detail elsewhere[223] and summarized briefly here, several workers[5,126–29,135–138,140,144–146,148,218,257,258] have contributed significantly to the experimental attack, which has consisted of comparing the spectra of Ce materials with those of La and Y isomorphs to infer the $4f$ part and to use the $h\nu$-dependence of the $4f$ cross section, primarily in the range 30 eV to 60 eV, and in the resonant region 110 eV to 130 eV. Figure 30 shows the first valence-band resonant photoemission data for γ-Ce.[5] The resonating peak at 2 eV was interpreted as showing the large $4f$ binding energy ε_f predicted by Johansson.[231] Figure 31 shows the γ-Ce valence band for photon energies between 40 eV and 60 eV.[233] In this energy range the $5d$ cross section decreases, while the $4f$ cross section increases,[145,259] so data like these are very important in showing that the resonating feature near E_F in Fig. 30 is also part of the $4f$ spectrum, rather than being of $5d$ origin as in the interpretation of the heavy rare-earth spectra described in subsection 4.2. The PES portions of Fig. 27 are the 60-eV spectra from these data. For compounds with transition metals having nearly full d shells, the resonance is essential in separating the $4f$ emission from that of the metal d electrons, although it is also necessary to take into account the strong energy dependence of the d cross section due to its Cooper minimum.[259] It is very important[223] to analyze energy-dependent data quantita-

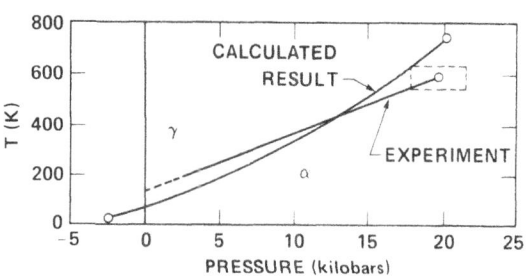

FIGURE 29. Phase diagram of Ce calculated from the Anderson Hamiltonian in the Kondo volume-collapse model, compared to experimental results. The dashed box shows the experimental uncertainty in the upper critical point. The lower critical point can be observed in Ce alloys, where it occurs at positive pressure. (From. Ref. 240.)

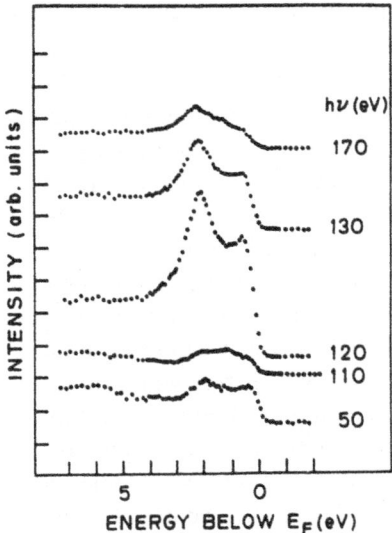

FIGURE 30. Valence-band photoemission EDCs of γ-Ce for photon energies around the $4d$ edge. The spectra have been normalized with respect to variations in photon flux, but not with respect to electron energy analyzer efficiency. (From Ref. 5.)

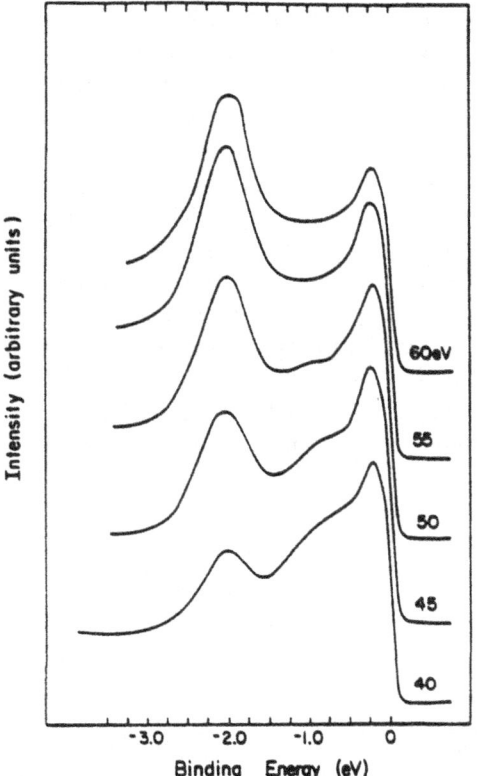

FIGURE 31. Photoemission EDCs for γ-Ce in the photon energy range of 40 V–60 eV. In this energy range the $4f$ cross section increases. The curves have been normalized to the intensity at an energy of -1.0 eV. The resolution is 120 meV. (From Ref. 233.)

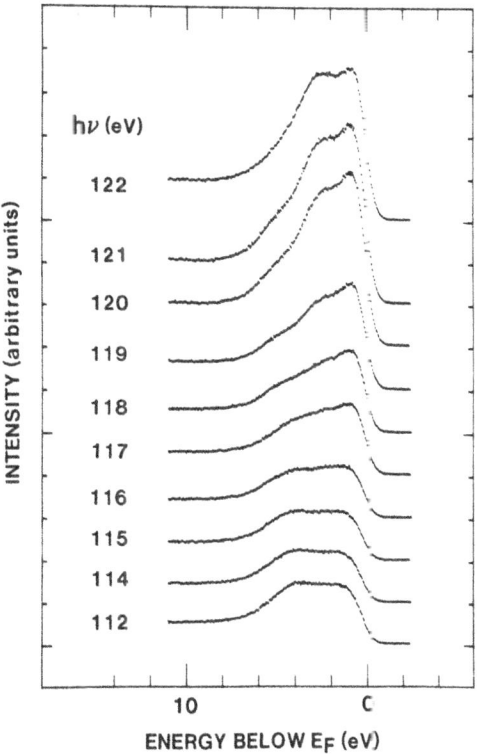

FIGURE 32. Photoemission EDCs for CeIr$_2$ in the photon energy range of the Ce 4d edge. (From Ref. 223.)

tively, rather than relying on the assumption that the 4f part will have a different shape. Especially for γ-like materials, analyses of the shape type have led to erroneous or conflicting results. Figure 32 shows a series of EDCs through the resonance region for CeIr$_2$, demonstrating the remarkable power of the resonance in enhancing the emission of one 4f electron over that from roughly 14 Ir 5d electrons. At present, the most reliable procedure is to subtract the Fano minimum spectrum ($h\nu$ about 112 eV to 114 eV) from the Fano maximum spectrum ($h\nu$ about 121 eV to 122 eV). This procedure relies on the information obtained from studies of Ce metal in the 40-eV-to-60-eV photon energy range, and from studies of La compounds, to the effect that the 5d emission resonates very weakly for the light rare earths, in contrast to the situation reported for the heavy rare earths.

A very interesting aspect[141,223] of the resonance behavior is shown in Figs. 33 and 34, which present for CeIr$_2$ and CeAl, respectively, the Fano minimum spectrum, the increase in emission for $h\nu$ about halfway through the resonance region, and the increase from this value of $h\nu$ to the Fano maximum. For both the γ- and the α-like systems, the increase in emission for the portion well below E_F is largest for the smaller values of $h\nu$ and that for emission near E_F is largest for the larger values of $h\nu$. Early interpretations[14] of this effect as reflecting a theoretical prediction[35] of different $h\nu$ dependency for the 5d and 4f resonances

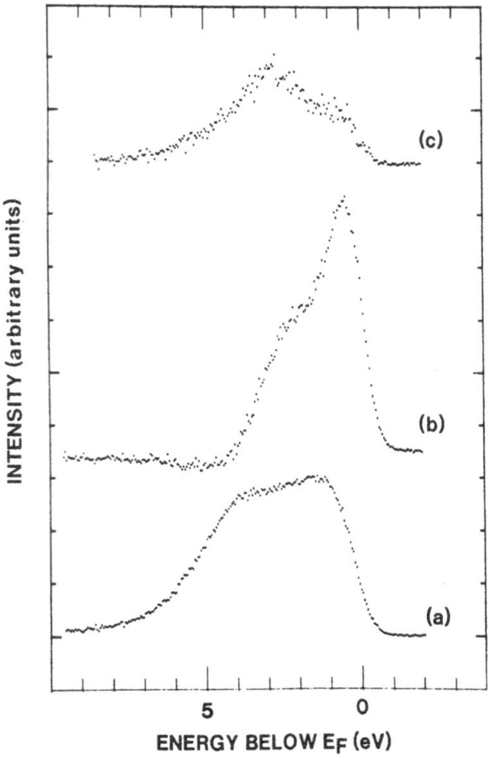

FIGURE 33. Details of resonant photo-emission behavior for CeIr$_2$: (a) the valence-band EDC at the Fano minimum, $h\nu_{min} = 112\,eV$; (b) the difference of the $h\nu_{mid} = 119\,eV$ spectrum and that for $h\nu_{min}$; (c) the difference of the $h\nu_{max} = 122\,eV$ spectrum and that for $h\nu_{mid}$. Note that the emission near E_F resonates most strongly at lower photon energies while the higher binding-energy emission resonates most strongly at higher photon energies (From Ref. 223.)

have been superseded by theories of the $4f$ resonance alone, based on the Anderson Hamiltonian.[17,18] In these theories, which describe the effect quite well, the different resonance behavior reflects the different phasings in the wave functions for the ionization and Kondo resonance parts of the spectrum and the hybridization between various intermediate states of the resonance. It is especially important to include the intermediate-state f^3 configuration, which occurs when U is not infinite, so that the ground state has f^2 configurations. Figure 35 illustrates the theoretical results for Hamiltonian parameters typical of cerium materials.

Yb is the one-hole analog to the one-electron case of Ce, so the theory worked out for Ce can be directly applied to Yb by a hole–electron mapping[244] in which the $4f$ BIS and PES spectra are interchanged. Features predicted for the Kondo resonance have been identified in the spectra of YbAl$_3$, and a fit of the spectra yields parameters consistent with the known ground-state properties of this material.[260]

4.6. Remaining Problems

In spite of the considerable progress described in this subsection on rare earths, there remain many experimental and conceptual challenges. First, the Kondo resonance should be studied with high resolution at low temperatures. Work of this type, with resolution of about 15 meV, has recently been done for certain cerium materials using a laboratory helium lamp with photon energies of

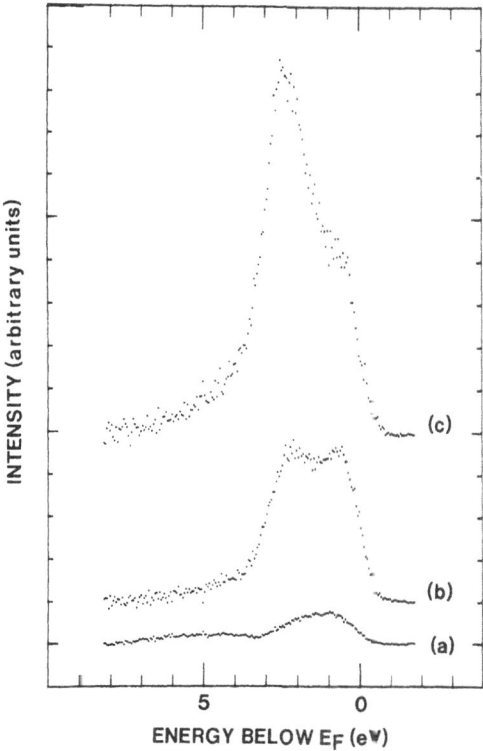

FIGURE 34. As for Fig. 33, but for CeAl and with $h\nu_{mid}$ = 118 eV. (From Ref. 223.)

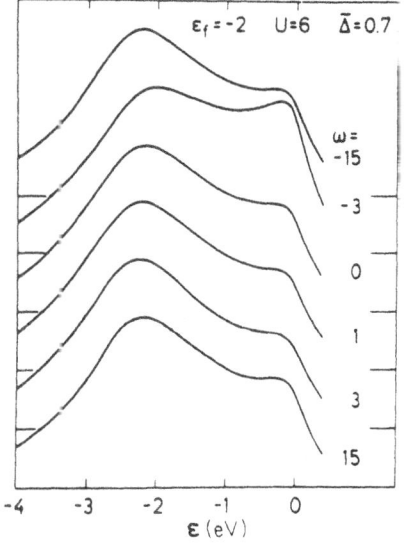

FIGURE 35. Theoretical photon energy (ω) dependence of the 4f spectrum calculated with the impurity Anderson Hamiltonian for the given parameters, typical for Ce materials; Δ is proportional to $N_f \rho V^2$. The photon energy zero is the energy where the Fermi energy structure has its maximum and the curves are normalized to the $\varepsilon = -2$ eV peak. For $\omega = -3$ eV, i.e., below the resonance, the $\varepsilon = 0$ peak is enhanced, as in the data of Figs. 33 and 34. (From Ref. 18.)

20 eV and 40 eV. For systems without transition metals, these spectra suffice for extracting the $4f$ emission of Ce materials. More generally, it would be highly desirable to perform such studies on other cerium materials with resonant photoemission. Second, the interpretations described above rest on a single-site theory for the Anderson Hamiltonian. Such theories cannot describe the transport phenomena of lattice systems, including, of course, superconductivity. It is expected[219] that the quantum states underlying the Fermi-level resonance have the coherence of the translational symmetry of the lattice. Experimental determination of k dependence in the resonance would have a major impact on the effort to understand coherence.

Third, the analysis of resonant-photoemission data thus far has been rather crude. Recent theoretical advances should be tested and used to obtain more detailed information from the complex resonance effects that have been observed.

Finally, the Anderson Hamiltonian parameter range of other rare-earth systems remains to be determined. For example, the theory for cerium strongly suggests that there must be, on some sufficiently small energy scale, relaxation effects for TmSe similar to those that give rise to the Kondo resonance in Ce materials; i.e., very near the Fermi level the states reached in PES and BIS should have the same f-level occupations as in the ground state, rather than having essentially integer occupations as seems to be observed in the spectra of subsection 5.4. It seems almost inescapable that models for the low-lying states of mixed-valent systems like TmSe must be augmented to include spin-fluctuation energy scales.

5. ACTINIDES

5.1. Introduction

The electronic structure of actinide materials is not well understood. Equivalently, the PES and BIS spectra of these materials have proved remarkably difficult to interpret, especially relative to the case of the rare earths. The usual perspective on this difficulty is that the $5f$ electrons of the actinides, while more localized than the transition-metal $3d$ electrons, are less localized than the rare-earth $4f$ electrons. While plausible, this perspective has yet to be given quantitative substance through a detailed interpretation of spectra.

By far the majority of electron spectroscopy work has been done for uranium or thorium since most workers are not equipped to handle highly radioactive materials, and in the case of resonant photoemission work,[261] only U and Th materials have been studied. Further, a recent thorough review on the subject, including resonant photoemission work through 1982, has been given by Baer,[262] so this section will concentrate on illustrating some general points and describing new developments for uranium materials. With the current uncertainties in interpreting $5f$ spectra, the role of certain reference materials which serve as examples of particular models is very important. The discussion here is focused largely around these reference materials.

RESPES studies for uranium have had less impact than those for the rare earths, in general, and for cerium in particular. This is partly because of the lack of understanding of the spectra, and partly, for uranium, because the emission from 3 $5f$ electrons is relatively easier to discern than that of 1 $4f$ electron. Nonetheless, for measuring the $5f$ spectrum in compounds with late transition metals, with their nearly filled d states, and in situations where the uranium is very dilute, resonance techniques have been very useful. RESPES studies also have provided a test of one of the few new theoretical efforts,[37] the application of the Anderson impurity Hamiltonian to UO_2.

One approach to interpreting the electron spectra is the single-site view taken with the rare earths. Even at the simplest level, in which hybridization is ignored, the situation is more difficult than with the rare earths because the fingerprint character of the transitions $f^n \rightarrow f^{n-1}$ is lost. This is because the spin–orbit splitting is larger than multiplet splittings for the $5f$ electrons, causing the predicted spectra[263] to be less sensitive to n. More seriously, various estimates[264] of the $5f$ Coulomb interaction suggest that it is much smaller than for rare earths, so that the assumption of a single valence becomes suspect as soon as hybridization is introduced into the picture. The spectra do not have the general appearance that might be expected for the Anderson Hamiltonian based on knowledge of the rare earths, and typically most of the $5f$ spectral weight is found in a broad band around the Fermi energy. There are only a few cases where the weight displays a gap around the Fermi energy, which could be taken as evidence of a single valence state stabilized by a Coulomb interaction, U. These are the only cases where a localized approach has been tried seriously, and they are discussed in section 5.2, especially a recent treatment of UO_2 using the Anderson impurity Hamiltonian.

Another approach is to compare measured PES/BIS spectra to the results of band theory. For some materials, discussed in section 5.3, there is reasonable agreement between theory and experiment, while in other cases, mentioned in section 5.4, spectral weight is found further above and below E_F than occurs in the band calculations. There is general agreement that this signals the presence of Coulomb interactions among the $5f$ electrons, but there is no detailed understanding as yet.

5.2. Materials Modeled as Ionic

UO_2 is a material frequently interpreted by an ionic model of a stable $5f^2$ valence state, corresponding to a U^{4+} ion, and a filled oxygen $2p$ band. That UO_2 is a magnetic insulator supports this picture, and the XPS/BIS spectrum,[265] shown in Fig. 36, has a clear gap around E_F. The traditional interpretation, that the XPS spectrum shows $5f^2 \rightarrow 5f^1$ transitions and the BIS spectrum shows $5f^2 \rightarrow 5f^3$ transitions, is indicated in the figure by the bar diagram for the two sets of final-state multiplets. A Coulomb energy U_{ff} of 4.6 eV is inferred from this interpretation. The oxygen $2p$ band lies at binding energies greater than 3.5 eV. The highest-binding-energy XPS peak does not correspond to any feature of the calculated oxygen band, and it is generally taken to be a satellite for which a variety of assignments have been given.

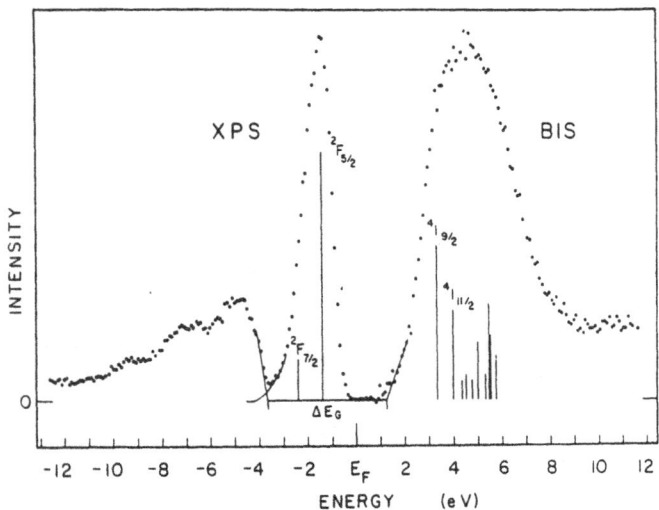

FIGURE 36. Combined XPS and BIS spectra of UO_2 for the energy range 12 eV below and above E_F. The bar diagram shows the final-state multiplet spectra expected for $5f^2 \rightarrow 5f^1$ and $5f^2 \rightarrow 5f^3$ transitions in *LS* coupling. (From Ref. 265.)

Resonant photoemission experiments[266,267] have been made for the valence band of UO_2. Figure 37 shows the valence-band spectra[267] for several photon energies, and Fig. 38 shows CIS spectra for the valence-band positions labelled in Fig. 37. Resonance effects are observed for all parts of the valence band, including the oxygen states and the satellite peak labelled 4. For the satellite peak, the CIS of the inelastic background, position 5 in Fig. 37, has been subtracted out. Earlier resonant photoemission studies by Reihl *et al.*[266] obtained the same behavior for the primary 5f peak, labelled 1, but concluded that any resonance of the rest of the band was too weak to observe. The resonance effects imply hybridization of uranium states into the oxygen band, as found in band calculations, and it is almost certain, though not proved, that the 5f admixture provides the largest resonance effects.

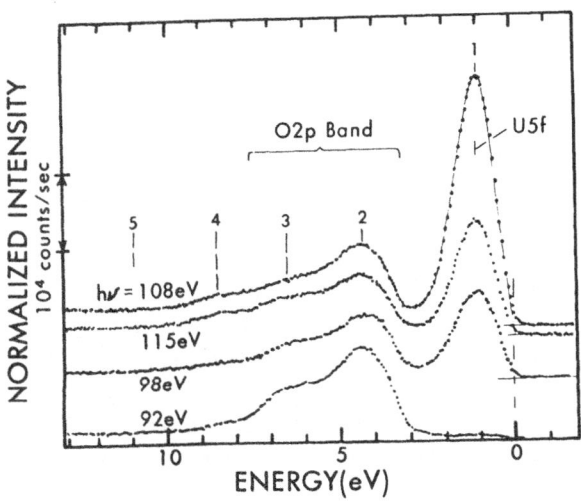

FIGURE 37. Valence-band EDC spectra of UO_2 (111) for various photon energies around the uranium 5d edge. The spectra are normalized to the photon flux. The $h\nu = 115$ eV spectrum shows enhancement of the 8.5-eV satellite. (From Ref. 267.)

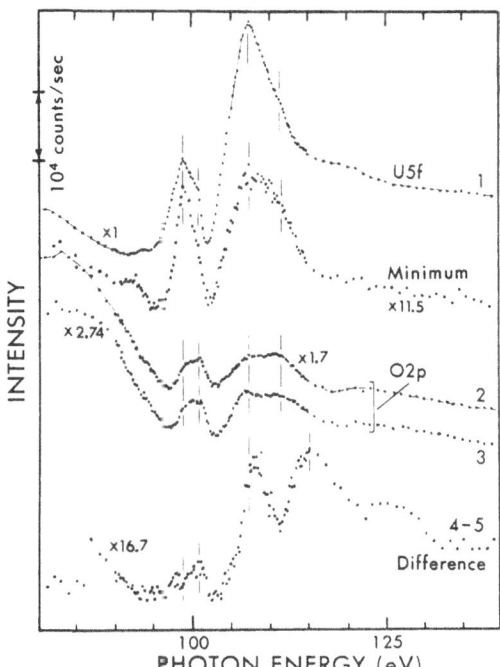

FIGURE 38. Valence-band CIS spectra of UO_2, (111) for photon energies around the uranium $5d$ edge obtained at the initial energies indicated in Fig. 37. (From Ref. 267.)

The $5f$ CIS of Fig. 38 is typical of the CIS spectra of uranium materials around the uranium $5d$ edge. There are resonant maxima at about 98 eV and 107 eV and resonant minima near 92 eV and 102 eV. The energies of these maxima and minima vary somewhat from material to material. The CFS spectrum, shown in Fig. 39, is similar except for showing additional structure at higher energies, near 115 eV, where there is also a peak in the satellite CIS. Figure 39 also compares the CFS spectrum with the absorption spectra calculated for the atomic transitions $5d^{10}5f^n \rightarrow 5d^95f^{n+1}$, for $n = 0, 1, 2$. The agreement is best for $n = 2$, as would be expected with the ionic model.

A realistic model for UO_2 must include both the hybridization between the U $4f$ and oxygen $2p$ electrons and the $5f$ Coulomb interaction U_{ff}. The impurity Anderson Hamiltonian has been applied[37] to the $4f$ core level and $5f$ spectra of UO_2, resulting in a conceptually unified and qualitatively successful description. Except for the improvement of treating the oxygen $2p$ states as a continuum, the model is basically the same as in the cluster model presented in section 3.2 for NiO, i.e., the ionic ground state is corrected by including configurations with electrons transferred from oxygen to uranium atoms. The ground state is an admixture of f^2p^6, f^3p^5, and f^4p^4; the $5f$ photoemission final states are admixtures of f^1p^6, f^2p^5, and f^3p^4; the $5f$ BIS final states are admixtures of f^3p^6, f^4p^5, and f^5p^4; and the $4f$ photoemission final states are admixtures of $4f^{13}5f^2p^6$, $4f^{13}5f^3p^5$, and $4f^{13}5f^4p^4$. Within each group, the states are coupled by the hybridization matrix element and have energies determined by the relative location of the f and p states, and by U. For the $4f$ photoemission, a Coulomb interaction $U_{fc} = 7.2$ eV is introduced to account for the attraction of the $5f$ electrons to the $4f$ hole. The effect of U_{fc} is to alter the relative energies of the $5f^2$

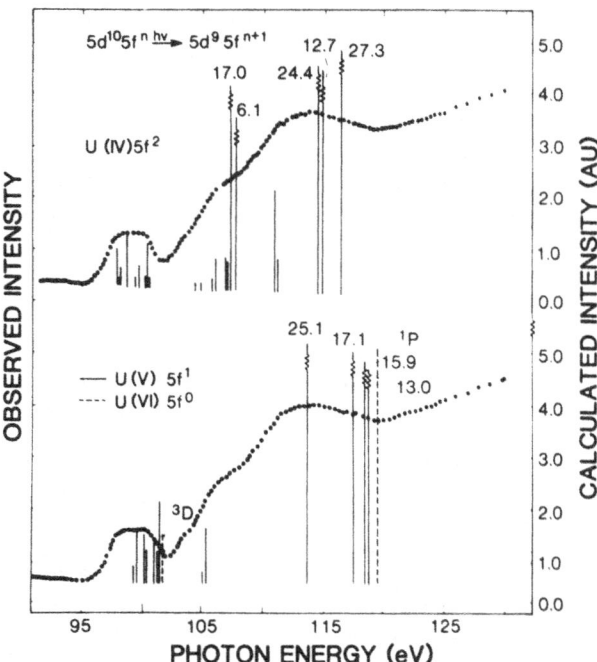

FIGURE 39. CFS yield spectrum of UO_2, for photon energies around the uranium $5d$ edge, compared to the oscillator strengths of the transitions from $5d^{10}5f^n$ to the various multiplet states of $5d^9 5f^{n+1}$ for $n = 0$, 1, and 2. (From Ref. 267.)

and $5f^3$ states such that $5f^3$ becomes more stable in the presence of the attractive potential of the core hole.

Figure 40 shows the calculated[37] $5f$ PES spectrum for UO_2, where the energy zero is that of Fig. 36. In the calculation, the oxygen $2p$ band $\rho(E)$ has been modeled by a semielliptical density of states, from -4 eV to -8 eV. For the parameters used, the unhybridized $5f^2 \rightarrow 5f^1$ transition lies 0.1 eV below the top of the $2p$ band, and $U_{ff} = 7.2$ eV, so the unhybridized BIS peak lies 7.1 eV above the top of the $2p$ band. The average value of the hybridization width $\Delta = \rho V^2$ is 0.76 eV, and as for Ce, it is $N_f \Delta$ that enters the PES theory, giving hybridization a larger effect than in a band calculation. Hybridization then

1. pushes the main $5f$ PES peak about 2.5 eV to smaller binding energy, in good agreement with experiment;
2. spreads some $5f$ spectral weight through the p-band region; and
3. pushes a satellite peak out of the bottom of the p band.

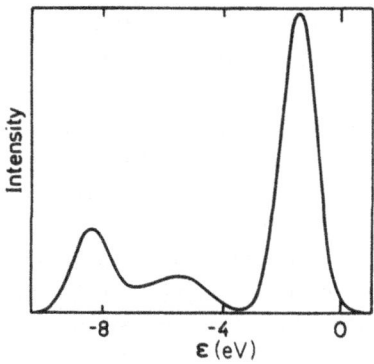

FIGURE 40. The theoretical $5f$ photoemission spectrum of UO_2 calculated for the impurity Anderson Hamiltonian as described in the text. (From Ref. 37.)

Thus the general features of the data are reproduced, but the weight of the satellite peak is too large. Hybridization also shifts the BIS peak (not shown) to higher energies, about 1 eV, to the center of the experimental peak. In the ground state the weights of $5f^2$, $5f^3$, and $5f^4$ are 0.85, 0.14, and 0.01, respectively, a modest departure from the ionic picture. However, the final state of the PES main peak is strongly mixed, about equally $5f^1$ and $5f^2$, because these configurations are degenerate in the PES final states.

Figure 41 shows experiment[265] and theory[37] for the 4f core level spectrum. All multiplet effects have been neglected. Each spin–orbit component has a main peak and a satellite. The occurrence of the satellite reflects the several possible valence-band configurations that can accompany the 4f core hole, similar to the situation for Ce described in section 4.5. In the presence of the core hole, the $5f^2$ and $5f^3$ states are nearly degenerate and the main and satellite lines are, respectively, bonding and antibonding mixtures of these. The model parameters were chosen to achieve the fit of Fig. 41, and then used to calculate the 5f spectrum. This interpretation of the satellite structure differs from previous ones either as to the configuration involved or as to the extent of mixing.

In principle the $5d \rightarrow 5f$ resonant photoemission CIS spectra can also be calculated using this impurity Anderson model. This calculation has not been done because in practice the situation is very complex. The intermediate states are admixtures of $5d^9 5f^3 p^6$, $5d^9 5f^4 p^5$, and $5d^9 5f^5 p^4$, which will produce phasing effects as described for cerium in section 4.5. Also, as pointed out in section 2, the large number of intermediate states associated with each configuration greatly complicates the resonance behavior, even without considering the hybridization mixing of the different configurations. It would be interesting to extend the comparison of Fig. 39 to include the multiplet spectra for $5d^9 5f^4$ and $5d^9 5f^5$, although with large hybridization mixing of the final states of the photon absorption it is not possible to deduce the uranium-ion ground state simply by analyzing the CFS spectrum into a sum of component spectra.

The theory has been applied to an unusual UO_2 resonant photoemission study,[268] that of the 4f core level emission using photon energies in the vicinity of the uranium 3d edge. This study adopted a purely ionic model and concluded from qualitative arguments that an assignment of $2p^5 5f^3$ for the satellite was consistent with its resonance behavior. However, the strongly admixed f^2-f^3 picture above leads to a good description of some details of the CIS spectra. The

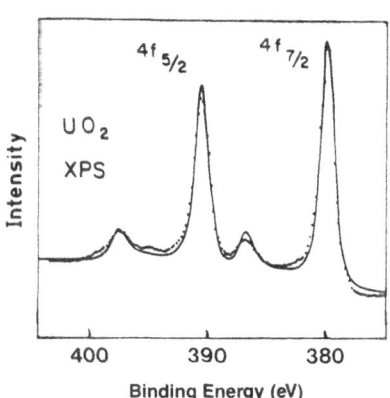

FIGURE 41. Theoretical (full line) and experimental (dots) 4f core level spectra of UO_2. (Experimental results from Ref. 265.) The spectra have a $4f_{5/2}$ and a $4f_{7/2}$ component, and each component has a main peak and a satellite, corresponding to a bonding and an antibonding $5f^2 - 5f^3$ state, respectively. (From Ref. 37.)

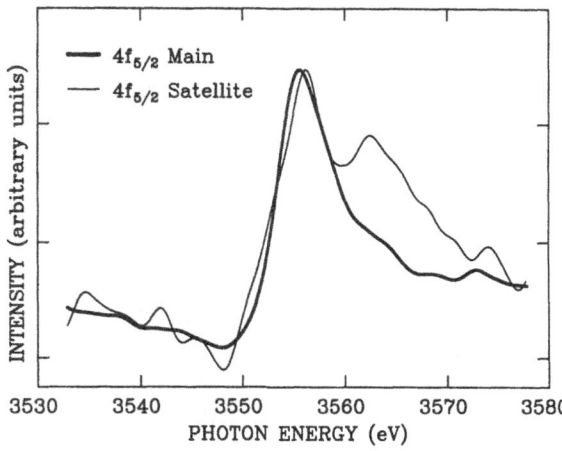

FIGURE 42. $4f$ CIS spectra of UO_2 at the $3d(M_5)$ threshold. (Previously unpublished results from the study of Ref. 268.)

resonance process involves the transitions

$$3d^{10}4f^{14}5f^n \rightarrow 3d^9 4f^{14}5f^{n+1} \rightarrow 3d^{10}4f^{13}5f^n\varepsilon_g \qquad (5.2.1)$$

for which multiplet splittings are sufficiently small that it may be justifiable to neglect them. The $4f$ spectra at various photon energies near the M_4 and M_5 edges have been published in Ref. 268. Figure 42 shows previously unpublished CIS spectra from the same study for the main and satellite $4f_{5/2}$ peaks at the M_5 threshold near $hv = 3554$ eV. Because the satellite emission is rather weak, its raw CIS spectrum reflects also the photon-energy dependence of the inelastic background that accompanies the lower binding-energy peak, so the spectrum shown in the figure has been corrected by subtracting an estimated inelastic CIS spectrum, taken to have the shape of the CIS spectrum of the $4f_{7/2}$ main peaks. The data have also been convolved with a Gaussian of 2.0-eV full-width-half-maximum, which suppresses noise without significant loss of spectral structure, and the two spectra have been scaled to have the same amplitude at the peak. It can be seen that the two spectra differ in that the satellite spectrum shows extra structure at higher energy. Figure 43 shows the theoretical CIS spectra calculated with the Anderson model. The theoretical spectra reproduce the essential features of the data, especially the additional structure of the satellite spectrum, for which the theory shows three structures corresponding to the $5f^3$, $5f^4$, and $5f^5$

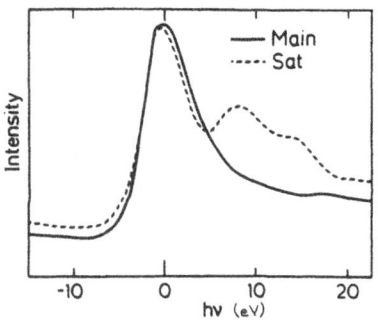

FIGURE 43. The theoretical CIS spectrum of the main peak (full curve) and the satellite (dashed curve) for the $4f$ spectrum of UO_2. The energy zero for hv is at the resonance threshold. The curves are normalized to their maximum values. (From Ref. 37.)

configurations of the intermediate states. The difference in the two CIS curves arises from differences in the phases of the various configurations in the wave functions of the initial, intermediate, and final states for the main and satellite lines. The parameters are the same as for the $4f$ XPS spectrum, except that U_{ff} was reduced to 4 eV, which apparently compensates for some aspect of the model that is oversimplified.

UPd$_3$ is the only metallic uranium material whose $5f$ PES/BIS spectra are universally interpreted on a localized model. Figure 44 shows the XPS/BIS spectrum.[269] By comparison with the XPS spectrum of ThPd$_3$, which shows the Pd $4d$ features B, C, and D, it was inferred that feature A is the $5f$ emission. As with cerium, the BIS spectrum is expected to be dominated by the $5f$ spectral weight. Various low-energy and low-temperature measurements have been interpreted on a model of a stable $5f^2$ valence state.[270–275] Accordingly, the small $5f$ weight around E_F, with its rather gap-like appearance, is usually interpreted as showing the Coulomb interaction U. The XPS feature A is ascribed to $5f^2 \rightarrow 5f^1$ transitions, while the broad BIS spectrum is ascribed to $5f^2 \rightarrow 5f^3$ transitions, its width being due to the final-state multiplets of $5f^3$. It must be remarked that the $5f$ spectrum provides no characteristic fingerprint features other than its gap-like appearance to support this interpretation, and the strongest evidence for the $5f^2$ ground state is from the low-energy data.

Figure 45 shows the CIS spectra[266] for features A and B of the valence band in the photon energy range around the uranium $5d$ edge. These data do not extend to high enough energy to show[276] that there is also a maximum around 110 eV, as occurs for UO$_2$ in Fig. 38. Figure 46 shows the valence band[266] at the photon energies $hv = 98$ eV and 92 eV where the CIS curves have a resonant maximum and minimum, respectively. The resonance of features A and A' confirms their assignment as $5f$ emissions, as does the fact that the $5f$ weight peaks at feature A, about 1 eV below E_F, with very little $5f$ weight at E_F. Although feature B resonates, it is not completely suppressed in the 92-

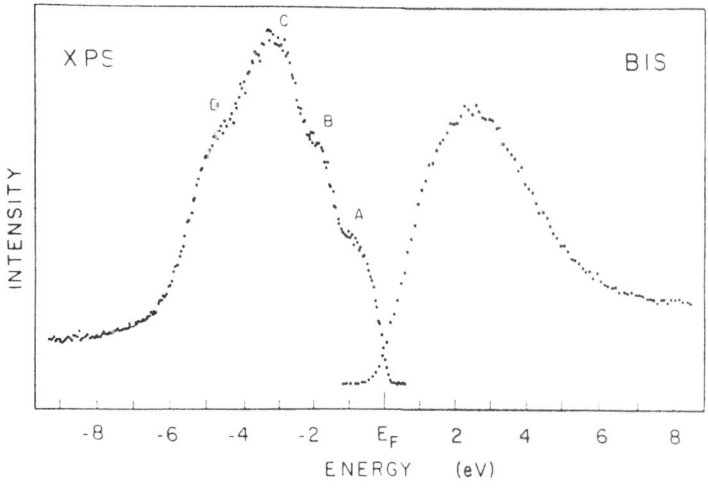

FIGURE 44. Combined XPS and BIS spectra of UPd$_3$ for 8 eV below and above E_F. (From Ref. 269.)

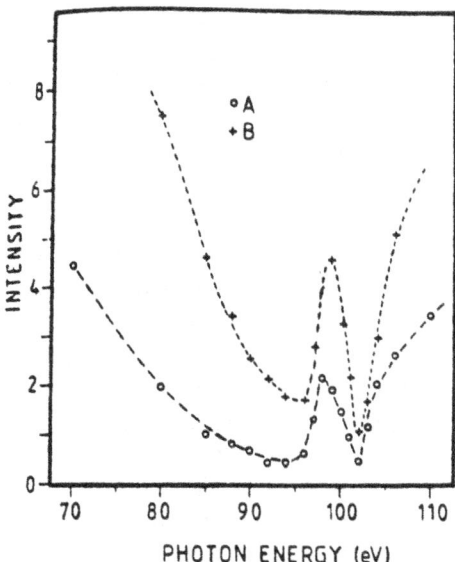

FIGURE 45. CIS spectra for UPd$_3$ for photon energies near the uranium 5d edge, for the valence band features labeled in Fig. 44. (From Ref. 266.)

eV spectrum, implying that at this binding energy there is overlapping uranium 5f and Pd 4d emission.

The effect of diluting the uranium in this material has been studied.[277] Figure 47 shows spectra for UPd$_3$ at the photon energies of $hv = 108$ eV and 102 eV where the CIS curves also have a resonant maximum and minimum, respectively. This energy range is more useful for the dilution experiments than that between 92 eV and 98 eV because the Pd states show over the latter range of hv a variation which, even though it is small, poses numerical problems for extracting the small 5f weight of the diluted material. The figure also shows spectra with $hv = 102$ eV for Y$_{1-x}$U$_x$Pd$_3$. It can be seen that the Pd 5d peaks move toward the Fermi level with dilution. Figure 48 presents for various x the uranium 5f spectra obtained as the difference of the spectra taken with $hv = 108$ eV and 102 eV. It is

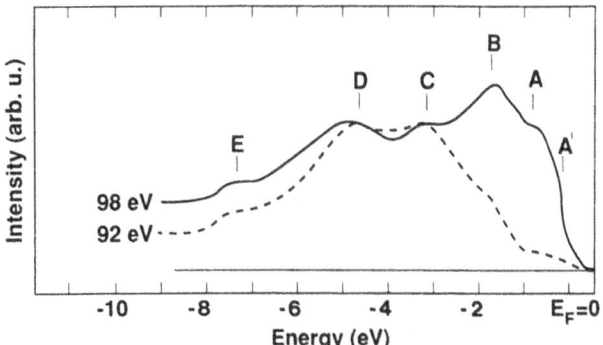

FIGURE 46. Valence-band spectra of UPd$_3$ (0001) at the uranium 5d resonance minimum ($hv = 92$ eV) and maximum ($hv = 98$ eV). The spectra are normalized at the point C. (From Ref. 266.)

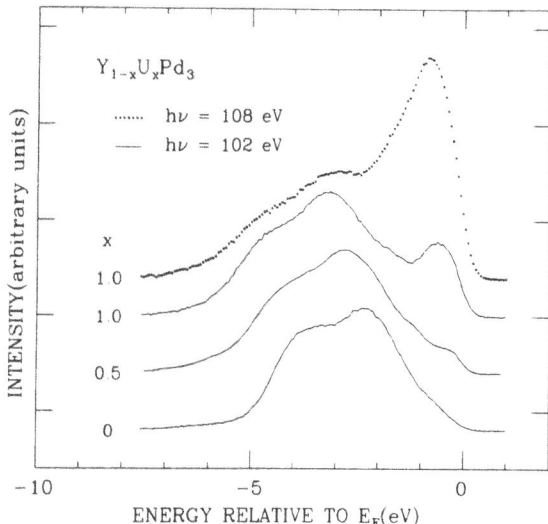

FIGURE 47. Valence-band spectra normalized to the incident photon flux for $Y_{1-x}U_xPd_3$ with $h\nu = 102$ eV for $x = 0$, 0.5, and 1.0 and $h\nu = 108$ eV for $x = 1.0$. Note the shift toward E_F of the Pd $4d$ emission between -2 eV and -6 eV as x decreases. (From Ref. 277.)

evident that the $5f$ peak at 1 eV narrows and moves to E_F with dilution. A crystal-structure change that occurs between $x = 0.5$ and $X = 0.9$ does not appear to be a major cause since half the movement of the $5f$ peak occurs between $x = 0.1$ and $x = 0.5$. The implications of this dilution result for the traditional interpretation of UPd$_3$ and the role of the shift of the Pd $4d$ peaks are not yet clear. Following the traditional interpretation, it has been argued[277] that trivalent yttrium substitutes for tetravalent uranium $4f^2$, so that the number of conduction electrons decreases with increasing yttrium concentration, causing the

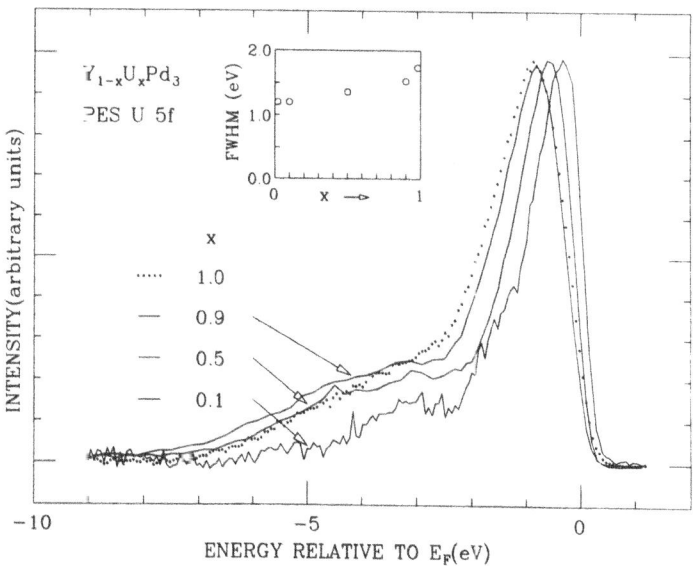

FIGURE 48. U $5f$ spectral weight for $x = 0.1$, 0.5, 0.9, and 1.0 in the $Y_{1-x}U_xPd_3$ system. Note that the $5f$ peak moves toward E_F as x decreases. Inset shows that the U $5f$ FWHM decreases as x decreases. For $x = 0.02$, the $5f$ spectrum (not shown) is identical to that for $x = 0.1$ within the experimental uncertainty due to statistically poor data from the highly diluted sample (From Ref. 277.)

Fermi energy to decrease. The fact that the unit cell volume changes little with x correlates well, within this picture, with the fact that the Goldschmidt radius is nearly the same for Y^{3+} and U^{4+}.[278]

A very interesting group of materials are compounds of uranium with the pnictides and the chalcogenides additional to oxygen, and resonant photoemission spectra have been measured for many of these materials. Many of these data are summarized in Ref. 262, and there has also been subsequent work.[279] There is little consensus as to the electronic structure of these materials, but several, including UTe and $U_{1-x}Th_xSb$, appear to have $5f$ peaks below E_F, which has been taken to imply for them an ionic model by the same criterion as for UPd_3.[266,280–282] The reader should be aware, however, that the ionic model has been disputed[283] for these materials.

5.3. Materials Modeled as Bandlike

Comparison with band theory results has been the most widely used, and widely useful, method of analyzing electron spectra of actinide materials. Unfortunately, the extent of the agreement is very much in the eye of the beholder. Alpha-uranium and thorium metals[284,285] are perhaps the most likely candidates for a successful band description of their $5f$ electrons. In both cases the shape, centroid, or width of the BIS spectra[284] differ considerably from calculations, and yet the agreement often has been termed satisfactory. There is much better agreement between experiment and band theory for the uranium valence-band photoemission spectrum,[284] although it is not as detailed as is found in simple metals. It was claimed,[286] on the basis of a resonant photoemission

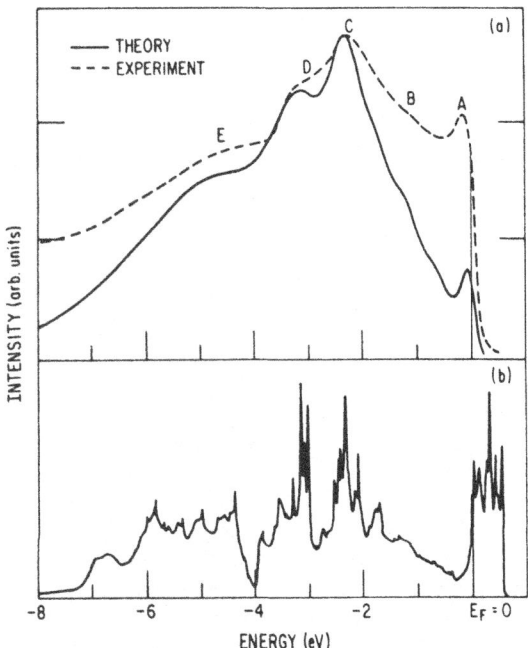

FIGURE 49. (a) Valence-band spectrum (dashed line) obtained from UI_3 (100) at $h\nu = 98$ eV, superposed on a calculated total density of states for UIr_3 (solid line) which has been broadened to account for instrumental and lifetime broadening. (b) The unbroadened density of states. (From Ref. 288.)

study, that α-uranium has a 5f valence band satellite somewhat like that of Ni metal, but more recent work[287] has shown that this feature, observed at the Fano minimum photon energy, is most probably due to the uranium 6d emission. Other materials often regarded as bandlike are uranium intermetallics such as UIr$_3$, discussed below; uranium pnictides and chalcogenides,[283] except for UO$_2$ and possibly UTe, and USb, as mentioned in the preceding subsection. Some of the uranium intermetallics, such as UPt$_3$, are heavy-fermion materials discussed in the next subsection, and for these the applicability of band theory is a delicate and fascinating question.

One material where the case for agreement between the 5f photoemission spectrum and band theory has been made with particular care is UIr$_3$.[288] The top panel of Fig. 49 compares the valence-band spectrum taken from a [100] surface at the Fano maximum photon energy 98 eV with the total calculated density of states broadened to account for instrumental resolution and lifetime effects. The bottom panel of the figure shows the unbroadened density of states. There is a good general agreement in the energies of the various features, with the magnitudes of features A and B enhanced in the experimental spectrum because they are largely of uranium 5f origin, while the other features are largely of Ir 5d origin. Figure 50 shows the valence-band spectra at the Fano minimum and maximum photon energies, normalized at the energy of feature E, and the difference spectrum, which is taken to be the 5f spectrum. Figure 51 compares this difference spectrum with the calculated 5f partial density of states (PDOS). The figure also shows an estimate[289] of the variation of the 5f photoemission cross section σ, and the product of σ and the 5f PDOS. The agreement between the latter and the experimental spectrum is as good as is ever found for uranium materials. Angle-resolved spectra showed dispersion of some of the valence-band features, in fair agreement with the calculated band dispersion, including that of one of the narrow f-like bands near E_F. The extent of the agreement is shown in Fig. 52. Figure 53 shows the BIS spectrum of UIr$_3$,[290] expected to show

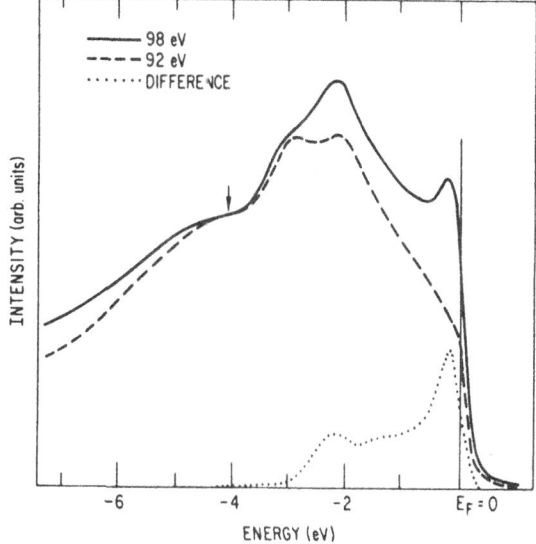

FIGURE 50. Photoemission spectra for UIr$_3$ (100) at the resonance maximum (solid line, $h\nu = 98$ eV) and at the resonance minimum (dashed line, $h\nu = 92$ eV). The dotted line is the difference of the two spectra and should represent primarily the 5f spectrum. (From Ref. 288.)

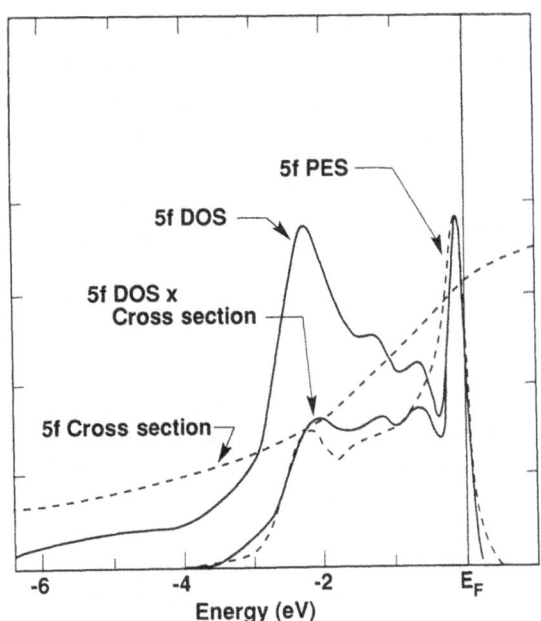

FIGURE 51. Experimentally extracted $5f$ spectrum of UIr$_3$ compared with broadened $5f$ PDOS and estimated $5f$ photoemission cross section σ for UIr$_3$, and the product of σ and the PDOS. (From Refs. 288 and 289.)

dominantly the $5f$ density of states, compared with the broadened band theoretic $5f$ PDOS. It is apparent that the experimental spectrum shows considerable extra weight in the energy range between 2 eV and 6 eV, and this weight has been interpreted[290] as indicating the effect of U_{ff}, analogous to the 6 eV satellite in the valence-band PES spectrum of Ni metal, as discussed in section 3.

5.4. Heavy-Fermion Materials

A number of materials containing uranium have very large low-temperature specific heats, leading to the descriptive term heavy fermion.[122,189,225] Although the term was invented at a time when attention was focused on uranium materials, it applies equally well to several Ce and Yb compounds, as mentioned in section 4. The ground states vary from magnetic to superconducting, and it is generally agreed that the $4f$ and $5f$ electrons give rise to these novel and variable properties. It has been speculated that the superconducting pairing in some of the materials is unusual and possibly of electronic rather than phononic origin. For cerium materials, as described in section 4.5, the PES/BIS $4f$ spectrum is dominated by single-site energetics, and the impurity Anderson Hamiltonian gives a unified description of the $4f$ spectrum and such low-energy properties as the specific heat and the magnetic susceptibility, which can be characterized by a spin fluctuation energy, the Kondo temperature T_K. As predicted theoretically, very heavy materials, which have small T_K, have very little $4f$ weight around E_F (although more than expected in the simplest models—see section 4.5). Such is not the case for the $5f$ spectrum of heavy-fermion uranium materials. For these materials, there is very much weight around E_F,[291–294] with a somewhat bandlike appearance, and the importance of the $5f$ Coloumb interaction can be inferred[295]

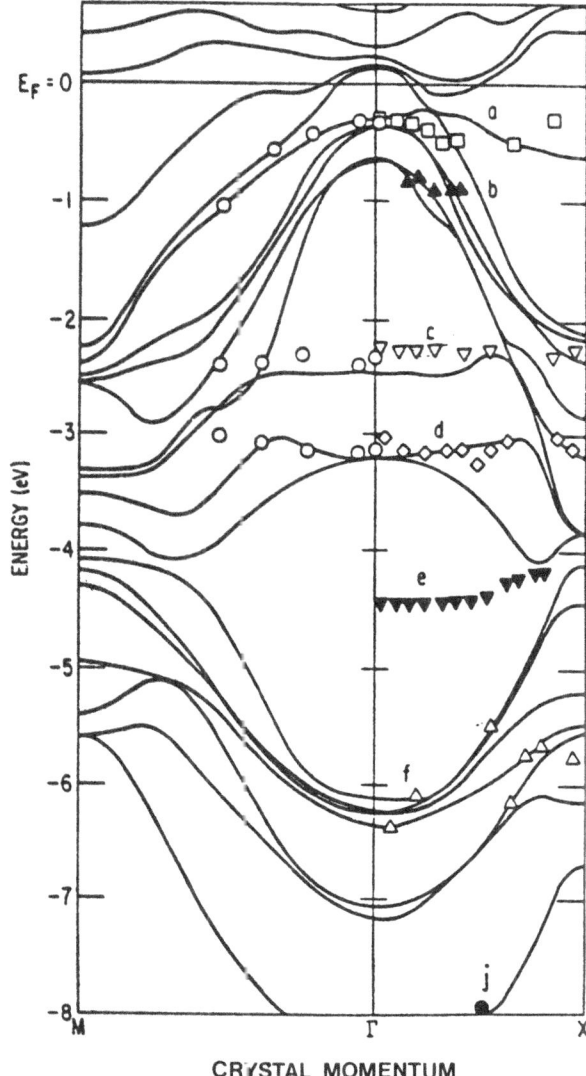

FIGURE 52. Experimental angle-resolved photoemission points for UIr_3 (100) superposed on a theoretical band structure. Data along the $\Gamma-X$ line are normal-emission data obtained at $hv = 26\,eV$ while varying the polar angle. (From Ref. 288.)

CRYSTAL MOMENTUM

only from differences between the spectrum and the results of density functional calculations, as described next.

Figures 54 and 55 show, for two heavy-fermion materials, UPt_3 and UAl_2, respectively, combined BIS and resonant photoemission spectra.[295] As usual, the PES spectra labeled U $5f$ results from subtracting the $hv = 92\,eV$ spectrum from that for $hv = 98\,eV$, and the 92 eV spectrum is taken to have the U $5f$ emission suppressed and thereby to reveal Al $3s-3p$ or Pt $5d$ emission, plus some U $5d$ emission that is suppressed only partly by the resonance. The BIS spectrum is dominated by the U $5f$ spectral weight. The figure also shows the results of band calculations[296,297] using the local density approximation (LDA) to the density functional theory. Many features of the LDA results are seen in the experimental spectra. For UPt_3, the $hv = 92\,eV$ spectrum shows a small Pt weight at E_F and the calculated width, if not the detailed shape, of the Pt states, while for UAl_2 the

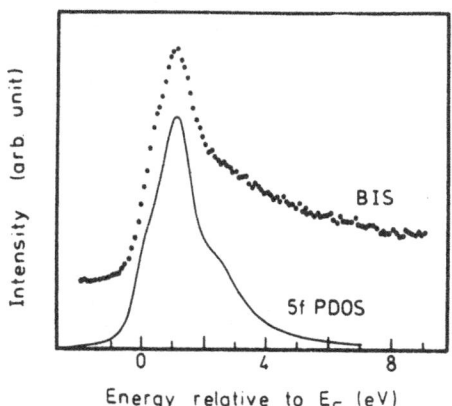

FIGURE 53. The BIS spectrum of UIr$_3$ compared to the calculated 5f PDOS after broadening. Note the extra experimental intensity from 2 eV to 7 eV. (From Ref. 290.)

Al s–p states can be seen, including the dip at $-5\,\mathrm{eV}$ and the U d states nearer E_F. It is interesting to note that the one-electron widths of the various theoretical subbands is small enough to observe a clear separation into two groups related to the atomic spin–orbit splitting. This separation suggests a spin–orbit origin of the 0.5-eV shoulder and the 1-eV peak found in the BIS spectra of both materials.

The most striking feature of the two spectra is that the measured widths of the 5f spectral weights greatly exceed the one-electron ones. Above E_F the BIS spectra show very much weight over an energy range as great as $6\,\mathrm{eV}$ above the 5f cutoffs, and below E_F the PES spectra show much more weight in the region from $-1\,\mathrm{eV}$ to $-2\,\mathrm{eV}$ than is predicted or can be explained by the experimental resolution, which is about 0.5 eV in both spectra. This difference in widths, also identified in UAu$_3$[298] and UBe$_{13}$,[299,300] signals the effects of the 5f Coloumb interactions, but as yet there is no detailed theoretical description. One proposal[289] is that the ligand–U 5f hybridization is sufficiently large that lattice effects dominate the 5f spectrum, so that one should use the type of model employed to describe the spectrum of nickel metal (see section 3). Alternatively,[293] it has been argued that there are important differences between cerium and uranium which, when included in a rigorous theory of the impurity. Anderson Hamiltonian, may lead to a description as good as that for the Ce 4f spectra.

Evidence of the importance of a single-site model has been provided[301] by studies of the diluted system $U_x Y_{1-x} Al_2$ for $x = 0.1$ and 0.02. The alloy system has the same crystal structure for all x. The U 5f PES spectrum has been extracted from valence-band spectra for $x = 0$, 0.02, and 0.1 taken at photon energies $h\nu = 92\,\mathrm{eV}$ and $108\,\mathrm{eV}$. The spectra have been normalized to the photon flux, and their inelastic backgrounds subtracted in a usual way. Figure 56 shows that the YAl$_2$ spectrum at $h\nu = 108\,\mathrm{eV}$ is almost identical in shape to that for $h\nu = 92\,\mathrm{eV}$, except that the total intensity has decreased by a factor of 0.7, consistent with the known photon-energy dependence of the Al 3s–3p states. Also, the $h\nu = 92\,\mathrm{eV}$ spectra for $x = 0.1$ and $x = 0.2$ (not shown) are essentially

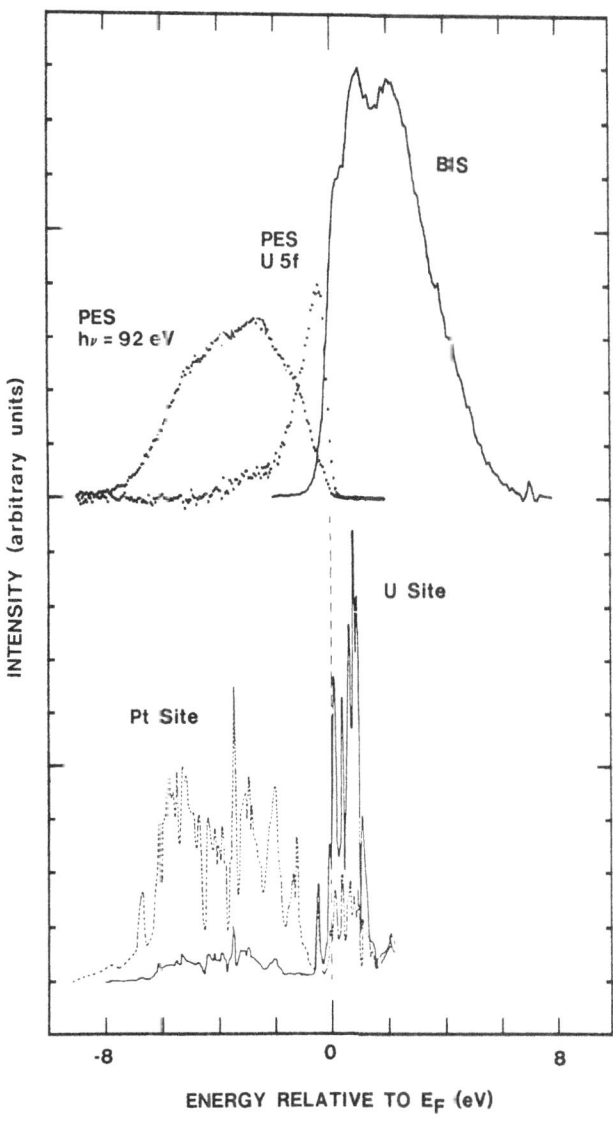

FIGURE 54. PES and BIS valence- and conduction-band spectra for UPt$_3$, as described in the text (top) and LDA density of states (bottom). Note extra width of the experimental PES/BIS 5f spectra relative to the LDA density of states. DOS—local density functional. (From Refs. 295 and 297.)

identical to the two YAl$_2$ spectra, confirming the usual assumption that the U 5f weight is suppressed at this photon energy. Figure 57a and b show normalized data with $hv = 108$ eV for $x = 0.1$ and 0.02, respectively. The $hv = 92$ eV spectra have been scaled by the factor 0.7 in order to account for the hv-dependence of the non-5f emission, and the two $hv = 92$-eV spectra have been given the same amplitude. The difference curves in each case give the 5f spectrum. The areas of the two extracted 5f spectra are in the ratio 0.18, very close to that of the two concentrations. Figure 58 compares the shapes of the 5f emission for $x = 0.02$, 0.1, and 1, and it is evident that they are essentially the

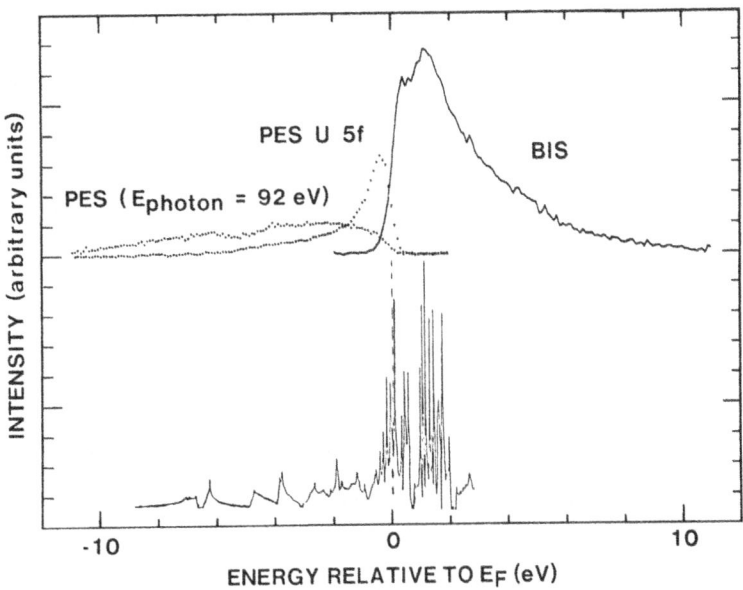

FIGURE 55. PES and BIS valence- and conduction-band spectra for UAl_2, as described in the text (top) and LDA density of states (bottom). Note extra width of the experimental PES/BIS $5f$ spectra relative to the LDA density of states. (From Refs. 295 and 296.)

same. This implies that single-site effects dominate even the $x = 1$ spectrum. This study also shows the power of the resonance technique for treating dilution systems, and the possibility for measuring for at least $x = 0.001$ when new insertion-device beam lines yielding perhaps 100 times the photon flux become available.

The valence bands of the heavy-fermion materials UPt_3, UBe_{13}, and U_2Zn_{17}

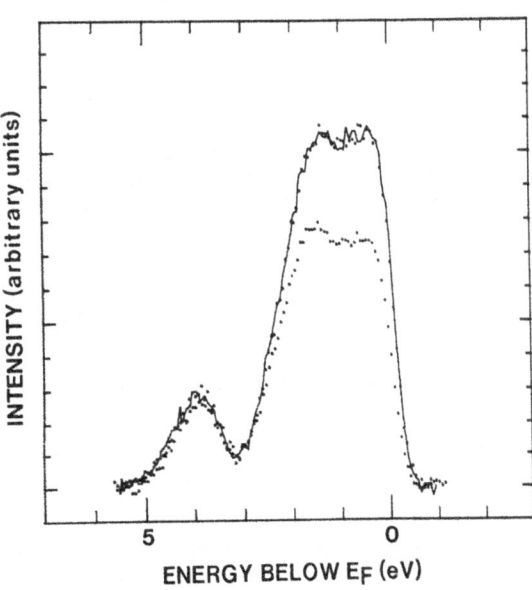

FIGURE 56. Normalized valence- band spectra with $hv = 92$ eV for $U_{0.1}Y_{0.9}Al_2$ (solid) and YAl_2 (dotted), and with $hv = 108$ eV for YAl_2 (dotted). The vertical scale for the two YAl_2 spectra is the same. All the spectra have essentially the same shape. (From Ref. 301.)

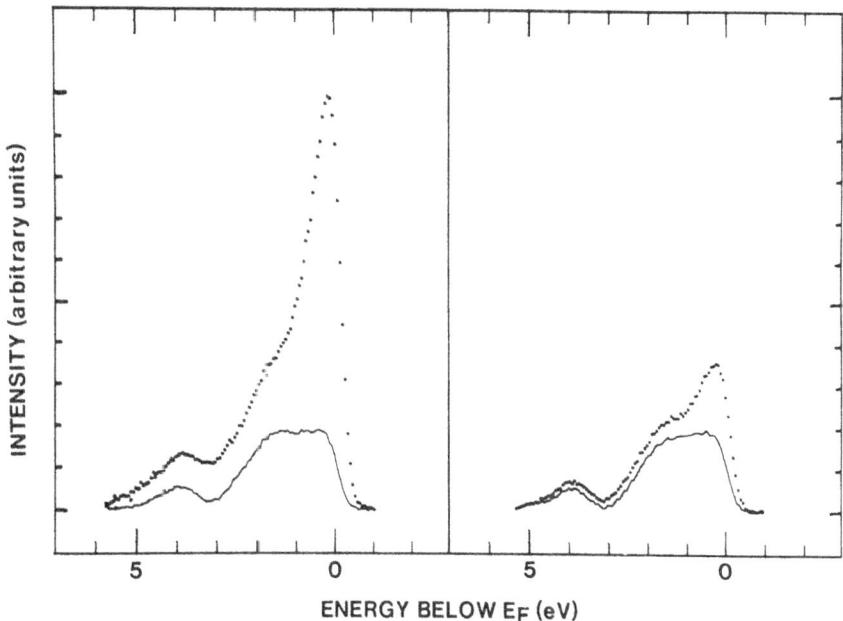

FIGURE 57. Normalized valence band spectra with $hv = 108\,eV$ (dotted) and with $hv = 92\,eV$ (solid) for (left) $Y_{0.9}\,U_{0.1}\,Al_2$ and (right) $Y_{0.98}\,U_{0.02}\,Al_2$. The difference of each pair gives the U $5f$ spectrum. Referenced to the $hv = 92\,eV$ spectra, the areas of the difference curves are in the ratio $1:0.18$, nearly that expected from the uranium concentration. (From Ref. 301.)

have been measured at temperatures down to 20 K with resolution about 100 meV using photons with energy of 40 eV.[302] For materials where the non-$5f$ emission is from sp states only, the $5f$ emission dominates at this photon energy.[259] Figure 59 shows[300] such data for UBe_{13}. There is a sharp feature near E_F, which has been found[502] to change very little for temperatures between 20 K

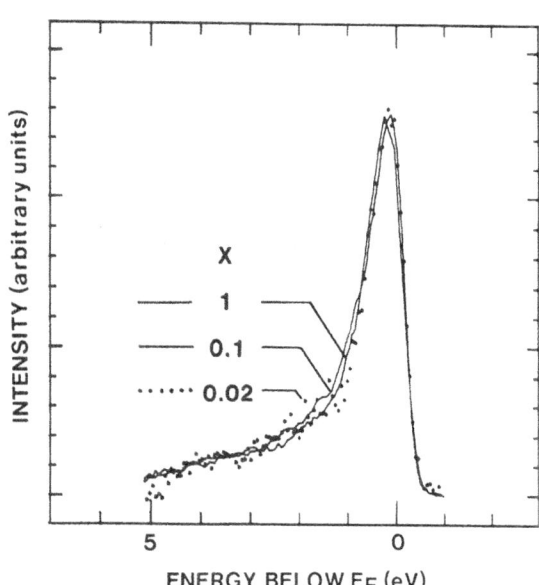

FIGURE 58. U $5f$ spectral weight for $U_xY_{1-x}Al_2$ with $x = 0.02$, 0.1, and 1.0. The three spectra are essentially identical. (From Ref. 301.)

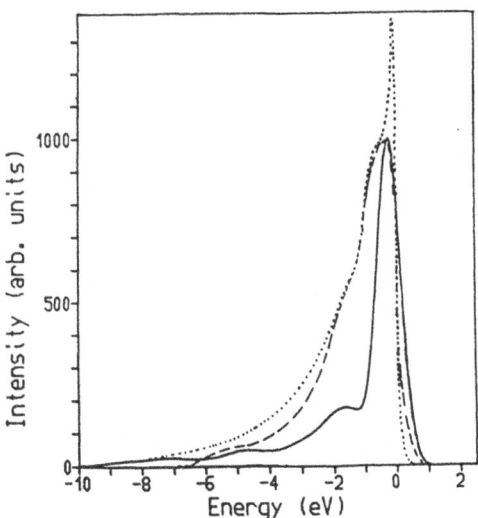

FIGURE 59. Valence-band spectra for UBe_{13} compared to a broadened theoretical density of states with resolution 800 meV (solid curve). Dashed curve is deduced from resonant photoemission at the uranium $5d$ edge, with resolution 300 meV, and the dotted curve is a composite $T = 20$ K spectrum for $h\nu = 40$ eV, taken with resolutions of 90 meV and 300 meV for the first 1 eV below E_F, and the rest of the spectrum, respectively. (From Ref. 300.)

and 300 K, except for smearing due to the temperature dependence of the Fermi function. A similar feature occurs for U_2Zn_{17} and UPt_3, although for the latter material it is somewhat obscured by Pt $5d$ emission. RESPES/BIS spectra for U_2Zn_{17} are given in Ref. 303.

Figure 59 also addresses[300] a claim[304] that in several uranium materials, including UBe_{13} and USi_3, the width of the $5f$ spectrum determined by resonant photoemission is substantially greater than that determined at other photon energies. One implication of this claim is that the excess width relative to that of the band calculations, pointed out above, is an artifact of the use of resonant photoemission. Figure 59 shows the uranium spectrum deduced from the difference of spectra measured at the resonance maximum and minimum photon energies of 99 eV and 92 eV, but with a lower resolution of 300 meV, set by the photon monochromator resolution. It can be seen that the two spectra are essentially identical, assuming that the sharp peak at E_F is absent from the resonance spectrum because of its lower resolution. On the basis of the data of Fig. 59 and similar data for USi_3, it is asserted[300] rather convincingly that there is not an extra width in a resonant photoemission spectrum and that the claim[304] of extra width resulted from not taking proper account of differing resolutions in the spectra that were analyzed. The figure also shows an LDA $5f$ PDOS, broadened by an amount larger than the experimental resolution, and it is clear that the PDOS is much narrower than the experimental spectra.

The difference between the $5f$ PDOS and the measured $5f$ spectrum has been analyzed[299] as indicating two components, corresponding to two screening channels, as shown in Fig. 60. One component is the PDOS itself, regarded as a well-screened or f-screened final state. The other component is the difference spectrum, regarded as a poorly screened or valence-screened final state in which a local $5f$ hole has been created. In this picture UPd_3 is interpreted as a material in which the entire $5f$ spectrum is poorly screened. A similar screening picture has been put forth for cerium materials, where the LDA spectrum is found to be concentrated around E_F and has no weight at the energies of the PES and BIS

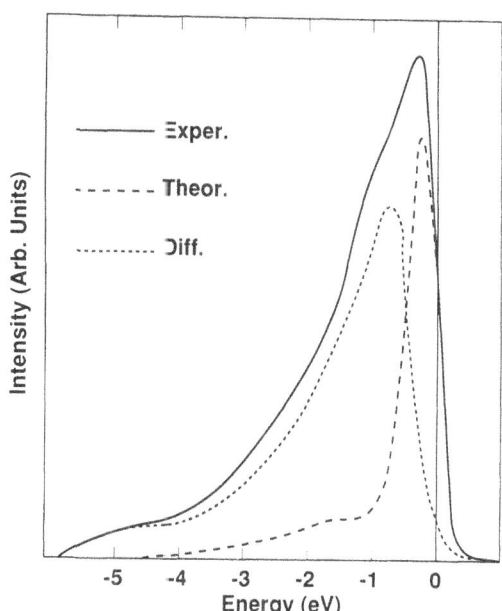

FIGURE 60. Analysis of UBe$_{13}$ 5f spectrum into well-screened (broadened LDA density of states) and poorly screened (difference of data and LDA) contributions. (From Ref. 299.)

peaks away from E_F. For cerium it has been possible[305] to use supercell LDA calculations to reproduce the energies of these peaks by constraining the 4f occupancy of a central site to be 0 or 2 for PES and BIS, respectively. The physical interpretation of the calculation is that the peaks away from E_F are poorly screened final states corresponding to a potential with an f hole or an added f electron, while the spectral weight near E_F is a well-screened final state because it corresponds to the ground-state charge density with the ground state number of 5f electrons. As described in subsection 4.5, the picture has a similarity to the Anderson model spectrum, in which the final states of the peaks away from E_F have roughly 0 and 2 electrons, while the final states of the Kondo resonance have a number of f electrons equal to that of the ground state. The important difference between the screening and Kondo pictures is that in the former the width of the spectrum around E_F is, by hypothesis, the LDA width, while in the latter, it is the small energy scale T_K. Thus the screening picture for either cerium or uranium materials inherently does not address the question of the heavy-fermion properties, but it does make a connection to an aspect of the Anderson Hamiltonian description that has been successful for cerium.

For one heavy-fermion material, URu$_2$Si$_2$, it has been observed[306] that the resonant photoemission behavior of the uranium 5f spectrum has a similarity to that of cerium 4f spectra. Figure 61 shows the valence-band spectrum for several photon energies. In the spectrum for $h\nu = 70$ eV the emission occurring between E_F and 4 eV is due to Ru 4d states. This emission decreases sharply as the photon energy is increased to that of the Cooper minimum[259] in the Ru 4d cross section around 100 eV. The uranium 5f emission resonates for $h\nu$ between 102 eV and 108 eV, and it can be seen that the emission away from E_F has a delayed resonance relative to the portion at E_F, very similar to the behavior of cerium 4f emission at the cerium 4d edge, described in subsection 4.5. Further work will be

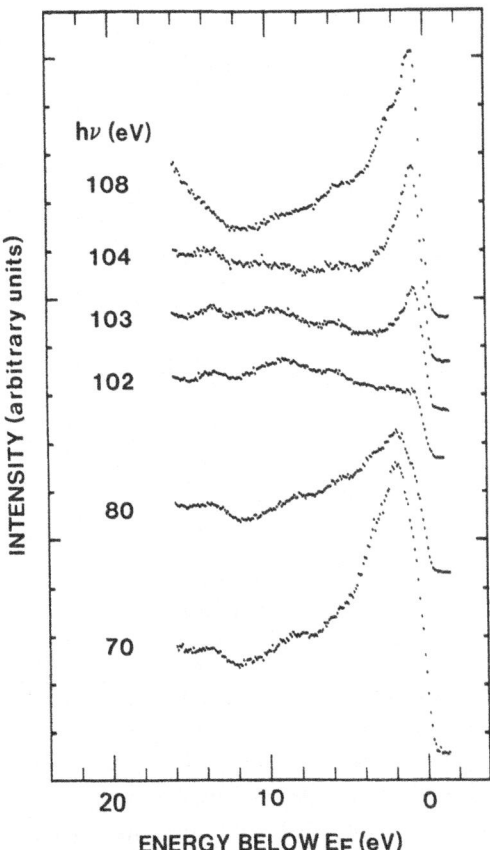

FIGURE 61. Valence-band spectra, normalized to the incident photon flux, for URu_2Si_2 at various photon energies, showing the effect of the Cooper minimum in the photoemission cross section for the Ru $4d$ states and the U $5f$ Fano resonance at the U $5d$ edge. The vertical scale is the same for all spectra. Note the delayed resonance of the U $5f$ weight away from E_F, similar to the behavior of Ce $4f$ spectra in Figs 32–35. (From Ref. 306.)

required to find out if this similarity to cerium occurs for other materials, and whether it is a clue to solving the puzzle of uranium spectra or merely a superficial resemblance.

6. SUMMARY

This chapter has described the current status of knowledge about the single-particle electronic structure of strongly interacting electrons in transition-metal, rare-earth, and actinide systems. Experimentally this knowledge is largely the result of combined PES and BIS experiments. Of the PES experiments, RESPES using synchrotron radiation has had a major impact by providing a spectroscopic tool for separating overlapping emissions, for identifying specific valence states, and for extracting that spectral weight of a dilute species. Theoretically the $1/N$ theory of the spectral weight of the impurity Anderson Hamiltonian, combined with theories of RESPES, has permitted the new spectroscopic data to be interpreted quantitatively and linked to the novel low-energy properties of the narrow-band electrons. The combination of experiment and theory has led to new paradigms for cerium and related rare-earth materials, and for transition-metal compounds.

Because many important and interesting problems remain, it is to be hoped that the recent surge of progress will continue These problems involve the theoretical and experimental properties of the lattice Anderson Hamiltonian and of deciding the correct basic model for uranium compounds. For the RESPES technique there is a need for theories that will allow a more quantitative analysis of data. It seems likely that the resonance line shapes contain detailed electronic-structure information not yet extracted. As synchrotron radiation sources improve, permitting higher resolution or the study or more dilute species, there is hope that the understanding of narrow-band systems can be brought to the same level of precision as exists for simple metals and semiconductors. The aim is no less than that a full many-body band theory of solids.

REFERENCES

1. J. P. Connerade, J.-M. Esteva, and R. C. Karnatak, eds, *Giant Resonances in Atoms, Molecules and Solids*, NATO ASI Series B: Physics, Vol. 151, Plenum, New York (1987).
2. C. Guillot, Y. Ballu. J. Paigné, J. Lecante, K. P. Jain, P. Thiry, R. Pinchaux, Y. Pètroff, and L. M. Falicov, *Phys. Rev. Lett.* **39**, 1632–1635 (1977).
3. W. Lenth, F. Lutz, J. Barth, G. Kalkoffen, and C. Kunz, *Phys. Rev. Lett.* **41**, 1185 (1978).
4. J. W. Allen, L. I. Johansson, R. S. Bauer, I. Lindau, and S. B. M. Hagström, *Phys. Rev. Lett.* **41**, 1499 (1978).
5. L. I. Johansson, J. W. Allen, T. Gustafsson, I. Lindau, and S. B. Hagström, *Solid State Commun.* **28**, 53–55 (1978).
6. W. Gudat, S. F. Alvarado, and M. Campagna, *Solid State Commun.* **28**, 943 (1978).
7. J. C. Fuggle, G. A. Sawatzky, and J. W. Allen, eds., *Narrow Band Phenomena*, Plenum, New York (1988).
8. L. I. Johansson, J. W. Allen, I. Lindau, M. H. Hecht, and S. B. M. Hagström, *Phys. Rev.* **B21**, 1408–1411 (1980).
9. J. Schmidt-May, F. Gerken, R. Nyholm, and L. C. Davis. *Phys. Rev.* **B30**, 5560 (1984).
10. U. Fano, *Phys. Rev.* **124**, 1866–1878 (1961); U. Fano and A. R. P. Rao, *Atomic Collisions and Spectra*, Academic, New York (1986), Chs. 7 and 8.
11. L. C. Davis and L. A. Feldcamp, *Phys. Rev.* **B23**, 6239–6253 (1981).
12. Y. Yafet, *Phys. Rev.* **B21**, 5023–5030 (1980).
13. Y. Yafet, *Phys. Rev.* **B23**, 3558–3559 (1981).
14. S.-J. Oh and S. Doniach, *Phys. Lett.* **81A**, 483–487 (1981).
15. S.-J. Oh and S. Doniach, *Phys. Rev.* **B26**, 1859–1872 (1982).
16. C.-O. Almbladh, in *Proceedings of X84-International Conference on X-ray and Innershell Processes in Atoms, Molecules and Solids* (Leipzig, 1984), pp. 435–445.
17. O. Gunnarsson and K. Schönhammer, p. 405 in Ref. 1.
18. O. Gunnarsson and T. C. Li, *Phys. Rev.* **B36**, 9488–9499 (1987).
19. A. Zangwill, p. 321 in Ref. 1.
20. G. Wendin, p. 171 in Ref. 1.
21. J.-I. Igarashi, *J. Phys. Soc. Jpn.* **54**, 2762–2773 (1985).
22. L. A. Feldkamp and L. C. Davis, *Phys. Rev. Lett.* **43**, 151–154 (1979).
23. D. R. Penn, *Phys. Rev. Lett.* **42**, 921–925 (1979).
24. L. C. Davis and L. A. Feldkamp, *Phys. Rev. Lett.* **44**, 673 (1980).
25. S. M. Girvin and D. R. Penn, *Phys. Rev. Lett.* **22**, 4081 (1980).
26. J. C. Parlebas, A. Kotani, and J. Kanamori, *Solid State Commun.* **41**, 439–443 (1982).
27. J. C. Parlebas, A. Kotani, and J. Kanamori, *J. Phys. Soc. Jpn.* **51**, 124 (1982).
28. T. Jo, A. Kotani, J.-C. Parlebas, and J. Kanamori, *J. Phys. Soc. Jpn.* **52**(7), 2581–2592 (1983).
29. G. P. Williams, G. J. Lapeyre, J. Anderson, F. Cerrina, R. E. Dietz, and Y. Yafet, *Surf. Sci.* **89**, 606–614 (1979).

30. L. C. Davis, *Phys. Rev.* **B25**, 2912 (1982).
31. G. v. d. Laan, *Solid State Commun.* **42**, 165 (1982).
32. A. Fujimori, F. Minami, and S. Sugano, *Phys. Rev.* **29**, 5225 (1984).
33. A. Fujimori and F. Minami, *Phys. Rev.* **B30**, 957 (1984).
34. J.-I. Igarashi and T. Nakano, *J. Phys. Soc. Jpn.* **55**, 1384–1391 (1986).
35. A. Zangwill and P. Soven, *Phys. Rev. Lett.* **45**, 204 (1980).
36. A. Sakuma, Y. Kuramoto, T. Watanabe, and C. Horie, *J. Magn. Magn. Mat.* **52**, 393 (1985).
37. O. Gunnarsson, D. D. Sarma, F. U. Hillebrecht, and K. Schönhammer, *J. Appl. Phys.* **63**, 3676–3679 (1988).
38. J. Hubbard, *Proc. R. Soc. London*, **A276**, 238–257 (1963); **A277**, 237–259 (1964); **A281**, 401–419 (1964).
39. P. W. Anderson, *Phys. Rev.* **124**, 41 (1961).
40. G. G. Tibbits and W. F. Egelhoff, Jr., *Phys. Rev. Lett.* **41**, 188–191 (1978).
41. M. Iwan, F. J. Himpsel, and D. E. Eastman, *Phys. Rev. Lett.* **43**, 1829 (1979).
42. J. Barth, G. Kalkoffen, and C. Kunz, *Phys. Lett.* **74A**, 360–362 (1979).
43. M. Iwan and E. E. Koch, *Solid State Commun.* **31**, 261 (1979).
44. M. Iwan, E. E. Koch, T. C. Chiang, and F.-J. Himpsel, *Phys. Lett.* **76A**, 177–180 (1980).
45. M. Iwan, E. E. Koch, T. C. Chiang, D. E. Eastman, and F.-J. Himpsel, *Solid State Commun.* **34**, 57–60 (1980).
46. D. Chandesris, G. Krill, G. Maire, J. Lecante, and Y. Petroff, *Solid State Commun.* **37**, 187–191 (1981).
47. R. Clauberg, W. Gudat, E. Kisker, E. Kuhlmann, and G. M. Rothberg, *Phys. Rev. Lett.* **47**, 1314 (1981).
48. D. Chandesris, C. Guillot, G. Chauvin, J. Lacante, and Y. Petroff, *Phys. Rev. Lett.* **47**, 1273–1276 (1981).
49. R. Clauberg, W. Gudat, W. Radlik, and W. Braun, *Phys. Rev.* **B31**, 1754–1758 (1983).
50. Y. Sakisaka, T. N. Rhodin, and P. A. Dowben, *Solid State Commun.* **49**, 563–565 (1984).
51. Y. Sakisaka, T. Komeda, M. Onchi, H. Kato, S. Masuda, and K. Yagi, *Phys. Rev.* **B36**, 6383–6389 (1987).
52. Y. Sakisaka, T. Komeda, M. Onchi, H. Kato, S. Masuda, and K. Yagi, *Phys. Rev. Lett.* **58**, 733–736 (1987).
53. R. E. Dietz, E. G. McRae, Y. Yafet, and C. W. Caldwell, *Phys. Rev. Lett.* **33**, 1372–1375 (1974).
54. L. C. Davis and L. A. Feldcamp, *J. Appl. Phys.* **50**, 1944–1949 (1979).
55. N. Mårtensson and B. Johansson, *Phys. Rev. Lett.* **45**, 482–485 (1980).
56. R. E. Dietz, E. G. McRae, and J. H. Weaver, *Phys. Rev.* **B21**, 2229–2247 (1980).
57. G. Treglia, M. C. Desjonqueres, F. Ducastelle, and D. Spanjaard, *J. Phys. C*: **14**, 4347–4355 (1981).
58. T. Aisaka, T. Kato, and E. Haga, *Phys. Rev.* **B28**, 1113 (1983).
59. V. I. Anisimov, E. Z. Kurmaev, and V. A. Gubanov, *J. Electron Spectros. Relat. Phenom.* **35**, 185–190 (1985).
60. A. Liebsch, *Phys. Rev. Lett.* **43**, 1431–1434 (1979).
61. A. Liebsch, *Phys. Rev.* **23**, 5203 (1981).
62. L. C. Davis and L. A. Feldkamp, *Solid State Commun.* **34**, 141 (1980).
63. J. Barth, F. Gerken, K. L. I. Kobayashi, J. H. Weaver, and B. Sonntag, *J. Phys. C.* **13**, 1369–1375 (1980).
64. D. Chandesris, J. Lecante, and Y. Petroff, *Phys. Rev.* **B27**, 2630 (1983).
65. J. Barth, F. Gerken, and C. Kunz, *Phys. Rev.* **B28**, 3608–3611 (1983).
66. H. Sugawara, K. Naito, T. Miya, A. Kakizaki, I. Nagakura, and T. Ishii, *J. Phys. Soc. Jpn.* **53**, 279–283 (1984).
67. H. Kato, T. Ishii, S. Masuda, Y. Harada, T. Miyano, Y. Komeda, M. Onchi, and Y. Sakisaka, *Phys. Rev.* **B32**, 1992 (1985).
68. J. Barth, F. Gerken, and C. Kunz, *Phys. Rev.* **B31**, 2022–2028 (1985).
69. S. Raaen and V. Murgai, *Phys. Rev.* **B36**, 887 (1987).
70. R. Bruhn, B. Sonntag, and H. W. Wolff, *Phys. Lett.* **69A**, 9–11 (1978).
71. R. Bruhn, E. Schmidt, H. Schröder, and B. Sonntag, *J. Phys. B* **15**, 2807–2817 (1982).

72. R. Bruhn, E. Schmidt, H. Schröder, and B. Sonntag, *Phys. Lett.* **90A,** 41 (1982).
73. E. Schmidt, H. Schröder, B. Sonntag, H. Voss, and H. E. Wetzel, *J. Phys. B.* **16,** 2961–2969 (1983).
74. P. H. Kobrin, U. Becker, C. M. Truesdale, D. W. Lindle, H. G. Kerkhoff, and D. A. Shirley, *J. Electron. Spectros. Relat. Phenom.* **34,** 129–139 (1984).
75. E. Schmidt, H. Schröder, B. Sonntag, H. Voss, and H. E. Wetzel, *J. Phys. B.* **18,** 79–93 (1985).
76. J. Barth, I. Chorkendorff, F. Gerken, C. Kunz, R. Nyholm, J. Schmidt-May, and G. Wendin, *Phys. Rev.* **B30,** 6251–6253 (1984).
77. A. Kakizaki, H. Sugawara, I. Nagakura, and T. Ishii, *J. Phys. Soc. Jpn.* **49,** 2183–2190 (1980).
78. I. T. McGovern, K. D. Childs, H. M. Clearfield, and R. H. Williams, *J. Phys. C.* **14,** L243 (1981).
79. S.-J. Oh, J. W. Allen, I. Lindau, and J. C. Mikkelsen, Jr., *Phys. Rev.* **B26,** 4845 (1982).
80. M. R. Thüler, R. L. Benbow, and Z. Hurych, *Phys. Rev.* **26,** 669 (1982).
81. A. Kakizaki, H. Sugawara, I. Nagakura, Y. Ishikawa, T. Komatsubara, and T. Ishii, *J. Phys. Soc. Jpn.* **51**(8), 2597–2603 (1982).
82. E. Schmidt, H. Schröder, B. Sonntag, H. Voss, and H. E. Wetzel, *J. Phys. B.* **16,** 2961–2969 (1983).
83. M. R. Thüler, R. L. Benbow, and Z. Hurych, *Phys. Rev.* **B27,** 2082–2088 (1983).
84. A. Kakizaki, K. Sugeno, T. Ishii, H. Sugawara, I. Nagakura, and S. Shin, *Phys. Rev.* **B28,** 1026–1035 (1983).
85. E. Bertel, R. Stockbauer, and T. E. Madey, *Phys. Rev.* **B27,** 1939 (1983).
86. C. Kunz, in *Vacuum Ultraviolet Radiation Physics: VUV VII,* (A. Weinreb and A. Ron, eds.) Adam Hilger, Bristol, and Israel Physical Society, Jerusalem (1983), p. 575.
87. H. Sakamoto, S. Suga, M. Taniguchi, H. Kanzaki, M. Yamamoto, M. Seki, M. Naito, and S. Tanaka, *Solid State Commun.* **52,** 721–724 (1984).
88. G. A. Sawatzky and J. W. Allen, *Phys. Rev. Lett.* **53,** 2339–2342 (1984).
89. H. Daimon, A. Ishizaka, K. L. I. Kobayashi, and Y. Murata, *J. Phys. Soc. Jpn.* **53,** 2130–2136 (1984).
90. R. D. Bringans and H. Höchst, *Phys. Rev.* **B30,** 5416 (1984).
91. K. Naito, A. Kakizaki, T. Komatsubara, H. Sugawara, I. Nagakura, and T. Ishii, *J. Phys. Soc. Jpn.* **54,** 416–423 (1985).
92. E. Bertel, R. Stockbauer, R. L. Kurtz, D. E. Ramaker, and T. E. Madey, *Phys. Rev.* **B31,** 5580–5583 (1985).
93. A. Kakizaki, T. Miya, K. Naito, A. Fukui, H. Sugawara, I. Nagakura, and T. Ishii, *J. Phys. Soc. Jpn.* **54,** 3638 (1985).
94. Y. Ueda, H. Negishi, M. Koyano, M. Inoue, K. Soda, H. Sakamoto, and S. Suga, *Solid State Commun.* **57,** 839–842 (1986).
95. A. Fujimori, M. Saeki, N. Kimizuka, M. Taniguchi, and S. Suga, *Phys. Rev.* **B34,** 7318–7328 (1986).
96. W. P. Ellis, R. C. Albers, J. W. Allen, Y. Lasailly, J.-S. Kang, B. B. Pate, and I. Lindau, *Solid State Commun.* **62,** 591–596 (1987).
97. L. Ley, M. Taniguchi, J. Ghijsen, R. L. Johnson, and A. Fujimori, *Phys. Rev.* **B35,** 2839–2843 (1987).
98. A. Fujimori, N. Kimizuka, M. Taniguchi, and S. Suga, *Phys. Rev.* **B36,** 6691–6694 (1987).
99. S. Masuda, Y. Harada, H. Kato, K. Yagi, T. Komeda, T. Miyano, M. Onchi, and Y. Sakisaka, *Phys. Rev.* **B37,** 8088 (1988).
100. K. E. Smith and V. E. Henrich, *Phys. Rev.* **B38,** 9571–9580 (1988).
101. R. J. Lad and V. E. Henrich, *Phys. Rev.* **B38,** 10860 (1988).
102. H. J. de Boer and E. J. Verwey, *Proc. Phys. Soc. London* **A49,** 59 (1937).
103. The most recent work is by K. Terakura, A. R. Williams, T. Oguchi and J. Kübler, *Phys. Rev. Lett.* **52,** 1830 (1984); K. Terakura, T. Oguchi, A. R. Williams, and J. Kübler, *Phys. Rev.* **B30,** 4734 (1984).
104. N. F. Mott, *Proc. Phys. Soc., London,* **A62,** 416 (1949).
105. P. W. Anderson, *Phys. Rev.* **115,** 2 (1959).
106. P. W. Anderson, in *Magnetism: A Treatise on Modern Theory and Materials,* (G. Rado and H. Suhl, ed.) Academic Press, New York (1963) Vol. 1, p. 25.

107. H. Scheidt, M. Glöbl, and V. Dose, *Surf. Sci.* **112,** 97 (1981).

108. F. J. Himpsel and Th. Fauster, *Phys. Rev. Lett.* **49,** 1583 (1982).

109. J. Zaanen, G. A. Sawatzky, and J. W. Allen, *Phys. Rev. Lett.* **55,** 418 (1985).

110. J. W. Allen, *J. Magn. Magn. Mat.* **47–48,** 168 (1985).

111. See footnote 25 of Ref. 88.

112. J. Zaanen and G. A. Sawatzky, *Can. J. Phys.* **65,** 1262 (1987).

113. J. W. Allen, p. 155 of Ref. 7.

114. G. van der Laan, C. Westra, C. Haas, and G. A. Sawatzky, *Phys. Rev.* **B23,** 4369 (1981).

115. G. van der Laan, J. Zaanen, G. A. Sawatzky, R. Karnatak, and J.-M. Esteva, *Phys. Rev.* **B33,** 4253 (1986).

116. J. Zaanen, C. Westra, and G. A. Sawatzky, *Phys. Rev.* **B33,** 8060 (1986).

117. Z.-X. Shen, J. W. Allen, J. J. Yeh, J.-S. Kang, W. Ellis, W. Spicer, I. Lindau, M. B. Maple, Y. D. Dalichaouch, M. S. Torikachvili, J. Z. Sun, and T. H. Geballe, *Phys. Rev.* **B36,** 8414 (1987).

118. P. W. Anderson, *Science* **235,** 1196 (1987); K. C. Haas, in *Solid State Physics,* (F. Seitz, D. Turnbull, and H. Ehrenreich, eds.) Academic Press, New York (1989), Vol. 42, pp. 213–270; C. G. Olson, R. Liu, A.-B. Yang, D. W. Lynch, A. J. Arko, R. S. List, B. W. Veal, Y. C. Chang, P. Z. Jiang, and A. P. Paulikas, *Science* **245,** 731 (1989); R. Manzke, T. Buslaps, R. Claessen, and J. Fink, *Europhys. Lett.* **9,** 477 (1989); J. W. Allen, C. B. Olson, M. B. Maple, J.-S. Kang, L.-Z. Liu, J.-H. Park, R. O. Anderson, W. P. Ellis, J. T. Markert, Y. Dalichaouch, and R. Liu, *Phys. Rev. Lett.* **64,** 595 (1990).

119. J. K. Lang, Y. Baer, and P. A. Cox, *J. Phys.* **F11,** 121 (1981).

120. P. A. Cox, *Struct. Bonding (Berlin)* **24,** 59–81 (1975); P. A. Cox, J. K. Lang, and Y. Baer, *J. Phys.* **F11,** 113–119 (1981).

121. J. F. Herst, D. N. Lowry, and R. E. Watson, *Phys. Rev.* **B6,** 1913 (1972); B. Johansson, *J. Phys.* **F4,** L169–L173 (1974); J. F. Herbst, R. E. Watson, and J. W. Wilkins, *Phys. Rev.* **B13,** 1439 (1976); J. F. Herbst, R. E. Watson, and J. W. Wilkins, *Phys. Rev.* **B17,** 3089 (1978); B. Johansson, *Phys. Rev.* **B20,** 1315 (1979).

122. See the Introduction to Section A, p. 15, in Ref. 7 for a historical overview of work on narrow-band materials.

123. F. Gerken, J. Barth, K. L. I. Kobayashi, and C. Kunz, *Solid State Commun.* **35,** 179 (1980).

124. J. W. Allen, L. I. Johansson, I. Lindau, and S. B. Hagstrom, *Phys. Rev.* **B21,** 1335–1343 (1980).

125. W. Gudat, S. F. Alvarado, M. Campagna, and Y. Petroff, *J. Phys. (Paris)* **41,** C5–1 (1980).

126. A. Franciosi, J. H. Weaver, N. Martensson, M. Croft, *Phys. Rev.* **24,** 3651 (1981).

127. M. Croft, A. Franciosi, J. H. Weaver, and A. Jayaraman, *Phys. Rev.* **B24,** 544–548 (1981).

128. W. Gudat, M. Campagna, R. Rosei, J. H. Weaver, W. Eberhardt, F. Hulliger, and E. Kaldis, *J. Appl. Phys.* **52,** 2123 (1981).

129. J. W. Allen, S.-J. Oh, I. Lindau, J. M. Lawrence, L. I. Johansson, and S. B. Hagström, *Phys. Rev. Lett.* **46,** 1100 (1981).

130. J. W. Allen, S.-J. Oh, J. Lawrence, I. Lindau, L. I. Johansson, and S. B. Hagström, in *Fluctuations in Solids,* (L. M. Falicov, W. Hanke, and M. B. Maple, eds.) North Holland, Amsterdam (1981), pp. 431–434.

131. L. I. Johansson, J. W. Allen, and I. Lindau, *Phys. Lett.* **86A,** 442–444 (1981).

132. F. Gerken, J. Barth, and C. Kunz, *Phys. Rev. Lett.* **47,** 933 (1981).

133. W. F. Egelhoff, Jr., G. G. Tibbetts, H. H. Hecht, and I. Lindau, *Phys. Rev. Lett.* **46,** 1071 (1981).

134. J. A. D. Matthew, G. Strasser, and F. P. Netzer, *J. Phys. C.* **15,** L1019–L1022 (1982).

135. H. Sugawara, A. Kakizaki, I. Nagakura, T. Ishii, T. Komatsubara, and T. Kasuya, *J. Phys. Soc. Jpn.* **51,** 915–921 (1982).

136. N. Mårtensson, B. Reihl, and R. D. Parks, *Solid State Commun.* **41,** 573–576 (1982).

137. D. Wieliczka, J. H. Weaver, D. W. Lynch, and C. G. Olson, *Phys. Rev.* **B26,** 7056 (1982).

138. W. Gudat, R. Rosei, J. H. Weaver, E. Kaldis, and F. Hulliger, *Solid State Commun.* **41,** 37–42 (1982).

139. J. W. Allen, R. J. Nemanich, and S.-J. Oh, *J. Appl. Phys.* **53,** 2145–2148 (1982).

140. J. W. Allen, S.-J. Oh, I. Lindau, M. B. Maple, J. F. Saussuna, S. B. Hagström, *Phys. Rev.* **B26,** 445 (1982).

141. J. M. Lawrence, J. W. Allen, S.-J. Oh, and I. Lindau, *Phys. Rev.* **B26**, 2362–2370 (1982).

142. F. Gerken, J. Barth, and C. Kunz, *AIP Conf. Proc.* **94**, 602–614 (1982).

143. M. Aono, T.-C. Chiang, F. J. Himpsel, and D. E. Eastman, *Solid State Commun.* **37**, 471 (1982).

144. D. J. Peterman, J. H. Weaver, and M. Croft, *Phys. Rev.* **B25**, 5530–5533 (1982).

145. R. D. Parks, N. Mårtensson, and B. Reihl, in *Valence Instabilities*, (P. Wachter and H. Boppart, eds.) North-Holland, Amsterdam (1982), pp. 239–247.

146. J. W. Allen, S.-J. Oh, M. B. Maple, and M. S. Torikachvili, *Phys. Rev.* **B28**, 5347–5349 (1983).

147. J. W. Allen, *J. Less-Common Met.* **93**, 183–187 (1983).

148. S. Suga, M. Taniguchi, M. Seki, H. Sakamoto, H. Kanzaki, M. Yamamoto, A. Kurita, Y. Kaneko, and T. Koda, *Solid State Commun.* **49**, 1005–1008 (1984).

149. R. D. Parks, S. Raaen, M. L. den Boer, Y.-S. Chang, and G. P. Williams, *Phys. Rev. Lett.* **52**, 2176 (1984).

150. S.-J. Oh, J. W. Allen, and I. Lindau, *Phys. Rev.* **B30**, 1937–1944 (1984).

151. K. Soda, S. Asaoka, T. Mori, M. Taniguchi, K. Naito, Y. Onuka, T. Komatsubara, T. Miyahara, S. Sato, and T. Ishii, *J. Magn. Magn. Mat.* **52**, 347 (1985).

152. S. Suga, M. Yamamoto, M. Taniguchi, M. Fuhisawa, A. Ochiai, T. Suzuki, and T. Kasuya, *J. Magn. Magn. Mat.* **52**, 293 (1985).

153. R. D. Parks, S. Raaen, M. L. den Boer, Y.-S. Chang, and G. P. Williams, *J. Magn. Magn. Mat.* **47–48**, 163–167 (1985).

154. U. Becker, H. G. Kerkhoff, T. A. Ferret, P. A. Heimann, P. H. Kobrin, D. W. Lindle, C. M. Truesdale, and D. A. Shirley, *Phys. Rev.* **A34**, 2858 (1986).

155. T. M. Zimkina, V. A. Fomichev, S. A. Gribovskii, and I. I. Zhukova, *Sov. Phys.-Solid State* **9**, 1128–1130 (1967).

156. V. A. Fomichev, T. M. Zimkina, S. A. Gribovskii and I. I. Zhukova, *Sov. Phys.-Solid State* **9**, 1163–1165 (1967).

157. R. Haensel, P. Rabe, and B. Sonntag, *Solid State Commun.* **8**, 1845–1848 (1970).

158. M. W. D. Mansfeld and J. P. Connerade, *Proc. R. Soc. London* **A352**, 125 (1976).

159. T. B. Lucatorio, T. J. McIlrath, J. Sugar, and S. M. Younger, *Phys. Rev. Lett.* **47**, 1124–1128 (1981).

160. S. M. Kazakov and O. V. Khristoforov, *Sov. Phys. JETP* **57**, 290 (1983).

161. A. F. Starace, *Phys. Rev.* **B5**, 1773–1784 (1972).

162. J. Sugar, *Phys. Rev.* **B5**, 1785–1796 (1972).

163. G. Wendin and A. F. Starace, *J. Phys. B.* **11**, 4119–4134 (1978).

164. A. F. Starace, *Handbuch der Physik*, Vol. XXXI: Corpuscles and Radiation in Matter I, (S. Flügge, ed.) Springer-Verlag, Berlin (1982) pp. 1–121.

165. C. W. Clark, *J. Opt. Soc. Am.* **B1**, 626–630 (1984).

166. U. Becker, p. 473 in Ref. 1.

167. J. P. Connerade and R. C. Karnatak, *J. Phys. F.* **11**, 1539–1544 (1981).

168. P. Motais, E. Belin, and C. Bonnelle, *J. Phys. F.* **11**, L169–L171 (1981).

169. J. A. D. Matthew, G. Strasser, and F. P. Netzer, *Phys. Rev.* **B30**, 4975–5979 (1984).

170. J. W. Allen, S.-J Oh, I. Lindau, and L. I. Johansson, *Phys. Rev.* **B29**, 5927 (1984).

171. J. J. Yeh, J. Nogami, G. Rossi, and I. Lindau, *J. Vac. Sci. Technol.* **A2**, 969–972 (1984).

172. G. Strasser, F. P. Netzer, and J. A. D. Matthew, *Solid State Commun.* **49**, 817–821 (1984).

173. G. Rossi and A. Barski, *Phys. Rev.* **B32**, 5492–5495 (1985).

174. G. Rossi and A. Barski, *Solid State Commun.* **57**, 277–281 (1986).

175. I. Abbati, L. Braicovich, U. del Pennino, C. Carbone, J. Nogami, J. J. Yeh, I. Lindau, A. Iandelli, G. L. Olcese, and A. Palenzona, *Phys. Rev.* **B34**, 4150–4154 (1986).

176. D. J. Friedman, C. Carbone, K. A. Bertness, and I. Lindau, *J. Electron Spectros. Relat. Phenom.* **41**, 59–66 (1986).

177. I. Lindau and W. E. Spicer, *J. Electron Spectrosc. Relat. Phenom.* **3**, 409–413 (1974).

178. G. Broden, S. B. M. Hagström, and C. Norris, *Phys. Kondens. Mater.* **15**, 327 (1973).

179. S. F. Alvarado, M. Campagna, and W. Gudat, *J. Electron. Spectros. Relat. Phenom.* **18**, 43–49 (1980).

180. M. H. Hecht, A. J. Viescas, I. Lindau, J. W. Allen, and L. I. Johansson, *J. Electron Spectros. Relat. Phenom.* **34**, 343–353 (1984).

181. Y. Takakuwa, S. Takahashi, S. Suzuki, S. Kono, T. Yokotsuka, T. Takahashi, and T. Sagawa, *J. Phys. Soc. Jpn.* **51**, 2045–2046 (1982).

182. Y. Takakuwa, S. Suzuki, T. Yokotsuka, and T. Sagawa, *J. Phys. Soc. Jpn.* **53**, 687–695 (1984).

183. W.-D. Schneider, C. Laubschat, and B. Reihl, *Phys. Rev.* **B27**, 6538–6541 (1983).

184. B. Johansson, *Phys. Rev.* **B19**, 6615–6619 (1979).

185. B. Johansson and N. Mårtensson, *Phys. Rev.* **B21**, 4427–4457 (1980).

186. F. Gerken, J. Barth, R. Kammerer, L. I. Johansson, and A. Flodström, *Surf. Sci.* **117**, 468–474 (1982); R. Kammerer, J. Barth, F. Gerken, A. Flodström, and L. I. Johansson, *Solid State. Commun.* **41**, 435–438 (1982).

187. L. I. Johansson, A. Flodström, S.-E. Hörnström, B. Johansson, J. Barth, and F. Gerken, *Surf. Sci.* **117**, 475–481 (1982); L. I. Johansson, A. Flodström, S.-E. Hörnström, B. Johansson, J. Barth, and F. Gerken, *Solid State Commun.* **41**, 427–430 (1982).

188. F. Gerken, A. S. Flodström, J. Barth, L. I. Johansson, and C. Kunz, *Phys. Scr.* **32**, 43–57 (1985).

189. The progress in research and thinking on the related topics of mixed valence, the Kondo lattice, and heavy fermions can be traced in the proceedings of a series of international conferences, as follows: *Valence Instabilities and Related Narrow-Band Phenomena* (R. D. Parks, ed.) Plenum, New York (1977); *Valence Fluctuations in Solids,* (L. M. Falicov, W. Hanke, and M. B. Maple, eds.) North-Holland, Amsterdam (1981); *Valence Instabilities,* (P. Wachter and H. Boppart, eds.) North-Holland, Amsterdam (1982); *Proc. 4th Int. Conf. Valence Fluctuations,* (E. Müller-Hartmann, B. Roden and D. Wohlleben, eds.) North-Holland, Amsterdam (1985); *Proc. Conf. Electronic Structure and Properties of Rare Earth and Actinide Intermetallics,* (G. Hilscher, G. Wiesinger, E. Gratz, C. Schmitzer, P., Weinberger and R. Grössinger, eds.) North-Holland, Amsterdam (1985); *Proc. 5th Int. Conf. Crystalline Field and Anomalous Mixing Effects in f-electron Systems* (T. Kasuya, ed.) North Holland, Amsterdam (1985); *Proc. Int. Conf. Anomalous Rare Earths and Actinides: Valence Fluctuations and Heavy-Fermions,* (J. X. Boucherle, J. Flouquet, C. Lacroix, and J. Rossat-Mignod, eds.) North-Holland, Amsterdam (1987); *Theoretical and Experimental Aspects of Valence Fluctuations and Heavy Fermions,* (L. C. Gupta and S. K. Malik, eds.) Plenum, New York (1987); *Proc. 6th Int. Conf. Crystal Field Effects and Heavy-Fermion Physics,* (W. Assmus, P. Fulde, B. Luthi, and F. Steglich, eds.) North-Holland, Amsterdam (1988).

190. For an early review, see J. M. Lawrence, P. S. Riseborough, and R. D. Parks, *Valence fluctuation phenomena, Rep. Prog. Phys.* **44**, 1–84 (1981).

191. G. K. Wertheim, W. Eib, E. Kaldis, and M. Campagna, *Phys. Rev.* **B22**, 6240 (1980).

192. S.-J. Oh and J. W. Allen, *Phys. Rev.* **B29**, 589–592 (1984).

193. G. Kaindl, C. Laubschat, B. Reihl, R. A. Pollack, N. Mårtensson, F. Holtzberg, and D. E. Eastman, *Phys. Rev.* **B26**, 1713–1727 (1982).

194. G. K. Wertheim and M. Campagna, *Chem. Phys. Lett.* **47**, 182–184 (1977).

195. G. K. Wertheim and G. Crecelius, *Phys. Rev. Lett.* **40**, 813–816 (1978).

196. J. K. Lang and Y. Baer, *Solid State Commun.* **31**, 945–947 (1979).

197. G. Krill, J. P. Senateur, and A. Amamou, *J. Phys. F,* **10**, 1889–1897 (1980).

198. G. Krill, J. Durand, A. Berrada, N. Hassanain, and M. F. Ravet, *Solid State Commun.* **35**, 547–550 (1980).

199. M. G. Mason, S. T. Lee, G. Apai, R. F. Davis, D. A. Shirley, A. Franciosi, and J. W. Weaver, *Phys. Rev. Lett.* **47**, 730 (1980).

200. Å. Fäldt and H. P. Myers, *Solid State Commun.* **48**, 253 (1983).

201. Å. Fäldt and H. P. Myers, *Phys. Rev.* **B30**, 5481 (1984).

202. Å. Fäldt and H. P. Myers, *Phys. Rev. Lett.* **52**, 1315 (1984).

203. A. Franciosi, P. Perfetti, A. D. Katnani, J. H. Weaver, and G. Margaritondo, *Phys. Rev.* **B29**, 5611 (1984).

204. Å. Fäldt and H. P. Myers, *Phys. Rev.* **B33**, 1424 (1986).

205. G. K. Wertheim, J. H. Wernick, and G. Crecelius, *Phys. Rev.* **B18**, 875–879 (1978).

206. G. Kaindl, B. Reihl, D. E. Eastman, R. A. Pollack, N. Mårtensson, B. Barbara, and T. Penney, *Solid State Commun.* **41**, 157–160 (1982).

207. W.-D. Schneider, C. Laubschat, G. Kalkowski, J. Haase, and A. Puschmann, *Phys. Rev.* **B28**, 2017 (1983).

208. G. Kaindl, W.-D. Schneider, C. Laubschat, R. Reihl, and N. Mårtensson, *Surf. Sci.* **126**, 105–111 (1983).

209. T. Penney, B. Reihl, R. A. Pollack, B. Barbara, and T. S. Plaskett, *J. Appl. Phys.* **55**, 1975–1977 (1984).

210. E. V. Sampathkumaran, K. H. Frank, G. Kalkowski, G. Kaindl, M. Domke, and G. Wortmann, *Phys. Rev.* **B29**, 5702–5707 (1984).

211. M. Domke, C. Laubschat, E. V. Sampathkumaran, M. Prietsch, T. Mandel, G. Kaindl, and H. U. Middelmann, *Phys. Rev.* **B32**, 8002 (1985).

212. M. Domke, C. Laubschat, M. Prietsch, T. Mandel, G. Kaindl, and W.-D. Schneider, *Phys. Rev. Lett.* **56**, 1287–1290 (1986).

213. I. Abbati, L. Braicovich, U. del Pennino, C. Carbone, J. Nogami, J. J. Yeh, I. Lindau, A. Iandelli, G. L. Olcese and A. Palenzona, *Phys. Rev.* **B34**, 4150–4154 (1986).

214. V. T. Rajan and L. M. Falicov, *Solid State Commun.* **21**, 347–351 (1977).

215. A. J. Martin, *Z. Phys.* **B36**, 235–243 (1980).

216. H.-J. Brocksch, D. Tománek, and K. H. Bennemann, *Phys. Rev.* **B25**, 7102–7109 (1982).

217. H.-J. Brocksch, D. Tománek, and K. H. Bennemann, *Phys. Rev.* **B27**, 7313–7317 (1983).

218. M. Croft, J. H. Weaver, D. J. Peterman, and A. Franciosi, *Phys. Rev. Lett.* **46**, 1104 (1981).

219. R. M. Martin, *Phys. Rev. Lett.* **48**, 362 (1982).

220. J. W. Allen and R. M. Martin, *Phys. Rev. Lett.* **49**, 1106 (1982).

221. O. Gunnarsson and K. Schönhammer, *Phys. Rev. Lett.* **50**, 604 (1983).

222. O. Gunnarsson and K. Schönhammer, *Phys. Rev.* **B28**, 4315 (1983).

223. J. W. Allen, S.-J. Oh, O. Gunnarsson, K. Schönhammer, M. B. Maple, M. S. Torikachvili, and I. Lindau, *Adv. Phys.* **35**, 275–316 (1986).

224. D. C. Koskenmaki and K. A. Gschneidner, in *Handbook of the Physics and Chemistry of Rare Earths*, (K. A. Gschneidner and L. Eyring, eds.) North-Holland, Amsterdam (1978), p. 337.

225. G. R. Stewart, *Rev. Mod. Phys.* **56**, 755 (1984).

226. F. Steglich, J. Aarts, C. D. Bredl, W. Lieke, D. Meschede, W. Franz, and H. Schäfer, *Phys. Rev. Lett.* **43**, 1892 (1979).

227. W. H. Zachariasen, quoted by A. W. Lawson and T. Y. Tang, *Phys. Rev.* **76**, 301 (1949).

228. L. Pauling, quoted by A. F. Schuck and J. H. Sturdivant, *J. Chem Phys.* **18**, 145 (1950).

229. R. Ramirez and L. M. Falicov, *Phys. Rev.* **B3**, 2425 (1971).

230. B. Coqblin and A. Blandin, *Adv. Phys.* **17**, 281 (1968).

231. B. Johansson, *Philos. Mag.* **30**, 469 (1974).

232. E. Wuilloud, H. R. Moser, W.-D. Schneider, and Y. Baer, *Phys. Rev.* **B28**, 7354 (1983).

233. D. M. Wieliczka, C. G. Olson, and D. W. Lynch, *Phys. Rev.* **B29**, 3028 (1984).

234. A detailed bibliography of work using all these techniques can be found in a recent review. N. E. Bickers, *Rev. Mod. Phys.* **59**, 845–939 (1987).

235. O. Gunnarsson and K. Schönhammer, *Phys. Rev.* **B31**, 4815 (1985).

236. O. Gunnarsson and K. Schönhammer, in *Handbook on the Physics and Chemistry of Rare Earths*, (K. A. Gschneidner, Jr., L. Eyring, and S. Hüfner, ed.) North-Holland, Amsterdam (1987) Vol. 10, p. 103.

237. J. W. Rasul and A. C. Hewson, *J. Phys. C* **17**, 2555 (1984).

238. O. Gunnarsson, K. Schönhammer, J. C. Fuggle, F. U. Hillebrecht, J.-M. Esteva, R. C. Karnatak, and B. Hillebrand. *Phys. Rev.* **B28**, 7330 (1985).

239. O. Gunnarsson and K. Schönhammer, *J. Magn. Magn. Mat.* **52**, 141 (1985); **52**, 227 (1985).

240. R. M. Martin and J. W. Allen, *J. Magn. Magn. Mat.* **47–48**, 257 (1985).

241. A. A. Abrikosov, *Physics (Long Island City, NY)* **2**, 5 (1965).

242. H. Suhl, *Phys. Rev.* **138**, A515 (1965).

243. J. W. Allen, *J. Magn. Magn. Mat.* **52**, 135–140 (1985).

244. N. E. Bickers, D. L. Cox, and J. W. Wilkins, *Phys. Rev. Lett.* **54**, 230 (1985).

245. T. Watanabe and A. Sukuma, *Phys. Rev.* **B31**, 6320–6323 (1985).

246. F. Patthey, B. Delley, W.-D. Schneider, and Y. Baer, *Phys. Rev. Lett.* **55**, 1518 (1985).

247. F. Patthey, W.-D. Schneider, Y. Baer, and B. Delley, *Phys. Rev.* **B34**, 2967 (1986).

248. F. Patthey, W.-D. Schneider, Y. Baer, and B. Delley, *Phys. Rev. Lett.* **58**, 2810 (1987).

249. J. C. Fuggle, F. U. Hillebrecht, Z. Zolnierek, R. Lässer, Ch. Freiburg, O. Gunnarsson, and K. Schönhammer, *Phys. Rev.* **B27**, 7330 (1983).

250. J. C. Fuggle, F. U. Hillebrecht, J.-M. Esteva, R. C. Karnatak, O. Gunnarsson, and K. Schönhammer, *Phys. Rev.* **B27**, 4637 (1983).

251. K. R. Bauschspiess, W. Kboksch, E. Holland-Moritz, H. Launois, R. Pott, and D. K. Wohlleben, in *Valence Fluctuations in Solids,* (L. M. Falicov, W. Hanke, and M. B. Maple, eds.) North-Holland, Amsterdam (1981), p. 417.

252. D. R. Gustafson, J. D. McNutt, and L. O. Roellig, *Phys. Rev.* **183**, 435 (1969).

253. U. Kornstädt, R. Lässer, and B. Lengeler, *Phys. Rev.* **B21**, 1898 (1980).

254. M. Lavagna, C. Lacroix, and M. Cyrot, *Phys. Lett.* **90**, 210 (1983); *J. Phys.* **F 13**, 1007 (1983).

255. O. K. Andersen, *Phys. Rev.* **B12**, 3060 (1975).

256. J. D. Thompson, Z. Fisk, J. Lawrence, J. L. Smith, and R. M. Martin, *Phys. Rev. Lett.* **50**, 1981 (1983).

257. A. Plautau and S.-E. Karlson, *Phys. Rev.* **B18**, 3820 (1978).

258. G. Crecelius, G. K. Wertheim, and D. N. E. Buchanan, *Phys. Rev.* **B18**, 6519 (1978).

259. J. J. Yeh and I. Lindau, *At. Data Nucl. Tables* **32**, 1 (1985).

260. S.-J. Oh, S. Suga, A. Kakizaki, M. Taniguchi, T. Ishii, J.-S. Kang, J. W. Allen, O. Gunnarsson, N. E. Christensen, A. Fujimori, T. Suzuki, T. Kasuya, T. Miyahara, H. Kato, K. Schönhammer, M. S. Torikachvili, and M. B. Maple, *Phys. Rev.* **B37**, 2861 (1988).

261. The first actinide RESPES study was reported by R. Baptist, M. Belakhovsky, M. S. S. Brooks, R. Pinchaux, Y. Baer, and O. Vogt, *Physica* **B102**, 63–65 (1980).

262. Y. Baer, in *Handbook on the Physics and Chemistry of the Actinides,* (A. J. Freeman and G. H. Lander, eds.) North-Holland, Amsterdam (1984), Ch. 4, p. 271.

263. G. Gerken and J. Schmidt-May, *J. Phys.* **F 13**, 1571 (1983).

264. J. F. Herbst, R. E. Watson, and I. Lindgren, *Phys. Rev.* **B14**, 3265 (1976).

265. Y. Baer and J. Schönes, *Solid State Commun.* **33**, 885 (1980).

266. B. Reihl, N. Mårtensson, D. E. Eastman, A. J. Arko, and O. Vogt, *Phys. Rev.* **B26**, 1842–1851 (1982).

267. L. E. Cox, W. P. Ellis, R. D. Cowan, J. W. Allen, S.-J. Oh, I. Lindau, B. B. Pate, and A. J. Arko, *Phys. Rev.* **B35**, 5761–5765 (1987).

268. L. E. Cox, W. P. Ellis, R. D. Cowan, J. W. Allen, and S.-J. Oh, *Phys. Rev.* **B31**, 2467–2471 (1987).

269. Y. Baer, H. R. Ott, and K. Andres, *Solid State Commun.* **36**, 387–391 (1980).

270. N. Shamir, M. Melamud, J. H. Shaked, and M. Weger, *Physica* **B94**, 225 (1978).

271. A. F. Murray and W. J. L. Buyers, in *Crystalline Electric Field and Structural Effects in f-Electron Systems,* (J. E. Crow, R. P. Guertin, and T. W. Mihalisin, eds.) Plenum, New York (1979), Vol. 1, p. 257.

272. W. J. L. Buyers, A. F. Murray, T. M. Holden, E. C. Svenson, P. de V. DuPlessis, G. H. Lander, and O. Vogt, *Physica* **B102**, 291 (1980).

273. W. J. L. Buyers, T. M. Holden, A. F. Murray, J. A. Jackman, P. R. Norton, P. de V. DuPlessis, and O. Vogt in *Valence Fluctuations in Solids,* (L. M. Falicov, W. Hanke, and M. B. Maple, eds.) North-Holland, Amsterdam (1981), p. 187.

274. W. Ubachs, A. P. J. van Deursen, A. R. de Vroomen, and A. J. Arko, *Solid State Commun.* **60**, 7 (1986).

275. K. Anders, D. Davidov, P. Dernier, F. Hsu, W. A. Reed, and G. J. Nieuwenhuys, *Solid State Commun.* **28**, 405 (1978).

276. J. W. Allen, unpublished data.

277. J.-S. Kang, J. W. Allen, M. B. Maple, M. S. Torikachvili, W. P. Ellis, B. B. Pate, Z.-X. Shen, J. J. Yeh, and I. Lindau, *Phys. Rev.* **B39**, 13529–13532 (1989).

278. V. M. Goldschmidt, *Trans. Faraday Soc.* **25**, 253 (1929).

279. See, e.g., S. Suga, M. Yamamoto, K. Soda, T. Mori, S. Takagi, N. Niitsuma, T. Suzuki, and T. Kasuya, *J. Magn. Magn. Mater.* **52**, 297 (1985).

280. R. Reihl, N. Mårtensson, P. Heiman, D. E. Eastman, and O. Vogt, *Phys. Rev. Lett.* **46**, 1480 (1981).

281. B. Reihl, N. Mårtensson, D. E. Eastman, and O. Vogt, *Phys. Rev.* **B24**, 406 (1981).

282. B. Reihl, N. Mårtensson, and O. Vogt, *J. Appl. Phys.* **53**, 2008–2013 (1982).

283. J. Schönes, in *Handbook on the Physics and Chemistry of the Actinides,* (A. J. Freeman and G. H. Lander, eds.) North-Holland, Amsterdam (1984) Ch. 5, p. 341.

284. Y. Baer and J. K. Lang, *Phys. Rev.* **B21**, 2060 (1980).
285. A. Fujimori and J. H. Weaver, *Phys. Rev.* **B31**, 6411 (1985).
286. M. Iwan, E. E. Koch, and F. J. Himpsel, *Phys. Rev.* **B24**, 613 (1981).
287. B. Reihl, M. Domke, G. Kaindl, G. Kalkowski, C. Laubschat, F. Hulliger, and W.-D. Schneider, *Phys. Rev.* **B32**, 3530–3533 (1985).
288. A. J. Arko, D. D. Koelling, and B. Reihl, *Phys. Rev.* **B27**, 3955 (1983).
289. A. J. Arko, private communication.
290. D. D. Sarma, F. U. Hillebrecht, W. Speier, N. Mårtensson, and D. D. Koelling, *Phys. Rev. Lett.* **57**, 2215 (1989).
291. G. Landgren, Y. Jugnet, J. F. Morar, A. J. Arko, Z. Fisk, J. L. Smith, H. R. Ott, and B. Reihl, *Phys. Rev.* **B29**, 493–496 (1984).
292. E. Wuilloud, Y. Baer, H. R. Ott, Z. Fisk, and J. L. Smith, *Phys. Rev.* **B29**, 5228–5231 (1984).
293. R. D. Parks, M. L. den Boer, S. Raaen, J. L. Smith, and G. P. Williams, *Phys. Rev.* **B30**, 1580 (1984).
294. J. Ghijsen, R. L. Johnson, J. C. Spirlet and J. J. M. Franse, *J. Electron Spectros. Relat. Phenom.* **37**, 165–169 (1985).
295. J. Allen, S.-J. Oh, L. E. Cox, W. P. Ellis, M. S. Wire, Z. Fisk, J. I. Smith, B. B. Pate, I. Lindau, and A. J. Arko, *Phys. Rev. Lett.* **54**, 2635–2638 (1985).
296. A. M. Boring, R. C. Albers, G. R. Stewart, and D. D. Koelling, *Phys. Rev.* **B31**, 3251 (1985).
297. P. Strange and B. L. Gyorffy, in *Proc. Conf. Electronic Structure and Properties of Rare Earth and Actinide Intermetallics,* (G. Hilscher, G. Wiesinger, E. Gratz, C. Schmitzer, P. Weinberger, and R. Grössinger eds.) North-Holland, Amsterdam (1985).
298. W.-D. Schneider, C. Laubschat, and B. Reihl, *Phys. Rev.* **B35**, 7922 (1987).
299. A. J. Arko, B. W. Yates, B. D. Dunlap, D. D. Koelling, A. W. Mitchell, D. J. Lam, Z. Zolnierek, C. G. Olson, Z. Fisk, J. L. Smith, and M. del Giudice, *J. Less-Common Met.* **133**, 87–97 (1987).
300. A. J. Arko, D. D. Koelling, C. Capasso, M. del Guidice, and C. G. Olson, *Phys. Rev.* **B38**, 1627 (1988).
301. J. S. Kang, J. W. Allen, M. B. Maple, M. S. Torikachvili, B. Pate, W. Ellis, and I. Lindau, *Phys. Rev. Lett.* **59**, 493–495 (1987).
302. A. J. Arko, C. G. Olson, D. M. Wieliczka, Z. Fisk, and J. L. Smith, *Phys. Rev. Lett.* **53**, 2050 (1984).
303. Y. Lasailly, J. W. Allen, W. Ellis, L. Cox, B. Pate, Z. Fisk, and I. Lindau, *J. Magn. Magn. Mat.* **63–64**, 512–514 (1987).
304. D. D. Sarma, F. U. Hillebrecht, C. Carbone, and A. Zangwill, *Phys. Rev.* **B36**, 2916 (1987).
305. A. J. Freeman, B. I. Min, and M. R. Norman, in *Handbook on the Physics and Chemistry of Rare Earths,* (K. A. Gschneidner, Jr., L. Eyring, and S. Hüfner, eds.) North-Holland, Amsterdam (1987) Vol. 10, p. 165.
306. J. W. Allen, J.-S. Kang, Y. Lasailly, M. B. Maple, M. S. Torikachvili, W. Ellis, B. Pate, and I. Lindau, *Solid State Commun.* **61**, 183–186 (1987).

Ion Spectroscopy

Photon-Stimulated Desorption

Victor Rehn and Richard A. Rosenberg

1. INTRODUCTION AND BACKGROUND

There is today an intense interest within the scientific and technological communities in surfaces and other low-dimensional structures. Many practical problems of importance in industry await scientific understanding of surfaces for their solution: Chemical catalysis would be high on this list. Scientifically, the understanding of surfaces has been an important goal for many decades, but especially within the past two decades, during which the advance of ultrahigh-vacuum technology has opened the way to more-reliable experimental studies of highly contamination-sensitive structures, such as chemically active surfaces. The advancement of ultrahigh-vacuum technology has made possible the high-current, high-energy electron storage ring, which produces synchrotron radiation (SR) in copious quantities for use in science and technology. Many surface-analysis techniques have been facilitated by SR. In this chapter, we describe a family of SR-based techniques utilizing photon-stimulated desorption (PSD) for the study of surfaces and other low-dimensional structures. It is the objective of this chapter to survey the unique scientific information concerning the physical and chemical structure of surfaces and adsorbate layers obtainable with PSD techniques. As we discuss in detail below, this uniqueness results partially from the unique properties of ultraviolet and soft X-ray photon probes and partially from the low kinetic energy of the ionic or neutral surface species desorbing from the surface.

PSD is defined here as the phenomenon of desorption of surface or adsorbed species in ionic, excited, or neutral form as a result of electronic excitation by a single-quantum photoabsorption event.[1,2] Customarily, PSD is mechanistically distinguished from photodesorption: PSD describes desorption induced by an

Victor Rehn • Physics Division, Research Department, Naval Weapons Center, China Lake, Caliornia 93555. *Present address*: Office of Naval Research—Adian Office, APO AP 96337-0007   *Richard A. Rosenberg* • Synchrotron-Radiation Center, University of Wisconsin, Stoughton, Wisconsin 53589 *Present address*: Advanced Photon Source, Argonne National Laboratory, Argonne, Illinois 60439.

Synchrotron Radiation Research: Advances in Surface and Interface Science, Volume 1: Techniques, edited by Robert Z. Bachrach. Plenum Press, New York, 1992.

electronic excitation[3-5] while photodesorption includes desorption induced by excitation of molecular or crystalline vibrations, phonons, or conduction-band electrons.[6] This chapter begins with an historical review of desorption experiments intended as a means to help understand surfaces, defects, adsorbates, and other low-dimensional systems. In later sections, we describe the physical processes involved in PSD phenomena from ionic, covalent, and metallic surfaces; review the experimental techniques and their applications, with examples; discuss the apparatus used; and close with a discussion of future research and applications. The discussion of PSD studies of molecular-solid surfaces is reserved for a chapter in Volume 2 of this set.[42]

In the study of low-dimensional structures, the ability to observe chemical interactions, atomic arrangements and electronic configurations in the region of a single type of atomic site would facilitate experimental tests of any theory of these structures. Experimental approaches to such observations may be categorized as type I, the measurement of such steady-state properties of the structure as infrared (IR) absorption or LEED or type II, measurement of transient properties of the structure following excitation, such as Auger electron spectroscopy (AES), secondary ion mass spectroscopy (SIMS), or photoelectron spectroscopy. In the type I experiment, the measurement is made via direct observation of the properties of an excitation from the ground or steady state of the system. In the type II experiment, the system is first excited to a transient (nonstationary) state (e.g., core-hole excitation in AES), and the properties of the transient state are observed subsequently and independently of the excitation (e.g., via the Auger electron emission spectrum). In either case, sensitivity to structure associated with a single type of atomic site (i.e., site selectivity) is hard to obtain. Intrinsic, steady-state properties of low-dimensional structures that do not overlap with the properties of nearby three-dimensional structures are rare. One such property is the vibrational spectrum of adsorbed molecular species, which can be observed in careful IR absorption or high-resolution, low-energy electron energy-loss spectroscopy. In these techniques, single-quantum events associated with isolated adsorbates or a monolayer of adsorbates are observable. By contrast, diffractive responses such as LEED are derived from the structure under study by way of an intrinsically nonlocal interference that averages over many sites of an ordered surface. Because of the nonlocal nature of the measurement, LEED elucidates long-range order in the system, including differences among types of sites in an ordered structure, but lacks local-site selectivity.

In addition to local-site selectivity, a second necessary feature for study of low-dimensional structures is sensitivity to the structure under study independent of the neighboring bulk matter. A third desirable feature for the study of low-dimensional structures is a relative freedom from damage or excessive disturbance of the structure under study. SIMS and AES cause far greater disturbance to the structure under study than is required by the quantum-mechanical disturbance of the observation, for example. Some of the most interesting physical properties of low-dimensional structures are destroyed or obscured by very small amounts of excitation energy. The experimenter needs to seek a gentle, selective, sensitive, and nondestructive probe for type II experiments on fragile low-dimensional structures: These three criteria are seldom satisfied in a single experiment.

Among the probes available for type II experiments on low-dimensional structures, photons have unique properties: Provided neither their energy nor their density is too large, they interact with matter by way of single-quantum events. Photons with energy $h\nu < 10^4$ eV carry too little momentum to displace even a hydrogen atom from a crystal-lattice site or a molecular bond. Thus, the interaction of photons with matter is gentler and less disturbing than those of other probes available for stimulating a low-dimensional structure for study of its transient response. For 0.2 eV $< h\nu < 0.2$ MeV, photoabsorption occurs in condensed matter primarily via electronic excitation; for $h\nu > 20$ eV, the electronic excitation occurs primarily in the highly localized atomic-core states. Thus, single-quantum, localized (site-specific) excitations of condensed systems and relative freedom from damage are characteristics of photon probes of energy 20 eV $< h\nu < 10$ keV. Excitation of the surface, relative to the underlying bulk, is maximized by utilizing lower-energy photons incident on the surface at low grazing angles. Still, however, photoexcitation occurs over a depth of several nanometers: The missing ingredient in the photon-probe type II excitation is sensitivity to the low-dimensional structure independent of underlying bulk matter.

Ions or neutral atoms or molecules desorbing at low kinetic energies from a low-dimensional structure are ideal complements to the photon-beam probe. Except for channeling-type desorption, low-energy ($E < 10$ eV) atomic or ionic particles usually do not pass through bulk condensed matter, although a contrary case has been reported for adsorbed films of CH_3Br recently.[7] Ions are susceptible to reneutralization by electrons from bulk matter, and neutral atoms or ions with kinetic energy less than 20 eV are unable to displace other atoms blocking their path. In this way, PSD of ions combines the unique properties of the photon-beam probe with the extreme surface sensitivity of the low-energy desorbing ion to satisfy all three of the general criteria for the type II experiment. Photon-stimulated ion desorption (PSID) is a unique tool for the study of low-dimensional structures. At this time, only a few experiments on photon-stimulated desorption of *neutral* atomic or molecular species have been reported: The impact of PSD of neutrals on the study of surfaces and other low-dimensional structures has not yet been felt.

Historical Background. Desorption of particles from surfaces has long been studied by a variety of techniques. Physisorbed surface species, bonded weakly to surfaces by van der Waals forces, may be desorbed by mild heating, since thermal energy may easily exceed the energy of the van der Waals bond.[8] Hydrogen-bonded species may be thermally desorbed as well. In some cases the energy of ionic or covalent surface bonds involved in chemisorption may be so large that simple thermal desorption may desorb as much from the underlying layers as from the surface. Rapid surface heating can supply more thermal energy in the surface without evaporating the underlying solid, and IR-laser heating is used in either single-(vibrational)-quantum or multiple-quantum versions of laser-assisted evaporation or thermal desorption. Thermal desorption has long been used for the study of surface-bonding energies or, with the use of a mass spectrometer, surface composition. If the heating is quick enough (as is that produced by a laser pulse) or if the substrate is a refractory material, thermal desorption can be sensitive primarily to the surface. Thermal desorption may cause little damage to the surface layer. If adsorption sites differ significantly in their surface-bond energies, they may be distinguished by their differing desorption temperatures, providing a degree of site selectivity in thermal-desorption experiments.

Desorption induced by ion impact is the mechanisms of SIMS. Ion impact or sputtering to remove surface layers has been used extensively in research. In SIMS experiments, the surface is eroded by heavy-ion impact and the desorbed or sputtered atoms and ions are mass-analyzed in a conventional way. Although SIMS is destructive of the surface under study, it offers high sensitivity to the composition of the outermost three or four atomic layers in the surface. However, site selectivity is not available, and the electronic structure of the surface is obliterated by the high kinetic energy of the impacting ions.

The term photodesorption has been restricted by Lichtman to refer to a specific mechanism of desorption from *semiconductor* surfaces following illumination by radiation of photon energy greater than the band gap.[6,9] Lichtman described the desorption of CO_2 from an oxygen-exposed semiconductor surface containing carbon impurities as the typical and most common example of photodesorption: Oxygen adsorbs, probably on the carbon impurity, attracting an electron from the conduction band and forming a tightly bound CO_2^- ionic adsorbate. Photoabsorption of photons with $h\nu > E_g$, the energy gap in the semiconductor, produces free holes. A free hole is attracted to the negative ion site and neutralizes the CO_2^- ion, greatly reducing its binding energy and allowing thermal desorption of neutral CO_2. Thus, the photoabsorption event need not occur at the desorption site, but only within range of free-hole Coulomb attraction to the CO_2^- ion. We note the contrast of this model of photodesorption with PSD, in which the substrate may be metal, insulator, or semiconductor; the desorbing particle may be an adsorbate or an intrinsic surface species; and the photon energy must be high enough to excite a deep-valence or core electron of either the adsorbate or the substrate.

The phenomenon of PSID was predicted by extrapolation from electron-stimulated ion desorption (ESID or, customarily, ESD) after it was realized that ESID thresholds correspond to core-level energies of either the desorbing particle or one of its nearest neighbors.[5] Experimental measurements of the electron-energy dependence of ESID yields showed that a threshold for ESID could be correlated with the core-excitation energy of a surface atom. A physical mechanism for desorption following core-hole excitation of ionic surfaces was proposed,[10] based on the concept of the Coulombic explosion observed in molecules.[11,12] Simply stated, the Coulombic explosion results from an abrupt reversal of the charge of an ion. On a site that is attractive to an ion of the original charge, such a charge reversal results in an instantaneous reversal of the Coulombic force of attraction; hence the Coulombic explosion. This early, innovative concept (illustrated in Fig. 1) was modeled on a fully valent ionic surface (e.g., TiO_2) in which the interatomic Auger transition following excitation of the $3p$ core hole in the Ti^{4+} ion would be highly probable and would act to expel two or three electrons from the desorbing atomic species (O^{--} becoming O^0 or O^+). (Note that in the fully valent ionic TiO_2, intraatomic Auger transition is not possible following excitation of a $Ti(3p)$ core hole because there are no

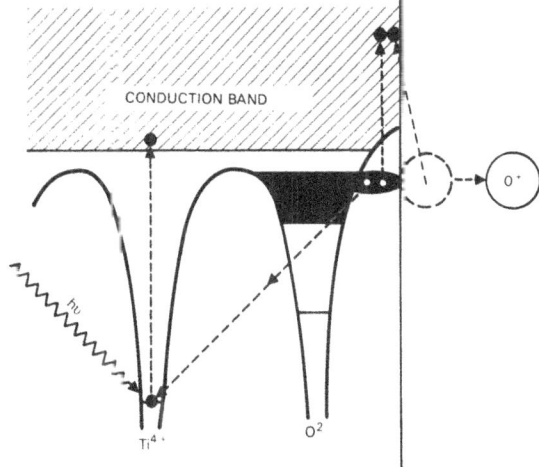

FIGURE 1. Conceptual diagram of the KF mechanism applied to TiO_2.

occupied electron states of the Ti^{4+} ion above $3p$.) This desorption mechanism, known as the Knotek–Feibelman (KF) mechanism,[5] was proposed to explain ESID threshold data. It suggested three temporally separate steps in desorption stimulated by core-hole excitation:

1. core-hole excitation, occurring in about 10^{-16} sec;
2. interatomic Auger decay of the core-hole excitation, requiring 10^{-15} to 10^{-13} sec;[13,19] and
3. Coulombic expulsion of a surface or adsorbate species at low kinetic energy ($<10\,eV$), requiring 10^{-13} to 10^{-12} sec.

As long as an adequate temporal separation exists, each step will be a separately identified physical process, with a distinct electronic "precursor" state from which it proceeds.

Three important properties became apparent from the KF model of the desorption process

1. Desorption should follow core-hole excitation by *any* stimulus, including vacuum-ultraviolet (VUV) or soft X-ray (SXR) photoexcitation;
2. photoexcitation of the core hole would be a single-quantum event,[14,15] free from the background of electrons produced in ESD by multiple scattering of the incident electron beam; and
3. the desorption–excitation spectrum would be controlled by the photoabsorption spectrum of the desorption-site locale.

The KF model for stimulated desorption has been extended successfully to other systems besides fully valent ionic systems and has been challenged in some cases by other mechanisms. In particular, the unique site specificity has been both confirmed and challenged in more recent results on various systems. A detailed discussion of indirect mechanisms for PSID, in which the unique site specificity is not observed, is given in section 2.4.

The first observation of PSID was facilitated by SR, in particular, by exploiting the tunability and the time structure of SR for spectrally resolved time-of-flight mass analyses of desorbing species.[16–18] Two types of spectra were recorded.

1. Using a constant (monochromatic) photon energy or a broad band of energies (e.g., central-image or zero-order radiation from the monochromator), the mass spectrum of desorbing ions was observed, as illustrated in Fig. 2a; and
2. monitoring the desorption rate for a particular ionic mass, the excitation-energy spectrum for a *particular desorbing ionic species* was observed, as illustrated in Fig. 2b.

For the first experiments, the TiO_2 (100) surface was chosen as a surface suitable for comparison with previously reported ESD excitation spectra and with the predictions of the KF model.

The first PSID experiments detected H^+, OH^+, and F^+ desorption from TiO_2 surfaces that had been heat-cleaned, lightly sputtered, and H_2O dosed. The basic concepts of the KF desorption mechanism were confirmed for photon stimulation.[1] It was clear that positive ions desorb with kinetic energy less than 10 eV—independently of the photoexcitation energy*—as predicted by the KF model. In addition, the desorption–excitation spectra above core-level thresholds showed near-edge and extended fine structure, again consistent with the KF model. This was particularly exciting because near-edge and extended fine structure observed in the PSD should represent the electronic structure of the local desorption site in the same way that near-edge (NEXAFS) and extended X-ray absorption fine structure (EXAFS) represent the local environment of an atom that absorbs an X-ray photon. (This is discussed in greater detail in sections 3.4 and 3.5.) Also, there was a rich mass spectrum of desorbing ions. H^+ desorption was most easily observed. This was particularly important because most surface analysis techniques are only indirectly sensitive to hydrogen, if at all. An additional feature that resulted from the use of time-of-flight mass analysis was the ability to observe the excitation-energy spectrum of all desorbing ions concurrently, thereby facilitating a quantitative comparison of relative desorption rates of various ions, as illustrated in Fig. 2.

Following publication of the first PSD experiments, workers in the field fanned out to address several questions, some of which were

- Can PSD be observed from metal surfaces? From insulator surfaces? From covalently bonded, hydrogen bonded, or van der Waals–bonded surface layers?
- How large are the desorption yields and cross sections for PSD processes in various systems, and can they be predicted? Might PSD possibily offer a practical method of surface cleaning?
- How do molecular adsorbates dissociate and desorb under VUV or SXR photoexcitation?

---

* In subsequent experiments, small variations in the kinetic-energy spectrum of the desorbing ions have been shown to correlate with the excited states of molecular adsorbates.[32]

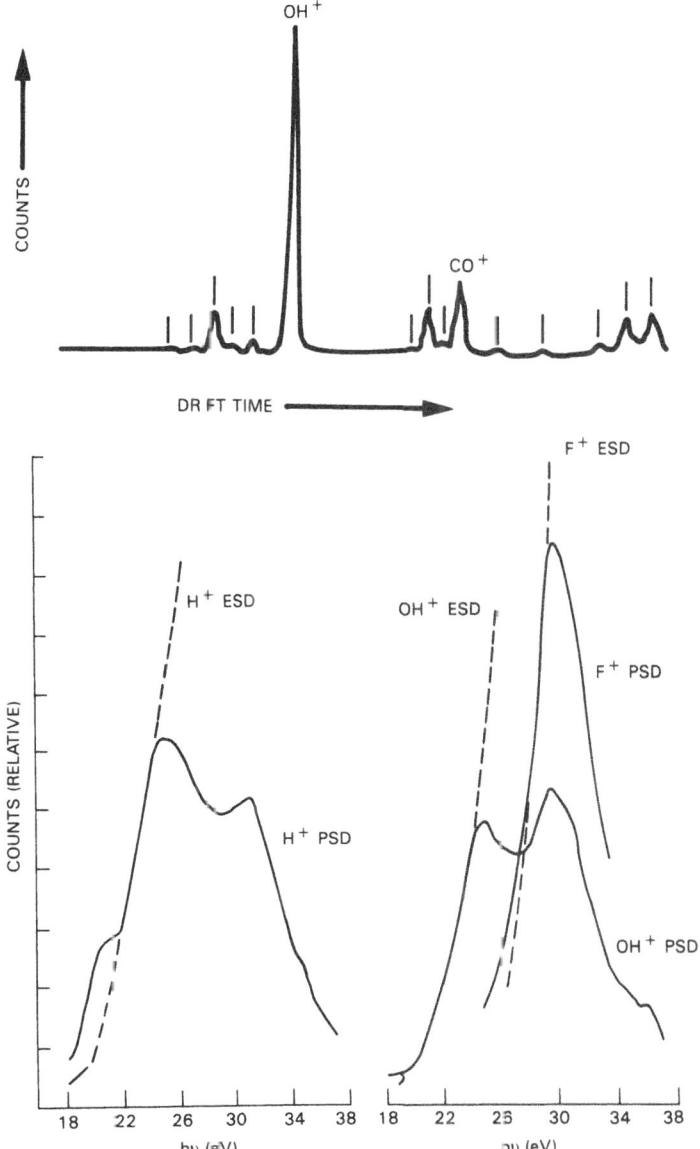

FIGURE 2. (a) Time-of-flight mass spectrum of desorbing ions and (b) excitation-energy spectrum of H^+, OH^+, and F^+ ions desorbing from TiO_2. In (a) the very strong H^+ peak is omitted. The dashed lines in (b) represent the excitation-energy dependence of ESD observed from TiO_2. Note that the threshold energies and the spectral features are different for each ion in these concurrently accumulated PSID spectra. (From Ref. 1.)

- What type of molecular states of adsorbates lead to ionic or neutral desorption?
- Can impurity and imperfection sites be studied on semiconductor, metal, or insulator surfaces?
- How does the kinetic-energy distribution of desorbing ions depend on photoexcitation?

- How do the desorption patterns of positive, negative, and neutral species compare under various conditions of excitation?
- How can ESD and PSD results be compared and contrasted?
- How do PSD ion-angular distributions (PSDIAD) compare and contrast with ESD ion-angular distribution (ESDIAD) results? How do the angular patterns change with excitation energy, polarization, or with atom excited?

Some theoretical questions were

- How can desorption mechanisms be studied, understood, generalized, and unified?
- How does the intermediary Auger process act to enhance the likelihood of ion desorption?
- How can PSD–NEXAFS data be used best to determine the electronic structure of desorption sites?

Significant progress has been reported by many investigators on these and other questions. The number of theoretical and experimental papers relating to PSD is so great that a comprehensive review is impractical. In the following sections, we will summarize some of the key research results that have led us to our current understanding of PSD processes and illustrate some of the applications of PSD to surface-science problems.

2. PHYSICS OF PSD

Stimulated desorption of surface species in ionic or neutral form has been studied for many years, and much has been learned about the physical processes involved. Simulated desorption is intrinsically a property of a transient state into which the desorbing-surface site is excited as a consequence of the stimulus. In developing an understanding of stimulated-desorption science, we try first to understand which transient, excited states are precursor states of desorption, and how the desorption proceeds from the moment of establishment of the system in a precursor state. Finally, we try to understand how the photon or electron stimulus establishes the site in a particular precursor state for desorption. Knowledge of both aspects, together with knowledge of other competing processes that quench or do not lead to desorption, is required before we will know how to predict and control desorption.

The separation of the physics of desorption into two distinct considerations makes sense only if the precursor state of the desorption site persists long enough to become an identifiable state of the electronic system. Chemically bonded systems with bond energies on the order of a few electron volts require times of order $\Delta t \sim h/\Delta E \sim 1$–$10$ fsec to relax sufficiently for precursor-state identification. Because desorbed-particle kinetic energies generally lie below 10 eV,[19] the acceleration of desorbing atomic or ionic particles by realistic potential gradients should be small enough that little particle motion takes place within 10 fsec, in accordance with the Franck–Condon principle. With this justification, we can proceed to examine possible desorption-precursor states at though they may be identified with steady states of the desorption site. It is within this approximation that much of the theoretical research on desorption mechanisms has been done.

2.1. Desorption Precursor States

Prior to the pioneering ESD research of Knotek and Feibelman, most interpretations of stimulated-desorption results were made on the basis of adiabatic potential curves of the desorbing particle in proximity to the surface, as described in the theories of Menzel and Gomer[3] and Redhead[4] (MGR). The basic idea of the MGR theory is that the precursor state is a nonstationary state described by one of the multitude of adiabatic potential curves of the desorbing particle and the surface. Following Gadzuk,[20] we illustrate in Fig. 3 several types of such excitations. Note that diagrams such as these are intended to represent schematically (with the single multidimensional variable **R**) the configuration-momentum-space or phase-space position of all the atomic nuclei of the system.

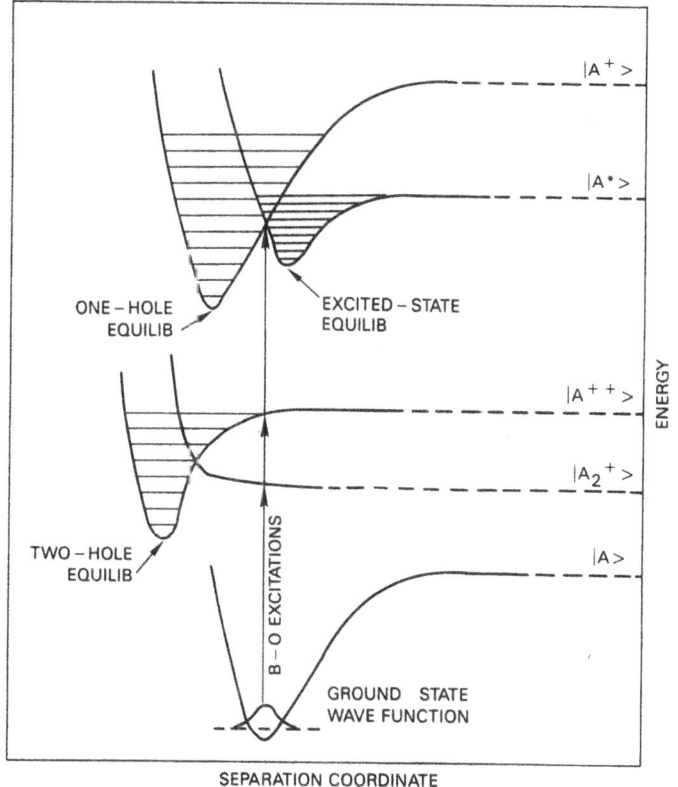

FIGURE 3. Diagram of adiabatic potential curves for an MGR-type desorption. Vibrational levels, desorption-threshold energies, and desorption products are illustrated for each excited potential curve. Changes in the equilibrium separation with ionization or excitation are illustrated as well. The vertical line labeled, "B–O excitation" illustrates various excitation energies within the Born–Oppenheimer approximation. Autoionization, curve crossing, predissociation and other molecular concepts important in MGR desorption may be illustrated on diagrams such as this.

When a vertical (Franck–Condon type) transition is hypothesized, it is assumed that the positions and momenta of all atomic nuclei in the system remain constant, while the energy of the system is changed by a change in the electronic configuration only. Within the Born–Oppenheimer approximation, an allowed electronic excitation from the ground state raises the system vertically, without change of R, to the adiabatic potential curve of one of the excited states of the system. As illustrated in Fig. 3, these states may correspond to bound or nonbound states of the desorbing ionic or neutral particle.

For bound states, the new equilibrium position in phase space, R_{eq} is located at the minimum in the adiabatic potential curve currently occupied. Under a driving force derived from the electronic-system energy, which provides a potential gradient accelerating nuclei, the nuclear coordinates of the system will change spontaneously toward their equilibrium positions. Although each potential curve is labeled with an electronic-state designation (usually corresponding to the stationary state or the asymptotic state for that curve), it must be remembered that the electron orbitals are a continuous function of R and may look quite different at various points of the same adiabatic potential curve. In desorption calculations, it is usually assumed that the nucleus of the desorbing particle only will move appreciably as the system seeks its equilibrium configuration, which gives it the lowest total energy. This assumption is discussed in more detail in section 2.3.

From the instant the system is placed in a nonequilibrium state, any of several processes may occur. The system may completely or partially de-excite radiatively or nonradiatively. The system may adjust its nuclear positions, following the adiabatic potential curve to its minimum energy (which may or may not remain bound). The system may follow a crossing adiabatic potential curve to obtain a lower final energy. If the system is positively ionized, it may recapture the electrons needed to return to neutral, or it may autoionize to a more highly ionized state. If the excitation energy is high enough, the system may make an Auger transition to a state with fewer electrons. An entirely new set of adiabatic potential curves must be used in either of the latter two cases, since the system now has a different number of electrons contributing to the energy of its electronic system. It is clear that very complex physical processes are involved and that only some of these lead to ionic or neutral desorption.

The main controversy raised by the work of KF centered around which precursor states are more likely to lead to desorption. Within the considerations presented by the MGR papers, it was generally assumed that a one-electron Franck–Condon-type ionization or excitation to an antibonding or possibly a nonbonding state would be sufficient to cause desorption. However, the KF papers presented an excitation process that removed two or more electrons from the desorption site. Because of the interatomic Auger transition, all electrons may be taken from the desorbing particle, converting the surface O^{--} into a desorbing O^0 or O^+ in the TiO_2 example illustrated in Fig. 1. Although the KF mechanism did not explicitly consider desorption precursor states, soon it was realized that such an excitation process might produce two or more holes in the valence or bonding orbitals, and possibly one or more electrons in nonbonding or antibonding excited orbitals of the desorption site. This was a point of contrast with the conventional Franck–Condon picture, which generally assumed a single valence hole with one or possibly more electrons in nonbonding or antibonding orbitals.

For the case of hydrogen adsorbed on a Ni (111) surface, the work of Melius, Stulen, and Noell[21] showed that indeed a precursor state involving two holes in the Ni–H bonding orbital facilitates H^+ desorption, while a single hole would yield H^- desorption, or no desorption. Ramaker, White, and Murday[22,23] showed that because of the hole–hole repulsion energy, the lifetime of the two-hole excitation may be much longer than that of the single-hole excitation. These results seemed to confirm the importance of the two-hole excitation and helped to account for the wide range of surfaces from which PSID yields had been observed following core-hole excitation. The original KF mechanism was applicable strictly to fully valent ionic surfaces only: With long-lived two-hole excitations, the mechanism became applicable to covalently bonded surfaces as well.

The importance of the hole–hole repulsion energy in determining whether a desorption threshold will accompany a core-hole-excitation threshold was emphasized by Ramaker, White, and Murday[22,23] by comparing stimulated desorption from SiO_2 and Si_3N_4 surfaces. In SiO_2, from which significant desorption of O^+ is observed, they found the hole–hole repulsion energy $U = 11\,eV$, which is greater than the energy width of the valence band, $E_{VB} = 4\,eV$, and calculated a desorption cross section of $10^{-21}\,cm^2$. In contrast, in Si_3N_4, from which no N^+ desorption is observed,[24] Ramaker, White, and Murday found $U < E_{VB}$ and calculated the cross section to be $10^{-39}\,cm^2$. The difference in U for these two cases is the result of the more diffuse character of the planar, trigonal sp^2-hybrid directed-valence bonding orbitals of Si_3N_4, compared to the more compact tetrahedral sp^3-hybrid bonding orbitals of SiO_2. For covalently bonded sites, the desorption branching ratio for two-hole precursor states is controlled by the magnitude and lifetime of the hole–hole repulsion energy: Two holes tightly packed in a compact valence orbital produces a large repulsive force. If the E_{VB} is small enough, the delocalization or separation of the two holes will be retarded and the repulsive force persists long enough to drive the desorbing particle out of range of recapture.

The lifetime of a local excitation at a surface site should be influenced by the concentration of free electrons in the surface. It was not clear *a priori* whether desorption precursor states would survive on metal surfaces long enough for desorption to occur. The first experiments to show PSID from a metal surface following core-hole excitation were those of Franchy and Menzel.[25] In this work, CO^+ desorption was observed following C (K) ionization and O^+ desorption was observed following O (K) ionization on the CO/W (100) surface. In order to ascertain that the observations were truly PSID and not photoelectron-induced ESD, Franchy and Menzel compared their PSID results with ESD observations obtained *in situ* on the same system. It was shown that the ratio of CO^+ desorption to O^+ desorption was different by a factor of seven for the same excitation energy, 1487 eV. This large difference negated the possibility of photoelectron-induced ESD. Although the interatomic Auger process of the KF mechanism was not invoked, these results showed that desorption precursor states can be excited by intra-atomic Auger decay of the core hole, and that such states can survive long enough for desorption to occur in spite of a metallic free-electron concentration present near the desorption site.

Another observation of PSID was reported from experiments on Nb and Nb_3Sn thin films,[26] in which both metal- and adsorbate-core excitations were shown to lead to desorption. In Fig. 4 the PSID excitation spectra of O^+ and H^+ are shown near C (K) and Nb ($M_{2,3}$) core energies for a thin film of Nb_3Sn after a light sputter treatment. These data were interpreted as indicating the presence of C–O and Nb–O bonds on the surface from which the O^+ issued following core-hole excitation of the C (K) and the Nb ($M_{2,3}$) states. Other early studies relevant to the question of desorption from metal surfaces were reported for F^+, O^+, and Cl^+ desorption from W (100);[27] for H^+, O^+, OH^+, and F^+ desorption from oxygen-exposed Mo (100)[28,29] for O^+ desorption from CO-exposed Ni (100)[30] for O^+ desorption from Ru (001)[31] and for H^+ desorption from Pd and Pt.[32] In the W (100) study, ESD and PSID results were compared, from which Woodruff and co-workers ruled out the possibility of ESD-intermediated PSID (see sec. 2.4.1). Also, they found considerable but not total support for desorption via intra- and interatomic Auger decay of core-hole excitations. The O^+ desorption from the surface α-oxygen phase on Mo (100) wax excited by the Mo (L_1) core-hole excitation and compared with the total-photoelectron-yield spectrum in the same energy range. In the study of CO chemisorbed on Ni (100), the O^+ desorption following O (K) excitation was reported to show an excitation spectrum distinct from the photoabsorption spectrum, a fact that was attributed to multielectron excitations occurring in the desorption process. In the Ru study, angle-resolved PSID was observed from the CO-coated (001) surface following Ru ($N_{2,3}$) core-hole excitation. The results were compared with ESDIAD and ultraviolet photoelectron spectra in this well-characterized system, in which the CO is covalently bonded to the Ru. It became clear from these and other experimental results that precursor states can be excited by either intra- or interatomic Auger decay of the core-hole excitation, and that they can survive even on metal surfaces long enough for desorption to occur.

Another question that was answered early in the development of PSID

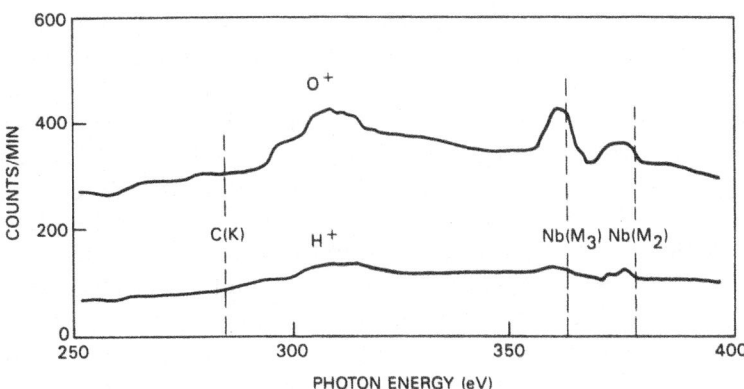

FIGURE 4. PSID excitation spectra of a Nb_3Sn thin film. The sample had been baked to 773 K and lightly sputtered. Note that the impurity carbon features are as prominant as those of the major component Nb.

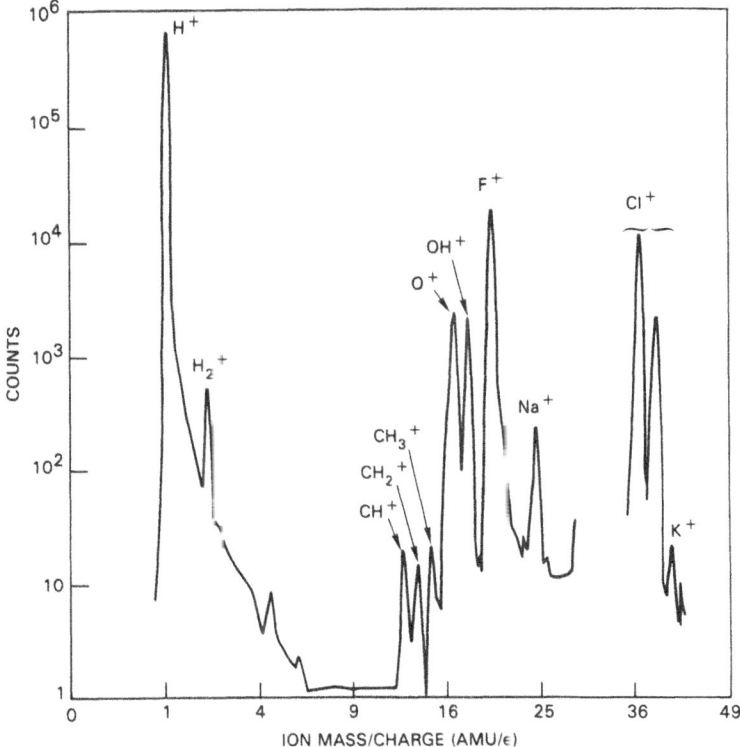

FIGURE 5. PSID mass spectrum of acid-cleaned GaAs (110). At least 12 impurity species were identified, but no Ga or As species. The sample had been optically polished previously, and baked *in situ* to about 750 K.

science was whether and how metal cations could be desorbed following core-hole excitation of either cation or anion. As shown in Fig. 5, both Na^+ and K^+ were among the many impurities observed desorbing from a polished, etched, and baked (110) surface of GaAs.[33] However, it was in the work of Parks and coworkers that a detailed explanation of the process for metal-cation desorption was first given.[34] They reported studies of cleaved NaF surfaces excited by X-ray photons near the Na (K) edge at 1075 eV. They argue that the most probable decay route for the Na (K) core hole is via the standard Na (KLL) intra-atomic Auger process, which should occur within 10^{-15} sec, producing Na^{3+}. Charge transfer from surrounding fluorine ions must follow. They suggest the process would be:

$$Na^{3+} + F^- \Rightarrow Na^{2+} + F^0. \tag{2.1.1}$$

This reaction is exothermic by 53 eV.

Either of two additional charge-exchange reactions would follow next:

$$Na^{2+} + F^0 \Rightarrow Na^+ + F^+ \tag{2.1.2}$$

on

$$Na^{2+} + F^- \Rightarrow Na^+ + F^0, \tag{2.1.3}$$

which are described as exothermic by 14 and 28 eV, respectively. The energy released in the charge-transfer steps may induce fluorescence or expulsion of electrons from the valence band (shakeup). The latter case has the net result of an interatomic Auger event. While the Na (L_3) F (L_3) F (L_3) Auger transition would be endothermic, the quasi-interatomic Na (L_3) F (L_3) F' (L_3) Auger transition is exothermic by about 7 eV, where F and F' are different neighboring fluorine atoms. Note that the Na atom, which had absorbed the original K-core excitation, is now again in its original Na^+ charge state, but now it has either two neutral F^0 neighbors or one unipositive F^+ neighbor. In either case, the total electrostatic (Madelung) energy of the Na^+ neighborhood is repulsive, leading to desorption of either Na^+ or F^+. Experimentally, Parks and coworkers found Na^+ desorption about twice as likely as F^+.

On the basis of results such as those summarized above, PSID of either anions or cations can be understood on the basis of precursor states induced by the Auger decay of a core-hole excitation in an either desorbing or neighboring anion or cation. Much remains to be understood, especially the role of surface imperfections and the quantitative interpretation of PSID spectral yields.

2.2. Antoniewicz Bounce

Another factor relating to the probability of desorption has come to be referred to as the "Antoniewicz bounce."[35] If the Franck–Condon excitation places the system on an ionic potential curve for which the bond strength is greater than that in the ground state, the equilibrium bond length will be shorter than in the ground state, as illustrated in Fig. 3. The potential gradient will drive the system toward the new, smaller value of R_{eq}, which may be so small that it crosses a lower-potential curve at an energy above its dissociation energy. Via curve crossing, the system would then desorb a particle along this lower-potential curve, hence seeming to "bounce" off a shortened bond length. It is evident that the Antoniewicz bounce should be observed mainly for weakly adsorbed species, such as rare gases on metals, but would not be likely for more strongly bound surface species.

2.3. Coulombic Explosion

The influence of the motion of neighboring nuclei on the desorption probability was addressed theoretically for the case of ionic surfaces for which a Coulombic explosion has been assumed. Using the fully valent, ionic NaF surface as an example, Walkup and Avouris[36,37] assumed an instantaneous charge change $F^- \Rightarrow F^+$ of a surface fluorine atom and performed classical-trajectory calculations for several near-neighbor atoms. They found that desorption via Coulombic explosion should not occur from the perfect NaF surface in this model, because the neighboring ions would move to positions that reduce the repulsive force before the F^+ gets out of recapture range.

On the other hand, both Na^+ and F^+ PSID from NaF have been reported,[34] and both desorptions were attributed to Auger-intermediated processes. As discussed in section 2.1, Parks and coworkers showed that charge transfer can occur rapidly between ions, producing environments in which either Na^+ or F^+ may be expelled from the surface via change in the Madelung potential (Coulombic expulsion). Thus the simple classical–trajectory calculation modeled with fixed ionic charges may not be representative of reality in this case. The results calculated by Walkup and Avouris should be used with caution. Adsorbate atoms, other types of ionic surfaces, covalent, metallic, or molecular surfaces all may behave differently, involving charge transfer or other physical processes ignored in the simple classical-trajectory calculation. However, it is apparent also that the Coulombic-explosion concept in its simplest form should be used with caution and that motions of neighboring ions, as well as interionic charge transfers, may be expected to play an important role in ion desorption.

2.4. Indirect Mechanisms of Stimulated Desorption

We have identified several types of desorption precursor states of the desorption site and have indicated that occupation of these states does occur following core-hole excitation of either the desorbing surface species or a nearest-neighbor species. The important and interesting question has been raised of whether the desorbing particle must issue from the site of the stimulated core-hole excitation, or whether this excitation energy could somehow be transferred to a remote site before inducing desorption. Such a remote-ion desorption is referred to as an "indirect desorption." If photon absorption at one site can stimulate desorption at a remote site, the unique site selectivity of the PSD excitation spectra is jeopardized.

Many previously reported PSID studies have shown features attributable only to direct PSID. In the original PSID report,[1] the difference in the optical-absorption and electron-scattering cross sections versus energy was pointed out in explaining the data of Fig. 2b. In the second PSID study reported, Franchy and Menzel showed PSID to occur by an intrinsic photoeffect from CO/W (100), and not from photo-induced ESD.[2] In Fig. 6 we show concurrently acquired excitation spectra of H^+, O^+, and F^+ desorption from a film of Nb_3Sn.[38] On the basis of an indirect mechanism, it would be difficult to understand the observations shown in Fig. 6 in which excitation of the Si ($L_{2,3}$) core hole induces additional F^+ desorption, but no additional O^+ or H^+ desorption. O^+ desorption from oxygen-exposed Mo (100)[28,29] and from Na_xWO_3[39] shows features that can be attributed only to direct PSID. PSID studies of monolayer CO and NO on Ni (100) near the Ni ($L_{2,3}$) edge[40] show an NO^+ yield spectrum similar to the total electron yield (TEY) spectrum, but an O^+ yield spectrum that was clearly different. In addition, the O^+ yield spectrum near the O (K) edge does not follow the absorption coefficient of the Ni.[40] In many cases, indirect mechanisms are contraindicated for covalently and ionically bonded systems.

Direct evidence for the direct Auger-intermediated desorption (AID) mechanism was reported by Knotek and Rabalais,[4] who observed the emitted ion and the Auger electron in coincidence. They studied an oxidized and fluorinated Ti (001) surface, from which desorption of O^+, OH^+, and F^+ was observed following excitation of a Ti-core hole or core holes in the desorbing ions. Using a cylindrical mirror analyzer for electron kinetic-energy analysis that

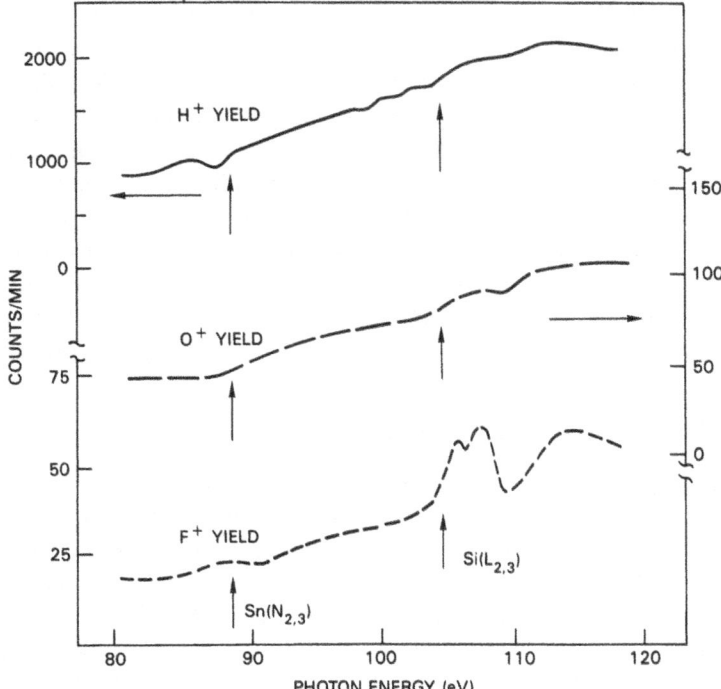

FIGURE 6. PSID excitation spectra for thin-film Nb_3Sn. The same sample as in Fig. 4, but before sputtering. Note the strong association of impurity F^+ desorption with impurity Si near 105 eV and the absence of threshold spectral features associated with Sn near 88.6 eV.

was modified to contain a coaxial time-of-flight ion-mass analyzer (see section 4), Knotek and Rabalais scanned the exciting-electron energy over the range (612–675 eV) for fluorine KLL-Auger excitation. They observed the coincidence count rate for the F^+ ion and the F (KLL) Auger electron. Although they accumulated data for 10,000 sec per point in their spectral scans, the number of coincidences was only 10^{-5}–10^{-6} of the ion count, leading to a very weak signal. Nevertheless, peaks in the coincidence spectrum were observed at energies of 615 and 641 eV, corresponding to F ($KL_1L_{2,3}$) and F ($KL_{2,3}L_{2,3}$) Auger processes, respectively. This direct evidence seems to leave no doubt that the direct AID process does occur, and we conclude that PSID via an AID process is the usual case for covalently or ionically bonded systems.

Nevertheless, evidence for PSD by an indirect process has been reported for some molecular surfaces, and indirect-mechanism models have been suggested. Many molecular surfaces have been studied and the PSID results interpreted using the direct AID process, as discussed in our later chapter.[42] Now we examine the evidence for indirect PSID from molecular surfaces and the model mechanisms.

2.4.1. X-Ray-Induced Electron-Stimulated Desorption

The first detailed mechanism for transfer of excitation energy to a remote desorption site was described by Jaeger, Stöhr, and Kendelewicz:[43,44] X-ray-induced electron-stimulated desorption (XESD). In this mechanism, energetic X-ray photoelectrons carry the excitation energy from the site of the photoabsorption event to the site of the ion expulsion. Thus, the XESD process combines

1. the internal X-ray photoemission characteristics of the photoabsorption site,
2. the characteristics of photoelectron transport between the sites, and
3. the ESD characteristics of the expulsion site.

Although only the photoabsorption step is directly dependent on the photon energy, the transport range and the electron scattering probability are dependent on the photoelectron energy, which is dependent in turn on the photon energy. The final ESD step is hypothesized to occur via the normal core-hole excitation (KF) mechanism.

The experimental support for the XESD mechanism resulted from a simple test for the localization of PSID by studying the desorption yield versus adsorbate thickness. Jaeger et al. reported on such studies using adsorbed layers of NH_3[43] or H_2O[44] on Ni (110). In the simple PSID model, desorption is predicted to result from core-hole excitation only within the surface layer. It has been shown that hydrogen-containing molecules such as NH_3 or H_2O condensed in thick or thin solid films yield mainly H^+ desorption,[45,42] for which there is no possibility of desorbed-ion core-hole excitation. The desorption of H^+ from NH_3/Ni should be stimulated by excitation of a nitrogen core hole or, if the NH_3 is bonded directly to the Ni surface, of a Ni core hole.* In a measurement of the H^+ yield as a function of the solid-NH_3 film thickness, the normal PSID model predicts that the yield resulting from nitrogen core-hole excitation will increase with film thickness until the Ni surface is completely covered, and then remain invariant with further increases in the NH_3-layer thickness. However, Jaeger, Stöhr, and Kendelewicz found that the H^+ yield at the N (K) edge continued to increase until the film became quite thick, as indicated by indirect estimates of its thickness.

Similarly, according to the normal PSID model, the H^+ yield resulting from Ni core-hole excitation should increase to a maximum when the surface is covered by a single NH_3 layer then decrease to zero when two or more molecular layers cover the entire surface. Jaeger et al. found that the decrease in the H^+ yield beyond the monolayer required a much thicker film. When they compared the H^+ yield behavior with that of the total photoelectron yield and the fluorescence yield, they found that the H^+ yield and the total photoelectron yield behaved similarly with film thickness, but quite differently from the fluorescence yield, in which fluorescence photons may issue from depths determined primarily by the penetration of the stimulating photon beam.

Next they analyzed the near-edge excitation spectra in these three cases.

---

* H^+ desorption via the normal PSID model from an NH_3 molecule adsorbed via weak van der Waals interaction on the Ni (110) surface was considered unlikely by Jaeger et al., due to the remoteness of the N–H bond from the Ni atom. However, the Ni^+ ionic charge created within 3 Å of an adsorbed NH_3 molecule may strongly perturb the $3s\,a_1$ and other large, bulky Rydberg orbitals of the NH_3 molecule; hence, desorption of the H^+ via shakeup excitation of a now-dissociative state may be quite likely. For a more complete discussion of PSID involving two-hole, one-electron excitations of solid H_2O or NH_3 films, see Ref. 42.

Near the N (K) edge, where there are no Auger photoelectrons energetic enough to induce ESD via N (K) excitation, the three spectra are different. In the H$^+$ yield spectra, there are two partially resolved near-edge peaks for which the relative intensities change with film thickness. Analysis led Jaeger et al. to conclude that H$^+$ yield originates from two mechanisms, XESD and PSID, in spite of the absence of photoelectrons of energy sufficient for N (K) ESD. In the "thick" NH$_3$ films, the ratio of the two spectral features was about 60:40, while in the "monolayer" film only the structure attributed to PSID was observed, as expected.

We note that it should be expected that two PSID near-edge spectra would be superimposed at the N (K) edge in the absence of an XESD contribution, because of the two types of NH$_3$ sites expected in the layer: the Ni–NH$_3$ site and the NH$_3$–NH$_3$ site characteristic of a thicker NH$_3$ film, or of a nonuniform NH$_3$ film containing thicker islands. It should also be expected that the surface densities of these two sites will be a function of the NH$_3$ layer thickness, and that the NH$_3$–NH$_3$ site will be rare in films of monolayer thickness or smaller. It is unexpected that the surface density of the NH$_3$–NH$_3$ site continues to increase beyond double-monolayer coverage, or that the Ni–NH$_3$ site does not disappear for the thicker layers studied, unless the layer morphology is snow-like or of island structure. The lack of morphological measurements and of direct thickness measurements in the data of Ref. 43 or 44 weakens the evidence for XESD. The method of varying film thickness (by warming the thicker film to thin it by evaporation) makes it likely that the film morphology varied (annealed) as the thickness was changed. If the NH$_3$ film grew with a snowflake or island structure, then the persistence of the H$^+$ desorption with Ni core-hole excitation would be expected as long as some Ni sites were not covered. Although Jaeger et al. concluded that XESD is probably important only in covalently bonded systems where ESD excitation-energy thresholds might be lower, the 23 eV threshold for NH$_3$ is as high as that for most ionic systems.[46]

In another study of indirect mechanisms of PSID, a thick solid film of a mixture of N$_2$ and O$_2$ (N$_2$/O$_2$ = 1.35) was condensed at a temperature below 20 K and studied with PSID in the spectral regions of the N (K) and O (K) core-hole excitation.[47] In Fig. 7, spectra of N$^+$ and O$^+$ PSID yields are compared with the total photoelectron yield in the photon-energy range 390–590 eV. A small increase in the N$^+$ yield at the excitation energy of the O (K) may be observed, as well as a large increase in the O$^+$ yield at the excitation energy of the N (K). Thus Rosenberg et al. found evidence for an intermolecular transfer of excitation from the photoexcited nitrogen molecule to an oxygen molecule that issues the desorbing O$^+$, and vice versa. Intermolecular Auger decay of the N (K) core-hole excitation, as would be required by the normal PSID mechanism in this case, again seems unlikely. However, Rosenberg et al. found that XESD seems unlikely also, because of the unusually high ESD cross section, $\sigma \geq 10^{-19}$ cm^{-2}, which would be required to account for the yield rates observed in this study. These experimental results can be understood, however, in terms of photoelectron-stimulated dissociation of surface molecular ions, a mechanism which offers an explanation of previous indirect-desorption observations as well.

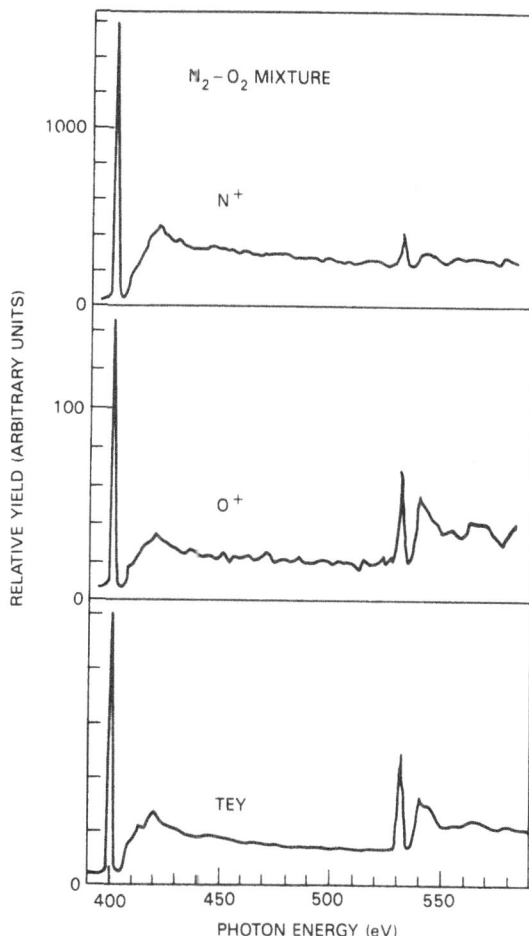

FIGURE 7. N^+ and O^+ PSID excitation spectra compared with the photo-electron total-yield spectrum for a mixed solid film of nitrogen (57%) and oxygen (43%) in the spectral range of the N (K) and O (K) edges. The resolution of the monochromator, using 20 μm slits, was 2.7 ev at 500 eV.

This new indirect mechanism is discussed in section 2.4.2, but first we consider the evidence against XESD for the case of the O_2–N_2 mixed solid.

Using previously published PSID spectra of N^+ from solid nitrogen and O^+ from solid oxygen,[45] the spectra of Fig. 7 were fitted to weighted-sum spectra of the pure solid films, which had been prepared in the same way. The fits are estimated to be accurate to only 20%, justifying the use of this simple additive model of desorption. The normalized "edge jumps," taken as the ratio of count rate 15–20 eV above the absorption edge to that just below the edge and normalized to incident photon flux, is given for all four cases in Table 1. Using calculated photoabsorption cross sections for gas-phase N_2[48] and O_2,[49] the yield spectra for both N^+ and O^+ ions above both the N (K) and O (K) absorption edges were renormalized to represent the number of desorbed ions per surface photo-ionization. These results are tabulated also in Table 1.

We adopt the viewpoint that N^+ desorbing as a consequence of O (K) excitation and O^+ desorbing as a result of N (K) excitation are purely indirect-desorption events, while N^+ desorption following N (K) excitation and

TABLE 1. Differential Increases in the Ion-Desorption Rates at the N (K) and O (K) Edges for a Mixed Solid Film of Nitrogen (57%) and Oxygen (43%)

Ion	Edge	Energy (eV)	Ions per incident photon[a]	Ions per surface ionization[b]
N^+	N (K)	420	2.5×10^{-6}	1.1×10^{-3}
O^+	N (K)	420	1.9×10^{-7}	8.1×10^{-5}
N^+	O (K)	550	8.5×10^{-7}	1.6×10^{-3}
O^+	O (K)	550	3.3×10^{-7}	6.6×10^{-4}

[a] Calculated assuming unity detector efficiency and using a photon-flux calibration taken subsequently with a National Bureau of Standards photodiode.[51]

[b] Calculated assuming 10^{15} molecules per cm^2 in a monomolecular layer, and using gas-phase photoabsorption cross-section data: 2.3×10^{-18} cm^2 for nitrogen at 420 eV,[48] 0.05×10^{-18} cm^2 for oxygen at 550 eV.[49]

O^+ desorption following O (K) excitation may result from either direct- or indirect-desorption processes. Above the O (K) absorption edge, the purely indirect N^+ desorption rate per surface ionization exceeds the total rate of O^+ desorption by a factor of 2.5, while above the N (K) absorption edge the purely indirect O^+ desorption rate is only about 7% of the total N^+ desorption rate. The O^+ desorption by indirect channels above the N (K) edge was only about 12% of the O^+ desorption by all channels above the O (K) edge. In contrast, the N^+ desorption was roughly the same at the two edges, a fact that cannot be explained entirely by appealing to a greater propensity for N^+ to desorb than for O^+.

If the indirect channel were attributable to XESD, then the ion-desorption yield Y would be calculable from the cross section for ESID, $\sigma_e(E)$, as a function of electron kinetic energy E and the photoelectron-energy-distribution function $N(E)$:

$$Y = C \int_0^\infty \sigma_e(E) N(E)\, dE, \qquad (2.4.1)$$

where C is a constant allowing for the efficiency of ion expulsion without reneutralization or recapture. $\sigma_e(E)$ data are not available for these molecular solids, but PSID results on these systems have been reported for energies less than 35 eV. Major thresholds in the PSID cross section $\sigma_p(E)$, for solid O_2, N_2, and NH_3 were found at energies of 20, 28, and 23 eV, respectively.[46,50] The maximum ion yields were on the order 2×10^{-8} ions per photon in each case.[45] The $N(E)$ spectrum resulting from photoabsorption usually peaks at about 5 eV and diminishes monotonically at higher electron energies, except for small peaks at higher energies corresponding to Auger electrons and primary photoelectrons. Therefore the lower O^+ desorption threshold from solid oxygen would favor O^+ yield via XESD in the O_2–N_2 mixed film, contrary to observation.

Having found evidence against XESD via low-energy photoelectrons, we will now discuss evidence against high-energy O (KLL) Auger electrons as a source of XESD in the O_2–N_2 mixed solid. From the measured photon flux, 7.2×10^7 photons/sec[51] and the molecular-oxygen K-shell photoabsorption cross section, 0.5×10^{-18} cm^2,[49] we estimate that 3.6×10^4 photoionizations occurred per second per molecular layer within the SR beam at $h\nu = 540$ eV. Each absorbed photon of energy above the O (K) edge generates one O (KLL) Auger electron, which has an energy, $E = 510$ eV, higher than the N (K) ionization energy, 400 eV. The N (K) electron-impact ionization cross section has been reported to be $<5 \times 10^{-20}$ cm^2.[52] If we assume that all Auger electrons created within ten molecular layers of the surface (about 35 Å) contribute to K-hole excitation of the surface nitrogen, then we obtain a hole-excitation rate for surface N (K) holes of <20 per/sec. Utilizing the largest ion-yield rate ever reported for PSID, 10^{-3} ions per surface excitation, the expected N$^+$ desorption rate from the O$_2$–N$_2$ mixed solid would be <0.02 ions/sec, corresponding to 3×10^{-10} ions per incident photon. In Table 1, the observed purely indirect desorption rate for N$^+$ aobe the O (K) edge is shown to be about 3000 times larger than this generous estimate. The evidence is against XESD as a significant indirect-desorption mechanism in the O$_2$–N$_2$ mixed solid.

2.4.2. Photoelectron-Stimulated Dissociation of Surface Molecular Ions

The evidence for the existence of an indirect mechanism for PSID is that, for some molecular surfaces, the PSID excitation spectra mimic the photoelectron-total-yield spectra and do not show the unique thickness dependence expected of direct PSID. We note that all molecular films from which an indirect PSID process has been suggested are insulating. In the SR environment, surface charging is likely to occur even though in the collection of positive ions, all emitted photoelectrons are driven back onto the surface by the strong, externally applied electric field. In a recent study of N$_2^+$ ions in a neon matrix, a surface charge of 25 to 35 V was observed.[53] If we assume a surface potential of 10 V on a film 1000 Å in thickness, then the surface-charge density must be about 10^{11} charges (ions)/cm^2. Molecular ions are easily dissociated by low-energy electron impact: For gas-phase N$_2^+$ and O$_2^+$, the cross section for dissociation by low-energy-electron impact has been reported as about 2×10^{-16} cm^2.[54] Combining these two numbers produces a predicted N$^+$ yield of 2×10^{-5} ions per incident electron, compared with 8.5×10^{-7} ions per photon from Table 1. Thus it seems likely that an appreciable ion yield should be expected from the surface of the mixed N$_2$–O$_2$ solid via photon-stimulated dissociation of surface molecular ions. (Avoiding acronyms such as PESDSMI, we refer to this mechanism as the "surface-ion mechanism" for indirect desorption.)

Another significant aspect of the data from the mixed N$_2$–O$_2$ solid was the larger indirect yield of N$^+$ at the O (K) edge compared to the indirect yield of O$^+$ at the N (K) edge. This fact can be accounted for by noting that, at the higher energy of the O (K) edge, there is still a significant photoionization cross section for the N (K) shell: about 0.6×10^{-18} cm^2.[55] N (K) hotoionization is expected to produce N$_2^+$ efficiency. However, the cross section for formation of O$_2^+$ ions at the N (K) edge via L-shell ionization is about ten times smaller.[56] Thus, the surface-ion mechanism accounts for all the facts observed for the indirect desorption from the mixed N$_2$–O$_2$ solid. In fact, the observations on the NH$_3$ and H$_2$O films on Ni can also be accounted for by similar reasoning with the surface-ion mechanism, as well as the observations of indirect H$^+$ PSID from water-dosed SrTiO$_3$.[57] In the latter report, the H$^+$ and OH$^+$ PSID was studied on fractured $SrTiO_3$ surfaces and on surfaces that had been ion sputtered and annealed. Again, spectral yields of PSID were compared with fluorescence and total photoelectron-yield spectra. In addition, arguments were made concerning the absence of $1s \Rightarrow 3d$ surface photo-excitations expected at the reduced-symmetry surface sites. All these observations appear to be consistent with the surface-ion mechanism of indirect PSID, although the surface resistivity of the specially treated $SrTiO_3$ was not reported.

In conclusion, indirect desorption or desorption from a site remote from the photoabsorption site, is indicated in some data on insulating solid molecular films or adsorbate layers such as ammonia, water, or atmospheric gases condensed at low temperatures. Of the physical mechanisms suggested to date, mechanisms involving photoelectron-stimulated dissociation of surface molecular ions fit the known facts best, although other variants of this process cannot be distinguished or ruled out from data available at this time.

2.5. Excitation Spectra and Photoabsorption

In previous sections, desorption has been shown to result from core-hole excitation either at the site of the desorbing ion or at a more remote site. It has been shown that specific excited states of the desorption site lead to ionic or neutral desorption of a surface or adsorbate species. In this section, we review briefly the core-hole photoexcitation process. The spectral dependence of PSD is determined by the photon-energy dependence of photoabsorption. The basic theory of electromagnetic absorption is reviewed in order to remind the reader of the important physical processes that control the PSD excitation spectra.

Photoabsorption by core electrons is a very complex physical process and has been studied for many years by many researchers. Several excellent reviews are available,[14,15,58,59] and we shall not present a full review here. For the understanding of PSD processes and applications, however, it is important that the physical principles of core-level photoabsorption be understood. Here we present a brief outline of relevant theory and refer the reader to more comprehensive reviews for additional details.

The response of matter to electromagnetic radiation is usually represented in the dipole approximation by the complex frequency-dependent refractive index

$$\bar{n}(\omega) = n(\omega) + ik(\omega), \tag{2.5.1}$$

or equivalently by the complex dielectric function

$$\varepsilon(\omega) = \varepsilon_1(\omega) + i\varepsilon_2(\omega) = [\bar{n}(\omega)]^2. \tag{2.5.2}$$

The optical-absorption coefficient, $\alpha(\omega)$, is linearly related to the extinction coefficient, $k(\omega)$;

$$\alpha(\omega) = \frac{4\pi k(\omega)}{\lambda}, \tag{2.5.3}$$

where $\lambda = nc/\omega$ Is the wavelength of the electromagnetic radiation within the material. In the dipole approximation, the response of bound, charged particles in matter to an electromagnetic field is taken to be a dipolar-polarizability density, while the response of free charged particles is taken to be a current density. Both can be complex, meaning that both energy storage and energy dissipation may be represented for either free or bound charges of matter: the former by the real part, $n(\omega)$ or $\varepsilon_1(\omega)$, and the latter by the imaginary part, $k(\omega)$ or $\varepsilon_2(\omega)$.

Our present interest is in the photoabsorption by matter, a process which dissipates electromagnetic-field energy within matter, and is included phenomenologically in $k(\omega)$ or $\varepsilon_2(\omega)$. In photoabsorption, transitions in the quantum state of the material system are induced by the electromagnetic field, with energy from the field being transferred to the material system. Because of the high frequencies associated with core-level excitations, the polarization response of materials is very small, meaning that $n(\omega) = 1 - \delta(\omega)$ is very nearly unity, and little energy storage takes place. Similarly, energy dissipation by free-charge currents is negligible, and most of the energy dissipation is via core-level excitation of matter.

According to Fermi's golden rule, the transition rate for core-level excitations, W_{if}, may be calculated from the matrix element of the electromagnetic Hamiltonian, H':

$$W_{if} = \frac{2\pi}{\hbar} \left| \langle f| H' |i \rangle^2 \, \delta(E_f - E_i - h\nu), \right. \tag{2.5.4}$$

where the material system is represented initially by the eigenstate $|i\rangle$ or eigenenergy E_i, and after the transition by the eigenstate $|f\rangle$ of eigenenergy E_f. $\hbar$ is Planck's constant divided by 2π. Note that the final state may differ from the initial state by single- or multiple-particle excitations. In the proper choice of gauge, H' may be written in terms of a vector potential, which in turn may be expressed in terms of the electron-momentum operator $\mathbf{p}$. The resulting expression for $\varepsilon_2(\omega)$ is

$$\varepsilon_2(\omega) = \frac{4\pi^2 e^2}{m^2\omega^2} \sum_{if} |M_{if}|^2 \, \delta(E_f - E_i - h\nu). \tag{2.5.5}$$

The matrix element of the transition may be written in the dipole approximation:

$$M_{if} \approx \langle f| \, \boldsymbol{\varepsilon} \cdot \mathbf{p} \, |i\rangle, \tag{2.5.6}$$

where $\boldsymbol{\varepsilon}$ represents the polarization (unit) vector of the electromagnetic field. Thus, the calculation of core-level photoabsorption spectra (and therefore PSD-excitation spectra) is reduced to the calculation of electric-dipole matrix elements for all possible core-level excitations and ionizations, with energy conservation as required by the delta function in Eq. (2.5.5).

In the energy range considered here, the momentum of the photon is very small compared to that of electrons in the material system. Momentum conservation requires that the momentum change in the material system be small; thus, in an energy-momentum diagram, the transition must be very nearly a vertical one. This fact is important in cases where the final state involves an itinerate electron in a crystal, for example.

Other conclusions of the general theory are important to note here. First, generally, one-electron excitations are dominant. The distinctive photoabsorption threshold associated with core spectra is usually very close to the threshold for one-electron excitations to the lowest unoccupied electronic quantum state, such as the conduction-band minimum is a semiconductor. Most multi-electron excitations require more energy than the fundamental one-electron core-hole excitation and, therefore, produce spectral features at higher energies than the one-electron threshold. Features at lower energies, caused by simultaneous phonon absorption, for example, are observed occasionally. Thus the initial interpretations of PSD-excitation spectra should be made, usually in terms of one-electron transitions.

Second, major features of core-level spectra can often be understood in terms of free atoms or molecules coupled weakly to the crystal or surface. Examples are the core-excitation threshold energies, the inner-well resonance behavior,[60-62] and the excitation of molecular excited states, including some Rydberg states.[42,46] Third, some of the most common multiple-electron excitations observed in core-level spectra are the core-exciton effects,[59] the shakeup effects,[63] "white" line absorptions,[59,64] high-angular-momentum coupling effects,[58,59] and electron-correlation[50,42] effects. Finally, a local approach to the calculation often provides a good description of itinerant photoelectron effects. This comment is discussed in more detail in sections 3.4 and 3.5 in connection with PSID-EXAFS and PSID-NEXAFS, respectively.

Based on the principles of the photoabsorption process, we may expect a rich physical content in PSD excitation spectra. This fact is well illustrated in the following section and in Ref. 42.

3. TECHNIQUES AND APPLICATIONS OF PSD

The major scientific questions of importance in the study of surfaces can be stated quite generally and simply:

- What atomic or molecular species reside on the surface?
- To what surface and subsurface species are the surface species bonded?
- What are the coordination numbers, bond lengths, and bond angles of surface species?
- What is the electronic structure of the surface site?

Indeed, if we knew the answers to these four questions, we would feel quite good about our understanding of the surface! In principle, PSD can answer each of these questions, providing an atom's-eye view of the surface. In the study of adsorbates, surfaces, and surface defects, PSD provides an unusual and surprising variety of experimental techniques. There are many other techniques, of both types I and II, which have been applied to answering the basic questions of surface science; they are reviewed elsewhere.[65–67] Among the most successful approaches has been photoelectron spectroscopy, which is treated in depth in this volume and other places.[68,69] In the following paragraphs, we describe and illustrate how the basic questions of surface science have been approached by PSD techniques and discuss problems that have occurred in experiment or interpretation.

3.1. Chemical Identification of Surfaces and Thin Films

One of the most definitive identifications of surface species is the mass spectrum of desorbing atomic and molecular species. An example is shown in Fig. 5. About a dozen atomic or molecular ions have been identified in that positive-ion PSID time-of-flight mass spectrum.[33] the sample is a polished and chemically etched GaAs (110) surface, which had been baked *in situ* to 500 °C. The mass spectrum was obtained during several minutes of illumination of the sample with broad-band, central-image radiation from the Grasshopper monochromator (Beam Line 1–1) at the Stanford Synchrotron Radiation Laboratory (SSRL).

The results shown in Fig. 5 are very similar to the early ESD results on "dirty GaAs."[70] Notably absent are the host species, Ga and As, and any ions heavier than potassium. PSID mass spectra of freshly cleaved GaAs, by contrast, show only H^+ desorption, which was attributed to dissolved hydrogen from the crystal-growth process and to ambient hydrogen in the ultrahigh-vacuum chamber.[33] Clearly, the PSID mass spectra do not represent everything on the surface. The sensitivity of PSID is not uniform over the periodic table. Among the ions identified in Fig. 5 are both molecular and atomic, both metal and ligand. However, species such as Ga or As that have multiple covalent bonds in the surface appear to desorb rarely if at all; species that are bonded to a single atom in the surface, such as singly valent covalently bonded or ionically bonded species, or isolated surface-impurity atoms (adatoms) or molecules are commonly desorbed. As illustrated in Fig. 6, F^+ is frequently observed, even from surfaces with fluorine concentrations far too low for observation with AES. Thus PSID mass spectra are not the ideal means to identify surface species; they are very sensitive to some types of surface species (especially hydrogen and fluorine) but quite insensitive to others. Although not many have been reported to date, PSD mass and excitation spectra of negative ions and neutral species may be expected to suffer a similar limitation. In section 2, we discussed some of the complicated physical processes of the excitation and expulsion of surface species by stimulated desorption. From that perspective, it may be unreasonable to expect that a quantitative assessment of surface populations might be possible using PSD data. Let us now examine what can be done.

With the use of the broad continuum spectrum available in SR beams, surfaces can be illuminated with radiation that will excite core holes in any atomic species of the periodic table. Subsequent Auger transitions are highly likely, leading to two-hole or higher excitations in most surface or adsorbate sites.

However, the probability of desorption from these excited precursor states is uneven. A simple mass spectrum of desorbing ions or neutrals will not facilitate the calculation of surface populations by any current theory. What can be said is that ionic desorption probabilities are higher for light atoms and lower for heavier atoms, higher for singly bonded species and much lower or zero for species with multiple bonding partners in the surface. The decreased sensitivity to heavy atoms is attributed to the increased probability of recapture or reneutralization for slowly moving desorbing ions.[71] With the intensity currently available in SR beams, it can be said that desorption of light ions, especially hydrogen and fluorine, should provide sensitivity to submonolayer coverages, perhaps to 10^{-5} of a monolayer in favorable cases. The sensitivity limits for desorption of neutral species are not yet well established, but appear promising.

In some cases, surface-species identification can be accomplished by the hydrogen-decoration method of PSID analysis. In this method, the surface under study is exposed to hydrogen in some form, and H^+ PSID excitation spectra are then observed. Hydrogen activated with an electron beam or a hot filament contains H and H_2^* in significant numbers,[72] greatly increasing its reactivity with many surface species. In this way, the high sensitivity to H^+ desorption is utilized for detection of surface species which otherwise might not desorb at measurable rates, as shown in Fig. 8.[38] The surface species are identified in the H^+

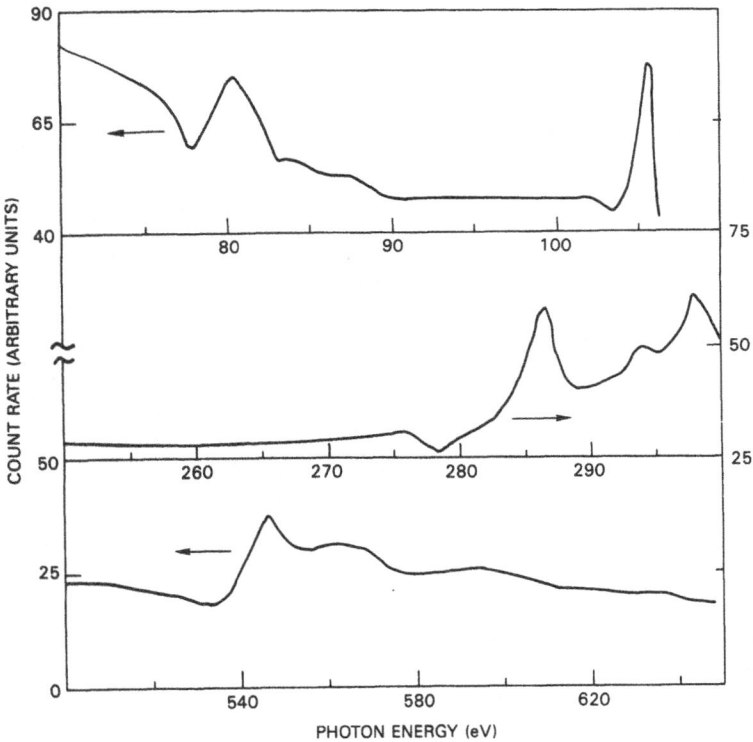

FIGURE 8. Characterization of a Nb surface by hydrogen decoration and H^+ PSID. Thresholds for H^+ desorption are observed near the Al ($L_{2,3}$) (chemically shifted to 77 eV), Si ($L_{2,3}$) (chemically shifted to 104 eV), C (K) (284 eV), and O (K) (532 eV). (From Ref. 38.)

PSID-excitation spectra via their characteristic core-excitation thresholds. Aluminum, silicon, carbon, and oxygen are identified as contaminants on the Nb surface of Fig. 8. Among these, only oxygen is observed directly in PSID mass spectra. Note that the hydrogen-decoration method still does not guarantee identification of all surface species: Species that do not adsorb or react with hydrogen remain undetected. Note also that exposure of a surface to hydrogen could induce chemical changes in the surface, and that the electronic configuration around the H^+-desorption site may not characterize the unhydrogenated surface. Of course, fluorine, oxygen, or other atoms or molecules may be used similarly to decorate the surface for the puprose of identification of surface species.

Hydrogen or oxygen decoration may be coupled with sputter-depth profiling to determine the thickness and chemical profile of overlayers that do not themselves desorb. This is illustrated in Fig. 9 for a Nb_3Ge surface which was coated with a 30-Å film of oxidized amorphous silicon (a-Si).[38] Although Si^+ desorption was not observed, H^+ and O^+ PSID were observed upon excitation of the Si ($2p$) core hole. In this research, the question to be answered was whether the oxidation process produced a uniformly oxidized SiO_2 layer suitable for a tunnel barrier. As sputtering proceeded, the Si ($2p$) edge features in the H^+ and O^+ PSID excitation spectra were monitored. As shown in Fig. 9, the ratio of PSID ion yields was not constant, indicating that the oxidation process, which involves water vapor, did not produce primarily a hydroxide of Si. However, the chemical shift of the Si ($2p$) edge was a constant +5 eV, a value normally associated with SiO_2. It was concluded that the desorption sites were the same throughout the sputtered thickness, and probably the entire SiO_2 layer.

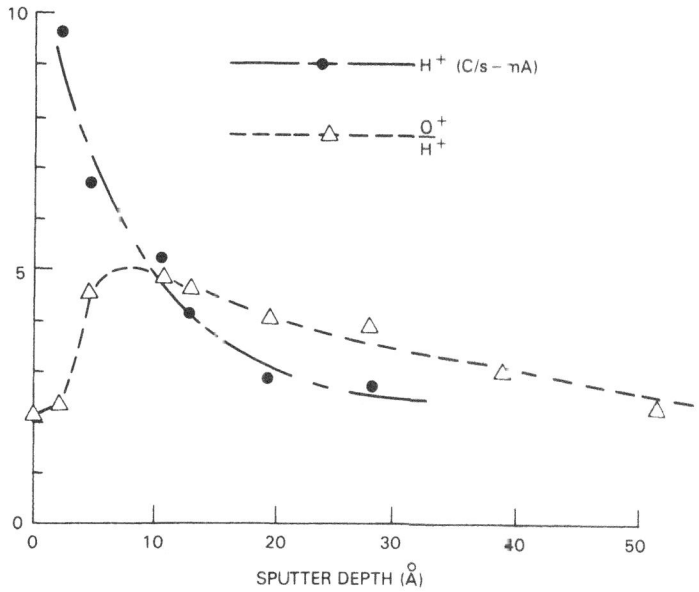

FIGURE 9. PSID depth profile in a 30-Å thick film of oxidized amorphous silicon on Nb_3Ge. Solid curve: normalized H^+ PSID yield. Dashed curve: ratio of O^+ count rate to the concurrent H^+ count rate. (From Ref. 38.)

In a study of frozen, amorphous H_2O, the PSID mass spectra showed only H^+ desorption: It was a mystery why no molecular ions such as OH^+ or H_2O^+ were observed in desorption.[73,74] Reasoning from the possible electronic states of the H_2O molecule and possible crystallographic site geometries, it was shown that only H^+ desorption is expected from crystalline or amorphous solid films of molecular H_2O.

From these examples, some of the power of the PSID mass spectra in determining the chemical nature of surfaces and thin films has been illustrated. Also, it was indicated that additional chemical information is obtainable from the PSID excitation spectra. This is discussed in more detail in the next section.

3.2. Surface-Bonding Partnerships and Chemical Associations

In PSID experiments, the identification of surface-bonding partners associated with desorbing ions is straightforward in cases where desorption occurs by the normal PSID model. If surface atom S is bonded to host atom H, then excitation of core holes in either atom S or H may stimulate desorption of S^+, as discussed in section 2. Hence, finding a desorption threshold for S^+ at a core-excitation threshold characteristic of atom H indicates that atom S was bonded to atom H before desorption. In Fig. 10, the identification of bonding partnerships is illustrated for a water-dosed $SrTiO_3$ surface.[75] The PSID mass spectrum shown in Fig. 10(a) was stimulated by 30 eV photons, and shows a strong peak at mass/charge = 17 (OH^+). In Fig. 10(b), the monochromator was set for $hv = 35$ eV, and the mass-17 peak is joined by a peak at mass/charge = 16 (O^+). Thus the O^+ desorption is shown not to occur below the threshold for Ti ($M_{2,3}$) excitation (34 eV), whereas OH^+ (and H^+) desorption follows excitation of either of two higher-lying core levels: Sr ($N_{2,3}$) at 22 eV or O (L_1) at 24 eV. From this study it was concluded that H_2O dissociates on the $SrTiO_3$ surface, with the hydroxyl radical bonding (oxygen down) to either Sr or O surface atoms. The fact that no O^+ desorption is observed following excitation of the O (L_1) is interpreted to indicate that no O atoms exist on the water-dosed surface except for those bonded to Ti, and excitation of OH species stimulates only H^+ desorption. (Note that if adequate normalization is built into the experiment, the comparison shown in Fig. 10 can be obtained automatically as the monochromator is scanned through various core-level energies.)

Other examples are shown in Fig. 5 of Chapter 2 and Figs. 1 and 2 of Chapter 3. The implicit assumption is that excitations of nearest-neighbor atoms only will stimulate desorption, with no likelihood of desorption by indirect mechanisms.[76] As discussed in section 2, among the PSID studies reported to date, this assumption appears questionable only for insulating molecular surfaces. To be safe, it must be considered in the interpretation of excitation spectra for surface-bonding partnerships.

Another factor to be considered in the interpretation of desorption–excitation spectra is the possibility of delayed onset of photoabsorption.[15,77,78] In photoabsorption spectra, delayed onset is associated with a severe mismatch of the oscillation periods of the wave functions of initial and final electronic states. In the calculation of the transition probability, Eq. (2.5.4), for the electronic

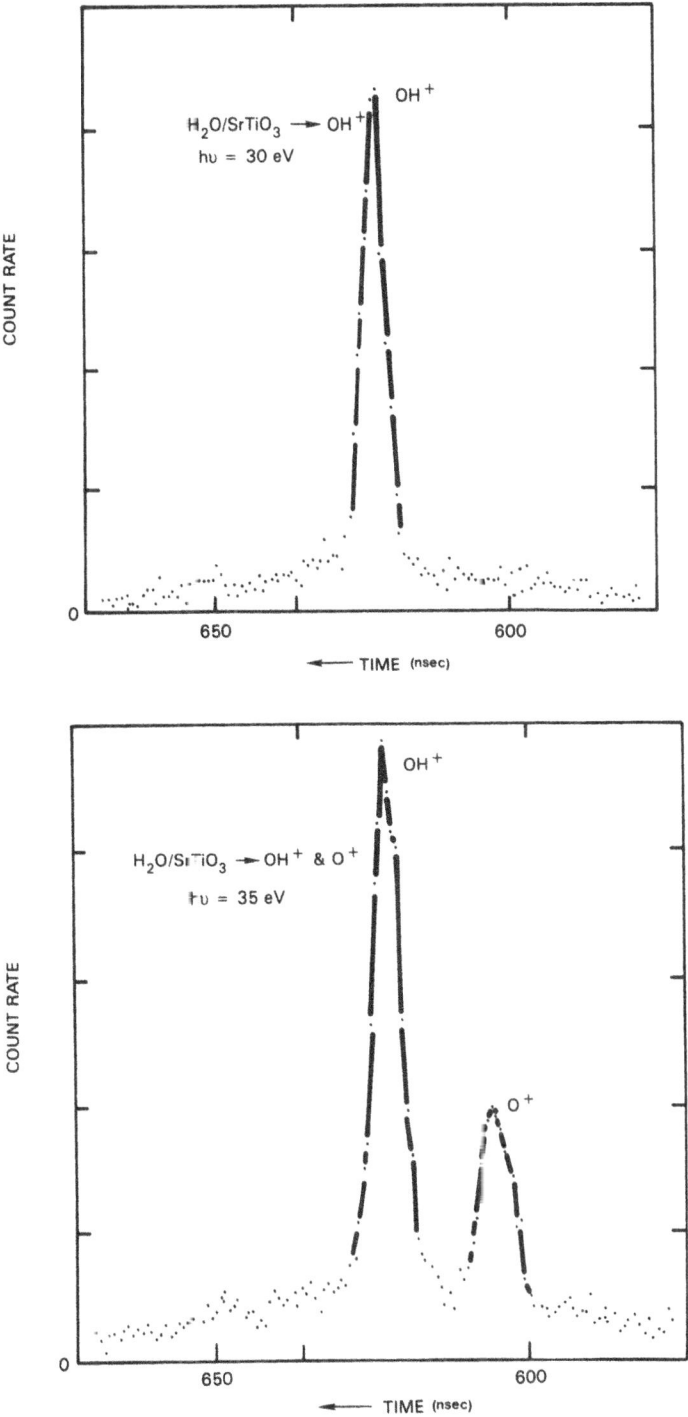

FIGURE 10. PSID mass spectra of H_2O on a $SrTiO_3$ surface: (a) 30-eV photon stimulation and (b) 35-eV stimulation. O^+ s desorbed from Ti ($M_{2,3}$), but not from Sr ($N_{2,3}$) or O (L_1) excitations. H^+ and OH^+ from the dissociated H_2O adsorbate molecules desorb from the lower-energy thresholds, however. (From Ref. 75.)

photoabsorption, the electric dipole matrix element, Eq. (2.5.6), will be small if the initial state has a much shorter spatial-oscillation period than the final state in the spatial region of greatest overlap. This occurs when the initial state is an atomic state of high angular momentum (d or f state) and the final state is a conduction-band state of nearly zero kinetic energy, for example. The net result is that photoabsorption (and therefore PSD) will be a weak at the threhsold for such a transition. At higher energies the photoabsorption constant, $\alpha(\omega)$, Eq. (2.5.3), increases gradually as the kinetic energy of the final state increases, thereby decreasing its spatial oscillation period to match more closely that of the initial state.

An example of PSD-threshold obscuring attributed to delayed onset of photoabsorption is shown in Fig. 11.[38] In these experiments, sputtering was used to remove a thin overlayer and desorption was monitored to determine the appearance of the underlying Nb surface. It was observed that spectral features associated with the Nb ($M_{2,3}$) edge, shown in Fig. 11a, began to appear at the same sputter depth as the delayed-onset O^+ and H^+ desorption above the Nb ($M_{4,5}$) edge, shown in Fig. 11b. It may be noted in Fig. 11 that the Nb edge features in the OH^+ PSID spectra are appreciably weaker than those in the O^+ and H^+ spectra. Because these data were concurrently taken in a single spectral

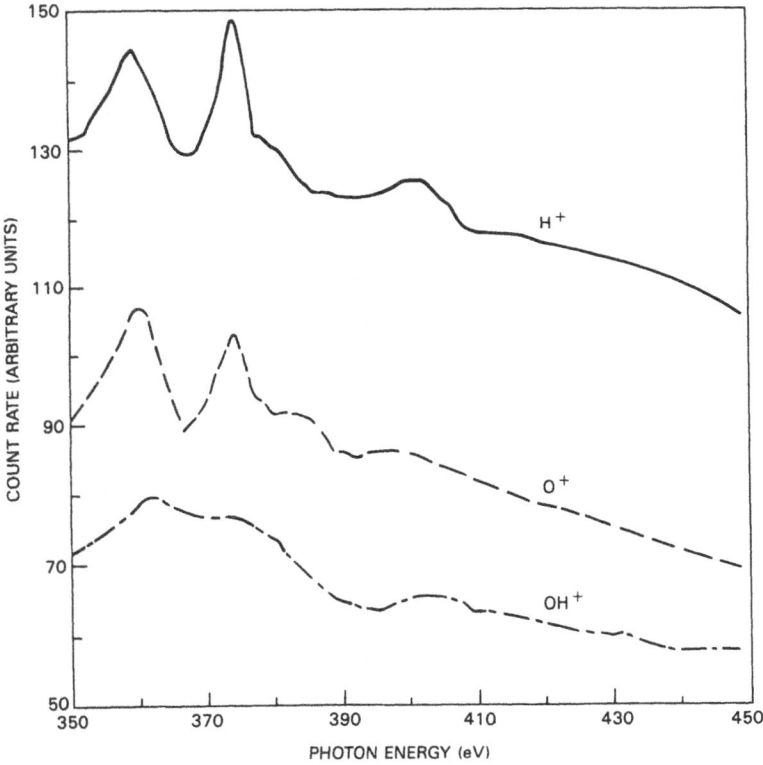

FIGURE 11. Delayed onset in PSID: H^+, O^+, and OH^+ PSID excitation spectra of a sputtered Nb film in the spectral region of (a) the Nb (M_2) and Nb (M_3) edges and (b) the Nb ($M_{4,5}$) edges. (From Ref. 38.)

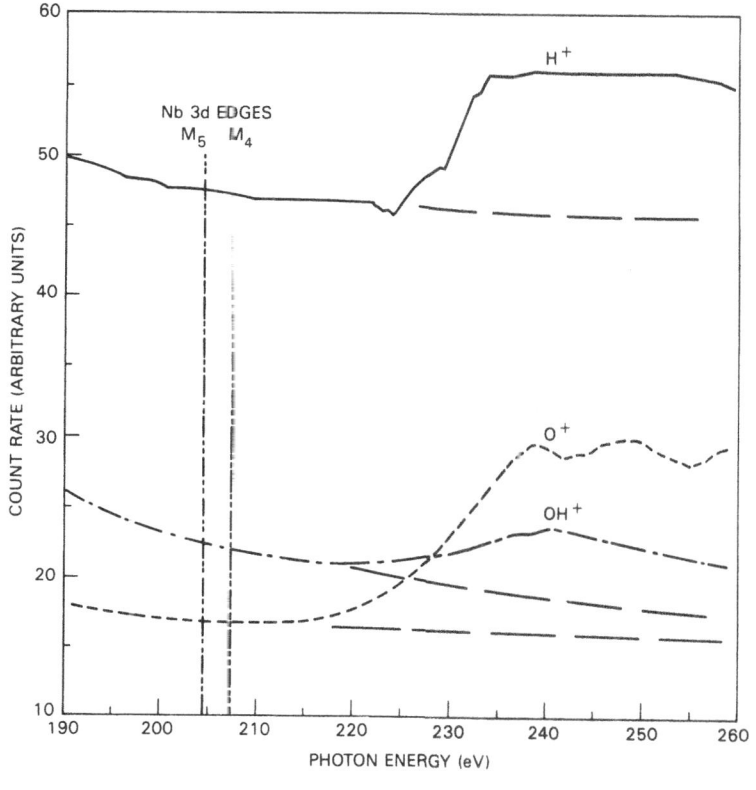

FIGURE 11. (*Continued*)

scan, many experimental uncertainites are cancelled in a comparison of ion yields, and these spectral differences are considered experimentally significant in spite of the weakness of the desorption signals. The evidence is that Nb–OH bonds were rarer than the Nb–O or Nb–H bonds on this sputtered thin-film surface, although the unknown branching ratio of OH$^+$ and H$^+$ desorption following Nb core-hole excitation at the Nb–OH site has not been accounted for.

An interesting study has been reported of the role of hydrogen in the surface reconstruction of diamond.[79–81] The H$^+$ PSID from excitation of the C ($2s$) or C ($1s$) core holes was monitored and compared with photoelectron spectroscopy and LEED. It was found that the $2 \times 2/2 \times 1$ reconstruction of the (111) surface, obtained after heating for several minutes at 950°–1000 °C, is free of hydrogen, while the 1×1 reconstruction observed at lower temperatures and on as-polished surfaces is hydrogen-terminated. The photoemission results showed that a surface state is formed about 2.5 eV below the valence-band maximum on the hydrogen-free $2 \times 2/2 \times 1$ reconstructed surface. Reproducibly, this surface state is destroyed and the H$^+$ yield is restored by exposure to activated hydrogen. Furthermore, the hydrogen is robustly bonded to the carbon, as heating to 600 °C has little effect on the H$^+$ PSID.

The study of the interaction of water with the surface of Si is another interesting application of PSID to the determination of surface-bonding partnerships.[82] In this work, an acid-cleaned Si (111) wafer was rinsed finally in

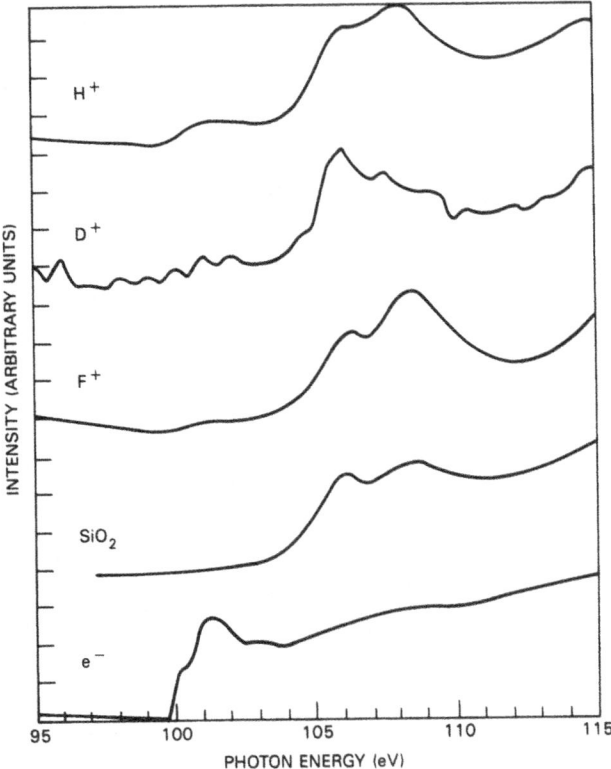

FIGURE 12. PSID from water-dosed Si (111). The acid-cleaned Si (111) sample was dipped in heavy water, D_2O, before introduction into the vacuum chamber: (a) shows the PSID spectra after a 500-K bake, compared with the total photoelectron yield (e^-) and SiO_2 spectra; (b) shows the spectra after heating to 800 K, compared with the 500-K data.

D_2O before introduction into the vacuum chamber and then was heat cleaned *in situ.* After mild heating to 500 K, PSID excitation spectra near the Si ($L_{2,3}$) core threshold stimulated desorption of H^+, D^+, and F^+ ions, and the chemical shifts (+5 eV for all) indicated that all three species had issued from SiO_2-type sites, as shown in Fig. 12a. (The total photoelectron-yield spectra of the sample, representing the underlying bulk Si, and of SiO_2 are shown for comparison.) The SiO_2 site type was confirmed by studying the H^+ desorption spectrum at the O (K) edge and comparing it with the total photoelectron-yield spectrum, from which a chemical shift of +2.6 eV was measured. In Fig. 12a, the spectra of the H^+ and D^+ desorption are different, indicating isotope-related differences between the Si–H and the Si–D sites. Upon further heating to 800 K, the H^+ and D^+ desorption spectra became quite similar, indicating similar site structures for both hydrogen isotopes, as shown in Fig. 12b. After further heat cleaning at 1150 K, which is known to produce a sharp 7 × 7 LEED pattern on Si (111), the desorption spectrum showed only H^+ desorption, and that spectrum showed no chemical shift from the Si ($L_{2,3}$) energy observed in the total photoelectron-yield spectrum. This residual hydrogen was shown to originate from the residual environmental gases in the vacuum chamber, even though the total pressure was

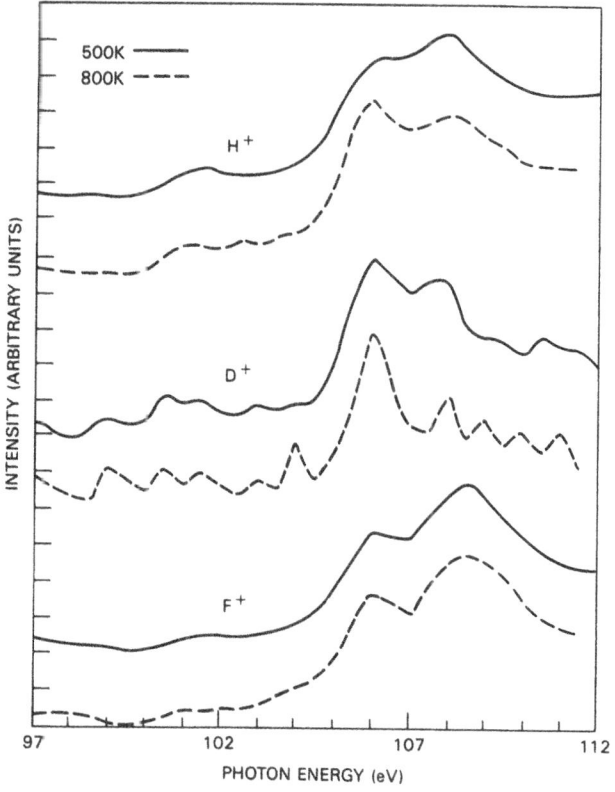

FIGURE 12. (*Continued*).

less than 2×10^{-10} Torr. Subsequent exposure of the clean-Si surface to heavy-water vapor showed that water vapor dissociates on the clean-Si surface, bonding as Si–H and Si–OH. Such studies as these show the utility and value of studies of surface-bonding partnerships and the high intrinsic sensitivity of PSID to surface hydrogen.

3.3. Surface-Bond Angles

The classical method of utilizing desorption for the study of bond angles is the ESDIAD technique.[83-86] In these experiments, an electron beam of variable energy is incident on the surface to produce ESD. Concentric around the surface point irradiated by the electron beam are two spherical grids. The first is at ground potential and the second at high potential, for the purpose of accelerating the desorbing ions radially outward along their desorption path. After passing the second grid, the ions are further accelerated onto a position-sensitive ion detector. (Further experimental details are given in section 4.1.) In this way, the polar and azimuthal angles of emission of the ion are obtained. According to the usual interpretation, the ion is expelled from the surface azimuthally along the direction of the surface bond broken by the electronic excitation, but the polar angle is increased somewhat from the surface-bond direction by image-charge force.[87,88] Many adsorbate systems have been studied this way, showing distinct temperature-dependent orientational patterns in some cases.[89] Comparison with the results of other techniques such as LEED, electron-energy loss, IR absorption, photoelectron diffraction, and angle-resolved ultraviolet photoemission have shown good agreement in most cases, allowing assignment of specific surface structure and site identification. Similar experiments have been reported using photons instead of electrons, a technique which may be called PSDIAD.[90–93] In cases where both ESDIAD and PSDIAD have been done on the same system, comparable results have been obtained. The main advantages of PSDIAD lie in the selectivity of the photon excitation and its polarization selection rules.[84] The selectivity of photon excitation allows ion angular distributions to be associated with specific desorption-precursor states. The polarization selection rules requires excitation of only directed-valence-bond orbitals oriented parallel to the polarization vector. No reports of surface-bond-angle measurements based on polarization-dependent PSDIAD have appeared to date.

3.4. Surface-Bond Lengths and Coordination Numbers

In bulk matter, local-bond lengths and coordination numbers may be obtained from the EXAFS spectrum if the photoabsorbing atom is distinct from its neighbors, such as the iron atom in a hemoglobin molecule, for example.[94] The technique involves analysis of the small modulations observed in photoabsorption spectra far above a core-absorption threshold. These modulations result from interference in the outgoing photoelectron wave function produced as it is scattered by the electrons on near-neighbor atoms.[63,69,95,96] Because the probability of multiple scattering of the outgoing electron diminishes rapidly with increasing kinetic energy, analyzing the X-ray absorption spectrum in the range 50–500 eV or more above the absorption edge emphasizes the nearest-neighbor single-atom scattering and permits deduction of the bond lengths and coordination numbers for the nearest atomic shells. Bond lengths may be deduced to an accuracy of ± 0.01 Å in favorable cases.

Techniques of EXAFS have been developed and reported for the study of surfaces also.[97–99] In the usual surface EXAFS (SEXAFS) technique, photoelectrons are detected as the photon energy is scanned above an X-ray absorption edge. The fine structure in the photoelectron-yield spectrum represents the X-ray absorption (and therefore the EXAFS) for the region from which the photoelectrons originate.[100] If photoelectrons with kinetic energy in the 50–150-eV range are detected, then the photoyield-SEXAFS spectrum represents the near-surface region within the minimum electron-escape depth of 0.5–1.0 nm, or two to four atomic layers.[101,61] Increased surface sensitivity can be obtained in SEXAFS experiments if adsorbate-core excitations are used.[102] In this case, the background of electrons from the bulk is subtracted by extrapolation of the photoelectron-yield spectrum from below the adsorbate-core threshold. Alternatively, two or more incidence angles might be used, and the contributions of bulk and surface regions to the photoelectron-excitation spectra might be deconvolved, using the known variation in the photon-penetration depth with incidence angle. Because of the weakness of the EXAFS undulations, however, deconvolution utilizing multiple spectra is difficult.

Other surface-absorption-dependent spectra have been used for SEXAFS detection. Both X-ray and visible luminescence-excitation spectra, as well as electron-energy loss spectra and reflectivity spectra, contain EXAFS information—each technique carrying varying degrees of surface and bulk sensitivity.[103] In the most surface-sensitive SEXAFS experiments, bond-length and coordination-number data representative of the outer few atomic layers may be obtained, utilizing well-established EXAFS data-processing and interpretation methods.

In the PSID excitation spectrum above a core-excitation threshold, fine structure is observed that originates in the photoabsorption of a specific desorbate-host bonding site within the outermost one or two atomic layers. Because the PSID excitation-energy dependence is controlled by the photoabsorption spectrum of the desorption site, the observed fine structure is again the EXAFS spectrum and contains bond-length and coordination-number information relevant to the photoabsorption-site locale.[1,104] Thus, in principle, an inhomogeneous surface (i.e., one consisting of various types of host atoms, impurities, adsorbates, and imperfections) may be examined experimentally by studying the PSID-EXAFS spectrum of each surface atom individually. For each site that is characterized by a unique desorbing ion (e.g., an impurity) or by a unique surface-bonding partner, a characteristic PSID-EXAFS spectrum should be obtainable and interpretable in terms of local-bond lengths and coordination numbers.

The first report of PSID-SEXAFS was a study of oxygen adsorbed on Mo (100). Using photon energies above the Mo L_1 core excitation at 2870 eV in the 3000-to-3600–eV range. Jaeger and coworkers compared the total-photoelectron-yield EXAFS spectrum (representative of the bulk Mo site) with the O^+ PSID-SEXAFS spectrum, representative of the Mo–O surface sites.[28] They showed that the two sites have the same Mo–Mo nearest-neighbor distance, 2.73 ± 0.03 Å, and differ only in coordination number (eight for the bulk and four for the surface). It was unexpected that the Mo–Mo nearest-neighbor distance in the surface should be the same as the bulk because of the oxide-like 2.9 eV chemical shift of the Mo ($M_{2,3}$) absorption edge previously reported for the oxygen-coordinated Mo surface. Because of the large difference in the atomic scattering factors of Mo and O, and because of the interference of the Mo (L_3) and Mo (L_2) EXAFS with the Mo (L_1) EXAFS spectrum, the Mo–O bond length was not determined. However, Jaeger and coworkers attribute the lack of reconstruction of the Mo surface to a highly ionic Mo–O bond, as opposed to a more covalent Mo–O bond found on the low-coverage surface phase, in which the O is believed to occupy a fourfold hollow site on the Mo (100) surface.

A study of amorphous H_2O and D_2O ice films utilized both PSID and photoelectron-yield EXAFS to obtain more details of the electronic structure of the photoexcited H_2O molecules.[74] In these molecular films, the difference between surface and bulk molecular sites is expected to be minimal, since the intermolecular perturbation is small even in the bulk solid. The EXAFS result first confirmed the theoretical value of the O–O phase shift,[105] which had not been reported for atoms with $Z < 11$. Next, a comparison of the near-edge PSID structure to the gas-phase EELS spectrum was used to identify differences between the PSID and the photoelectron total-yield spectra: PSID spectra, representing the surface sites, show some structure related to the Rydberg states that are observed in gas-phase H_2O, but the photoelectron total-yield spectra showed no hint of the Rydberg states at bulk sites. The conclusion was that some of the Rydberg states (which have large orbitals) survive on the surface H_2O molecules, while all Rydberg states are quenched in the bulk H_2O film by intermolecular perturbations, which will be much stronger for the larger Rydberg orbitals than for other, smaller molecular orbitals.

A recent study of water-dosed Si (100) reported the application of PSID-SEXAFS for the purpose of resolving several previously suggested surface-site models.[106] The excitation spectra of H^+ desorption following Si (K) core-hole excitation were recorded and compared with photoelectron total-yield spectra, which represent the bulk Si. Bond lengths for the three nearest-neighbor shells were deduced for surface and bulk sites, along with coordination numbers, for comparison with theoretical models. It was found that the model of Chadi[107] fit the data within experimental error, while four other models did not. In these examples, PSID-SEXAFS has been shown to be a valuable technique for the deduction of local-site bond lengths and coordination numbers. Additional site details, including electronic structure, are obtainable by studying the near-edge spectra, which are ignored in SEXAFS.

3.5. The Complete Local-Surface-Site Geometry

In the previous section, PSID-SEXAFS was discussed as a means of determining the bond lengths and coordination numbers for specific sites on surfaces, such as surface-impurity sites. There it was mentioned that the portion of PSD-excitation spectra within 40–50 eV of the core-excitation threshold is disregarded in EXAFS interpretations because of the effects of multiple scattering of the outgoing photoelectron among neighboring atoms. Although more difficult to model or calculate, multiple scattering creates in the photoabsorption spectra the positions of all near-neighbor atoms involved in the scattering of the photo-electron. In principle, photoabsorption spectra near the core-excitation threshold (and therefore PSID-NEXAFS spectra) contain information not only on the local bond lengths and coordination numbers but on the complete geometry of the photoabsorption locale. Added to this prospect is the experimental fact that spectral features in this near-edge region are much stronger than the weak, fine-structure undulations in the EXAFS spectral range.

An example of PSID-NEXAFS spectra in comparsion with total photoelectron yield (PEY) is shown in Fig. 13 for InP (110).[108,109] At the P (K) edge, the surface-sensitive PSID-NEXAFS spectrum is quite similar to the bulk-derived PEY spectrum. In contrast, at the In (L) edges, the two spectra are quite different. In the conventional model of the (110) surface construction of III–V semiconductors, the surface In–P bond is tilted by a considerable angle (22°–28°) out of the (110) plane (in which it lies in the bulk crystal) such that the surface P atom gains electron density and becomes more like a free P atom.[107] At the same time, the surface In atom loses electron density, thereby becoming more like the free In

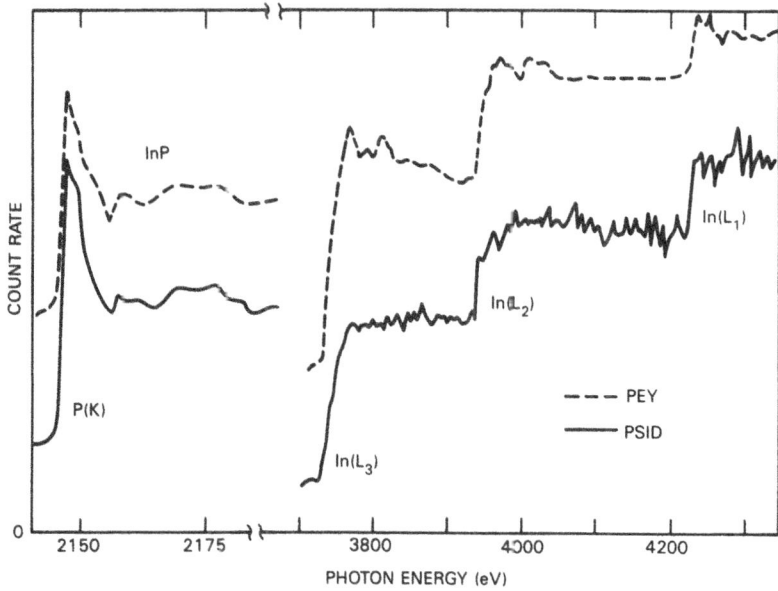

FIGURE 13. PSID-NEXAFS spectra of cleaved InP (110) at the In (L) and P (K) edges, compared with photoelectron total-yield (PEY) spectra. (From Refs. 108 and 109.)

atom. Presumably, these differences in the surface crystal structure and the associated differences in electron distribution among atomic sites generate the differences in the surface and bulk NEXAFS spectra illustrated in Fig. 13. If it is possible to deconvolve the NEXAFS spectra for both surface and bulk sites, the comparison should yield a highly detailed description of the desorption site's geometric and electronic structure.[110]

The study of near-edge spectra is referred to as XANES or NEXAFS spectroscopy. Early applications were directed toward developing "fingerprints" of representative local structures to be used in interpreting NEXAFS spectra.[61] These were applied to the PSID excitation spectra of BeO, Al_2O_3, and SiO_2 with only partial success.[60] Calculations that include multiple scattering among enough nearby atoms to reproduce experimentally observed spectra have been reported[111] and applied to the study of adsorbed-oxygen sites on Ni (100).[112] One of the hollow sites calculated fits the experimental NEXAFS spectrum quite well, while the other sites calculated do not fit at all. Thus NEXAFS spectra seem to be interpretable in a site-specific way for atomic adsorbates on metal surfaces, although convergence of the calculation requires inclusion of as many as 30 near-neighbor Ni atoms.

Considerable success has been reported in the NEXAFS studies of molecular adsorbates, such as CO on Ni (100) or on Cu (100). In molecular systems, often the NEXAFS spectra are dominated by intramolecular scattering and even, in some cases, by atomic features.[13] These strong atomic and molecular effects in the NEXAFS spectra may obscure adsorbate–substrate interactions.[113] Possible additional difficulties may lie in the adequate inclusion of many-electron excitations in the calculations (see section 2.5). With the identification of sources of difficulty in the NEXAFS calculations, further progress should be forthcoming in realizing the full potential of NEXAFS spectroscopy. No combined experimental and theoretical PSD-NEXAFS studies have been reported as yet. It can be expected, however, that both the promise and the problems discussed above will be found also in the calculation of site-specific PSID-NEXAFS spectra.

3.6. Surface-Defect Studies

In the original report of PSID, it was noted that the TiO_2 surface had been heated and lightly sputtered (with 500-eV argon ions for a few minutes) before observation of PSID, a technique found effective previously in ESD studies.[1] This surface treatment generates a small density of surface defects, as well as activating electronically active (dopant) impurity atoms to increase the conductivity of the sample. Later studies of cleaved GaAs (110) showed little desorption without sputtering or other "activation" of the surface, as mentioned in section 3.1.

On the other hand, the sputter-profiling study discussed in section 3.1 showed that the PSID ion yields from a thin, 30-Å layer of oxidized amorphous silicon (a-Si) diminished with sputter time, although the surface-defect density is assumed to increase with sputter time. This is illustrated in Fig. 14.[38] Both the Si ($L_{2,3}$) edge strength and the (subtracted) background decreased by more than a factor of two within the first fifteen seconds of sputtering and continued to diminish monotonically up to the total sputter time of 360 sec for the experiment. The sputtering process utilized a large-area rastered beam of 500 eV Ar^+ ions. By all measures, the PSID yields diminished significantly with sputter time: evidence that the desorbing ions were not issuing primarily from sputter-induced surface-defect sites. The role of surface imperfections in PSID and ESD continues to be poorly understood.

Recently a systematic study of the role of surface defects in PSID has been reported. Utilizing both PSID and ESDIAD, Kurtz studied TiO_2 (001) surfaces which had been sputtered or heat-treated to temperatures as high as 1200 K.[89] It was shown that low-temperature annealing, ≤ 700 K, increases the positive-ion yield for both surface studied, but higher-temperature annealing produces marked differences in the behavior of the two surfaces. The ion yield from the (001) surface became independent of further annealing, but that from the (110) surface decreased with higher-temperature annealing. ESDIAD patterns are different on the two surfaces: The (110)-surface pattern shows a complex annealing behavior. Kurtz interprets his results as indicating that desorption occurs primarily from facet edges on the (001) surface and from step edges on the (110) surface. High-temperature annealing of the (110) surface reduces the step density on the surface, resulting in a reduced ion yield as observed.

For the case of sputtering defects, a comparison of NEXAFS spectra from the oxidized a-Si surface discussed above is shown in Fig. 14.[38] H^+ PSID spectra from the Si ($L_{2,3}$) edge taken at various times during the sputtering process are shown, with a constant background subtracted from each curve. Within the limits of the signal-to-noise ratio of the data, the Si NEXAFS spectrum remains unchanged as sputtering continues (except for amplitude), a somewhat surprising result. It

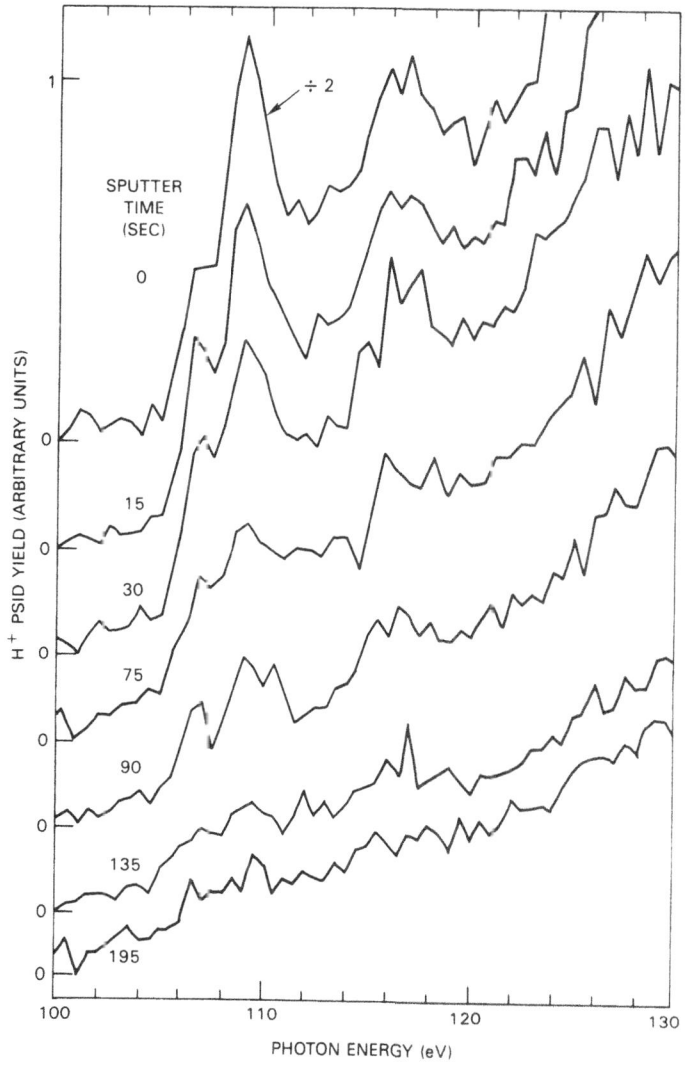

FIGURE 14. Changes in the PSID-NEXAFS spectra while sputtering. The sample was an oxidized film of amorphous silicon, 30 Å thick, on Nb_3Ge. Sputtering utilized 500-eV Ar^+ ions in a large-area rastered beam. (From Ref. 38.)

might be expected that sputtering would induce a variety of modified Si–H surface sites, each of which would show its own characteristic NEXAFS spectrum. A superposition of possibly several different NEXAFS spectra observed from the sputter-damaged surface might be expected to show less definition and other spectral changes. It must be concluded that the surface erosion via 500 eV Ar^+ ions does not significantly change the structure of the Si–H sites from which H^+ ions are desorbed.

Studies of defective surfaces may provide significant advances in the understanding of surface-defect structure, both geometrical and electronic, in future years. Defect geometrical structures are closely related to the electronic structure of the defect site. In systems where desorption from a single type of defect site can be obtained, the PSID-NEXAFS spectra will yield electronic-state information, represented in the photoabsorption spectrum of each site as discussed in section 3.5. The ESDIAD or PSDIAD patterns will give confirming geometrical information on which model calculations may be based. In this way, the chemical bonding arrangement and electronic structure of the imperfection site may be obtained.

3.7. Desorption of Neutral Species

It is generally agreed that the desorption of neutral species from surfaces as a result of photoabsorption is more probable than ion desorption; estimates range from one to three orders of magnitude more probable.[114] Although sensitive detection of desorbing neutral species is appreciably more difficult than for ions, the complexity of the physical process involved foretells a productive future for studies of PSD of neutrals.

It has long been known that alkali halide surfaces erode under photon or electron irradiation, but the physical mechanisms remain poorly understood. Excited neutral species desorbing from surfaces have been studied by observing their characteristic (gas-phase) resonance-radiation spectra[114,115] as a function of the exciting photon or electron energy and fluence. Results of such studies have shown clearly that core-level excitations are involved in the energy-transfer mechanism, and that previous theories involving the production and migration of defects and the ejection of ions are inadequate to describe the observations.

Neutral species issuing from a standard copper sample have been analyzed using a technique called surface analysis by laser ionization (SALI).[116] Although the initial experiments utilized a 2.7-keV Ar^+ beam to sputter small quantities of neutral species, ESD or PSD could be used as well. The desorbing neutral species were ionized nonresonantly by multiphoton absorption or KrF* excimer radiation (photon energy = 5 eV), 10–15 nsec pulses of about 10 mJ per pulse. The resulting ions were accelerated into a reflection time-of-flight mass analyzer. Both high sensitivity and high resolution were reported: The mass-resolving power of the detector was reported to be $M/\Delta M = 500$ at $M/e = 200$, and its sensitivity to parts per million with 10^{-10} g of sputtered neutral material.

The use of time-of-flight kinetic-energy analysis of the desorbed neutral species followed by laser-resonance ionization and ion detection has been reported,[117,118] and applied to the study of NO or CO adsorbed on Pt (111).[119] These studies have revealed that neutral NO species can be desorbed via a process of freeing the hindered rotation of a surface NO. In contrast to the case of CO/Pt (111), the kinetic-energy distribution of the desorbing NO has a peak at a low 50 meV due to this desorption mechanism, in which the exciting electron beam induces a transition from a hindered-rotor to a free-rotor state of the adsorbed NO. This is followed by a second transition to the ground continuum state.

Another recent result that illustrates the scientific importance of studying neutral-species desorption is excited-atom production by electron bombardment of alkali halides.[120] In this work, laser-induced fluorescence was used to detect desorbing neutral species and to analyze their velocity distribution via Doppler shifts. Na atoms excited to the $^3P_{1/2}$ state by the incident electron beam are further excited resonantly to the $^4D_{3/2}$ state by a dye laser. The cascade $4p \Rightarrow 3s$ fluorescence emission at a wavelength of 3303 Å is detected. By varying the dye laser's frequency, the particular Doppler-selected velocity of the emitting Na atom is varied, permitting measurement of the velocity spectrum of desorbing Na atoms.

Although the latter two examples used ESD techniques, it is clear that similar studies of neutral desorption via PSD would provide new views of the physics of neutral-species desorption with the added features of the photon probe discussed in sections 1 and 2.

4. EXPERIMENTAL DETAILS

In this section we discuss the apparatus of PSD and some of the experimental concerns, most of which are shared with other surface-analysis techniques. In the previous sections, we have indicated the importance of attributes of SR photon beams to the techniques of PSD. These attributes are well known and have been reviewed several times previously. For example, see in various articles in the books listed as WD80 and Ko83 in the reference list. The SR continuum spectral coverage makes it possible to excite core-holes in any atom of the periodic table, either selectively using monochromatized SR or nonselectively using broadband SR. Because of the small cross sections for PSID from most surfaces, intensity as high as that of the SR source is a necessity. New higher-intensity wiggler,[121] undulator,[122,123] or free-electron laser[124] beams may increase the PSD sensitivity for the study of rare (low-concentration) surface sites. The ultraclean-vacuum environment of storage-ring SR sources is an essential feature for any optical study of surfaces, including PSD, in spectral ranges characterized by a scarcity or absence of transparent materials for windows. Although not essential in PSD, utilization of time-of-flight detection permits highly efficient, concurrent detection of several desorbing species. Time-of-flight detection is facilitated by the narrow-pulse character of SR and its long interpulse periods.[1,16–18] The high degree of collimation of SR makes angular studies convenient and precise without sacrificing intensity. Finally, the high degree of polarization of the SR beam can be utilized in the study of surface-bond characteristics. On the other hand, the disadvantages of SR for PSD experiments include the difficulty of focusing to micron or smaller spot sizes for spatially resolved surface studies and the difficulties associated with conducting experiments at national facilities far from one's own laboratory. It may be hoped that small, table-top SR sources may become available at a price affordable by individual research institutions.

Apparatus. Figure 15 illustrates in block diagram the apparatus used for PSID. It consists of an ultrahigh-vacuum chamber, open to the SR beamline, with sample-holder suitable for positioning the sample in the SR beam, and an ion detector. Note that the photon beam strikes the sample at glancing incidence so that reflected photons do not find their way into the ion detector, which is

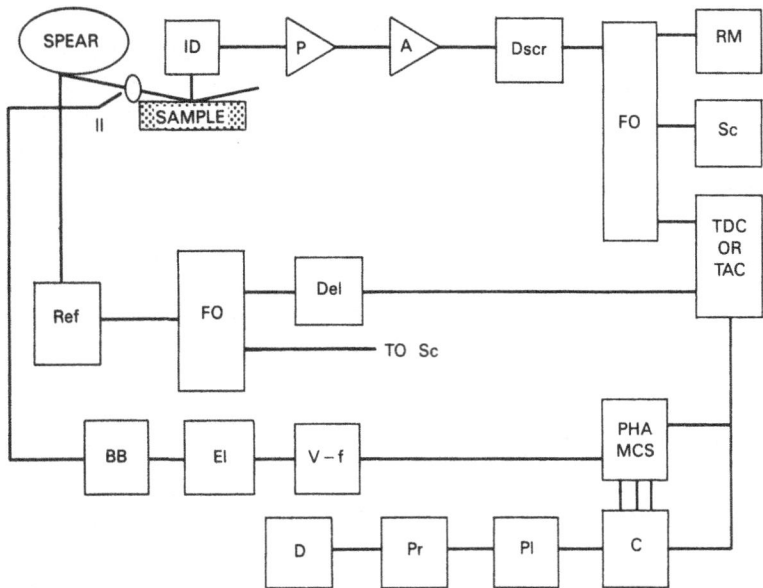

FIGURE 15. Apparatus for PSD in block diagram. SPEAR represents SR source, ID is the ion detector, P is a preamp, A is an adjustable-gain amplifier, Dscr is a threshold discriminator, FO are NIM fan-out units, RM is a counting-rate meter, Sc is a 250-MHz oscilloscope, and TDC and TAC are time-domain converter and time-to-amplitude converter, respectively (either may be used). PHA and MCS are pulse-height analysis and multichannel-scaling modes of the multichannel pulse-height analyzer; C is the control computer. II is the incident-intensity monitor, BB Is a battery box, EI is an electrometer, V–f is a voltage-to-frequency converter, Ref is the SR-source timing reference, Del is an adjustable delay, PI is a plotter, pr is a printer, and D is a disc drive for data storage.

sensitive to soft X-ray photons. Ions emitted from the sample are accelerated normally away from the sample surface and into the ion detector by a uniform electric field.

The time-of-flight (TOF) ion detector shown in Fig. 16 utilizes an accelera-tion screen to increase the ion kinetic energy high enough (1–2 keV) so that the variations in initial kinetic energy (0–10 eV) become negligible. Next a field-free drift region allows ions of differing velocity to drift apart, and a dual microchannel-plate assembly detects ions with high efficiency. The microchannel plate converts ion pulses to pulses of electrons and multiplies the number of electrons in the microchannels. The electron pulses are collected by an anode and conducted out of the vacuum chamber by a bakable 50-Ω coaxial cable to a low-noise, fast (>250 MHz) preamplifier. In order to permit control of the drift time without affecting the ion-detection efficiency, an additional acceleration screen is added between the drift region and the microchannel-plate surface. By holding the sum of the two acceleration potentials constant, the ion kinetic energy incident on the microchannel-plate surface is independent of the initial accelera-tion potential, which controls the ion TOF. In this way, the TOF can be set to any convenient value without affecting the ion-detection efficiency. For efficient ion detection using microchannel plates, the ion kinetic energy has a broad optimum range, 1–3 keV. The time resolution obtained is illustrated in Fig. 17, as well as in the spectra of Fig. 10.

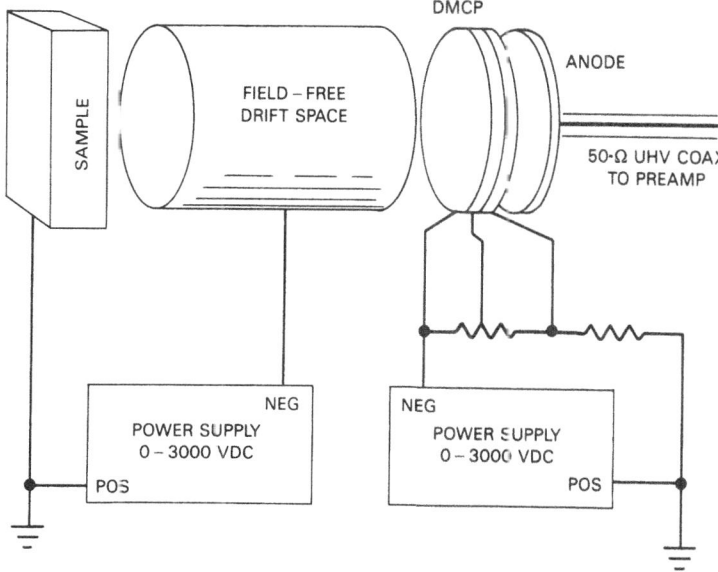

FIGURE 16. Time-of-flight ion detector. The drift-space enclosure is biased to attract positive ions through the fine mesh screens on each end. The dual microchannel-plate (DMCP) assembly is biased still higher to ensure efficient detection of impinging ions. Electrons freed by the impinging ions are accelerated through the capillary holes in the MCP, with multiplication of up to 1,000,000, and are collected by the anode. The coaxial cable leads through a vacuum feedthrough to the preamp and other electronics shown in Fig. 15. In order to eliminate interference from stray photons or charged particles, a grounded metal shield, not shown, encloses the entire in-vacuum assembly. To detect negative ions, the polarity of the drift-space power supply is reversed and the DMCP power supply is floated on the drift-spaced supply. The preamp must then be floated at high positive potential.

After amplification in the electronic system, the pulses pass through a discriminator that rejects small noise pulses and converts signal pulses to standard fast-NIM pulses (-1 V, with a rise time of 1 nsec) and convenient pulse width, typically 10–20 nsec.[16,17] At that point the signal pulses are ready for TOF analysis. This can be accomplished in three ways. One, the traditional method illustrated in Fig. 15, utilizes a time-to-pulse-height (TPHC) or time-to-amplitude (TAC) converter, in which a suitably timed reference pulse from the storage ring is used as a stop pulse. Then the pulse-height-coded time information is analyzed in a multichannel pulse-height analyzer, which plots ion count rate versus ion arrival time directly, although on a right-to-left time scale. A limitation of this scheme is the length of pulses (typically 10 μsec) utilized by the TAC, during which no detection takes place. Another method utilizes a gated amplifier in which the gate time is set appropriately for the ion mass-to-charge ratio sought. Then a fast-NIM counter is used to count that species directly. In this arrangement, separate gated amplifiers are required for each ionic species to be counted concurrently. However, the arrangement avoids the long deadtime associated with the time-to-pulse-height conversion. A third method now available is the time-domain converter (TDC), such as the one made by LeCroy Research Systems. These instruments put out a digitally encoded time interval for each pulse received and can handle throughputs as high as 1 MHz.

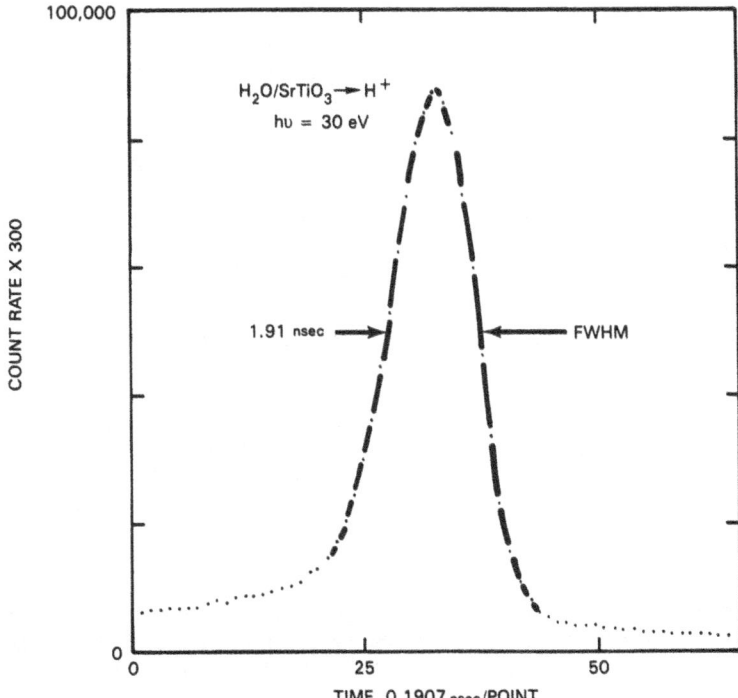

FIGURE 17. TOF ion detector performance.

The timing-reference pulse needed in any time-resolved experiment is a vital concern. Reference pulses available from the storage-ring electronics accurately clock the SR pulses, but the timing of the SR pulse, relative to the storage-ring clock at any experiment station, must be set by the experimenter using a variable delay (Del in Fig. 15). In PSD experiments, it is fortunate that a prompt pulse is usually observed due to fluorescence or scattered X-ray photons from the sample, which are detected by the ion detector. These pulses can be distinguished by their lack of dependence on ion-accelerating voltage and can be used to adjust the delay of the storage-ring-clock pulse for use as the needed timing reference. Because relatively short flight times are used, it is necessary that the timing-reference pulse from the storage ring be free of jitter on the nanosecond time scale and of fast rise time. Because of the long transmission distances and possible interference with other uses of these pulses, the experimenter must be cautious in accepting a timing reference from the storage ring.

An important consideration in PSD experiments is the normalization of the ion count rate to the intensity of SR incident on the sample. This has been accomplished in various ways but the most satisfactory generally has been the use of the real-time photoelectron total-yield signal from an appropriately coated fine-mesh screen in the incident beam as near to the sample as is practical (II in Fig. 15). (Alternatively, a coated grazing-incidence mirror may be used as a source of photoelectrons for normalization). Proper normalization is very valuable in PSD experiments. Use of freshly evaporated (*in situ*) coatings of copper, graphite, gold, sodium salicylate, or other appropriate material allows deduction of cross-section data, as well as allowing quantitative comparison of measurements made in different laboratories or under different SR beam conditions. The TEY spectra of the coating materials should be well characterized, reproducible, and lacking in sharp spectral features over the spectral range of interest.

Errors in this normalization scheme may come from several sources. The beam may be partially obscured after passing the screen, and that obscurity may vary with time, for example. The spectral dependence of the normalization may have unwanted structure due to absorption edges in the screen coating, such as those of oxygen. These spectral features may vary with time as the coating-surface changes, for example. If the mesh is too coarse compared to the cross-sectional size of the beam, variations in beam position will be converted into unwanted variations in the normalization signal. The detector used for the normalization-signal photoelectrons may also collect electrons from other sources, such as ion pumps or gauges, or even collect and detect ultraviolet photons from sources within the chamber. Finally, leakage current from the battery box needed for applying proper bias voltages to the normalization-signal detector (e.g., Spiraltron^{TM}) may grow slowly to give an unacceptable DC offset in the normalization-channel signal. In the electronic system needed to operate the normalization signal, problems in matching time constants between the two channels to be later ratioed can be obviated by use of a voltage-to-frequency (V–f) converter and scalar in the normalization channel. In this case, long time constants are not needed in the normalization channel, and the V–f output is counted by the scalar over precisely the same period as ions are counted in the PSID-signal channel. The response time of this arrangement is determined by the minimum frequency used in the V–f converter and can be less than 1 msec—a much faster response time than can be obtained by using the electrometer current directly.

The path length required for TOF-ion detection of low-energy, heavy ions may be reduced by using a reflection in the TOF detector, as illustrated in Fig. 18.[116] In this TOF ion detector, a lens is used to increase the collection aperature

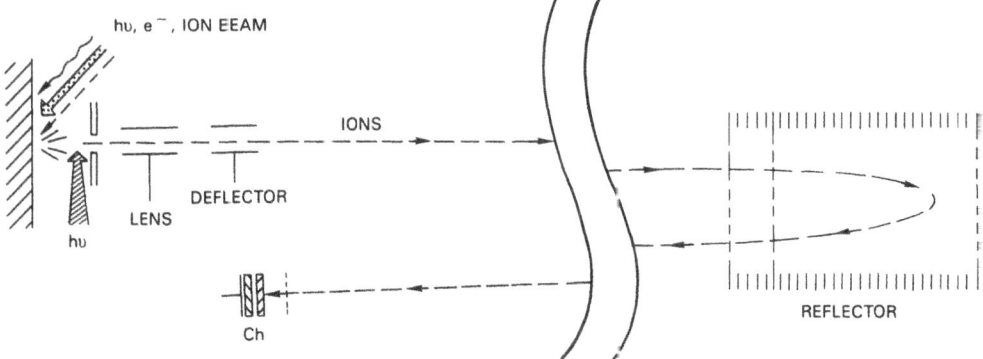

FIGURE 18. The reflection TOF ion detector. The photons indicated by lower "hv" beam are the laser photons used by Becker and Gillam to ionize neutral atoms emitted by the sample. "Ch" indicates the channel plate ion detectors. (From Ref. 116.)

and to minimize the variation of ion path lengths through the detector. A deflector is used to add a small transverse kinetic energy so that the channel plates can be offset laterally from the sample normal. The reflector is a retarding potential, which reverses the longitudinal kinetic energy. Even for heavy masses, excellent mass resolution was reported by Gillem and Becker for this TOF ion detector.[116]

Other types of detectors have been used for PSID in order to achieve specific objectives. One innovative example is the ellipsoidal-mirror display-analyzer system. This sytem was built for angle-resolved-photoemission experiments and adapted for angle-resolved PSID experiments.[125] A schematic illustration is shown in Fig. 19. The heart of the detector is the ellipsoidal mirror for reflecting and focusing charged particles emitted from the sample through a small aperature onto a multichannel plate.[126] Within the angular acceptance of the ellipsoidal mirror, angles are preserved, so that the angular information contained in the emission direction is faithfully represented on the multichannel plates. Hence, the pattern of amplified electron-current density impinging on the phosphor screen represents the angular distribution of the charged particles emitted from the sample and is analyzed externally to the vacuum system via the video cameras or photomultiplier tube, as illustrated.

Another innovative concept of the TOF detector is the combination cylindrical-mirror electron detector and TOF ion detector.[92] Knotek and Rabalais utilized this detector for the observation of ion-electron coincidences.[41] In this detector, shown in Fig. 20, the TOF ion detector is coaxially located within the cylindrical-mirror electron-energy analyser (CMA). If both electrons and positive

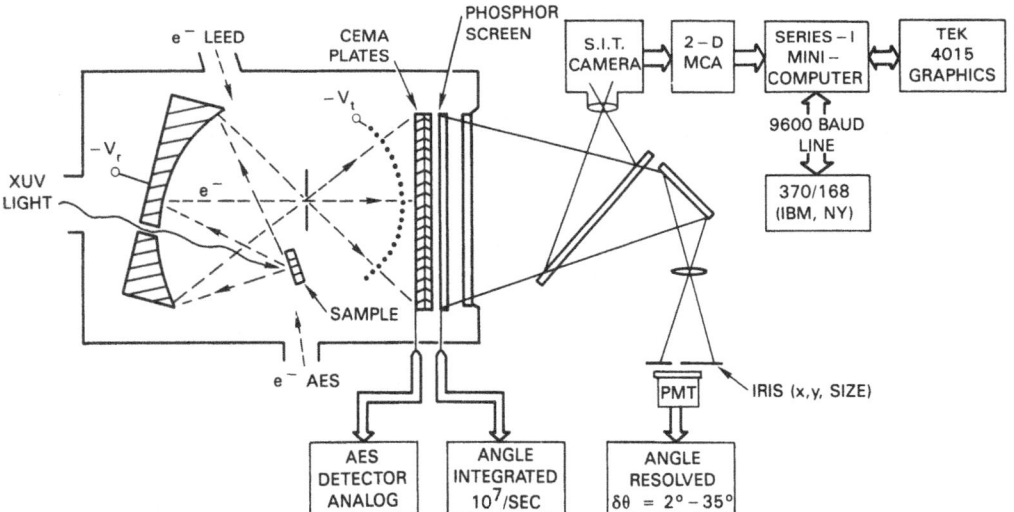

FIGURE 19. An imaging detector used for electrons or ions. The ellipsoidal mirror and sample are positioned to collect ions or electrons emitted into only half the available emission cone, thereby allowing the stimulating photon beam to be reflected away from the collection aperture. Fluorescence photons are always collected, of course. In this chamber, electron beams for LEED or AWS are shown, and the sample manipulator allows orienting the sample appropriately for these experiments. (From Ref. 125.)

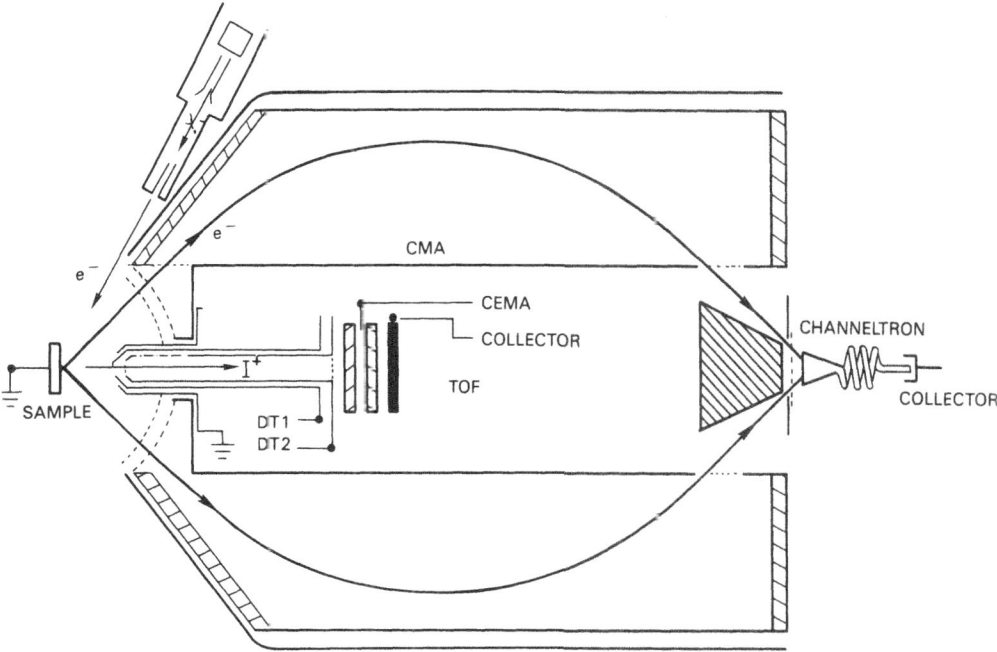

FIGURE 20. An electron- on coincidence detector. A standard cylindrical-mirror analyzer (CMA) was modified by insertion of a well-shielded TOF ion detector coaxially. CEMA denotes the channel-electron multiplier assembly for detection of ions. DT1 and DT2 represent electrodes for the ion-accelerating potentials for the drift region and the CEMA, respectively. The external electron gun was used for ESD experiments. (From Ref. 41.)

ions are to be allowed to leave the surface, then the electric field needed to accelerate positive ions into the TOF detector must be carefully contained, lest the electron energy be disturbed. Hence, both positive ions and electrons must be allowed to drift in field-free space until entering a shielded region, where an appropriate amount of kinetic energy can be given to ions independently of the electrons. A coincidence gate can be set electronically to separate the electron signal that is coincident with the ion from that originating at other times. As with all timing experiments, care must be taken to compensate for any time offsets in the relative particle paths or in the separate electronic units in the chain.

The examples above are just a few of the concepts used for PSID experiments in the recent past. They are presented as concepts on which to build new experiments of the future. It is hoped that these examples will stimulate the development of new experimental methods and new innovative concepts for future PSD experiments.

5. FUTURE RESEARCH AND APPLICATIONS OF PSD

Any discussion of the future of scientific activity is fraught with the risk of failure to predict the most important future activities. We will not attempt to predict what will be important in science in the future, but only to comment on opportunities apparent to us at the present time. Mainly these are extrapolations of research currently underway or planned. Perhaps they may serve as a guide for those who may have thought about this field a little less than we have. We offer them with no apologies, and are confident that our list of future applications will surely omit applications which will turn out to be most important in science and technology.

Theoretical Understanding of Adsorption and Desorption Processes. Although these processes are much better understood now than a few years ago, the time is ripe for more specific calculations on model systems, which can now be compared quantitatively with experiment. Adsorption by surface defects will be better understood when the electronic and geometric structures of surface defects are better known. Model calculations now can serve as a guide for experimentalists exploiting the techniques discussed above.

Following up on the work of Walkup and Avouris,[36,37] a study of charge exchange during the desorption event will help in understanding both the dynamics of desorption and the probability of recapture or reneutralization.

Theoretical studies of multi-electron effects, such as shakeup excitations, innerwell resonances, and core-exciton effects, could be of great value in interpreting NEXAFS spectra, which is clearly a rich source of physical information, and one largely unused at present. Model calculations of NEXAFS spectra that include as many of the one-electron and multi-electron interactions as possible would be timely, since uninterpreted or partially interpreted experimental NEXAFS spectra abound in the literature.

Surface Hydrogen. The role of hydrogen in surface relaxation or reconstruction, in surface catalysis or reactivity, in surface diffusion and nucleation, etc., might be expected to be of predominant importance over the years to come. Because of the especially high sensitivity of PSID compared to the especially low sensitivity of other surface-analysis techniques to surface hydrogen, it would seem likely that PSID techniques will be increasingly used for these important studies.

Surface-Impurity Distributions and Depth Profiling. In the experimental determination of impurity distributions on surfaces and within thin layers, AES or SIMS are the most common techniques used at present. PSID can be used in a complementary way to increase the sensitivity to light-atom impurities on the surface, especially hydrogen, as has been illustrated in several places above (see Fig. 2, 3, 4, 6, 8, 11, and 12). Although little has been reported to date, depth profiling of light-impurity atoms can be done with PSID by analogy with the more common Auger or SIMS sputter-depth-profiling techniques. An example is given in Figs. 9 and 14.[38] In this work, the role of hydrogen and oxygen in the formation of effective tunnel barriers of oxidized amorphous silicon was studied by sputter profiling slowly through the 30-Å-thick tunnel-barrier film formed on superconductors such as Nb, Nb_3Sn, or Nb_3Ge. chemical bonding is indicated by the chemical shift of the Si ($L_{2,3}$) edge, and PSID-NEXAFS spectra may be interpretable in terms of the site geometries.

Surface-Imperfection Studies, or Studies of Rare Surface Sites. Following the recent work of Kurtz,[89] studies of the geometrical and electronic structure of isolated impurities or imperfections, dislocations, and topographical and morphological features such as steps, grainboundaries, and the like can be undertaken with PSID in combination with other techniques such as photoemission, ESDIAD, or PSDIAD. By combining multitechnique experiments with theoretical efforts, it may be possible to obtain a fairly detailed description of both geometric and electronic structure of surface-imperfection sites. These surface features control thin-film and epitaxial growth by serving as sites for inhomogeneous nucleation. Also, they are possibly catalytically active sites, a detailed understanding of which has evaded scientists for so long. Their study would be important for scientific, technological, and economic reasons.

Adsorbate–Surface Interactions: Catalysis, Etching, and Nucleation. In the ongoing study of surface bonds between substrate and adsorbate species, PSID excitation spectra have shown that much can be learned. Recent studies of the PSD of neutrals have shown that combining the selectivity of the excitation beam stimulating PSD with an equally selective desorbate-detection laser-photon beam will provide an incredibly detailed description of adsorbate–surface interactions. these are hard experiments because they are conducted in a multidimensional parameter space. However, for selected model systems, the level of detailed understanding obtainable seems to justify the effort. Simpler experiments to identify species involved in surface chemical, pyrolytic, or photolytic reactions, for example, may be much more common in future years.

Surface Structural Studies. Although PSID is not the technique of preference for long-range surface structure, details of the surface structure may be easily checked with PSID SEXAFS or PSID-NEXAFS. This would be particularly true of surfaces involving both heavy and light atoms. In these surfaces, it is difficult to obtain the light-atom position with diffraction techniques, but PSID-SEXAFS, utilizing the light-atom core hole, should have both the surface sensitivity and the signal strength necessary for the task.

PSD of Neutrals. Studies of the desorption of neutral species is only beginning, but already it appears that new desorption physics is involved and high sensitivites are obtainable.

Spatially Resolved PSD. Up to the present, the focusing of the exciting SR on very small spots has been difficult. SR tomography with resolution of about 1 μ is the state of the art. With the advent of improved SR sources such as long undulators and soft X-ray (SXR) free-electron lasers, the possibility arises to focus SXR radiation to considerably smaller spots. By detecting the PSID from such small spots, spatially resolved surface analysis comparable with scanning-Auger microscope (SAM) systems may become available, and along with it comes all the advantages of the photon-probe beam, as discussed in section 1.

These ideas about future applications are indicative of the richness of the future we expect for scientists studying nature with the aid of various PSD techniques. The scientist is limited only by his or her limited imagination, for nature seems to offer an unlimited variety of scientific problems related to surfaces and other low-dimensional structures. Also, the society of humankind develops a similarly large variety of technological problems and applications for the dedicated and imaginative surface scientist to ponder.

Acknowledgments. It is a pleasure to acknowledge the collaboration and assistance of several people in various synchrotron experiments reviewed here; Peter Love, Christofer Parks, Geoffrey Thornton, Ian Owen, Arold Green, Teresa Cole, Ty Kobal, David Katz, James Bethke, and Carey Schwartz. The exciting collaboration with Michael Knotek is acknowledged with pleasure. The financial support for the preparation of this manuscript came from the N.W.C. Independent Research Fund and the Office of Naval Research. All our synchrotron experiments were conducted at the Stanford Synchrotron Radiation Laboratory, which is supported by the Department of Energy in cooperation with the Stanford Linear Accelerator Center.

REFERENCES

Many of the references for this chapter are located in the four books listed below. Abbreviated references to these four books are indicated below.

DIET I: N. H. Tolk, M. M. Traum, J. C. Tully, and T. E. Madey, eds, *Desorption Induced by Electronic Transitions, DIET, I* Springer-Verlag, Berlin (1983).

DIET II: W. Brenig and D. Menzel, eds., *Desorption Induced by Electronic Transitions, DIET II,* Springer-Verlag, Berlin (1985).

WD80:   H. Winick and S. Doniach, eds., *Synchrotron Radiation Research,* Plenum, New York (1980).

Ko83:   E. E. Koch, ed. *Handbook on Synchrotron Radiation,* Vols. 1A, and 1B (D. E. Eastman and Y. Farge, gen. eds.) North Holland, Amsterdam (1983).

1. M. L. Knotek, V. O. Jones, and Victor Rehn, Photon-stimulated desorption of ions, *Phys. Rev. Lett.* **43**, 300–303 (1979).
2. Search and Discovery News, *Phys. Today* **33**, 17–18 (1980); Victor Rehn, Surface science with photon-stimulated ion desorption, *Nav. Res. Rev.* **35**, 36–41 (1983); **35**, M. L. Knotek, Stimulated desorption from surfaces, *Phys. Today* **37**, 24–32 (1984).
3. D. Menzel and R. Gomer, Desorption from metal surfaces by low-energy electrons, *J. Chem. Phys.* **41**, 3311–3328 (1964).
4. P. A. Redhead, Interaction of slow electrons with chemisorbed oxygen, *Can. J. Phys.* **42**, 886–905 (1964).
5. M. L. Knotek and P. J. Feibelman, Ion desorption by core-hole Auger decay, *Phys. Rev. Lett.* **40**, 964–967 (1978) *Surf. Sci.* **90**, 78–87 (1979).
6. D. Lichtman and Y. Shapira, Photodesorption: A critical review, *CRC Crit. Rev. Solid State Mater. Sci.* **8**, 93–118 (1978).
7. F. L. Tabares, E. P. Marsh, G. A. Bach, and J. P. Cowin, Laser photofragmentation and photodesorption of physisorbed CH_3Br on lithium fluoride, *J. Chem. Phys.* **86**, 738–744 (1987).
8. D. A. King, in *Chemistry and Physics of Solid Surfaces,* Vol II (R. Vanselow ed.) Chemical Rubber Co., Cleveland (1979), pp. 87–128.
9. D. Lichtman, Photodesorption and Negative-Ion ESD in *DIET I*, pp. 117–119.

10. P. J. Feibelman, and M. L Knotek, Reinterpretation of electron-stimulated desorption data from chemisorbed systems, *Phys. Rev.* **B18**, 6531 (1978).

11. T. A. Carlson and M. O. Krause, Relative abundances and recoil energies of fragment ions formed from the X-ray photoionization of N_2, O_2, CO, NO, CO_2, and CF_4, *J. Chem. Phys.* **56**, 3206–3209 (1972).

12. R. B. Kay, Ph. E. Van der Leeuw, and M. J. Van der Weil, Ion fragmentation and postcollection interaction in the Auger decay of $C(K)$-ionized CO, *J. Phys.* **B10**, 2521–259 (1977).

13. L. G. Parratt, Electron band structure of solids by X-ray spectroscopy, *Rev. Mod. Phys.* **31**, 616–645 (1959).

14. F. Bassani and M. Alterelli, Interaction of Radiation with Condensed Matter in *Ko83* Ch. 7, pp. 463–605.

15. F. C. Brown, "Ultraviolet Spectroscopy of Solids with the Use of Synchrotron Radiation," in *Solid State Physics* Vol. 29 (H. Ehrenreich, F. Seitz, and D. Turnbull, eds.), Academic Press, New York (1974), pp. 1–73.

16. K. M. Monahan and V. Rehr, Exploiting the unique time structure of synchrotron radiation at SSRL, *Nucl. Instrum. Methods* **152**, 255–259 (1978).

17. V. Rehn, Time-resolved spectroscopy in synchrotron radiation, *Nucl. Instrum. Methods* **177**, 193–205 (1980).

18. I. H. Munro and A. P. Sabersky, "Synchrotron Radiation as a Modulated Source for Fluorescence-Lifetime Measurements and for Time-Resolved Spectroscopy" in *WD80*, pp. 323–352.

19. S. Prigge, H. Niehus, and E. Bauer, Electron stimulated desorption ion energy distribution (ESDIED) and surface structure: O on W (100), *Surf. Sci.* **75**, 635–656 (1978).

20. J. W. Gadzuk, "Fundamental Excitations in Solids Pertinent to Desorption Induced by Electronic Transitions" In *DIET I*, pp. 4–25.

21. C. F. Melius, R. H Stulen, and J. O. Noell, Mechanism of near-threshold stimulated desorption of protons from transition-metal surfaces, *Phys. Rev. Lett.* **48**, 1429–1430 (1982).

22. D. E. Ramaker, C Y. White and J. S. Murday, On Auger induced decomposition/desorption of covalent and ionic systems, *J. Vac. Sci. Technol.* **18**, 748–753 (1981).

23. D. E. Ramaker, C T. White, and J. S. Murday, On Auger-induced decomposition/desorption of covalent and ionic systems, *Phys. Lett.* **89A**, 211–214 (1982).

24. M. L. Knotek and J. E. Houston, Study of stepwise oxidation and nitridation of Si (111) by ESD and Auger spectroscopy, *J. Vac. Sci. Technol.* **20**, 544–547 (1982).

25. R. Franchy and D. Menzel, Adsorbate-core ionization as primary process in electron- and photon-stimulated desorption from metal surfaces, *Phys. Rev. Lett.* **43**, 865–867 (1979).

26. Victor Rehn, A. K. Green, R. A. Rosenberg, G. Loubriel, and C. C. Parks, The chemical makeup of Nb and Nb_3Sn films, *Physica (The Hague)* **107B**, 533–534 (1981).

27. D. P. Woodruff, M M. Traum, H. H. Farrell, N. V. Smsith. P. D. Johnson, D. A. King, B. L. Benbow, and Z. Hurych, Photon- and electron-stimulated desorption from a metal surface, *Phys. Rev.* **B21**, 5642–5645.

28. R. Jaeger, J. Feldhaus, J. Haase, J. Stöhr, Z. Hussain, D. Menzel, and D. Norman, Surface EXAFS by means of photon-stimulated ion desorption: O on Mo (100), *Phys. Rev. Lett.* **45**, 1870–1873 (1980); *Phys. Rev. Lett.* **49**, 1264–1267 (1982).

29. R. Jaeger, J. Stöhr, J. Feldhaus, S. Brennan, and D. Menzel, Photon-stimulated desorption following deep-core excitation: O on Mo (100), *Phys. Rev.* **B23**, 2102–2110 (1981).

30. R. Jaeger, J. Stöhr, R. Treichler, and K. Baberschke, Photon-stimulated desorption due to multielectron excitations in chemisorbed molecules: CO on Ni (100), *Phys. Rev. Lett.* **47**, 1300–1303 (1981).

31. T. E. Madey, R. Stockbauer, S. A. Flodström, J. F. van der Veen, F. J. Himpsel, and D. E. Eastman, Photon-stimulated desorption from covalently bonded species: CO adsorbed on Ru (001), *Phys. Rev.* **B23**, 6847–6350 (1981).

32. R. H. Stulen and R. A. Rosenberg, High-resolution photon-stimulated desorption of H^+ from H_2O on Pd and Pt, *J. Vac. Sci Technol.* **A2**, 1051–1052 (1984).

33. V. Rehn, R. A. Rosenberg, and G. Thornton, "PSID studies of chemically cleaned, cleaved and H_2O-dosed GaAs (110) surfaces" (1980), unpublished.

34. C. C. Parks, Z. Hussain, D. A. Shirley, M. L. Knotek, G. Loubriel, and R. A. Rosenberg, Auger decay mechanism in photon-stimulated desorption from sodium fluoride, *Phys. Rev.* **B28,** 4793–4798 (1983).

35. P. R. Antoniewicz, Model for electron- and photon-stimulated desorption, *Phys. Rev.* **B21,** 3811–3815 (1980).

36. R. E. Walkup and Ph. Avouris, Classical-trajectory studies of electron- or photon-stimulated desorption from ionic solids, *Phys. Rev. Lett.* **56,** 524–527 (1986).

37. R. E. Walkup & Ph. Avouris, Summary abstract: Classical-trajectory studies of photon-stimulated desorption of ions from alkali halides, *J. Vac. Sci. Technol.* **A4,** 1247 (1986).

38. A. K. Green, Victor Rehn, R. A. Rosenberg, C. C. Parks, R. H. Hammond, and M. R. Beasley, The role of hydrogen and oxygen in oxidized-amorphous-silicon tunnel-barrier films studied by PSID and AES (1983) unpublished.

39. R. L. Benbow, M. R. Thuler, and Z. Hurych, PSD of O^+ from Na_xWo_3: Demonstration of bonding-site selectivity, *Phys. Rev. Lett.* **49,** 1264–1267 (1982).

40. R. Jaeger, R. Treichler, and J. Stöhr, Evidence for multielectron excitations in PSID, *Surf. Sci.* **117,** 533–548 (1982).

41. M. L. Knotek and J. W. Rabalais, "Measurement of Auger Electron Electron-Ion Coincidence Events from Surfaces" In *DIET II,* pp. 77–83.

42. R. A. Rosenberg and Victor Rehn, "Surface studies of molecular solid films," in *Synchrotron Radiation Research, Vol. 2,* (R. Z. Bachrach, ed.) Plenum, New York (1992).

43. R. Jaeger, J. Stöhr, and T. Kendelewicz, X-ray induced electron-stimulated desorption versus photon-stimulated desorption: H_2O on Ni (110), *Phys. Rev.* **B28,** 7145–7148 (1983).

44. R. Jaeger, J. Stöhr, and T. Kendelewicz, X-ray induced electron-stimulated desorption versus photon-stimulated desorption: NH_3 on Ni (110), *Surf. Sci.* **134,** 547–565 (1983).

45. R. A. Rosenberg, V. Rehn, A. K. Green, P. R. LaRoe, and C. C. Parks, "PSID from Condensed Molecules: N_2, CO, C_2H_2, CH_3OH, N_2O, D_2O, and NH_3" in *DIET-I,* pp. 247–261.

46. R. A. Rosenberg, P. J. Love, P. R. LaRoe, V. Rehn, and C. C. Parks, K-shell photoexcitation of solid N_2, CO, NO, O_2 and N_2, *Phys. Rev.* **B31,** 2634–2642 (1985).

47. R. A. Rosenberg, P. J. Love, P. R. LaRoe, V. Rehn, and C. C. Parks, Indirect mechanisms in ion desorption from solid molecular films (1982) unpublished.

48. H. Peterson, A. Bianconi, F. C. Brown, and R. Z. Bachrarch, Absolute N_2 K-photoionization cross section up to $hv = 450$ eV, *Chem. Phys. Lett.* **58,** 263–266 (1978).

49. R. E. LaVilla, The O (K_a) and C (K_a) emission and O (K) absorption spectra for O_2 and CO_2, *J. Chem. Phys.* **63,** 2733–2737 (1975).

50. R. A. Rosenberg, V. Rehn, A. K. Green, P. R. LaRoe, and C. C. Parks, Electron correlation and hole localization in photon-stimulated ion desorption from solid N_2, CO and C_2H_2, *Phys. Rev. Lett.* **51,** 915–18 (1983).

51. D. Charleston, private communication (1982).

52. G. Glupe and W. Mehlhorn, A new method for measuring electron-impact-ionization cross sections of inner shells, *Phys. Lett.* **25A,** 274–275 (1967).

53. L. B. Knight, Jr., J. M. Bostick, R. W. Woodward, and J. Steadman, An electron-bombardment procedure for generating cations and neutral radicals in solid-Ne matrices at 4 K: ESR study of $^{14}N_2^+$ and $^{14}N_2^+$, *J. Chem. Phys.* **78,** 6415–6421 (1982).

54. B. Van Zyl and G. H. Dunn, Dissociation of N_2^+ and O_2^+ by electron impact, *Phys. Rev.* **163,** 43–45 (1967).

55. R. B. Kay, Ph. E. Van der Leuw, and M. J. Van der Wiel, Absolute oscillator strengths for the shape resonances near the K edges of N_2 and CO, *J. Phys.* **B10** 2513–2519 (1977).

56. F. M. Goldberg, C. S. Fadley, and S. Kono, Photoionization cross sections for atomic orbitals with random and fixed spacial orientations, *J. Electron Spectros. Relat. Phenom.* **21,** 285–363 (1981).

57. I. W. Owen, N. B. Grookes, C. H. Richardson, D. R. Warburton, F. M. Quinn, D. Norman, and G. Thornton, On the dominance of an indirect mechanism for PISD from $SrTiO_2$–H_2O," *Sur. Sci.* **178,** 897–900 (1986).

58. C. O. Almbladh and L. Hedin, "Beyond the One-electron Model: Many-body Effects in Atoms, Molecules and Solids" in *WD80,* 607–904.

59. F. C. Brown, "Inner-Shell Threshold Spectra" in *WD80,* pp. 61–100.

60. M. L. Knotek, R. H. Stulen, G. M. Loubriel, V. Rehn, R. A. Rosenberg, and C. C. Parks, Photon-stimulated desorption of H^+ and F^+ from BeO, Al_2O_3 and SiO_2: Comparison of near-edge structure to photoelectron yield, *Surf. Sci.* **133**, 291–304 (1983).

61. A. Bianconi, Surface x-ray absorption spectroscopy: surface EXAFS and surface XANES, *App. Surf. Sci.* **6**, 392–418 (1980).

62. J. L. Dehmer and D. Dill, Molecular effects on inner-shell photoionization: K-shell spectra of N_2, *J. Chem. Phys.* **65**, 5327–5334 (1976).

63. G. S. Brown and S. Doniach, "The Principles of X-Ray Absorption Spectroscopy" in *WD8C*, 353–385.

64. M. Brown, R. E. Peierls, and E. A. Stern, White lines in X-ray absorption, *Phys. Rev.* **B15**, 738–744 (1977).

65. G. A. Sommerjai and M. A. Van Hove, *Adsorbed Monolayers on Solid Surfaces*, Vol. 38, *Structure and Bonding*, Springer-Verlag, Berlin (1979).

66. H. D. Hagstrom, "Surface Physics," in *Physics Vade Mecum* (H. L. Anderson, ed.) Am. Inst. Phys., New York (1981), pp. 300–313.

67. R. Vanselow, ed., *Chemistry and Physics of Solid Surfaces*, Vol. II and III, Chemical Rubber Company, Cleveland, 1979 and 1982.

68. I. Lindau and W. E. Spicer, "Photoemission as a Tool to Study Solids and Surfaces," Ch. 6 in *WD80*, pp. 159–221.

69. N. V. Smith and F. J. Himpsel, "Photoelectron Spectroscopy," Ch. 9 in *Ko83*, pp. 905–954.

70. T. E. Madey and J. T. Yates, Jr., Electron-stimulated desorption and work function studies of clean and cesiated (110) GaAs, *J. Vac. Sci. Technol.* **8**, 39–44 (1971).

71. T. E. Madey and J. T. Yates, Jr., *J. Chem. Phys.* **52**, 5215 (1970).

72. P. Pianetta, I. Lindau, C. M. Carner, and W. E. Spicer, *Phys. Rev. Lett.* **37**, 1166–1169 (1976).

73. R. A. Rosenberg, V. Rehn, V. O. Jones, A. K. Green, C. C. Parks, G. Loubriel, and R. H. Stulen, The photodissociative ionization of amorphous ice, *Chem. Phys. Lett.* **80**, 488–494 (1981).

74. R. A. Rosenberg, P. R. LaRoe, V. Rehn, J. Stohr, R. Jaeger, and C. C. Parks, K-shell excitation of D_2O and H_2O ice: photoion and photoelectron yields, *Phys. Rev.* **B28**, 3036–3030 (1983).

75. G. Thornton, R. A. Rosenberg, V. O. Jones, and V. Rehn, "Studies of freshly fractured and H_2O-dosed $SrTiO_3$ using PSID and TEY spectroscopies" (1982), unpublished.

76. R. L. Kurtz, R. Stockbauer, and T. E. Madey, "Site Specificity in Stimulated Desorption from TiO_2" in *DIET II*, pp. 89–93.

77. J. P. Connerade, The non-Rydberg spectroscopy of atoms, *Contemp. Phys.* **19**, 415–447 (1978).

78. S. T. Manson and J. W. Cooper, Photoionization in the soft X-ray range: Z dependence in a central-potential model, *Phys. Rev.* **165**, 126–138 (1968).

79. B. B. Pate, P. M. Stefan, C. Binns, P. J. Jupiter, M. L. Shek, I. Lindau and W. E. Spicer, Formation of surface states on the (111) surface of diamond, *J. Vac. Sci. Technol.* **19**, 349–354 (1981).

80. B. B. Pate, M. H. Hecht, C. Binns, I. Lindau, and W. Spicer, Photoemission and photon stimulated ion desorption studies of diamond (111): hydrogen, *J. Vac. Sci. Technol.* **21**, 364–367 (1982).

81. B. B. Pate, Ph.D Thesis, *The Diamond Surface: Atomic and Electronic Structure*, Stanford U. (1984), unpublished.

82. R. A. Rosenberg, P. J. Love, V. Rehn, I. Owen, and G. Thornton, Bonding of hydrogen on water-dosed Si (111), *J. Vac. Sci. Technol.* **A4**, 1451–1454 (1986).

83. T. E. Madey, in *Inelastic Particle-Surface Collisions*, Springer Series in Chemical Physics Vol. 17, (E. Taglauer and W. Heiland, eds.), Springer-Verlag, Berlin (1981), 80.

84. T. E. Madey, F. P. Netzer, F. P. Houston, D. M. Hanson, and R. Stockbauer, "Determination of Molecular Structure at Surfaces Using Angle-Resolved Electron- and Photon-Stimulated Desorption" in *DIET I*, pp. 120–138.

85. T. E. Madey, D. Ramaker, and R. Stockbauer, *Ann. Rev. Phys. Chem.* **35**, 215 (1984).

86. T. E. Madey, C. Benndorf, N. D. Shinn, Z. Miskovic, and J. Vukanic, "Recent advances using ESDIAD: Applications to surface chemistry" in *DIET II*, pp. 104–115.

87. W. L. Clinton, ESDIAD of H_2O and NH_3 on Ru (001), *Surf. Sci.* **112**, L791–L796 (1981).

88. R. A. Gibbs, S. P. Holland, K. E. Foley, B. J. Garrison, and N. Winograd, Image potential and ion trajectories in SIMS, *Phys. Rev.* **B24,** 6178–6181 (1981).

89. R. L. Kurtz, Stimulated desorption studies of defect structures on TiO_2, *Surf. Sci.* **177,** 526–552 (1986).

90. J. F. van der Veen, F. J. Himpsel, D. E. Eastman, and P. Heimann, *Solid State Commun.* **36,** 99 (1980).

91. T. E. Madey, R. L. Stockbauer, J. F. van der Veen, and D. E. Eastman, Angle-resolved photon-stimulated desorption of oxygen ions from a W (111) surface, *Phys. Rev. Lett.* **45,** 187–190 (1980).

92. M. M. Traum and D. P. Woodruff, Time-of-flight measurements with a CMA for simultaneous energy and mass determinations of desorbed ions, *J. Vac. Sci. Technol.* **17,** 1202–1207 (1980).

93. D. M. Hanson, R. Stockbauer, and T. E. Madey, PSD and other spectroscopic studies of the interaction of oxygen with a Ti (001) surface, *Phys. Rev.* **B24,** 5513–5521 (1981).

94. B. K. Teo and D. C. Joy, eds, *EXAFS Spectroscopy: Techniques and Applications,* Plenum, New York (1981).

95. E. A. Stern, Theory of the extended X-ray absorption fine structure, *Phys. Rev.* **B10,** 3027–3037 (1974).

96. G. S. Brown, "Extended X-Ray Absorption Fine Structure in Condensed Materials" in *WD80,* pp. 387–400.

97. P. H. Citrin, P. Eisenberger, and R. C. Hewitt, EXAFS of surface atoms on single-crystal substrates: Iodine adsorbed on Ag (111), *Phys. Rev. Lett.* **41,** 309–312 (1978).

98. J. Stöhr, D., Denley, and P. Perfetti, SEXAFS in the soft X-ray region: Study of an oxidized Al Surface, *Phys. Rev.* **B18,** 4132–4135 (1978).

99. J. Stöhr, *Jpn. J. Appl. Phys.* **17-2,** 217 (1978).

100. P. A Lee, P. H. Citrin, P. Eisenberger, and B. M. Kincaid, Extended X-ray absorption fine structure—its strengths and limitations as a structural tool, *Rev. Mod. Phys.* **53,** 769–806 (1981).

101. P. Pianetta, I. Lindau, C. M. Gardner, and W. E. Spicer, Chemisorption and oxidation studies of the (110) surfaces of GaAs, GaSb and InP, *Phys. Rev.* **B18,** 2792–2806 (1978).

102. J. Stöhr, L. I. Johnsson, S. Brennan, M. Hecht, and J. N. Miller, Surface extended X-ray-absorption-fine-structure study of oxygen interaction with Al (111) surfaces, *Phys. Rev.* **B22,** 4052–4065 (1980).

103. R. G. Jones and D. P. Woodruff, Sampling depths in total yield and reflectivity SEXAFS studies in the soft X-ray region, *Surf. Sci.* **114,** 38–46 (1982).

104. M. L. Knotek, V. O. Jones, and Victor Rehn, Photon-stimulated desorption by core-level excitation: Density of states and extended fine structures, *Surf. Sci.* **102,** 566–577 (1981).

105. B. K. Teo and P. A. Lee, *J. Am. Chem. Soc.* **101,** 2815 (1979).

106. R. McGrath, I. T. McGovern, D. R. Warburton, G. Thornton, and D. Norman, A PSID SEXAFS study of H_2O adsorption on Si (100), *Surf. Sci.* **178,** 101–107 (1986).

107. D. J. Chadi, Energy-minimization approach to the atomic geometry of semiconductor surfaces, *Phys. Rev. Lett.* **41,** 1062–1065 (1978), Erratum, 1332.

108. Victor Rehn, R. A. Rosenberg, P. J. Love, and C. C. Parks, "PSID from clean and hydrogen-dosed InP (110) and GaP (110) in the spectral range 2–4 keV" (1982), unpublished. Some of these results were presented in Ref. 109.

109. Victor Rehn, "Synchrotron-radiation induced desorption from semiconductors: recent results," presented at the U.S.–Japan Seminar on Recombination-Induced Defect Formation in Crystals," University of Illinois, Urbana, 2–5 June 1982, unpublished.

110. E. E. Koch, C. Kunz, and B. Sonntag, Electronic states in solids investigated by means of SR, *Phys. Rep.* **C29c,** 153–231, (1977).

111. P. J. Durham, J. B. Pendry, and C. H. Hodes, *Contemp. Phys. Commun* **25,** 193–206 (1982).

112. D. Norman, J. Stöhr, R. Jaeger, P. J. Durham, and J. B. Pendry, Determination of local atomic arrangements at surfaces from NEXAFS studies: O on Ni (100), *Phys. Rev. Lett.* **51,** 2052–2055 (1983).

113. D. Norman, X-ray absorption spectroscopy (EXAFS and XANES) at surfaces, *J. Phys.* **C19,** 3273–3311 (1986).

114. N. H. Tolk, M. M. Traum, J. S. Kraus, T. R. Pian, W. E. Collins, N. G. Stoffel, and G. Margaritondo, Optical radiation from photon-stimulated desorption of excited atoms, *Phys. Rev. Lett.* **49,** 812–815 (1982).

115. N. H. Tolk, W. E. Collins, J. S. Kraus, R. J. Morris, T. R. Pian, N. M. Traum, N. G. Stoffel, and G. Margaritondo, "The Electronic Desorption of Excited Alkali Atoms from Alkali Halide Surfaces" in *DIET I*, pp. 156–162.

116. C. H. Becker and K. T. Gillen, Surface analysis by nonresonant multiphoton ionization of desorbed or sputtered species, *Appl. Phys. Lett.* **45**, 1063 (1984).

117. A. R. Burns, ESD of neutral NO from Ni: Site-specific resonance-ionization detection, *Phys. Rev. Lett.* **55**, 525–528 (1985).

118. A. R. Burns, Site-selective ionization detection of neutral NO following ESD, *J. Vac. Sci. Technol* **A4**, 1499–1500 (1986).

119. A. R. Burns, E. B. Stechel, and D. R. Jennison, Desorption by electronically stimulated adsorbate rotation, *Phys. Rev. Lett.* **58**, 250–253 (1987).

120. R. E. Walkup, Ph. Avouris, and A. P. Ghosh, Excited-atom production by electron bombardment of alkali halides, *Phys. Rev. Lett.* **57**, 2227–3320 (1986).

121. J. E. Spenser and H. Winick, "Wiggler Systems as Sources of Electromagnetic Radiation," in *WD80*, pp. 636–715.

122. G. Brown, G. K. Halbach, J. Harris, and H. Winick, Wiggler and undulator magnets: A review, *Nucl. Instrum. Methods* **208**, 65–78 (1983).

123. R. Z. Bachrach, R. D. Bringans, B. B. Pate, and R. G. Carr, The SSRL Insertion-Device Beam Line "WUNDER," *Proc. SPIE—Int. Soc. Opt. Eng.* **582**, 238–250 (1986).

124. J. M. J. Madey and C. Pellegrini, eds, *Free-Electron Generation of Extreme Ultraviolet Coherent Radiation*, Am. Inst. Phys. Conf. Proc. **118**, American Institute of Physics, New York (1984).

125. D. E. Eastman, J. J. Donnelon, N. C. Hein, and F. J. Himpsel, "An ellipsoidal mirror display analyzer system for electron and angular measurements" (preprint, 1979).

126. R. Stockbauer, Instrumentation for PSD, *Nucl. Instrum. Methods* **222**, 284–290 (1984).

Diffraction and Scattering

Grazing Incidence X-Ray Scattering

P. H. Fuoss, K. S. Liang, and P. Eisenberger

1. INTRODUCTION

The structural challenges presented by the two-dimensional world of surfaces have proven formicable. In spite of the critical role that surfaces and interfaces play in such diverse sciences as catalysis, electrolysis, tribology, metallurgy, and the study of electronic devices and in spite of the expected richness of two-dimensional (2-D) physics of melting, magnetism, and their related phase transitions, only a few surface structures are known and most of those only semiquantitatively (e.g., their symmetry).[1] Our inability in many cases to understand atomic structure and to make the structure–function connection in the 2-D region of surfaces and interfaces has significantly inhibited progress in understanding this potentially rich area of science. In the more thoroughly explored three-dimensional (3-D) world of materials, the use of X-ray scattering has provided us with most of our knowledge of the structure of crystals, amorphous solids, and liquids; from simple materials like silicon to complex materials like DNA.

Of course, the reason for this difference is that 3-D materials have enough scattering power to be successfully studied using conventional X-ray sources despite the relatively low scattering cross section of X-ray photons. For surfaces and interfaces, the relatively small number of atoms has made structural experiments difficult at best and impossible in most cases. Because of this low X-ray cross section, previous investigators used other probes like electrons or atoms, which have much larger cross sections, to study surfaces. The complexity of surface and interface problems together with the limitations of various types of techniques using photons, atoms, ions, electrons, or neutrons strongly supports the need for a variety of approaches. Listed in Table 1 are the various techniques currently used to investigate surfaces.[2] The most prominent of these is LEED,[3] which has been invaluable in exposing the structural richness of surfaces as well as providing important structural information. However, the accuracy of that information has been limited by interpretive problems associated with multiple

P. H. Fuoss • AT & T Bell Laboratories, Murray Hill, New Jersey 07974   *K. S. Liang* • EXXON Corporate Research Laboratories, Annandale, New Jersey 08801   *P. Eisenberger* • Princeton University, Princeton, New Jersey 08544.

TABLE 1. Current Surface-Structure Probes and Their Acronyms

Probe	Technique	Acronym or descriptive phase	Monolayers sampled
Electrons	Elastic scattering	LEED	1–5
	Inelastic scattering	ELS	1–5
	Electron stimulated desorption		1
Photons	Absorption spectroscopy	SEXAFS	1–5
	Photoemission	XPS, UPS	1–5
	Elastic scattering	GIXS	30
	Interferometry (standing waves)		1000
	Inelastic light scattering	Raman	1000
	Photon stimulated desorption	PSD	1
Ions	Elastic scattering (Shadowing and blocking)		1–10
	Secondary ion emission	SIMS	~ 1
Atoms	Elastic scattering		1
	Inelastic scattering		1
Neutrons	Elastic scattering		Bulk
	Inelastic scattering		Bulk
Positrons	Elastic scattering		1–5
	Inelastic scattering		

scattering due to the high cross section of the low-energy electrons. High cross sections also limited the use of particle probes in studying interfaces because they cannot in most cases penetrate through the overlayer to the interface without experiencing an unacceptable attenuation. Finally, the need to conduct the measurements in a vacuum limits the types of systems that can be studied.

Of the many recent advances based on different techniques for the study of surfaces and interfaces, this chapter will focus on one: the use of high-brightness X-ray synchrotron sources coupled with the grazing incidence X-ray scattering (GIXS) geometry to extend X-ray scattering techniques to surface and interface studies. Following the recent publication of the first GIXS expement (on structural relaxations at the Al–GaAs interface) in 1979[4] and subsequent demonstration of its monolayer surface sensitivity in 1981,[5] the potential of this technique has been demonstrated in numerous studies. In view of the rapidly growing literature in this area, we shall limit this review to the work published before early 1986.

In the next section, some theoretical aspects of GIXS will be discussed, followed by a discussion of experimental techniques and synchrotron radiation sources. Experimental results obtained so far in such diverse applications as reconstructed and adsorbate structures, liquid surfaces, amorphous material surfaces, and surface phase transitions will then be described, followed by a brief discussion of potential future applications.

2. THEORETICAL BACKGROUND

The goal of structural studies is to understand correlations between atoms and molecules and their relationship to the physical properties of the system.

These studies can be straightforwardly performed for bulk systems using the classic techniques of X-ray scattering. Much of the surface-sensitive X-ray scattering results are similar to bulk experiments in basic technique and differ only in details which enhance surface sensitivity and signal rates. This section reviews the basic ideas of X-ray scattering with a formalism that makes introduction of the surface-sensitive details straightforward.

2.1. Basic Scattering Theory

The basic positional correlations of atoms in an arbitrary object can be represented by a function $\rho(\mathbf{R})$

$$\rho(\mathbf{R}) = \sum_{i=1}^{N} \delta(\mathbf{R}_i - \mathbf{R}), \tag{1}$$

where $\mathbf{R}_i$ is the location of the ith atom and the sum runs over all of the N atoms in the sample.[6]

From this correlation function, relationships for the scattered intensity can be straightforwardly derived if kinematic scattering is assumed.[6] First, the structure factor is defined as

$$S_\alpha(\mathbf{Q}) = \sum_{i=1}^{N_\alpha} \Omega(\mathbf{R}_i)e^{i\mathbf{Q}\cdot\mathbf{R}_i}$$

$$= \int_v \Omega(\mathbf{R})\rho_\alpha(\mathbf{R})e^{i\mathbf{Q}\cdot\mathbf{R}}\,dV, \tag{2}$$

where $\mathbf{Q}$ is the momentum transfer, $\rho_\alpha(\mathbf{R})$ is the quantity defined in Eq. (1) except for α-type atoms and Ω is a geometric factor which gives the effective contribution of each atom to the total scattering. The standard assumption in kinematic scattering theory is that Ω is dominated by absorption and is a slowly varying function of position (e.g., $\Omega = e^{-\mu z}$ where μ is an inverse absorption length of order $10^{-4}\,\text{Å}^{-1}$). In this case, Ω can be removed from the integral to become an overall multiplication factor. As will be seen, in GIXS, Ω can have length scales of order $10\,\text{Å}$ and hence must be explicitly retained at this stage. The scattered intensity $I(\mathbf{Q})$ is related in the standard way to the structure factor by.[6]

$$I(\mathbf{Q}) \alpha \sum_{\alpha} \sum_{\beta} f_\alpha(\mathbf{Q})f_\beta^*(\mathbf{Q})S_\alpha(\mathbf{Q})S_\beta^*(\mathbf{Q}), \tag{3}$$

where f_i is an atomic scattering factor and the sums run over all types of atoms in the sample.

It is customary in considering crystalline systems to rewrite Eq. (3) in terms of the product of two scattering amplitudes, the scattering from the regular array of unit cells (yielding the Laue conditions) and the scattering from the contents of each unit cell. In that case, the scattered intensity is given by

$$I \propto F^{*}F \frac{\sin^2 N_1 \mathbf{Q} \cdot \mathbf{a}_1}{\sin^2 \mathbf{Q} \cdot \mathbf{a}_1} \frac{\sin^2 N_2 \mathbf{Q} \cdot \mathbf{a}_2}{\sin^2 \mathbf{Q} \cdot \mathbf{a}_2} \frac{\sin^2 N_3 \mathbf{Q} \cdot \mathbf{a}_3}{\sin^2 \mathbf{Q} \cdot \mathbf{a}_3}, \tag{4}$$

where $\mathbf{a}_i$ is the unit cell size and N_i is the number of unit cells in the sample, both in the ith direction, and F, the unit cell structure factor, is given by:

$$F = \sum_{j=1}^{n} f_j e^{i\mathbf{Q} \cdot \mathbf{r}_j}, \tag{5}$$

where f_j is an atomic scattering factor and $\mathbf{r}_j$ is an atomic position within the unit cell and the sum runs over the contents of the unit cell. Since N_i is usually a large number, Eq. (4) implies that most of the scattering from a crystalline system is at very sharp peaks where $\mathbf{Q} \cdot \mathbf{a}_i = \pi$. For the quasi-two-dimensional systems which we are discussing, the number of unit cells in the third direction is small and the scattering is diffuse in that direction.

We should also note that this customary equation of X-ray diffraction does not contain a dependence on geometry (e.g., extinction) and is for that reason not strictly correct. For example, it is commonly thought that scattering is forbidden for certain Bragg reflections [e.g., the Si (200) reflection] because successive atomic planes scatter out of phase and cancel. However, when absorption is added to the sum in Eq. (5) the resultant structure factors at forbidden anti-Bragg reflections appear like the scattering from a single layer of atoms. Thus, on the scale of the scattering from a single atomic layer, forbidden reflections are always present.* Another source of monolayer-level scattering arises from two Laue conditions being satisfied while the third is between

---

* This can be illustrated by writing the scattered intensity as:

$$I \propto \left[\sum_{j=0}^{\infty} \Gamma(e^{-\mu\Delta j} - e^{-\mu\Delta(j+1/2)}) \right]^2,$$

where Γ is the scattering per layer, μ is the linear absorption coefficient, and Δ is the thickness of the atomic bilayer. Using geometric series, this equation reduces to:

$$I \propto \Gamma^2 \left(\frac{1 - e^{-\mu\Delta/2}}{1 - e^{-\mu\Delta}} \right)^2,$$

which has the limits

$$I = \left(\frac{\Gamma}{2} \right)^2 \quad \text{for small } \mu\Delta$$

and

$$I = \Gamma^2 \quad \text{for large } \mu\Delta.$$

Thus, the scattered intensity from an anti-Bragg reflection is similar in magnitude to the scattering from a single layer.

FIGURE 1. The grazing incidence X-ray scattering (GIXS) geometry. The X-rays impinge on the sample at a grazing angle ϕ and the scattered radiation is detected at a second (not necessarily equivalent) grazing angle ϕ' and at an azimuthal angle 2θ.

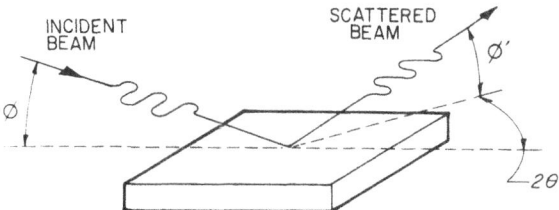

reciprocal lattice points.* Interference between these two sources of weak scattering from the bulk and from overlayers can, in principle, be used to determine the registry between the substrate and overlayer.

The discussion to this point has considered classic X-ray scattering theory. Many of the details enhancing surface sensitivity are incorporated in Ω, which we now consider. Normally, Ω is a slowly varying function of $\mathbf{r}$ (as compared to the atomic correlations involved) and is dominated by absorption. In this case, Ω can be removed from the sum and treated as an overall absorption correction (within the limits previously discussed). The situation is much more complicated for the grazing incidence case, however, because Ω can have length scales that are short compared to the atomic correlations and is anisotropic. The following section will obtain an expression for Ω for the simplest grazing incidence–total reflection case.

2.2. Effects of Grazing Incidence

The GIXS geometry is shown in Fig. 1. The X-rays are incident on the surface at an angle ϕ and are specularly reflected at the same angle ϕ. The scattered X-rays are detected at a grazing angle ϕ' with respect to the sample surface and at an angle 2θ with respect to the transmitted beam. The intensity of scattering from a layer can be calculated following the distorted-wave approximation of Vineyard.[8] In this method, the fields inside the sample (the distorted wave) are calculated followed by a calculation of the scattered intensity from this distorted wave. The final step, as pointed out by Dietrich and Wagner,[9] is calculating the observed field outside the material, including refractive-index effects on the scattered fields.

---

* It is interesting to compare the intensity of scattering from a monolayer to that from a three-dimensional crystal off of the exact Bragg condition. If two of the Laue conditions are satisfied, the scattered intensity will be proportional to

$$I \propto F * FN_1 N_2 \frac{\sin^2 N_3 \mathbf{Q}_3 d_3}{\sin^2 \mathbf{Q}_3 d_3},$$

which has the limit

$$\lim_{\mathbf{Q} d \to \pi/2} I = \frac{N_1 N_2 F * F}{2}.$$

Thus, if two of the Laue conditions are satisfied, the intensity of scattering from a bulk crystal will be at least 1/2 of that due to a monolayer.[7]

The basic features of the total reflection process are illustrated by a simple sharp interface between a vacuum and a continuous medium with a refractive index n. For X-rays, the refractive index is given by[10]

$$n = 1 - \delta - i\beta, \tag{6}$$

where δ is given by

$$\delta = \frac{\lambda^2 e^2}{2\pi m c^2} \sum_\alpha N_\alpha (Z_\alpha + f'_\alpha). \tag{7}$$

β is given by

$$\beta = \frac{\lambda^2 e^2}{2\pi m c^2} \sum_\alpha N_\alpha f''_\alpha, \tag{8}$$

and Z_α is the atomic number of species α, $f'_\alpha(k, E)$ and $f''_\alpha(k, E)$ are real and imaginary resonant (anomalous) scattering terms.

The reflectivity of an ideally flat interface can be calculated from the Fresnel formulas (e.g., see Born and Wolf[11]). We assume a simple plane wave:

$$\varepsilon = A e^{i(\mathbf{k}\cdot\mathbf{r} - \omega t)}, \tag{9}$$

where ε is the electric field, A is the amplitude, $\mathbf{k}$ is a complex wave vector, $\mathbf{r}$ is a position vector, ω is the wavenumber, and t is time. We also assume that A is parallel to the plane of the surface and that the interface is located perpendicular to the z axis at $z = 0$ (see Fig. 2). For small glancing angles (i.e., $\sin \phi \sim \phi$), the theoretical reflectivity is given by

$$I_r(\phi) = \frac{(\phi - p)^2 + q^2}{(\phi + p)^2 + q^2} I_0. \tag{10}$$

Here I_r is the reflected intensity, I_o is the incident intensity,

$$p^2 = \tfrac{1}{2}((\phi^2 - 2\delta)^2 + 4\beta)^{1/2} - \phi^2 + 2\delta, \tag{11}$$

and

$$q^2 = \tfrac{1}{2}((\phi^2 - 2\delta)^2 + 4\beta)^{1/2} + \phi^2 - 2\delta. \tag{12}$$

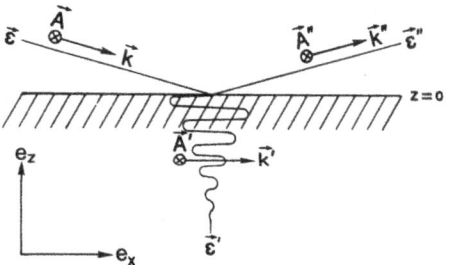

FIGURE 2. The total reflection geometry used for the calculation of the reflectivity and scattered intensity at grazing angle. The polarization vectors are normal to the page.

Since the refractive index for X-rays is slightly less than one ($\delta \sim 10^{-6}$), (almost) total external reflection occurs below a critical angle. Under these conditions, there is no energy flow into the reflector (sample) and only evanescent fields which decay over tens of angstroms are present. It is the scattering from those fields that we now discuss.

The evanescent field, calculated following Born and Wolf[11] is given by

$$\mathbf{\varepsilon}' = \mathbf{A}' e^{i(\mathbf{k}' \cdot \mathbf{r} - \omega t)}, \tag{13}$$

where the primed quantities are analogous to the quantities in Eq. (13). The complex amplitude is given by

$$\mathbf{A}' = \frac{\phi}{\phi + u_2 + i v_2} \mathbf{A}, \tag{14}$$

where

$$2u_2^2 = ((1 - \delta)^2 - \beta^2 - 1)^{1/2} + ((1 - \delta)^2 - \beta^2 - 1) \tag{15}$$

and

$$2v_2^2 = ((1 - \delta)^2 - \beta^2 - 1)^{1/2} - ((1 - \delta)^2 - \beta^2 - 1). \tag{16}$$

The vector $\mathbf{k}'$ in this solution is given by

$$\mathbf{k}' = k \cos \phi \hat{\mathbf{e}}_x - ik(\phi_c^2 - \phi^2)^{1/2} \hat{\mathbf{e}}_x, \tag{17}$$

where $\hat{\mathbf{e}}_i$ is a unit vector in the ith direction. Note that the k_z' term results in an exponentially decaying field in the $\hat{\mathbf{e}}_x$ direction (normal to the surface) with a $1/e$ depth ξ given by

$$\xi = \frac{1}{k(\phi_c^2 - \phi^2)^{1/2}}. \tag{18}$$

We can now write down a function for Ω, the effective contribution of each atom to the scattering, as[12]

$$\Omega(\mathbf{r}) = |\mathbf{A}'| e^{-k(\phi_c^2 - \phi^2)^{1/2}}. \tag{19}$$

This evanescent field penetrates from $\sim 25 \, \text{Å}$ to several microns into the reflector. We note that the surface contribution to the scattered intensity, $\Omega^*(0)\Omega(0)$, as a function of the grazing angle ϕ peaks at the critical angle. The physical interpretation of this peak is that the path length of each X-ray in the material is maximized at the critical angle, where each ray travels parallel to the surface. Thus, for monolayers, we expect a significant enhancement in signal rate at the critical angle.

Knowing the fields inside the material, the scattered intensity from this wave can be calculated within the limitations of the kinematic theory of diffraction by following Eqs. (2) and (3), with $\mathbf{Q}$ given by

$$\mathbf{Q} = (k_x'' - k \cos \phi)\hat{\mathbf{e}}_x + k_y''\hat{\mathbf{e}}_y + k_z''\hat{\mathbf{e}}_z, \qquad (20)$$

where k'' is defined in Fig. 2.

The formalism discussed so far is valid for output angles ϕ' much larger than the critical angle. For small ϕ' (near or below the critical angle), refraction effects on the diffracted beam must be included. The dominant effect in such a case is to increase the depth sensitivities. For a more complete derivation for a plane interface, the interested reader should consult the paper by Dietrich and Wagner.[9]

The above theory shows that X-rays can be confined to the surface of ideally flat samples. However, for practical cases this theory is not complete because it neglects microscopic details about the surface (including the structural information we are trying to determine). Early tests of total reflection theory were performed by Parratt.[13] He found that the reflected intensity could be adequately described using simple Fresnel theory. However, the important effects for surface science are the penetration depth and effective evanescent field intensity as a function of the grazing angle ϕ'. A fairly sensitive method of examining these effects is to study the fluorescence yield as a function of grazing angle. This quantity should be proportional to

$$\Gamma(t) = \int_0^t \Omega * (z)\Omega(z) \, dz = \xi\Omega \circledast (0)\Omega(0) \qquad \text{for } t \gg \xi, \qquad (21)$$

where t is the thickness of the layer fluorescing. This relationship was tested by Becker et al.[14] who measured the fluorescence from polished Ge (111) surfaces, as shown in Fig. 3. Clearly, theory and experiment agree exceptionally well for this case.

Another probe of the grazing-incidence effect is the intensity of scattering from a sample. A particularly simple case involves the scattering from an amorphous film, because this scattering is generally very diffuse and isotropic and oscillates about the free-atom scattering intensity. Fischer-Colbrie and Fuoss[15] have studied the scattering from a 250 Å-thick $GeSe_2$ film deposited on a Si wafer, as well as a variety of other thin amorphous samples. The intensity of scattering at the principal maximum of the 250-Å a-$GeSe_2$ sample ($|\mathbf{Q}| \sim 2.2 \, \text{Å}^{-1}$) is plotted as a function of ϕ in Fig. 4. Also shown in Fig. 4 are the ideal reflectivity and the integrated intensity. The integrated intensity agrees fairly well with the data except in the region of the critical angle, the discrepancies at which are due mainly to macroscopic bowing of the surface. This bowing results in a variety of critical angles and substantial errors near the critical angle. Although other effects such as surface roughness cannot be ruled out, the very abrupt drop in scattered intensity at low angle is certainly due to limited penetration of the X-rays resulting from refractive-index effects.

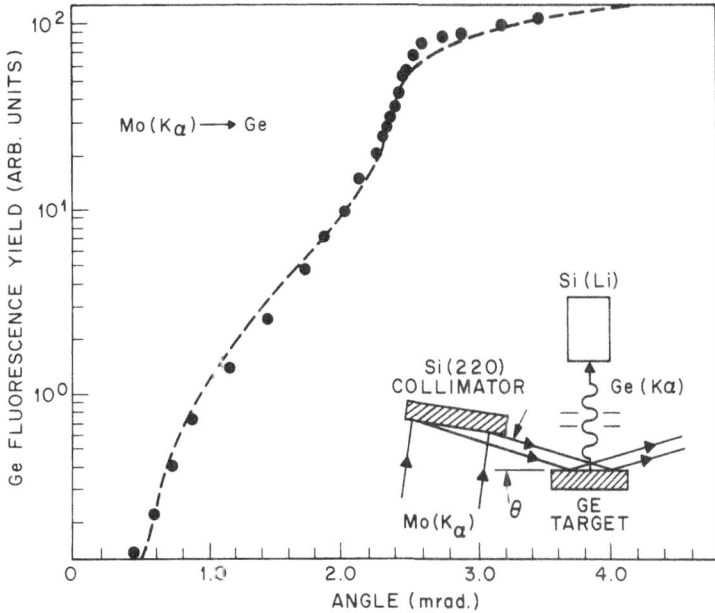

FIGURE 3. The theoretical reflectivity, integrated intensity, and measured intensity as a function of grazing angle from a polished Ge surface.

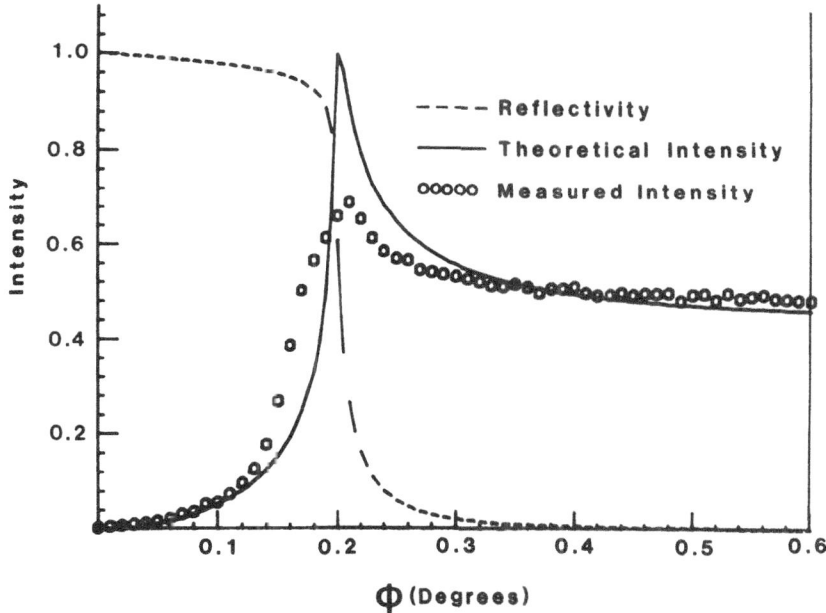

FIGURE 4. The scattered intensity from an amorphous $GeSe_2$ sample using 11-KeV photons at a momentum transfer of $k = 2.2\,\text{Å}^{-1}$. Note the sharp peak at 0.2°. The extremely rapid falloff of the intensity below 0.2° is due to a convolution of the rapidly decreasing penetration depth with the also decreasing electric field intensity in the sample.

2.3. Dynamic Scattering Considerations

It should be pointed out that most GIXS studies performed so far have been analyzed using kinematic scattering theory. The kinematic approach breaks down for some physical problems. For example, diffraction at the integer-order reflections from Ge must be described using a dynamical diffraction theory. Such a theory is essentially a special case of a three-beam dynamic diffraction theory since the totally reflected beam can be considered as the (000) beam. Theories for these situations have been developed by Afanas'ev and Melkonyan[16] and by Cowan,[17] and experiments performed by Cowan and coworkers[18] yield a fairly complicated energy flow between the specularly reflected wave and the diffracted beam.

2.4. Surface Structural Determination

From the above discussions, we know how the scattered intensities of X-rays from a surface layer can be obtained. We now give some brief background on how the surface structure can be determined from the measured scattered intensities. The study often includes the determination of the atomic coordinates and charge distribution of the 2-D layer of both the in-plane and perpendicular correlations, the registry of this layer (commensuration) with the underlying 3-D structure, and the defects in the 2-D layer. These goals can usually be achieved by straightforward extensions of standard X-ray analysis techniques employed for the 3-D structures.

2.4.1. Ordered Structures

The most well-developed theories for X-ray structural analysis naturally are in the area of crystallography, which involves the determination of atomic coordinates and electron densities of an ordered structure. Two approaches are commonly applied. The first uses direct Fourier inversion of the scattering data to calculate a Patterson map of the electron density.[19] From the Patterson map, approximate atomic coordinates can usually be determined. These approximate coordinates are then least-squares refined to determine a more precise structure. For simple systems it is possible to skip the Patterson step and proceed directly to the modeling refinement. Another set of techniques consists of the so-called direct methods.[20] For detailed discussions of the direct methods, readers are referred to the standard texts.

The starting point for the Patterson technique is to rewrite the scattered intensity of Eq. (3) in terms of the electron density:

$$I(\mathbf{Q}) \sim |S(\mathbf{Q})|^2 = \int\int \rho(\mathbf{u})\rho(\mathbf{x} + \mathbf{u}) \exp\left(-2\pi i \mathbf{Q} \cdot \mathbf{x}\right) d\mathbf{u}_x \, d\mathbf{x}_u \qquad (22)$$

$$= \int P(\mathbf{x}) \exp\left(-2\pi i \mathbf{Q} \cdot \mathbf{x}\right) d\mathbf{x}_x,$$

where

$$P(\mathbf{x}) = \int \rho(\mathbf{u})\rho(\mathbf{x} + \mathbf{u})\, d\mathbf{u}. \tag{23}$$

The function $P(\mathbf{x})$ in real space is called the Patterson function, which is the Fourier transform of the intensity function $I(\mathbf{Q})$ in reciprocal space,

$$P(\mathbf{x}) = \int I(\mathbf{Q}) \exp(2\pi i \mathbf{Q} \cdot \mathbf{x})\, d\mathbf{Q}. \tag{24}$$

Experimentally, after $I(\mathbf{Q})$ is measured, the question is reduced to the determination of the Patterson function. However, it should be clear that $P(\mathbf{x})$ is the average value of the product of the electron densities at two points separated by the vector $\mathbf{x}$. X-ray scattering therefore directly reveals only a highly complex function of the distribution of the electron cloud in the object, and it is often difficult to interpret this function directly in terms of a particular structure.

With respect to surface analysis, it should also be clear that if a set of reflections at some small $\mathbf{Q}_z$ normal to the 2-D layer is measured, the subsequent Patterson map is a projection of the 3-D structure onto a plane. This limited information is often sufficient to determine the 2-D structure, but ambiguities are likely to remain. In general, it is necessary to measure the integrated intensity along the rod of scattering to remove this ambiguity. This is a much larger data-collection task and has not been completely implemented for any system (although limited data have been collected in some cases, these are discussed in section 4).

After a preliminary structure has been determined, either by generating a Patterson map or by postulation, the structure can be refined by a fitting analysis. In general, this is accomplished by systematically varying the atomic coordinates to minimize a function which gives a "goodness of fit." Two common choices are the R factor, given by[3]

$$R = \frac{\sum |I_{\text{measured}}(h, k) - I_{\text{calculated}}(h, k)|}{[I_{\text{measured}}(h, k)]^{1/2}}, \tag{25}$$

and chi-squared, given by[21]

$$\chi^2 = \frac{1}{N_{\text{points}}} \sum \frac{S_{\text{measured}}(h, k) - |S_{\text{calculated}}(h, k)|}{\sigma^2(h, k)}, \tag{26}$$

where $\sigma(h, k)$ is the uncertainty associated with that reflection. The R factor has the advantage of ease of calculation and a general transferability. (However, we should note that there are numerous definitions of the R factor, and that they are not equivalent.) The χ^2 analysis incorporates a measure of the statistical accuracy of each peak into the analysis and has the advantage that the solution is statistically valid if $\chi^2 < 1$.

Following a calculation of the 2-D structures' atomic coordinates, it is desirable to determine the registry of the substrate with the overlayer (if the structures are commensurate). The immediate thought is to look for systematic variations in the Bragg reflections where the scattering from the bulk and the layer overlap. Unfortunately, the Bragg reflections from the 3-D substrate are many orders of magnitude more intense than the overlayer reflections, and the interference effects are negligible. Thus, the use of interference between the extremely weak scattering at the forbidden reflections of the bulk (see section 2.1) and the overlayer to determine the substrate–overlayer registration has been suggested.[7] Care needs to be exercised, as this weak scattering from the substrate is dependent on interface details (e.g., surface roughness).

2.4.2. Surface Defects

While it is often possible to deduce the actual physical arrangement of defects at interfaces, the development of scanning tunneling microscopes[22] and atomic-resolution UHV electron microscopes[23] suggests that they will be the dominant techniques for studying the nature of defects, at least for highly ordered films. However, statistical information about the nature and distribution of extended and point defects in a 2-D overlayer is often desired. Statistical information about extended defects is contained in the lineshape of the Bragg reflections, as is information about point defects in diffuse scattering at both Bragg and anti-Bragg reflections. The analysis of this information is again heavily dependent on Fourier analysis techniques. There seems to be little difficulty in adapting classical X-ray analysis techniques to the analysis of surface defects, and the reader is referred to the classic texts[6] for detailed discussions. Specific examples such as determining correlations between atomic steps on surfaces and between antiphase domains are included in section 4.

2.4.3. Amorphous and Liquid Structures

The structure of amorphous and liquid surfaces is potentially a very rich area for research since current structural probes have failed to yield information about these systems. The traditional radial distribution function (RDF) analysis needs considerable modification for these problems, primarily because RDFs are typically calculated by Fourier inversion of the scattered intensity given by (for a one-component system)[6]

$$I(k) = Nf^2 + Nf^2 \int_0^\infty 4\pi r^2 [\rho(r) - \rho_0] \frac{\sin(kr)}{kr} \, dr. \qquad (27)$$

Here $\rho(r)$ is the radial distribution function, ρ_0 is the average electron density, and $I(k)$ is the scattered intensity. The derivation of this relationship explicitly assumes that the material is spherically symmetric when averaged over the sample.

For a 2-D amorphous or liquid system, this assumption is not necessarily valid and a cylindrical correlation function is more appropriate (although it may turn out that the a 2-D amorphous layer does not even have complete cylindrical symmetry). For cylindrical symmetry, the appropriate correlation function for a one-component system is $\rho(R, z)$ and is related to the scattering by[24]

$$I(k_R, k_x) = 2\pi \int_{z_{min}}^{z_{max}} \int_0^\infty (\rho(R, z) - \rho_0)J_0(k_R, R)e^{ik_z z}R \, dR \, dz, \qquad (28)$$

where k_R is the in-plane wavevector, k_z is the normal wavevector, R is a distance in the plane of the surface, z is normal to the surface, and J_0 is the zero-order spherical Bessel function. Clearly, this function is much more difficult to invert than for the spherical case. If k_z is zero, a transform yields the projection of the amorphous layer onto the surface. This projection is, however, much different from a RDF. For example, a common test of the quality of an RDF is to look for nonzero contributions below the shortest interatomic distance. For this cylindrical projection, however, low R ripples are allowed since there can be projections of bond lengths that are shorter than actual bond distances.

A detailed understanding of the collection and analysis of scattering from 2-D amorphous systems is still being developed. One expects that the analysis will be much more complex and model-dependent than similar scattering from 3-D amorphous systems. On the other hand, observing transitions from 2-D to 3-D behavior may greatly improve our understanding of both regimes of amorphous structure.

3. INSTRUMENTATION

This section will discuss the instrumentation required to perform GIXS measurements following the principles discussed in the previous section. Because the field is so diverse in application, this is a formidable task. We will therefore concentrate on three major aspects. First, we will discuss the synchrotron X-ray sources available for these measurements and their near-term development. Second, we will examine the basic requirements of surface-sensitive X-ray scattering apparatus. Finally, we will discuss the integration of surface science and X-ray scattering apparatus.

3.1. X-Ray Sources

The two primary difficulties presented in measurements of X-ray scattering from surfaces are achieving a detectable signal rate (i.e. the incident photon beam must be sufficiently intense) and separating the signal from the substrate background. To accomplish these two goals, synchrotron radiation sources are generally required, due to their high intensity and extremely small divergences. Although certain classes of problems can be studied with conventional rotating-anode X-ray tubes, the number of those experiments is limited. Thus, this section will concentrate only on synchrotron radiation sources and, in particular, on their properties which make them useful for surface-sensitive X-ray diffraction sources.

At the present time, synchrotron sources produce 7-to-8-KeV photon fluxes of, for example, $\sim 1 \times 10^{13}$ photons/sec/mrad at the NSLS[25] and $\sim 1 \times 10^{14}$

photons/sec/mrad from the SSRL 54-pole wiggler.[26] These are ideal numbers, and the realities of X-ray optics reduce them substantially, but they are within an order of magnitude of the experimentally determined values. Estimates of total scattering from a monolayer can easily be obtained. The scattering cross section of an atom of Si, for example, is 3.7×10^{-23} cm^2.[27] Thus, a single amorphous layer of atoms would scatter on the order of 10^{-7} of the incident photons resulting in total scattering rates of between 10^5 and 10^7 photons/sec (assuming that all the photons impinge on the surface). While this signal rate is distributed over 4π sr, it should be easily observed.

Normally, synchrotron scattering experiments are performed with both the incident and scattered beams lying in a vertical scattering plane. This geometry is used because of the high polarization of the source in the horizontal plane and its small vertical divergence. The electron source size in typical synchrotrons ranges from 5 mm × 0.38 mm at SSRL to 0.5 mm × 0.2 mm at the NSLS, with the first number being the horizontal beam size. Hence, a 1-cm sample at a 3-mrad grazing angle will accept only 0.5 percent of the beam at SSRL and 6 percent of the beam at the NSLS (assuming one-to-one imaging of the source). It is possible, of course, to demagnify the image, but that increases the horizontal divergence. To achieve surface sensitivity and/or reasonably aberration-free scattering, the horizontal divergence should be limited to <1 mrad, with 0.1 mrad being desirable.

Adding the constraint of horizontal divergence, the relevant figure of merit becomes the horizontal brilliance, photons/sec/mm/rad (which is a conserved quantity). Clearly, based on the previous argument, the product of the divergence and the horizontal source size should be $<3 \times 10^{-3}$ mm mrad. Limiting an experiment to an emittance of this amount reduces the useful fluxes discussed above to $\sim 6 \times 10^{10}$ for the NSLS bending magnet and the SSRL 54-pole wiggler. In the near future, undulators on existing and planned storage rings should improve these figures to $\sim 10^{12}$ photons/sec.

As discussed, the horizontal size of the synchrotron beam is not yet ideal for GIXS studies where the scattering plane is vertical. However, for looking at systems with short correlation lengths where high resolution is not required, a horizontal scattering plane is preferable. For example, studies of amorphous thin films at the NSLS with the conditions described above would result in a ratio of signals between the horizontal and vertical scattering geometries of ~ 25. On the other hand, for high-resolution scattering experiments, the ratio of the divergence in horizontal and vertical planes essentially cancels the ratio of the sizes, and the vertical scattering plane is preferred because of polarization corrections.

3.2. Surface Scattering Diffractometers

Early GIXS experiments were mainly performed using special sample chambers attached to the existing four-circle diffractometers. As generalized scattering instruments, these diffractometers have the advantage that they are readily adapted to a variety of experiments and hence make maximum use of limited synchrotron resources. However, they are not constructed to fit the special needs of GIXS experiments (e.g., the large weight and size of a UHV chamber). Recently, specialized instruments have been developed which allow for much more robust instruments incorporating different geometries. In the case of UHV surface experiments, two types of diffractometers have been developed which include complete sets of traditional surface-science diagnostic probes.

The first type of instrument was designed by Brennan and Eisenberger[28] and is based on the z-axis four-circle geometry. This instrument, shown in Fig. 5, is mounted on a large goniometer with a vertical axis. This goniometer is used to set the incident angle ϕ. On this goniometer is mounted a two-circle diffractometer, which is used to set θ and 2θ. Through the axis of this two-circle diffractometer is mounted the vacuum chamber with a LEED/Auger system, sputter system, and various sample dosers. The detector moves on a fourth circle which is used to set

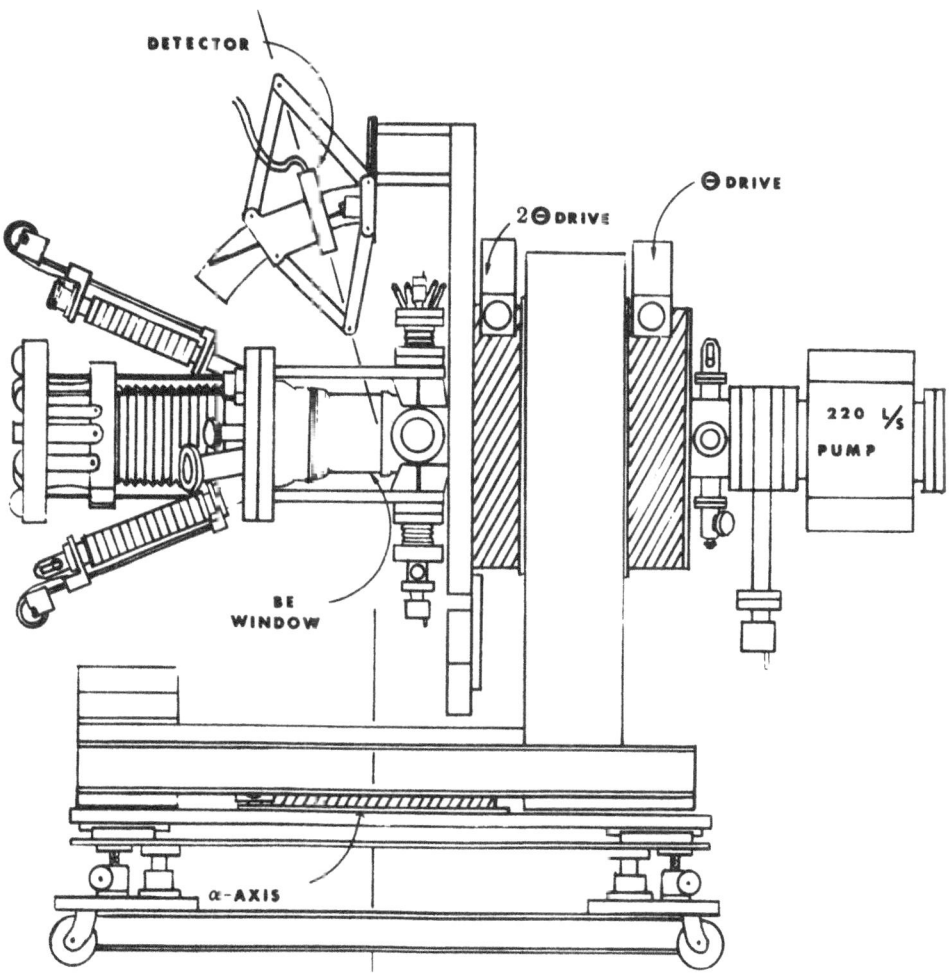

FIGURE 5. Illustration of the Brennan and Eisenberger scattering diffractometer, pointing out the important components. The three goniometers affect angles ϕ, θ and 2θ, while the detector angle ϕ, is independently controlled. The vacuum chamber extends from the pump through the two goniometers to the window and the bellows where the LEED unit, ion sputter gun, and metal evaporator are located.

the output angle ϕ'. X-rays enter and leave the chamber through a 360° Be window which allows ϕ values up to 45°. The advantages of this arrangement are

1. the sample is never moved from the scattering position, so that simultaneous surface-science and X-ray analysis can be performed;
2. the mechanical arrangement is very robust, so that large weights can be carried on the diffractometer; and
3. evolution from 2-D to 3-D physics can be easily followed, since the diffraction perpendicular to the sample surface can be studied easily.

The system has the disadvantage that the range of surface-science tools that can be used at any one time is limited by access to the sample.

Another type of instrument, which can function in either the z-axis or a limited conventional four-circle mode, has been constructed by Fuoss and Robinson.[29] This instrument, shown in Fig. 6, has a number of features. First, the surface-science section and the X-ray diffractometer are relatively separate entities connected only by a differentially pumped rotary seal with teflon O-rings. The X-ray diffractometer is similar to a conventional four-circle diffractometer, except that the range of the x motion is ±10° instead of the 360° of the standard instrument. This limited range is dictated by a long sample manipulator that transfers the sample from an X-ray scattering position in the Be window section to a surface-science portion and, if desired, through the standard surface-science section into a specialized vacuum chamber (not shown). This entire instrument is supported by air pads floating on an optical table and can be rotated about a

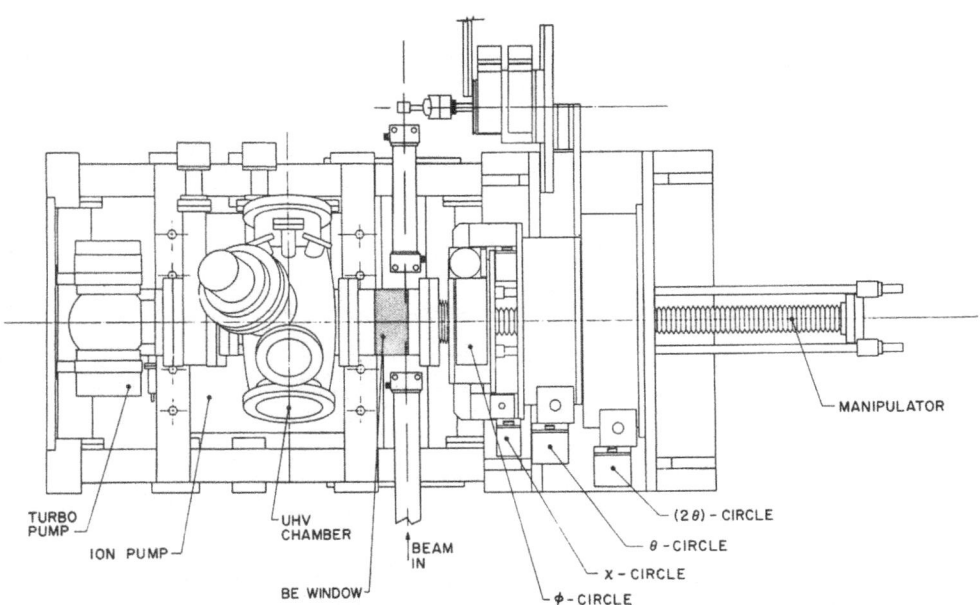

FIGURE 6. A top view of the Fuoss and Robinson instrument. This entire instrument floats on airpads that allow the incident angle ϕ to be independently adjusted.

vertical axis through the sample, thus implementing the z-axis configuration. The advantages of this instruments are that

1. an extremely wide range of surface-science equipment can be incorporated into the design;
2. provision can be made for alternate surface-science chambers so that non-UHV materials can be studied;
3. the modes of X-ray diffraction are extremely flexible; and
4. there are no blind spots due to Be-window supports.

The difficulties are that

1. the sample must be moved to perform traditional surface-science experiments (which may compromise X-ray diffraction reproducibility and certainly makes simultaneous experiments difficult),
2. the differentially pumped seal is a potential vacuum problem, and
3. the very long manipulator may cause outgassing problems.

These two designs are representative of current efforts in generalized UHV surface diffraction chambers. A number of these chambers are being built, and all have unique features. Continuous improvements are expected in the near future to reach an optimum design. In addition to generalized instruments, there are much-more-specialized chambers optimized for specific experiments. For these problems, a full chamber is not necessary and more rapid progress can be made with a simple system.

One of the great promises of X-ray diffraction from surfaces is that it may be possible to study surfaces under normal conditions of pressure, temperature, etc. For example, chemical reactions are normally carried out at atmospheric pressure or above. UHV studies using the same reactants do not necessarily give the correct answer, because different reaction channels may open up at high pressure. Surface diffraction chambers are being built to study such problems. An approach being pursued by Liang and co-workers is to construct a UHV chamber for surface preparation and a high pressure cell to study chemical reactions.[30] Other studies of, for example, liquid surfaces by Als-Nielson and Pershan[31] (see section 4.3.1) do not require UHV conditions at any step and can be performed on relatively simple instruments.

From this quick summary of the existing equipment, it should be obvious that there are a great many ways of incorporating surface-sensitive X-ray scattering into specific experimental problems. We expect that a similar review made in a few years will yield a wide variety of innovative ideas. For example, a number of researchers are constructing extremely compact UHV systems, which will mount on conventional four-circle diffractometers.

3.3. Diffractometer Design Considerations

Before leaving the subject of instrumentation, we should explore briefly the optical design principles of the two type of UHV surface diffractometers discussed above. We will first examine the detector-acceptance function in reciprocal space of a z-axis diffractometer, which implements the geometry described in the previous section in a very straightforward way. First, the incident grazing angle is fixed by rotating the plane of the sample about a vertical axis (i.e., keeping the surface normal in a horizontal plane). Second, the θ and 2θ angles of the incident and scattered wave vectors projected onto the sample surface are set and, finally, the output grazing angle is fixed. This geometry keeps the broad horizontal divergence of the synchrotron along the surface normal. Mathematically, this can be expressed as[32]

$$\mathbf{k} = \frac{2\pi}{\lambda}\{((\cos\theta\cos\phi - \cos\theta'\cos\phi')\hat{\mathbf{e}}_x$$
$$+ (\sin\theta\cos\phi + \sin\theta'\cos\phi')\hat{\mathbf{e}}_y + (\sin\phi + \sin\phi')\hat{\mathbf{e}}_z\}, \quad (29)$$

where the y direction is along the in-plane momentum transfer, the x direction is normal to the y direction and in the plane of the sample, and z is normal to the plane of the sample. The detector-acceptance function can be calculated from these equations by differentiating with respect to the appropriate angles and is:

$$d\mathbf{k}_x = \frac{2\pi}{\lambda}(\cos\theta\sin\phi\,d\phi + \cos\theta\sin\phi'\,d\phi'$$
$$+ \sin\theta\cos\phi\,d\theta + \sin\theta'\cos\phi'\,d\theta') \quad (30a)$$

$$d\mathbf{k}_y = \frac{2\pi}{\lambda}(\sin\theta\sin\phi\,d\phi + \sin\theta'\sin\phi'\,d\phi'$$
$$+ \cos\theta\cos\phi\,d\theta + \cos\theta'\cos\phi'\,d\theta') \quad (30b)$$

$$d\mathbf{k}_z = \frac{2\pi}{\lambda}(\cos\phi\,d\phi + \cos\phi'\,d\phi'). \quad (30c)$$

The same function for the conventional four-circle diffractometer can be obtained from these equations for the $\theta = 0$ mode by writing the transform:

$$\begin{pmatrix} k'_x \\ k'_y \\ k'_z \end{pmatrix} = \begin{pmatrix} 1 & 0 & 0 \\ 0 & \cos x & -\sin x \\ 0 & \sin x & \cos x \end{pmatrix}\begin{pmatrix} k_x \\ k_y \\ k_z \end{pmatrix}, \quad (31)$$

where the primed quantities are in the four-circle frame of reference and ϕ and ϕ' are approximately zero. Thus, the detector-acceptance function for this system is given by

$$d\mathbf{k}'_x = \frac{2\pi}{\lambda}\sin\theta(d\theta + d\theta')\hat{\mathbf{e}}_x \quad (32a)$$

$$d\mathbf{k}'_y = \frac{2\pi}{\lambda}\{\cos x\cos\theta(d\theta + d\theta') - \sin x(d\phi + d\phi')\}\hat{\mathbf{e}}_y \quad (32b)$$

$$d\mathbf{k}_z = \frac{2\pi}{\lambda}\{\sin x\cos\theta(d\theta + d\theta') + \cos x(d\theta + d\theta')\}\hat{\mathbf{e}}_z. \quad (32c)$$

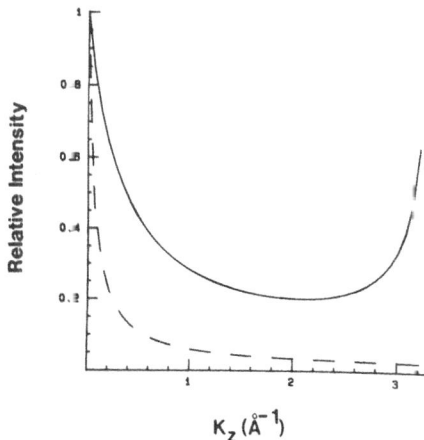

FIGURE 7. Calculated peak intensity from a resolution-limited rod for (broken line) a conventional four-circle diffractometer and (solid line) a z-axis diffractometer. The in-plane momentum transfer is correct for a Si (110) reflection. $\Delta\Theta = \Delta\Theta' = 0.0001$; $\Delta\phi = \Delta\phi' = 0.005$.

These functions have a strong practical impact on the measured intensities because the scattering is relatively diffuse normal to the surface and, typically, very sharp in the plane. For example, Fig. 7 shows the calculated peak intensity along a resolution-limited diffraction rod. Clearly, the z-axis arrangement yields a stronger and less distorted *peak* intensity. However, the integrated intensity would not show these effects, and if the signal-to-background ratio is high there is little practical difference between the two instruments.

4. CURRENT APPLICATIONS

Many two-dimensional systems have been the focus of research for consensed-matter scientists in recent years. These include examples in diversified scientific disciplines such as the study of single-crystal surfaces, adsorbed monolayers, membranes, and interfaces between materials. In these areas, surface X-ray scattering techniques have been rapidly emerging as a vital atomic and molecular probe to examine the nature of these systems. At present, the application of the GIXS technique to surface problems is still quite new, but substantial progress has already been made and the future potential of this technique is clear. In the following sections, we group the review of GIXS experimental results into three major areas: the structure of solid surfaces and interfaces, surface phase transitions, and liquid and liquid-crystal surfaces.

4.1. Structure of Solid Surfaces and Interfaces

Since the development of the LEED technique in the early 1960s, a large number of single-crystal surfaces have been observed to undergo structural rearrangements of the top most atomic layers of the truncated surfaces.[1] These rearrangements are driven by a lowering of the free energy at the bulk–vacuum interface and are sensitively dependent on the exact properties of the material (e.g., Ge and Si have much different reconstructions). Although a variety of theories have been developed to explain the stabilization of many surface phases, definitive explanations are not expected before detailed surface structures become known. In the following, we will review four structural areas where GIXS has been applied: semiconductor surfaces, metal surfaces, buried interfaces, and amorphous films.

4.1.1. Reconstructed Semiconductor Surfaces

Semiconductor surfaces undergo a wide variety of surface structural rearrangements and were the first surfaces studied with GIXS. Since the study of Ge (001) (2 × 1) surface by Eisenberger and Marra,[5] a number of other surfaces have been examined including InSb (111) (2 × 2),[33] InSb (111) (3 × 3),[34] and Si (111) (7 × 7).[35]

The study of Ge (001) (2 × 1) surfaces demonstrated for the first time the monolayer sensitivity of the GIXS technique. Prior to the X-ray study, LEED had been used to study the Ge (001) (2 × 1) surface, but there was little agreement between theory and experiment. Using a 60-Kw rotating-anode X-ray source and a standard four-circle diffractometer equipped with a transferable X-ray cell (section 3.2) at Bell Laboratories, Eisenberger and Marra were able to measure Bragg intensities from five half-order reflections. The results of a least-squares fitting analysis of these data agree with a paired-atom model whose

TABLE 2. Observed and Calculated Surface-Structure Factors for InSb (111)-2 × 2[a]

h	k	F_{hk}^{obs}	$F_{hk}^{calc\ b}$	$F_{hk}^{calc\ c}$
1/2	0	3.5 ± 0.2	9.0	4.0
1/2	1/2	6.2 ± 0.5	9.4	6.4
1	1/2	21.1 ± 0.4	10.0	21.0
3/2	0	32.1 ± 0.6	30.0	31.6
3/2	1/2	8.6 ± 0.5	3.1	9.0
3/2	1	7.6 ± 0.5	13.8	8.0
3/2	3/2	9.3 ± 1.0	11.3	11.6
2	1/2	18.5 ± 9.0	15.9	19.0
2	3/2	17.5 ± 0.8	9.3	17.3
5/2	0	26.2 ± 1.1	14.6	22.0
5/2	1/2	16.6 ± 1.0	5.2	17.5
5/2	1	29.4 ± 1.2	26.5	29.2
5/2	3/2	14.1 ± 0.8	9.0	14.7
3	1/2	3.8 ± 2.6	4.1	2.6
7/2	0	8.9 ± 2.2	12.2	9.1
7/2	1/2	12.8 ± 2.0	12.6	13.7

[a] The indices (h, k) used to label the reflections refer to the hexagonal coordinate frame defined in terms of the cubic lattice by $[1, 0]_{hex} = [1/2, -1/2, 0]_{bulk}$ and $[0, 1]_{hex} = [0, 1/2, -1/2]_{bulk}$.
[b] Values calculated for the six-atom model derived directly from the Patterson function after adjustment of a scale factor only.
[c] Final four-parameter model.

displacements are close to Chadi's[36] silicon calculations when scaled for Ge by their relative lattice parameters. However, their results also revealed the need for second-layer displacements consistent with Appelbaum and Hamann's[37] prediction that these reconstructions induce subsurface strain.

The study of InSb (111) (2 × 2) was performed at Deutsches Elektronen-Synchrotron from the storage ring DORIS in Hamburg by Bohr *et al.*[33] The study is a prototypical example of surface reconstruction of compound semiconductors. In this work, sixteen Bragg reflections were collected with a position-sensitive detector with the incident X-rays kept near the critical angle of InSb (0.236°). It was found that the rod profiles were flat except near the critical angle. The authors therefore concluded that the surface diffraction arose from a single plane of atoms. Using the total intensity under each rod corrected for polarization, Lorentz factor, and sample area, the experimental-structure factors were determined (Table 2). Since a large number of reflections were obtained, it is of particular significance that a Patterson analysis was applied to reveal novel structural features of this surface directly.

Figure 8a shows the Patterson function calculated from the observed structure factors in Table 2. The figure reveals three clear nonorigin peaks that must correspond to interatomic vectors in the surface structure. Such peaks can not come from a truncated bulk structure in which the repeating unit is a covalently bonded hexagon, such as the ones shown as open circles in Fig. 8b. The authors examined a distorted hexagon structure, the shaded circles in Fig. 8b, which has a simple arrangement of atoms with 3m symmetry that agrees with the Patterson function. But in the structural refinement step, a structure-factor

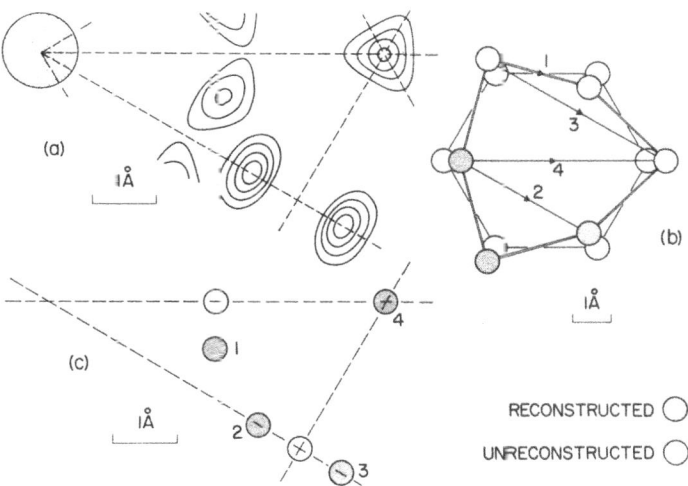

FIGURE 8. The results of Bohr *et al.* on the InSb (111) surface: a) Repeating unit of the Patterson function calculated for the F^{obs} in Table 1. Positive contour levels above zero are shown. Dashed mirror lines surround the asymmetric repeating unit. The shaded circle is the origin peak that rises 17 contour levels. (b) Distortion of a hexagonal arrangement of atoms taken from the projected unreconstructed InSb (111) surface. (c) Pair-correlation peaks 1 to 4 derived from vectors 1 to 4 in (b).

calculation performed with these six atoms of the distorted hexagon in the (2 times 2) unit cell showed poor agreement ($\chi^2 = 125$) with this model. The authors further examined the Patterson map of the difference electron density to reveal a seventh atom sitting on one of the threefold axes. The seven-atom structure had a residual of $\chi^2 = 25$, which dropped to 2.1 when the positions of the atoms were least-squares refined and a thermal-vibration parameter was included. Finally, by allowing the atomic number of the outer three atoms of the distorted hexagon to vary, the authors concluded that the structure had three In and four Sb atoms in the (2×2) unit cell. That is, the projection of a single layer of InSb has a vacant In site. Such a structure is remarkably similar to that of GaAs (111) (2×2) determined by LEED analysis.[38]

The results of this InSb work clearly demonstrate the power of the GIXS technique in surface crystallography. More recently, work was done by Robinson *et al*[35] on an even more complicated surface, Si (111) (7×7). The complication of this surface is partially due to the large unit cell of reconstruction and to the involvement of at least four layers of atoms. This surface has been extensively studied using various techniques, including LEED,[39] low-energy ion scattering,[40] scanning tunneling microscopy,[41] ion channeling[42–44] and transmission electron diffraction.[45] The X-ray measurements were performed on an eight-pole wiggler source at the Stanford Synchrotron Radiation Laboratory (SSRL). The results of this X-ray work are consistent with a stacking-fault model originally revealed by transmission electron diffraction. This work shows that, even for a complex reconstructed surface, surface structures can be determined in great detail using GIXS in conjunction with other surface techniques.

4.1.2. Reconstructed Metal Surfaces

Metal surfaces can exhibit a variety of reconstructions and interlayer relaxations. These structural transformations can take place on clean surfaces or can be induced by chemisorption of foreign species. Prototypical systems of *fcc* (110) surfaces of Au and Cu have been studied using GIXS.

The (110)-terminated *fcc* structure consists of atoms in close-packed rows along surface [$\bar{1}$10] directions spaced apart by one unit cell along the perpendicular [001] direction. The clean surface of Au (110) reconstructs and doubles the unit-cell size along the [001] direction. Several models have been proposed for the observed (1×2) reconstruction of Au (110) to explain this doubling of the unit cell. Most experimental evidence favors the class of missing-row models[46] in which every second [$\bar{1}$10] row of top-layer atoms is omitted. Other structures like paired-row and buckled models with alternating top-layer row displacements (parallel or perpendicular to the surface) were also suggested. A GIXS study of the Au (110) surface was performed by Robinson[46] using the 60-Kw rotating anode at Bell Laboratories. A direct Patterson synthesis of the X-ray diffraction intensities yielded a missing-row structure for the Au (110) surfaces with lateral second-layer displacements.

The unique power of high-resolution X-ray scattering was demonstrated in the new salient features of Au (110) surfaces revealed by examining the observed intensity profiles of the surface reflections. First, it was found that the intensity profiles along the rods of the superlattice reflections were not constant (Fig. 9). Second, the peak positions of superlattice reflections are systematically displaced for different samples studied (Fig. 10). The nonuniform rod profiles indicate the prescence of extended structural rearrangements normal to the top layer of surface. By least-square fitting of the profiles, the results indicated that the spacing between the top two layers of the (1 × 2) reconstructed surface was 2.06 ± 0.44 Å, i.e., an expansion of 40 ± 30%. This expansion of the outer layer is contrary to most cases known on clean metal surfaces and is still a subject of controversy. On the other hand, explanation of the observed displacements of the superlattice reflections is more straightforward. The results can be reconciled with those of ion scattering and tunneling electron microscopy by a domain structure which has [001] monoatomic steps with the exposed {111} surfaces as domain walls.[47] These results give an illustration of how X-ray diffraction profiles obtained in GIXS can be used for the study of surface defects. For 3-D systems, the employment of X-ray diffraction in defects (point, line, or plane) has been well established in both theory and experiment.[5] One can expect that high-resolution GIXS studies using synchrotron radiation should also yield important knowledge about surface and subsurface defects.

Another example is a chemisorption-induced surface reconstruction. A clean Cu (110) surface maintains the in-plane structure of the terminated bulk with an oscillatory relaxation of the surface layers. Oxygen adsorbed on this surface is known to form a (2 × 1) structure near a saturation coverage of 0.5 monolayer with the doubling of the periodicity along the [$\bar{1}$10] direction. Even for such a simple system, the exact atomic arrangement of this surface is still a subject of dispute since LEED, low-energy ion scattering, high-energy ion scattering, and helium diffraction results supported different models. Analogously

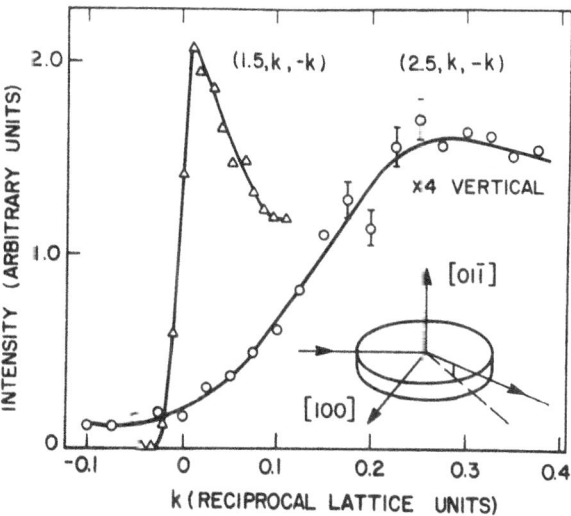

FIGURE 9. Rod profiles parallel to [01$\bar{1}$] to two superlattice reflections performed at constant (glancing) incidence angle. The profile has intensity above the surface only; the resolution-limited edge at $k = 0$ corresponds to eclipsing of the exit beam by the sample.

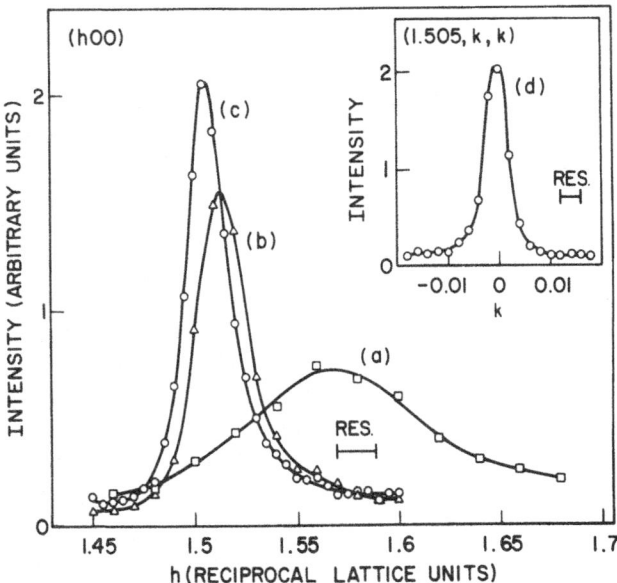

FIGURE 10. Scans of the superlattice $\frac{3}{2}00$ Bragg diffraction peaks from a Au (110) (2 × 1) reconstructed surface: (a)–(c) Radial scans along [100], resolution (full width at half maximum) indicated. (a) A radial scan along the [100] of a sample with the surface inclined by 1.5° from the [100] (maximum counting rate 0.6 count/sec). (b) A radial scan of a second sample along an incline of 0.1° (maximum counting rate 2 count/sec). (c) A radial scan on the sample in (b) after preparing the surface for a second time (maximum counting rate 10 counts/sec). (d) Transverse scan, parallel to [011], of the peak in (c).

to the Au (110) case, commonly proposed models include the missing-row model, in which alternating [001] rows of Cu atoms are missing, and the buckled-row model, in which the nonoxygen-containing [001] Cu rows of the top layer are displaced significantly outwards from their bulk position.

GIXS was measured from the O–Cu (110) (2 × 1) surface at SSRL using the intense 54-pole wiggler source. The results demonstrated the feasibility of using GIXS for the study of low Z adsorbates.[48] However, the complexity of this surface was shown in a later study of the coverage dependence of the superlattice structure.[49] Using a bulk Cu crystal with a small mosaic width (0.12°) and after extensive routines of surface preparation by mechanical polishing, electrochemical polishing, and UHV surface cleaning, a spread of surface mosaics were found with some having exceedingly good widths (≈0.5°). The GIXS scans of the half-order superlattice reflection along the (110) direction following oxygen exposure are shown in Fig. 11. A very sharp peak (resolution-limited) was observed following exposure to ≈50 L of oxygen. This peak had the same mosaic (≈0.05°) as that of the starting Cu surface. At higher oxygen exposures, this reflection became broad with an additional shoulder at an incommensurate ($Q \approx 0.515$ unit) position. These results suggest that this reconstruction probably takes place in different stages which can not be observed by LEED because of limited resolution.

The above discussions demonstrate that, on single-crystal metal surfaces,

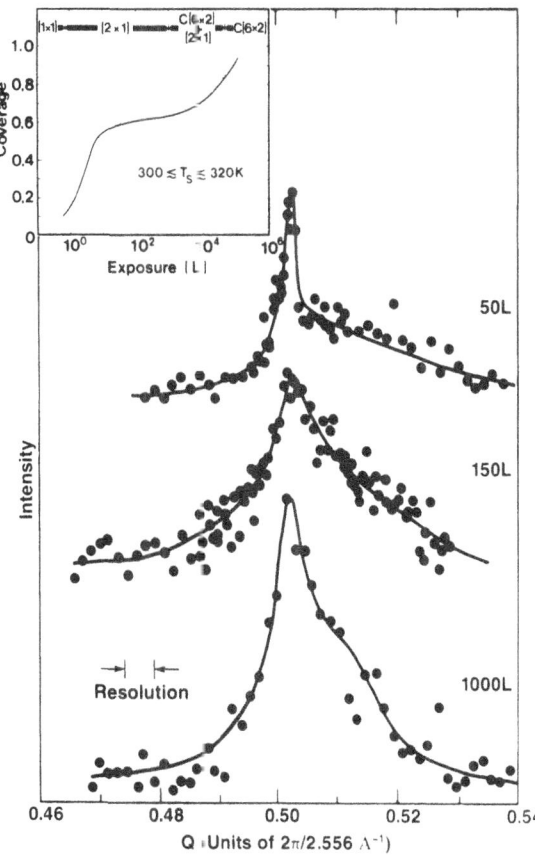

FIGURE 11. GIXS scans of the half-order superlattice $(-\frac{1}{2}, \frac{1}{2}, 0)$ reflection along the (110) directions from an O–Cu (110) (2 × 1) surface as a function of oxygen exposure. The inset shows the coverage dependence of LEED patterns. (From Ref. 67.)

defects and domain structures are often present. GIXS techniques provide us a tool not only to study the surfaces but also to study now to prepare the right surface to begin with. We should emphasize that high resolution is the key to these new capabilities, as will be further demonstrated in later sections.

4.1.3. Buried Interfaces

The structure of interfaces represents another area where GIXS can be directly applied. The first demonstration of the GIXS technique by Marra, Eisenberger, and Cho was in fact performed on an epitaxially grown Al–GaAs (110) surface.[4] Since the index of refraction for Al is closer to unity than is that of GaAs, it was possible to choose a grazing angle such that total reflection occurred at the interface between the two materials rather than at the Al surface. Their results revealed a tetragonal distortion at the interface that compensates for the 1% mismatch between the Al and GaAs lattices. Recently, Segmuller and co-workers have studied lattice mismatches and strain fields of other epitaxial films using GIXS.[50] Since GIXS is nondestructive, more applications of this technique to study various interfaces can be expected, especially for liquid–solid interfaces where UHV surface probes cannot be applied.

Another example of buried interfaces is a synthetic multilayer structure fabricated by vapor-deposition techniques. Besides practical interests, such layered structures provide good models for the study of grazing-angle X-ray effects. Most X-ray studies reported so far have been limited to the measurements of interference patterns with the scattering wave vector normal to the surface. One example is the high-resolution measurement of the X-ray reflectivity of a-Si/a-Ge superlattice films by Ruppert *et al.*,[51] where the layer spacings were chosen to exhibit interference with Bragg and anti-Bragg reflections due to dynamical effects. Dynamical modeling of reflection and interference near the critical angle shows such interference effects, but the calculated Fresnel reflectivity deviates from the measured curve (Fig. 12). These deviations can be attributed to the interfacial roughness of the layered composites if the substrate is smooth (see section 4.3.2). For a complete study, measurement of the diffuse scattering with the scattering wave vector parallel to the surface is particularly important.

4.1.4. Amorphous Thin Films

Our discussions show that the sensitivity of the GIXS technique is well established for ordered monolayers, even with low-Z adsorbates such as oxygen. It will be necessary to have much more intense synchrotron sources than those presently available to routinely measure scattering from disordered monolayers. In this context, it is interesting to consider the results of a GIXS study on amorphous thin films by Fuoss and Fischer-Colbrie.[52]

Fuoss and Fischer-Colbrie studied evaporated and sputtered films of amorphous GeSe$_2$ to determine the structure and the orientational dependence of this structure as a function of film thickness down to a minimum thickness of 250 Å. Currently, there is considerable uncertainty about whether these films have intermediate-range order in the form of layers parallel to the substrate surface, or

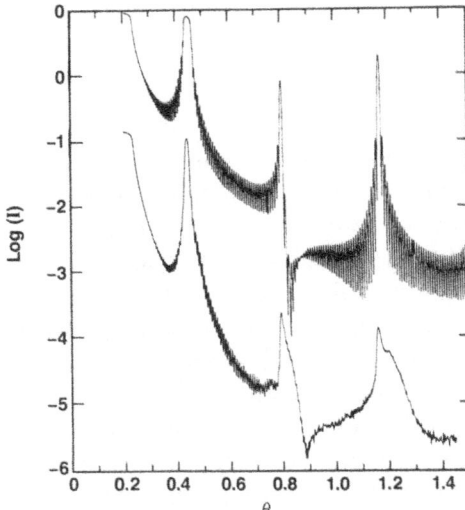

FIGURE 12. The X-ray interference pattern from an amorphous 55-Å Si/49-Å Ge multilayer shows Bragg and anti-Bragg peaks superpositioned with Fresnel reflectivity (lower curve, experimental; upper curve, theoretical).

whether each is a chemically ordered random network with three-dimensional connectivity. Results for a variety of films prepared by different techniques are shown in Fig. 13. Radial distribution analysis (section 2.4.3) has been performed to yield good RDFs from the thicker films. Difficulties in normalizing the data prevented high-quality RDFs from the thicker films. Difficulties in normalizing the data prevented high-quality RDFs from being calculated for the thinnest films. The results showed that there was no evidence for a layered structure derived from the crystal structure of GeSe₂. Extrapolating from this study, it is clear that it is possible to study amorphous thin films with GIXS techniques but that brighter X-ray sources are needed to make these experiments practical.

4.2. Surface Phase Transitions

So far, we have discussed the static aspects of the structure of surfaces. Surfaces, of course, can undergo various structural transformations as a function of temperature, coverage, and the like. For the surfaces discussed above, the Au (110) surface undergoes a $(1 \times 2) \rightarrow (1 \times 1)$ transition at ~923 K. The nature of this transition was shown to follow the universal behavior of the 2-D Ising class,[53] but questions remain.[54] Other interesting 2-D systems beyond the scope of this review include physisorbed inert gases on graphite[55] and intercalation layers.[56]

The first GIXS experiment performed in this area is the study of the melting of Pb overlayers on a Cu (110) surface by Marra, Fuoss, and Eisenberger at SSRL.[57] Pb monolayers on Cu surfaces have the structure neither of Pb nor of

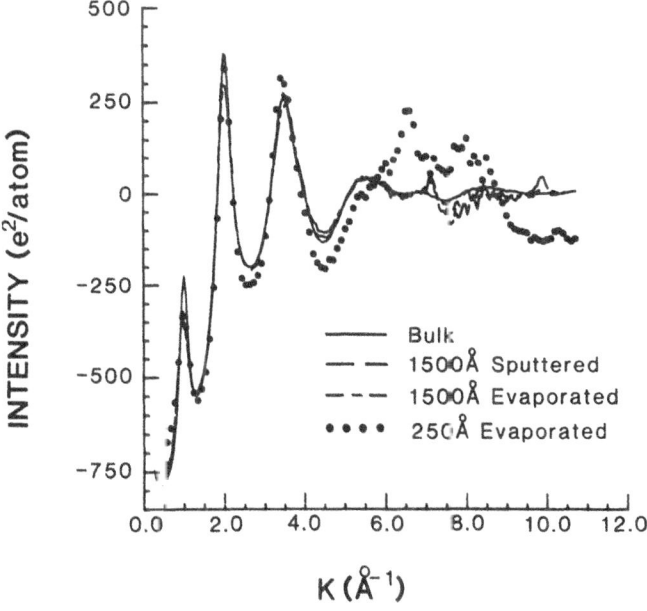

FIGURE 13. The corrected, normalized scattered intensity from amorphous GeSe₂ films of various thicknesses. Note the similarity of the curves as thickness is changed. This indicates that, contrary to some theories, there are no orientational effects induced by the substrate.

Cu.[58] The GIXS study found that Pb deposited at 120 °C formed a commensurate (5 × 1) structure with five times the Cu periodicity along the [1̄10] direction, in agreement with a previous LEED study. This Pb overlayer melted near 320 °C, as monitored by the shape of the diffraction spots. Upon cooling, the Pb unexpectedly assumed an incommensurate structure, which reappeared after every subsequent melting of the surface (Fig. 14). Further study of the behavior of melting of the incommensurate structure revealed that the incommensurate Pb phase melted first along the troughs of the Cu (110) surface at ~240 °C and perpendicular to the troughs at a higher temperature of ~325 °C. (Bulk Pb melts at 327 °C).

Brennan, Fuoss, and Eisenberger continued the investigation of this phase diagram and melting behavior of Pb on the Cu (110) surface (Fig. 15).[59] They found three distinct regions of the phase diagram:

1. an incommensurate region at low coverage,
2. a commensurate (5 × 1) structure at higher coverage, and
3. an island structure at high coverage (which had previously been found by the LEED studies of Henrion and Rhead).

All of these phases were found to melt at up to 80 °C below the bulk Pb melting point. The incommensurate structure was found to be a distortion of the commensurate (5 × 1) structure, which allows for a relaxation of Pb atoms

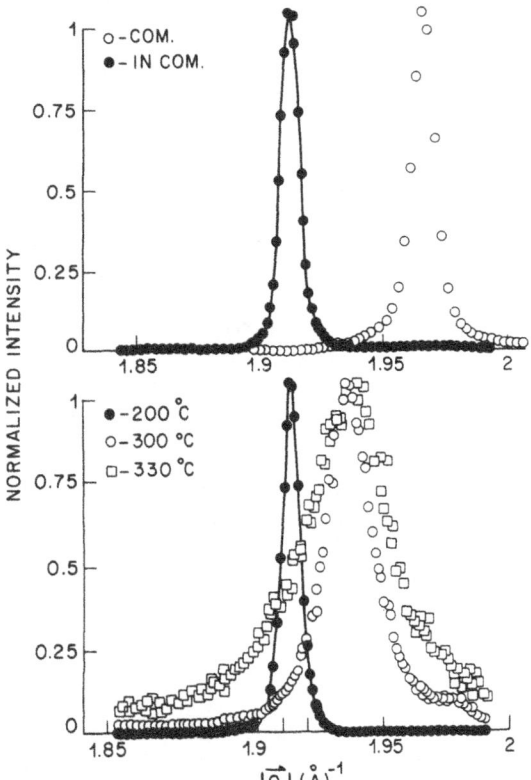

FIGURE 14. Radial scans of the (0.8, −0.8, 0) Bragg reflection from a Pb–Cu (110) (5 × 1) surface. The top panel shows scans indicating the shift to incommensurability: commensurate (open circles) and incommensurate (filled circles). The bottom panel shows radial scans as a function of temperature. Note the changing line shapes as the solid melts and the shift of the peak towards commensurability as the Pb layer melts.

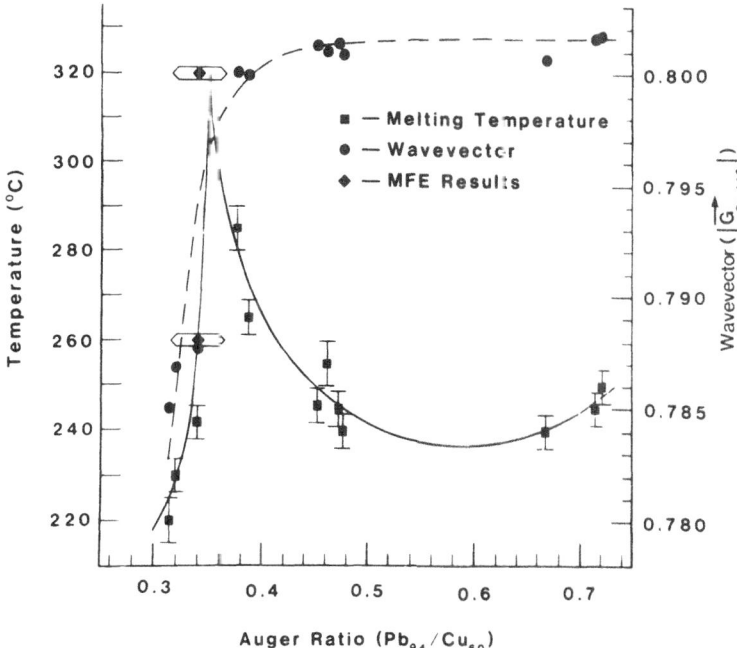

FIGURE 15. The melting behavior of Pb monolayers on the Cu (110) surface as a function of coverage. The coverage units are in terms of the ratio of the Pb Auger line at 94 eV and the Cu Auger line at 60 eV. A coverage of ~0.33 corresponds to 1 monolayer, while 0.46 corresponds to several layers.

toward their bulk metallic distances. The Pb islands had a bulk Pb-like structure with (111) faces interfacing to the Cu (110) surface.

4.3. Surfaces of Liquids and Liquid Crystals

The surfaces of liquids and liquid-crystalline materials represent another area in which GIXS can significantly affect our understanding. Most studies discussed above, on solid surfaces, centered primarily on the in-plane structures of ordered surfaces by employment of X-ray scans with scattering wave vector nearly parallel to the surface. The incidence angle of X-rays is kept near Θ_c on the basis of the existence of an evanescent wave near the surface to enhance surface sensitivity. The measurements and interpretation of the X-ray diffraction pattern in terms of the in-plane distribution of molecules in a liquid interface are more difficult than for a solid, due to both the diffuse nature of the scattering and the removal of the contribution to the X-ray scattering from atoms below the surface, which requires knowledge of their distribution as a function of density. Currently, there have been no GIXS studies of liquid surfaces. However, in the case of liquids, it it is also interesting to study the variation in distribution normal to the surface by keeping the scattering wave vector near normal to the surface. Essentially, this requires measuring the X-ray reflectivity as a function of incidence angle.[60] The principle and examples of such measurements have been discussed in sections 2.2

and 4.1.3 for solid systems. The following sections will discuss reflectivity measurements from liquid surfaces.

4.3.1. Liquid Scattering Spectrometer

The requirement that the liquid surface remain horizontal places severe requirements on diffractometer design for these studies. A high-resolution scattering spectrometer, specially constructed for experiments at the synchrotron radiation laboratory HASYLAB at DESY, was recently described by Als-Nielsen and Pershan.[31] A schematic of the spectrometer is shown in Fig. 16. This instrument uses a crystal reflection to deflect the beam downward onto the liquid sample and a second, identical crystal to deflect the reflected beam back into the horizontal plane. By rotating these crystals about the incident beam, this arrangement allows large incident angles to be reached. Readers are referred to the reference for further operational details of this instrument.

4.3.2. Liquid–Vapor Interfaces

The liquid–vapor interface involves an inhomogeneous distribution with a thickness of a few molecular layers. The first application of X-ray reflectivity measurements to study the liquid surface were performed by Lu and Rice on liquid mercury using X-rays from a conventional sealed tube.[60] Later, high-resolution measurements were carried out by Pershan, Als-Nielsen, and their coworkers.[61]

More recently, Braslau et al.[61] reported X-ray reflectivity measurements in the study of the surface roughness of water. The required dynamic range on the order of 10^7 to 10^8 was obtained in this experiment by using a synchrotron source (Fig. 17). Reflectivity data of such quality can be interpreted in terms of the mean-square roughness of the surface without prescribing the molecular mechanism by which the roughness is generated. The value they obtain for this measure of the surface-density profile is 3.24 ± 0.05 Å, which is close to the characteristic width of 3.56 Å obtained by a molecular-dynamics simulation.[62]

Another aspect of the total external reflection of X-rays is the accompanying fluorescence from evanescent waves. This has recently been exploited to study the density distribution perpendicular to the surface (see section 2.2) by Bloch et al.[63] They used this technique to infer the concentration profile of a polymer solution–vapor interface of sulfonated polystyrene dissolved in dimethylsulfoxide

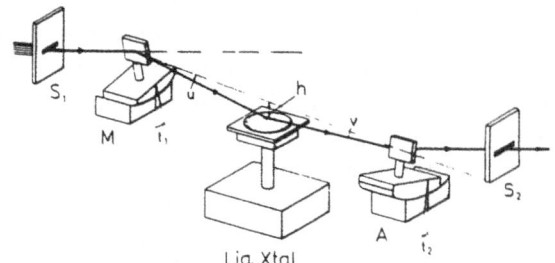

FIGURE 16. The liquid scattering spectrometer of Als-Nielson and Pershan. The use of the crystal monochromator to change the grazing angle allows very large angles to be reached (and hence large momentum transfer perpendicular to the surface) while allowing the liquid surface to remain horizontal.

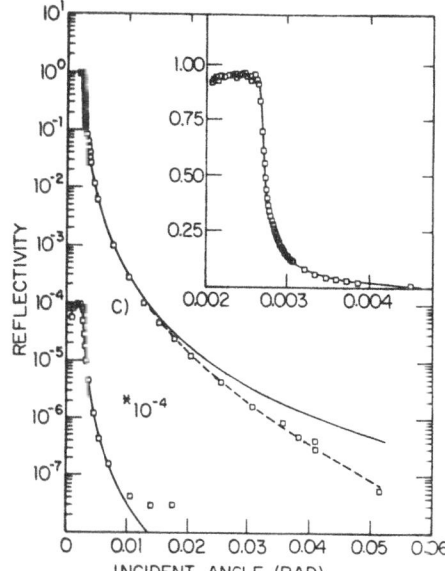

FIGURE 17. The reflectivity of water, showing the effects of surface roughness measured by Braslau *et al.* The solid curves were calculated assuming a sharp interface while the dashed curves include the effect of a root-mean-square roughness of 3.2 Å. The inset shows the reflectivity at low angle and curve (c) is data obtained with a rotating anode source (for comparison).

(DMSO). Fluorescence from two types of atoms was monitored: the sulfur $K\alpha$ line from the solvent DMSO and the $K\alpha$ line of the Mn ion attached to the sulfonated polystyrene chain. The result, when fitted with recent mean-field theories, clearly implies a strong attraction of the polymer to the interface with a concentration enhancement factor of about 100 over the bulk solution concentration value.

4.3.3. Liquid-Crystal Surfaces

Liquid crystals exhibit fascinating structural behavior between that of liquids and 3-D solids. For example, rod-like molecules are known to form nematic phases in which molecules show one preferred orientation but without long-range order and smectic phases in which molecules are aligned into a layered structure.

A first study of liquid-crystal surfaces was recently performed by Pershan and Als-Nielsen[64] to observe the surface effect of the nematic-to-smectic phase transition at HASYLAB using the liquid-scattering spectrometer described above. They observed Bragg reflections in bulk nematic and isotropic phases, which arose from a surface layer of the smectic phase propagating from the free surface into the bulk. This propagation length was found to be equal to the correlation length of a bulk smectic sample. The measured X-ray reflectivity from this layered surface showed a coherent superposition of Fresnel reflection from the free surface and Bragg reflection from the smectic order induced by the surface (section 4.1.3). By studying the temperature dependence of the diffuse Bragg reflection near the nematic–smectic transition, Pershan and Als-Nielson were able to investigate the critical fluctuation at the liquid-crystal surface.

Another example closely related to the liquid crystal is lipids of the membrane structure. Seul, Eisenberger, and McConnell performed a GIXS study on phospholip monolayers deposited on silicon substrates at SSRL. Phospholips are lyotropic liquid crystals that are stable in water solution. The diffraction experiments were carried out in a moist helium atmosphere. The results demonstrated 2-D order in these monolayers and water intercalcation between the layers. Although many details remain to be studied, the results point to the possibility of applying GIXS to the biological membrane systems.

5. FUTURE APPLICATIONS

As is indicated in this review, the last six years (1981–1986) have seen the application of GIXS to the solution of many surface–interface problems. While significant structural information has been obtained in these initial studies, they also have the general character of exploring the possible applications of the technique and testing the various theoretical and experimental approaches required to make GIXS a versatile and quantitative structural technique. As a generality, it is fair to say that this period has uncovered no obstacles to this objective. It is natural, therefore, that the immediate future holds more quantitative and substantiative structural studies by an increasing number of practitioners.

While the potential for quantitative crystallography has been demonstrated, it has not yet been completely systematized nor developed to the point of 3-D crystallography. This should occur in the coming years. GIXS crystallography will be applied to the full range of reconstructed and adsorbate systems, as well as interfaces, resulting in as complete a structural knowledge of those systems as has been obtained for bulk 3-D systems.

While structural phase transitions have been investigated, they have not yet been pursued with the vigor exhibited in studies of either 3-D systems or the inert gases on graphite. This should also occur in the coming years and is expected to be a very rich area scientifically.

The first studies of nonordered interfaces, either amorphous or liquid, have been made, but one awaits the next generation of sources and monochromator–mirror systems to be able to study liquid or amorphous monolayers.

The first steps *have been taken* in using GIXS to study biological structures that cannot be crystallized in 3-D. With the recent success of using electron microscopes to see such ordered 2-D biological crystals, there is every reason to be optimistic that a more comprehensive structural study will be performed by GIXS.

The first application of GIXS to study surface roughness has demonstrated its utility for such studies. Given the current theoretical and technological interest in roughening one would expect this application to also grow into a robust enterprise.

However, beyond linear extrapolations of past work and improved instrumentation and analysis, the future will also hold many new firsts. Examples include the investigation of surface vs. bulk melting at the solid–vacuum interface, the aforementioned determination of 2-D biological structures, microscopic understanding of roughening and wetting, the evolution of interfaces from 2-D to 3-D structures, and the development of the capability for studying time-dependent phenomena on surfaces. In addition, adaptation of many of these studies to *in situ* conditions by the development of specialized cells and chambers will obviously be a strong feature in the future.

Although the discussions in this chapter are limited to the area of elastic scattering in the determination of surface atomic structures, the areas of inelastic X-ray scattering are beginning to explore 3D magnetic[65] and vibrational[66] structures. Applications of inelastic X-ray scattering on surfaces will be explored, in particular using the next generation of synchrotron radiation sources.

In spite of the very high potential for GIXS, there are significant barriers to its future growth. Most important is the fact that many of the more interesting studies require the intensity of synchrotron radiation sources. These sources are sparsely distributed and require people to work away from their home institution. A second barrier is the cost of the experimental apparatus. For many surface–interface studies the chamber required is very costly, certainly so for those studies requiring UHV. A final barrier is that, even more so than modern crystallography, this technique in many cases requires a team of scientists with different skills. In general, one needs an experimentalist to design the apparatus, a materials scientist to prepare the samples, and the team of people required to run experiments at a synchrotron facility. With the dominant role played by X-ray scattering in determining 3-D bulk structural phenomena as our guide, one can feel confident that it is only a matter of time before these problems are overcome and X-ray scattering studies of surfaces and interfaces occupy a similar role.

REFERENCES

1. G. A. Somorjai, *Chemistry in Two Dimensions: Surfaces,* Cornell University Press, Ithaca, NY, (1981).
2. P. Eisenberger and L. C. Feldman, *Science* **214,** 300 (1981).
3. M. A. Van Hove and S. Y. Tong, *Surface Crystallography by LEED,* Springer-Verlag, Berlin (1979).
4. W. C. Marra, P. Eisenberger, and A. Y. Cho, *J. Appl. Phys.* **50,** 6927 (1979).
5. P. Eisenberger and W. C. Marra, *Phys. Rev. Lett.* **46,** 1081 (1981).
6. A. Guinier, *X-Ray Diffraction in Crystals, Imperfect Crystals, and Amorphous Bodies,* Freeman, San Francisco (1963); B. E. Warren, *X-Ray Diffraction,* Addison-Wesley, Reading, MA (1969).
7. I. K. Robinson, *Phys. Rev.* **B33,** 3830 (1986); S. R. Andrews and R. A. Cowley, *J. Phys. C.* **18,** 6427 (1985).
8. G. H. Vineyard, *Phys. Rev.* **B26,** 4145 (1982).
9. S. Dietrich and H. Wagner, *Phys. Rev. Lett.* **51,** 1469 (1983).
10. R. W. James, *The Optical Principles of the Diffraction of X-Rays,* Cornell University Press, Ithaca, NY (1955).
11. M. Born and E. Wolf, *Principles of Optics,* Pergamon, New York (1983).
12. P. H. Fuoss, in *Synchrotron Radiation and Surface Science* (D. Norman and J. E. Inglesfield, eds.), Daresbury Laboratory, Daresbury (1983).
13. L. G. Parratt, *Phys. Rev.* **95,** 359 (1954).
14. R. S. Becker, J. A. Golovchenko, and J. R. Patel, *Phys. Rev. Lett.* **50,** 153 (1983).
15. A. Fischer-Colbrie and P. H. Fuoss, in preparation; A. Fischer-Colbrie, Ph.D. Thesis, Stanford University (1986).
16. A. M. Afanas'ev and M. K. Melkonyan, *Acta Crystallog.* **A39,** 376 (1984).
17. P. L. Cowan, *Phys. Rev.* **B32,** 5437 (1985).
18. P. L. Cowan, S. Brennan, T. Jach, M. J. Bedzyk, and G. Materlik, *Phys. Rev. Lett.* **57,** 2399 (1986).

19. H. Lipson and W. Cochran, *The Determination of Crystal Structures,* G. Bell and Sons, London (1953).

20. J. Karle, *Science* **232,** 837 (1986); H. Hauptman, *Science* **233,** 178 (1986).

21. I. K. Robinson, in *Handbook on Synchrotron Radiation,* Vol. 3 (D. Moncton, ed.), North-Holland, Amsterdam, to be published.

22. J. A. Golovchenko, *Science* **232,** 48 (1986).

23. L. D. Marks, *Phys. Rev. Lett.* **51,** 1000 (1983); L. D. Marks and D. J. Smith, *Nature* **303,** 316 (1983); L. D. Marks and D. J. Smith, *Surf. Sci.* **143,** 587 (1984).

24. N. Norman, *Acta Crystallog.* **7,** 462 (1954) and references therein.

25. A. van Steenbergen and NSLS Staff, *Nucl. Instrum. Methods* **172,** 25 (1980).

26. E. Hoyer, *Nucl. Instrum. Methods* **208,** 117 (1983).

27. C. H. MacGillavry and G. D. Rieck, *International Tables for X-Ray Crystallography,* Vol. III, D. Reidel (1983).

28. S. Brennan and P. Eisenberger, *Nucl. Instrum. Methods* **222,** 164 (1984).

29. P. H. Fuoss and I. K. Robinson, *Nucl. Instrum. Methods* **222,** 171 (1984).

30. K. L. Liang, G. J. Hughes, and R. C. Hewitt, unpublished.

31. J. Als-Nielson and P. S. Pershan, *Nucl. Instrum. Methods* **208,** 545 (1983).

32. P. H. Fuoss, unpublished.

33. J. Bohr, R. Feidenhans'l, M. Nielsen, M. Toney, R. L. Johnson, and I. Robinson, *Phys. Rev. Lett* **54,** 1275 (1985).

34. R. L. Johnson, J. H. Fock, I. K. Robinson, J. Bohr, R. Feidenhans'l, J. Als-Nielson, M. Nielsen, and M. Toney, in *The Structure of Surfaces,* (M. A. Van Hove and S. Y. Tong eds.), Springer-Verlag, Berlin (1985).

35. I. K. Robinson, W. K. Waskiewicz, P. H. Fuoss, J. B. Stark, and P. A. Bennett, *Phys. Rev.* **B33,** 7013 (1986).

36. J. Chadi, *Phys. Rev. Lett.* **43,** 43 (1979).

37. J. A. Appelbaum and D. R. Hamann, *Surf. Sci.* **74,** 21 (1978).

38. S. Y. Tong, G. Xu, and W. N. Mei, *Phys. Rev. Lett.* **52,** 21 (1978).

39. F. Jona, H. D. Shih, D. W. Jepsen, and P. M. Marcus, *J. Phys. C* **12,** L455 (1979).

40. M. Aono, R. Souda, C. Oshima, and Y. Ishizawa, *Phys. Rev. Lett.* **51,** 801 (1983).

41. G. Binnig, H. Rohrer, Ch. Gerber, and E. Weibel, *Phys. Rev. Lett.* **50,** 120 (1983).

42. R. J. Culbertson, L. C. Feldman, and P. J. Silverman, *Phys. Rev. Lett.* **45,** 2043 (1980).

43. P. A. Bennett, L. C. Feldman, Y. Kuk, E. G. McRae, and J. E. Rowe, *Phys. Rev.* **B28,** 3656 (1983).

44. E. G. McRae, *Phys. Rev.* **B28,** 2305 (1983).

45. K. Takayanagi, Y. Tarishiro, M. Takahashi, and S. Takahashi, *J. Vac. Sci. Technol.* **3,** 1502 (1985).

46. I. K. Robinson, *Phys. Rev. Lett.* **50,** 1145 (1983).

47. I. K. Robinson, Y. Kuk, and L. C. Feldman, *Phys. Rev.* **B29,** 4762 (1984).

48. K. S. Liang, P. H. Fuoss, G. J. Hughes, and P. Eisenberger, in *The Structure of Surfaces,* (M. A. Van Hove and S. Y. Tong, Eds.), Springer-Verlag, Berlin (1985).

49. K. S. Liang and P. Eisenberger, in "Characterization of Defects in Materials," MRS Symp. Proc., Vol. 82, p. 493 (1986).

50. A. Segmuller, in *Advances in X-Ray Analysis,* Vol. 29, (C. S. Barrett, ed.), Plenum, New York (1986).

51. A. F. Ruppert, P. D. Persans, G. J. Hughes, K. S. Liang, B. Ables and W. Lanford, *Phys. Rev.* **B44,** 11381 (1991).

52. P. H. Fuoss and A. Fischer-Colbrie, *Phys. Rev.* **B38,** 1875 (1988); A. Fischer-Colbrie and P. H. Fuoss, *J. Non-Cryst. Solids* **59–60,** 859 (1983).

53. J. C. Campuzano, M. S. Foster, G. Jennings, R. F. Willis, and W. Unertl, *Phys. Rev. Lett.* **54,** 2684 (1985).

54. W. N. Unertl, *Comments Cond. Mat. Phys.* **12,** 289 (1986).

55. R. J. Birgeneau and P. M. Horn, *Science* **232,** 329 (1986).

56. A. Erbil, A. R. Kortan, R. J. Birgeneau, and M. S. Dresselhaus, *Phys. Rev.* **B28,** 6329 (1983).

57. W. C. Marra, P. H. Fuoss, and P. Eisenberger, *Phys. Rev. Lett.* **49,** 1169 (1982).

58. J. Henrion and G. E. Rhead, *Surf. Sci.* **29,** 20 (1972).

59. S. Brennan, P. H. Fuoss, and P. Eisenberger, in *The Structure of Surfaces,* (M. A. Van Hove and S. Y. Tong, eds.), Springer-Verlag, Berlin (1985).

60. B. C. Lu and S. A. Rice, *J. Chem. Phys.* **68,** 5558 (1978).
61. A. Braslau, M. Deutsch, P. S. Pershan, A. H. Waeiss, J. Als-Nielsen, and J. Bohr, *Phys. Rev. Lett.* **54,** 114 (1985).
62. M. R. Townsend, J. Gryko, and S. A. Rice, *J. Chem. Phys.* **82,** 4391 (1985).
63. J. M. Bloch, M. Sansone, F Rondelez, D. G. Peiffer, P. Pincus, J. W. Kim, and P. Eisenberger, *Phys. Rev. Lett.* **54,** 1039 (1985).
64. P. S. Pershan and J. Als-Nielsen, *Phys. Rev. Lett,* **52,** 759 (1984); J. Als-Nielson, F. Christensen, and P. S. Pershan, *Phys. Rev. Lett.* **48,** 1107 (1982).
65. D. Gibbs, D. E. Moncton, K. L. D'Amico, J. Bohr, and B. H. Grier, *Phys. Rev. Lett.* **55,** 234 (1985).
66. D. E. Moncton, private communication.
67. G. R. Gruzalski, D. M. Zehner and J. F. Wendelken, *Surf. Sci.* **147,** L623 (1984).

9

The Study of Surface Structures by Photoelectron Diffraction and Auger Electron Diffraction

Charles S. Fadley

ABBREVIATIONS AND ACRONYMS

AED	Auger electron diffraction
APD	azimuthal photoelectron diffraction
ARPEFS	angle-resolved photoemission fine structure (acronym for scanned-energy photoelectron diffraction)
CMA	cylindrical mirror analyzer
DL	double-layer model
EELS	electron energy loss spectroscopy
ESDIAD	electron stimulated desorption ion angular distributions
EXAFS	extended X-ray absorption fine structure
FT	Fourier transform
FWHM	full width at half maximum intensity
GIXS	grazing incidence X-ray scattering
HT	high temperature limit (in SPPD experiment)
LEED	low energy electron diffraction
LT	lower temperature of measurement (in SPPD experiment)
ML	monolayer
MEIS	medium-energy ion scattering
MQNE	magnetic quantum number expansion
MS	multiple scattering
MSC	multiple scattering cluster
MTL	missing-top-layer model
NEXAFS	near edge X-ray absorption fine structure = XANES

Charles S. Fadley • Department of Chemistry, University of Hawaii, Honolulu, Hawaii 96822. *Present address:* Department of Physics, University of California–Davis, Davis, California 95616, *and* Materials Science Division, Lawrence Berkeley Laboratory, Berkeley, California 94720.

Synchrotron Radiation Research: Advances in Surface and Interface Science, Volume 1: Techniques, edited by Robert Z. Bachrach. Plenum Press, New York, 1992.

NPD   scanned-energy photoelectron diffraction with normal emission
ODAC   one-dimensional alkali-chain model
OPD   scanned-energy photoelectron diffraction with off-normal emission
PD, PhD   photoelectron diffraction
PLD   path-length difference
PPD   polar photoelectron diffraction
PW   plane-wave scattering
RBS   Rutherford back scattering
SEXAFS   surface extended X-ray absorption fine structure
SMSI   strong metal support interaction
SPAED   spin polarized Auger electron diffraction
SPPD   spin polarized photoelectron diffraction
SRMO   short-range magnetic order
SS   single scattering
SSC   single scattering cluster
STM   scanning tunneling microscopy
SW   spherical-wave scattering
XANES   X-ray absorption near-edge structure = NEXAFS
XPD   X-ray photoelectron diffraction, typically at energies of 500–1400 eV
XPS   X-ray photoelectron spectroscopy

1. INTRODUCTION

A knowledge of the atomic identities, positions, and bonding mechanisms within the first 3–5 layers of a surface is essential to any quantitative microscopic understanding of surface phenomena. This implies knowing bond directions, bond distances, site symmetries, coordination numbers, and the degree of both short-range and long-range order present in this selvedge region. A number of surface-structure probes have thus been developed in recent years in an attempt to provide this information.[1] Each of these methods has certain unique advantages and disadvantages, and they are often complementary to one another.

We will here concentrate on the basic experimental and theoretical aspects of photoelectron diffraction (PD or PhD) and its close relative, Auger electron diffraction (AED). Although the first observations of strong diffraction effects in X-ray photoelectron emission from single-crystal substrates by Siegbahn et al.[2] and by Fadley and Bergstrom[3] took place almost 20 years ago, and the use of such effects at lower energies to determine surface structures was proposed by Liebsch[4] 15 years ago, it was not until about 10 years ago that quantitative experimental surface-structure studies were initiated by Kono et al.,[5] Woodruff et al.,[6] and Kevan et al.[7] By now both photoelectron diffraction and Auger electron diffraction are becoming more widely used to study surface atomic geometries.[8-13] We will thus consider here both the present status and future prospects of these methods, and then return at the conclusion of this chapter to make a critical comparison of them with several other surface-structure probes such as LEED, grazing incidence X-ray scattering (GIXS), and scanning tunneling microscopy (STM).

The basic experiment in PD or AED involves exciting a core photoelectron or a relatively simple core-like Auger transition from an atom in a single-crystal environment and then observing modulations in the resulting peak intensities that are due to final-state scattering from atoms neighboring the emitter. For a general Auger peak of the type XYZ, it is thus important that the upper levels Y and Z involved are not so strongly influenced by chemical bonding as to induce an anisotropy in emission that is more associated with initial-state electronic structure. The directly emitted photoelectron- or Auger electron–wave exhibits interference with various scattered waves, and this interference pattern is analyzed to derive structural information. Peak intensities can be monitored as a function either of the emission direction or, in the case of photoelectron diffraction, of the exciting photon energy. In AED, excitation can also derive from anything producing core holes: an electron beam, VUV/soft-X-ray radiation, or even an ion beam.

The three basic types of measurement possible are as shown in Fig. 1: an azimuthal or ϕ scan, a polar or θ scan, and, for photoelectron diffraction, a scan of energy in a normal or off-normal geometry. Several abbreviations and acronyms have arisen in connection with such measurements. With soft X-ray excitation at about 1.2–1.5 keV at the typical X-ray photoelectron spectroscopy (XPS) limit, scanned-angle measurements have been termed X-ray photoelectron diffraction (XPD).[5,9] Scanned-energy photoelectron measurements spanning the VUV-to-soft-X-ray regime have also been called normal photoelectron diffraction (NPD),[7,8,14] off-normal photoelectron diffraction (OPD),[15] or angle-resolved photoemission fine structure (ARPEFS)[8] to emphasize their similarity to the more familiar surface extended X-ray absorption fine structure (SEXAFS).[16] Both standard X-ray sources and synchrotron radiation can be used for excitation, with photon energies being as low as 60 eV[6,17,18] and as high as a few keV.[7,8,19] Synchrotron radiation adds the capability of varying the photon energy continuously and of studying the dependence of the diffraction on polarization.

The degree of modulation of intensity observed in PD or AED experiments can be very large, with overall values of anisotropy as high as $(I_{max}-I_{min})/I_{max} = \Delta I/I_{max} = 0.5$–$0.7$. Thus, it is not uncommon to observe 30–50% changes in the peak intensity as a function of direction or energy, and such effects are relatively easy to measure. This is by contrast with the related surface-structure technique

FIGURE 1. The three basic types of photoelectron or Auger electron diffraction measurement: an azimuthal (ϕ) scan at constant polar angle, sometimes referred to as azimuthal photoelectron diffraction or APD; a polar (θ) scan at constant azimuthal angle, referred to as polar photoelectron diffraction or PPD; and a scan of $h\nu$ in fixed geometry that can be done only in photoelectron diffraction and for emission either normal or off-normal to the surface (denoted NPD or OPD, respectively). The scanned-energy type has also been referred to as angle-resolved photoemission fine structure or ARPEFS. Note that θ is measured with respect to the surface.

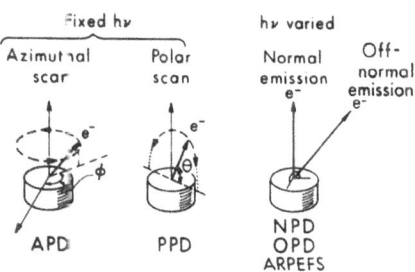

of SEXAFS,[16] in which typical modulations are about one tenth as large. This difference arises from the fact that SEXAFS effectively measures an angle-integrated photo electron diffraction pattern as a function of energy, and it is not surprising that this integration averages over various phases and leads to considerably lower relative effects.

We shall consider both scanned-angle photoelectron and Auger results and scanned-energy photoelectron results here. To date, scanned-angle studies are much more numerous; this is due to their greater simplicity, since scanned-energy work has several requirements in addition: the sweeping of photon energy with a synchrotron radiation source, the correct normalization of photon fluxes and electron-analyzer transmissions as a function of energy, and the possibility of allowing for interference between Auger peaks and photoelectron peaks in certain kinetic-energy ranges. Finally, it requires more-complex theoretical calculations in that scattering phase shifts and other nonstructural parameters have to be generated for all of the energies in a scan.[8,19-22] However, an advantage in scanned-energy work is that Fourier transform (FT) methods can be used to estimate the path-length differences for various strong scatterers.[8,20-22]

A key element in either photoelectron or Auger electron diffraction is the energy dependence of the relevant elastic-scattering factors. Figure 2 illustrates this for the case of atomic Ni with curves of the plane-wave scattering amplitude $|f_{Ni}|$ as a function of both the scattering angle θ_{Ni} and the electron energy. For low energies of 50–200 eV, it is clear that there is a high amplitude for scattering into all angles. For the intermediate range of about 200–500 eV, it is a reasonable approximation to think of only forward scattering ($\theta_{Ni} = 0°$) and backscattering ($\theta_{Ni} = 180°$) as being important. However, at energies above 500 eV, we see that the scattering amplitude is significant only in the forward direction, in which it is strongly peaked. The degree of forward peaking increases as the energy is increased. The utility of such forward scattering at higher energies in surface-structural studies was noted in very early XPD investigations,[5,23] and it has more recently been termed a "searchlight effect"[11] or "forward focusing"[24] in connection with XPD analyses of epitaxial overlayers. This effect turns out to be one of the most useful and simply interpretable aspects of higher-energy photoelectron or Auger electron diffraction, and we will make reference to it in several of the examples considered in following sections. These qualitative observations con-

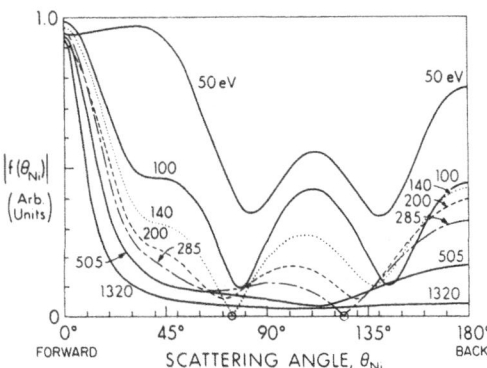

FIGURE 2. Nickel plane-wave scattering factor amplitudes $|f_{Ni}|$ as a function of both the scattering angle θ_{Ni} and the photoelectron kinetic energy. Note the zeroes occurring for both 140 eV and 285 eV, which have been termed a generalized Ramsauer–Townsend effect. (From Ref. 21.)

cerning the energy dependence of the scattering factor will later also assist in explaining which multiple scattering effects may be the most significant. A special aspect of such scattering factors is that they may exhibit zeroes for certain angles and energies; this has been termed a generalized Ramsauer–Townsend effect, and its influence on the analysis of ARPEFS data is considered elsewhere.[21,25]

A final important aspect of either photoelectron or Auger electron diffraction is that both are *atom-specific* probes of *short-range order*. Thus, each type of atom in a sample can in principle be studied, and each will have a unique diffraction signature associated with the neighbors around it. Previous work shows that the principal features of diffraction curves are due to the geometry of the first 3–5 spheres of scatterers around a given emitter, although data may exhibit useful fine structure that is associated with scatterers as far as 20 Å away.[20,26] This short-range sensitivity is thus shared with SEXAFS. We will later point out the potential uses of PD and AED in studying the degree of order present in the near neighbors of the emitter.

The remainder of this chapter begins by briefly reviewing the experimental requirements of these methods and considering both the simplest single-scattering model and other more accurate models that have been used to analyze both PD and AED data. The bulk of the text discusses several illustrative cases to which these techniques have been applied. This is not intended to be an exhaustive listing of all such studies to date, but the examples have been chosen to demonstrate certain basic phenomena, to illustrate the range of structural information that can be obtained, and to provide some idea of the different classes of systems that can be fruitfully studied. In certain cases, the limitations of the analysis or the need for future improvements are pointed out. Finally, some particularly interesting new directions for the future are discussed, and comparisons to other currently used structural probes are also made.

The studies discussed represent a mixture of work utilizing both standard X-ray or electron excitation sources and synchrotron radiation, with the number of investigations using standard sources certainly being greater to date. Thus, the methods discussed here are not limited to synchrotron radiation, by contrast with several others discussed in this volume.[27,28] However, both PD and AED will benefit greatly by the use of the higher-intensity facilities in the vacuum ultraviolet/soft X-ray range that are now becoming more available, and we return to this point toward the end of the chapter.

2. EXPERIMENTAL CONSIDERATIONS

The basic experimental requirements for carrying out photoelectron or Auger electron diffraction measurements are relatively simple. A minimal experiment can consist of the excitation source, a specimen holder with only one axis of angular motion (usually the polar angle as defined in Fig. 1), and an electron energy analyzer with an angular resolution of at least approximately ±5°. Thus, most of the commercially available hemispherical analyzers are suitable, and even a cylindrical mirror analyzer (CMA) with some sort of baffle at its entry slit can be used. Peak intensities can be measured very simply as the difference in

height between some point at the maximum and a point in the high-energy
background. Measurements at this level are thus quite easy to take, and
interesting surface-structural information has been obtained from them.[11]

Going beyond this minimal experiment to be able to tap all of the
information available in the diffraction pattern involves several possible
elaborations:

• The specimen holder should have both polar and azimuthal axes of
rotation (cf. Fig. 1) so that the electron emission direction can be oriented
arbitrarily with respect to the surface. The optimal scanning capabilities in this
case are to be able to vary θ from grazing excitation incidence to grazing electron
exit and to vary ϕ over a full 360° or more. The latter is very useful for
establishing the symmetry of the surface and for verifying the reproducibility of
features from one symmetry-equivalent azimuthal direction to another. Scanning
ϕ over its full range is the most difficult to achieve in practice if there are
electrical or mechanical connections to the sample for heating, cooling, or
measuring temperature, but designs of this type have been in use for some time.[9]
The reproducibility and accuracy of both of these motions should be at least
±0.5°, with even smaller values on the order of ±0.1° being required for very
high angular resolution work.
• Automated scanning of spectra, determining of peak intensities by more
accurate area-integration and/or peak-fitting procedures, and stepping of angles
under computer control are also essential for efficiently obtaining the most
reliable data. Systems for doing this are discussed elsewhere.[9,29]
• It also may be desirable to rotate both the specimen and the analyzer (or
excitation source) on two axes so as to be able to orient the excitation source at
various positions with respect to the electron emission direction. In photoelectron
diffraction, this permits making use of the radiation polarization to preferentially
excite the direct wave toward different scatterers while at the same time observing
the electron intensity along a special direction.[8,19] This is particularly important in
studies utilizing synchrotron radiation. In Auger electron diffraction, it can also
be useful for assessing the degree to which the penetration of the exciting flux
along different incidence directions influences the outgoing diffraction pattern,
even though results to date indicate that such effects are minor.[30] (Similar
anisotropic penetration might also be expected with X-rays due to Bragg
reflections,[2,3] but such effects have so far not been found to be significant in
photoelectron diffraction patterns.)
• Improving the angular resolution of the analyzer to the order of ±1.0° has
also been found to yield data at higher energies with considerably more fine
structure.[10,26] Achieving this may involve specially designed entrance optics,[31,32]
or more simply the use of movable tube-array baffles at the entry to a more
standard analyzer.[33] High-resolution results of this kind will be discussed in more
detail in sections 4.2.1, 4.2.2, and 5.1.
• Improving the energy resolution of the system to on the order of 0.1 eV is
also desirable, because it permits resolving small chemical shifts or surface shifts
of core levels and studying the diffraction patterns of these species separately.[18]
• Scanning angle or energy obviously involves an added cost in time for any

study, and so it is desirable to have the highest overall count rates. This can be achieved by using a more-intense excitation source (as, for example, from insertion-device-generated synchrotron radiation) and/or the most efficient and highest-speed electron analyzer and detection system. Making the latter as effective as possible is important, since there are always potentially deleterious effects of radiation damage as the excitation intensity is increased. Analyzer improvements include the use of multichannel energy-detection systems involving several single-channel electron multipliers or a microchannel plate[32,34] and the use of special spectrometer geometries in which spectra at several angles can be recorded at the same time.[35-37] However, a potential disadvantage of systems recording several angles at once is that the angular resolution may be limited, particularly if it is desired to scan kinetic energies to several hundred eV. A final method for increasing data acquisition rates with a pulsed synchrotron radiation source is to use a time-of-flight analysis system;[38] a logistical problem with such systems however, is that they may require running the storage ring in a less frequently used "timing" mode with fewer electron bunches. Leckey[39] has recently reviewed many of the more novel proposals for analyzers with high energy resolution, high angular resolution, and/or high data acquisition rates.

• Finally, if scanned-energy photoelectron diffraction is to be performed, it is essential to use a reasonably stable synchrotron radiation source and to have an analyzer system whose transmission properties as a function of energy are well understood. This is because photon energies must be scanned in small steps over a total period on the order of hours in present experiments, and the influences of both the decay of photon flux with time and the change of the analyzer's sensitivity with kinetic energy must be corrected out of the final intensity data so as to yield something that is truly proportional to the energy-dependent photoelectric cross section in a given emission direction. Methods for making these corrections are discussed elsewhere.[8,20]

3. THEORETICAL MODELING

3.1. Single-Scattering Theory

3.1.1. Overview of Model

Since the first theoretical paper on low-energy photoelectron diffraction by Liebsch,[4] several detailed discussions of the modeling of photoelectron and Auger electron diffraction have appeared in the literature.[9,15,21,24,25,40-45] Thus, we will begin here by presenting only the essential ingredients of the simplest approach, the single-scattering cluster (SSC) model, and then comment toward the end of this section on several improvements that can be made to it, as well as on some effects expected due to multiple scattering (MS) events.

The basic elements of this single-scattering cluster model are shown schematically in Fig. 3(a). The fundamental assumptions are essentially identical to those used in describing extended X-ray absorption fine structure (EXAFS),[41] and a similar model has also been applied some time ago to angle-resolved Auger emission at very low energies of ≤100 eV.[40] We consider photoelectron emission first and then discuss the modifications required to describe Auger emission.

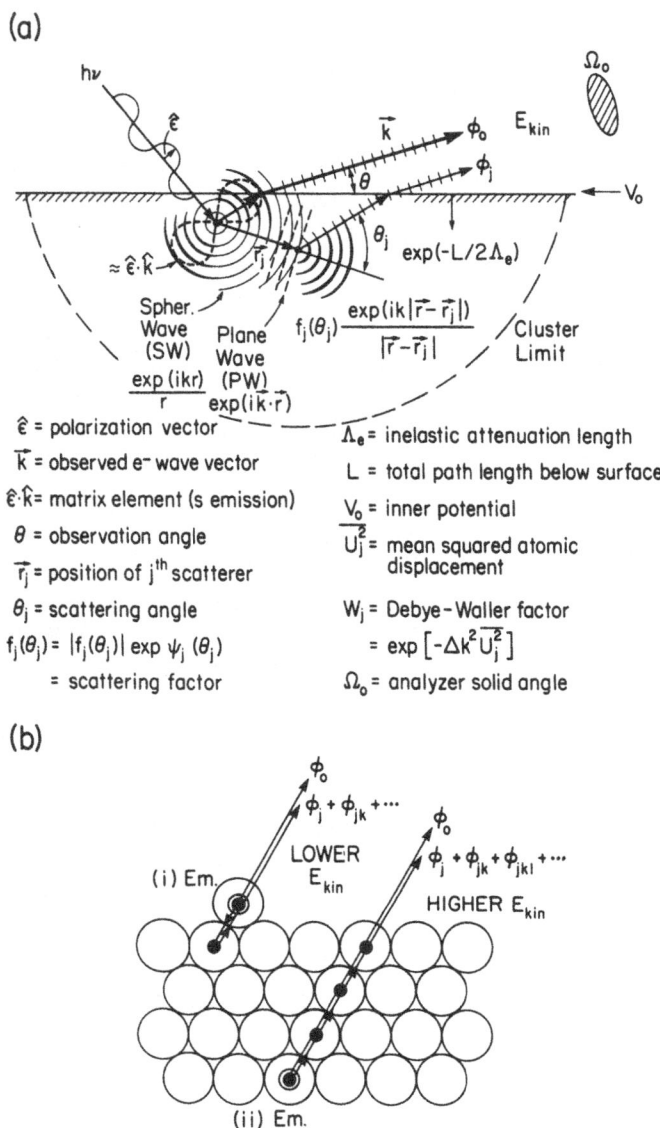

FIGURE 3. (a) Illustration of the assumptions used in the single-scattering cluster (SSC) model, with various important quantities indicated. (b) Two types of multiple-scattering corrections to the SSC model that may be significant for certain energies and geometries: (i) at lower energies of <200 eV, backscattering from a nearest-neighbor atom behind the emitter and then forward scattering by the emitter (from Ref. 25); (ii) multiple forward scattering along lines of closely spaced atoms that leads to a reduction of the expected intensity enhancement, particularly at higher energies of >500 eV (from Refs. 24 and 73).

Radiation with polarization $\hat{\epsilon}$ is incident on some atom in a cluster, from which it ejects a core-level photoelectron. (In Fig. 3a, the emitting atom is shown near the surface, but it could as well be any atom in the substrate.) If the initial core-electron wave function is denoted by $\psi_c(\mathbf{r})$ and the final photoelectron wave function corresponding to emission with wave vector $\mathbf{k}$ by $\psi(\mathbf{r}, \mathbf{k})$, then the observed intensity will be given in the dipole approximation by

$$I(\mathbf{k}) \propto |\langle \psi(\mathbf{r}, \mathbf{k})| \, \hat{\boldsymbol{\varepsilon}} \cdot \mathbf{r} \, |\psi_c(\mathbf{r})\rangle|^2. \tag{1}$$

The final-state wave function in single scattering is further described as being the superposition of a direct wave $\phi_0(\mathbf{r}, \mathbf{k})$ and all singly scattered waves $\phi_j(\mathbf{r}, \mathbf{r}_j \rightarrow \mathbf{k})$ that result from initial ϕ_0 emission toward a scatterer j at $\mathbf{r}_j$ and then subsequent scattering so as to emerge from the surface in the direction of $\mathbf{k}$. Thus, the overall wave function can be written as[21,41]

$$\psi(\mathbf{r}, \mathbf{k}) = \phi_0(\mathbf{r}, \mathbf{k}) + \sum_j \phi_j(\mathbf{r}, \mathbf{r}_j \rightarrow \mathbf{k}). \tag{2}$$

Because the detector is situated at essentially infinity along $\mathbf{k}$, all of the waves in Eq. (2) can finally be taken to have the limiting spherical forms $\phi_0 \propto \exp{(ikr)}/r$ or $\phi_j \propto \exp{(ik\,|\mathbf{r} - \mathbf{r}_j|)}/|\mathbf{r} - \mathbf{r}_j|$, although the effective amplitudes and phases of each type in a given direction will be modulated by the photoexcitation matrix element and, for each ϕ_j, also $\exp{(ikr_j)}/r_j$ and the scattering factor. Flux conservation also dictates that the portion of ϕ_0 which passes to the scatterer j to produce ϕ_j decays in amplitude as a spherical wave, or as $1/r_j$. This decay is a principal reason why PD and AED are short-range probes, although the effects of inelastic scattering contribute additionally to this. If the scattering angle is θ_j, the overall path length difference (PLD) between ϕ_0 and any ϕ_j is $r_j(1 - \cos{\theta_j})$, and it is these PLDs that provide most of the bond-length information in photoelectron or Auger electron diffraction.

3.1.2. Matrix Elements and Final-State Interference

When this model has been applied to photoelectron emission, the dipole matrix element has usually been treated as involving a p-wave final state (that is, the case that is appropriate for emission from an s subshell). This yields a matrix-element modulation of the form $\hat{\boldsymbol{\varepsilon}} \cdot \hat{\mathbf{k}}$ for an arbitrary direction of emission $\hat{\mathbf{k}}$.[21,41] For emission from other subshells with l not equal to zero, more complex expressions including both of the interfering $l + 1$ and $l - 1$ channels are involved,[43,45–47] and we return below to consider how important these effects can be. However, at higher energies, the assumption of a p-wave final state has been found to be reasonably adequate in several prior studies of non-s emission.[9,10,48–50]

Since the differential photoelectric cross section $d\sigma_{nl}(\hat{\boldsymbol{\varepsilon}}, \mathbf{k})/d\Omega$ is proportional to intensity rather than amplitude, another possible approximation might be to use a ϕ_0 modulation of $[d\sigma_{nl}(\hat{\boldsymbol{\varepsilon}}, \mathbf{k})/d\Omega]^{1/2}$.[51] Although this is not strictly correct and it also does not account for possible sign changes in the matrix element with direction due to the photoelectron parity,[15,52] it may be a reasonably adequate approximation for higher-energy XPD in which the forward-dominated electron-scattering process selects out $\mathbf{r}_j$ choices very nearly parallel to $\mathbf{k}$. That is, for the range of $\mathbf{r}_j$ directions near the $\mathbf{k}$ direction that produce significant scattering, the matrix element varies little, so that a very precise description of it is not required. In fact, predicted XPD patterns have not been found to be very sensitive to the exact way in which the matrix-element modulation is included. At lower energies, such simplifications are not generally possible, however, and Treglia[53] has, for example, recently shown that not using the correct final-state angular momenta can have a strong effect on predicted azimuthal diffraction patterns at energies of about 30 eV.

Such final-state momentum and interference effects have been studied in more detail recently by Friedman and Fadley,[47] who have made use of a newly developed Green's function matrix approach due to Rehr and Albers.[54] Representative results as a function of electron kinetic energy are presented in Fig. 4. Here, a Cu emitter is 3.5 Å away from a single Cu scatterer, and three different electron kinetic energies of 100, 300, and 1000 eV are considered. Scattering is in all cases full spherical wave. The intensity fluctuations as a function of scattering angle are normalized to the unscattered intensity I_0 as $\chi = [I - I_0]/I_0$. In order to illustrate in these calculations only the effects of changing the final-state angular momenta that are involved, emission from a Cu $2p$ orbital was taken as a reference. For this p-emission case, the correct final-state interference involves s

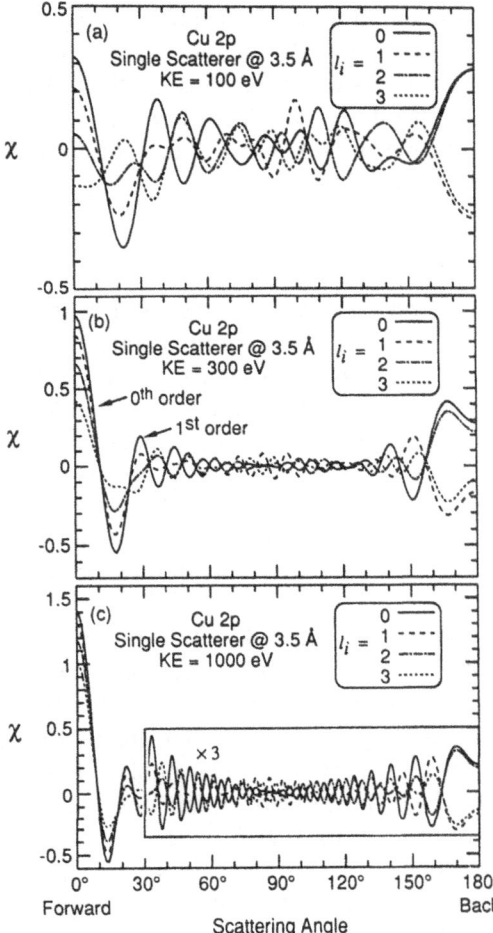

FIGURE 4. Theoretical calculations of electron scattering from a single Cu atom at a distance of 3.5 Å from the emitter and for energies of (a) 100 eV, (b) 300 eV, and (c) 1000 eV. Intensity is shown as the normalized function $\chi = (I - I_0)/I_0$. Full spherical-wave (SW) scattering is used, and different final-state assumptions are compared: $l_i = 0$ (s to a single p channel), $l_i = 1$ (p to interfering $s + d$ channels), $l_i = 2$ (d to $p + f$), and $l_i = 3$ (f to $d + g$). The radiation is taken to be unpolarized, with the plane of polarization lying in the plane of r_{Cu} and k. Note the sign reversals due to photoelectron parity in the backscattering direction. (From Ref. 47.)

and d waves, and includes the radial matrix elements R_s and R_d and the phase shifts δ_s and δ_d. These have been calculated using an atomic cross-section program due to Manson.[55] The ratio R_d/R_s changes relatively little, from 4.62 to 3.91, as we go from 100 eV to 1000 eV. The curves shown for $l_i = 0$ are the simple limit, discussed previously, of an s initial state and single p final state with no interference. The results for $l_i = 1$ are the correct description of Cu 2p emission. For the other two cases of $l_i = 2$ and $l_i = 3$ shown, emission into final waves at $l_f = 1$ and 3 and $l_f = 2$ and 4, respectively, is allowed, and the same radial matrix elements R_s and R_d and phase shifts δ_s and δ_d were used for the $l_f = l_i + 1$ and $l_f = l_i - 1$ channels in both cases. These sets of four curves thus permit systematically observing only the effect of the different final-state character and interference associated with the dipole matrix element.

Several general conclusions can be drawn from the curves of Fig. 4:

- Increasing the angular momenta in the final state from 1 to 0 + 2 to 1 + 3 to 2 + 4 is found to decrease systematically the amplitude of forward scattering, thus constituting a reason for which calculations using the p final state may overpredict the degree of anisotropy for emission from subshells with $l_i \geq 1$.
- In the backscattering direction, the parity of the photoelectron waves is evident, since the odd waves from $l_i = 0$ and 2 exhibit the same sign of χ, and the opposite sign is seen for the even waves from $l_i = 1$ and 3. The previously discussed approximation of using the square root of the differential cross section neglects these sign differences. It implicitly assumes photoelectron waves of even character unless an *ad hoc* sign change is introduced as appropriate for emission angles greater than 90° with respect to the polarization vector.[15]
- The smallest differences between different final-state angular momenta are for the highest energy, where, in the dominant forward direction, the main effect is a reduction of amplitudes in the forward scattering direction, but little change occurs in the shapes of the '0th-order' peak at a scattering angle of 0° and, for $l_i < 3$, also in the 1st-order peak at about 22°. However, as energy is decreased to 100 eV, the differences between the curves become increasingly more significant, and they begin also to involve phase changes in the regions of both of these peaks nearest forward scattering.
- At the highest energy typical of the XPS limit, one thus expects the general shape of the 0th order or forward scattering peak to be the same regardless of final-state angular momenta, and to see a general suppression of the relative importance of the higher-order features.

Overall, these results indicate that the use of the correct final-state angular momenta with interference will probably be important for energies below about 500 eV. For higher energies of 1000 eV or move, forward scattering should be reasonably well treated by the simple p final state (as has been verified in prior XPD studies), although both overall anisotropies and the relative intensities of higher-order features may be overestimated. Similar conclusions concerning the suppression of higher-order diffraction features have been reached by both Parry[46] and Sagurton[56] using more approximate calculations based upon plane-wave scattering and/or plane-wave final states.

Keeping in mind the discussion of the last paragraphs, we shall for simplicity and heuristic reasons in what follows still use the p final state and its factor $\hat{\boldsymbol{\varepsilon}} \cdot \hat{\mathbf{k}}$ in describing photoelectron emission.

3.1.3. Electron–Atom Scattering

The electron–atom scattering that produces ϕ_j is most simply described by a complex plane-wave (PW) scattering factor

$$f_j(\theta_j) = |f_j(\theta_j)| \exp[i\psi_j(\theta_j)]. \tag{3}$$

where $\psi_j(\theta_j)$ is the phase shift associated with the scattering. The scattering factor is in turn calculated from partial-wave phase shifts δ_l according to the usual expression:

$$f(\theta) = (2ik)^{-1}\sum_{l=0}^{\infty} (2l + 1)[\exp(2i\delta_l) - 1]P_l(\cos\theta), \tag{3'}$$

where the P_l are Legendre polynomials. For large r, the scattered wave ϕ_j is thus proportional to $f_j(\theta_j)\exp(ik|\mathbf{r} - \mathbf{r}_j|)/|\mathbf{r} - \mathbf{r}_j|$, with an overall phase shift relative to ϕ_0 of $kr_j(1 - \cos\theta_j) + \psi_j(\theta_j)$ that is due to both path-length difference and scattering. The use of this form for ϕ_j implicitly assumes that the portion of ϕ_0 incident on the jth scatter has sufficiently low curvature compared to the scattering potential dimensions to be treated as a plane wave. This is the so-called small-atom approximation,[57] and its limitations in comparison to the more accurate spherical-wave (SW) scattering[21,25,58,59] of Fig. 4 are discussed below.

The PW scattering factor $f_j(\theta_j)$ is thus determined by applying the partial-wave method to a suitable spherically symmetric scattering potential for each atomic type in the cluster. The number of partial-wave phase shifts needed for convergence goes up with energy, and for a typical scattering potential of effective radius 1.5 Å would be $\gtrsim 8$ for $E_{kin} = 500\,eV$ and $\gtrsim 24$ for 1500 eV. Tabulations of free-atom scattering factors at energies going up to the XPS regime also exist.[60] Alternatively, scattering potentials more appropriate to a cluster of atoms with overlapping charge densities and potentials can be constructed via the muffin-tin model employed, for example, in LEED theory.[61] The free-atom f_j is generally larger in magnitude in the forward direction than its muffin-tin counterparts due to the neglect of charge and potential overlap.[42] Both types of f_j have been employed in higher-energy PD and AED calculations, and they usually do not yield markedly different $I(\mathbf{k})$ curves, although the use of the free-atom f_j is expected to predict slightly higher peak intensities due to its larger amplitudes in the forward direction. The PW scattering factor amplitudes in Fig. 2 were calculated using the more accurate muffin-tin procedure. Whatever procedure is used to calculate these scattering factors, there are two useful generalizations concerning their behavior as atomic number is varied:

• The forward scattering amplitude $|f_j|$ at higher energy is found to be primarily sensitive to the radius of the atom (or muffin tin) involved. It is for this reason that free-atom forward scattering amplitudes are always larger than those for a muffin tin in which the potential is effectively truncated at the tin radius. This behavior can be rationalized by a classsical argument in which it is noted that forward scattering trajectories graze the outer reaches of the scattering potential and so are only deflected slightly; these trajectories are thus primarily sensitive to the outer regions of the potential.

• The backscattering amplitude at higher energy is by contrast found to increase monotonically with atomic number. This also is expected from a classical argument in which backscattering involves strongly deflected trajectories that pass close to the nucleus.

3.1.4. Inelastic Scattering

The effects of *inelastic* scattering on wave amplitudes during propagation below the surface must also be included. If intensity falls off as $\exp(-L/\Lambda_e)$, where L is an arbitrary path length below the surface and Λ_e is the inelastic attenuation length, then *amplitude* is expected phenomenologically to fall off as the square root of this or $\exp(-L/2\Lambda_e) \equiv \exp(-\gamma L)$. Each wave ϕ_0 or ϕ_j is multiplied by such an exponential factor involving an L value which includes the total path length below some surface cutoff point (cf. Fig. 3a). This surface cutoff is often chosen to be the substrate surface as defined by hard-sphere atoms,[42] although this choice should not influence the diffraction patterns unless some atoms are positioned above the cutoff. Thus, the attenuation coefficient $\gamma = 1/2\Lambda_e$, although γ values up to 1.3–2 times this have been suggested in prior EXAFS,[41,62] AED,[12] and PD[42,48,63] analyses. That is, the effective inelastic attenuation length Λ_e in these diffraction experiments is suggested to be about 0.50–0.75 times literature values based upon intensity-attenuation measurements or theoretical calculations.[64] In fact, some inelastic attenuation lengths derived from EXAFS measurements do not appear to take account of the difference between amplitude and intensity mentioned above.[62]

These reduced values of Λ_e are not surprising in view of several factors: Uncertainties of at least ±20% are common in measurements of attenuation lengths,[64,65] and some recent measurements in fact yield values that are significantly lower than others in the literature.[65] The effects of elastic scattering and diffraction on intensities can introduce additional uncertainties of this order,[66,67] and it is, for example, now well recognized that the actual mean free path between inelastic scattering events is about 1.4 times the attenuation length discussed above. Finally, the effective attenuation length in a diffraction measurement should be shorter than in a simple intensity-attenuation experiment, because quasielastic scattering events of small energy (e.g., from phonons) that leave the electron kinetic energy within the peak being measured[68] can still introduce direction changes and phase shifts that effectively remove such electrons from the coherent intensity for diffraction. In addition, multiple elastic-scattering events similarly cause a reduction of the effective coherent intensity in a single-scattering theory. Thus, one overall expects effective attenuation lengths related as $\Lambda_e(\text{intensity}) > \Lambda_e(\text{multiple-scattering diffraction}) > \Lambda_e(\text{single-scattering diffraction})$.

Fortunately, electron diffraction features for most cases are not strongly affected by varying Λ_e over its plausible range, and so its choice is in general not crucial to final structural conclusions. Nonetheless, it is desirable to verify this insensitivity by varying Λ_e in model calculations.[42,48,63]

3.1.5. Vibrational Effects

Vibrational attenuation of interference effects is furthermore potentially important and can be included in the simplest way by multiplying each ϕ_j by its associated temperature-dependent Debye–Waller factor:

$$W_j(T) = \exp\left[-\Delta k_j^2 \overline{U_j^2}(T)\right] = \exp\left[-2k^2(1 - \cos\theta_j)\overline{U_j^2}(T)\right], \qquad (4)$$

where Δk_j is the magnitude of the change in wave vector produced by the scattering, and $\overline{U_j^2}(T)$ is the temperature-dependent one-dimensional mean-squared vibrational displacement of atom j. At this level of approximation, $\overline{U_j^2}$ is assumed to be isotropic in space, and any correlations in the movements of near-neighbor atoms are neglected. (The importance of *correlated* vibrational motion in certain types of lower-energy diffraction experiments is considered below.) Suitable bulk and surface $\overline{U_j^2}$ values or Debye temperatures can be obtained from the literature. At high energy, the electron scattering is significant only when θ_j is rather close to zero, and this acts through the $(1 - \cos\theta_j)$ factor in the argument of Eq. (4) to yield W_j very close to unity for all important scattered waves. So vibrational effects are to first order not very significant in forward-scattering-dominated XPD or AED, although they can be very important in LEED, EXAFS, and lower-energy PD and AED, where backscattering is the dominant diffraction mode and thus $1 - \cos\theta_j$ is a maximum.

An alternate method for allowing for vibrational effects is to assume some probability distribution of atomic positions due to vibration (as, for example, a harmonic-oscillator envelope) and then to numerically sum separate weighted diffraction intensities for all possible combinations of atomic positions. This is cumbersome, but it has been used to quantitatively look at the effects of specific types of wagging molecular vibrations at surface.[23,69]

3.1.6. Single-Scattering Cluster Model

With these assumptions, the simplest SSC–PW expression for photoelectron intensity $I(\mathbf{k})$ can now be written down from Eqs. (1–3) as

$$I(\mathbf{k}) \propto \int \left| \hat{\boldsymbol{\varepsilon}} \cdot \hat{\mathbf{k}}e^{-\gamma L} + \sum_j \frac{\hat{\boldsymbol{\varepsilon}} \cdot \hat{\mathbf{r}}_j}{r_j}|f_j(\theta_j)|\, W_j e^{-\gamma L_j}\{\exp i[kr_j(1 - \cos\theta_j) + \psi_j(\theta_j)]\} \right|^2 d\hat{\boldsymbol{\varepsilon}}$$

$$+ \sum_j \int (\hat{\boldsymbol{\varepsilon}} \cdot \hat{\mathbf{r}}_j)^2 \frac{|f_j(\theta_j)|^2}{r_j^2}(1 - W_j^2)e^{-2\gamma L_j}\, d\hat{\boldsymbol{\varepsilon}}. \qquad (5)$$

Here, $\hat{\boldsymbol{\varepsilon}} \cdot \hat{\mathbf{k}}$ and $\hat{\boldsymbol{\varepsilon}} \cdot \hat{\mathbf{r}}_j$ represent p-wave photoemission matrix-element modulations along the unit vectors $\hat{\mathbf{k}}$ and $\hat{\mathbf{r}}_j$, respectively, and $\exp(-\gamma L)$ and $\exp(-\gamma L_j)$ are appropriate inelastic attenuation factors. Thus, $(\hat{\boldsymbol{\epsilon}} \cdot \hat{\mathbf{k}}) \exp(-\gamma L)$ is the amplitude of the direct wave $\phi_0(\mathbf{r}, \mathbf{k})$ and $(\hat{\boldsymbol{\epsilon}} \cdot \hat{\mathbf{r}}_j) |f_j(\theta_j)| W_j \exp(-\gamma L_j)/r_j$ is the effective amplitude of $\phi_j(\mathbf{r}, \mathbf{r}_j \rightarrow \mathbf{k})$ after allowance for both inelastic scattering and vibrational attenuation of interference. The complex exponential allows for the total final phase difference between ϕ_0 and each ϕ_j.

The integrals on $\hat{\boldsymbol{\epsilon}}$ simply sum over the different polarizations perpendicular to the radiation progagation direction, as appropriate to the particular case at hand. Closed-form expressions for a totally unpolarized source that are applicable to high-energy work are given elsewhere;[42] however, the simplest way to carry out this integration for a general case is just to sum the intensities for two perpendicular polarizations of convenient orientation.

The second Σ_j corrects the first absolute value squared for the incorrect inclusion of Debye–Waller attenuations in terms involving a product of a scattered wave with itself. That is, in expanding the absolute value squared, only products involving unlike waves like $\phi_0 \phi_j^*$ or $\phi_j \phi_l (j \neq l)$ should include Debye–Waller products of W_j or $W_j W_l$, respectively. The $(1 - W_j^2)$ factor in the second summation is thus necessary to yield overall correct products of the form $\phi_j \phi_j^*$ without any W_j^2 factor. The second sum has been called thermal diffuse scattering,[40] and it is often quite small with respect to the overall modulations. Equation (5) is thus the basic starting point of the single-scattering cluster model.

In modifying this model to describe Auger emission, the usual assumption is that the much freer mixing of angular momenta in the final state overall leads to an outgoing wave with s character.[12,40,70–73] Although selection rules do limit the allowed final angular momentum states in Auger emission,[72] for certain cases, the $l = 0$ channel is dominant. Also, if filled subshells are involved in both the initial and final levels of the transition, the implicit sums over all initial and final m_l values would be expected to produce an overall distribution of emitted primary intensity that could be approximated as an s wave. Although it is possible for higher-l components to be present in the final state that could affect the scattering,[72,73] these are often found at higher energies to be minor effects.[12,70,71] For Auger emission into such an assumed s final state, we thus simply remove all factors involving $\hat{\boldsymbol{\epsilon}} \cdot \hat{\mathbf{k}}$ and $\hat{\boldsymbol{\epsilon}} \cdot \hat{\mathbf{r}}_j$ in Eq. (5). Non-s character in Auger final states deserves further study however.

It is also worth noting here that the cluster sum on j in Eq. (5) makes no explicit use of the 2- or 3-dimensional translational periodicities that may be present, even though the atomic coordinates $\mathbf{r}_j$ used as inputs may incorporate such periodicities. Thus, neither surface- nor bulk-reciprocal lattice vectors $\mathbf{g}$ are explicitly involved, and it is not appropriate at this level of description to speak of diffraction "beams" associated with certain $\mathbf{g}$ vectors as in LEED. However, in section 5.1 we will consider the relationship of this model to an alternative Kikuchi-band picture that does involve $\mathbf{g}$ vectors and the idea of Bragg reflections from sets of planes.

The last parameter of importance in actually using Eq. (5) is the range of j or the choice of a suitable cluster of atoms. This is done empirically so as to include all significant scatterers by verifying that the predicted diffraction patterns do not change in any significant way with the addition of further atoms at the periphery of the cluster. Clusters can range from a few atoms for near-normal high-energy emission from a vertically oriented diatomic molecule on a surface[23] to as many as several hundred atoms for substrate emission in which both the emission and the scattering must be summed over several layers into the bulk.[42] In the latter case, each structurally unique type of atom emits incoherently with respect to the other, so that intensities from each must be added layer by layer. However, even for the largest clusters so far considered, the inherent simplicity of Eq. (5) still yields calculations which do not consume excessive amounts of computer time, especially by comparison with those necessary for such procedures as multiple-scattering LEED simulations.

A further physical effect of importance in making comparisons to experiment is the possibility of electron refraction at the surface in crossing the surface barrier or inner potential of height V_0. Even at the relatively high energies of XPS, for emission angles near grazing, refraction effects of a few degrees can be produced (cf. Fig. 14 in Ref. 9). Thus, for lower takeoff angles relative to the surface and/or lower kinetic energies, a proper allowance for refraction is necessary. This is accomplished most simply by using a suitable inner potential V_0 derived from experiment and/or theory to predict the internal angle of emission θ' for a given external propagation direction θ.[9] The resulting expression for an electron energy of $E'_{kin} = E_{kin} + V_0$ inside the surface is

$$\theta' = \cos^{-1}\left\{\left[\frac{E_{kin}}{E_{kin} + V_0)}\right]^{1/2} \cos\theta\right\}, \qquad (6)$$

where, as before, θ and θ' are measured with respect to the surface. In the presence of an adsorbate, the exact form of the surface potential barrier thus becomes important, as it may not then be possible to assume an abrupt rise to the vacuum level at the substrate surface. Also, the presence of adsorbate atoms may alter V_0 through changes in the work function, and these atoms also may occupy positions above the surface in which only a fraction of V_0 is appropriate. In some photoelectron diffraction studies, V_0 has also been treated as an adjustable parameter.[20,25,63] Although prior studies indicate that structural conclusions are not particularly sensitive to the choice of V_0,[25,42] it is important to realize that not allowing for it properly may shift theoretical diffraction patterns by as much as a few degrees with respect to the actual θ values at which they will be observed. The precise method of allowing for inner potential and related image–force effects has also been considered in more detail theoretically.[25]

We stress also at this point that any uncertainties in final structures associated with the choices of nonstructural parameters such as the scattering phase shifts, the attenuation length for inelastic scattering, vibrational attenuation, and the inner potential are equally well shared with the techniques of LEED, EXAFS, and SEXAFS, although in EXAFS/SEXAFS, empirical phase shifts from known structures can sometimes be used.

A final step in any realistic calculation based upon this model is to integrate the direction of emission **k** over the solid angle Ω_0 accepted into the electron analyzer. For most of the calculations reported here, this has been over a cone of ± 3.0–$3.5°$ half angle, although for certain high-resolution cases a smaller cone of ± 1.0–$1.5°$ has been used.

3.1.7. Improvements to the Model

We now consider some possible improvements to this simple SSC–PW model:

• A first possible correction is to choose a *more correct form for the primary wave as it leaves the emitter*. The SSC–PW result of Eq. (5) assumes a simple outgoing plane wave from the emitter which then scatters to produce an outgoing spherical wave from each scatterer. In fact, the correct primary wave should be of the type used in free-atom photoelectric cross sections and should consist of an ingoing spherical wave plus the outgoing plane wave.[21,41,45,74] Such a primary wave experiences the emitter potential and represents the correct solution to the Schrödinger equation inside of a muffin-tin-like region centered on the emitter. If this form of the primary wave is used, the equivalent of Eq. (5) with neglect of effects due to vibrations is:[21,41]

$$I(\mathbf{k}) \propto \int \left| \hat{\boldsymbol{\varepsilon}} \cdot \hat{\mathbf{k}} e^{-\gamma L} + \sum_j \frac{\hat{\boldsymbol{\varepsilon}} \cdot \hat{\mathbf{r}}_j}{r_j} |f_j(\theta_j)| \, e^{-\gamma L_j} \{ \exp i[kr_j(1 - \cos\theta_j) + \psi_j(\theta_j)] \} \right.$$
$$\left. + \sum_j \frac{\hat{\boldsymbol{\varepsilon}} \cdot \hat{\mathbf{r}}_j}{r_j^2} f_j(\pi) f_{em}(\pi - \theta_j) e^{-\gamma(L + 2r_j)} \{ \exp i[2kr_j] \} \right|^2 d\hat{\boldsymbol{\varepsilon}}. \quad (7)$$

This result, although still single scattering in assumption, now contains, through the scattering of the incoming wave, a second sum of terms that are the classic double scattering events of the type emitter → scatterer → emitter → detector discussed in EXAFS theory.[4] Because these added terms are in effect double scattering and also exhibit stronger attenuation due to both $1/r_j^2$ and $e_j^{-2\gamma r}$, this sum is expected for many cases to be a small correction to Eq. (5). This should be especially true for higher energies where backscattering is negligible. In fact, the inclusion of this sum can be shown to lead to the central-atom (emitter) phase shift that is always present in EXAFS theory, and we comment further on this later in this section.

• A next important correction is the use of *spherical-wave (SW) scattering* instead of the asymptotic and much simpler plane-wave (PW) scattering. The nature of such SW corrections in reducing forward scattering amplitudes in XPD was first pointed out some time ago,[23] but more recent studies have presented detailed comparisons of PW and SW results for different systems.[58,59] For example, Fig. 5 compares PW and SW scattering at energies from 50 eV to 950 eV,[58] with the results being displayed in a format identical to that of Fig. 4. Emission from an s level ($l_i = 0$, $l_f = 1$) to a single Ni scatterer 2.49 Å away is considered. For larger scattering angles ($\gtrsim 40°$) and higher energies ($\gtrsim 200$ eV), the PW and SW results are essentially identical. However, for lower energies and in the forward scattering direction, there are significant differences. In particular, for energies $\gtrsim 100$ eV, the forward scattering peak is significantly reduced in amplitude by a factor that can be as low as 0.5. As expected, the differences between PW and SW curves also decrease as the scatterer is moved away from the emitter,[58] because in the limit of a scatterer at infinity, the incident wave is

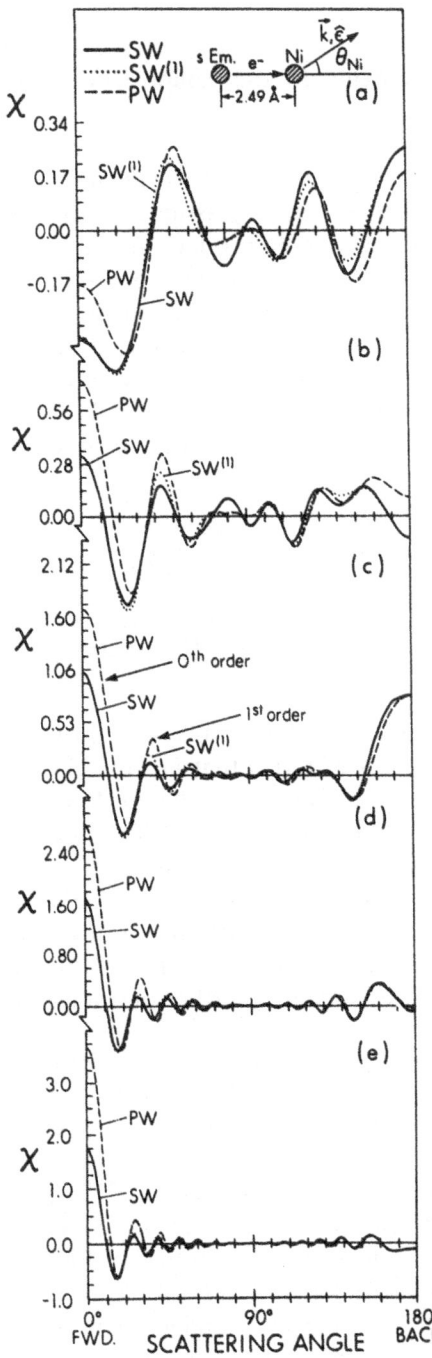

FIGURE 5. As in Fig. 4, but comparing plane-wave (PW) and spherical-wave (SW) scattering from a single Ni scatterer at a distance of 2.5 Å from the emitter with energies of (a) 50 eV, (b) 100 eV, (c) 200 eV, (d) 500 eV, and (e) 950 eV. Here, PW results are compared to SW results for the case of $l_i = 0$ (s emission to a single p channel). Polarization is parallel to the emission direction. The curves labelled SW$^{(1)}$ represent a first-order approximation to the full SW scattering. (From Refs. 21 and 58.)

planar. One general conclusion from these results is thus that, at higher energies, the primary effect of including curvature in ϕ_0 is to reduce the amplitudes of the forward-scattering peaks in $I(\mathbf{k})$ for near-neighbor atoms as compared to those predicted from Eq. (3).

Fortunately, such SW corrections can now be very simply and accurately incorporated into the SSC framework via effective SW scattering factors developed by Barton and Shirley using a Taylor-series magnetic quantum number expansion (MQNE)[25] and by Rehr *et al.* using separable Green's function approaches.[45,54] For example, Rehr *et al.*[45] derive an equation identical to Eq. (7) in form, but in which the plane-wave scattering factors $f_j(\theta)$ are replaced by three effective spherical-wave scattering factors $f_{j,\text{eff}}^{(1)}(\theta, r_j)$, $f_{j,\text{eff}}^{(2)}(\pi, r_j)$, and $f_{\text{cm,eff}}^{(3)}(\pi - \theta, r_j)$ that are used to describe the three types of scattering events present. These effective scattering factors depend on r_j, as they must converge to the PW result as r_j goes to infinity. They are also very simply calculable, involving expressions closely related to that in Eq. (3′).

However, particularly at higher energies, the much simpler PW approximation is still found to yield results very similar in form to those with SW scattering, and it has been found possible to draw useful structural conclusions with it. Sometimes, PW scattering at high energy has been used together with an empirical reduction factor of forward scattering amplitudes by a factor of 0.4–0.5[42] that can be largely justified as being due to SW effects (cf. Fig. 5).

• An additional important correction for some cases is the use of *correlated vibrational motion* in which atoms that are near neighbors of the emitter have lower vibrational amplitudes relative to the emitter, and thus Debye–Waller factors for diffraction that are nearer unity. This correction is more important in special geometries and at lower energies for which large-angle or, particularly, backscattering events become more important, as first pointed out in connection with the interpretation of scanned-energy data by Sagurton *et al.*[21] and also discussed by Barton and Shirley.[25] This more correct form for vibrational attenuation involves a factor W_j^{corr} of the form:[21]

$$W_j^{\text{corr}}(T) = \exp\left[\frac{-\Delta k_j^2 \sigma_j^2(T)}{2}\right] = \exp\left[-k^2(1 - \cos \theta_j)\sigma_j^2(T)\right], \qquad (8)$$

where $\sigma_j^2(T) = \langle(\Delta\hat{\mathbf{k}}_j \cdot \mathbf{u}_j)^2\rangle$ is a thermal average of the projection of the atomic displacement $\mathbf{u}_j$ as measured with respect to the emitter onto the direction of the change in wave vector produced by the scattering $\Delta\hat{\mathbf{k}}_j$. Thus, each scatterer in a photoelectron diffraction experiment is sensitive to a different type of vibrational displacement, varying from no effects for forward scattering, to small effects for small-angle scattering associated with components of $\mathbf{u}_j$ perpendicular to the emitter–scatterer axis, to maximum effects for backscattering associated with components of $\mathbf{u}_j$ along this axis. By contrast, in SEXAFS, it is only the along-axis components that contribute. Correlation effects are also expected to be largest for atoms that are backscatterers, because along-axis vibrations will be reduced more than those perpendicular to this axis. Ultimately, this might make it possible to measure anisotropies in vibration in a more precise way with temperature-dependent photoelectron diffraction, for example, by looking at the variation of different peaks in Fourier transforms of scanned-energy data. A first attempt at this has recently been made by Wang *et al.*[75] Also, even forward scattering features at high energy contain vibrational information because of peak broadening by motion perpendicular to a bond,[23,69] and this has permitted

Wesner *et al.*[76] to determine the vibrational amplitude anisotropy for an adsorbed molecule, as discussed further in section 4.1.3.

• A final aspect of the model which might be improved but which has only been discussed in a limited way to date is *more accurate allowance for both surface refraction and attenuation due to inelastic scattering.* Refraction has been treated differently from the phenomenological approach indicated here both by Lee[41] and by Tong and Poon,[77] who have considered the proper matching of the attenuated photoelectron wave inside the surface to the free electron wave outside the surface. However, the latter have found that, if refraction is allowed for in the way described here in calculating the path length for inelastic scattering in approaching the surface, the net result is very little different from the correct treatment of the wave matching. Another more complex problem is choosing the proper value for the inelastic attenuation length: As we have noted above, these lengths in electron-diffraction problems appear empirically to be only about 0.5–0.75 times the typical literature values based upon intensity attenuation. It would be desirable to understand these attenuation lengths more quantitatively, including both elastic and inelastic effects, for example, within the framework of more accurate methods of measuring peak intensities developed by Tougaard.[78] Finally, it might be useful to consider the possibility of *nonuniform or anisotropic inelastic scattering.* Such effects have been considered in both LEED[79a] and EXAFS,[79b] where the use of complex scattering phase shifts is proposed; but the influence of such effects on predicted diffraction patterns in PD or AED has not been assessed. More recently, Treglia *et al.*[80a] have used SSC–SW calculations to describe very low energy photoelectron diffraction at about 30 eV from different surfaces of W. They see evidence for a significantly different inelastic attenuation length in emission from W (001) and W (110). This could well be possible, but at this low energy, it would also be useful to carry out full MS calculations to eliminate such effects as another cause of effective anisotropic attenuation. In another recent paper, Frank *et al.*[80b] have discussed Auger electron diffraction data from Pt(111) with various adsorbates and for energies varying from about 65 eV to 420 eV. They have analyzed these results in terms of a classical model of anisotropic inelastic attenuation which totally neglects all wave inteferences and diffraction phenomena. Unfortunately, there is no basis in prior experiment or theory for this extreme model, even though it seems to fortuitously fit some of the features in the experimental data. Thus, this classical analysis by Frank *et al.* provides neither a useful method for analyzing AED data, nor any new information concerning the possibility of anisotropic inelastic attenuation. Such attenuation is in any case expected to produce only small corrections to the strong anisotropies associated with diffraction effects.

3.1.8. Relationship to EXAFS/SEXAFS Theory

As a further aspect of the SSC model, we note that it can be directly reduced to an expression very close to that used in EXAFS/SEXAFS analyses if it is assumed that all scattered waves ϕ_j are small in magnitude in comparison to ϕ_0.[15] Then, if we begin at Eq. (5) (for simplicity neglecting any averaging over $\hat{\varepsilon}$), we see that all terms such as $\phi_j\phi_i^*$ and $\phi_i\phi_j^*$ can be neglected in expanding the absolute value squared. The thermal diffuse scattering term can also be neglected. After some simple algebra, it can then be shown that

$$I(\mathbf{k}) \propto (\hat{\boldsymbol{\varepsilon}} \cdot \hat{\mathbf{k}})^2 e^{-2\gamma L} + 2(\hat{\boldsymbol{\varepsilon}} \cdot \hat{\mathbf{k}}) e^{-\gamma L} \sum_j \frac{\hat{\boldsymbol{\varepsilon}} \cdot \hat{\mathbf{r}}_j}{r_j} |f_j(\theta_j)| \, W_j e^{-\gamma L_j}$$
$$\times \cos[kr_j(1 - \cos\theta_j) + \psi_j(\theta_j)], \quad (9)$$

and that this can be converted to a normalized function $\chi(\mathbf{k})$ if we take the unscattered intensity to be $I_0 = (\hat{\boldsymbol{\varepsilon}} \cdot \hat{\mathbf{k}})^2 e^{-2\gamma L}$ and finally write

$$\chi(\mathbf{k}) \equiv \frac{I(\mathbf{k}) - I_0}{I_0} \propto \frac{2}{(\hat{\boldsymbol{\varepsilon}} \cdot \hat{\mathbf{k}})e^{-\gamma L}} \sum_j \frac{\hat{\boldsymbol{\varepsilon}} \cdot \hat{\mathbf{r}}_j}{r_j} |f_j(\theta_j)| \, W_j e^{-\gamma L_j}$$
$$\times \cos[kr_j(1 - \cos\theta_j) + \psi_j(\theta_j)]. \quad (10)$$

This last equation thus has a form very close to the standard kinematical expression for EXAFS/SEXAFS, with the only differences being that double scattering events of the type emitter → scatterer → emitter → detector in Eq. (7) are included in the integration over direction in EXAFS to better describe the primary wave,[41] with these producing the central-atom phase shift; and the integration over direction changes the cosine function here finally to a sine function for EXAFS/SEXAFS. Equations (9) and (10) were first used in connection with the interpretation of ARPEFS data by Orders and Fadley,[15] and they have later been refined in this context by Sagurton et al.[21] Their form also suggests the possibility of using Fourier transform methods in scanned-energy PD to derive information concerning the set of path-length differences associated with a given structure, as discussed first by Hussain et al.[14] and now in active use by Shirley and co-workers[20,25] as a preliminary step of ARPEFS analysis.

As a final comment concerning this level of the diffraction theory, we consider the conservation of photoelectron flux. In the small-atom (or large r_j) limit, where PW scattering is adequate, the usual optical theorem assures that flux will be conserved if it is integrated over 4π.[81] Thus, even if high-energy scattering produces forward-scattering peaks, there will be, somewhere else, sufficient phase space with reduced intensity to exactly cancel them. However, in using the SSC–PW model for cases in which some scatterer distances require SW corrections, it is doubtful that flux will be conserved properly.[40] Nonetheless, with SW scattering correctly included, Rehr et al.[45] have shown that their SW equivalent of Eq. (7) does conserve flux and lead to a generalized optical theorem on each l channel involved.

In subsequent sections, we will consider several applications of this SSC model to the interpretation of experimental data, including especially several substrate and adsorbate systems of known geometry to test the degree of its validity.

3.2 Effects beyond Single Scattering

Finally, the possible importance of multiple scattering (MS), particularly along rows of atoms in a multilayer substrate, has been discussed qualitatively for some time,[9,42] and more recent papers have presented quantitative estimates of such effects and suggested improved methods for including MS corrections if they are needed.[24,25,54,82–84] In general, the MS analogue of Eq. (2) can be written as

$$\psi(\mathbf{r}, \mathbf{k}) = \phi_0(\mathbf{r}, \mathbf{k}) + \sum_j \phi_j(\mathbf{r}, \mathbf{r}_j \to \mathbf{k}) + \sum_j \sum_k \phi_{jk}(\mathbf{r}, \mathbf{r}_j \to \mathbf{r}_k \to \mathbf{k})$$

$$+ \sum_j \sum_k \sum_l \phi_{jkl}(\mathbf{r}, \mathbf{r}_j \to \mathbf{r}_k \to \mathbf{r}_l \to \mathbf{k})$$

$$+ \sum_j \sum_k \sum_l \sum_m \phi_{jklm}(\mathbf{r}, \mathbf{r}_j \to \mathbf{r}_k \to \mathbf{r}_l \to \mathbf{r}_m \to \mathbf{k}) + \text{higher orders},$$

$$(11)$$

where events up to fourth order are shown here and, in the multiple scattering sums, the combinations of j, k, l, and m are limited only in that they do not involve consecutive scattering by the same scatterer. Such MS calculations have been done in two basic ways: first by Tong and co-workers using LEED-type methods that require full translational symmetry along the surface,[24] and more

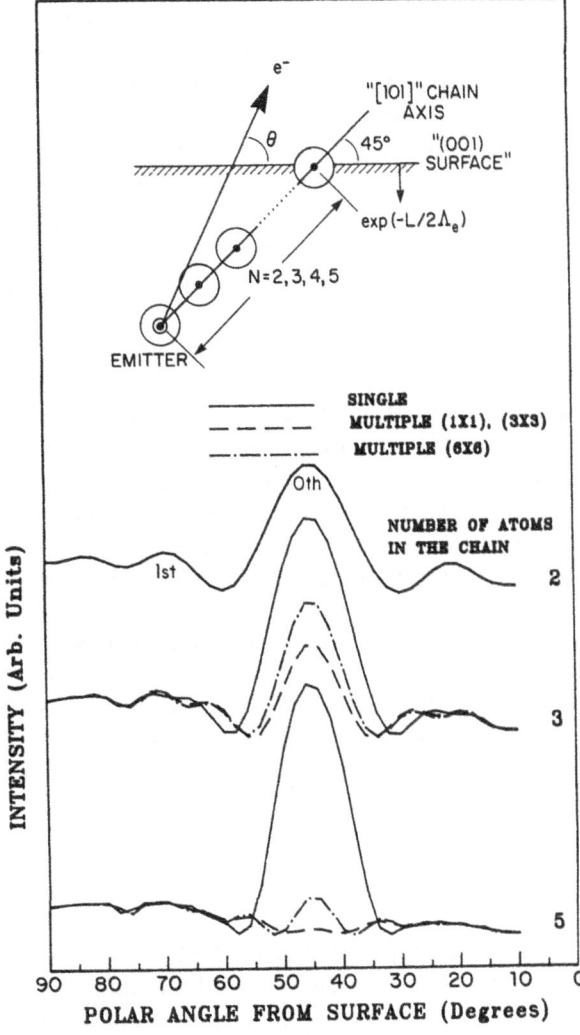

FIGURE 6. Calculated Auger electron diffraction patterns at 917 eV from linear chains of Cu atoms in single and multiple scattering. The geometry of the calculation with the emitter at the base of the chain is shown at the top. The primary outgoing Auger wave is treated as having s character. The multiple scattering results are shown at three levels of the matrix used to describe the scattering: (1 × 1), (3 × 3), and the most accurate, (6 × 6). The (1 × 1) and (3 × 3) cases are found to be superimposable for this case. (From Ref. 84, with similar results also appearing in Refs. 73 and 83(b).)

recently by Barton and co-workers using a cluster approach with SW scattering and the Taylor series MQNE method to simplify the calculations.[25,82,83] The cluster method is really more appropriate to the physics of such a short-range order probe, and we will term it MSC–SW. More recently, Rehr and Albers[54] have proposed a Green's-function matrix method for such MSC–SW calculations that shows promise as an alternate approach in extensive applications by Kaduwela *et al.*[84]

One effect of MS first discussed by Poon and Tong[24] is a defocusing of intensity occurring in multiple forward scattering at higher energies along a dense row of atoms, such that an SSC–PW or SSC–SW calculation along such a row may overestimate the intensity by a factor of two or more. This is illustrated schematically in Fig. 3b(ii). For an embedded species at some distance from the surface but again emitting along such a row, it has more recently been shown that these defocusing effects may be even more dramatic.[73,82,83]

Such defocusing effects have been very nicely illustrated in recent MSC–SW calculations by Barton, Xu, and van Hove[73,82,83] and by Kaduwela *et al.*[84] for emission from chains of Cu atoms of variable length. Some recent results of this type are shown in Figs. 6 and 7. In both figures, chains of 2, 3, or 5 atoms with the emitter at their base are tilted at 45° with respect to the surface of a medium of uniform density that simply serves to attenuate the emitted waves inelastically (see inset in Fig. 6). This geometry thus simulates the intensity distribution expected for emission from the 2nd, 3rd, and 5th layers along a low-index [110]

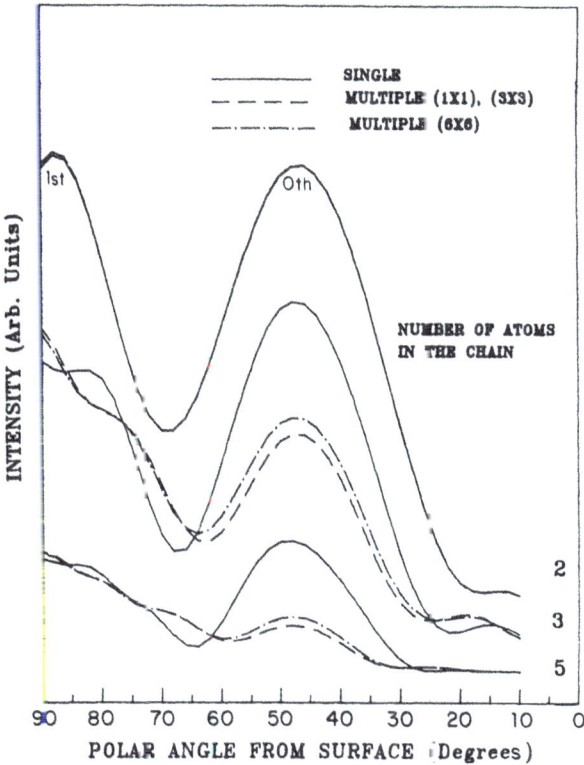

FIGURE 7. As in Fig. 6 (bottom), but for an energy of 100 eV.

row of Cu with (001) orientation, but without any diffraction effects due to scatterers adjacent to the row. Emission into a simple s-wave final state approximating Auger emission is treated. Both single-scattering and fully converged (6×6) multiple-scattering calculations are shown for each case.

In Fig. 6, for an emission energy of 917 eV, it is clear that the single- and multiple-scattering curves are identical for the two-atom case (as appropriate to a diatomic adsorbate, for example), but they diverge more and more as additional scatterers are added between the emitter and the detector. For the five-atom chain, the forward scattering peak is suppressed to only about 10–15% of its value for single scattering. There is also a systematic narrowing of the width of this peak as more defocussing due to multiple scattering comes into play. For scattering angles more than about ±15° from the chain axis, the differences between single and multiple scattering are much more subtle, as is to be expected since strong multiple forward scattering is no longer possible directly in the emission direction. At the much lower energy of 100 eV in Fig. 7, one expects less strongly peaked forward scattering, as shown by the wider peaks along a polar angle of 45°. Here again, the single-scattering and multiple-scattering results are identical for a two-atom chain, but one sees a suppression and narrowing of the forward scattering peak with increasing chain length that is qualitatively similar to, but less severe than, that observed at the higher energy.

Overall, these and other recently published results by Xu and van Hove[73] indicate that, for emitters in the first one or two layers of a surface and/or for which the emission direction does not involve near parallelism with a dense row of scatterers, a single scattering model should be quite accurate. For atoms further below the surface and/or for emission directions along such high-density rows, certain forward scattering features are expected to be suppressed by multiple scattering, but single-scattering calculations should nonetheless predict their positions with good accuracy.

An additional important multiple scattering effect pointed out by Barton et al.[25] is due to strong nearest-neighbor backscattering at lower energies. This they find in certain scanned-energy cases to significantly increase intensity due to events of the type emitter → neighbor → emitter → detector, as illustrated in Fig. 3b(i).

A further important point in connection with such multiple-scattering calculations is that events up to at least the fifth order have to be included to assure reasonable convergence.[25,84] In fact, it is found that including only second-order events can often lead to curves which are in much poorer agreement with experiment than the corresponding first-order calculation![84] This is similar to the experience in EXAFS theory, in which including only lower-order multiple-scattering corrections can yield worse results than those of single scattering.[57,85] A more reasonable procedure is to include events up to, say, the fifth order if the total path length $r_j + r_{jk} + r_{kl} + \cdots$ is less than some cutoff value of 10–20 Å,[20,25,57,85] although an inproved cutoff criterion has been suggested by Kaduwela et al.[84]

As noted previously, there is by now a considerable body of data which indicates that useful structural information can be derived at the SSC–SW or even SSC–PW level, and we will show illustrations of this in subsequent sections. Nonetheless, MS effects such as those described above can cause discrepancies between experiment and theory for certain classes of system, and full MS treatments of both photoelectron and Auger electron diffraction are beginning to be more often used. Several advances in the simplification of these methods, as well as rapid improvements in computer technology, should lead to a greater reliance on MS approaches in future work. In the examples which follow, a variety of theoretical models have been used, and the specific approach followed will be indicated with each set of results to permit the reader to draw his or her own conclusions.

4. ILLUSTRATIVE STUDIES OF DIFFERENT TYPES

4.1. Small-Molecule Adsorption and Orientation

We here consider primarily the case of small-molecule adsorption as studied by higher-energy XPD. The cases treated are thus of considerable interest in studies of surface chemistry and catalysis, and they provide the first simple illustrations of the utility of the forward-scattering peaks discussed in the preceding section. Auger peaks at similar energies of about 1000 eV could also in principle be used for such studies, but all of the cases to date involve photoelectron diffraction.

4.1.1. CO/Ni (001)

We begin with the first system of this type studied by Petersson et al.[23] and Orders et al.:[69] c(2 × 2) CO on Ni (001). Figure 8 compares experimental C 1s

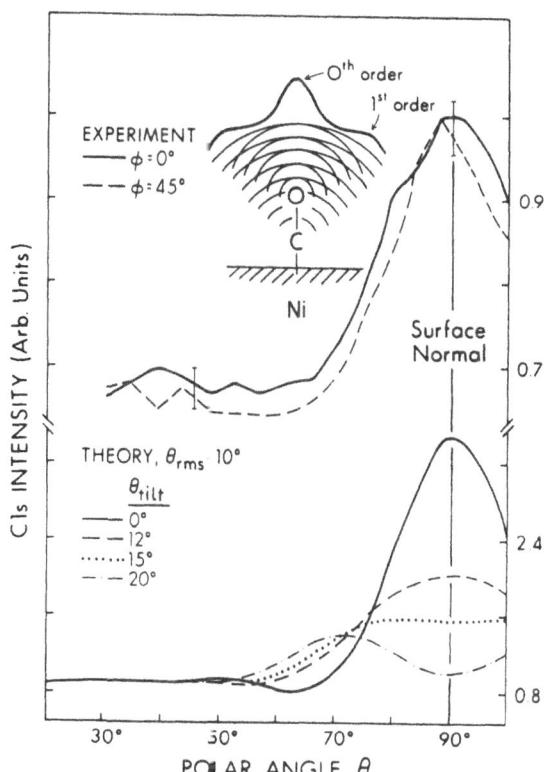

FIGURE 8. Comparison of polar-scan C 1s XPD data from c(2 × 2) CO on Ni (001) at a kinetic energy of 1202 eV with SSC-PW theory. The inset indicates the type of intramolecular forward scattering and interference involved. Note definitions of zeroth-order and first-order effects, as shown also in Figs. 4 and 5. (From Ref. 69.)

polar scans in two high-symmetry azimuths (normalized by dividing by the O $1s$ intensity to eliminate the θ-dependent instrument-response function) to SSC–PW calculations for varying degrees of CO tilt relative to the surface normal.[69] The theoretical model also includes a wagging or "frustrated-rotation" molecular vibration with an rms displacement of 10 Å. The experimental curves are essentially identical along both azimuths and show a strong peak along the surface normal that represents about a 35% anisotropy. Comparing experiment and theory furthermore permits concluding very conservatively that CO is within 10° of normal for this overlayer and that it has no preferential azimuthal orientation.

The inset in this figure also indicates that, in addition to the forward scattering or zeroth-order diffraction peak, one expects higher-order features such as the first-order peak indicated. (These also appear in the single-scatterer calculations of Figs. 4 and 5, where higher orders also are shown.) The first-order peak corresponds to a 2π phase difference between the direct wave and the scattered wave, or a path length difference of approximately one deBroglie wavelength. We will further consider such higher-order features in the next case and subsequent examples.

4.1.2. CO/Fe (001)

A more recent and more complex case of CO adsorption is that on Fe (001). In Figs. 9a and 9b, we show both polar and azimuthal C $1s$ data obtained by Saiki et al.[86] from CO adsorbed at room temperature on Fe (001) so as to form predominantly the so-called α_3 state. This rather unusual species has been the subject of prior studies by several techniques, including EELS, ESDIAD, and NEXAFS.[87] Its structure is of considerable interest because it is thought to be bound in a highly tilted geometry with a significantly weakened C–O bond and thus to be a possible intermediate state for the dissociation of the molecule. However, the best that the tilt angle could be determined from NEXAFS data was 45 ± 10°, and no information was obtained on the most likely azimuthal orientation(s) of the molecules. It is thus of interest to see what more can be learned about such a species from XPD.

The strong peak in the normalized C $1s$ polar-scan results for the [100] azimuth shown in Fig. 9a immediately permits a direct estimate of the tilt angle with respect to the surface normal as $\theta_{\text{tilt}} = 55 \pm 2°$ (that is, with the molecule oriented 35° from the surface). Also, the fact that this forward scattering peak is not seen in polar scans along the [1$\bar{1}$0] azimuth indicates that the preferred tilt is along $\langle 100 \rangle$ directions, or into the open sides of the fourfold-hollow sites that are the sterically most reasonable choices for the bonding location. Complementary evidence confirming this structure comes from the azimuthal data at a polar angle with respect to the surface of $\theta = 35°$ in Fig. 9b. These results again show the preferred tilt in the $\langle 100 \rangle$ azimuths via strong peaks along $\phi = 0°$ and 90°. It is thus concluded that the CO molecules are tilted along the four $\langle 100 \rangle$ axes, perhaps in separate but equally populated domains, as illustrated schematically for one fourfold-hollow site in Fig. 9c.

As a self-consistency check of these data, it is also of interest that the overall effects seen in both parts a and b of Fig. 9 are of very nearly the same magnitude.

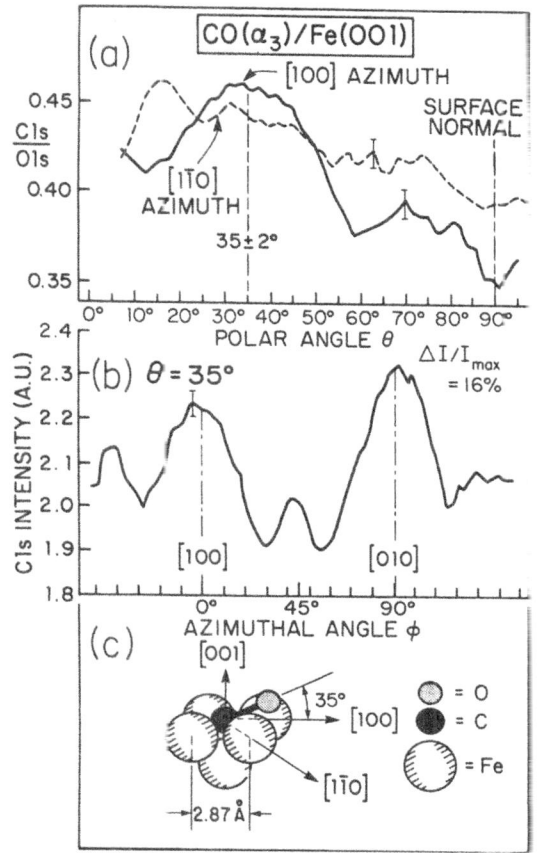

FIGURE 9. (a) Experimental polar scans of the C $1s$/O $1s$ intensity ratio for the α_3 state of CO on Fe (100). The C $1s$ kinetic energy is 1202 eV. Curves are shown for two azimuths: [100] (solid curve) and $[1, \bar{1}, 0]$ (dashed curve). (b) Experimental azimuthal scan of C $1s$ intensity for the α_3 state of CO at a polar angle of 35° chosen to coincide with the peak in the [100] data of part (a). (c) The bonding geometry as deduced from these data. (From Ref. 86.)

That is, if the overall anisotropy as mentioned previously is measured as a percentage by $\Delta I/I_{max}$, we find about a 14% effect in Fig. 9a and a 16% effect in Fig. 9b. Thus, it is possible to reliably measure rather small diffraction effects with XPD, particularly in the azimuthal data, which do not need to be corrected for any systematic instrumental changes in intensity. By contrast, polar scans will always be influenced by a θ-dependent instrument-response function[9] and must somehow be corrected for this. Since the O $1s$ intensity is not expected to be very much affected by final-state scattering and diffraction, using the C $1s$/O $1s$ ratio in Fig. 9a acts to normalize out any such instrumental effects.

Another useful observation from Figs. 9a and b is that the main peaks exhibit very similar full widths at half-maximum intensity (FWHM) of 30–35°. Thus, the resolutions for determining both the polar and the azimuthal senses of the tilt are about the same.

The results in Fig. 9b also exhibit much smaller but quite reproducible peaks along the $\langle 110 \rangle$ azimuths (that is, at $\phi = 45°$) that could be due to scattering from Fe atoms in the $\langle 110 \rangle$ corners of the hollow. A more detailed theoretical analysis of these azimuthal results using the SSC–SW model in fact shows that these peaks are due to constructive addition of first-order scattering from oxygen

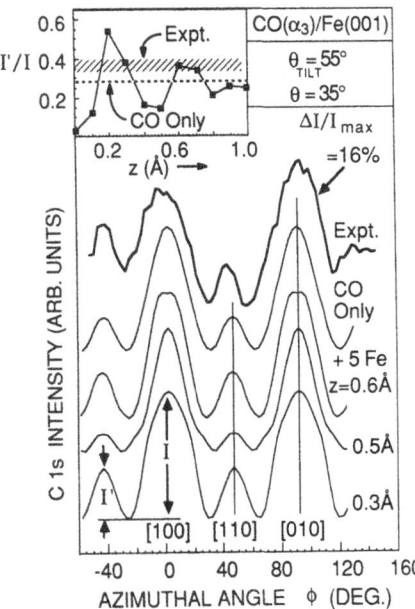

FIGURE 10. Experimental results of Fig. 9(b) are compared with theoretical SSC-SW calculations of C $1s$ azimuthal scans for CO on Fe (001) tilted at 35° with respect to the surface along the $\langle 100 \rangle$ directions and assumed to be in four equally populated domains. In the top theoretical curve, no Fe scatterers are included. In the lower theoretical curves, five Fe scatterers are added, as in Fig. 9(c). The C atom is centered in the fourfold hollow, and the distance z with respect to the Fe surface plane is varied. The inset shows the ratio of the two main peak intensities I'/I as a function of z. (From Ref. 86.)

(see inset in Fig. 8) and second- or third-order scattering from the corner Fe atoms, depending upon the distance z of the tilted CO from the Fe surface. Some results of these calculations of the azimuthal scan of Fig. 9b are compared to experiment in Fig. 10. The top theoretical curve is from a calculation in which only CO molecules are present; these are assumed to be present in four equally populated domains tilted at $\theta = 35°$. This very simple calculation correctly predicts the positions and approximate widths of the strong forward scattering peaks along $\langle 100 \rangle$ azimuths, as well as the additional weaker first-order features seen along $\langle 110 \rangle$ at $\theta = 45°$. However, if the five Fe nearest neighbors are also included as scatterers (as shown in Fig. 9c) and the C atom is further assumed to be centered in the fourfold hollow but with variable vertical distance z relative to the first Fe layer, we arrive at what should be a more realistic set of curves. These are striking in that the small peaks along $\langle 110 \rangle$ are predicted to oscillate in intensity, as shown in the figure inset. Comparing experiment and theory for the ratio I'/I as indicated yields z values of both about 0.22 ± 0.10 Å and 0.63 ± 0.10 Å that agree best; these z values also correspond to very reasonable C–Fe distances of 1.6–2.0 Å. Multiple-scattering calculations for this system by Kaduwela et al.[84] also quantitatively confirm the single-scattering results shown here; this is as expected in view of the high energy and high takeoff angles relative to the surface.

Figure 11 shows a further aspect of this analysis in which the experimental polar scans of Fig. 9a are compared to SSC–SW theory for the two azimuths involved and for several z distances. Polar scans are also seen to be sensitive to both azimuth and vertical distance, with in particular the results for the [110] azimuth favoring a z value nearer 0.3 Å. This study thus indicates the significant advantage of having both polar and azimuthal XPD data for such systems.

The theoretical anisotropies $\Delta I/I_{max}$ in Figs. 8, 10, and 11 are found to be about 2–3 times larger than those of experiment. This kind of discrepancy has

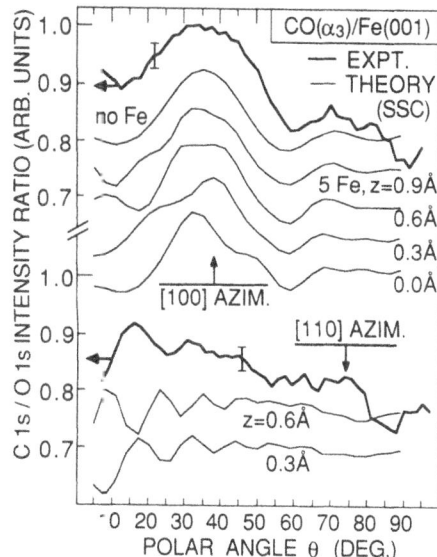

FIGURE 11. Comparison of experimental C 1s polar scans for α_3 CO on Fe (001) to SSC-SW theory for two azimuths and different vertical distances z. (From Ref. 86.)

been found in most previous XPD studies of adsorbates[9,10,23,69] and can be explained by the combined effects of the following:

• Molecular vibration. This has not been included in the calculations for CO/Fe (001) shown here, but is considered in prior work for CO/Ni (001).[23,69]
• The presence of more than one type of emitter on the surface. For the present case, this could be due either to the method of formation of the α_3 state or to adsorption at defects. There could also be additional C-containing impurities beyond those associated with CO and its dissociation products on the surface. All of these act to diminish diffraction features relative to background and thus to reduce the experimental anisotropy. Such effects will tend to be present in any adsorbate system to some degree.

4.1.3. CO/Ni (110)

A final example of a molecular adsorbate system is that of CO on Ni (110), as studied with polar-scan measurements by Wesner, Coenen, and Bonzel.[76,88] For this case, Fig. 12 shows a comparison of normalized C 1s polar scans from CO adsorbed to saturation on Ni (110) at two different temperatures of 300 K and 120 K. The polar scans are markedly different, with the high-temperature results being very similar to those of CO on Ni (001) (cf. Fig. 8), and thus suggestive of a simple vertical adsorption of the CO, and the low-temperature results being widely split into a doublet along the [001] azimuth, but retaining a weaker peak along the normal for the [1$\bar{1}$0] azimuth. The low-tempeature, higher-coverage results have been explained by a structure in which the CO molecules are tilted by $\pm21°$ along the [001] azimuth, as shown in Fig. 12d.[89] This structure is nicely confirmed in Fig. 12c, where SSC–PW calculations with an rms vibrational amplitude of 8° are found to yield excellent agreement with experiment.

Wesner et al.[88] have also considered the effect of adsorbing CO on a Ni (110)

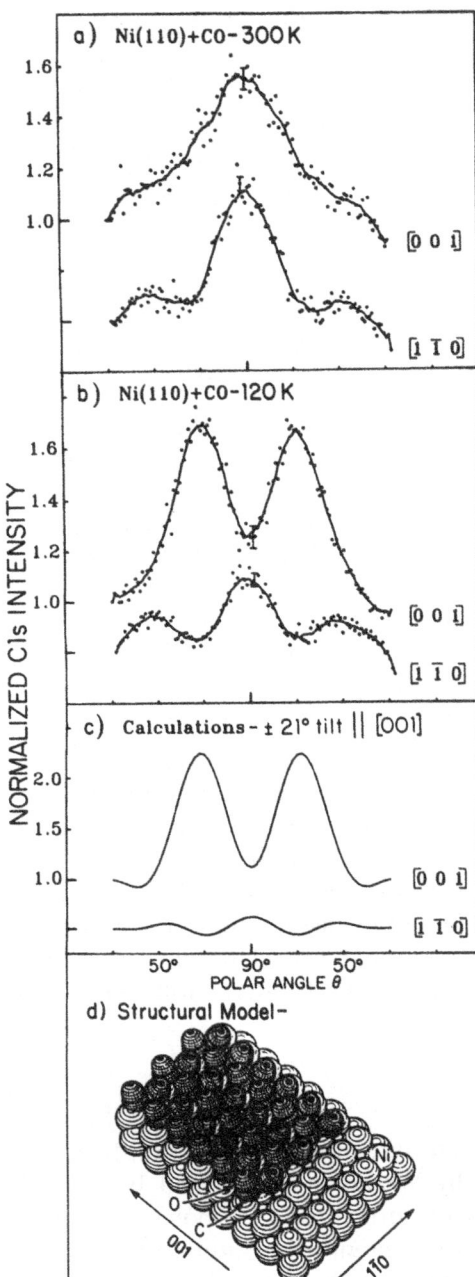

NORMALIZED C1s INTENSITY

a) Ni(110)+CO–300 K

[0 0 1̄]

[1̄ 1̄ 0]

b) Ni(110)+CO–120 K

[0 0 1̄]

[1̄ 1̄ 0]

c) Calculations – ± 21° tilt ∥ [001]

[0 0 1̄]

[1̄ 1̄ 0]

50°   90°   50°
POLAR ANGLE θ

d) Structural Model –

Ni

O

C

001

110

FIGURE 12. (a) Experimental polar scans in two azimuths of the C 1s intensity from CO adsorbed on Ni (110) at 300 K. The kinetic energy is 970 eV. (b) As in (a), but for adsorption at 120 K. (c) SSC-PW calculations modeling the data in (b), with an assumed tilt of ±21° along [001] and an rms vibrational amplitude of 8°. (d) The geometric model assumed for the calculations of (c). [(a)–(c) from Ref. 88, (d) from Ref. 89.]

surface pretreated with K, which is known to act as a promoter in many catalytic reactions. This system is found to have both vertical and more highly tilted CO species present. Finally, the same group has made use of the temperature dependence of the widths of peaks such as those in Fig. 12 for CO on Ni (011) to study the anisotropy of wagging vibrational amplitudes in different azimuths.[76]

4.1.4. Other Systems and Other Techniques

These simple examples thus show that XPD (or in principle also higher-energy AED) is a very powerful tool for studying the orientations and bonding of small molecules on surfaces, and that it is well suited to even very highly tilted species that may exhibit enhanced reactivity and thus be important in such phenomena as catalysis. Each of the cases discussed here is also significant in that other surface structural probes have been applied to the same problem without being capable of a clean resolution of the structure. Similar XPD measurements and theoretical analyses have also recently been applied to several other systems: CO and CH_3O on Cu (110) by Prince et al.[90] and CO on Pt (111) treated with K as a promoter by Wesner et al.[91]

Similar forward-scattering effects have also been seen by Thompson and Fadley[92] in emission from an atomic adsorbate on stepped surfaces: oxygen on Cu (410) and Cu (211). For this case, scattering by near-neighbor atoms up the step face from the emitter is found to be particularly strong. Stepped surfaces in fact represent a particularly attractive kind of system for study by this technique, since any atomic or molecular adsorbate that bonds preferentially at the base of the step has atoms on the step face as nearest-neighbor forward scatterers in the upstep direction.

The use of intramolecular forward scattering also appears to have several advantages for determining molecular or fragment orientations on surfaces in comparison to other techniques such as high-resolution electron energy loss spectroscopy (EELS),[93] electron stimulated desorption-ion angular distributions (ESDIAD),[94] and NEXAFS[95] or SEXAFS.[16] In EELS, the presence of a tilted species can be detected by which vibrational modes are excited, but estimating the magnitude of the tilt is difficult.[87a,93] In ESDIAD, the ion angular distributions for bond tilts away from normal can be significantly distorted by image forces and ion-neutralization effects,[87b,94] and tilts further away from normal than 25–30° therefore cannot in practice be measured accurately, if at all. In NEXAFS[95] and SEXAFS,[16] the experimental intensities of different features vary only relatively slowly with polarization, as $\sin^2 \alpha$ or $\cos^2 \alpha$, if α is the angle between the radiation polarization and the appropriate molecular symmetry axis. In forward-scattering XPD or AED, by contrast, it is the much narrower peak in the scattering amplitude $|f|$ near 0° (cf. Fig. 2 and Fig. 8) that controls the precision of orientation determinations, leading to FMWHs of 25–35° for all molecules studied to date. Comparing these values to the effective widths of $\sin^2 \alpha$ or $\cos^2 \alpha$ thus leads to the conclusion that forward scattering in XPD or AED should be about 3–4 times more precise in determining bond directions. An additional problem in NEXAFS is that a correct assignment of the peak(s) to be studied is necessary.

We close this section by noting that scanned-energy photoelectron diffraction or ARPEFS also has been applied recently to the study of small-molecule fragments such as formate (HCOO) and methoxy (CH_3O) adsorbed on Cu (100). The lower energies involved in this work imply that information on bond distances to backscattering neighbors below the adsorbate are also derivable. Such studies are described in more detail in the chapter by Haase and Bradshaw in Volume 2 of this set.

4.2. Atomic Adsorption and the Oxidation of Metals

4.2.1. Oxygen/Ni (001)

Saiki and co-workers[26] have carried out an XPD/LEED investigation of the interaction of oxygen with Ni (001) over the broad exposure range from $c(2 \times 2)$ O at 30 Langmuirs (L) to saturated oxide at 1200 L. Scanned-angle measurements were performed with Al $K\alpha$ radiation at 1486.6 eV for excitation. Although this system has been extensively studied in the past by various structural and spectroscopic probes,[96–98] several questions remain as to the exact structures formed. The combined use of XPD and LEED proves capable of answering several of these, as well as pointing out some new features of XPD that should be generally useful in surface-structure studies.

For example, in Fig. 13a, we show azimuthal scans of O $1s$ intensity at a relatively high polar angle θ of 46° with respect to the surface for four oxygen exposures from the onset of sharp $c(2 \times 2)$ LEED spots (30 L) to full oxide saturation (1200 L). The experimental curves are compared to SSC–SW calculations for a $c(2 \times 2)$ overlayer in simple fourfold sites with a vertical oxygen distance of $z = 0.85$ Å above the first Ni layer (the by now generally accepted structure), for two monolayers (ML) of NiO (001) with ideal long-range order, and for two monolayers of NiO (111) with long-range order. The dominant peaks at $\phi = 0°$ and 90° for the highest two exposures of 150 L and 1200 L are correctly predicted by theory and are due to simple forward scattering of photoelectrons emitted from oxygen atoms below the surface by oxygen atoms situated in the upper layers of the oxide, as indicated by the arrows in Fig. 13b. These peaks furthermore persist as the strongest features down to 30 L, indicating very clearly the existence of buried oxygen emitters, probably in small nuclei of NiO (001), over the full region of observation of the $c(2 \times 2)$ overlayer. The presence of such oxide nuclei in varying degrees on Ni (001) surfaces prepared in different laboratories is thus a likely cause of some of the previous controversy surrounding the vertical positions of both $c(2 \times 2)$ and $p(2 \times 2)$ oxygen on this surface,[96,97] but XPD provides a sensitive probe of the presence of any sort of buried species via such forward-scattering effects.

Comparing the 1200-L experimental curve and the theoretical curve for 2 ML of ideal Ni (001) in Fig. 13a for the region near $\phi = 45°$ shows qualitative agreement as to the existence of a region of enhanced intensity for $30° < \phi < 60°$, but disagreement as to exact fine structure, with theory showing a doublet where experiment shows a single broad peak. However, annealing this saturated oxide to approximately 250 °C for ≈ 10 minutes to increase its degree of long-range order parallel to the surface (as well as perhaps its thickness)[97] is found to yield a significantly altered XPD curve, with a doublet centered at $\phi = 45°$ that is in very good agreement with theory for NiO (001), as shown in the higher-resolution results of Fig. 14. It is also striking that the annealed oxide overlayer shows much more fine structure and generally narrower features, even though the dominant peaks in both the unannealed and annealed data are still those for simple forward scattering along $\langle 101 \rangle$ directions (i.e., at $\phi = 0°$ and 90°). The theoretical curves for 2 ML or 3 ML of ideal NiO (001) in Fig. 14 are

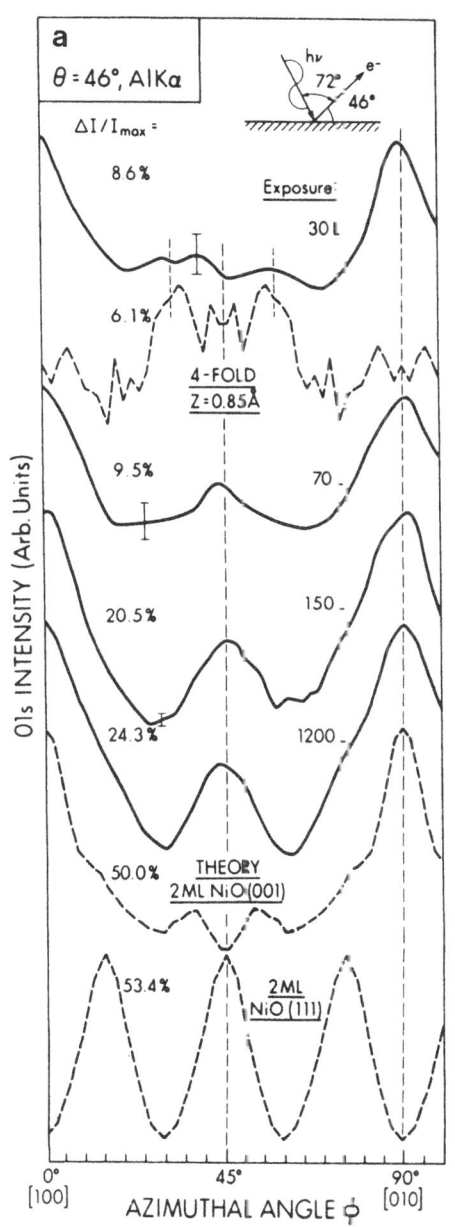

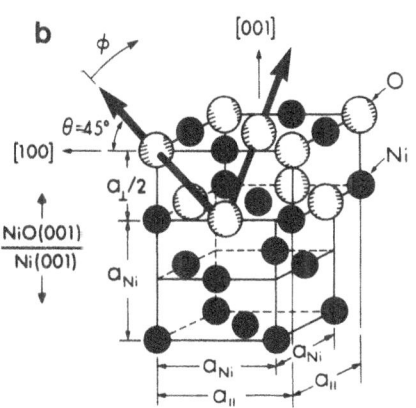

FIGURE 13. (a) O $1s$ azimuthal XPD data from oxygen on Ni (001) at four exposures from the $c(2 \times 2)$ regime (30 L) to saturated oxide (1200 L). The kinetic energy is 945 eV. The polar angle of 46° involves scanning very close to the $\langle 110 \rangle$ directions at $\theta = 45°$ and $\phi = 0°$ and 90°. Also shown are SSC-SW calculations for fourfold $c(2 \times 2)$ oxygen at $z = 0.85$ Å and fully ordered 2 ML overlayers of NiO with both (001) and (111) orientations. (b) An approximate representation of the structure of the dominant Ni (001) formed at high exposures, indicating the oxide lattice expansion and strain involved. Also shown as arrows are the directions of the strong forward-scattering peaks observed at $\phi = 0°$, 90° in (a). (From Ref. 26.)

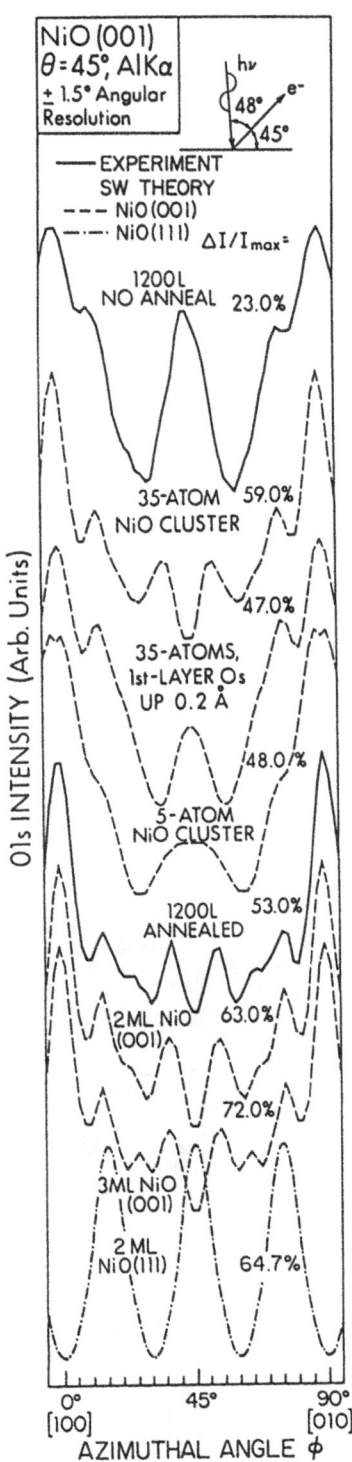

FIGURE 14. O 1s azimuthal XPD data from the saturated oxide formed at 1200 L exposure on Ni (001) obtained at a high angular resolution of ±1.5° with an emission angle of 45° with respect to the surface. Experimental curves are shown for both the ambient-temperature oxide and the same overlayer after a brief low-temperature anneal. SSC-SW calculations are also shown for several cases: smaller five-atom and 35-atom clusters to simulate loss of long-range order and strain and large fully converged clusters to simulate ideal NiO growing in either the (001) orientation (with 2 ML or 3 ML thickness) or the (111) orientation (with 2 ML thickness). (From Ref. 26.)

also in remarkably good agreement with the annealed data, verifying that annealing has produced a very highly ordered overlayer, and suggesting that the unannealed oxide exhibits diffraction effects due to strain and disorder.

The data shown in Fig. 14 are different from all results presented up to this point in being obtained at a very high angular resolution of $\pm 1.5°$ or less; precise angular resolution has in this case been obtained by using interchangeable tube arrays of the proper length-to-diameter ratio, as discussed in detail by White et al.[33] Note the additional fine structure in the unannealed 1200-L curve of Fig. 14 as compared to that of Fig. 13a.

The bottom theoretical curves in Figs. 13a and 14 are for 2 ML of NiO (111), an orientation of oxide growth which is also thought from LEED to coexist with NiO (001) on this surface.[97] The total lack of agreement of the NiO (111) curve with experiment makes it clear that this is only a minority species affecting no more than 5% of the NiO present.

In order to better understand the unannealed oxide data in Figs. 13a and 14, we also show in Fig. 14 theoretical curves for smaller 35-atom and 5-atom clusters of NiO (001). The previous calculations discussed involved much larger clusters with about 100 atoms per layer to insure full convergence. The 35-atom cluster includes atoms in about the first $1\frac{1}{2}$ unit cells around a given oxygen emitter; the 5-atom cluster is minimal and represents only nearest-neighbor and next-nearest-neighbor scatterers. The results for the full 2-ML cluster and the 35-atom cluster are found to be very close except for somewhat more fine structure in the full-cluster curve. This is consistent with prior XPD studies which have concluded that near-neighbor scatterers dominate in producing the observed patterns. However, much better agreement with the unannealed oxide results is seen if either the first-layer oxygen atoms (but not the nickel atoms) in the 35-atom cluster are relaxed upward by 0.2 Å or the effective cluster size is reduced to five atoms. Both of these models are consistent with a highly strained unannealed oxide overlayer of (001) orientation in which the long-range order is severely disturbed. The LEED spots for NiO (001) in fact indicate a lattice expanded by very nearly $\frac{1}{6}$ relative to the underlying Ni (001) surface, as indicated schematically in Fig. 13b. Although these results do not permit choosing between these two possibilities for stress relief in such a disordered system, they are significant in that both the experimental and theoretical XPD curves are quite sensitive to these more subtle deviations from an ideal NiO (001) overlayer with long-range order. This suggests a broad range of applications of XPD or higher-energy AED to studies of epitaxy and overlayer growth.

It is also significant in the comparisons of experimental data for annealed oxide with theory for 2–3 ML of NiO (001) in Fig. 14 that the agreement extends even to the overall degree of anisotropy, as judged again by $\Delta I/I_{max}$. The theoretical anisotropies are only about 1.2–1.3 times those of experiment. As noted previously, theory is in general expected to overestimate these anisotropies, in some previous cases by as much as factors of 2–3. One important reason for this kind of discrepancy is the lack of allowance in the calculations for atoms bound at various defect or impurity sites along or below the surface, as these are expected to produce a rather diffuse background of intensity, thus lowering the overall anisotropy. However, for the present case, the very good agreement suggests that the annealed oxide overlayer consists of oxygen atoms that are almost completely bound in a highly ordered NiO (001) structure.

At lower exposures, XPD has also been used to determine the $c(2 \times 2)$ *oxygen structure on Ni* (001).[26] The high θ values of Figs. 13 and 14 minimize the effects of any forward-scattering events in emission from oxygen in the $c(2 \times 2)$ overlayer (cf. Fig. 2), so that the 30-L curves here are dominated by the

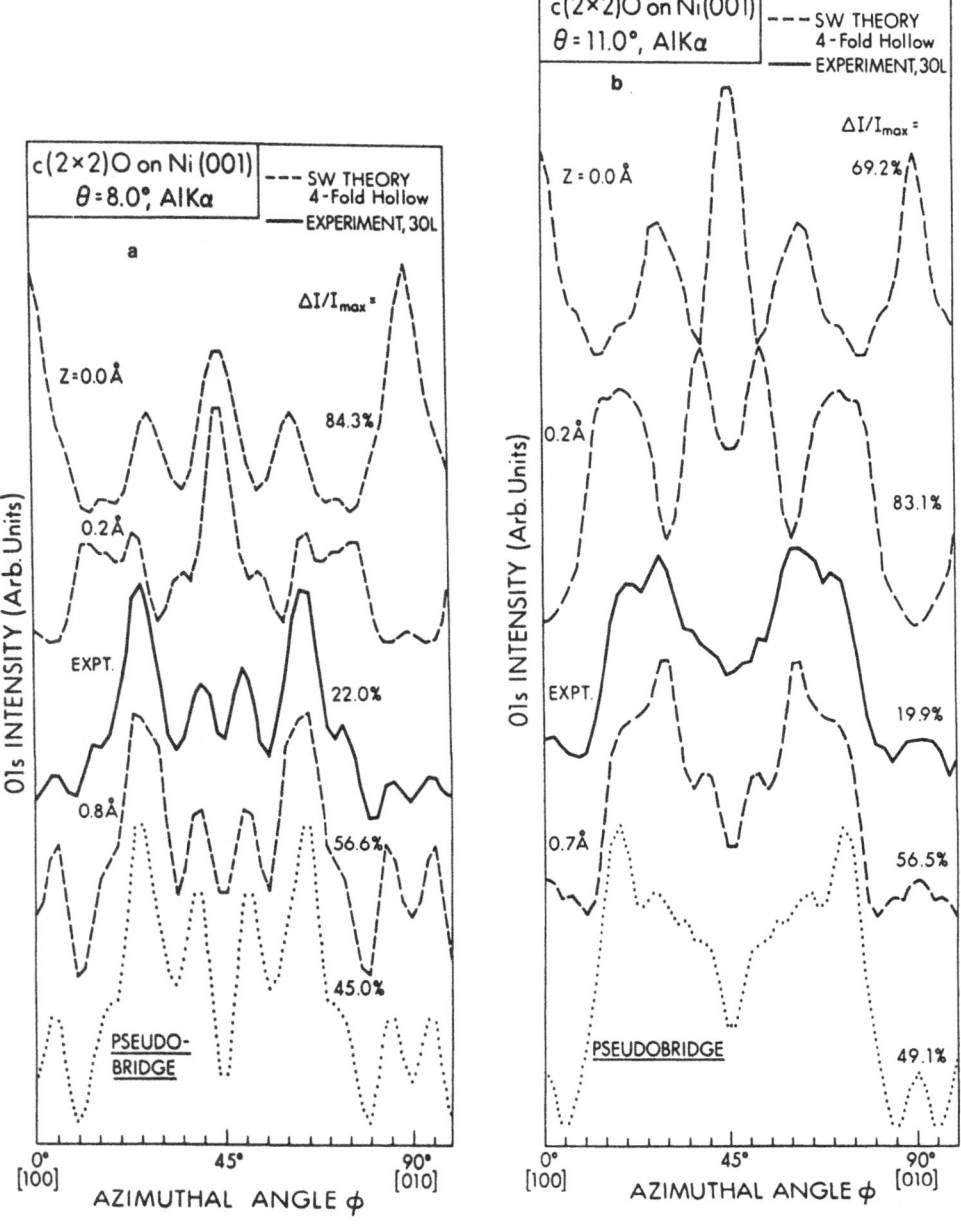

FIGURE 15. (a) Grazing-emission O 1s azimuthal data from $c(2 \times 2)$ O on Ni (001) at $\theta = 8°$. The experimental data are compared to SSC-SW curves for four possible fourfold-hollow $c(2 \times 2)$ structures, including the pseudobridge geometry of Ref. 98. (b) As in (a), but for $\theta = 11°$. (From Ref. 26.)

presence of a certain fraction of buried oxygen, probably in oxide nuclei. However, at very low takeoff angles with respect to the surface of approximately $8°-15°$, forward elastic scattering from adsorbed oxygen becomes much stronger, and the signal from buried oxygen is also suppressed by enhanced inelastic scattering.[9] Thus, the diffraction patterns at such low θ values are expected to be more strongly associated with overlayer effects.

Figure 15 shows such experimental and theoretical results for two representative θ values, $8°$ and $11,°$ of the four angles studied (data were also obtained for $14°$ and $17°$). Experiment is here compared with SSC–SW theoretical curves for four possible $c(2 \times 2)$ structures: in-plane fourfold bonding ($z = 0.0 \text{ Å}$); slightly-above-plane fourfold bonding ($z = 0.2 \text{ Å}$); the vertical distance in four-fold bonding yielding the empirical best fit to experiment at that θ value as judged both visually and by R factors;[26d] and the so-called pseudobridge geometry suggested by Demuth et al. on the basis of a LEED analysis.[98] For this last geometry, $z = 0.8 \text{ Å}$ and the oxygen atoms are offset horizontally by 0.3 Å in the fourfold hollow toward any of the four symmetry-equivalent $\langle 110 \rangle$ directions.

In Fig. 15a for $\theta = 8°$, it is very clear that $c(2 \times 2)$ oxygen does not occupy a position in the 0.0-to-0.2-Å range, although certain prior studies have suggested this as the most likely bonding position.[96,97] Simple fourfold bonding at $z = 0.80 \text{ Å}$, by contrast, yields excellent agreement with experiment, with all observed features being present in the theoretical curve. The only points of disagreement are the relative intensity of the weak doublets centered at $\phi = 0°$ and $90°$, which is too strong in theory; and the degree of anisotropy $\Delta I / I_{max}$, which is predicted to be too high by approximately a factor of 2.6. The latter discrepancy could be due to a significant fraction of oxygen atoms occupying defect or buried sites, e.g., in the oxide nuclei mentioned previously. Also, for such a low takeoff angle that begins to be within the forward scattering cone at this kinetic energy ($\approx 954 \text{ eV}$), there may be some defocusing and reduction of peak heights due to multiple scattering effects; in fact, $\phi = 0°$ and $90°$ are the directions of nearest-neighbor oxygen scatterers in the $c(2 \times 2)$ structure, as shown in Fig. 16a. The pseudobridge geometry does not fit experiment as well, since the relative intensity of the doublet centered at $\phi = 45°$ is too high.

In Fig. 15b, for $\theta = 11°$ the two geometries close to being in plane again do not agree at all with experiment, which is very well described by simple fourfold bonding at an optimum z of 0.70 Å. The pseudobridge geometry in this case also differs considerably from experiment as to the shape of the two main peaks. When these results are combined with those at the other two θ values studied,[26] it can overall be concluded that $c(2 \times 2)$ oxygen does not bond in either simple fourfold positions at $0.0 < z < 0.3 \text{ Å}$ or in the pseudobridge geometry, but does occupy simple fourfold positions at $z = 0.80 \pm 0.10 \text{ Å}$. This choice of structure is also confirmed by an R-factor comparison of experiment and various theoretical curves. The z distance found here also agrees very well with several more recent structural studies of this system.[96,97]

A final point in connection with the results of Fig. 15 is that, in order for theory to adequately reflect all of the fine structure seen in experiment, the cluster used in the calculations must include all O and Ni atoms within the first few layers of the surface (adsorbate plus two layers of Ni) and out to a relatively large radius of about 20 Å from the emitter. The rate of convergence with cluster size is illustrated in Fig. 16. Due to the rotational symmetry of the surface, calculations need be performed only over the 45° wedge indicated in Fig. 16a, but it is important to include sufficient atoms at the edge of this wedge. It is clear from the diffraction curves in Fig. 16b that going out to only 10 Å in radius does not yield the correct diffraction fine structure. This indicates sensitivity in forward scattering at grazing emission to well beyond the first 3–5 spheres of neighbors. The effective *diameter* of the cluster is thus about 40 Å.

Thus, these results for a prototypical surface oxidation over a broad exposure range, from ordered overlayers at partial monolayer coverage to saturated oxide, indicate several very useful types of structural information that can be derived from XPD (or by implication also by high-energy AED) in conjunction with SSC calculations.

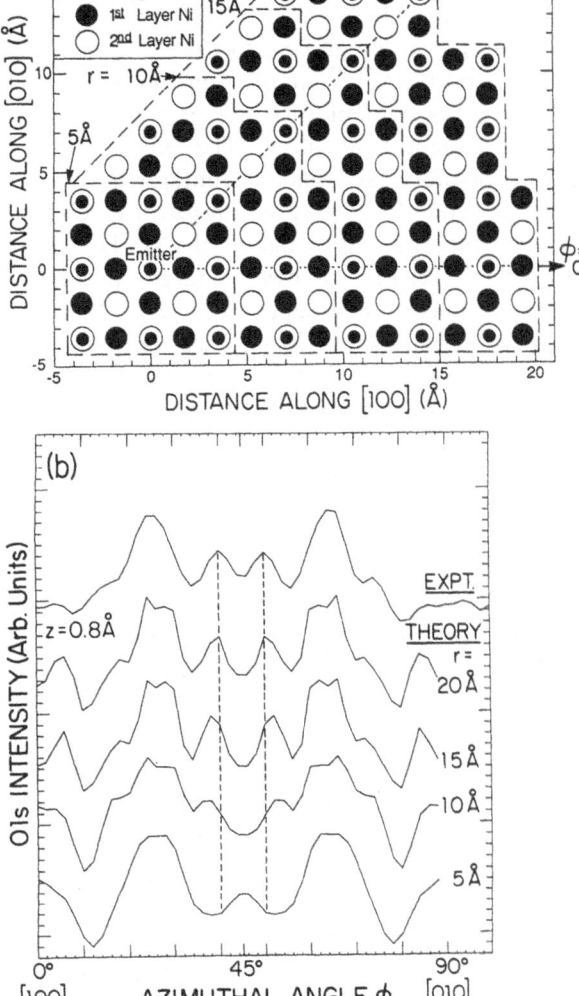

FIGURE 16. (a) Choices of cluster. Different-sized clusters used in testing the convergence of the SSC-SW curves for $\theta = 8°$ in Fig. 15(a), labelled with the approximate radii outward from the emitter that they represent. (b) Convergence with cluster size. Calculated curves for the clusters of (a) with $z = 0.8$ Å are compared to experiment. Note that a radius of at least 20 Å is required to yield optimum agreement. (From Ref. 26.)

4.2.2. Sulfur/Ni (001)

The sulfur/Ni (001) system has been much used as a test case for surface-structure techniques because it represents a rather unique example of a system for which there is a general consensus on a structure: the $c(2 \times 2)$ sulfur overlayer is bound with atomic S in fourfold sites at a distance z of 1.3–1.4 Å above the first Ni plane.[99] Several photoelectron diffraction studies have been made of this system,[14,15,19,21,25,99] including both scanned-angle and scanned-energy measurements, and we will consider a few of these.

Higher-energy scanned-angle XPD measurements have been made for this system by Connelly et al. (Fig. 44 in Ref. 9), and experimental azimuthal scans of S $2p$ emission at grazing takeoff angles are found to be in good agreement with SSC–PW calculations for the known structure. However, for a structure with this high a distance above the Ni surface, the effects of forward scattering become weaker, since the scattering angle from any near-neighbor Ni atom becomes larger. For example, for the Ni nearest neighbors in the fourfold hollow, a very low emission angle of 5° with respect to the surface still corresponds to a minimum scattering angle of approximately 43° that is well outside of the forward scattering cone at high energy (cf. Fig. 2). Thus, the strongest contribution to azimuthal anisotropy is scattering from the other (coplanar) S adsorbate atoms, for which the scattering angle is simply the emission angle with respect to the surface. The sensitivity of such XPD measurements to the vertical S–Ni distance is thus expected to be lower than for more nearly in-plane or below-plane adsorption, and it has been questioned as to whether such measurements will be sensitive enough to determine structures for any adsorbate sitting well above the surface.[9] Several possibilities appear to exist for improving the positional sensitivity for such cases: working at higher angular resolutions and taking advantage of additional diffraction fine structure, using lower energies for which large-angle and backscattering are stronger, and/or using special polarization geometries to enhance certain substrate scatterers. Some of these possibilities thus involve synchrotron radiation, and we consider now their application to the S/Ni case in both the scanned-angle and scanned-energy modes.

We first look at the influence of higher angular resolution. S $2p$ azimuthal XPD data at a polar angle of 13° obtained by Saiki et al.[100] with a high angular resolution of about ±1.0° are shown in Fig. 17. The data were obtained in scans over 100° in ϕ and then mirror-averaged across [110] to improve statistical accuracy, but all of the features shown were reproduced in the full scan. These results exhibit considerably more fine structure than similar data obtained with a ±3.0° resolution, and the anisotropy is found to go up from 31% to a very high 40% with increased resolution. Also, when these data are compared with the SSC–SW curves shown in this figure for different z positions of S above the fourfold hollow, they exhibit a high sensitivity to position. A more quantitative analysis of these high-resolution results by Saiki et al.[100] using R factors for comparing experiment and theory[26d] in fact yields a z value of 1.39 Å for this structure that is in excellent agreement with prior work. This analysis furthermore permits estimating the first nickel–nickel interplanar distance (d_{12}), which is found to be expanded to about 1.86 Å from the bulk value of 1.76 Å. Thus, there

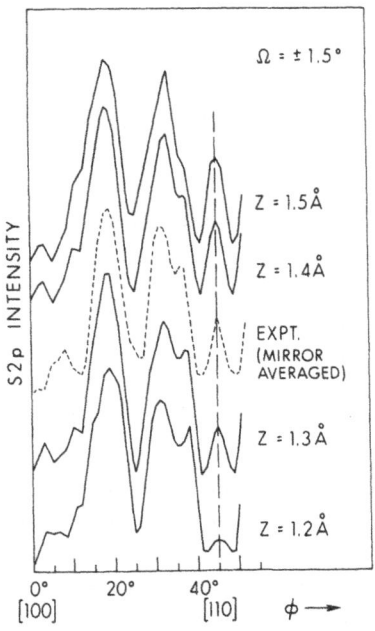

FIGURE 17. Azimuthal XPD data for S $2p$ emission from $c(2 \times 2)$ S on Ni (001) at a kinetic energy of 1085 eV obtained with a high angular resolution of approximately $\pm 1.5°$. The polar angle is 13° with respect to the surface. The anisotropy $\Delta I/I_{max}$ is a high 40% for these results, compared to only 31% for the same measurement with a $\pm 3.0°$ angular resolution; the fine structure is also considerably enhanced with higher resolution. SSC-SW calculations are shown for various distances z of the S above the Ni surface. (From Ref. 100.)

is considerable potential in using high-energy measurements with high angular resolution, even for adsorption at large z distances above approximately 1.0Å.

Going to lower energies with synchrotron radiation in such azimuthal measurements also has potential for such studies. We show in Fig. 18 results for S $1s$ emission from the $c(2 \times 2)$S overlayer on Ni (001) obtained by Orders et al.[19b] Here, the experimental geometry was chosen so that the polarization vector

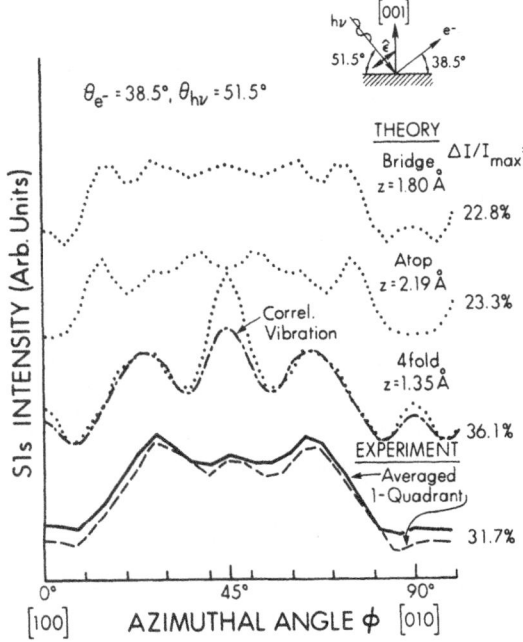

FIGURE 18. Synchrotron radiation excited S $1s$ intensity from $c(2 \times 2)$ S on Ni (001) at a kinetic energy of 282 eV. The geometry chosen emphasized nearest-neighbor backscattering because the polarization vector was oriented directly toward the relevant Ni nearest neighbor, as shown in the inset at upper right. SSC-PW calculations for three possible adsorption sites of bridge, atop, and fourfold are shown as dotted curves. The dashed-dotted fourfold curve involves a more correct inclusion of correlated vibrational effects. (From Refs. 19(b) and 101.)

was directed rather precisely toward nearest-neighbor Ni atoms for certain azimuthal positions in a ϕ scan. Backscattering from this type of Ni atom should also be rather strong at the photoelectron energy of 282 eV chosen (cf. Fig. 2). This energy is nonetheless high enough that a single-scattering model should still be reasonably quantitative. The experimental data is here compared with SSC–PW calculations for three different bonding sites (bridge, atop, and fourfold) with reasonable S–Ni bond distances, and the correct fourfold site is clearly in better agreement with experiment. The agreement is also significantly improved if a more accurate allowance for correlated vibrations is included in the SSC calculations, as shown by the dashed–dotted curve.[101]

However, a note of caution is in order concerning the use of different polarization orientations, since experimental and theoretical work on S/Ni by Sinkovic et al.[19a] indicates that a geometry in which the polarization is nearly perpendicular to the electron emission direction (instead of parallel, as in Fig. 18) increases the importance of multiple-scattering events and causes more significant deviations from a simple theoretical model. This is thought to occur through a weakening of that portion of the photoelectron wave emitted directly in the detection direction in comparison to the various scattered waves that can interfere with it. The intensity distribution is thus produced by the interference of direct and scattered waves that are all of the same magnitude, a situation rather like that in LEED where all contributions to intensity are those due to relatively weak backscattering; thus, MS effects might be expected to be more important. In most photoelectron and Auger experiments, the direct-wave amplitudes are stronger than those of the scattered waves, and it can be argued that this is a fundamental reason for the higher degree of applicability of a single-scattering approach.

Finally, we consider scanned-energy or ARPEFS measurements on S/Ni (001) of the type pioneered by Shirley and co-workers.[8,25] In this type of experiment, an adsorbate core intensity is measured as a function of $h\nu$ in a fixed θ, ϕ geometry, and the resulting EXAFS-like oscillations are analyzed in order to derive the adsorbate position. The data are usually analyzed as a normalized $\chi(E)$ or $\chi(k)$ function. Figure 19 shows typical experimental data of this type in a normal-emission geometry, for S $1s$ emission from $c(2 \times 2)$ S/Ni (001).[8] Allowance has been made here for the interference between the S Auger peak at 155–160 eV and the S $1s$ photoelectron peak. These results are compared to both MSC–SW calculations by Barton and Shirley[25] in Fig. 19a and SSC–SW calculations by Sagurton et al.[21] in Fig. 19b. The agreement is very good for both sets of theoretical curves, provided that the first nickel–nickel interlayer distance (d_{12}) is relaxed outward from the bulk value of 1.76 Å to 1.84 Å (cf. the two theory curves in Fig. 19b). This interlayer relaxation, as first pointed out by Barton and Shirley, thus illustrates the high sensitivity of photoelectron diffraction to subtle structural changes on the order of 0.10 Å or less.

It is also clear from this figure and other work on the S/Ni system[21,22] that both the single-scattering and multiple-scattering approaches describe the experimental results well and that they also lead to very similar structural conclusions, with only the perpendicular distance for S being different by 0.05 Å between the two analyses. Thus, although the MSC–SW approach is certainly in principle more accurate and does lead to $\chi(k)$ amplitudes in better agreement

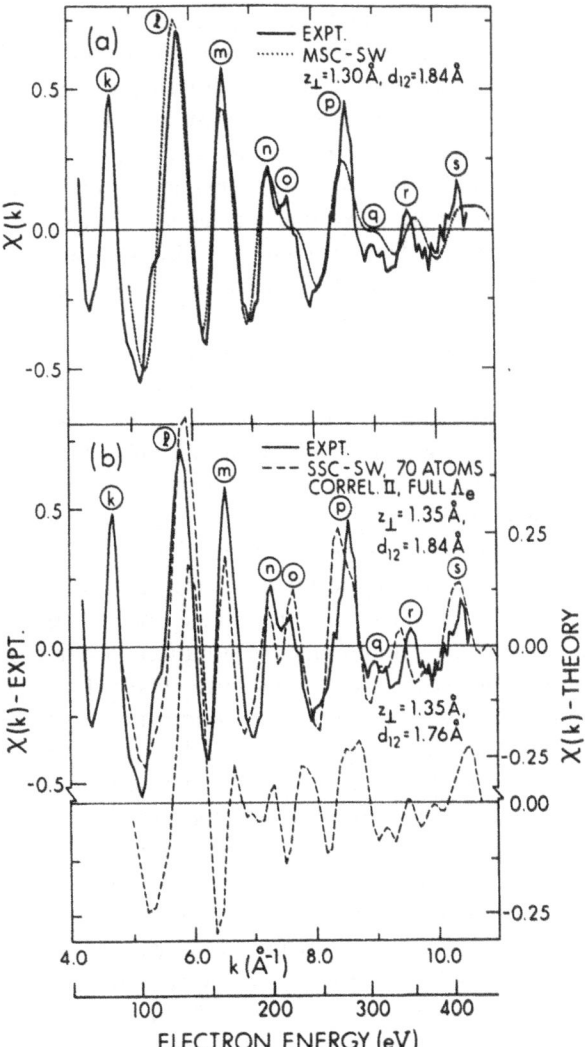

FIGURE 19. Comparison of scanned-energy S $1s$ data for $c(2 \times 2)$ S on Ni (001), $\theta_{hv} = 70°$, $\theta_{e^-} = 0°$ (From Ref. 8) with: (a) a multiple-scattering cluster spherical-wave (MSC-SW) calculation due to Barton and Shirley (From Ref. 25), and (b) single-scattering cluster spherical wave (SSC-SW) calculations due to Sagurton et al. (From Ref. 21). Both the sulfur vertical distance z and the first Ni–Ni interplanar distance d_{12} are specified. (Fig. from Ref. 21.)

with experiment, the SSC–SW method appears capable of a usefully quantitative description of the observed oscillations and fine structure.

Another aspect of this analysis noted by Barton and Shirley[25] is that nearest-neighbor backscattering followed by emitter forward scattering (cf. Fig. 3b–i) can be an important factor in producing the full amplitude of the ARPEFS oscillations at low energies. This may be the reason why the single-scattering curves in Fig. 19b have lower amplitudes, although a different allowance for vibrational effects also could play a role.[21]

An additional useful aspect of such ARPEFS data is in being able to Fourier transform $\chi(k)$ curves to yield peaks which are for some (but not necessarily all) of the strongest scatterers rather directly related to interatomic distances via the path-length difference and the scattering angle [cf. Eq. (10)]. The degree to which Fourier transforms can be used in this way is discussed in detail elsewhere.[21,25] However, ARPEFS Fourier transforms (FTs) need not be as simply associated with certain spheres of neighbors as are those of EXAFS and SEXAFS; the reason for this is the potentially large number of scattering events and various possible scattering angles that can be associated with a given region in the transform.[21] Nonetheless, such FTs have been used to rule out certain structures as part of a more detailed structure determination; we consider such an example in the next section.

4.2.3. Sulfur/Cr (001)

We now turn to a recent study of $c(2 \times 2)$ S/Cr (001) by Terminello *et al.*[20] that serves to represent a state-of-the-art analysis of scanned-energy or ARPEFS data. In this work, S $1s$ intensities were scanned as a function of energy up to about 475 eV above threshold; two different emission directions were studied; [001] and [011], with polarizations oriented in general along the emission direction (35° off normal toward [011] for [001] emission and along [011] for [011] emission). Special care was taken to avoid spurious energy-dependent effects in the measuring of intensities, with normalization being needed for both the incident photon flux and the transmission function of the electron-energy analyzer. As for S/Ni (001), the interference between the S Auger peak at 155–160 eV and the S $1s$ photoelectron peak was allowed for by carefully subtracting out the former. Fourier transforms of the data were made, with the inner potential being treated as an adjustable parameter and the $\chi(k)$ data being multiplied by a Gaussian window function to reduce ringing effects in the final FTs. The strongest peaks in these transforms were then taken to be semiquantitatively indicative of certain near-neighbor path-length differences; this analysis thus implicitly assumes that the single-scattering Eq. (10) represents a good first-order description of the diffraction and that there are no significant interferences between the effects of different near-neighbor scatterers. The approximate geometric information from the FT peak positions was found to point to the fourfold-hollow site as the adsorption position.

The final quantitative determination of the site type and the structure was made by directly comparing the experimental $\chi(k)$ curves (Fourier filtered to remove effects due to path-length differences beyond about 20 Å) with multiple-scattering cluster calculations using spherical-wave scattering. As one example of these results, Fig. 20a compares experimental curves along the two directions with curves calculated for S adsorbed on three types of sites. It is very clear here that the fit is best for the fourfold site (cf. similar comparison for the scanned-angle S/Ni results in Fig. 18).

Pursuing the fourfold site further by means of an R-factor comparison of experiment and MSC–SW theory, the authors derive a geometry that includes a determination of S–Cr distances down to the fifth layer of the substrate. Some of the results of this R-factor analysis are shown in Fig. 21. It is interesting here that the two sets of data for emission along [001] and [011] azimuths and with polarization nearly parallel to each emission direction are complementary in their sensitivities to different structural parameters. The [001] results are much more sensitive to the Cr_2-atop position because strong single and multiple backscattering can be involved (cf. Fig. 3b–i). By contrast, the [011] data is much more sensitive to the Cr_2-open position for the same reason. The polarization orientations enhance these effects by preferentially directing the initial photoelectron wave toward these scatterers (cf. Fig. 3a). The final results of this

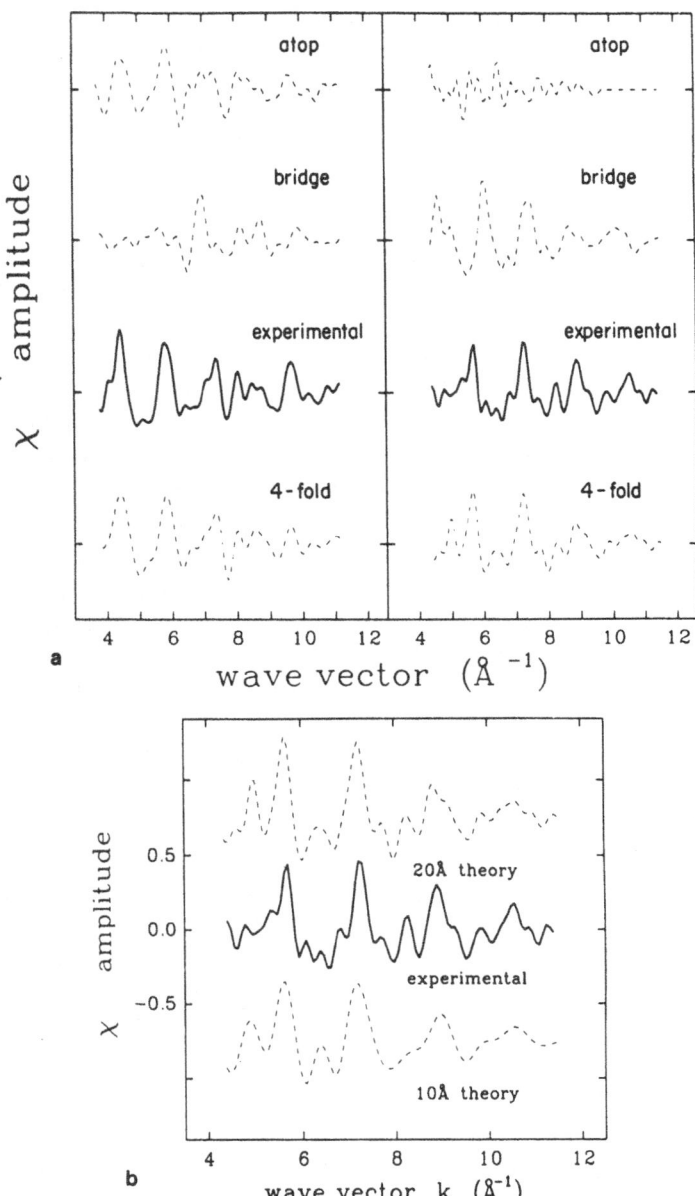

FIGURE 20. (a) Comparison of scanned-energy S 1s experimental data for emission from $c(2 \times 2)$ S on Cr (001) along the [001] direction (left panel) and [110] direction (right panel) with MSC-SW calculations for different adsorption sites of atop, bridge, and fourfold. (b) As in (a), but comparing the data obtained in the [011] azimuth to MSC-SW theory for the final optimized fourfold-hollow structure with different path-length cutoffs of 20 Å and 10 Å. (From Ref. 20.)

R-factor analysis show an 8% reduction of the mean separation of the first and second Cr layers (compare the 3% expansion in similar S/Ni results in Fig. 19) and further suggest a slight corrugation of the second layer and a slight expansion of the separation of the second and third layers, although the latter are not fully conclusive within the error limits of 0.02–0.03 Å estimated by the authors.

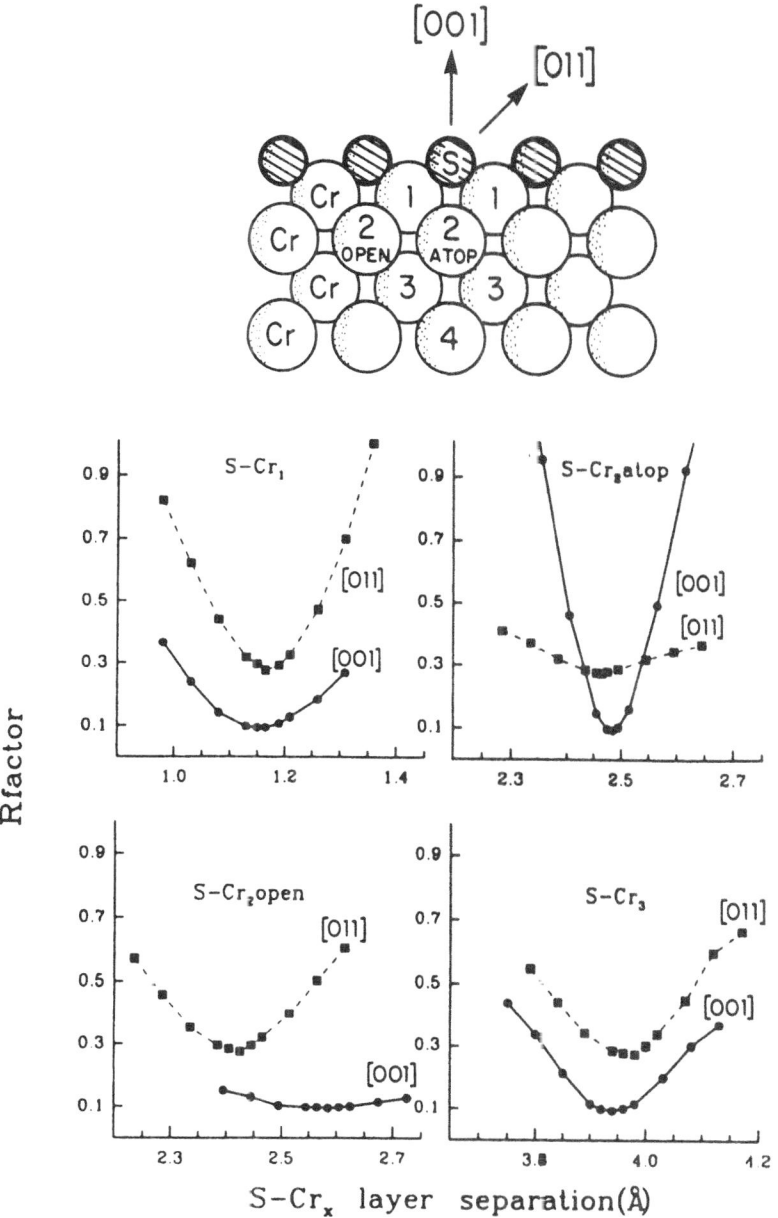

FIGURE 21. *R*-factor analysis of the scanned-energy results of Fig. 20, showing the geometry involved and the variation of the *R* factor with various S–Cr layer separations. (From Ref. 20.)

A further important point made in this work is that the $\chi(k)$ curves exhibit fine structure associated with path-length differences out to about 20 Å. Such fine structure in ARPEFS data and the need to use rather large clusters of up 50–100 atoms to adequately model S/Ni data have also been discussed previously (see Fig. 19 and Ref. 21). The work by Terminello *et al.* shows this explicitly by comparing experimental $\chi(k)$ curves for S/Cr with MSC–SW curves that have been cut off at both 10 Å and 20 Å total scattering lengths; these results are presented in Fig. 20b, where it is clear that the fine structure in experiment is better modeled by the 20-Å curve, especially for wave vectors above about 7 Å^{-1}. This sensitivity permitted a final determination of Cr layer spacings down to that between the fourth and fifth layers, although the accuracy decreases from an estimated ± 0.02–0.03 Å for the first three spacings to ± 0.07 Å for the fourth spacing measured. It is, finally, worth noting that the approximately 20 Å limit noted here is in the same range as that found in the higher-energy scanned-angle O/Ni results presented in Fig. 16. Thus, both methods seem to have similar sensitivity to more-distant neighbors.

This work demonstrates the full power of the scanned-energy approach, provided that the initial intensities are measured carefully and that the final results are analyzed by means of a quantitative comparison of experimental $\chi(k)$ curves with calculations for a range of choices of geometrical parameters. A very similar analysis has been carried out for the system $c(2 \times 2)$ S/Fe (001) by Zhang et al.[102] Although much more time-consuming multiple-scattering calculations were used for all of the geometries tried in these cases, it should be possible in general to do a much more rapid search for promising geometries in single scattering, with only fine tuning of the parameters then being required in multiple scattering.

4.3. Epitaxial Oxide, Metal, and Semiconductor Overlayers

4.3.1. NiO/Ni (001)

Although the case of NiO grown on Ni (001) considered in the previous section does not represent perfect epitaxy, the degree of agreement between experiment at 1200 L and theory in Figs. 13a and 14 clearly shows that the predominant form of NiO present is of (001) orientation. Certain structural conclusions concerning the form of this oxide and its degree of long-range order before and after annealing have also been made (section 4.2.1 and Refs. 26b,c). An analysis of the LEED spot patterns (including a splitting of the NiO (001) spots and corresponding XPD data in fact suggests a two-dimensional super-lattice growth of NiO (001) with a lattice constant expanded by exactly $\frac{1}{6}$ with respect to the underlying Ni substrate (cf. Fig. 13b). Although LEED patterns for the unannealed oxide also exhibit a 12-spot ring throught to be due to NiO (111),[97] the XPD results of Figs. 13a and 14 indicate that it is at most a minority species of the total NiO present, since NiO (111) would produce 12-fold symmetric XPD patterns (bottom theory curves in Figs. 13a and 14) that are not seen experimentally. This example thus indicates a very useful sensitivity of high-energy XPD to the orientation of an epitaxial overlayer and its degree of short-range order under various conditions of annealing and deposition.

4.3.2. Cu/Ni (001) and Fe/Cu (001)

We now consider two very different limits of metal-on-metal epitaxial growth taken from some of the first experimental studies in this field, those by Egelhoff and co-workers and Chambers and co-workers: pseudomorphic epitaxial growth of Cu on Ni[11,103] and island formation by Fe on Cu (001).[104]

Figure 22 illustrates high-energy AED for the first case of Cu on Ni (001). The different near-neighbor forward scattering events allowed as each new Cu layer is added are illustrated by the arrows in Fig. 22a. In Fig. 22b, experimental data from Egelhoff[11] are compared to theoretical SSC–PW curves from Bullock and Fadley.[71] In Fig. 22c, some of the same experimental data are compared to very recent multiple-scattering calculations by Xu and van Hove.[73]

In Fig. 22b, the relatively abrupt appearance at certain overlayer thicknesses of forward-scattering features such as those at $\theta = 45°$ and 90° (normal emission) can be used as a direct measure of the number of overlayers in the range of about 0–3 ML. Comparison with Fig. 22a also shows that the appearance of each of these two peaks corresponds to the onset of forward scattering by the two nearest neighbors encountered in this polar scan from [100] to [001]. The simple origin of these two peaks has also been directly verified by comparing SSC calculations with and without these important scatterers present.[71]

Thus, simple forward scattering peaks from nearest and next-nearest neighbors are very useful in studies of epitaxy, as we have also discussed for the oxide case in the last section. However, the interpretation of weaker features such as those at $\phi \approx 20°$ and 70° in Fig. 22b need not be so simple. Calculations with various atoms removed from the cluster show that these have more complex origins which require at least a full SSC calculation for their explanation.[71] For example, the peak near 70° is a superposition of simple forward scattering by atoms along [103] and [102] and, more importantly, first-order effects (cf. the inset of Fig. 8) from the atoms along [001] and [101]. Thus, for atoms that are further away than the first three or four spheres of neighbors, a mixed origin in forward scattering and higher-order interference effects is generally to be expected. This conclusion has also been confirmed in a recent analysis by Osterwalder et al.[48] of an extensive set of high-resolution Ni $2p_{3/2}$ data from bulk Ni (001) that we discuss further in section 5.1.

Figure 22a also makes it clear that, in pseudomorphic growth with the lateral lattice constants locked to those of Ni, the vertical spacing of the Cu layers will determine the θ position of the peak near 45°. A ±1° change in this peak position from 45° would correspond to a ±0.12-Å change in the vertical lattice parameter or a ±0.06-Å change in the interplanar spacing. This sensitivity has in fact recently been used by Chambers et al.[12a,b] to measure the degree of outward vertical relaxation in thin Cu overlayers on Ni (001). It should thus be possible to measure interlayer spacings with accuracies of better than 0.1 Å in this way[24,71,73] although doing some sort of theoretical modeling at least at the SSC–PW or SSC–SW level (as Chambers et al. have done[12]) is advisable to verify peak origins, shapes, and predicted shifts with relaxation. Using higher angular resolution also should be beneficial for such studies by making it possible to determine forward-scattering peak positions more precisely.

The main point of discrepancy between experiment and SSC–PW theory in Fig. 22b is that the peak for forward scattering along the nearest-neighbor [101] direction has a relative intensity too high for thicker overlayers by about a factor of about 2. As expected from the prior discussion of Fig. 5, using spherical-wave

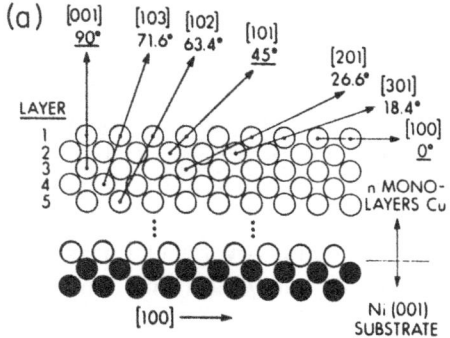

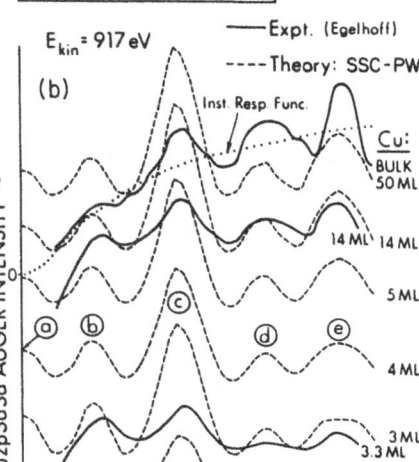

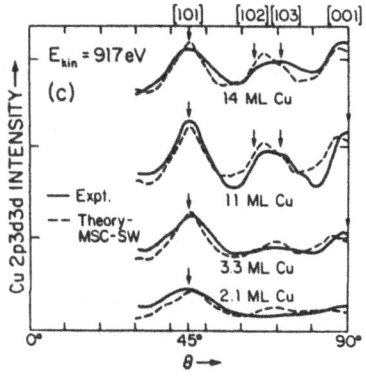

FIGURE 22. (a) Illustration of possible near-neighbor forward scattering events in the [001]–[100] plane for Cu grown in pseudomorphic epitaxy on Ni (001). Only those at 45° and 90° are fully explained by the simple one-event interpretation suggested here. (b) Experimental Cu Auger polar scans at 917 eV (from Ref. 11) are compared to SSC-PW calculations for successive layers of epitaxial growth of Cu on Ni (001) (from Ref. 71). Although the Cu *LMM* Auger intensity is monitored here, very similar results are obtained from the Cu 3*p* photoelectron intensity. (c) The same experimental data are compared to multiple-scattering cluster calculations. (From Ref. 73.)

scattering in the SSC model is found to significantly improve agreement for this relative intensity by reducing it to about $\frac{2}{3}$ of the magnitudes seen in Fig. 22b for thicknesses >3.0 ML[50b]; it is nonetheless still too high by 1.3–1.5 times in comparison with experiment. The remaining discrepancy is due to multiple scattering effects, and the calculations of Fig. 22c include the additional defocusing of intensity along the [101] direction. Much more quantitative agreement with experiment is obtained here. However, even though certain forward-scattering peaks may have their relative intensities decreased by multiple scattering, it should nonetheless still be possible to use the peaks along [001] and [101] in the simple way described in the preceding paragraphs to monitor overlayer thicknesses and determine interlayer relaxations.[71,73]

A more recent paper by Egelhoff[103] has also looked experimentally at a single pseudomorphic Cu (001) layer on Ni (001) buried under various numbers of Ni (001) overlayers. In this work, the attenuation and broadening of certain features with increasing layer thickness is interpreted as evidence of stronger multiple-scattering effects in emission from greater depths. Although the defocusing effects seen in the MS results of Fig. 6 make this a plausible conclusion, Herman et al.[105] have made SSC–SW predictions for the cases studied, and these are found to show very similar attenuation to the experimental data. As one example of this comparison of experiment and SSC–SW theory, Fig. 23 shows results for the 917 Auger peak; the experimental data have been corrected for the θ-dependent instrument response by dividing by the curve for a single Cu

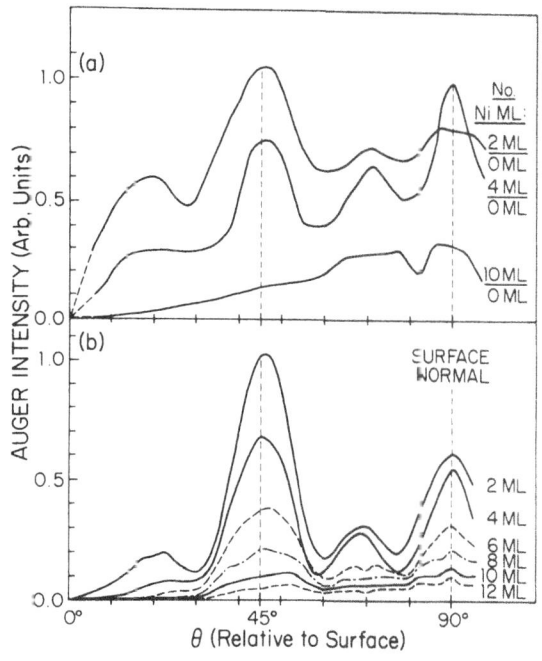

FIGURE 23. (a) Experimental data for Auger emission from a single pseudomorphic Cu (001) layer on top of Ni (001) buried underneath different numbers of layers of epitaxial Ni, also in (001) orientation. (From Ref. 103) (b) Theoretical calculations within the SSC-SW approximation of the results in (a), including curves for other overlayer thicknesses. (From Ref. 105.)

monolayer with no overlying Ni (shown as "0 ML"). Although the relative intensity of the peak at 45° compared to that at 90° is again predicted in theory to be too high, the trends in experiment as the Ni overlayer is increased in thickness are surprisingly well reproduced by the SSC calculations. In particular, the change in the absolute intensity of the peak at 45° with thickness is well reproduced by the calculations, and its final broadening out and diminution of importance in comparison to the peak at 90° is also correctly predicted. Discrepancies noted are that the broad, flat feature seen in experiment at about 70° is not fully developed in the single scattering theory and that an initial narrowing of the peak along 45° that may be due to multiple-scattering effects (cf. the discussion of Fig. 6 and Fig. 22c) is not seen. Experimental errors of as much as ±10–20% in measuring the number of monolayers (cf. calculated curves at other thicknesses), as well as the possible presence of defects in the growing Ni layer,[12a,b] could also affect the agreement between experiment and theory. More recent multiple-scattering calculations for this buried-monolayer system by Xu and van Hove[73] and by Kaduwela et al.[84] yield a more quantitative description of the decrease in intensity of the peak at $\theta = 45°$, although the experimental overlayer thicknesses have to be decreased by from 0.6 to 1.5 ML in the calculations to yield optimum agreement. However, on going to thicker over-layers on the order of ten layers, there is still a stronger peak in MS theory than in experiment near $\theta = 70°$.

Thus, although such a deeply imbedded emitter layer clearly represents an extreme case of the type shown in Fig. 3b–ii, for which multiple-scattering effects ought to be maximized, the case for these data definitely exhibiting such effects is not as strong as might be expected, and the SSC approach still yields at least a semiquantitative description of the data.

A final note of caution in connection with this study[103] concerns the idea that classical trajectories can be used to predict when and how multiple scattering will be important in AED or XPD. Although classical arguments can be didactically useful once the correct answer is known, taking them further seems to be very risky, particularly when the quite simple and wave-mechanical SSC model is already available for comparisons to experiment and to more-accurate calculations including higher-order multiple scattering.

We now turn to the second system: Fe/Cu (001) as studied by Chambers, Wagener, and Weaver[104a] and by Steigerwald and Egelhoff.[104b] Figure 24 shows a similar set of AED data from the latter study for the case of Fe deposited on Cu (001) at ambient temperature and compares it to results like those in Fig. 22b. It is striking here that coverages of one monolayer or less (even down to 0.1 ML) already exhibit the strong forward-scattering peak at 45° characteristic of *fcc* Fe in islands or clusters at least two layers thick, as well as the beginning of the peak along the surface normal associated with three-layer structures. In fact, the 1-ML Fe curve looks very similar to that for 3.3 ML of pseudomorphic Cu in Fig. 22b. These results[104b] and a more detailed set of polar and azimuthal data discussed by Chambers et al.[104a] thus show that at least the first one or two layers of Fe grown under these conditions have a strong tendency to agglomerate on Cu (001), a conclusion that has important implications for the magnetic properties of such overlayers.[104] This work nicely demonstrates the general usefulness of such

FIGURE 24. Experimental polar scan data for Fe $2p_{3/2}$ emission at 780 eV from Fe deposited at ambient temperature on Cu (001). Data for both 0.1 ML and 1 ML total coverages are compared to similar results for Auger emission from Cu deposited up to 1 ML and 2 ML on Ni (001); cf. Fig. 22(b). Note the presence of strong forward scattering peaks at 45° in both Fe curves and the beginning of a peak along normal for the 1-ML Fe data. [From Ref. 104(b), with more detailed polar and azimuthal data appearing in Ref. 104(a).]

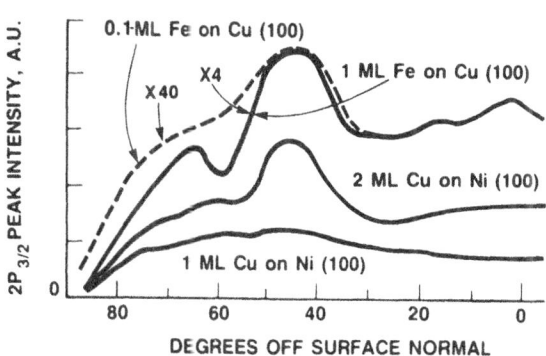

scanned-angle measurements for detecting the presence of island or cluster formation, as discussed further in section 4.5.

4.3.3. Fe/GaAs (001)

We now consider another example from the work of Chambers *et al.*[12a,b] in which Auger electron diffraction has been applied to the growth of epitaxial layers of Fe on GaAs (001). This system has been studied extensively because of its interesting magnetic anisotropies in the surface plane, as first discussed by Krebs, Jonker, and Prinz.[106] It is complicated by the fact that outward diffusion of As is thought to occur, even though at the same time the Fe atoms appear in LEED to be growing in (001) epitaxy. A polar scan in the [100] azimuth of the $L_3M_{4,5}M_{4,5}$ Fe Auger peak at approximately 710 eV kinetic energy provides further information on how this might be occurring, as illustrated in Fig. 25. Here, the experimental AED curve of Chambers *et al.* for a 10-ML Fe overlayer on GaAs is compared to an analogous experimental Fe $2p_{3/2}$ XPD curve for a clean *bcc* Fe (001) surface due to Herman *et al.*[107]; the XPD peak furthermore has a kinetic energy of about 780 eV, very close to that of the Auger peak, so that the two diffraction patterns would be expected to be very similar for a given crystal structure. In fact, the two experimental curves are very different, with the *bcc* Fe (001) showing a much lower intensity for the peak along [101] and different fine structure at polar angles of about 15–30° and 60–75°.

Also shown in Fig. 25 are SSC–PW theoretical curves for three overlayer crystal structures: *bcc* Fe with $a = 2.82$ Å [the bulk-lattice constant which also gives a very good match to the GaAs (001)], primitive cubic (*pc*) Fe with $a = 2.82$ Å, and *fcc* Fe with $a = 2.82$ Å. It is clear that the *fcc* calculation gives the best agreement with the Fe/GaAs experimental data as to both the relative intensity of the [101] peak and the fine structure. The calculations for the other two structures seriously underestimate the intensity of the peak along the [101] direction. The *bcc* calculation also agrees best with the XPD curve from clean Fe (001), particularly as to the relative intensities of the weaker features from $\theta = 15°$ to 75°, even if all of the fine structure is not correctly predicted. All

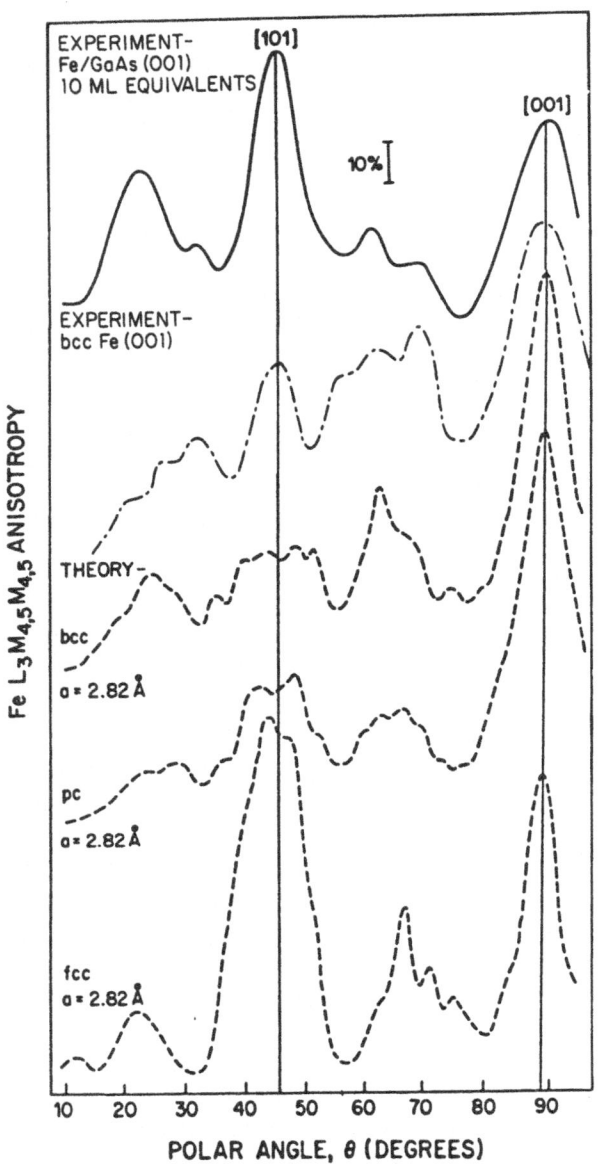

FIGURE 25. Experimental polar scan of the Fe *LMM* Auger intensity at 703 eV from 10 ML of Fe deposited on GaAs (001) (solid curve) is compared to theoretical calculations for various Fe lattices (dashed curves). (From Ref. 12(b).) The scans are in the [100] azimuth ($\phi = 0°$), with the directions [101] and [001] indicated. The calculations are at the SSC-PW level, and they are shown for Fe in three crystal structures: *bcc*, *pc* (primitive cubic), and *fcc* (which is proposed to be *bcc* Fe with As atoms outwardly diffused into the *fcc* interstitial sites). Also shown for comparison is an experimental *polar scan for bulk* Fe (001) in the same azimuth (dot-dash curve) from a separate study. (From *ref.* 107.)

calculations predict a strong peak along the normal or [001] direction; this is due to forward scattering from atoms with a closest spacing of $1.000a$ for all three structures. Along the [101] direction, by contrast, the *fcc* structure has nearest-neighbor scatterers at a distance of $a/\sqrt{2} = 0.707a$ (cf. Fig. 22a) whereas, in the *bcc* and *pc* structures, the nearest scatterers are twice that distance away at $\sqrt{2}a = 1.414a$. This explains the stronger forward-scattering peak along [101] in the *fcc* theory.

The combined experimental and theoretical results in Fig. 25 thus suggest that the local structure in Fe/GaAs has scatterers that are at the *fcc* positions. These results have been explained by the interesting proposal[12a,b] that the outward-diffusing As atoms occupy the face-centered positions in a *bcc* Fe lattice so as to yield an overall AED pattern that is essentially *fcc* in nature. Although Fe and As are slightly separated in atomic number (26 and 33, respectively) so that the all-Fe calculations of Fig. 25 are not in that case strictly correct, the forward-scattering strength that is dominant at these energies is not a strong function of atomic number (but rather of atomic size, as noted in section 3.1.3), and thus these theoretical simulations should be reasonably accurate for the hypothesized structure as well.

This work thus illustrates another aspect of higher-energy AED and XPD that should be generally useful in studying the detailed structures of complex epitaxial overlayers that may have impurities present, such as atoms diffusing outward from the substrate or inward from the surface. An obvious complementary and useful type of data that could be derived for such a system would be to look at the AED or the XPD of the impurity. For the example of Fe/GaAs, if the hypothesized structure is correct, As also should show an *fcc* type of diffraction pattern, although perhaps weaker or with less fine structure if it is preferentially segregated to the surface of the Fe overlayer. Another recent example of this type is a combined AED/XPD study of dopant P and Sb atoms in Ge epitaxial layers on GaAs (001) by Chambers and Irwin;[12c] here P was found to occupy lattice sites, whereas Sb was segregated to the surface.

4.3.4. $Hg_{1-x}Cd_xTe$ (111)

As a final example of an epitaxial system, we consider a recent scanned-angle XPD study by Granozzi, Herman *et al.*[108] of $Hg_{1-x}Cd_xTe$ (111) grown by liquid-phase epitaxy. This sample underwent transport at atmospheric pressure before being studied and was minimally ion-bombarded so as to remove a thin oxide layer from the surface. It was not subjected to bakeout or annealing after ion bombardment, to avoid depleting Hg from the surface region. At the time of measurement, the value of x was approximately 0.4. In spite of the less-than-ideal surface expected to remain after such a treatment, XPD modulations of $\Delta I/I_{max} \approx 15\text{--}25\%$ were seen in all of the major photoelectron peaks observable (Hg $4f_{7/2}$ at a kinetic energy of 1383 eV, Cd $d_{5/2}$ at 1078 eV, and Te $3d_{5/2}$ at 910 eV). Qualitatively comparing Hg, Cd, and Te diffractions curves immediately indicated that the Hg and Cd atoms were occupying similar lattice sites, as expected.

As another more subtle structural problem resolvable from this data, the question of the nature of the termination of the surface also was addressed. That is, was the surface terminated preferentially with double layers having cationic Cd (or Hg) on top and anionic Te on the bottom (termed Model *A*) or with the reverse (termed Model *B*)? Comparing the azimuthal XPD patterns for Cd and Te obtained at several polar angles with SSC–SW calculations for both Models *A* and *B* permits determining the dominant type of termination, even for a surface that probably has a reasonable amount of damage on it. Some of this data is shown in Fig. 26, where Cd emission at $\theta = 19°$ and $35°$ (both chosen to pass through near-neighbor scattering directions) is considered. It is clear that, for both angles of emission, the agreement between experiment and theory as to both visual fit and *R* factor[26d] is much better for a Model *A* termination; peak relative intensities, positions, and fine structure are much better predicted. Similar conclusions can be drawn from analogous Te azimuthal scans.

As one further aspect of this study, we consider the forward scattering origin of the various major peaks observed in Fig. 26 with the aid of Fig. 27, which indicates the several near-neighbor forward-scattering events possible in a surface terminated as in Model *A*. For the data at $\theta = 19°$, the effects of the event labelled as $\theta = 19°$, $\phi = 0°$ are clear in both experiment and theory. For the data at $\theta = 35°$, the principal peaks are due to events of the types labelled $\theta = 35°$, $\phi = 60°$ and $\theta = 30°$, $\phi = 30°$, $90°$.

The analogous Te curves at these polar angles are very different from those of Cd in both experiment and theory, with peak shifts and relative intensity

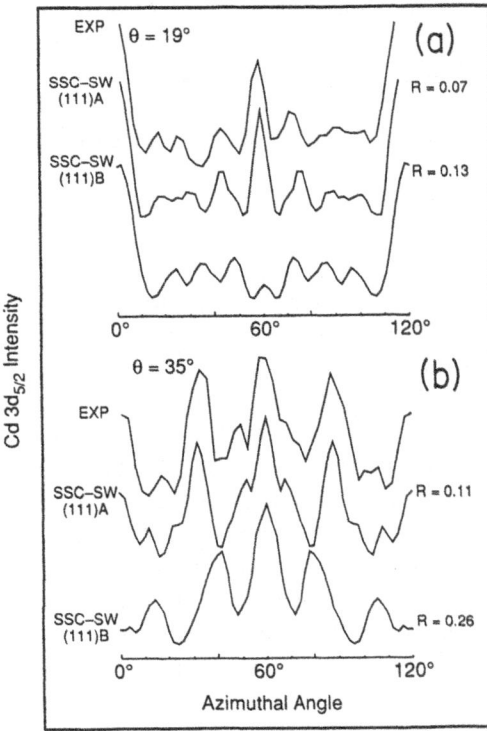

FIGURE 26. Al $K\alpha$-excited azimuthal scans of Cd $3d_{5/2}$ intensities from $Hg_{1-x}Cd_xTe\,(111)$ $(x \approx 0.4)$ at polar angles of (a) 19° and (b) 35° passing through or very close to forward-scattering low-index directions shown in Fig. 27 as $\theta = 19°$, $\phi = 0°$, $\theta = 35°$, $\phi = 60°$, and $\theta = 30°$, $\phi = 30°$, $90°$. Also shown are SSC–SW curves for the two possible surface terminations (Model A = Cd or Hg on top, Model B = Te on top), together with R factors comparing experiment and theory. (From Ref. 108.)

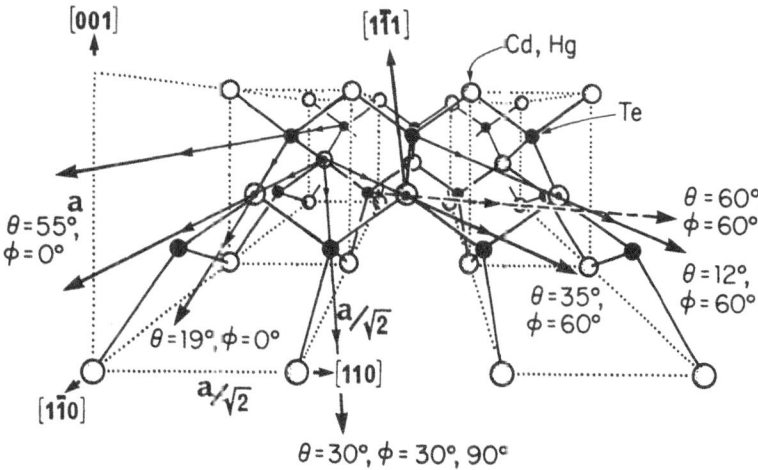

FIGURE 27. Perspective view of the unreconstructed (111) surface of $Hg_{1-x}Cd_xTe$ (111) in the Model A surface termination of Fig. 26, with the θ, ϕ coordinates of various near-neighbor/low-index directions along which forward scattering might be expected to be strong. These directions would be the same for the unreconstructed (111) surfaces of any material with the zincblende or diamond structure, as will be used later in discussing Fig. 36.

changes. In particular, the peaks at $\theta = 35°$, $\phi = 30,°$ and $90°$ for Cd disappear in Te and are replaced by two weaker features at $\theta = 35°$, $\phi \approx 38°$ and $80°$. This is easily explained, since Fig. 27 shows that, in an A-type termination, the peaks that disappear are only strong forward-scattering events in the first double layer for Cd emission; thus, they are not expected to be seen for Te.

Inspection of other azimuthal data of this type shows that most of the strong features can be assigned an origin in the various simple near-neighbor forward-scattering effects illustrated in Fig. 27, although it is again important to realize that higher-order interference effects can significantly influence the intensities due to forward scattering by atoms further from the emitter (cf. the discussion of Fig. 22 and, below, Figs. 37 and 38).

This study thus illustrates the further use of higher-energy XPD for epitaxial systems, for which bonding sites of substitutional atoms and the type of surface termination of a compound semiconductor can be determined.

4.3.5. Diffraction Effects in Quantitative Analysis and Photoelectron-detected EXAFS

We conclude this discussion of epitaxial systems with two notes of caution concerning the strong diffraction effects that are expected in either photoelectron or Auger emission from well-ordered lattices.

Diffraction Effects Must Be Carefully Allowed for in Any Attempt to Do Quantitative Analyses of Surface Composition. Methods of correcting for such effects have been considered by both Connelly *et al*, for simple adsorption on a metal,[109] and more recently for semiconductor surfaces by Alnot *et al*.[110] Not adequately allowing for such effects can lead to errors of as high as $\pm50\%$ in measured stoichiometries! Some of the methods for such corrections are averaging over diffraction curves obtained in more than one polar or azimuthal scan, taking advantage of the crystal-structure symmetry to find scans in which different constituents will have nearly identical diffraction patterns (e.g., this is possible in the zincblende structure[110]), or using theoretical calculations to try to determine directions in which diffraction effects can be neglected.

By contrast, a potentially useful aspect of diffraction effects for surface analysis is in monitoring intensities along different directions as a function of coverage during epitaxial growth, as suggested by Idzerda et al.[67] Model calculations of such curves in the SSC–PW model suggest that it should be possible to resolve the completion of the first few layers of growth.

The Use of Photoelectron Intensities to Monitor EXAFS-like Oscillations Requires Sufficient Angular Averaging. The idea of using photoelectron intensities to measure EXAFS oscillations for near-surface species has recently been proposed by Rothberg et al.[111] and applied to semiconductor systems by Choudhary et al.[112] It is clear from the strong oscillations of up to 70% seen in scanned-energy photoelectron diffraction and their dramatic dependence on emission direction (cf. Fig. 20) that an adequate averaging over direction must be undertaken to yield something related to the 4π-averaged EXAFS signal. Although this is automatic for disordered or polycrystalline systems,[111] it is problematic in single-crystal studies. Lee[41] has in fact questioned on theoretical grounds whether even the maximum 2π averaging possible in photoemission for such cases is sufficient to yield the EXAFS limit. Nonetheless, preliminary experimental results of this type[112] using the modest type of averaging inherent in the conical solid angle of a cylindrical mirror analyzer (CMA) appear to yield EXAFS-like data. However, it is the author's opinion that a single-geometry CMA measurement does not represent sufficient angular averaging to reliably yield the EXAFS limit and that the close similarity of these result to EXAFS data may have a fortuitous component. Perhaps measuring intensities for several different orientations of the specimen with respect to the analyzer would improve the reliability of this approach, but it is not clear that this has been done to date. The solid-angle averaging of a particular analyzer could also be checked by carrying out SSC calculations over the directions involved and summing these intensities, as was done recently by Idzerda et al. in another context.[67]

Overall, both XPD and AED thus have considerable potential for the study of the morphology of the first 1–5 layers of an epitaxial system. The strongest peaks are expected to be directly connected with simple forward scattering from the first few spheres of neighbors around a given emitter. Weaker features may involve a superposition of several types of scattering events. Thus, a quantitative analysis of the full intensity profile will require calculations at least at the SSC level. Predicting peak relative intensities correctly if emission along a dense row of atoms is involved may also require the inclusion of multiple scattering. However, much useful information about the surface structure, layer thickness, morphology, impurity-site type, and surface termination should be derivable from a consideration of the possible strong forward-scattering peaks due to the nearest neighbors (cf. Figs. 22 and 27) combined with theoretical modeling at the single scattering level.

4.4. Metal–Semiconductor Interface Formation

We now consider two recent examples of the application of higher-energy XPD to the study of metal–semiconductor interface formation. This kind of XPD study was pioneered by Kono and co-workers, and more detailed discussions appear elsewhere, including work on other metal–semiconductor combinations.[49,113,114] The examples chosen here both involve the initial stage of metal reaction with Si surfaces and represent structures over which controversy still exists. The examples differ in the final structure proposed. The first case, K/Si (001), is a metal overlayer relatively far above the Si surface. The second case, Ag/Si (111), is a metal layer nearly coplanar with the first Si layer. This strongly affects the degree and manner in which forward scattering by Si or metal atoms influences the observed diffraction patterns.

4.4.1. K/Si (001)

In this study by Abukawa and Kono,[114] azimuthal K $2p$ XPD data have been obtained for the structure formed by depositing K to saturation onto the Si (001) (2×1) reconstructed surface. The substrate surface is thought from a number of previous studies to consist of rows of dimers, as shown by the small open circles in Fig. 28. The most-often-discussed model for the potassium structure on this surface is the so-called one-dimensional-alkali-chain (ODAC) model illustrated in Fig. 28a; it corresponds to a $\frac{1}{2}$ ML coverage, and leaves open grooves adjacent to each high-lying row. However, there is still considerable controversy surrounding the structure of K adsorbed on Si (001), and this geometry has not been directly determined.[115] There is also disagreement as to what constitutes the saturation coverage of K on the surface.[114,115b]

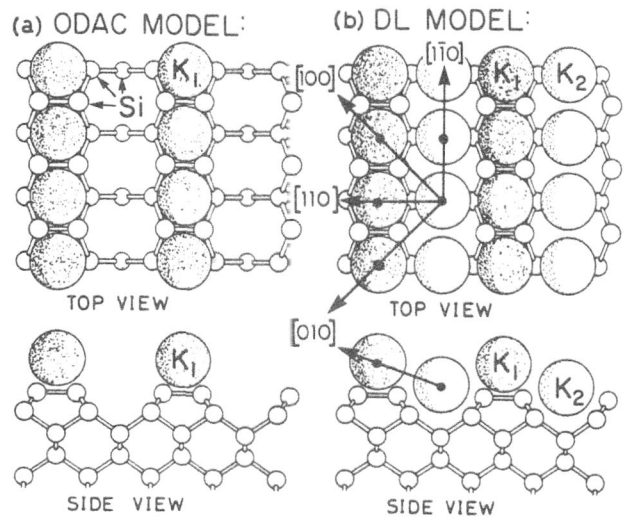

FIGURE 28. Schematic illustration of two structural models for the Si (001) (2×1) surface saturated with K: (a) one-dimensional-alkali-chain (ODAC) model, (b) double-layer (DL) model proposed from an analysis of azimuthal XPD data (see Fig. 29). Silicon dimers appear along the $[1, -1, 0]$ rows in both models. Each model can exist in two domains rotated by 90° with respect to one another. Some strong forward-scattering directions in the DL model are shown by arrows. (From Ref. 114.)

Even before considering the actual XPD data, we note that, if only atoms of type K_1 in the ODAC structure are present, the diffraction patterns would be dominated by forward scattering from other K_1 atoms, and this would further-more be strong only for very low θ and along the $\langle 1, -1, 0 \rangle$ rows for which the interatomic distances are shortest. The Si atoms should play only a minor role, perhaps producing fine structure in the azimuthal curves for very low takeoff angles.

A set of azimuthal experimental data for this system with emission angles relative to the surface of $14°$–$22°$ is shown as the points in Fig. 29. The strongest peak is seen along $\langle 100 \rangle$ for a relatively high value of $\theta = 14°$, an observation which already seems at odds with the ODAC model. Considering also the experimental anisotropy $\Delta I / I_{max}$ (scale along left of figure), we see that it can be as high as about 30%, a value which is significantly above those expected in general for such higher-θ scattering from neighbor atoms that are either all in-plane or all below-plane relative to the emitter (cf. Fig. 15 for $c(2 \times 2)$ O/Ni (001) as a typical example).

These results suggest trying in addition to the ODAC model another structure in which there are scatterers well above some K emitters. One such model is the obvious one of putting rows of atoms of type K_2 in all of the grooves to yield a 1-ML coverage, as illustrated in Fig. 28b. For this double-layer (DL) model, strong forward scattering can occur for higher takeoff angles, as indicated by the arrows along both $\langle 110 \rangle$ and $\langle 100 \rangle$ directions. For very low takeoff angles approaching zero, either model is expected to show strong forward scattering for emission along the K rows parallel to $\langle 1, -1, 0 \rangle$. The presence of two equivalent domains of either structure rotated by 90° with respect to one another also implies

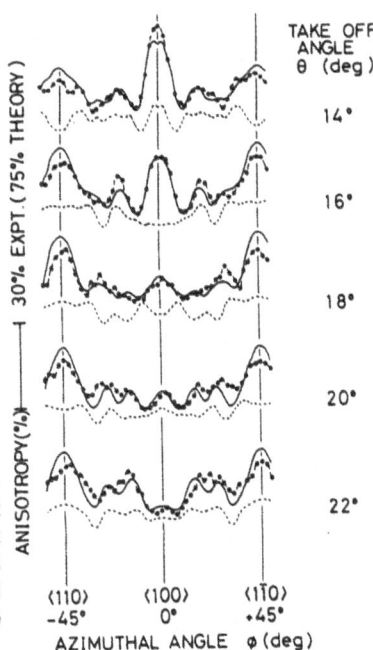

FIGURE 29. Azimuthal data for Al $K\alpha$-excited K $2p$ emission from the Si (001) (2×1) surface saturated with K at polar angles from 14° to 22° above the surface. Experiment is compared with SSC-PW calculations for the two models shown in Fig. 28: ODAC = dashed curves and DL model best fitting data = solid curves. Very similar curves were also obtained with SSC-SW calculations. (From Ref. 114.)

summing two diffraction patterns in the analysis and overall C_{4v} symmetry in both the observed and calculated patterns.

Comparing these experimental data to SSC–PW (or very similar SSC–SW) calculations for the two models[114] is now found to yield clearly superior agreement for the DL model (solid curves in Fig. 29). The strong peak at $\phi = 0°$ which grows in for θ approaching 14° can be explained as being due to emission from K_2 atoms and scattering by their second-nearest K_1 neighbors along $\langle 100 \rangle$. The peaks along $\langle 110 \rangle$ and $\langle 1, -1, 0 \rangle$ are due to K_2 emission again, but now involve scattering from nearest-neighbor K_1 atoms (and a sum over domains 90° apart). Additional azimuthal data for θ as low as 4°[114] show strong peaks for $\phi = \pm 45°$ that can be ascribed to the expected forward scattering along $\langle 1, -1, 0 \rangle$ directions within either K_1 or K_2 rows. Not surprisingly, these latter peaks are also present for very low θ in the theoretical curves for both models, and they are the most significant features in calculations for the ODAC model. Comparing experiment and theory for these lower-θ data also is found to support the DL model. By testing various vertical placements of the two K row types, the authors were able to determine a 1.1 Å vertical separation between the two K rows, and less accurately to determine that the bottom K row was not lower than about 0.5 Å above the first Si layer. For such a 1.1-Å separation, the K_1–K_2 distance is 3.99 Å and slightly larger than the K–K distance of 3.84 Å along either the K_1 or K_2 rows. It is also interesting that, for this structure, the $K_2 \to K_1$ forward scattering peaks should occur at $\theta \approx 16°$ along [110] and $\theta \approx 11°$ along [100]; this explains the strong peaks seen in the data over this range of polar angles. The registry of the DL along $\langle 1, -1, 0 \rangle$ with respect to the underlying Si surface was not determined, but the six-coordinate site shown in Fig. 28b for atoms of type K_1 is that predicted by theory to be the lowest energy.[115a,d]

In a more recent theoretical study of this system by Ramirez,[115d] it is found that adsorption in groove sites (including type K_2 in Fig. 28b) is significantly lower in energy than the six-coordinate site shown for K_1 atoms. Thus, adsorption in the grooves is supported by theory as well. However, the 1-ML structure proposed in this study is different from Fig. 28b in that the atoms of type K_2 are shifted along the $\langle 1, -1, 0 \rangle$ direction so as to be directly opposite the Si dimers. The K_2 atoms in this model are also predicted to be approximately in-plane with respect to the Si dimers. However, it is doubtful that this structure would yield the strong forward scattering peak seen in XPD along $\phi = 0°$ for relatively high theta values of 12–16°. Thus, even though these calculations[115d] indicate that a double layer with such shifted K_2 atoms is lower in energy than the structure shown in Fig. 28b, the latter structure still represents a better choice based upon the XPD data.

Overall, these XPD results thus provide important new insights into the bonding of K on Si (001) and illustrate several aspects of the use of this technique for metal–semiconductor studies.

4.4.2. Ag/Si (111)

The Ag/Si (111) system has been studied by almost every modern surface-science technique and is known to exhibit, among other things, a well-ordered

($\sqrt{3} \times \sqrt{3}$) Ag structure and the formation of *fcc* Ag clusters or islands with (111) orientation for exposures that go above the 0.7–1.0 ML needed just to form the ($\sqrt{3} \times \sqrt{3}$) structure.[49,50,116,117] In the following section, we consider the use of XPD in studying such clusters; here, we concentrate on a recent XPD study by Bullock *et al.* of the ($\sqrt{3} \times \sqrt{3}$) structure.[50a,b]

In this study, polar and azimuthal Ag $3d_{5/2}$ XPD data were obtained for a well-ordered and very stable ($\sqrt{3} \times \sqrt{3}$)Ag structure, and these experimental results are summarized in Figs. 30a and 31. The smooth and structureless nature of the polar scans in Fig. 30a indicates an absence of strong forward scattering effects, except perhaps at very low takeoff angles of $\theta \approx 4$–$8°$ where a four-peak structure is seen in Fig. 31. A simple geometric calculation then permits the conclusion that the Ag cannot be more than approximately 0.5 Å below the surface Si layer. This is also consistent with the lower anisotropy values of no more than 21% that are found for the azimuthal scans of Fig. 31. It can thus be concluded that there are no strong forward scatterers above the Ag. The azimuthal data are also fully consistent with an earlier XPD study of this system by Kono *et al.*,[49] but they are more detailed in involving full 360° ϕ scans and more θ values.

It is beyond the scope of this review to discuss the many models that have been and are being proposed for this structure, but all known structures have been tested against this azimuthal data by Bullock *et al.*, using R factors[26d] as the final quantitative measure of goodness of fit. The calculations were carried out at the SSC–SW level, and in final optimizations also with the full final-state interference of $3d$ emission into p and f channels. (This latter correction was not found to alter the structural conclusions, a result which is expected to be true in general for higher-energy XPD, but certainly not for work at less than a few hundred eV, as discussed in section 3.1.2).

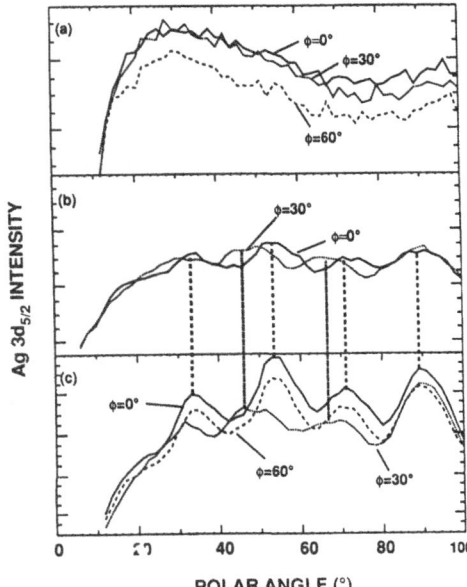

FIGURE 30. Polar XPD scans of Ag $3d_{5/2}$ intensity at 1120 eV from: (a) the ($\sqrt{3} \times \sqrt{3}$) Ag structure on Si (111) formed after annealing an ≈1.3-ML Ag overlayer to 550°C; (b) a Ag overlayer of approximately 2 ML average thickness at 450°C; and (c) a thick Ag overlayer of approximately 6 ML thickness at ambient temperature. (From Ref. 50.)

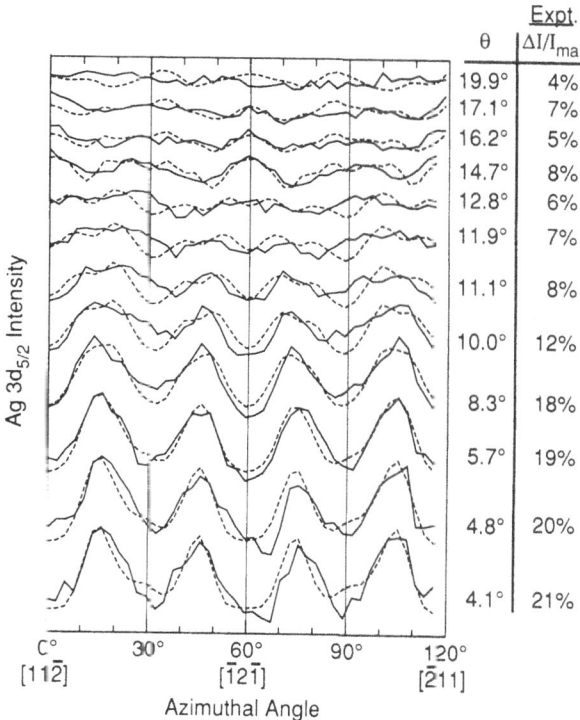

FIGURE 31. Azimuthal XPD scans of Ag $3d_{5/2}$ intensity from ($\sqrt{3} \times \sqrt{3}$) Ag/Si (111) at polar angles from 4° to 20° (solid lines) are compared to SSC-SW calculations for the optimized two-domain model of Fig. 32 (broken lines), for which $s_1 = s_2 = 0.86$ Å; $z_1 = -0.10$ Å, $z_2 = -0.30$ Å, and a 50:50 mixture of the two domains. Full final-state interference in the d- to $-p + f$ emission process has been included. This comparison yields an R factor of 0.14 (cf. values in Figs. 21 and 26). (From Ref. 50.)

The final model proposed on the basis of this work is for two nearly equivalent domains of Ag in a honeycomb array on a Si surface that has had the top layer of the first Si double layer removed. This two-domain missing-top-layer (MTL) model is illustrated in top view in Fig. 32. The optimized structural parameters are a contraction of the Si trimers toward one another in both domains of $s_1 = s_2 = 0.86$ Å, vertical distances of the Ag relative to the Si layer of $z_1 = -0.1$ Å for Domain 1 and $z_2 = -0.3$ Å for Domain 2 (that is, the Ag is very nearly coplanar with the Si in both domains, but just slightly below it), and a mixture of the two domain types that is between 50:50 and 40:60, with Domain 2 perhaps being slightly more predominant. The fits between experiment and theory for this fully optimized structure are shown in Fig. 31. All other models that have been tried yield significantly worse agreement as judged both visually and by R factors. This two-domain model is also closely related to one derived in a prior XPD study by Kono *et al.*: a single-domain MTL Ag honeycomb structure of type 1 with $s = 0.66$ Å and a vertical distance of -0.15 Å. The presence of Domain 2 is suggested to explain the four-peak structure at low θ values in Fig. 31, as illustrated by the nearest-neighbor forward-scattering peaks for the two domains shown at the bottom of Fig. 32. For

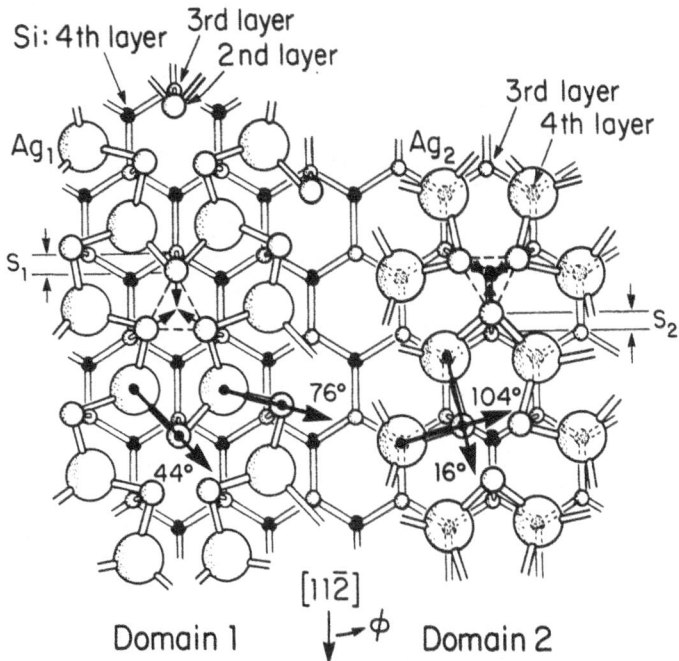

FIGURE 32. The two-domain missing-top-layer (MTL) honeycomb model proposed for ($\sqrt{3} \times \sqrt{3}$) Ag/Si (111). The parameters characterizing it are: vertical positions $z_1 = -0.1$ Å and $z_2 = -0.3$ Å, Si trimer contractions of $s_1 = s_2 = 0.86$ Å, and a 50:50 mixture of Domains 1 and 2. The lower half of the figure shows the two sets of nearest-neighbor Si forward-scattering peaks that produce the four-peak structure seen at low θ values in Fig. 31. (From Ref. 50.)

the lowest θ values near 4°, an additional correction of possible importance is the reduction of nearest-neighbor Si forward-scattering strengths due to multiple scattering effects along the nearly linear rows of atoms that can be labelled Ag emitter → Si first-neighbor scatterer → Si second-neighbor scatterer (cf. Figs. 32, 3b, and 6); very recent MS calculations by Herman *et al.*[50c] show that this reduces the absolute peak intensities for $\theta = 4°$ and $\phi = 16°$, 44°, 76°, and 104° by about 30%, thus improving the agreement of theoretical and experimental anisotropies.

A further interesting point in connection with this structure is that a recent LEED study of the clean Si (111) surface by Fan *et al.*[117] concludes that a little-studied ($\sqrt{3} \times \sqrt{3}$) Si reconstruction has very nearly the same geometry as Domain 1 in Fig. 32 if Ag adatoms are replaced by Si adatoms. Although these authors do not consider the possibility of a second domain of type 2 for ($\sqrt{3} \times \sqrt{3}$) Si, it might be expected to have approximately the same energy (due to weak fourth-layer interactions) and thus also to exist on the clean surface. This work thus lends support to the two-domain model for ($\sqrt{3} \times \sqrt{3}$) Ag, since one can imagine its growth simply by replacing the Si adatoms with Ag atoms.

This structure is still very controversial, and these results thus cannot be called conclusive, but they further illustrate the way XPD can be used for such metal–semiconductor studies. This study is also state-of-the-art for XPD in that it involves a large azimuthal data set, SSC–SW calculations with correct final-state interference, and the use of R factors[26d] to judge goodness of fit. As one qualitative figure of merit in connection with this study, the minimum R factors of 0.14 found are about $\frac{1}{2}$ of those found in recent LEED studies of the same system.[117,118]

4.5. Supported Clusters

In this section, we briefly consider two examples of how higher-energy XPD has been used to study the formation of three-dimensional clusters on surfaces. (A third example has already been considered in the data for Fe deposited on Cu (001) shown in Fig. 24, where agglomeration effects are visible even for very low coverages.)

4.5.1. Ag/Si (111)

We have noted in the last section that Ag readily forms islands and three-dimensional clusters on the Si (111) surface if the coverage exceeds the 0.7–1.0 ML needed for the ($\sqrt{3} \times \sqrt{3}$) Ag structure. If these clusters are more than one atomic layer in thickness, then strong forward-scattering effects are expected for emitters in the lower layer(s) of the cluster. Such effects are illustrated in Fig. 30b,c, where polar scans of Ag $3d_{5/2}$ intensity have been measured first in Fig. 30c for a thick Ag reference layer of approximately 6 ML thickness, and then after heating to 450 °C so as to desorb all but an average coverage of about 2 ML. In Fig. 30c, a LEED pattern characteristic of the epitaxial Ag (111) that is known to grow on Si (111) is seen, and strong diffraction peaks due to buried-atom emission from this thick overlayer are found. In Fig. 30b, the Ag (111) LEED pattern is weakly present and there are still clear remnants of the photoelectron-diffraction features seen in the thick overlayer. Thus, such XPD patterns are very sensitive to the presence of three-dimensional islands.

The previous discussion of Figs. 22 and 23 also suggests that it might be possible to estimate the average thickness of such clusters up to about 5 ML, where the XPD features begin to converge to the bulk pattern. An additional type of information that could be very useful for some systems is the orientation of the cluster crystal axes with respect to the surface normal. In fact, even if clusters grow in a textured way (that is, without preferred azimuthal orientation), polar scans of the type shown here should permit determining whether there is any preferred vertical axis. Bullock and Fadley[50b, 19] have also recently pointed out that, even for two-dimensional islands, it should be possible to use low-θ azimuthal scans to determine the island orientation and, for smaller islands, the average number of atoms present.

4.5.2. Pt/TiO₂

As a second example of cluster studies using XPD, Tamura *et al.*[120] have considered the interaction of Pt with three low-index faces of TiO$_2$, a system of interest in catalysis and for which the so-called strong metal-support interaction

(SMSI) can occur. In this study, Pt was deposited at room temperature to a mean thickness of about 10 ML onto the (110), (100), and (001) surfaces of TiO_2, and azimuthal XPD measurements were made at different polar angles for the Ti 2p and O 1s photoelectron peaks before deposition and for the Pt 4f peaks after deposition. Similar Pt 4f measurements were made after annealing the samples up to 800 K.

Some of these results are shown in Fig. 33a for the (110) surface at $\theta = 40°$ and Fig. 33b for (100) at $\theta = 45°$. Considering first Fig. 33a, we see that curves (i) and (ii) show weak diffraction features for both Ti 2p (clean) and Pt 4f (just after the deposition). The nonconstant background under these curves, particularly for (ii), is thought to be due to a nonuniform deposition over the region of the sample seen by the electron analyzer; thus, with changes in ϕ, a slightly different area and average Pt thickness might be seen. After the high-temperature anneal, the Pt 4f features in (iii) are strongly enhanced, with a concomitant increase in the anisotropy $\Delta I/I_{max}$ from 16% to 29%. This is consistent with the growth of thicker or larger clusters upon annealing, although (ii) indicates that some sort of ordering must be present even without annealing. Finally, (iv) shows a theoretical calculation based upon PW-cluster calculations with the effects of double scattering included. (The possible risk of including only double-scattering events has been mentioned already in section 3.2). The Pt clusters assumed had (111) orientation and contained 13 atoms in three planes; two symmetry-equivalent orientations with respect to the substrate 180° apart were considered. The resulting curve in (iv) is found to agree rather well with the annealed Pt 4f experimental results, suggesting that the clusters are growing with preferred (111) orientation.

A similar set of data for the (100) surface are shown in Fig. 33b. Here, (i) and (ii) exhibit strong diffraction from the O 1s and Ti 2p peaks of the substrate. Curve (iii) shows the strong diffraction of Pt 4f after the anneal. (A more uniform deposition of Pt has here made the background levels very flat.) Finally, curve (iv) is calculated for the same type of two-domain, three-layer Pt cluster [but with different assumed registry with the (100) surface], and it again shows good agreement with experiment, suggesting (111) orientation for the clusters on this surface as well.

For the third (001) surface studied, it is interesting that the Pt 4f oscillations were weak both before and after annealing, indicating a different kind of overlayer growth and/or a lower degree of cluster formation.

Together the three studies related to clusters that have been considered up to this point illustrate the utility of both polar and azimuthal XPD or AED data for studying the amount of cluster formation present and the average orientation and morphology of the aggregates formed. Two possible limitations of this kind of study are that XPD and AED average over all of the clusters present and so cannot easily be used to estimate the cluster-size distribution. In certain cases, it might even be difficult to detect the difference between, for example, a full 4-ML epitaxial overlayer and a collection of independent clusters with an average thickness of 4 ML, even if the crystallographic orientation could be easily determined. Although with careful measurements of both substrate and deposited-atom intensities before and after deposition and/or heat treatment, the

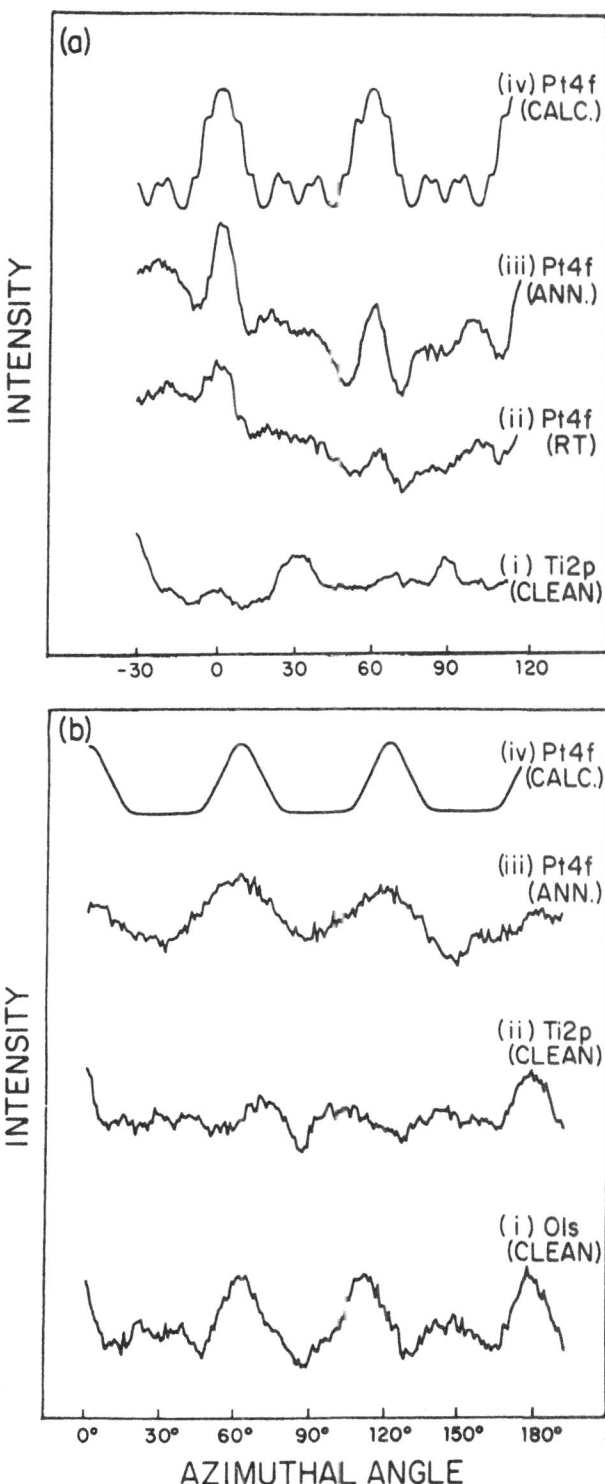

FIGURE 33. (a) Azimuthal XPD data for Pt $4f$ and Ti $2p$ emission from Pt on TiO_2 (110) at $\theta = 40°$ are compared to PW cluster calculations including double scattering for Pt emission from a (111)-oriented metal cluster of about 15 Å diameter. Al $K\alpha$ radiation was used for excitation. (b) As in (a) but for Pt $4f$, Ti $2p$, and O $1s$ emission from Pt on TiO_2 (100) at $\theta = 45°$. (From Ref. 120.)

implicit effects of "patching" in cluster growth should be evident in deposited-atom–substrate relative intensities. Simple formulas for analyzing such patched-overlayer relative intensities appear elsewhere.[9] It is also clear that combining XPD or higher-energy AED with scanning tunneling microscopy (STM) would yield a particularly powerful set of data for cluster and epitaxial growth studies. This is because STM can be used to measure directly both the cluster size distribution and the step and defect densities that are averaged over in XPD/AED. But it may be difficult or impossible with STM to see into a cluster or overlayer so as to determine its crystallographic orientation or thickness. This is because STM cannot probe below the surface density of states and also is not atom-specific.

4.6. Core Level Surface Shifts and Chemical Shifts

A further type of problem that has been studied by low-energy photoelectron diffraction using synchrotron radiation for excitation is metal core level surface shifts.[18,53,80,121,122] In particular, Sebilleau, Treglia et al.[18,53,80] have tuned the photoelectron energy to low values to achieve high surface sensitivity and have looked with high energy resolution at photoelectron diffraction from such surface-shifted core levels.

Some of their results for tungsten $4f$ emission from W (100) are illustrated in Fig. 34, where both the surface and bulk peaks are shown, together with their individual azimuthal diffraction patterns and corresponding SSC–PW theoretical curves. The two types of peaks clearly exhibit very different diffraction patterns, and both of these are rather well predicted by the SSC model, even at this quite low photoelectron energy of approximately 30 eV. It is remarkable that a single-scattering approach is so quantitative at such a low energy, and this may to some degree be fortuitous. However, later work by Treglia et al.[18c,53,80] has reached similar conclusions, with the only qualification being that it is necessary at such energies to use the correct final-state angular momenta, as expected from the discussion of Fig. 4 in section 3.1.2. For the low energy of this case, the $4f$-to-εd channel is assumed to be dominant.

This work thus illustrates the added ability of photoelectron diffraction to carry out independent structure determinations of physically or chemically different species of the same atom through core level shifts. These shifts are not limited to the clean-surface type considered above, but may also involve the well-known chemical shifts commonly seen when different chemical bonding or oxidation states are present. Such state-specific structure studies should be a very powerful probe of surface reactions, overlayer growth, and interface formation. They will, however, require very high energy resolutions of 0.3 eV or better to be fully effective in resolving small shifts.

As an obvious example for future work, it should also be possible to do state-specific diffraction studies on semiconductor surfaces, since both clean surfaces[123a,b] and chemically reacted surfaces[123c] exhibit shifted core levels characteristic of the different bonding sites and/or oxidation states.

One technologically important example of a semiconductor system for which more structural information concerning different chemical species would be useful

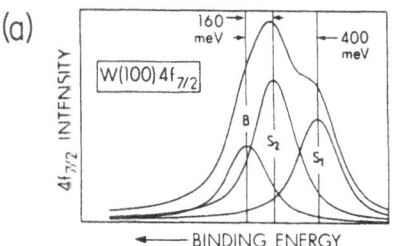

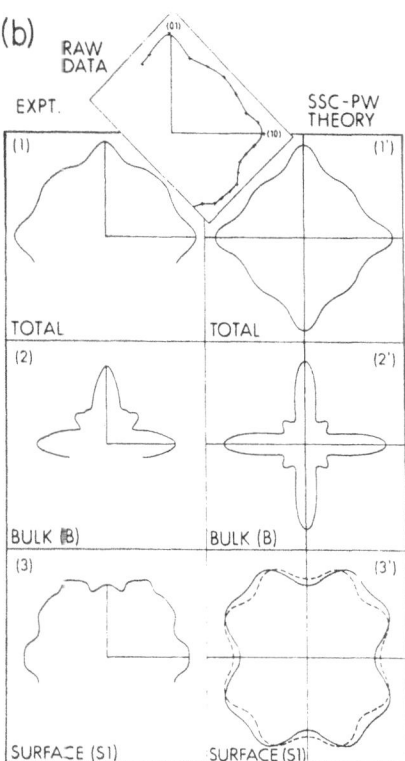

FIGURE 34. (a) A W $4f_{7/2}$ spectrum from W (001) at a kinetic energy of ≈ 30 eV, showing two surface-shifted core levels (S_1 and S_2) as well as a bulk peak (B). (b) The azimuthal dependences of these intensities at a polar angle of 60° above the surface: (1) represents the total $4f_{7/2}$ intensity, (2) the bulk intensity (B), and (3) the surface intensity (S_1). SSC-PW calculations of these intensities are shown in (1'), (2'), and (3'), respectively. The inset represents raw data for the total intensity. (From Ref. 18(a).)

is the formation of the interface between SiO_2 and Si. Figure 35 shows high-resolution Si $2p_{3/2}$ core spectra obtained by Himpsel *et al.*[123c] from Si (100) and Si (111) surfaces that were thermally oxidized in UHV conditions (2.5 Torr O_2, 750° C, 20 sec) so as to produce a very thin 5-Å oxide film. The overall resolution here was 0.3 eV, and it is striking that all of the oxidation states of Si are clearly seen, from the elemental substrate to the 4+ dioxide. The different nature of the oxidizing surface for Si (111) is further found to lead to a suppression of the Si^{2+} state. These intermediate oxidation states are thought to be associated with the interface, and, from quantitative estimates of the different depth distributions of these states, it is concluded that an extended rather than abrupt interface is involved. Models of such an extended interface have been proposed by Himpsel *et al.*, but these cannot be tested in detail without additional data. It seems clear that separately measuring the scanned-angle photoelectron diffraction patterns of the different oxidation states would provide some very useful information in this

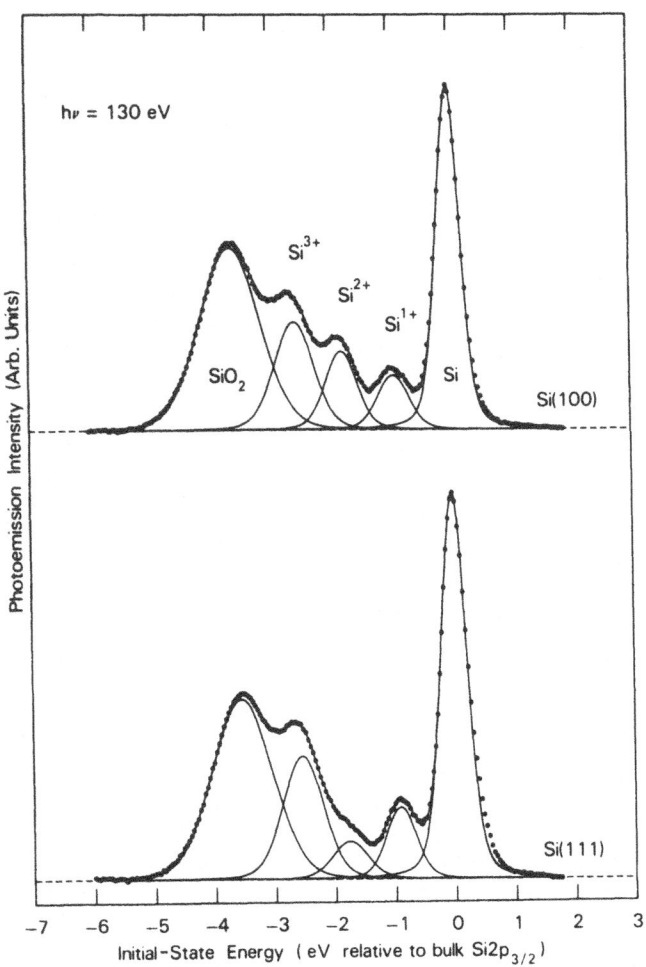

FIGURE 35. The Si $2p_{3/2}$ components of Si $2p$ spectra from thin oxide films of approximately 5 Å thickness thermally grown on Si (100) and Si (111) surfaces. Note the reduced intensity of Si^{2+} for Si (111), assumed to be due to structural differences in the interface. [From Ref. 123(c).]

direction, since each state is hypothesized to occupy one or at most a few distinct site types relative to the substrate lattice.

Although these are difficult experiments at present, the detailed state-by-state information derivable should help in unraveling the microscopic structures of many surface and interface systems. Being able to tune photon energy so as to vary surface sensitivity or to move on or off of resonant photoemission conditions would also be an advantage, as noted in prior studies.[123] Going to higher photon energy not only permits looking deeper into the material and assessing the relative depth distributions of the different species, but should also lead to more simply interpretable forward-scattering peaks for emission from interface-associated atoms. A disadvantage of higher energies is that the substrate signal tends to dominate the spectrum, but with high enough resolution and suitable reference spectra for subtracting the substrate signal, such high-energy measurements should be possible. Synchrotron radiation will thus be necessary to fully exploit this potential for studying interface growth by state-specific photoelectron diffraction.

4.7. Surface Phase Transitions

We conclude this discussion of applications of photoelectron diffraction and Auger electron diffraction by considering briefly their possible use in studying various types of surface phase transitions such as surface premelting, roughening, or disordering at a temperature below the bulk melting temperature,[124] as well as surface reconstructions that are temperature-dependent.[123a,b] The short-range order and directional sensitivity of both PD and AED suggest that they should be useful probes of such surface phase transitions, which may involve changes in near-neighbor atom positions and/or the introduction of considerably more disorder in these positions. The number of such studies is still very small, but the most recent are quite promising.

An unsuccessful attempt at observing surface premelting for Cu (001) in grazing-emission XPD was made some time ago by Trehan and Fadley.[63a] For this surface, roughening and possibly faceting was observed before any evidence was seen in the XPD anisotropies of the extra disorder associated with surface melting. However, much more recently, evidence for surface phase transitions involving surface disordering and perhaps premelting has been seen in XPD from two separate systems: Pb (110) by Breuer, Knauff, and Bonzel[125] and Ge (111) by Friedman, Tran, and Fadley.[126]

For the case of Ge (111), prior LEED studies and theoretical modeling by McRae and co-workers[127] indicate that there is a reversible surface order–disorder transition at a temperature of 1060 K that is 0.88 times the bulk melting temperature. Is this transition visible in XPD? In Fig. 36, we show such XPD data in which the Ge $3d$ azimuthal anisotropy was monitored as a function of temperature. The polar angle of 19° chosen here causes the emission direction to sweep through nearest-neighbor forward-scattering directions in the unreconstructed surface, as shown in Fig. 27. This relatively low θ value also leads to higher surface sensitivity.

Figure 36a shows four azimuthal scans taken at temperatures from ambient to about 50 K above the transition. (Note the expected similarity of the azimuthal scan at ambient temperature to that for $Hg_{1-x}Cd_xTe$ (111) in Fig. 26a.) As the temperature is increased, the azimuthal curves gradually lose much of their fine structure, and upon passing above the transition point, only two main peaks remain in the azimuths $[1, 1, -2]$ ($\phi = 0°$) and $[-1, 2, -1]$ ($\phi = 60°$). In Fig. 36b, the intensity of the $[1, 1, -2]$ peak corresponding to nearest-neighbor scattering is plotted against temperature, and it is clear that an abrupt drop occurs over the interval 850–1050 K. This drop furthermore cannot be explained by simple Debye–Waller modeling.

McRae et al.[127a] have measured the intensities of several LEED beams for the same system as a function of temperature, and their data is similar to Fig. 36b in that the intensities drop sharply toward 1060 K and level off thereafter. Some of the LEED intensities drop more rapidly than the curve of Fig. 36b near 1060 K;

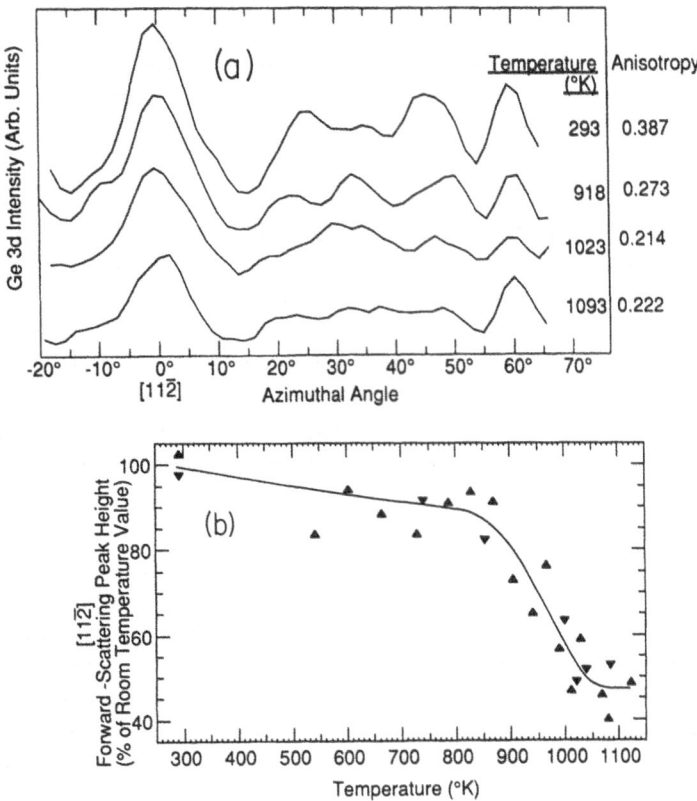

FIGURE 36. Temperature-dependent azimuthal XPD data for Ge $3d$ emission at 1458 eV from Ge (111) at low takeoff angle of $\theta = 19°$. This θ value corresponds to scanning through nearest-neighbor scattering directions for $\phi = 0°$, as shown in Fig. 27. (a) Four azimuthal scans at temperatures from ambient to above the order-disorder transition. (b) The detailed temperature dependence of the height of the peak along $\phi = 0°$. 1060 K is where a prior LEED study (Ref. 127(a)) has seen evidence for a surface-disordering transition. Upright triangles represent increasing temperature; inverted triangles, decreasing temperature. (From Ref. 126.)

some have a form very similar to this curve. Thus, it can be concluded that the same transition is observed in both sets of data, even though the LEED measurement is expected to be sensitive to longer-range order on a scale of approximately 100 Å, whereas XPD should probe distances on the order of 10–20 Å.

Although these XPD results have not as yet been analyzed in detail so as to derive additional structural information, it is clear that obtaining both polar and azimuthal data at temperatures below and above the transition temperature and comparing the diffraction structures seen with calculations for different types of disorder models should yield a better understanding of this and other surface phase transitions.

Similar abrupt changes in polar-scan diffraction anisotropies have also been seen by Breuer *et al.*[125] for the surface disordering of Pb (110), which has been observed previously with Rutherford backscattering and low-energy electron diffraction.[128]

As one interesting future direction for such work, the study of surface phase transitions should also benefit greatly from doing separate diffraction measurements on the various core peaks observed. For example, the Ge (111) surface exhibits one bulk peak and two surface peaks[123a,b] that could all be studied separately. However, the small shifts of only about 0.3–0.7 eV involved here would require very-high-resolution data and the use of curve-deconvolution procedures.

5. FUTURE DIRECTIONS

5.1. Measurements with High Angular Resolution and Bragg-like Reflections

As noted previously, most prior PD and AED measurements have been carried out with resolutions of at best a few degrees in half angle. In many systems, the acceptance solid angle is also not a simple cone, but may have different dimensions along two perpendicular axes.[29] For future work, the question thus arises as to what additional information might be gained by going to much better conic resolutions of, for example, $\pm 1.0°$.

As discussed in section 2, various methods exist for limiting angular spreads upon entry into the analyzer, but one which has the advantages of being very certain in its limits and operationally very convenient is the insertion of externally selectable angle-defining tube or channel arrays between sample and analyzer entry. The use of such channel arrays has been discussed by White et al.,[33] and they have been used to precisely limit angles to $\pm 1.5°$ or better (that is, $< \frac{1}{4}$ of typical prior solid angles).

We have already discussed two examples of this kind of data: for NiO grown on Ni (001) in Fig. 14 and for $c(2 \times 2)$ S on Ni (001) in Fig. 17. For these cases, we have pointed out the greater sensitivity to the degree of short-range order and the adsorbate position, respectively.

As a final example of the dramatic effects seen in going to high angular resolution, we compare in Figs. 37a and b low- and high-resolution XPD data obtained by Osterwalder, Stewart et al.[48] for Ni $2p_{3/2}$ emission from a clean Ni (001) surface at $\theta = 47°$. A great deal more fine structure is seen in the data with $\pm 1.5°$ resolution, and the form of the fine structure for $\phi \approx 25°–65°$ is in fact completely changed due to a lower degree of angular averaging over such structures. Very narrow features of only a few degrees at FWHM are also seen in the results at high resolution.

Figure 37c summarizes a more complete set of such high-resolution azimuthal data for Ni $2p_{3/2}$ that represents the most detailed investigation of XPD fine structure to date. Here, the polar angle of emission was varied in 1° steps from $\theta = 40°$ to 50°, passing through the high-symmetry value of $\theta = 45°$ which contains the $\langle 110 \rangle$ directions of nearest-neighbor scattering in its ϕ scan. Full 360° scans were used to generate each curve, and fourfold averages of this data into one quadrant shown elsewhere[10,48] agree excellently with the single-quadrant results presented here. This three-dimensional plot makes it clear that high-energy electron diffraction features can change extremely rapidly with either θ or

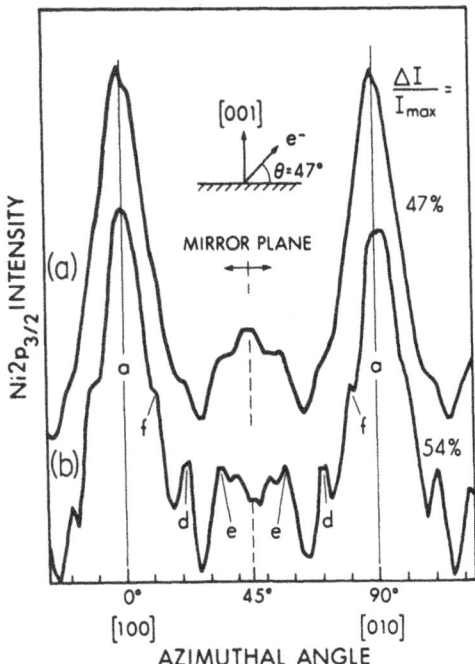

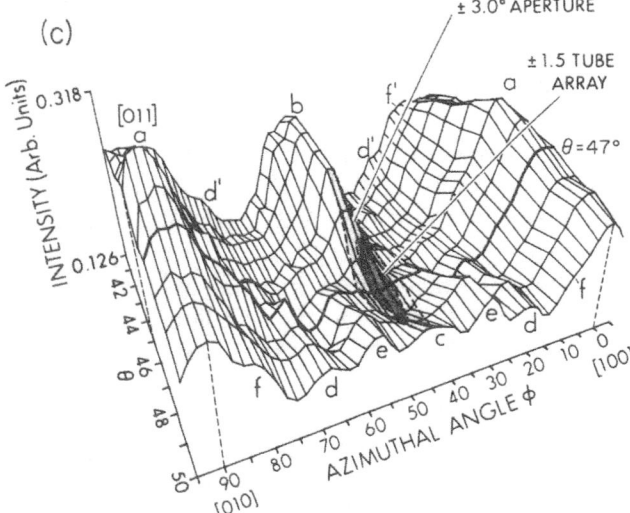

FIGURE 37. Effect of increasing angular resolution on Ni $2p_{3/2}$ azimuthal XPD data from a clean Ni (001) surface at 632 eV. (a) and (b) show single scans at $\theta = 47°$ with resolutions defined by a single aperture of nominal ±3.0° acceptance and a tube array yielding ±1.5° or less, respectively. (c) a three-dimensional summary of a series of single-quadrant high-resolution Ni $2p_{3/2}$ scans with a ϕ step of only 1°. The regions averaged over with the two different angle-defining devices in (a) and (b) are shown as shaded. (From Ref. 48.)

ϕ. These results also qualitatively explain how the approximately ±3.0° averaging in Fig. 37a yields features for $\phi \approx 25°-65°$ that are so different from those for the high-angular-resolution curve in Fig. 37b. That is, Fig. 37a represents an average over all of the curves in Fig. 37c from $\theta = 44°$ to $\theta = 50°$, as bounded by the lighter-shaded elliptical area, and the steeply rising ridge toward $\theta = 44°$ thus accounts for the peak seen at $\phi = 45°$ with lower resolution. The results in Fig. 37b, by contrast, represent an average over only the darker-shaded area in Fig. 37c, and so retain a minimum at $\phi = 45°$.

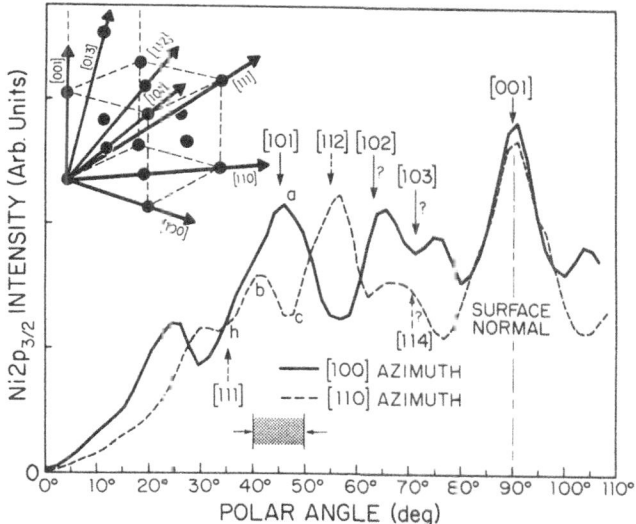

FIGURE 38. High-resolution Al $K\alpha$ polar scans of Ni $2p_{3/2}$ intensity above Ni (001) in two different azimuths, with certain low-index directions and special points noted [cf. lower-case letter labeling in Fig. 37(c)]. The region covered by Fig. 37(c) is shaded. The inset shows the near-neighbor/low-index directions within an *fcc* unit cell. (From Ref. 48.)

Figure 38 shows two high-resolution polar scans from the same study of Ni (001). The unit cell of the metal and various near-neighbor scatterers along low-index directions is also indicated to permit judging how well various strong features correlate with them (cf. also Fig. 22a). These polar scans also show considerable extra fine structure, for example, as compared to the same sort of [100] polar scan for higher-energy Auger emission from bulk Cu (001) shown in Fig. 22b. These high-resolution data are found to exhibit peaks for emission along some, but not all, of the near-neighbor directions shown. Peaks are found at positions corresponding closely to the nearest neighbors (and fourth-nearest neighbors) along [101], the second neighbors along [001], and the third neighbors along [112]. However, minima and/or significant peak shifts are seen for the fifth neighbors along [103] and the sixth neighbors along [111]. Neighbors even further away along [102] and [114] are also found to show significant shifts compared with the observed peaks. In particular, the [111] direction corresponds to a local minimum (indicated as point *h*), with enhanced intensity on either side of the minimum; a ϕ scan through [111] at $\theta = 35°$ shows the same sort of profile. As noted previously in the discussion of Fig. 22, this is due to the influence of higher orders of interference[71] and perhaps multiple scattering effects.[73] Thus, we conclude that the first 3–4 spheres of neighbors in any lattice will probably produce strong and simply interpretable forward scattering peaks. Beyond these spheres, more-complex origins will require modelling at least at the SSC–PW level for interpretation.

Three-dimensional data of the type shown in Fig. 37c have also been obtained at lower angular resolution by Baird, Fadley and Wagner for XPD from

Au $(001)^{129}$ and by Li and Tonner for high-energy AED from Cu $(001).^{29}$ These two data sets span a high fraction of the 2π solid angle above these two surfaces, and they exhibit very similar intensity contours, as expected since they both represent high-energy emission from the same *fcc* crystal structure. The more recent data of Li and Tonner serves as a more accurate reference for the overall features of such *fcc* XPD/AED patterns at lower angular resolution. These studies also agree with the preceding paragraph and the discussion of section 4.3.2 in seeing simple correlations of peaks with near-neighbor forward-scattering directions out only to the fourth shell, with directions such as [111], [114], [102], and [103] showing more complex behavior.

The Ni data discussed here and the other high-resolution results discussed previously thus make it clear that, at least in higher-energy XPD and AED, using resolutions that are much worse than $\pm 1.0°$ will blur out some features and lead to a loss of structural information. Such sharp features are generally the result of superpositions of several scattering events, since the relevant scattering factor by itself exhibits nothing narrower than the forward scattering peak of some 20–25° FWHM. These features also tend to involve scatterers further away from the emitter and thus to be associated with the degree of short-range order around the emitter. (This is nicely illustrated by the NiO/Ni (001) results of Fig. 14.) Thus, there is little doubt that XPD or AED with high resolution will contain more fine details of the structure under study.

At lower energies, by contrast, one expects generally wider features due to the broader, more diffuse scattering factors involved (cf. Fig. 2) and the larger de Broglie wavelengths that spread out different orders of interference (cf. the curves in Figs. 4 and 5). However, even for such energies, it is possible for superpositions of multiple events to produce rather narrow features, and high resolution might also be a benefit in this case.

The most obvious disadvantage of working at high angular resolution is the longer data-acquisition times, which may be 10–30 times those of typical low-resolution operation.[33] A second disadvantage is that it is likely that the effects of multiple scattering will tend to be averaged out somewhat in lower-resolution data because of cancellations of phases in the many events involved.[21] Conversely, in high-resolution data, such MS effects may be more important, even though the information content is inherently greater.

A further aspect of the relationship of such high-resolution data to more complex interference effects and more distant neighbors is the influence of Bragg-like diffraction effects from planes in multilayer substrate emission. In the presence of the strong inelastic damping characteristic of both PD and AED, such Bragg-like events lead to what has been termed a Kikuchi-band model of these phenomena.[2,3,63b,129,130] Although a fully quantitative Kikuchi-band theory of higher-energy PD or AED based upon the superposition of many Bragg-like scattered waves is lacking, simple model calculations have been carried out by Baird et al.,[129] by Goldberg et al.,[130] and more recently also by Trehan et al.,[63b] and they are found to semiquantitatively reproduce the results of XPD measurements on both Au (001) and Cu (001). In particular, the superposition of several Kikuchi bands along low-index directions yields the forward-scattering peaks seen in both experiment and SSC calculations.

More interestingly, there are features in experimental data at high angular resolution that appear to be associated with specific Bragg events from low-index planes (such as features d and f in Fig. 37b here and as discussed in connection with Fig. 31 of Ref. 9). This suggestion has been given more quantitative support in a recent high-resolution study of Ni (001) by Osterwalder et al.[48a] Furthermore, calculations with the SSC model exhibit these same Bragg-like features if the cluster size is permitted to be large enough and/or the inelastic damping is sufficiently reduced,[48a,63b] thus verifying that a cluster-based theory can be used for problems varying from short-range order to long-range order.

This formal equivalence of the SSC model and the Kikuchi-band picture for describing bulk-like multilayer emission was first pointed out some time ago,[9,42,130] but additional clarification seems appropriate in view of misleading statements concerning the role of the Kikuchi model in the interpretation of XPD and AED that have nonetheless appeared in the more recent literature.[11] From an experimental point of view, the essentially identical intensity profiles for LMM Auger electron diffraction and backscattered LEED "Kikuchi patterns" from Ni (001) at 850 eV observed by Hilferink et al.[70e] provide a particularly clear verification of this equivalence. From a theoretical point of view, the relationship of the two approaches, if both are carried to comparable quantitative accuracy, is analogous to the equivalence of the so-called short-range-order and long-range-order theories of EXAFS, as discussed elsewhere.[63b,131] It is clear, however, that the SSC and MSC approaches are of greater generality in that they can be applied to both surface- and bulk- emission and to problems of differing degrees of order. The Kikuchi-band picture is, by contrast, formulated on a basis of inelastically attenuated Bloch states that reflect long-range translational order. Thus, the cluster-based theories are inherently more rapidly convergent and are more appropriate ways to look at near-surface diffraction from adsorbates and thin overlayers, as noted previously.[42,130] But it is absolutely incorrect to say that the ability of the cluster approach to explain forward-scattering features makes the Kikuchi-band model invalid for describing substrate emission.[11]

In summary, the use of high angular resolutions on the order of $\pm 1.0°$ should permit even more precise structural conclusions to be derivable from both photoelectron diffraction and Auger electron diffraction, especially at energies of >500 eV. Such data should contain information on neighbors further away from the emitter, including features related to Bragg-like scattering events. It is also clear that the use of resolutions of $\pm 3.0°$ or worse may conceal a great deal of fine structure inherent in the experimental curves.

5.2. Spin-Polarized Photoelectron and Auger Electron Diffraction

Beyond increasing both the energy resolution and the angular resolution in PD and AED as means of deriving more detailed structural information, we can also ask what is to be gained if the last property of the electron, its spin, is also somehow resolved in the experiment. This prospect has so far been considered quantitatively and observed experimentally only in the case of photoelectron diffraction, but we return at the end of this section to comment on how it might also be possible in Auger electron diffraction.

In the first attempts at what has been termed spin-polarized photoelectron diffraction (SPPD), the fundamental idea has been to use core-level multiplet splittings to produce internally referenced spin-polarized sources of photoelectrons that can subsequently scatter from arrays of ordered magnetic moments in magnetic materials. Figure 39a illustrates how such a splitting can give rise to spin-polarized photoelectrons for $3s$ emission from high-spin Mn^{2+}. The splitting is intra-atomic in origin and arises from the simple LS terms of ^5S and ^7S in the final ionic state of Mn^{+3} with a $3s$ hole.[132] The net effect is to cause the peaks in the doublet to be very highly spin-polarized, with ^5S predicted to be 100% spin-up and ^7S to be 71% spin-down relative to the net $3d$ spin of the emitting atom.[133,134] The relatively large exchange interaction between the highly overlapping $3s$ and $3d$ electrons is responsible for the easily resolvable splitting of 6.7 eV between the ^5S and ^7S final states of the photoemission process.

The basic experiment in SPPD thus involves looking for spin-dependent scattering effects that make two such peaks behave slightly differently in the presence of a magnetically ordered set of scatterers. Such effects were first discussed theoretically by Sinkovic and Fadley,[134a] and they have several special properties:[135,136]

- There is no need for any kind of external spin detector beyond an electron spectrometer capable of resolving the two peaks in energy.

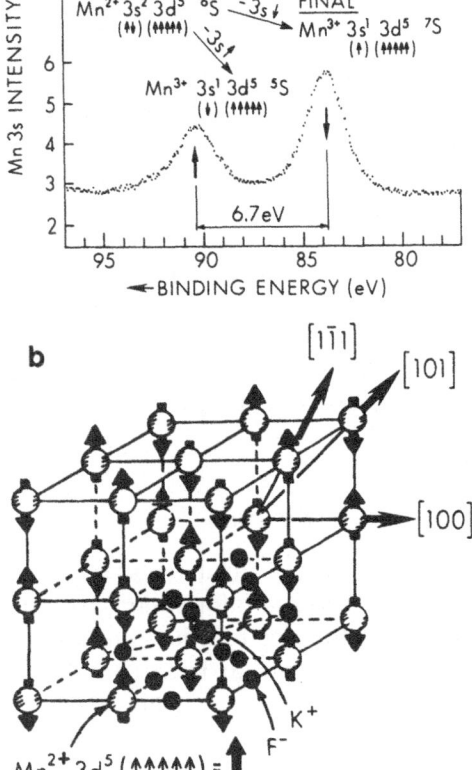

FIGURE 39. (a) The Al $K\alpha$-excited Mn $3s$ spectrum of $KMnF_3$, with the initial and final states leading to the multiplet splitting indicated, together with the predominant photoelectron spin expected in each peak. (b) The crystal structure of $KMnF_3$, with the antiferromagnetic ordering of the Mn^{2+} spins also indicated. [From Ref. 134(a).]

• The fact that the photoelectron spins are referenced to that of the emitting atom or ion means that SPPD should be capable of sensing magnetically ordered scatterers even when the specimen has no net magnetization. Thus, studies of both ferromagnetic and antiferromagnetic materials should be possible, and meaningful measurements should also be feasible above the relevant macroscopic transition temperatures (Curie or Neel temperatures, respectively). For the latter case, the photoelectrons in each peak would be unpolarized with respect to any external axis of measurement but still polarized relative to the emitting atom.

• The photoelectron emission process is also very fast, with a time scale of only about 10^{-16} to 10^{-17} seconds; thus, such measurements should provide an instantaneous picture of the spin configuration around each emitter, with no averaging due to spin–flip processes, which are much slower at roughly 10^{-12} seconds.

• Finally, the previously discussed strong sensitivity of any form of photo-electron diffraction to the first few spheres of neighboring atoms means that SPPD should be a probe of short-range magnetic order (SRMO) in the first 10–20 Å around a given emitter. Thus, provided that a sufficiently well-characterized and resolved multiplet exists for a given material, this technique has considerable potential as a rather unique probe of SRMO for a broad variety of materials and temperatures.

Before discussing the first observations of such spin-dependent scattering and diffraction effects, it is appropriate to ask to what degree final-state effects such as core-hole screening may alter or obscure these multiplets. We note first that the cases of principal interest in SPPD are outer core holes, which are more diffuse spatially than inner core holes and for which the interaction with the surrounding valence electrons is thus not as strongly polarizing as for inner core holes (which can often be very well described in the equivalent-core approximation). Nonetheless, it has been suggested by Veal and Paulikas[137] that both screened and unscreened multiplets corresponding to $3d^{n+1}$ and $3d^n$ configurations, respectively, are present in the $3s$ spectra of even highly ionic compounds such as MnF_2.

As such effects would make the carrying out of SPPD measurements more difficult (although still certainly not impossible) due to the potential overlap of peaks of different spin polarization, Hermsmeier et al.[138] have explored this problem in a study of Mn $3s$ and $3p$ multiplets for which the experimental spectra from several reasonably ionic solid compounds have been directly compared to the analogous spectra from gaseous Mn, a simple free-atom system in which no extra-atomic screening can occur. In Fig. 40, we show their compilation of $3s$ spectra for the diluted magnetic semiconductor $Cd_{0.3}Mn_{0.7}Te$ (a), single-crystal MnO with (001) orientation (b), polycrystalline MnF_2 as obtained some time ago by Kowalczyk et al.[139] (c), gaseous atomic Mn (d), and a free-ion theoretical calculation of these multiplets by Bagus et al. including configuration interaction, but totally neglecting extra-atomic screening (e).[140] From a consideration of the experimental data only, it is striking that for both $3s$ multiplets and $3p$ multiplets (not shown here, but discussed in Ref. 138) the solid-state spectra are very similar to the gas-phase spectra, with the only differences being some extra broadening in the solid state and some small changes in peak positions that are not at all surprising. Thus, even without resorting to theory, it seems clear that these

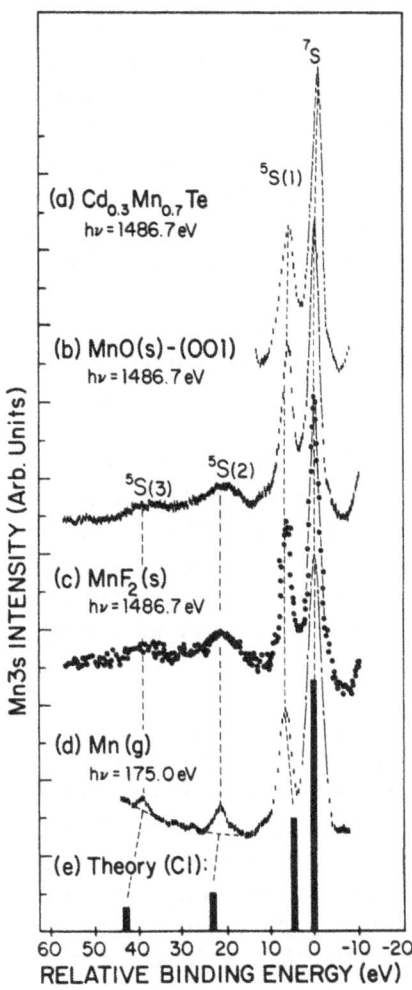

FIGURE 40. Experimental Mn 3s spectra for (a) the diluted magnetic semiconductor $Cd_{0.3}Mn_{0.7}Te$, (b) MnO (001), (c) polycrystalline MnF_2 (Ref. 139), and (d) gaseous atomic Mn are compared to (e) theoretical calculations for emission from a free Mn^{2+} ion including final-state configuration interaction (Ref. 140). (From Ref. 138.)

spectra are very free-atom–free-ion like, and that a simple multiplet interpretation such as that in Fig. 39a should rather accurately describe the spin polarizations of the photoelectrons involved.

If we consider now the best available free-ion theoretical prediction for the 3s spectra, this conclusion becomes even more convincing. In Fig. 40e, the results of a calculation by Bagus, Freeman, and Sasaki[140] for Mn^{3+} with a 3s hole and limited configuration interaction (CI) are shown. There is excellent agreement with experiment not only for the two dominant members of the multiplet that would be most useful in SPPD, but also for the two much weaker satellites that directly result from including CI. Similar conclusions are reached in a comparison of experiment and theory for analogous 3p spectra.[138] We thus conclude that extra-atomic screening does not cause a major perturbation of these multiplet splittings and thus also that outer core holes such as 3s and 3p should exhibit relatively free-atom–free-ion like multiplets for a variety of high-spin systems. Such multiplets in turn should be useful as spin-resolved sources in SPPD.

Direct experimental evidence of spin polarization in core spectra also exists. A recent measurement with an external spin detector of the spin polarization over the $3p$ peak from ferromagnetic Fe by Kisker and Carbone[141] yields significant spin-up polarization at lower kinetic energy and spin-down polarization at higher kinetic energy that are in the same sense as those expected for a simple $3p$ multiplet.[138] These results thus suggest that SPPD should be possible with ferromagnetic metals as well, particularly on the simpler and more widely split $3s$ peaks.

Returning now to a consideration of the SPPD experiments carried out to date, we have shown in Fig. 39b the crystal structure of the first material for which such effects were observed: a (110)-oriented sample of the simple antiferromagnet $KMnF_3$. It is clear from this that the relative spins of the emitter and the first scatterer encountered can be different for different directions of emission, as for example, between [100] and [101]. Spin-dependent scattering effects were first observed for this system by Sinkovic, Hermsmeier, and Fadley[142] as small changes of up to about 15% in the ratios of the $^5S(1)$ (spin-up) and ^7S (spin-down) peaks in the dominant doublet shown in Fig. 39a. For this study, a lower energy of excitation of 192.6 eV (Mo Mζ radiation) was used in order to yield lower-energy photoelectrons at approximately 100 eV, which are expected to exhibit significant spin-dependent effects in scattering.[134a] This requirement of low kinetic energies thus makes SPPD inherently well suited to synchrotron radiation with its tunable energy.

The $^5S(1):{}^7S = I(\uparrow)/I(\downarrow)$ intensity ratio was found to be sensitive to both direction of emission (as qualitatively expected from Fig. 39a) and temperature. Its variation with temperature is furthermore found to exhibit a surprisingly sharp transition at a point considerably above the Neel temperature (T_N), as shown in Fig. 41a. Here, we plot a normalized intensity ratio or "spin asymmetry" S_{expt} that is measured relative to the value of $I(\uparrow)/I(\downarrow)$ at a limiting high-temperature (HT) paramagnetic limit. This asymmetry is defined in the inset of Fig. 41a; it goes to zero at high temperature.

The abrupt high-temperature change observed in S_{expt} has been suggested to be due to the final destruction of the short-range magnetic order that is expected to dominate in producing such spin-polarized photoelectron diffraction effects. Note also that the short-range-order transition temperature T_{SR} at which this occurs is approximately $2.7T_N$.

In an important confirmation and extension of this earlier work, very similar SPPD effects have also more recently been observed by Hermsmeier et al. for (100)-oriented MnO,[143] and two of their curves for the temperature dependence of the spin asymmetry are shown in Fig. 41b. As for $KMnF_3$, there is a relatively sharp change in the $^5S(1):{}^7S$ ratio at a temperature that is again well above the long-range-order transition temperature at $T_{SR} = 4.5T_N$. For both $KMnF_3$ and MnO, it is also interesting that the form of the short-range order transition is very sensitive to emission direction, being steepest for the nearest-neighbor scattering direction in Fig. 41a and changing sign with only a 15° shift of emission direction in Fig. 41b. This sensitivity to direction is qualitatively consistent with single-scattering calculations of the spin-dependent exchange-scattering processes that may be involved.[134,143,144]

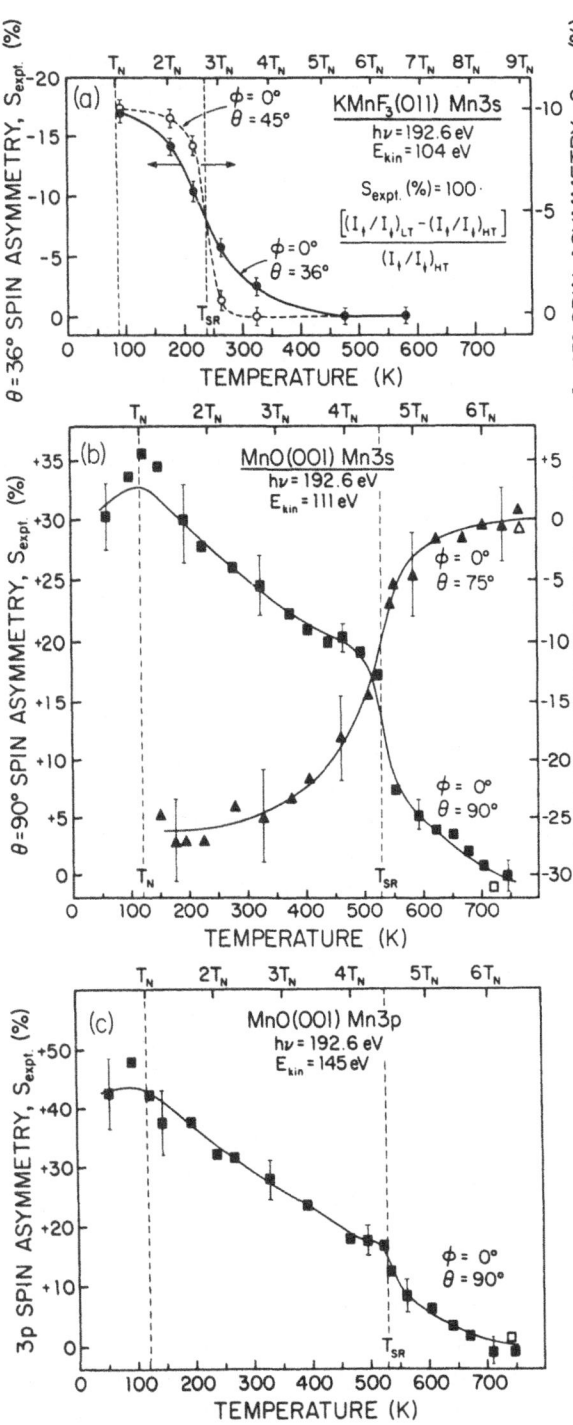

FIGURE 41. Spin-polarized photo-electron diffraction data indicating the presence of a high-temperature transition in antiferromagnetic short-range order. (a) Experimental spin asymmetries for the Mn 3s doublet from KMnF$_3$ with (110) orientation, as a function of temperature. Mo $M\zeta$ radiation at 192.6 eV was used for excitation to photoelectron energies of approximately 100 eV. The spin asymmetry is defined in the inset, where $I(\uparrow)/I(\downarrow)$ is the ratio of spin-up (^5S) to spin-down (^7S) intensities, HT refers to the highest temperature of measurement. Data are shown for two emission directions, one of which is along the [100] nearest-neighbor direction and the other 9° away from this. (From Ref. 142.) (b) As in (a), but for Mn 3s emission from MnO with (001) orientation. Note the different signs of the spin asymmetries for this case. (From Ref. 143.) (c) As in (b), but for Mn 3p emission from MnO with (001) orientation and treating the spin-up (^5P) and spin-down (^7P) peaks. (From Ref. 143.)

Although we have discussed only $3s$ emission thus far, the more complex $3p$ multiplets also should be spin polarized.[138,143] And in fact, a very similar transition has also been seen for MnO in the more widely split $^5P(1):^7P = I(\uparrow)/I(\downarrow)$ doublet at the same temperature T_{SR},[143] as shown in Fig. 41c. The fact that the same sort of transition is seen for these two peaks in spite of the fact that they are different from $^5S(1):^7S$ in both energy separation and mean kinetic energy provides strong support for the conclusion that this is a new type of magnetic transition.

It is also interesting that the T_{SR} values are, for both cases, approximately equal to the Curie–Weiss temperatures of the two materials, a connection which may be associated with the fact that this constant is proportional in mean-field theory to the sum of the short-range magnetic interactions.[145]

A final observation concerning this data is that the results for MnO in Figs. 41b and c show a possible indication of sensitivity to the long-range-order transition at T_N, as both curves possess a weak peak at T_N which is just outside of the estimated-error bar of the ratio measurement. If this is true, it is perhaps not surprising in view of the longer-range sensitivity of PD to neighbors that may be 20 Å from the emitter, as discussed in connection with both Figs. 16 and 20b.

A number of questions are thus raised by these results concerning the nature of short-range order above the long-range-order transition temperature and the way in which such effects can be incorporated in a spin-polarized variant of photoelectron diffraction theory. Although a quantitative theory of all aspects of the short-range-order transition and its inclusion in a spin-dependent modeling of the diffraction process does not yet exist, results in qualitative or semiquantitative agreement with experiment have been obtained in a few previous studies.[134,136,143,144]

The observation that Auger spectra from ferromagnetic materials exhibit strong spin polarization from one part of the manifold of features to another by Landolt and co-workers[146] also suggests that spin-polarized Auger electron diffraction (SPAED) should be possible. The more complex nature of Auger spectra in general will make the *a priori* prediction of the type of spin polarization more difficult, but for ferromagnets with net magnetization, an external spin detector could be used to first calibrate the spectrum for polarization.[146] Then, measurements of spin-up–spin-down ratios as functions of direction and/or temperature could be taken in the same way as for the spin-split core multiplets in SPPD. Even in antiferromagnetic systems with equal numbers of up and down $3d$ moments so that external calibration is impossible, any transition involving the polarized $3d$ valence electrons might be expected to show a net polarization that would again be internally referenced to the emitter.

A final aspect of such spin-polarized studies is to make use of left or right circularly polarized radiation, in conjunction with spin–orbit interaction in the energy levels involved, to preferentially excite one or the other spin polarization, as discussed recently by both Schuetz and co-workers[147] and Schoenhense and co-workers.[148] The use of such radiation already has produced very interesting spin-polarized NEXAFS and EXAFS structure from ferromagnets and ferrimagnets[147] and circular dichroism angular distributions (CDAD) from nonmagnetic surfaces and adsorbates.[148] In CDAD for light elements with negligible spin–orbit effects, no net spin polarization of the photoelectron flux is involved, but such measurements provide the interesting possibility of measuring the contributions of individual m_1 components to photoemission and photoelectron diffraction.[148] The CDAD studies require lifting the degeneracy of the m_1 sublevels, and so have been carried out on valence levels; however, with very high energy resolution, it might be possible to do similar measurements on outer core levels with, for example, small crystal-field and/or spin–orbit splittings present.

With the availability of higher-intensity sources of circularly polarized radiation from next-generation insertion devices, it should be possible to greatly expand both of these kinds of study so as to look in more detail at both the angle and the energy dependence of the photoelectron intensities. For example, spin-polarized EXAFS requires measuring very accurately the differences in absorption for right and left polarizations, because the overall effects may be as small as a few times 10^{-4} in K-shell absorption.[147] However, studying L_2- and L_3 absorption for heavier elements with $Z \geq 60$ leads to considerably larger effects that can be on the order of 10^{-3}–10^{-2}. Extending this to do SPPD would thus imply measuring similarly accurate ratios or differences of photoelectron intensities. In this case, the magnitudes of the photoelectron spin polarizations are only on the order of 1% for K-shell emission, but for heavier elements, they can be up to 40–50% in L_2 emission and 20–25% in L_3 emission.[147] The latter two cases are thus about $\frac{1}{3}$–$\frac{1}{2}$ as highly polarized sources as a high-spin multiplet such as that in Fig. 39a. One advantage of such an approach would be to expand such studies to cases for which a suitable high-spin multiplet is not available. A disadvantage is that an external axis of polarization is involved, so that only ferro- or ferrimagnetic specimens could be studied. However, in CDAD experiments, this last restriction is not present.[148]

SPPD is thus a very new area of photoelectron diffraction, but it has considerable potential for providing information on the short-range spin order and spin–spin correlation functions around a given type of emitter site in the near-surface region of magnetic materials. Other antiferromagnetic and also ferromagnetic materials are currently being studied in order to better establish the systematics of the short-range-order transition and the range of utility of this method. Spin-polarized Auger electron diffraction and other measurements making use of circularly polarized radiation for excitation also should be possible.

5.3. Synchrotron Radiation–Based Experiments

Looking ahead to the much more intense and/or much brighter synchrotron radiation sources in the VUV/soft X-ray region that are currently either coming into operation or being conceived as next-generation devices based upon undulators or wigglers, one can see much-expanded possibilities for all of the types of photoelectron diffraction measurements discussed up to this point.

Measurements with both high-energy resolution (to distinguish different surface layers or chemical states as shown in Figs. 34 and 35, respectively) and high angular resolution (to enhance fine structure and thus structural sensitivity) should be possible. For some types of experiments (e.g., with maximum surface sensitivity and/or with spin-polarized diffraction in mind), lower photoelectron energies of approximately 50–100 eV may be necessary, but for much structural work, energies of 1000 eV or even higher will be beneficial in yielding strongly peaked forward scattering and more nearly single-scattering phenomena. Being able to go to much higher photoelectron energies of up to 5000–10,000 eV may also be of interest in yielding even narrower forward-scattering peaks (as considered from a theoretical viewpoint by Thompson and Fadley[149]), more true bulk sensitivity via the longer electron attenuation lengths, and simpler theoretical interpretation. Being able to tune energy is also essential for the scanned-energy or ARPEFS experiments; it should be possible to carry these out much more rapidly and over a broader energy range above threshold. The polarization vector can also be oriented in either scanned-angle or scanned-energy measurements so as to enhance the contributions of various important scatterers (cf. Figs. 3a and 18). And we have already considered in the last section the possibility of using circularly polarized radiation. Finally, photoelectron *microscopy* with resolutions on the order of 500 Å or less is currently being developed,[150] and the additional dimension of using simultaneous photoelectron diffraction to probe the local atomic structure in such a small spot is quite exciting.

Auger electron diffraction may not benefit as much from synchrotron radiation, because excitation can be achieved with either photons or electrons and because the spectral form is not dependent on the excitation utilized if the initial hole is formed well above threshold. However, even for this case, synchrotron radiation could provide a more intense and less destructive excitation source than, for example, an electron beam or a standard X-ray tube. Also, it would be interesting to look at the diffraction process as the excitation energy is swept through threshold, so as to yield a purer one-hole initial state.

5.4. Combined Methods and Novel Data-Analysis Procedures: Photoelectron Holography?

It is clear from the foregoing examples that both scanned-angle and scanned-energy photoelectron diffraction measurements can provide useful information concerning surface structures, but that scanned-angle measurements are simpler in general to perform. Going to higher energies leads to easily interpretable forward-scattering features for many systems, but at the same time provides little information on the atoms that are below or behind the emitting atom as viewed from the detection direction. Thus, there are clear advantages to using lower energies as well, even if these lead to a potentially greater influence of multiple scattering. In the scanned-energy ARPEFS work discussed in sections 4.2.2 and 4.2.3, a major reason why interlayer spacings down into the bulk were derivable is that these lower energies exhibit the strongest backscattering effects and provide the largest oscillations in the $\chi(k)$ curves (cf. Fig. 20).

It is thus easy to suggest that the ideal photoelectron diffraction experiment based upon present methodology would consist of carrying out both high-energy measurements at kinetic energies greater than approximately 500 eV and low-energy measurements at approximately 50–100 eV. Being able to scan $h\nu$ would also be desirable, but not essential. A typical structure could then be analyzed by first making scanned-angle measurements at high energy and using the real-space aspects of any forward-scattering effects to narrow down the range of possible structures (cf., for example the discussion of Figs. 29 and 30). Combining scanned-angle measurements at high and low energies then should permit determining structures in detail, including atomic positions both below and above the emitter in the sense mentioned above. Or scanned-energy measurements could be performed as a second step as well, leading to the utility of Fourier transform methods for narrowing down the number of structures. Using an electron spectrometer that can simultaneously analyze and detect electrons over a range of emission directions[35–38] would also clearly speed up such studies, with the only likely drawback being that angular resolution is often lower in such systems, particularly when working at higher energies. In all of these methods, the final precise structural determination would require comparison of experimental diffraction curves or χ functions with calculated curves for a number of geometries, with the most quantitative method of comparison being via some sort of R factor.[20,26d] This is thus exactly the same methodology employed in LEED, except that in photoelectron diffraction, a single scattering approach should already provide useful information for many cases and there is additional readily available structural information concerning the type of local bonding site that can assist in ruling out structures.

As a final new direction in the analysis of scanned-angle data, we consider the recent interesting proposal by Barton,[151] based on an earlier suggestion by Szoeke,[152] that it should be possible to directly determine atomic positions via photoelectron holography. According to this idea, the photoelectron leaving the emitter is treated in first approximation as a spherical outgoing wave that, by virtue of the scattering and diffraction from its neighbors, produces an intensity modulation outside of the surface that can be considered a hologram. This hologram is then simply the intensity distribution of a given peak over a two-dimensional range in θ, ϕ (or, equivalently, some two-dimensional range in k_x, k_y). This intensity distribution can then be described by a formula of exactly the same type as Eq. (10), but with some important generalizations. These generalizations are that the scattering amplitude $|f_j|$ and phase shift ψ_j, together with the factors for the excitation matrix element and attenuation due to spherical wave, inelastic, and vibrational effects, must be replaced by an overall wave amplitude $|F_j|$ and phase ψ_j for each scatterer that sums over all single- and multiple-scattering events which terminate in atom j as the last scatterer before the detector. It can then be shown[151] that inverting this two-dimensional hologram mathematically to produce a real image is equivalent to a double Fourier integral in k_x and k_y, in which the desired z plane of the image is a variable paremeter within the integral. Thus, two dimensional x–y cross sections at different z positions are in principle possible with this method.

Barton has carried out a theoretical simulation of this new method using MSC–SW intensity distributions in θ, ϕ for the $c(2 \times 2)$ S/Ni (001) system at a kinetic energy of 548 eV and with a width of angular detection in both the k_x and k_y directions of $\pm 40°$. The inversion of this hologram is found to have maxima that can be directly related to different near-neighbor Ni atoms, with an estimated resolution in x and y of 0.5 Å and in z of a much higher 2.3 Å. The x and y resolutions are ultimately limited by the Rayleigh criterion for a lens (hologram) of a given opening angle. For the maximum reasonable detection-angle ranges in a spectrometer of ±40° to ±60°, this in turn yields resolution limits Δx and Δy that are very close to the de Broglie wavelength of the electron (i.e., 0.52 Å at 548 eV). This is a likely reason why a rather high kinetic energy in the typical XPS range was used for this simulation.

As noted by Barton, some limitations and/or problems that need to be addressed in the further development of this technique are the relatively low position-resolution obtainable, particularly in z; the presence of twin images at $\pm z$ for each atom (a universal effect in holography), which could cause serious overlap problems for bonding geometries involving atoms that are below-plane; the fact that multiple scattering effects on the F_j may cause deviations of the image positions from the actual sites, thus requiring an iterative correction via theoretical calculations of these generalized scattering amplitudes for an assumed geometry; the fact that several images at different energies, or even an additional Fourier transform of energy-dependent data at each θ, ϕ, may be necessary to effect this correction; and the added experimental difficulty in requiring some sort of high-speed multichannel electron analyzer that can obtain such large data sets in a reasonable amount of time.[35]

Another limitation not mentioned in connection with this theoretical simulation is that the high energy used implies relatively weak backscattering effects of only 15% or so compared to forward scattering (cf. Fig. 2); thus, the actual degree of modulation in intensity observed may be quite small, making the measurements rather difficult. Going to higher energies to improve resolution via shorter de Broglie wavelengths will make this problem worse due to even weaker backscattering. Thus, for an adsorbate or surface atom that has not significantly penetrated a surface, there will always be a tradeoff between resolution and ease of measurement in photoelectron holography. Of course, if the emitter is found below the surface, then strong forward scattering of the type discussed previously here can take place, and the resulting hologram should then show larger intensity modulations; however, forward scattering effects by themselves contain bond direction information, but not bond length information, so that the weaker modulations due to higher-order features would still need to be accurately measured in order for the inversion of the hologram to yield the full structure.

As a potentially more convenient experimental alternative for holography, a suitable Auger peak involving three filled levels might be useful as a source of a more nearly spherical wave as assumed in the image reconstruction, although the poorly understood mixing in of other l components could complicate a precise theoretical analysis of the effective amplitudes $|F_j|$ and phases ψ_j. Also, using Auger peaks that are too broad in energy would reduce the degree of monochromaticity (i.e., coherence) required in the source.[21b]

No matter how these problems are dealt with, even low-resolution three-dimensional images from such holography could be useful in ruling out certain bonding geometries in a semiquantitative way, much as Fourier transforms in ARPEFS can be useful through the approximate path-length differences they provide. It will be interesting to see what the first inversion of an experimental photoelectron hologram brings. (Please see the added note on holographic methods at the end of this chapter.)

6. COMPARISONS TO OTHER TECHNIQUES AND CONCLUDING REMARKS

We begin this concluding section by comparing photoelectron and Auger electron diffraction to several other current probes of surface structure in order to assess their relative strengths and weaknesses. As a first overall comment, it is clear from any perusal of the current literature (e.g., Ref. 1) that no one surface-structure probe directly and unambiguously provides all of the desired information on atomic identities, relative numbers, chemical states, positions, bond distances and bond directions in the first 3–5 layers of the surface. The very small number of surface structures for which there is a general consensus in spite of several decades of careful study of some of them testifies to the need for using complementary information from several methods.

To provide some idea of this complementarity of approaches, we show in Table 1 several techniques assessed according to a number of characteristics: photoelectron diffraction (PD) in both scanned-angle and scanned-energy forms, Auger electron diffraction (AED), surface extended X-ray absorption fine structure (SEXAFS),[16] near-edge X-ray absorption fine structure NEXAFS,[95,153,154] low-energy electron diffraction (LEED),[61] surface-sensitive grazing incidence X-ray scattering (GIXS),[27] scanning tunneling microscopy (STM),[155] and Rutherford backscattering (RBS) or medium-energy ion scattering (MEIS).[156] This is not intended to be a complete list of modern structure probes, but it roughly represents the group most used at present.

These techniques are rated, first, according to whether they directly provide information on atomic identity (a positive feature of all techniques except for LEED, GIXS, and STM) and chemical state (possible only with PD, AED, and NEXAFS). Atom identification is possible in GIXS only if use is made of anomalous dispersion near a certain absorption edge. State-specific information is not derivable in typical SEXAFS measurements because of the overlap of different oscillatory absorption structures above a given edge.

Also, we assess whether other subsidiary types of structural and bonding information can be obtained in a straightforward manner. Of course, once a structure has been determined and optimized to fit the data of any one of these methods, it has implicit in it bond directions, bond distances, site symmetries, and coordination numbers, but the table entries have been chosen to reflect the directness with which these can be extracted from the raw data with a minimum of data analysis. The types of information considered are valence electronic levels or excitations (directly accessible only in NEXAFS and STM), bond directions (particularly easy to determine in high-energy PD/AED with forward scattering—as discussed in comparison to other techniques in section 4.1.4—and RBS/MEIS with shadowing and blocking), bond distances (very direct in SEXAFS Fourier transforms), local bonding-site symmetries (easiest to determine with PD, AED, SEXAFS, and RBS/MEIS), and coordination numbers (derivable directly from high-energy PD and AED and less directly from the amplitudes of SEXAFS oscillations). STM can also directly image surface atoms and thus provide coordination numbers, but it is limited to looking at only the outermost surface density of states, and so does not probe the bonding below this level in a direct way. Distinguishing between structures that are related to atomic positions and protrusions in the density of states can also be a problem in STM. It has been suggested that NEXAFS resonance energies can be used to measure bond distances,[153] but this approach may be limited to well-calibrated series of homologous molecules, and has been called into question.[154]

The estimated accuracies of finally determining atomic positions with the current state of these techniques is also indicated. Numbers smaller or larger than these will be found for some cases in the literature, but it is the author's opinion that the numbers in the table are a better representation of the true absolute accuracies if all of the various uncertainties in both experimental parameters and the modelling or treatment of the data are taken into account. Surface X-ray diffraction is the most accurate, but its principal sensitivity is to horizontal positions, with vertical positions being derivable only via the more difficult method of measuring rod profiles normal to the surface. PD in any of its forms and AED should be inherently as accurate as SEXAFS, if not more so, particularly if the latter has been analyzed only with transform and back-transform methods without any final theoretical modeling. PD should also ultimately be as accurate as LEED,[73] particularly for a given amount of input to the theoretical analysis.

The degree to which these techniques probe short-range order in the first 10–20 Å around a given site versus longer-range order over 100 Å or more is also considered. Except for LEED and X-ray diffraction, all of the techniques are primarily sensitive to short-range order, although we have also pointed out that PD and AED actually have sensitivity extending over a region of diameter as large as 40 Å. Although inherently larger-scale probes, LEED and X-ray diffraction can with spot profile analysis be used to study the breakdown of long-range order in such phenomena as surface phase transitions.

Next, several characteristics relating to the ease of obtaining data and analyzing it theoretically are indicated: the overall percentage change in intensity as one measure of the ease of determining the signal (which is particularly large for PD, AED, and LEED); the possibility of using a simple, usually kinematical, theory to analyze the results; and the feasibility of using Fourier transform methods to more directly derive structural parameters. The overall figures for percentage effect should be assessed carefully, however, since the inelastic background under some photoelectron and Auger spectra can be high, thus making even a 50% modulation of the peak intensity difficult to measure. By contrast, for some applications of SEXAFS, background effects can be much reduced by using X-ray fluorescence detection,[157] although surface specificity is then lost. Problematic background effects can also arise in SEXAFS scans such as Auger-photoelectron interferences if either type of peak is being used to monitor the absorption and sharp spikes or glitches of intensity due to Bragg reflection of X-rays from very well-ordered crystals such as semiconductors. Auger-photo-electron interferences can also make the use of scanned-energy photoelectron diffraction more difficult if there are any Auger peaks from the sample that lie in the kinetic-energy range from about 100 eV to 400 eV. Standard Auger tabulations show that this could yield difficult background subtraction problems for the atomic number ranges 4–7, 14–22, and 37 upward. As examples of this, sulfur at 16 involves such an interference, as noted previously in sections 4.2.2 and 4.2.3,

TABLE 1. Comparison of Several Surface-Structure Techniques by Different Criteria ("yes" and "no" responses based on deriving a given type of information with minimal data analysis)

	Scanned-angle PD	Scanned-energy PD	AED	SEXAFS	NEXAFS	LEED	GIXS	STM	RBS/MEIS
Atom specific?	Yes	Yes	Yes	Yes	Yes	No	No[a]	No	Yes
Chemical-state specific?	Yes[b]	Yes[b]	Yes[b]?	No	Yes	No	No	No	No
Bond directions?	Yes[c]	Yes	Yes[c]	Yes[c]	Yes[c]	No	No	Yes[d]	Yes[e]
Bond distances?	Yes	Yes[f]	Yes	Yes[f]	No[g]	No	No	Yes[d]	Yes[e]
Adsorption site symmetries?	Yes[h]	Yes?	Yes[h]	Yes?	No	No	No	Yes[d]	Yes
Coordination numbers?	Yes[c]	Yes	Yes[c]	Yes	No	No	No	Yes[d]	Yes
Position accuracies?	~0.02–0.05 Å	~0.02–0.05 Å	≤0.05 Å	~0.02–0.05 Å	?[g]	~0.01–0.05 Å	~0.001 Å (Horizontal) ?[i] (Vertical)	~0.3 Å (Horizontal) ~0.05 Å (Vertical)	~0.01 Å
Valence-electron states?	No	No	No	No	Yes	No	No	Yes	No
Short-range order (~10–20 Å)?	Yes	Yes	Yes	Yes	Yes	No	No	Yes[d]	Yes

Long-range order (>100 Å)?	No	No	No	No	No	Yes	Yes	Yes[d]	No
Overall % effects?	20–70%	20–70%	~5%	20–70%	Variable	70–90% in I–V	Weaker surface peaks	Large	~10%[e]
Kinematical theory?	Yes[j]	Yes[j]	Yes[j]	Yes[j]	No	No	Yes	No	Yes
Fourier transform analysis?	No[k]	Yes[f]	Yes[f]	No	No	No	Yes	Yes[l]	No
Requires synchrotron radiation?	No	Yes	Yes	No	Yes	No	Yes	No	No

[a] GIXS only atom-specific if anomolous dispersion used

[b] If core shifts/fine structure can be resolved in PD/AED, but more difficult in Auger spectra

[c] Forward scattering in high-energy PD and AED yields ~3 times better resolution tor bond directions than in polarization-dependent NEXAFS and SEXAFS. Also directly gives coordination numbers for neighbors between emitter and detector.

[d] STM senses only surface density of state contours, so that structural parameters are for outermost atoms only.

[e] Via shadowing and blocking in RBS/MEIS.

[f] Fourier transforms to yield path-length differences in scanned-energy PD may be complicated by overlap of close-lying differences; SEXAFS Fourier transforms are simpler in this respect.

[g] Only via correlation of bond length with NEXAFS resonance energies that may be limited in application (see Refs. 153 and 154).

[h] May be very direct from azimuthal PD or AED data.

[i] Vertical position information in GIXS less accurate and available only by measuring vertical rod profiles.

[j] Some multiple scattering effects may have to be considered in all of PD, AED, and SEXAFS (see Refs. 57 and 85), and dynamical effects also may have to be included in the analysis of GIXS data for some scattering geometries (see Ref. 27).

[k] Only if holographic imaging is possible.

[l] Fourier transforms of STM images are useful for detecting lateral symmetries present.

and the Ag/Si system considered in section 4.4.2 was found in a recent scanned-energy experiment[50(c)] to exhibit extensive interferences over the full 90–350 eV range due to the various peaks in both the Si *KLL* and the Ag *MNN* spectra.

As a last and important criterion for the present volume, we indicate whether a given technique requires synchrotron radiation, as about half of them do.

The ideal structural probe would have "yes" for all of the nonquantitative characteristics in this table except the last one, which for reasons of broadest utility would be "no." It is clear that each method has positive features, but none constitutes this ideal probe. Thus, complementary information from several methods is in general desirable for fully resolving any structure. PD and AED are positive on sufficient points to be attractive additions to this list. AED is easier to excite (e.g., with photons or electrons), but the more complex nature of Auger spectra will prevent doing state-specific diffraction measurements for many cases, and an accurate theory, especially for lower energies, will be more difficult. Not being able to use radiation polarization to selectively excite towards a given scatterer is also a disadvantage of AED. As one disadvantage of scanned-angle PD and AED, we note the present lack of being able to use Fourier transform methods to determine structure directly (although photoelectron holography is a proposal to do this); thus it may be necessary to carry out a number of calculations for various structures, a procedure analogous to that used in LEED. However, some aspects of the data (e.g., forward scattering peaks at high energy) provide structural information very directly, and a good deal of any analysis should be possible within the framework of a simple single scattering picture. And in any case, the final test of any structural model derived in PD, AED, or SEXAFS should be to compare experiment to a diffraction calculation on a cluster of atoms of sufficient size to adequately include all significant scatterers.

Thus, although photoelectron diffraction and its close relative Auger electron diffraction are relatively new additions to the array of tools for studying surface structures, they have already proven to be useful for a broad variety of systems. Even at the present stage of development of both techniques with, for example, standard X-ray tubes or electron guns as excitation sources, and theory at the single-scattering-cluster–spherical-wave level, structurally useful and unique information can be derived for a range of problems including adsorption, molecular orientation, oxidation, epitaxial growth, metal–semiconductor interface formation, cluster growth, surface phase transitions, and short-range magnetic order. The use of higher angular resolutions promises to provide more precise structural information, particularly concerning longer-range order. The wider availability of synchrotron radiation, especially from the next generation of high-brightness insertion devices, will enormously increase the speed of both scanned-angle and scanned-energy measurements, thus permitting more studies of surface dynamics. The accurate-intensity ratio measurements at low kinetic energy required in spin-polarized photoelectron diffraction will also become easier. Some degree of lateral-resolution photoelectron microscopy-plus-diffraction should also become possible. And with focused electron beams, Auger electron microscopy-plus-diffraction is also feasible. Also, high-brightness radiation sources should permit increased energy resolutions of the order of 0.3 eV even at the higher photon energies of 1.0–2.0 keV that are optimum for taking advantage of forward scattering and a single-scattering approach. Separate diffraction patterns will be obtainable for the various peaks in a given spectral region that are produced by chemical shifts, multiplet splittings, or more complex final-state effects. Using both linearly and circularly polarized radiation will also permit the selection of specific scatterers and spin-polarized final states, respectively. State-specific structural parameters should thus be derivable in a way that is not possible with other methods.

NOTE ON HOLOGRAPHIC METHODS

Since the original writing of this review, the use of holographically motivated Fourier-transform inversion methods for deriving surface structural information from both photoelectron- and Auger electron-diffraction data (cf. discussion in section 5.4) has advanced considerably. Some of these developments are discussed below.

The first experimental data have successfully been inverted to yield direct images of atomic positions near Cu surfaces by Tonner et al.[158] More recently, the same types of images have been observed for the semiconductors Si and Ge by Herman et al.[159] and for the simple adsorbate system $c(2 \times 2)$ S/Ni (001) by Saiki et al.[100] In general, these images are accurate to within about ±0.2–0.3 Å in planes parallel to the surface and more or less perpendicular to strong forward scattering directions, but only to within about ±0.5–1.0 Å in planes perpendicular to the surface or containing forward scattering directions.

Methods have been proposed for eliminating the observed distortions in atomic images due to both the anisotropic nature of the electron–atom scattering and the phase shift associated with the scattering by Saldin et al.,[160] Tong et al.,[161] and Thevuthasan et al.[162] Preliminary tests of these methods are encouraging, but more applications to experimental data are needed to assess them fully. Further image distortions due to anisotropies in the electron emission process have been discussed,[160,161] and corrections for these also appear to be useful. Additional spurious features that may arise in images due to the strength of the electron–atom scattering and resultant self-interference effects have been pointed out by Thevuthasan et al.[162] By contrast, the multiple scattering defocusing illustrated in Figs. 3b–ii and 6 has been shown to reduce the image distortions for the special case of buried emitters that are separated by several atoms from the detector.[162]

Finally, Barton[163] has shown in theoretical simulations that the simultaneous analysis of photoelectron holographic data obtained at several different photon energies, involving in effect an additional Fourier sum on energy, should act to reduce the influence of both twin images and multiple scattering on atomic images.

Thus, the holographic analysis of both photoelectron and Auger electron data is in an intense period of evaluation, with several indications already that it may ultimately provide reasonably good starting-point structures which can then be refined by the more classic trial-and-error methods discussed previously in this review, but in much reduced time.

ACKNOWLEDGMENTS

Some of the work reported here has been supported by the Office of Naval Research under Contract N00014-87-K-0512, the National Science Foundation under Grant CHE83-20200, and the New Energy Development Organization (NEDO) of Japan. Some of the studies discussed here have also benefitted by grants of Cray XMP/48 time at the San Diego Supercomputer Center. The author is also very grateful to his various former and present co-workers at the University of Hawaii for their significant contributions to the development of these techniques. J. J. Rehr has also provided various ideas concerning efficient methods of treating both spherical-wave effects and multiple scattering. Special thanks also go to D. J. Friedman for critically reading this chapter, and to S. A. Chambers and G. Schoenhense for providing helpful comments concerning it. The fine work of M. Prins on many of the figures is also gratefully acknowledged. I also thank J. Hanatani for assistance with preparing this manuscript.

REFERENCES

1. Excellent summaries of the current status of most of these techniques can be found in the proceedings of three International Conferences on the Structure of Surfaces: (a) *The Structure of Surfaces* (M. A. Van Hove and S. Y. Tong, eds.), Springer Verlag, Berlin (1985); (b) *The Structure of Surfaces II* (J. F. van der Veen and M. A. van Hove, eds.), Springer-Verlag, Berlin (1988); (c) *The Structure of Surfaces III* (S. Y. Tong, M. A. Van Hove, K. Takayanagi, and X. D. Xie, eds.), Springer-Verlag, Berlin (1991).
2. K. Siegbahn, U. Gelius, H. Siegbahn, and E. Olsen, *Phys. Lett.* **32A,** 221 (1970).
3. C. S. Fadley and S. A. L. Bergstrom, *Phys. Lett.* **35A,** 375 (1971).
4. A. Liebsch, *Phys. Rev. Lett.* **32,** 1203 (1974); *Phys. Rev.* **B13,** 544 (1976).
5. S. Kono, C. S. Fadley, N. F. T. Hall, and Z. Hussain, *Phys. Rev. Lett.* **41,** 117 (1978); S. Kono, S. M. Goldberg, N. F. T. Hall, and C. S. Fadley, *Phys. Rev. Lett.* **41,** 1831 (1978).
6. D. P. Woodruff, D. Norman, B. W. Holland, N. V. Smith, H. H. Farrell, and M. M. Traum, *Phys. Rev. Lett.* **41,** 1130 (1978).
7. S. D. Kevan, D. H. Rosenblatt, D. Denley, B.-C. Lu, and D. A. Shirley, *Phys. Rev. Lett.* **41,** 1565 (1978).
8. J. J. Barton, C. C. Bahr, Z. Hussain, S. W. Robey, J. G. Tobin, L. E. Klebanoff, and D. A. Shirley, *Phys. Rev. Lett.* **51,** 272 (1983); J. J. Barton, C. C. Bahr, Z. Hussain, S. W. Robey, L. E. Klebanoff, and D. A. Shirley, *J. Vac. Sci. Technol.* **A2,** 847 (1984); J. J. Barton, S. W. Robey, C. C. Bahr, and D. A. Shirley in *The Structure of Surfaces* (M. A. van Hove and S. Y. Tong, eds.), Springer Verlag, Berlin (1985) p. 191.
9. C. S. Fadley, *Prog. in Surf. Sci.* **16,** 275 (1984).
10. C. S. Fadley, Phys. Scr. **T17,** 39 (1987) and earlier references therein.
11. (a) W. F. Egelhoff, *Phys. Rev.* **B30,** 1052 (1984); (b) R. A. Armstrong and W. F. Egelhoff, *Surf. Sci.* **154,** L225 (1985); (c) W. F. Egelhoff, *Phys. Rev. Lett.* **59,** 559 (1987).
12. (a) S. A. Chambers, H. W. Chen, I. M. Vitomirov, S. B. Anderson, and J. H. Weaver, *Phys. Rev.* **B33,** 8810 (1986); (b) S. A. Chambers, I. M. Vitomirov, S. B. Anderson, H. W. Chen, T. J. Wagener, and J. H. Weaver, *Superlattices and Microstructures* 3, 563 (1987); (c) S. A. Chambers and T. J. Irwin, *Phys. Rev.* **B38,** 7858 (1988); and earlier references therein.
13. C. S. Fadley, in *Core-Level Spectroscopy in Condensed Systems* (J. Kanamori and A. Kotani, eds.), Springer Verlag, Berlin (1988), p. 236.
14. Z. Hussain, D. A. Shirley, C. H. Li, and S. Y. Tong, *Proc. Natl. Acad. Sci. USA.* **78,** 5293 (1981). The use of Fourier transforms was here first thought to yield interlayer distances but later shown to provide path-length difference information, as discussed in Refs. 15 and 21.

15. P. J. Orders and C. S. Fadley, *Phys. Rev.* **B27**, 781 (1983).

16. (a) P. Citrin, in *Springer Series on Surface Science*, Vol. II. Springer Verlag, Berlin (1985), p. 49; (b) D. Norman, *J. Phys.* **C19**, 3273 (1986); (c) J. Stohr, in *X-Ray Absorption: Principles, Applications, Techniques of EXAFS, SEXAFS, and XANES* (R. Prins and D. Konigsberger, eds.), Wiley, New York (1987).

17. N. V. Smith, H. H. Farrell, M. M. Traum, D. P. Woodruff, D. Norman, M. S. Woolfson, and B. W. Holland, *Phys. Rev.* **B21**, 3119 (1980); H. H. Farrell, M. M. Traum, N. V. Smith, W. A. Royer, D. P. Woodruff, and F. D. Johnson, *Surf. Sci.* **102**, 527 (1981).

18. (a) D. Sebilleau, M. C. Desjonqueres, D. Chauveau, C. Guillot, J. Lecante, G. Treglia, and D. Spanjaard, *Surf. Sci. Lett.* **185**, L527 (1987); (b) M. C. Desjonqueres, D. Sebilleau, G. Treglia, D. Spanjaard, C. Guillot, D. Chauveau, and J. Lecante, *Scanning Electron Microsc.* **1** 1557 (1987); (c) D. Sebilleau, G. Treglia, M. C. Desjonqueres, D. Spanjaard, C. Guillot, D. Chauveau, and J. Lecante, *J. Phys. (Paris)* **49**, 227 (1988).

19. (a) B. Sinkovic, P. J. Orders, C. S. Fadley, R. Trehan, Z. Hussain, and J. Lecante, *Phys. Rev.* **B30**, 1833 (1984); (b) P. J. Orders, B. Sinkovic, C. S. Fadley, Z. Hussain, and J. Lecante, *Phys. Rev.* **B30**, 1838 (1984).

20. L. J. Terminello, X. S. Zhang, Z. Q. Huang, S. Kim, A. E. Schach von Wittenau, K. T. Leung, and D. A. Shirley, *Phys. Rev.* **B38**, 3879 (1988).

21. (a) M. Sagurton, E. L. Bullock, and C. S. Fadley, *Phys. Rev.* **B30**, 7332 (1984); and (b) *Surf. Sci.* **182**, 287 (1987).

22. D. P. Woodruff, *Surf. Sci.* **166**, 377 (1986).

23. L. G. Petersson, S. Kono, N. F. T. Hall, C. S. Fadley, and J. B. Pendry, *Phys. Rev. Lett.* **42**, 1545 (1979).

24. H. C. Poon and S. Y. Tong, *Phys. Rev.* **B30**, 6211 (1984); S. Y. Tong, H. C. Poon, and D. R. Snider, *Phys. Rev.* **B32**, 2096 (1985).

25. J. J. Barton and D. A. Shirley, (a) *Phys. Rev.* **B32**, 1892 (1985) and (b) *Phys. Rev.* **B32**, 1906 (1985); (c) J. J. Barton, Ph.D. Thesis, U. Cal., Berkeley (1985); J. J. Barton, S. W. Robey, and D. A. Shirley, *Phys. Rev.* **B34**, 778 (1986).

26. (a) R. S. Saiki, A. P. Kaduwela, J. Osterwalder, M. Sagurton, C. S. Fadley, and C. R. Brundle, *J. Vac. Sci. Technol* **A5**, 932 (1987); (b) R. S. Saiki, A. P. Kaduwela, J. Osterwalder, C. S. Fadley, and C. R. Brundle, *Phys. Rev.* **B40**, 1586 (1989); (c) R. S. Saiki, A. P. Kaduwela, J. Osterwalder, C. S. Fadley, and C. R. Brundle, to be published; (d) The R factors used in this study are generally the R1 defined in M. A. van Hove, S. Y. Tong, and M. H. Elconin, *Surf. Sci.* **64**, 85 (1977), although final checks have also been made with the other possibilities R2–R5 discussed there.

27. Chapter 8 in this volume by P. H. Fuoss, K. S. Liang, and P. Eisenberger.

28. Chapter 3 of Vol. 2 of this set by J. Haase and A. M. Bradshaw.

29. H. Li and B. P. Tonner, *Phys Rev.* **B37**, 3959 (1988).

30. S. A. Chambers, H. W. Chen, S. B. Anderson, and J. H. Weaver, *Phys. Rev.* **B34**, 3055 (1986).

31. S. D. Kevan, *Rev. Sci. Instrum.* **54**, 1441 (1983).

32. S. D. Kevan, Ph.D. thesis, U. Cal., Berkeley (1980).

33. R. C. White, C. S. Fadley, and R. Trehan, *J. Electron Spectros. Relat. Phenom.* **41**, 95 (1986).

34. J. Osterwalder, M. Sagurton, P. J. Orders, C. S. Fadley, B. D. Hermsmeier, and D. J. Friedman, *J. Electron. Spectros. Relat. Phenom.* **48**, 55 (1989).

35. D. E. Eastman, J. J. Donelon, N. C. Hien, and F. J. Himpsel, *Nucl. Instrum Methods* **172**, 327 (1980).

36. R. C. G. Leckey and J. D. Riley, *Appl. Surf. Sci.* **22/23**, 196 (1985).

37. H. A. Engelhardt, W. Back, and D. Menzel, *Rev. Sci. Instrum.* **52**, 835 (1981); H. A. Engelhardt, A. Zartner, and D. Menzel, *Rev. Sci. Instrum.* **52**, 1161 (1981).

38. M. G. White, R. A. Rosenberg, G. Gabor, E. D. Poliakoff, G. Thornton, S. H. Southworth, and D. A. Shirley, *Rev. Sci. Instrum.* **50**, 1268 (1979); Z. Hussain and D. A. Shirley, private communication.

39. R. C. G. Leckey, *J. Electron Spectros. Relat. Phenom.* **43**, 183 (1987).

40. L. McDonnell, D. P. Woodruff, and B. W. Holland, *Surf. Sci.* **51**, 249 (1975).

41. P. A. Lee, *Phys. Rev.* **B13**, 5261 (1976).

42. S. Kono, S. M. Goldberg, N. F. T. Hall, and C. S. Fadley, *Phys. Rev. Lett.* **41**, 1831 (1978); *Phys. Rev.* **B22**, 6085 (1980).

43. T. Fujikawa, *J. Phys. Soc. Jpn.* **50,** 1321 (1981); **51,** 251 (1982); **54,** 2747 (1985); *J. Electron. Spectros. Relat. Phenom.* **26,** 79 (1982).

44. H. Daimon, H. Ito, S. Shin, and Y. Murata, *J. Phys. Soc. Jpn.* **54,** 3488 (1984).

45. J. J. Rehr, R. Albers, C. Natoli, and E. A. Stern, *Phys. Rev.* **B34,** 4350 (1986); J. J. Rehr, J. Mustre de Leon, C. R. Natoli, and C. S. Fadley, *J. Phys. (Paris)* **C8,** Suppl. 12, 213 (1986); J. J. Rehr, J. Mustre de Leon, C. R. Natoli, C. S. Fadley, and J. Osterwalder, *Phys. Rev.* **B39,** 5632 (1989).

46. D. E. Parry, *J. Electron. Spectros. Relat. Phenom.* **49,** 23 (1989).

47. D. J. Friedman and C. S. Fadley, *J. Electron Spectrosc.* **51,** 689 (1990).

48. (a) J. Osterwalder, E. A. Stewart, D. Cyr, C. S. Fadley, J. Mustre de Leon, and J. J. Rehr, *Phys. Rev.* **B35,** 9859 (1987); (b) J. Osterwalder, A. Stuck, D. J. Friedman, A. P. Kaduwela, C. S. Fadley, J. Mustre de Leon, and J. J. Rehr, *Phys. Scr.* **41,** 990 (1990); (c) J. Osterwalder, E. A. Stewart, D. J. Friedman, A. P. Kaduwela, C. S. Fadley, J. Mustre de Leon, and J. J. Rehr, unpublished results.

49. S. Kono, K. Higashiyama, and T. Sagawa, *Surf. Sci.* **165,** 21 (1986) and references to prior structure studies therein.

50. (a) E. L. Bullock, G. S. Herman, M. Yamada, D. J. Friedman, and C. S. Fadley, *Phys. Rev.* **B, 41,** 1703 (1990); (b) E. L. Bullock, Ph.D. thesis, U. Hawaii (1988); (c) A. P. Kaduwela, D. J. Friedman, M. Yamada, E. L. Bullock, C. S. Fadley, Th. Lindner, D. Ricker, A. W. Robinson, and A. M. Bradshaw in *Structure of Surfaces III* (S. Y. Tong, M. A. Van Hove, K. Takayanagi, and X. D. Xie, eds.), Springer-Verlag, Berlin (1991), p. 600; and to be published.

51. M. Owari, M. Kudo, Y. Nihei, and H. Kamada, *J. Electron. Spectros. Relat. Phenom.* **21,** 131 (1981).

52. D. H. Rosenblatt, S. D. Kevan, J. G. Tobin, R. F. Davis, M. G. Mason, D. A. Shirley, J. C. Tang, and S. Y. Tong, *Phys. Rev.* **B26,** 3181 (1982).

53. G. Treglia, in *Core-Level Spectroscopy in Condensed Systems* (J. Kanamori and A. Kotani, eds.), Springer Verlag, Berlin (1988), p. 281.

54. J. J. Rehr and E. A. Albers, *Phys. Rev.* **B41,** 8139 (1990).

55. S. T. Manson, *Adv. Electron. Electron. Phys.* **41,** 73 (1976).

56. M. Sagurton, private communication.

57. P. A. Lee and J. B. Pendry, *Phys. Rev.* **B11,** 2795 (1975).

58. M. Sagurton, E. L. Bullock, R. Saiki, A. P. Kaduwela, C. R. Brundle, C. S. Fadley, and J. J. Rehr, *Phys. Rev.* **B33,** 2207 (1986).

59. H. C. Poon, D. Snider, and S. Y. Tong, *Phys. Rev.* **B33,** 2198 (1986).

60. M. Fink and A. C. Yates, *At. Data Nucl. Data Tables* **1,** 385 (1970); M. Fink and J. Ingram, *At. Data Nucl. Data Tables* **4,** 129 (1972); D. Gregory and M. Fink, *At. Data Nucl. Data Tables* **14,** 39 (1974).

61. J. B. Pendry, *Low Energy Electron Diffraction,* Academic Press, London (1974); M. A. van Hove and S. Y. Tong, *Surface Crystallography by LEED,* Springer-Verlag, New York (1979).

62. E. A. Stern, B. A. Bunker, and S. M. Heald, *Phys. Rev.* **B21,** 5521 (1980).

63. (a) R. Trehan and C. S. Fadley, *Phys. Rev.* **B34,** 6784 (1986); (b) R. Trehan, J. Osterwalder, and C. S. Fadley, *J. Electron. Spectros. Relat. Phenom.* **42,** 187 (1987).

64. M. P. Seah and W. A. Dench, *SIA, Surf. Interface Anal.* **1,** 2 (1979); C. J. Powell, *Scanning Electron Micros.* **4,** 1649 (1984); S. Tanuma, C. J. Powell, and D. R. Penn, *SIA Surf. Interface Anal.* **11,** 577 (1988).

65. P. M. Cadman and G. M. Gossedge, *J. Electron. Spectros. Relat. Phenom.* **18,** 161 (1980); C. M. Schneider, J. J. de Miguel, P. Bressler, J. Garbe, S. Ferrer, R. Miranda, and J. Kirscher, *J. Phys. (Paris) C,* Suppl. 12, **49,** 1657 (1988).

66. O. A. Baschenko and V. I. Nefedov, *J. Electron. Spectros. Relat. Phenom.* **21,** 153 (1980) and **27,** 109 (1982) and earlier references therein.

67. Y. U. Idzerda, D. M. Lind, and G. A. Prinz, *J. Vac. Sci. Technol.* **A7,** 1341 (1989).

68. N. E. Erickson and C. J. Powell, *Phys. Rev.* **B40,** 7284 (1989).

69. P. J. Orders, S. Kono, C. S. Fadley, R. Trehan, and J. T. Lloyd, *Surf. Sci.* **119,** 371 (1981).

70. (a) H. Helferink, E. Lang, and K. Heinz, *Surf. Sci.* **93,** 398 (1980); (b) P. J. Orders, R. E. Connelly, N. F. T. Hall, and C. S. Fadley, *Phys. Rev.* **B24,** 6161 (1981).

71. E. L. Bullock and C. S. Fadley, *Phys. Rev.* **B31,** 1212 (1985).

72. R. G. Weissman and K. Muller, *Surf. Sci. Rep.* **1**, 251 (1981).

73. M. L. Xu and M. A. van Hove, *Surf. Sci.* **207**, 215 (1989).

74. G. Breit and H. Bethe, *Phys. Rev.* **94**, 888 (1954).

75. L.-Q. Wang, A. E. Schach von Wittenau, Z. G. Ji, L. S. Wang, Z. Q. Huang, and D. A. Shirley, *Phys. Rev.* **B44**, 1292 (1991); and unpublished results.

76. D. A. Wesner, F. P. Coenen, and H. P. Bonzel, *Surf. Sci.* **199**, L419 (1988).

77. S. Y. Tong and H. C. Poon, *Phys. Rev.* **B37**, 2884 (1988).

78. S. Tougaard, *SIA, Surf. Interface Anal.* **11**, 453 (1988) and earlier references therein.

79. (a) J. B. Pendry and B. S. Ing, *J. Phys.* **C8**, 1087 (1975); (b) B. K. Teo and P. Lee, *J. Am. Chem. Soc.* **1979**, 101 (1979).

80. (a) G. Treglia, M. C. Desjonqueres, D. Spanjaard, D. Sebilleau, C. Guillot, D. Chauveau, and J. Lecante, *J. Phys. (Paris)* **1**, 1879 (1989); (b) D. G. Frank, N. Batina, T. Golden, F. Lu, and A. T. Hubbard, *Science* **247**, 182 (1990) and references therein.

81. L. I. Schiff, *Quantum Mechanics*, 3rd ed., McGraw-Hill, New York (1968), p. 136.

82. J. J. Barton, M.-L. Xu, and M. A. van Hove, *Phys. Rev.* **B37**, 10475 (1988).

83. (a) M.-L. Xu, J. J. Barton, and M. A. van Hove, *J. Vac. Sci. Technol* **A6**, 2093 (1988) and (b) *Phys. Rev.* **B39**, 8275 (1989)

84. A. P. Kaduwela, G. S. Herman, D. J. Friedman, C. S. Fadley, and J. J. Rehr, *Phys. Scr.* **41**, 948 (1990); A. P. Kaduwela, G. S. Herman, D. J. Friedman, and C. S. Fadley, *J. Electron Spectrosc. Relat. Phenom.* **57**, 223 (1991).

85. (a) R. C. Albers and J. J. Rehr, to be published; (b) K. Baberschke, in *The Structure of Surfaces II* (J. F. van der Veen and M. A. van Hove, eds.), Springer-Verlag, Berlin (1988), p. 174; (c) P. Rennert and N. V. Hung, *Phys. Status Solid* **148**, 49 (1988).

86. R. S. Saiki, G. S. Herman, M. Yamada, J. Osterwalder, and C. S. Fadley, *Phys. Rev. Lett.* **63**, 283 (1989).

87. (a) D. W. Moon, S. L. Bernasek, D. J. Dwyer, and J. L. Gland, *J. Am. Chem. Soc.* **107**, 4363 (1985); (b) C. Benndorf, B. Kruger, and F. Thieme, *Surf. Sci.* **163**, L675 (1985); (c) D. W. Moon, S. Cameron, F. Zaera, W. Eberhardt, R. Carr, S. L. Bernasek, J. L. Gland, and D. J. Dwyer, *Surf. Sci.* **180**, L123 (1987).

88. (a) D. A. Wesner, F. P. Coenen, and H. P. Bonzel, *Phys. Rev. Lett.* **60**, 1045 (1988); (b) *Phys. Rev.* **B39**, 10770 (1989).

89. H. Kuhlenbeck, M. Neumann, and H.-J. Freund, *Surf. Sci.* **173**, 194 (1986).

90. K. C. Prince, E. Holub-Krappe, K. Horn, and D. P. Woodruff, *Phys. Rev.* **B32**, 4249 (1985): E. Holub-Krabbe, K. C. Prince, K. Horn, and D. P. Woodruff, *Surf. Sci.* **173**, 176 (1986).

91. D. R. Wesner, F. P. Coenen, and H. P. Bonzel, *Phys. Rev.* **B33**, 8837 (1986).

92. K. A. Thompson and C. S. Fadley, *Surf. Sci.* **146**, 281 (1984).

93. H. Ibach and D. L. Mills, *Electron Energy Loss Spectroscopy*, Academic Press, New York (1982).

94. T. E. Madey, David E. Ramaker, and R. Stockbauer, *Ann. Rev. Phys. Chem.* **35**, 215 (1984).

95. J. Stohr and R. Jaeger, *Phys. Rev.* **B26**, 4111 (1982); A. L. Johnson, E. L. Muetterties, J. Stohr, and F. Sette, *J. Phys. Chem.* **89**, 4071 (1985); J. Stohr and D. A. Outka, *Phys. Rev.* **B36**, 7891 (1987).

96. See discussion and references in I. P. Batra and J. A. Barker, *Phys. Rev.* **B29**, 5286 (1984) and R. L. Strong and J. L. Erskine, *Phys. Rev.* **B31**, 6305 (1985).

97. An excellent review of the O/Ni (001) system is by C. R. Brundle and J. Q. Broughton in *The Chemical Physics of Solid Surfaces and Heterogeneous Catalysis* (D. A. King and D. P. Woodruff, eds.), Elsevier, Amsterdam (1991), Vol. 3a.

98. J. E. Demuth, N. J. DiNardo, and C. S. Cargill, *Phys. Rev. Lett.* **50**, 1373 (1983).

99. J. J. Barton, C. C. Bahr, Z. Hussain, S. W. Robey, J. G. Tobin, L. E. Klebanoff, and D. A. Shirley, *Phys. Rev. Lett.* **51**, 272 (1983) and earlier references therein.

100. R. S. Saiki, A. P. Kaduwela, Y. J. Kim, D. J. Friedman, J. Osterwalder, S. Thevuthasan, E. Tober, R. Ynzunza and C. S. Fadley, to be published.

101. M. Sagurton, B. Sinkovic, and C. S. Fadley, unpublished results.

102. X. S. Zhang, L. J. Terminello, S. Kim, Z. Q. Huang, A. E. Schach von Wittenau, and D. A. Shirley, *J. Chem. Phys.* **89**, 6583 (1988).

103. W. F. Egelhoff, *Phys. Rev. Lett.* **59**, 559 (1987).

104. (a) S. A. Chambers, T. J. Wagener, and J. H. Weaver, *Phys. Rev.* **B36,** 8992 (1987); (b) D. A. Steigerwald and W. F. Egelhoff, *Phys. Rev. Lett.* **60,** 2558 (1988).

105. G. Herman, J. Osterwalder, and C. S. Fadley, unpublished results.

106. J. J. Krebs, B. T. Jonker, and G. A. Prinz, *J. Appl. Phys.* **61,** 2596 (1987).

107. G. S. Herman, T. Lindner, R. S. Saiki, and C. S. Fadley, unpublished results.

108. G. Granozzi, A. Rizzi, G. S. Herman, D. J. Friedman, C. S. Fadley, J. Osterwalder, and S. Bernardi, *Phys. Scr.,* **41,** 913 (1990); G. S. Herman, D. J. Friedman, C. S. Fadley, G. Granozzi, G. A. Rizzi, J. Osterwalder, and S. Bernardi, *J. Vac. Sci. Technol.* **B9,** 1870 (1991).

109. R. E. Connelly, C. S. Fadley, and P. J. Orders, *J. Vac. Sci. Technol.* **A2,** 1333 (1984).

110. P. Alnot, J. Olivier, F. Wyczisk, and C. S. Fadley, *J. Electron. Spectros. Relat. Phenom.* **43,** 263 (1987); P. Alnot, J. Olivier, and C. S. Fadley, *J. Electron. Spectros. Relat. Phenom.* **49,** 159 (1989).

111. G. M. Rothberg, K. M. Choudhary, M. L. denBoer, G. P. Williams, M. H. Hecht, and I. Lindau, *Phys. Rev. Lett.* **53,** 1183 (1984).

112. K. M. Choudhary, P. S. Mangat, A. E. Miller, D. Kilday, A. Filipponi, and G. Margaritondo, *Phys. Rev.* **B38,** 1566 (1988).

113. S. Kono in *Core-Level Spectroscopy in Condensed Systems* (J. Kanamori and A. Kotani, eds.), Springer-Verlag, Berlin (1988), p. 253.

114. T. Abukawa and S. Kono, *Phys. Rev.* **B37,** 9097 (1988).

115. (a) S. Ciraci and I. P. Batra, *Phys. Rev. Lett.* **56,** 877 (1986); (b) E. M. Oellig and R. Miranda, *Surf. Sci.* **177,** L947 (1986); (c) T. Kendelewicz, P. Soukassian, R. S. List, J. C. Woicik, P. Pianetta, I. Lindau, and W. E. Spicer, *Phys. Rev.* **B37,** 7115 (1988); (d) R. Ramirez, *Phys. Rev.* **B40,** 3962 (1989).

116. E. J. van Loenen, J. E. Demuth, R. M. Tromp, and R. J. Hamers, *Phys. Rev. Lett.* **58,** 373 (1987); R. J. Wilson and S. Chiang, *Phys. Rev. Lett.* **59,** 2329 (1987) and *J. Vac. Sci. Technol* **A6,** 800 (1988); T. L. Porter, C. S. Chang, and I. S. T. Tsong, *Phys. Rev. Lett.* **60,** 1739 (1988); and earlier references therein.

117. W. C. Fan, A. Ignatiev, H. Huang, and S. Y. Tong, *Phys. Rev. Lett.* **62,** 1516 (1989).

118. S. Y. Tong, H. Huang, C. M. Wei, W. E. Packard, F. K. Men, G. Glander, and M. B. Webb, *J. Vac. Sci. Technol* **A6,** 615 (1988).

119. E. L. Bullock and C. S. Fadley, unpublished results.

120. K. Tamura, U. Bardi, M. Owari, and Y. Nihei, in *The Structure of Surfaces II* (J. F. van der Veen and M. A. van Hove, eds.), Springer-Verlag, Berlin (1988), p. 404.

121. D. Spanjaard, C. Guillot, M. C. Desjonqueres, G. Treglia, and J. Lecante, *Surf. Sci. Rep.* **5,** 1 (1985).

122. Y. Jugnet, N. S. Prakash, L. Porte, T. M. Duc, T. T. A. Nyugen, R. Cinti, H. C. Poon, and G. Grenet, *Phys. Rev.* **B37,** 8066 (1988).

123. (a) T. Miller, T. C. Hsieh, and T.-C. Chiang, *Phys. Rev.* **B33,** 6983 (1986); (b) J. Aarts, A.-J. Hoeven, and P. K. Larsen, *Phys. Rev.* **B38,** 3925 (1988); (c) F. J. Himpsel, F. R. McFeely, A. Taleb-Ibrahimim, J. A. Yarmoff, and G. Hollinger, *Phys. Rev.* **B38,** 6084 (1988).

124. E. Tosatti, in *The Structure of Surfaces II* (J. F. van der Veen and M. A. van Hove, eds.), Springer-Verlag, Berlin (1988), p. 535.

125. U. Breuer, O. Knauff, and H. P. Bonzel, *J. Vac. Sci. Tech.,* **A8,** 2489 (1990).

126. T. Tran, D. J. Friedman, Y. J. Kim, G. A. Rizzi, and C. S. Fadley, *Structure of Surfaces III* (S. Y. Tong, M. A. Van Hove, K. Takayanagi, and X. D. Xie, eds.) Springer-Verlag (1991), p. 522; T. Tran, S. Thevuthasan, Y. J. Kim, G. S. Herman, D. J. Friedman and C. S. Fadley, to be published.

127. (a) E. G. McRae and R. A. Malic, *Phys. Rev. Lett.* **58,** 1437 (1987) and *Phys. Rev.* **B38,** 13183 (1988); (b) E. G. McRae, J. M. Landwehr, J. E. McRae, G. H. Gilmer, and M. H. Grabow, *Phys. Rev.* **B38,** 13178 (1988).

128. J. W. M. Frenken, J. P. Toennies, Ch. Woell, B. Pluis, A. W. Denier van der Gon, and J. F. van der Veen, in *The Structure of Surfaces II* (J. F. van der Veen and M. A. van Hove, eds.), Springer-Verlag, Berlin (1988), p. 547.

129. R. J. Baird, C. S. Fadley, and L. F. Wagner, *Phys. Rev.* **B15,** 666 (1977).

130. S. M. Goldberg, R. J. Baird, S. Kono, N. F. T. Hall, and C. S. Fadley, *J. Electron. Spectros. Relat. Phenom.* **21,** 1 (1980).

131. W. Schaich, *Phys. Rev.* **B8**, 4078 (1973); E. A. Stern, *Phys. Rev.* **B10**, 3027 (1974).

132. See, for example, discussion of multiplets by C. S. Fadley in *Electron Spectroscopy: Theory, Techniques, and Applications* (C. R. Brundle and A. D. Baker, eds.), Vol. 2, Academic Press, London (1978), Ch. 1.

133. G. M. Rothberg, *J. Magn. Magn. Mater.* **15-18**, 323 (1980).

134. (a) B. Sinkovic and C. S. Fadley, *Phys. Rev.* **B31**, 4665 (1985); (b) B. Sinkovic, Ph.D. thesis, U. Hawaii (1988); (c) B. Sinkovic, D. J. Friedman, and C. S. Fadley, *J. Magn. Magn. Mater.* **92**, 301 (1991).

135. B. Hermsmeier, B. Sinkovic, J. Osterwalder, and C. S. Fadley, *J. Vac. Sci. Technol* **A5**, 1082 (1987).

136. C. S. Fadley in (*Magnetic Properties of Low-Dimensional Systems II: New Developments* (L. R. Falicov, F. Meija-Lira, and J.-L. Moran-Lopez, eds.), Springer-Verlag, Berlin (1990), p. 36; See especially calculations by J. M. Sanchez and J.-L. Moran-Lopez for a frustrated antiferromagnetic system discussed therein.

137. B. W. Veal and A. P. Paulikas, *Phys. Rev. Lett.* **51**, 1995 (1983).

138. B. Hermsmeier, C. S. Fadley, B. Sinkovic, M. O. Krause, J. Jiminez-Mier, P. Gerard, and S. T. Manson, *Phys. Rev. Lett.* **61**, 2592 (1988).

139. S. P. Kowalczyk, L. Ley, R. A. Pollak, F. R. McFeely, and D. A. Shirley, *Phys. Rev.* **B7**, 4009 (1973).

140. P. S. Bagus, A. J. Freeman, and F. Sasaki, *Phys. Rev. Lett.* **30**, 850 (1973).

141. E. Kisker and C. Carbone, *Solid State Commun.* **65**, 1107 (1988).

142. B. Sinkovic, B. Hermsmeier, and C. S. Fadley, *Phys. Rev. Lett.* **55**, 1227 (1985).

143. B. Hermsmeier, J. Osterwalder, D. J. Friedman, and C. S. Fadley, *Phys. Rev. Lett.* **62**, 478 (1989); B. D. Hermsmeier, Ph.D. thesis, U. Hawaii (1989); B. Hermsmeier, J. Osterwalder, D. J. Friedman, B. Sinkovic, T. Tran, and C. S. Fadley, *Phys. Rev.* **B42**, 11895 (1990).

144. D. J. Friedman, B. Sinkovic, C. S. Fadley, *Phys. Scr.* **41**, 909 (1990).

145. J. E. Smart, *Effective Field Theories of Magnetism*, Saunders, New York (1966), Ch. 4.

146. M. Landolt and D. Mauri, *Phys. Rev. Lett.* **49**, 1783 (1982); D. Mauri, R. Allenspach, and M. Landolt, *Phys. Rev. Lett.* **52**, 152 (1984); M. Taborelli, R. Allenspach, and M. Landolt, *Phys. Rev.* **B34**, 6112 (1996); R. Allenspach, D. Mauri, M. Taborelli, and M. Landolt, *Phys. Rev.* **B35**, 4801 (1987).

147. G. Schuetz, W. Wagner, W. Wilhelm, P. Keinle, R. Zeller, R. Frahm, and G. Materlik, *Phys. Rev. Lett.* **58**, 737 (1987); G. Schuetz, R. Frahm, P. Mautner, R. Wienke, W. Wagner, W. Wilhelm, and P. Kenle, *Phys. Rev. Lett.* **62**, 2620 (1989); G. Schuetz, *Phys. Scr.* **T29**, 172 (1989); G. Schuetz and R. Wienke, *Hyperfine Interactions* **50** 457 (1989).

148. G. Schoenhense, *Appl. Phys.* **A41**, 39 (1986); C. Westphal, J. Bansmann, M. Getzlaff, and G. Schoenhense, *Phys. Rev. Lett.* **63**, 151 (1989); G. Schoenhense, *Phys. Scr.*, **T31**, 255 (1990).

149. K. A. Thompson and C. S. Fadley, *J. Electron. Spectros. Relat. Phenom.* **33**, 29 (1984).

150. P. L. King, A. Borg, P. Pianetta, I. Lindau, G. Knapp, and M. Keenlyside, *Phys. Scr.* **41**, 413 (1990); B. P. Tonner, *Phys. Scr.* **T31**, (1990); W. Ng, A. K. Ray-Chaudhuri, R. K. Cole, S. Crossley, D. Crossley, C. Gong, M. Green, J. Guo, R. W. C. Hansen, F. Cerrina, G. Margaritondo, J. H. Underwood, J. Kortright, and R. C. C. Perera, *Phys. Scr.* **41**, (1990); H. Ade, J. Kirz, S. L. Hulbert, E. D. Johnson, E. Anderson, and D. Kern, *Appl. Phys. Lett.* **56**, 1841 (1990).

151. J. J. Barton, *Phys. Rev. Lett.* **61**, 1356 (1988) and private communication.

152. A. Szoeke, in *Short Wavelength Coherent Radiation: Generation and Applications* (D. T. Attwood and J. Bokor, eds.), AIP Conf. Proc. No. 147, American Institute of Physics, New York (1986).

153. J. Stohr, J. L. Gland, W. Eberhardt, D. A. Outka, R. J. Madix, F. Sette, R. J. Koestner, and U. Doebler, *Phys. Rev. Lett.* **51**, 2414 (1983); J. Stohr, F. Sette, and A. L. Johnson, *Phys. Rev. Lett.* **53**, 1684 (1984); A. P. Hitchcock and J. Stohr, *J. Chem. Phys.* **87**, 3253 (1986).

154. M. N. Piancastelli, D. W. Lincle, T. A. Ferrett, and D. A. Shirley, *J. Chem. Phys.* **86**, 2765 (1987) and **87**, 3255 (1987).

155. P. K. Hansma and J. Tersoff. *J. Appl. Phys.* **61**, R1 (1987); J. Tersoff, in *The Structure of Surfaces II* (J. F. van der Veen and M. A. van Hove, eds.) Springer-Verlag, Berlin (1988), p. 4; and references therein.

156. T. Gustafsson, M. Copel, and P. Fenter, in *The Structure of Surfaces II* (J. F. van der Veen and M. A. van Hove, eds.) Springer-Verlag, Berlin (1988), p. 110; and references therein.

157. F. Sette, S. J. Pearton, J. M. Poate, J. E. Rowe, and J. Stohr, *Phys. Rev. Lett.* **56**, 2637 (1986).

158. G. R. Harp, D. K. Saldin, and B. P. Tonner, *Phys. Rev. Lett.* **65**, 1012 (1990), and *Phys. Rev.* **B42**, 9199 (1990).

159. G. S. Herman, S. Thevuthasan, T. T. Tran, Y. J. Kim, and C. S. Fadley, *Phys. Rev. Lett.* **3** Feb., 1991.

160. D. K. Saldin, G. R. Harp, B. L. Chen, and B. P. Tonner, *Phys. Rev.* **B44**, 2480 (1991) and earlier references therein.

161. S. Y. Tong, C. M. Wei, T. C. Zhao, H. Huang, and H. Li, *Phys. Rev. Lett.* **66**, 60 (1991); S. Y. Tong, H. Li, and H. Huang, *Phys. Rev. Lett.* **67**, 3102 (1991).

162. S. Thevuthasan, G. S. Herman, A. P. Kaduwela, R. S. Saiki, Y. J. Kim, W. Niemczura, M. Burger, and C. S. Fadley, *Phys. Rev. Lett.* **67**, 469 (1991).

163. J. J. Barton, *Phys. Rev. Lett.* **67**, 3406 (1991).

Index

Boldface indicates volume number.

The manufacturer's authorised representative in the EU is Springer
Nature Customer Service Centre GmbH, Europaplatz 3, 69115 Heidelberg,
Germany. If you have any concerns regarding our products, please
contact ProductSafety@springernature.com

Printed and bound by CPI Group (UK) Ltd, Croydon, CR0 4YY
29/04/2026
02099318-0001